Developments in Precambrian Geology 11

ARCHEAN CRUSTAL EVOLUTION

DEVELOPMENTS IN PRECAMBRIAN GEOLOGY

Advisory Editor B.F. Windley

Further titles in this series

1. B.F. WINDLEY and S.M. NAQVI (Editors)
   Archaean Geochemistry
2. D.R. HUNTER (Editor)
   Precambrian of the Southern Hemisphere
3. K.C. CONDIE
   Archean Greenstone Belts
4. A. KRÖNER (Editor)
   Precambrian Plate Tectonics
5. Y.P. MEL'NIK
   Precambrian Banded Iron-formations. Physicochemical Conditions of Formation
6. A.F. TRENDALL and R.C. MORRIS (Editors)
   Iron-Formation: Facts and Problems
7. B. NAGY, R. WEBER, J.C. GUERRERO and M. SCHIDLOWSKI (Editors)
   Developments and Interactions of the Precambrian Atmosphere, Lithosphere and Biosphere
8. S.M. NAQVI (Editor)
   Precambrian Continental Crust and its Economic Resources
9. D.V. RUNDQVIST and F.P. MITROFANOV (Editors)
   Precambrian Geology of the USSR
10. K.C. CONDIE (Editor)
    Proterozoic Crustal Evolution

DEVELOPMENTS IN PRECAMBRIAN GEOLOGY 11

ARCHEAN CRUSTAL EVOLUTION

Edited by
K.C. CONDIE
New Mexico Institute of Mining & Technology, Department of Geoscience, Socorro, NM 87801, U.S.A.

1994
ELSEVIER
Amsterdam — Lausanne — New York — Oxford — Shannon — Tokyo

ELSEVIER SCIENCE B.V.
Sara Burgerhartstraat 25
P.O. Box 211, 1000 AE Amsterdam, The Netherlands

ISBN: 0-444-81621-6

This book is printed on acid-free paper.

Printed in The Netherlands

CONTRIBUTING AUTHORS

NICHOLAS T. ARNDT

Institute de Geologie, Université de Rennes, Avenue de General Leclerc, 35042 Rennes Cedex, France

LEWIS D. ASHWAL

Department of Geology, Rand Afrikaans University, P.O. Box 524, Auckland Park 2006, South Africa

MARK E. BARLEY

Key Centre for Strategic Mineral Deposits, Department of Geology, University of Western Australia, Nedlands, WA 6009, Australia

KENT C. CONDIE

Department of Geoscience, New Mexico Institute of Mining and Technology, Socorro, NM 87801, USA

DAVID J. DES MARAIS

NASA, Ames Research Center, Mail Stop 239-4, Moffett Field, CA 94035, USA

KENNETH A. ERIKSSON

Department of Geological Sciences, Virginia Polytechnic Institute, Blacksburg, VA 24061-0420, USA

CHRISTOPHER M. FEDO

Department of Geological Sciences, Virginia Polytechnic Institute, Blacksburg, VA 24061-0420, USA

DAVID I. GROVES

Key Centre for Strategic Mineral Deposits, Department of Geology, University of Western Australia, Nedlands, WA 6009, Australia

DONALD R. LOWE

Department of Geology, Stanford University, Stanford, CA 94305-2115, USA

HERVE MARTIN

CNRS – URA 10, Université Blaise Pascal, 5 Rue Kessler, 63038 Clermont-Ferrand Cedex, France

JOHN S. MYERS
Geological Survey of Western Australia, 100 Plain Street, East Perth, WA 6004, Australia

LEV M. NATAPOV
State Geology Company Aerogeologia, Lenin Prospect 35, Moscow 117171, Russia

A.D. NOZHKIN
United Institute of Geology, Geophysics and Mineralogy, Siberian Branch, Academy of Sciences of Russia, Universitetsky Prospect 3, Novosibirsk 630090, Russia

JOHN PERCIVAL
Canadian Geological Survey, 601 Booth Street, Ottawa, Ontario K1A OE8, Canada

OLEG M. ROSEN
Institute of the Lithosphere, Academy of Sciences of Russia, Staromeonetny per. 22, Moscow 109180, Russia

PAUL SYLVESTER
Research School of Earth Sciences, Australian National University, P.O. Box 378, Canberra, ACT 2601, Australia

PHILIP C. THURSTON
Ontario Geological Survey, 933 Ramsey Lake Road, Sudbury, Ontario P3E 6B5, Canada

CONTENTS

INTRODUCTION

K.C. CONDIE

As geochronology has progressed in the last 20 years, the Archean has continued to attract interest. Advancements in our understanding of Archean crustal and mantle evolution can be traced through a series of meetings and published papers beginning in 1970 with the first International Archean Symposium in Perth in Western Australia (Glover, 1970). Among the exciting problems at that meeting were the relationship between low and high grade Archean terrains, the origin of Archean ultramafic rocks and, in particular, the recently discovered komatiites in South Africa, and the origin of various Archean mineral deposits. Only about 30% of the papers dealt with geochronology and Archean crustal evolution. If one were to identify a landmark meeting for the Archean, it would almost certainly be the NATO Advanced Study Institute convened by Brian Windley in April of 1975 at Leicester (Windley, 1976). At this meeting the Archean truly "came of age". Investigators from many different disciplines focused their expertise on the early history of the earth. Exciting debate and discussion centered on such topics as the role of plate tectonics in the Archean, the relative importances of compressive versus vertical tectonics, and the use of trace elements in understanding both the origin and tectonic setting of Archean rocks. Detailed accounts of Archean high grade terrains from SW Greenland were presented and rigorous comparisons were made between granite-greenstone and high grade terrains. For the first time, the nature of the atmosphere, oceans, and life during the Archean was an important part of an Archean symposium.

When the Archean International Symposium was again convened in Perth in 1980 (Glover and Groves, 1980), high levels of interest continued on the recently dated 3.8–3.5 Ga rocks in SW Greenland and South Africa. Discussion focused around the widespread development of continental crust in the Late Archean, the significance of komatiites, whether Archean greenstones were deposited on ensialic or ensimatic basement, the role of plate tectonics in greenstone formation (back arc basins and rifts being the leading contenders), the relation of greenstones to surrounding gray gneisses, the use of trace elements to constrain the tectonic settings of Archean volcanics and sediments, and differences in deformational styles between Archean low and high grade terrains. This was the first international meeting at which Chinese geologists appeared and gave the rest of the world a glimpse of the Archean in China. Although some U/Pb zircon ages were presented (including the first ion microprobe data), dating was still largely by

Rb/Sr and Sm/Nd methods. In contrast to the 1970 Archean Symposium, almost 50% of the papers dealt with Archean crustal evolution.

In 1981, two important volumes were published in this same Elsevier series: *Precambrian Plate Tectonics* edited by Alfred Kroner (Kroner, 1981) and *Archean Greenstone Belts* by Kent Condie (Condie, 1981). In both books it was clear that plate tectonics in some form was important in the Archean. Detailed descriptions of Early Archean greenstone belts and discussions of Archean cratons and continental growth, the use of REE in monitoring crustal evolution, and paleomagnetic constraints on Archean plate motions are among but a few of the important topics presented by various authors in Kroner's book. In my book, I attempted to bring together factual and interpretational data related to origin of Archean granite-greenstone terrains, including a summary of the various models that had been published for their tectonic development.

During the most recent Archean Symposium in Perth in 1990, there was a shift in interest from field and trace element data to the new rapidly evolving high-precision U/Pb geochronology of Archean rocks and to detailed structural studies of both low and high grade Archean terrains (Glover and Ho, 1992). The terrane concept so widely applied to the Phanerozoic was proposed for the Archean Yilgarn Province in Western Australia by John Myers, and now is widely accepted for the Archean as evident by the articles in this book. The importance of Nd model ages, epsilon-Nd data, and SHRIMP U/Pb zircon ages to understanding the Archean was clearly apparent at this meeting. Other topics of continuing or new interest included tracing P-T histories of segments of Archean crust, changes at the Archean/Proterozoic boundary, and tracing of the tectonic history of greenstones with provenance information from clastic sediments.

In 1975, I made a list of what appeared to be the most important questions related to Archean crustal evolution. Among these were the following:

(1) Why are there no Archean ophiolites?

(2) What are the differences between Archean and Early Proterozoic greenstones and what do they tell about changing tectonic regimes?

(3) What are the characteristics of Archean high grade terrains and in what tectonic settings did they form? How are they related to greenstones?

(4) Of what significance are the quartzites found in some Archean greenstone belts?

(5) How many different tectonic settings are represented by Archean greenstones?

(6) Are the felsic volcanics in greenstones the extrusive equivalents of surrounding TTG complexes?

(7) What does the cyclicity in greenstone stratigraphy reflect?

(8) What are the protoliths of the Archean TTG complexes and are they the same in high and low grade terrains?

It is encouraging that although most or all of these questions are still with us in some form, we have a much greater understanding of Archean crustal evolution today than we did in 1975. Moreover, we have testable models for many of these questions today that we did not have in 1975.

Plate tectonics is now widely accepted as the principal process that controls the history of continents and oceans. How far back in time plate tectonic extends, however, is still a subject of active debate among geologists. If we accept a strictly uniformitarian approach to earth history, then we must accept plate tectonics from at least 4 Ga onwards. Although it is agreed that for the earth to cool with time heat must be lost at the surface, it is not clear if the formation of new lithosphere at ocean ridges has always been the major mechanism by which this heat is dissipated, as it is today. There is considerable interest in the possibility that mantle plumes may have been more important is dissipating terrestrial heat in the Archean than they are today.

One way of assessing the role of plate tectonics in the Archean has been to examine the rocks formed at that time. Do we see the same rock assemblages we have today, and are their age relationships, tectonic histories, and chemical compositions similar? If so, plate tectonics would seem to be acceptable in the earliest part of earth history. Figure 1 is an updated version of a graph I originally published in 1989 showing the distribution of major rock assemblages with time. The greenstone association, except for ophiolites, is recognized in the oldest known rocks at 4.0–3.6 Ga. The oldest well preserved cratonic sediments are in

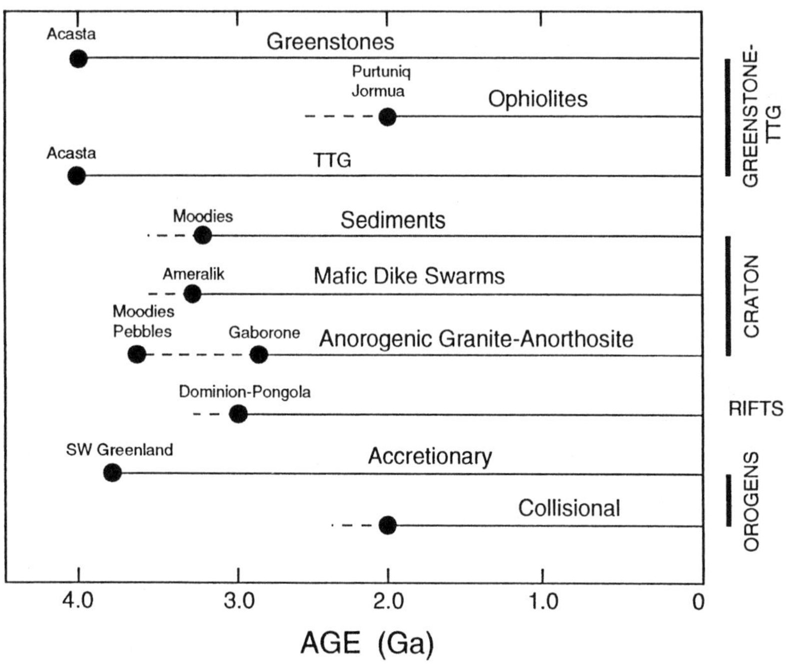

Fig. 1. Time distribution of rock assemblages.

the Moodies Group in South Africa deposited about 3.2 Ga, and by 3.1–2.9 Ga cratonic sedimentary successions were deposited in both intracratonic basins and on stable continental shelves. The oldest preserved dike swarms occur in deformed TTG complexes, such as the Ameralik dikes in SW Greenland intruded at about 3.25 Ga (Friend et al., 1988). Although the oldest anorogenic granite–anorthosite complex whose age is tightly constrained by U/Pb zircon ages is the Gaborone suite in Botswana at 2875 Ma (Moore et al., 1993), it is likely that such granites formed much earlier in southern Africa as recorded by granite conglomerate clasts in the Moodies Group, which have been dated at about 3.6 Ga from their zircons (Kroner and Compston, 1988). The oldest known rift assemblages are about 3.0 Ga in the Kaapvaal craton in southern Africa, where both the Dominion Supergroup and parts of the Pongola Supergroup developed in rifts (Eriksson and Fedo, Chapter 5). Two types of orogens have been recognized in the geologic record (Windley, 1992). A collisional orogen involves the collision of two or more large continental fragments, whereas an accretionary orogen forms by the growth and amalgamation of arcs, submarine plateaus, oceanic islands, ophiolites, and small continental blocks. Although not well known, the oldest reported accretionary orogen is in SW Greenland at about 3.8 Ga (Nutman et al., 1989), and the oldest well described accretionary orogen is the Birimian of West Africa and equivalent rocks in Guiana at 2.1 Ga. The oldest recognized collisional orogens are Early Proterozoic in age, such as the Kola–Karelian orogen in the northern Baltic Shield, the Wopmay and Thelon orogens in NW Canada, and the Capricorn orogen in Western Australia.

The fact that greenstones, sodic TTG, anorogenic granites, and accretionary orogens appear in the very earliest vestiges of our preserved geologic record at 4.0 to 3.5 Ga, strongly supports some sort of plate tectonics operating on the earth at this time. By 3.3 to 3.0 Ga mafic dike swarms, cratonic sediments, and continental rifts had appeared recording the development of the earliest cratons. Only ophiolites do not appear in the geologic record until about 2.0 Ga. The absence of pre-2-Ga ophiolites, however, does not mean the absence of remnants of Archean oceanic crust. Archean oceanic crust may not have looked like an idealized ophiolite, and thus has not yet been identified in Archean supracrustals. Although plate tectonics appears to have been with us since at least 4 Ga, there are now well substantiated differences between Archean and post-Archean rocks that indicate that Archean tectonic regimes must have differed in some respects from modern ones. The best documented of these differences are the abundance of komatiites (Chapter 1) and heavy-REE depleted TTG in the Archean (Chapter 6), differences in the composition of Archean shales and some igneous rocks (Taylor and McLennan, 1985; Condie, 1992), and the unique composition of Archean anorthosites (Chapter 8). These differences have lead to the concept of having our cake and eating it to, or in other words, plate tectonics operated in the Archean, but it differed in some ways from modern plate tectonics. This view appears to be widely accepted and we are now faced with the question of how and to what degree did Archean plate tectonics differ from modern plate tectonics, a question which is treated in many of the chapters of this book.

In Chapter 1, Nick Arndt discusses komatiites: how they form, what their textures and composition tell us, and what they can and cannot tell us about the mantle. Nick elegantly reviews the experimental and theoretical studies related to the origin of spinifex texture, and how it is related to cooling and nucleation rate. It seems clear that the cooling history of komatiitic liquids varies and depends critically on the initial composition of the magma and its phenocryst content, both of which are controlled by the initial melting conditions in the mantle, and by the path followed by the komatiitic magma on its way to the surface. He shows how it is possible to separate the effects of alteration from olivine fractionation using the MgO contents of komatiites, and discusses element mobility in altered and metamorphosed komatiites. He also suggests that trace element ratios such as Th/Nb and La/Nb are sensitive to the amount of crustal contamination in a komatiitic melt. As others have also concluded, the author indicates that komatiitic magmas must be produced in anomalously hot mantle, such as mantle plumes. Clearly, komatiites cannot be used to constrain the composition of the depleted mantle reservoir during the Archean.

In the next chapter Phil Thurston reviews the principal characteristics of Archean greenstone belts and relates them to modern tectonic settings. From lithologic assemblages in greenstone belts from the Superior Province in Canada, he identifies five assemblages: (1) the platform assemblage consisting of clastic, chemical, and biochemical sediments deposited on shallow platforms; (2) mafic assemblages comprised chiefly of submarine basalts and komatiites occurring in extensive mafic plains; (3) mafic to felsic sequences including submarine volcanics and hyaloclastic sediments; (4) late sequences of stream, deltaic, and submarine fan sediments deposited in overlying unconformable basins (pull-apart basins); and (5) and continental style volcanics composed chiefly of calc-alkaline volcanics and associated ash-flow tuffs. He suggests all of these fit within the framework of modern plate tectonics, mostly in arc-related environments. He also discusses the apparent absence of Archean ophiolites, which he suggests may be due to our not recognizing them. Thurston sees the complex histories of Archean greenstones as the product of accretionary tectonics, and proposes that individual greenstone belts are really collages of various rock assemblages formed in different tectonic environments. He does conclude, however, that there are differences between Archean and modern volcanics (such as the relative abundance of komatiites in the Archean), establishing some unique tectonic features for the Archean.

In Chapter 3 I examine several features of greenstone evolution through time, comparing stratigraphic, lithologic, isotopic ages, and geochemistry of greenstones that appear to have formed in oceanic tectonic settings. One rather interesting observation is that earth history can be divided into three greenstone time periods: >2.7 Ga, when greenstones appear to have formed and collided continuously, although probably not forming supercontinents; 2.7–1.0 Ga where a clear episodisity is apparent in greenstone eruption and collision age, with two greenstone time gaps (2450–2200 and 1650–1350 Ma) corresponding to supercontinent fragmentation; and <1.0 Ga, when again greenstones appear to have formed and collided rather continuously, perhaps in response to overlap in the times of dispersal and assembly of supercontinents. Greenstone lithologic associations suggest that oceanic tectonic settings were less diverse in the Archean than afterwards. A greater proportion of submarine plateaus in Archean greenstones probably reflects an enhanced preservation rate of Archean greenstones due to felsic magma underplating or/and a greater frequency of mantle plumes.

Don Lowe, in Chapter 4, presents a thorough discussion of greenstone-related sedimentary rocks. He recognizes five lithofacies associations that reflect different tectonic settings and depositional environments: (1) a mafic volcaniclastic–orthochemical biogenic association representing oceanic islands or ocean ridges; (2) an orthochemical–biogenic association that commonly overlies the first association and represents an oceanic intraplate setting; (3) a felsic volcaniclastic and terrigenous association, and (4) an orogenic terrigenous association, both of which are characteristic of the upper parts of greenstone successions and reflect deformation and accretion at convergent plate margins; and (5) a polycyclic terrigenous association, which develops along mature continental margins adjacent to active greenstones. Some data suggest that preserved pre-3 Gy-old greenstones developed at ocean ridges or at hotspots, whereas later Archean greenstones formed chiefly at subduction zones.

In Chapter 5, Ken Eriksson and Chris Fedo summarize our current understanding of Archean cratonic and rift successions, focusing primarily on southern Africa, where the best preserved successions occur. The earliest sediments deposited in these environments occur in the Kaapvaal and Zimbabwe cratons at 3.2–2.9 Ga, and clearly document the existence of at least a few stable cratons by this time. These early cratonic sediments were deposited in synrift basins associated with bimodal volcanics, as well as in stable-shelf basins. Sedimentary facies have been well described by Eriksson and others and include alluvial, lacustrine, and shallow-marine environments. Eriksson and Fedo relate the large increase in volume of cratonic sediments in the Late Archean to the growth of continental crust at this time.

Herve Martin presents an exhaustive summary on the occurrence, composition, and origin Archean gray gneisses (TTG) in Chapter 6. Included are descriptions of their mineralogy, major element composition, and a discussion of the differences between the tonalite–trondhjemite and calc-alkaline suites. The now well established depletion in heavy REE in Archean TTG is compared to similar depletion in a few young granitoids and suggests a relatively high temperature of the source. Herve thoroughly reviews geochemical and experimental results for the origin of these rocks and compares them to modern granitoids. As previous investigators have suggested, Archean TTG magmas appear to have been produced in anomalously hot mafic crust, probably garnet amphibolite, in which garnet is left in the melting residue, thus retaining the heavy REE. The author favors a convergent margin plate tectonic model in which relatively warm descending mafic crust partially melts, probably by dehydration melting. This in turn, implies steeper geotherms in Archean subduction zones. He supports his model with a modern analog where the Chile rise is being subducted beneath the Andes and partially melted to produce heavy REE depleted volcanics.

In the next chapter Paul Sylvester reviews the occurrence and origin of Archean granite plutons, including and in-depth geochemical comparison with Phanerozoic granites. He suggests a three-fold classification of these plutons into calc-alkaline, strongly peraluminous, and alkaline types with all three types emplaced over broad areas of Archean cratons following several million years or more after greenstone volcanism, TTG plutonism, and deformation. Geochemical modeling coupled with experimental petrologic results suggest all three types of plutons are the products of partial melting of igneous and sedimentary rocks in the middle and lower crust. Differences in chemical composition between Archean and Phanerozoic plutons reflect either or both changes in crustal source compositions through time, or steeper geothermal gradients in Archean continents.

Only in the last few years have anorthosites been recognized as important minor components in the preserved Archean crust. In Chapter 8, Lew Ashwal and John Myers describe Archean anorthosites and show that they differ from most post-Archean anorthosites by their high An content (ca. An_{80}) and their occurrence as megacrystic sheet-like bodies. Ashwal and Myers also show that Archean anorthosites are cumulates probably derived by fractional crystallization of Fe-rich basalts. They appear to have been emplaced at shallow depths and many of them later metamorphosed at high grades. It is difficult to propose a tectonic setting for Archean anorthosites because of the lack of any modern analogs. However, their close association with pillow basalts suggests an oceanic environment. Why they did not continue to be produced after the Archean remains an unsolved mystery.

In Chapter 9 John Percival reviews Archean high grade rocks and their relationship to low grade rocks. He surveys results from P-T studies in Archean high grade crustal provinces and discusses these in light of processes involved and possible tectonic settings. He concludes that almost all Archean high grade terrains can be explained in the context of one of two tectonic settings: deep crustal portions of magmatic arcs or collisional orogens. In particular, collisional orogens can account for widespread reworking of older rocks, clockwise P-T paths, and syntectonic exhumation. The fact, however, that some Archean granulites are characterized by isobaric cooling paths suggests that their uplift and exhumation may be unrelated to the tectonic processes that formed them. Widespread Archean granulite provinces like the Minto block in eastern Canada appear to have no young counterparts exposed at the surface and they may represent the root zones of foreland or hinterland areas associated with Archean collisional orogens. Underplated basaltic magmas may have been the heat source for many Archean granulites and also they may have promoted deep crustal melting producing felsic magmas emplaced at deep levels.

Chapter 10 by Oleg Rosen, Kent Condie, Lev Natapov and A.D. Nozhkin focuses on the Siberian craton. In this chapter the authors bring together data from geologic mapping, gravity and magnetic anomaly distributions, thermobarometric studies, and U/Pb zircon chronology from the Siberian craton. From the available database, the authors define seven Archean crustal provinces and constituent terranes within some of the provinces. Major periods of deformation and plutonism are recognized at ≥ 3.5, 3.4, 3.0 and 2.75 Ga, and together with Nd isotopic data, results suggest that 90% of the Precambrian basement of the Siberian craton was extracted from the mantle during the Archean. Seismic sounding results are consistent with thin Archean lithosphere (100–150 km thick). The Archean basement in Siberia was widely remobilized at 2.0–19 Ga, when many the Archean provinces appear to have collided to form the Siberian craton. Only the Atitkan orogenic belt may comprise mostly juvenile Early Proterozoic crust.

In Chapter 11 Dave Groves and Mark Barley present an in-depth summary of Archean mineral deposits, emphasizing the important deposits in greenstone belts. The authors consider mineral deposits an important parameter together with lithologic association and structural style in constraining the tectonic setting of greenstone belts. They show that the distribution of volcanic-associated mineral deposits varies both with age and lithologic association. For instance the Cu-Zn massive sulfides are associated with mafic to felsic greenstone successions, analogous to arc systems, whereas Ni sulfides are associated with komatiites in mafic plain successions, similar perhaps to submarine plateaus. Archean gold deposits, so important in some greenstones, are interpreted as being deposited along convergent margins based on the nature of controlling structures and associated igneous rocks. The authors also re-emphasize the interesting yet still poorly understood observation that most Archean mineralization occurred in the Late Archean, with only a few anomalous sulfate deposits in the Early Archean. The authors suggest a parallel between the tectonic and metallogenic evolution of Phanerozoic arcs, including the breakup and aggregation of supercontinents, and Archean greenstones, again supporting a plate tectonic Archean regime.

In the final chapter, Dave Des Marais reviews the origin and Archean evolution of the earth's atmosphere. He suggests that the early prebiotic atmosphere was composed principally of CO_2, N_2, and H_2O and small amounts of H_2 and CO. As microorganisms developed, H_2 was captured in organic matter, and the amount of organic matter buried was controlled by a redox balance with volcanic emanations. He also suggests that because the earliest continents should have been largely submerged beneath seawater, and because the earth's heat flow was greater, that hydrothermal activity on the seafloor was more important in controlling the composition of the seawater–atmosphere system than erosion from the continents. During the Archean, atmospheric O_2 levels were maintained at low values because the burial rate of photosynthetic organic matter was small and O_2 was consumed largely by volcanic emanations. In the Late Archean, however, increases in land area and increasing burial rates of photosynthetic organic matter caused CO_2 levels to begin to drop and O_2 levels to rise.

Numerous individuals helped review the chapters for completeness and accuracy. In particular, I want to thank the following people for the time and effort they gave to reviewing one or more chapters: Fred Barker, Mike Brown, Kevin Burke, Mark Drummond, Ken Eriksson, Simon Harley, Claude Herzberg, Dick Hutchinson, Rob Kerrich, Case Klein, Mike Lesher, Don Lowe, Tony Morse, Wulf Mueller, Euan Nisbet, Allen Nutman, Phil Thurston, Robert Wiebe, Brian Windley, and Hank dela R. Winter.

REFERENCES

Condie, K.C., 1981. Archean Greenstone Belts. Elsevier, Amsterdam, 434 pp.

Condie, K.C., 1989. Plate Tectonics and Crustal Evolution. Third Edition, Pergamon Press, New York, N.Y., 476 pp.

Condie, K.C., 1993. Chemical composition and evolution of the upper continental crust: contrasting results from surface samples and shales. Chem. Geol., 104: 1–37.

Friend, C.R.L., Nutman, A.P. and McGregor, V.R., 1988. Late Archean terrane accretion in the Godthaab region, southern West Greenland. Nature, 335: 535–538.

Glover, J.E. (Ed.), 1971. Symposium on Archean Rocks. Perth, 23–26 May 1970, Geol. Soc. Australia, Spec. Public. No. 3, 469 pp.

Glover, J.E. and Groves, D.I. (Eds.), 1981. Archaean Geology. Second International Symposium, Perth, 1980. Geol. Soc. Australia, Spec. Public. No. 7, 515 pp.

Glover, J.E. and Ho, S.E. (Eds.), 1992. The Archaean: Terrains, Processes and Metallogeny. Proceeding Volume for the Third International Archaean Symposium, 17–21 September, 1990, Perth, Western Australia. Geology Dept. and University Extension, Univ. Western Australia, Public. No. 22, 436 pp.

Kroner, A. (Ed.), 1981, Precambrian Plate Tectonics. Elsevier, Amsterdam, 781 pp.

Kroner, A. and Compston, W., 1988. Ion microprobe ages of zircons from Early Archean pebbles in graywacke, Barberton greenstone belt, southern Africa. Precamb. Res., 38: 367-380.

Moore, M., Davis, D.W., Robb, L.J., Jackson, M.C., and Grobler, D.F., 1993. Archean rapakivi granite–anorthosite–rhyolite complex in the Witwatersrand basin hinterland, southern Africa. Geology, 21: 1031–1034.

Nutman, A.P., Friend, C.R., Kinny, P.D., and McGregor, V.R., 1993. Anatomy of an Early Archean gneiss complex: 3900 to 3600 Ma crustal evolution in southern West Greenland. Geology, 21: 425–418.

Taylor, S.R. and McLennan, S.M., 1985. The Continental Crust: Its Composition and Evolution. London, Blackwell Scientif. Pub., 312 pp.

Windley, B.F., 1992. Proterozoic collisional and accretionary orogens. In: K.C. Condie (Ed.), Proterozoic Crustal Evolution, Elsevier, Amsterdam, pp. 419–445.

Windley, B.W. (Ed.), 1976, The Early History of the Earth. John Wiley & Sons, New York, N.Y., 619 pp.

Chapter 1

ARCHEAN KOMATIITES

N.T. ARNDT

INTRODUCTION

Twenty-five years have passed since komatiites were first recognized in South Africa by Richard and Morris Viljoen (Viljoen and Viljoen, 1969a,b; 1982) and the existence of ultramafic volcanic rocks is now generally accepted. We know of occurrences in all continents except Antarctica, and the principal mineralogical, petrological and chemical characteristics of the rocks are well documented. Investigations have moved on to specific questions of petrogenesis or have focused on the information these rocks provide about compositions and physical conditions in the Archean mantle.

In this chapter I will not attempt to summarize all the information available on komatiites but will focus on three broad subjects. The first is the study of komatiite as an object of petrological curiosity. Here I discuss the origin of those textures and structures of komatiites that set them apart from other volcanic rocks: their spinifex textures and their layering. The second subject is the vexing problem of alteration of komatiites, and the extent to which the present compositions of komatiite samples reflect those of the original lavas. The third is the use of komatiites as *precious witnesses* (to use the French term) of the composition and physical conditions in the Archean mantle. There are two main issues, one being the question of how well the composition of the Archean mantle is represented by the rocks we now sample, bearing in mind that these rocks are metamorphosed and hydrothermally altered, and are the solidification products of variably fractionated and perhaps contaminated lavas. The other is what part of the mantle is sampled by komatiites and whether this is representative of any large mantle reservoir.

SPINIFEX TEXTURE

Definition

Spinifex is the most spectacular and most characteristic texture in komatiites (Fig. 1a–c). A slightly reworded version of the definition presented by Arndt and Nisbet (1982) is:

Fig. 1. Photos of spinifex-textured komatiites. (a) Upper part of a thin komatiite flow from Barberton showing the transition from chilled flow top through fine random spinifex to coarse plate spinifex (photo provided by A. Kröner); (b) coarse plate spinifex from Pyke Hill in Munro Township; (c) spinifex-textured vein in the cumulate layer of a thin flow from Pyke Hill, Munro Township.

"Spinifex is a texture characterized by large, skeletal, platy, bladed or acicular grains of olivine or pyroxene, found in the upper parts of komatiitic flows, or, less commonly, at the margins of sills and dikes. The texture is believed to form during relatively rapid, *in situ* crystallization of ultramafic or highly mafic liquids.

In platy olivine spinifex texture, olivine has a plate or lattice habit and forms complex grains made up of many individual plates arranged roughly parallel to one another. The 'books' of parallel grains are oriented approximately perpendicular to flow or intrusive margins. Composite olivine plates may be as long as 1 m but are only 0.5 to 2 mm thick. Interstitial material is fine skeletal pyroxene, cruciform or dendritic chromite, and devitrified glass. Random olivine spinifex texture contains smaller, less elongate, randomly oriented olivine plates.

Pyroxene spinifex texture contains pigeonite or augite or both pyroxenes in complex skeletal needles that are arranged in sheaths perpendicular to flow margins. The pyroxene needles typically are 1–5 cm long but only 0.5 mm wide and lie in a matrix of much finer augite needles and devitrified glass, or augite, plagioclase and quartz.

Primary phases usually are replaced by secondary minerals such as serpentine, chlorite, tremolite, talc, epidote and albite."

In general the term seems to be used in a manner consistent with this definition, but some problems exist. "Spinifex" continues to be applied to various textures produced during the growth of metamorphic olivines and other bladed or spiky minerals in metamorphosed ultramafic rocks. As has been pointed out by several authors (e.g. Collerson et al., 1976; Donaldson, 1982; Oliver et al., 1972), these textures are readily distinguishable from true spinifex textures: metamorphic olivine grains are bladed, non-skeletal, with no evidence of parallel growth; they commonly lie in matrices of secondary hydrous minerals such as talc, amphibole, chlorite or serpentine and commonly cut across the metamorphic fabric. In some cases their compositions are similar to those of spinifex olivines, in others they are richer in Fe.

Another problem lies in the use of the term spinifex for textures visible only with hand lens or microscope. It is true that some textures in the chilled margins of komatiite flows (e.g. Fig. 2b) are simply finer-grained versions of the macroscopic random olivine spinifex textures developed deeper in the flows (Figs. 1 and 2c,d); and that the acicular pyroxenes in flow margins, pillows or fragments of komatiitic basalt differ only in grain size from macroscopic pyroxene spinifex textures. Yet to accept these microscopic textures as spinifex seems to provide license for the term to be used for all manner of textures characterized by acicular minerals, be they in tholeiitic basalts, alkaline lavas, ocelli, or even skarns. To avoid ambiguity and potential confusion, it seems safer to reserve the term spinifex for the spectacular macroscopic textures found in komatiitic lavas.

Yet another problem stems from the use of terms such as "amphibole spinifex" to describe textures in which pyroxene needles are replaced by secondary tremolite or actinolite (de Wit et al., 1983, 1987). Ever since the texture was formally described by Viljoen and Viljoen (1969a,b) and Nesbitt (1971), the name of the dominant *igneous* mineral has been specified (as in "olivine spinifex" or "pyroxene spinifex") and that the types of secondary minerals have been mentioned separately. Although this is frowned upon by some petrologists accustomed to working with fresher rocks (Thompson (1983) coined the term "komatispeak" for the practice), it is practical for most komatiites in which good textural preservation

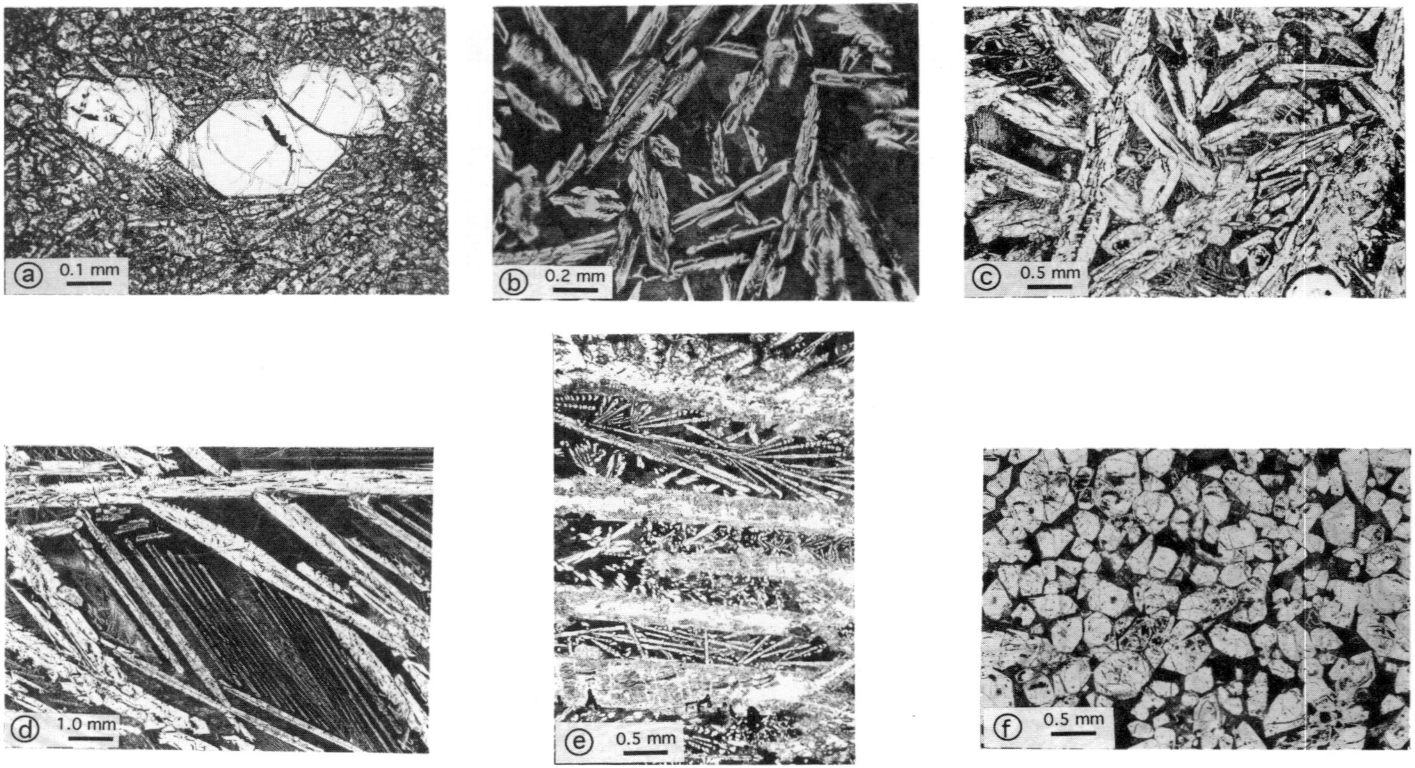

Fig. 2. Photomicrographs of komatiites. (a) Olivine phenocrysts in the chilled flow top of the Alexo komatiite; (b) fine random spinifex texture from the top of a flow from Munro Township; (c) random spinifex from the Alexo komatiite. This sample contains high MgO due in part to the presence of excess (accumulated) olivine; (d) plate spinifex from komatiite from Munro Township; (e) B1 layer from a komatiite flow from the Ottawa Islands; (f) olivine cumulate of komatiite from Munro Township. The white crystals are olivine, virtually unaltered in (a), partially replaced by serpentine or chlorite in (c) and (d) and completely replaced in (b), (e) and (f). The matrix contains acicular augite and devitrified glass.

allows easy recognition of the primary minerals. With this background, the use of the term "amphibole spinifex" is misleading and suggests yet another type of spinifex in which an igneous amphibole is the dominant component.

Occurrence

During past years some earlier ideas about the nature and origin of spinifex texture have been shown to be incorrect. The concept that spinifex lavas directly inherit the compositions of silicate liquids seems no longer valid. Detailed petrological and chemical studies indicate that olivine or pyroxene accumulate during spinifex growth, and that the resultant spinifex rocks commonly have a higher abundance of these minerals than the original liquids. At Alexo in the Abitibi belt of Canada, for example, Barnes (1983) showed that pyroxene spinifex lavas contained excess pyroxene, and Arndt (1986a) estimated that both the plate and random spinifex section of a komatiite flow contained 5–25% excess olivine.

Another concept requiring examination is the idea that spinifex textures are restricted to volcanic rocks, and are typically found in the upper layers of thin lava flows. This picture comes mainly from the komatiites in Munro Township (Pyke et al., 1973; Arndt et al. 1977), where many flows display conspicuous layering with upper spinifex zones and lower cumulate zones (Fig. 3a). The spinifex texture has a marked polarity defined by a downward increase in the size of the spinifex crystals. The same type of flow is known in many other regions, such as Marshall Pool in Western Australia (Barnes et al., 1973), Barberton in South Africa (de Wit et al., 1987; Smith et al., 1980; Viljoen and Viljoen, 1969b), the Belingwe belt in Zimbabwe (Nisbet et al., 1977; Renner et al., 1993), and the Crixas belt in Brazil (Arndt et al., 1989). On the other hand, spinifex textures with a quite different character and mode of occurrence have been revealed by studies of komatiites in Canada and Australia. In the upper crusts of the thick fossil lava lakes and sills, spinifex-textured veins are common (Arndt, 1986b). These are similar to veins in thin komatiite flows (Fig. 1c and Arndt et al., 1977; de Wit et al., 1987; Pyke et al., 1973; Viljoen and Viljoen, 1969b), but those in the lava lakes are much thicker and often contain spectacular textures (Fig. 3b). Examples from the Texmont mine region, south of Timmins in the Ontario part of the Abitibi belt, display olivine blades up to 80 cm long, commonly with a symmetrical decrease in grain size from the center towards both margins of the veins. Spinifex veins are also found in the lower border zones of certain thick flows.

A recent re-examination by Paul Davis and the author of komatiitic units in Dundonald Township, also in the Abitibi belt, provides firm evidence of spinifex textures in intrusive rocks. A certain proportion of these komatiitic units turn out to be thin sills, not flows, as had earlier been thought (Muir and Comba, 1979). The sills are 50 cm to several meters thick, and many have an upper spinifex layer and a lower olivine cumulate layer (Fig. 3c). The upper contacts differ from those of similarly-layered flows in that there is an abrupt, but on a fine scale gradational,

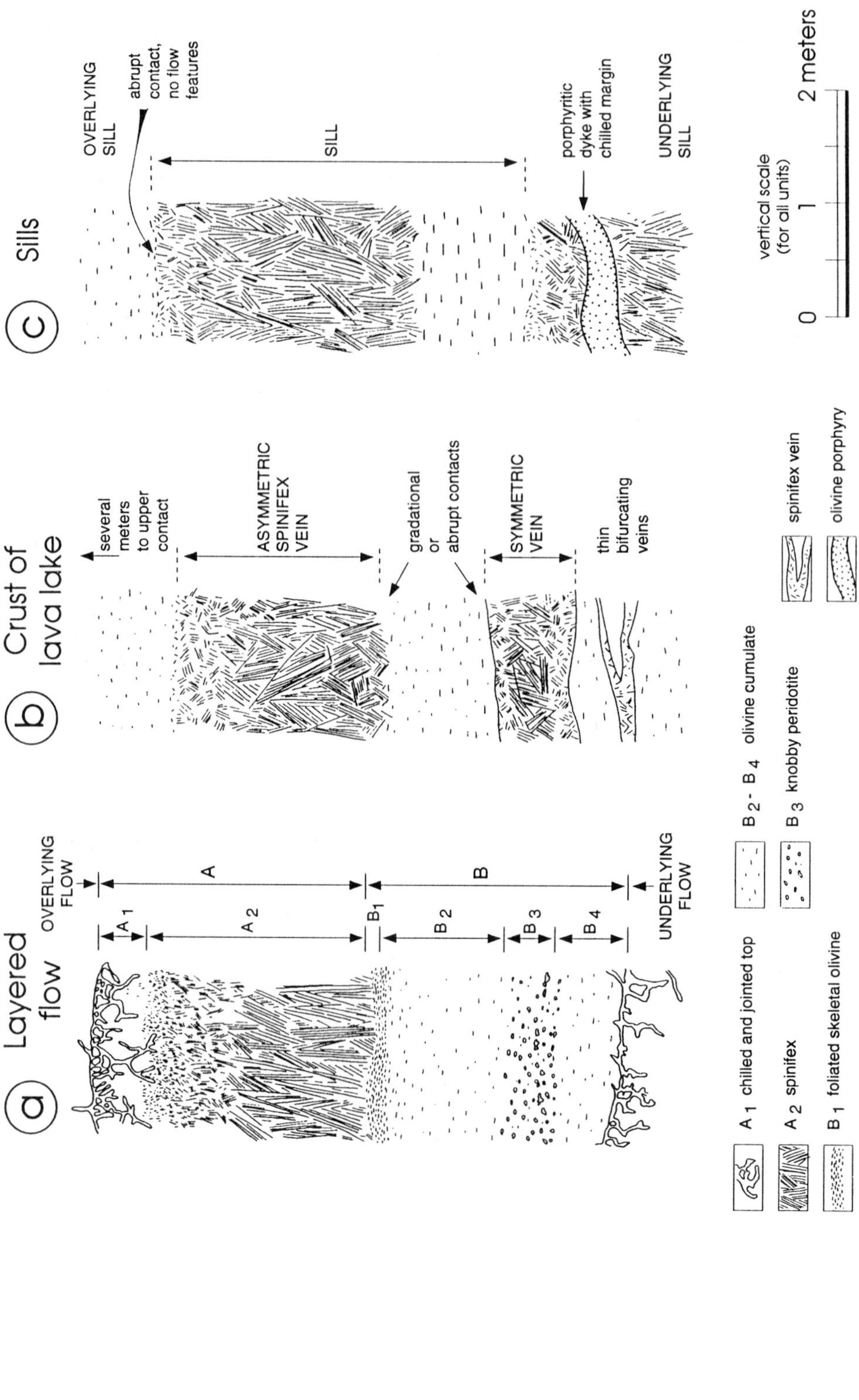
a Layered flow
OVERLYING FLOW
A 1
A 2
A
B 1
B 2
B 3
B 4
B
UNDERLYING FLOW
A 1 chilled and jointed top
A 2 spinifex
B 1 foliated skeletal olivine
B 2 - B 4 olivine cumulate
B 3 knobby peridotite
b Crust of lava lake
several meters to upper contact
ASYMMETRIC SPINIFEX VEIN
gradational or abrupt contacts
SYMMETRIC VEIN
thin bifurcating veins
spinifex vein
olivine porphyry
c Sills
OVERLYING SILL
abrupt contact, no flow features
SILL
porphyritic dyke with chilled margin
UNDERLYING SILL
vertical scale (for all units)
0
1
2 meters

transition from fine random spinifex to overlying olivine cumulate, and no sign of the brecciated or fractured zones found at normal flow tops. These units alternate with, or are cut by, sills of massive, porphyritic komatiite.

It is now clear that the simple picture of spinifex in the upper parts of layered flows requires revision. As shown in Fig. 3, this is but one mode of occurrence of the texture. Only when in this form can the asymmetry of the texture — an upward decrease in crystal size — be used as a reliable indicator of polarity.

Why spinifex texture?

The earlier ideas of Viljoen and Viljoen (1969b) and Nesbitt (1971), who proposed that spinifex texture forms as a result of quenching of ultramafic komatiite liquids, were criticized by Donaldson (1982) who pointed out that cooling rates during solidification of komatiite flows would be relatively low. Donaldson quoted cooling rates of 1–0.5°C h^{-1} for lava 0.5–1 m beneath the surface of a 5–20 m thick flow. These figures were questioned by Turner et al. (1986) who argued that these estimates failed to take into account convection, which would lead to greatly enhanced cooling rates. The latest treatments of Renner (1989) indicate, however, that cooling rates in the upper parts of komatiite flows were generally less than 1°C h^{-1}, a figure higher than for normal basaltic flows, but perhaps not high enough to explain the highly unusual textures.

Further information can be obtained by considering in more detail the occurrence of spinifex. It is striking that the texture is restricted to the type of flow we call komatiitic, but is absent from tholeiitic units with similar thickness and compositions. Consider two examples.

(a) *Fred's Flow vs Theo's Flow*. These thick layered mafic-ultramafic flows in Munro Township, Ontario, were described by Arndt (1977) and Arndt et al. (1977). Both are about 120 m thick, both are relatively continuous along strike, and both are layered with mafic upper sections and ultramafic cumulate lower

---

Opposite: Fig. 3. Sketch showing three difference situations in which komatiite can crystallize spinifex texture. (a) Layered komatiite flows (from Pyke et al. (1973). (b) Veins within the crust of thick lava lakes (from Arndt (1986b; Barnes et al. (1988) and personal observations in other parts of the Abitibi belt). Two types of vein are shown. The first has diffuse contacts with the surrounding olivine porphyritic rock that comprises most the upper border of the lava lake, and probable formed by segregation, into a fracture, of interstitial liquid from incompletely crystallized surrounding rock. This vein is asymmetric and displays a downward increase in the size of the spinifex olivine crystals. The second type of vein has sharp contacts with the surrounding rock and probably formed by injection of liquid into largely or completely solid rock. (c) Spinifex-textured sills (from observations in Dundonald Township, Ontario). A complete spinifex-textured sill and the base and top of two others are shown. These sills are layered, with spinifex upper sections and cumulate lower sections, but they lack the fractured upper contacts that are used to identify lava flows. An olivine porphyritic sill is shown intruding the lowermost sill.

sections. Although there are minor differences in mineralogy and trace element chemistry, their major-element compositions are similar: both units apparently crystallized from liquids with 15–20% MgO. Fred's Flow has a remarkable spinifex layer, some 20 m thick and changing from olivine spinifex at the top to pyroxene spinifex at the base. In Theo's Flow we see no such zone: at the top, below a breccia and in a position equivalent to that of Fred's Flow's spinifex, there is just a thick layer of aphanitic, pyroxene-rich rock.

(b) *The basalts of Gilmour Island, Hudson Bay, Canada.* The petrology and chemical compositions of these Proterozoic rocks have been described by Baragar and Lamontagne (1980) and Arndt (1982). Two types of magnesian basalt are found on the island. The komatiitic basalts have spinifex textures: the tholeiitic basalts, which are recognized on the basis of slightly different trace-element and isotopic compositions but similar bulk compositions, do not. The presence or absence of spinifex can neither be correlated with eruptive environment nor with the morphological features of the flows. The two types of flow are interlayered, and have similar thicknesses, lateral continuity, internal layering and grain sizes (except for the spinifex sections).

There are other puzzling aspects of spinifex textures. Why, when a fracture forms in the interior of a komatiitic flow and fills with komatiite liquid, does it crystallize to spinifex texture? As explained above, the veins formed in this manner are more-or-less symmetrical, and show evidence of nucleation at both margins and inward growth of spinifex crystals (Fig. 1c). Similar veins in tholeiitic units crystallize to porphyritic or aphanitic rock. Even more intriguing are the flows or sills that have skeletal or spinifex textures throughout, and apparently are non-differentiated. Examples include the remarkable ~100 m flow or sill from Murphy Well, in the Yilgarn Craton of Australia, which has a uniform composition from top to bottom and skeletal olivine textures throughout (Lewis and Williams, 1973), and certain komatiite flows from Gorgona island, which consist almost entirely of spinifex texture beneath a thin flow top breccia (Aitken and Echeverría, 1984; Echeverría, 1980).

Donaldson (1982) stated that two conditions are necessary for the formation of plate spinifex texture "(a) a thermal and/or compositional gradient in the magma along which the elongate crystals grow competitively (and hence mutually parallel) in constrained fashion, and (b) absence of new nuclei". He noted that the first condition is likely to be met close to the margins of komatiitic flows, but he was unable to satisfactorily explain an absence of nuclei.

Perhaps the key lies in the earlier history of the komatiite magmas. Donaldson (1979) and Lofgren (1980, 1983) have shown experimentally that a period of superheating strongly influences the subsequent crystallization history of a silicate liquid. The process of heating a silicate melt well above its liquidus apparently breaks down the structure of the liquid, depolymerizing it, and destroying the chains and networks that act as nuclei during crystallization on subsequent cooling. A liquid subjected to this type of treatment crystallizes quite differently from one that was never superheated. Superheated liquids display a reluctance to nucleate when cooled and the crystals that do form tend to be few, large and skeletal. With rapid cooling, phenocrysts may form, but textures in the matrix are very different from those resulting from the cooling of non-superheated liquids.

Komatiitic liquids show a disinclination to nucleate, and a strong tendency towards heterogeneous nucleation. When phenocrysts are present, they act as the sites of olivine growth: in parts of flows without phenocrysts, as in the upper parts of layered flows or in veins, olivine or pyroxene grows into the liquid from quenched or solid margins. There can be no doubt that much of this behavior is due to the inherent characteristics of the Mg-rich, Si- and Al-poor komatiite liquids, particularly their low viscosities and non-polymerized structure that allow rapid diffusion and rapid growth of olivine or pyroxene. However, the particular features of the komatiitic units mentioned above suggests that this may not be the whole story, and, as suggested earlier by several authors (e.g. Nesbitt, 1971; Donaldson, 1979; Lofgren, 1983; Aitken and Echeverría, 1984; Lesher and Groves, 1986), an earlier period of superheating may be important. It will be shown later in this chapter that komatiite magma follows a path through the mantle that takes it to temperatures well above the liquidus. As the magma approaches or reaches the surface, the cooling rate accelerates and the magma starts to crystallize. The period of superheating has the effect that nucleation is inhibited, relatively few phenocrysts form, and heterogeneous nucleation on quenched margins is favored. Spinifex texture is the consequence.

LAYERING

According to Donaldson (1982) "perhaps the most appealing aspect of spinifex-textured (komatiitic) cooling units is that they are petrographically and chemically layered". Many, though not all, komatiite flows have upper spinifex-textured layers and lower layers of olivine cumulate (Fig. 3a). Viljoen and Viljoen (1969a,b) recognized the two main divisions in their classic papers on the Barberton Mountain Land, but the relationship of these layers to one another only became clear when Pyke et al. (1973) studied the better-exposed examples from Munro Township.

The origin of layering is a subject of great prominence in the brief history of komatiite studies. Summaries of ideas have been presented by Donaldson (1982) and Bédard (1987), and it is from these sources that much of the following material is drawn. The issue was first discussed by Pyke et al. (1973) and later by Arndt et al. (1977) who proposed that the lower cumulate layer formed by gravitative settling of phenocryst olivine that had been transported to the site of crystallization in the lava, or had crystallized after emplacement. This left the upper part of the flow free of phenocrysts, and this portion of the flow then crystallized to spinifex texture. Lajoie and Gélinas (1978) proposed that the two units solidified in the opposite order. They argued that the spinifex layer of flows from Lamotte Township in Quebec solidified first and that the olivines of the cumulate layer were later deposited from liquid that flowed beneath the thickened crust. Donaldson (1982) followed his review of these early models with one of his own. He was impressed by a correlation between the habits of olivine grains and their position in the flow. Olivine with a habit corresponding to high cooling rates was restricted to the flow margins. Deeper in the flow interior the habit changes, from skeletal near the top to solid polyhedral in the interior, corresponded to a systematic decline in cooling rate. He proposed that olivine crystallized simultaneously throughout the flow, and that the marked difference in textures of the spinifex and cumulate units was largely due to contrasting cooling rates.

A new phase came with the publication of a paper by Turner et al. (1986) in which it was proposed that the liquid within a komatiite flow should convect turbulently with a vigor sufficient to inhibit the settling of olivine grains. This led to the models of Turner et al. (1986) and Arndt (1986a) in which the spinifex layer was said to grow downwards from a crust, concentrating the suspended olivine phenocrysts in the lower part of the flow. When the viscosity of the phenocryst-charged liquid exceeded a critical level, convection ceased and the flow solidified as a layered unit.

The key to the Turner et al. model is the vigor of convection within the flow. This is an important parameter because it also controls the rate at which the flow will cool, and thus has a strong influence on the extent of possible differentiation and the types of textures that might form. The first estimates of cooling rates by Donaldson (1982) were based on conductive cooling of basaltic units. He quoted 1–0.5°C h^{-1} for lava just beneath the surface of a 5–20 m thick flow, a figure similar to that obtained by Usselman et al. (1979) in a more sophisticated treatment. These values were questioned by Turner et al. (1986) who argued that the treatments failed to take into account convection, which would be particularly vigorous in hot, low-viscosity komatiite liquids, and would lead to greatly enhanced cooling rates. They calculated rates of 1–100°C h^{-1} soon after emplacement of a 1600°C komatiite. Most recently Renner (1989) and Cheadle et al. (in prep.) reconsidered the role of convection. These authors studied in detail the komatiite flows of the Reliance Formation, Belingwe, Zimbabwe, and showed that textural, mineralogical and chemical variations within these flows were best explained by cooling of an stagnant, non-convecting liquid. Their calculated cooling rates during the growth of the spinifex-textured crust were less than 1°C h^{-1}.

A critical parameter in this discussion is the temperature variation immediately below the crust of the flow, because it is this variation (δT) that is responsible for the density difference that drives convection. If the crust is thin and the temperature gradient across it steep, the temperature in the liquid immediately below the crust will be far lower than in the interior of the flow, and this will lead to vigorous convection in a low-viscosity liquid like komatiite. One of many special features of komatiite magmas is the very large interval between liquidus and solidus. For typical basalts the interval is about 100°C, but for a komatiite with 28% MgO, the figure is closer to 500°C (Arndt, 1976). This is important, because there is a direct relationship between this temperature interval and the thickness of a partially-liquid zone at the top of a cooling lava flow, between the completely solid crust and the completely liquid interior. In komatiites the partially-liquid zone will be exceptionally thick. In their quantitative model Turner et al. assumed that crystallization occurs only at the eutectic temperature. Although this simplified the calculations, it probably is not appropriate for komatiite. In a eutectic system there will be no partially-liquid zone: the crust thickness will be small and the temperature difference that drives convection large. Turner et al. discussed the effect of a zone of dendritic crystals growing downwards from a crust, but did not model this situation quantitatively.

In the Renner–Cheadle treatments, the partially-liquid zone is thick enough to diminish radically the temperature gradient across the crust. In her quantitative treatment of the solidification of Belingwe komatiites, Renner (1989) used very low values for δT (0–50), which led to cooling sequences in which convection was sluggish or absent. Under these conditions all olivine phenocrysts that were present in the lava at the time of emplacement or which grew following ponding settled to the base of the flow to form the cumulate lower while spinifex olivines grew down from the crust. The last part of the flow to solidify was the base of the spinifex layer. The model is thus very similar, in all essential aspects, to the original scheme of Pyke et al. (1973): our ideas have turned full circle.

But is this the last word? There is little to criticize in the application of the Renner–Cheadle modeling of the Zimbabwe flows, but it has to be asked whether their conclusions apply to komatiites in general. In addition to their remarkable freshness (Nisbet et al., 1987), the Zvishevane flows are peculiar in two important respects: (a) they contained an unusually large proportion of phenocryst olivine at the time of emplacement, and (b) they formed from relatively low-MgO liquids. Work is currently underway to establish how great an influence these differences might have on the solidification history. Preliminary results suggest during the solidification of the more magnesian Alexo flow, convection may have been important. Cooling rates were high, and olivine phenocrysts may have been suspended in the interior of the flow for part of the cooling history. The last part of flow to remain fluid was the B1 layer (Figs. 1a and 2e), as evidenced by flow textures illustrated by Arndt (1986a).

It can be concluded that the cooling history of komatiite flows is variable, and depends critically on the initial composition of the liquid and phenocryst content, factors that probably are controlled by conditions during mantle melting and on the path followed by the komatiite on its way to the site of emplacement.

CHEMICAL COMPOSITIONS

Effects of alteration

Komatiites are extinct. The last eruption we know of was some 80 Ma ago, on Gorgona Island (Echeverría, 1980; Aitken and Echeverría, 1984), and although recent sightings are claimed with some regularity in the literature, none is convincing. When dealing with the chemical compositions of modern basalts, andesites, phonolites, even carbonatites, a natural step is to obtain samples of newly erupted lavas which are little altered and have compositions like those of the original magmas. We know of no newly-erupted komatiites, and even the youngest examples from Gorgona Island have been affected by the circulation of hydrothermal fluids. They contain secondary hydrous minerals in veins and as pseudomorphic replacement of olivine and glass. Chemical analyses typically have 5% or more H_2O, and variable ferrous/ferric iron ratios. Such features would be grounds for rejection from most chemical data banks of modern igneous rocks. The freshest Archean komatiites, those from the Zvishavane area in Zimbabwe and Alexo in Abitibi, are only slightly more altered than the Gorgona rocks, but most other Precambrian komatiites are less well preserved. It must be accepted that all komatiites are altered to a greater or lesser extent, and any investigation of their geochemistry has to penetrate a veil of alteration.

A special characteristic of komatiites allows this to be done, and indeed by using this feature, the chemistry of komatiites can often be interpreted more confidently than that of all other Precambrian or many Phanerozoic volcanic rocks. In komatiites the only important liquidus mineral is olivine. (Chromite also crystallizes, but in amounts so small that it influences only the Cr concentrations and has no significant effect on other elements). Because of the low viscosities of komatiites and the large interval between the liquidus and the temperatures at which other silicate phases appear, olivine crystallizes alone and readily segregates, and most komatiite flows display a wide range of compositions produced by olivine fractionation and accumulation. Thermal erosion and wall-rock contamination can upset such trends, but the effects of this process normally are minor, as discussed in a later section.

The olivine control lines produced by fractionation and accumulation provide a valuable tool for monitoring the effects of alteration: if an element plots on a control line it probably was immobile; if it scatters or plots on a trend oblique to the control line, the element is likely to have been influenced by alteration. In the following section I discuss the effects of olivine fractionation and the use of the olivine-control-line tool, and then, having established which elements are immobile and reliable, go on to a more general discussion of komatiite geochemistry.

Olivine fractionation in komatiites

Olivine has fractionated or accumulated in all komatiite samples, and chemical variations due to these processes normally far outweigh the effects of mantle melting, contamination or alteration. MgO contents in samples from individual differentiated, spinifex-textured komatiite flows vary extremely widely, by as much as 34% in some cases. In the Alexo komatiite flow of Ontario which was studied by Barnes (1983) and Arndt (1986a), olivine-rich cumulates have up to 42% MgO and the most evolved spinifex-textured lavas only 19% (Fig. 4). The lowermost flow in the Kambalda komatiite sequence (Lesher, 1983; Arndt and Lesher, 1992) has MgO contents ranging from 16 to near 50% MgO, and thinner differentiated flows exposed in many greenstone belts have 20–40% MgO. Because of these large variations, it is usually a simple matter to collect a suite of samples across a flow or suite of flows and use them as a basis for monitoring the extent of olivine fractionation and the effects of alteration.

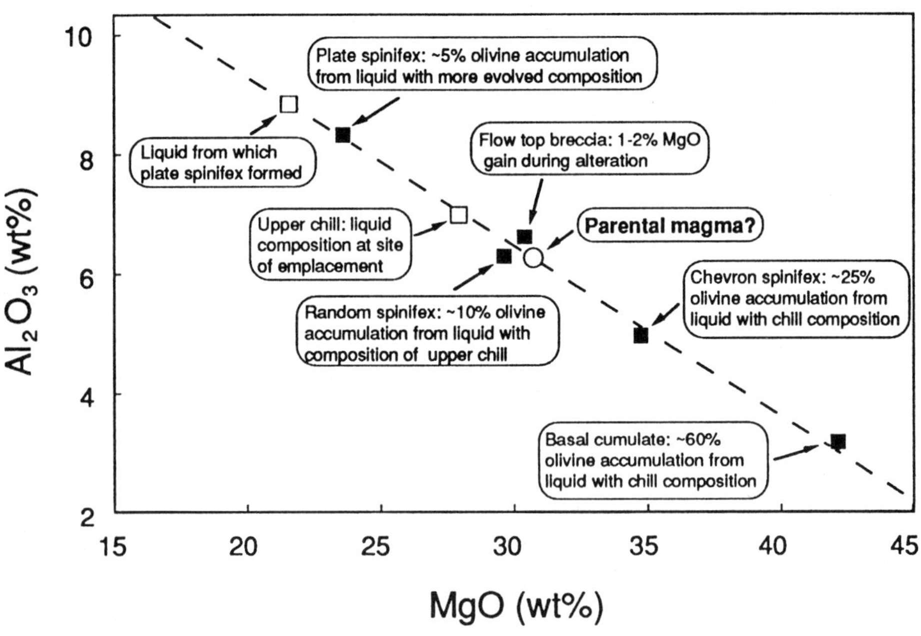

Fig. 4. Diagram illustrating the variations in MgO composition produced by the accumulation and fractional crystallization of olivine in the Alexo komatiite flow. The filled squares indicate the compositions of samples from the flow, the open symbols give calculated or estimated compositions of the parental liquid, the liquid at the site of emplacement and the evolved liquid from which the lower plate-spinifex-textured rock crystallized. The analytical data and the procedure used to calculate liquid compositions are given by Arndt (1986).

Mobile and immobile elements: the olivine control line criterion

The most convenient way to represent chemical data from komatiites is to plot variation diagrams with MgO on the x-axis. MgO, which is a relatively immobile element in komatiites (as demonstrated below), varies widely and systematically with olivine fractionation and accumulation (Figs. 4–6), and competing petrogenetic processes such as the fractionation or accumulation of other minerals, or contamination and alteration, produce trends that are oblique to olivine fractionation trends and easily recognized. Figure 4 is a plot of modal olivine vs. MgO for a suite of well-preserved Zvishavane komatiites. The data correlate reasonably well and demonstrate the relationship between MgO and olivine content. In this example, the initial liquid had 19–20% MgO (Renner, 1989; Renner et al., 1994): about 15% olivine fractionated from the liquid to produce the most evolved spinifex-textured lavas, and about 25% olivine accumulated in the most magnesian olivine cumulates. The scatter about the trend is partly due to biases in the modal analyses arising from the difficulties in counting coarse-grained spinifex-textured samples, but is also influenced by variations in cooling rate that affected the amount of occult olivine, and hence the amount of MgO, in the original glass.

In Fig. 6, variation diagrams are plotted for the same sample suite. Consider the plot of MgO vs Al_2O_3. The data fall on a well-defined linear trend ($r^2 = 0.998$) which intercepts the MgO axis at 48.6±0.5%. Because olivine contains insignificant Al_2O_3 (<0.3 wt.), the MgO intercept should give the MgO content of the olivine that fractionated and accumulated. The value, which can be read off Fig. 6,

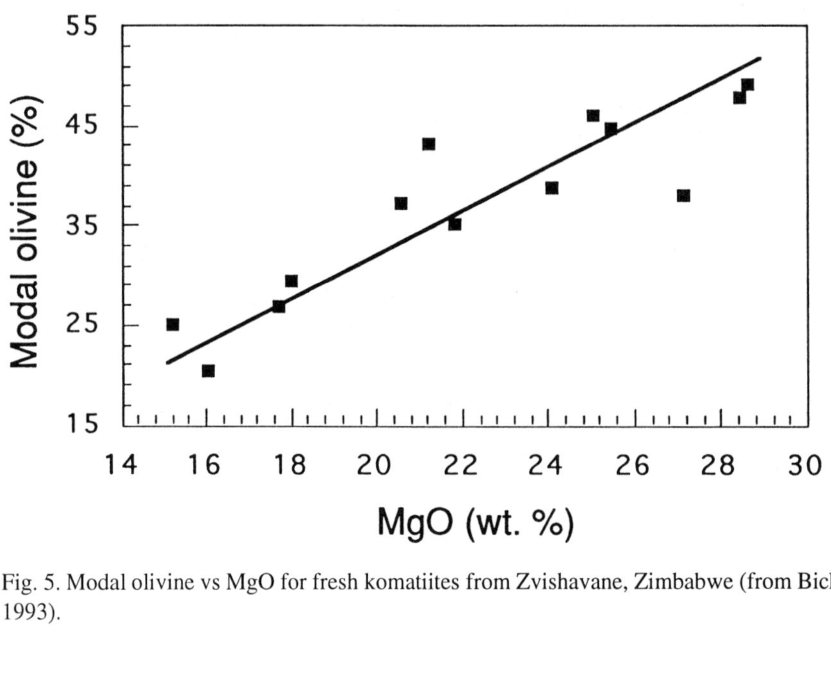

Fig. 5. Modal olivine vs MgO for fresh komatiites from Zvishavane, Zimbabwe (from Bickle et al., 1993).

agrees within error with the average MgO content of olivine (Fo_{89}) that crystallized in the flow. The amounts of fractionated and accumulated olivine, which can be calculated assuming incompatible behavior of Al, are consistent with the measured variations in modal olivine given in Fig. 6. In all respects the variations of MgO and Al_2O_3 contents are consistent with olivine control: it can be concluded that in these komatiite flows *both* these elements were essentially unaffected by hydrothermal alteration, metamorphism and other secondary processes.

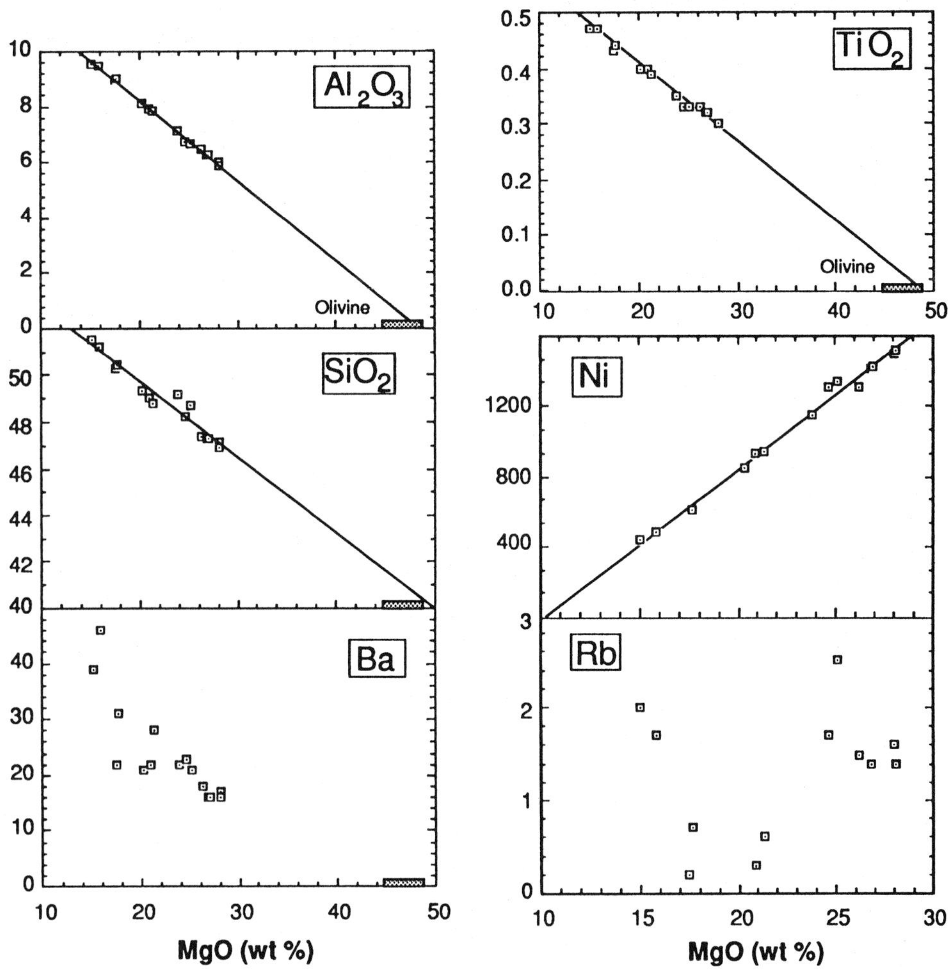

Fig. 6. Variation diagrams for fresh komatiites from Zvishavane, Zimbabwe (data from Bickle et al., 1993 and Nisbet et al., 1987). Al_2O_3, TiO_2, SiO_2 and Ni behave as immobile elements and plot on tightly constrained trends that intersect the olivine composition. Ba and Rb are mobile and scatter broadly.

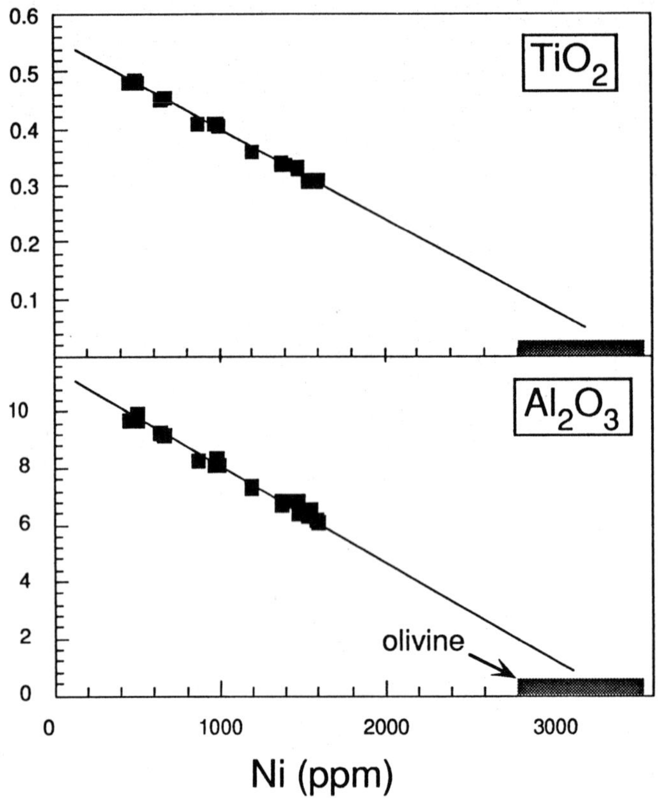

Fig. 7. Ni vs Al_2O_3 and TiO_2 diagrams for fresh komatiites from Zvishavane, Zimbabwe (data from Bickle et al., 1993 and Nisbet et al., 1987). Further evidence of immobile behavior of these elements.

Similar conclusions can be reached for other incompatible elements plotted in Fig. 6. These comprise elements normally thought of as impervious to hydrothermal processes — Al and Ti amongst the major elements, and trace elements such as the rare earth (REE) and high-field-strength elements (HFSE).

Immobility for an element compatible with olivine can be demonstrated by substituting it for MgO and comparing the X-axis intercept with concentration of the element measured in olivine. Data from the Zvishavane flows plot on tight trends in the Ni vs Al_2O_3 and Ni vs TiO_2 diagrams (Fig. 7) with an intercept at 3400 ppm, a Ni concentration typical of those measured in olivine phenocrysts. This approach can only be used for elements abundant enough to be analyzed with the microprobe. For others such as the platinum group elements (PGE), immobile behavior can only be inferred on the basis of limited scatter in variation diagrams. In the case of Cr, simple olivine control can only be assumed in the most magnesian komatiites. In others chromite coprecipitates with olivine in sufficient quantities to control the behavior of Cr, and possible affects of alteration are not easily monitored.

Mobile elements

Element mobility takes two forms. Many elements scatter widely in variation diagrams and show no relationship to olivine control lines. A good example is Rb whose concentrations range from 0.1 to 2.5 ppm and do not correlate with MgO (Fig. 6). Isotopic studies by E. Hegner (reported in Nisbet et al., 1987) demonstrate that Rb was lost from many samples from the Zvishavane komatiites. Potassium is similarly mobile. In contrast to the situation in other more altered komatiites (see below), Na_2O plots on a fairly tight trend (Fig. 8) with an intercept on the MgO axis at about 48%: the trend also passes through the composition of fresh glass trapped in inclusions in olivine grains (Nisbet et al., 1987; McDonough and Ireland, 1993). Some other mobile elements define tight linear trends but have MgO intercepts far from the appropriate olivine composition. CaO, for example, has an anomalously high intercept at about 56% MgO (not shown), and Ba and Sr have intercepts that are too low (Fig. 6). A peculiarity of the Zvishevane data is the mobile behavior of Ce. When plotted as part of a chondrite-normalized REE spectrum, pronounced positive and negative Ce anomalies are observed. Bickle et al. (1993) and Renner et al. (1994) suggest that these komatiites may have been subaerial, a characteristic that might explain their unusual freshness. Recent circulation of groundwaters through these unusually fresh rocks then caused some redistribution of Ce.

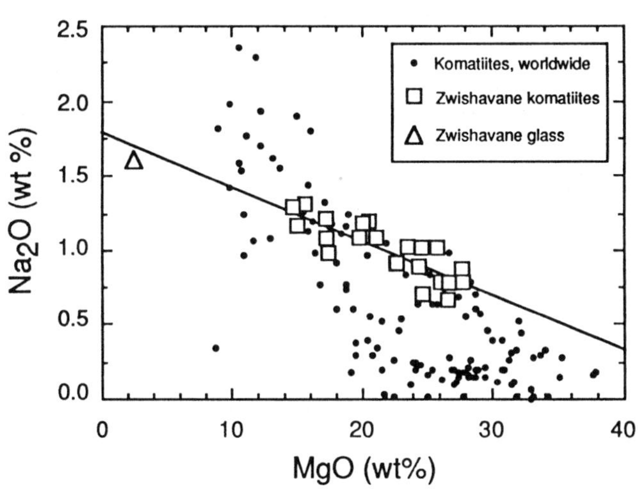

Fig. 8: Diagram of MgO vs Na_2O. The data from the relatively well preserved Zvishavane flows form a trend that intersects the olivine composition (not shown) and passes close to the composition of fresh glass in inclusions in olivine phenocrysts (data from Bickle et al., 1993; Nisbet et al., 1987; McDonough and Ireland, 1993). The majority of komatiites from other localities plot below this trend: they appear to have lost 50–80% of their original Na_2O content.

More altered komatiites

The behavior of elements in more altered komatiites is similar in many respects to that in the Zvishavane komatiites. As illustrated in Fig. 9 and discussed by Beswick (1982), Smith and Erlank (1982), Arndt and Nesbitt (1982), Barnes (1983), Bickle et al. (1993), elements such as Al, Ti, Zr, most REE and Ni are usually immobile and Rb, Ba, Ca, Si and Eu are mobile. Conspicuous differences

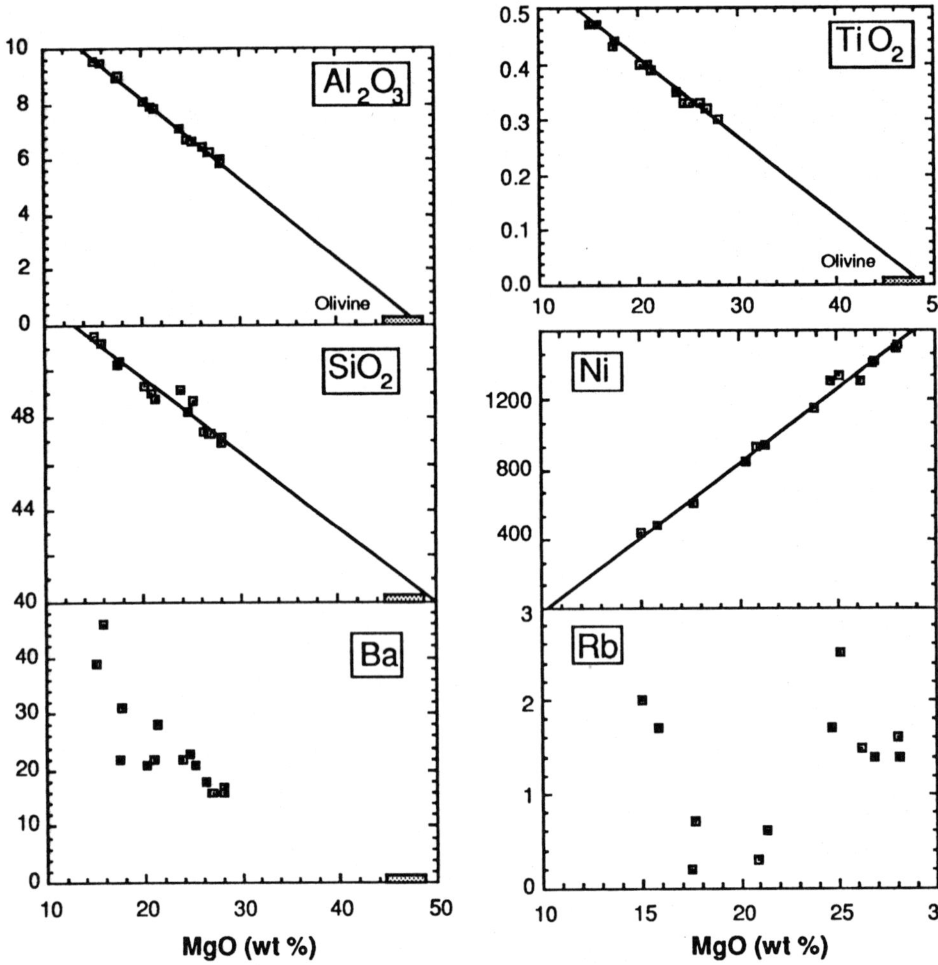

Fig. 9. Variation diagrams for more altered komatiites from the Abitibi Belt. Al_2O_3 and TiO_2 exhibit immobile behavior but SiO_2, Na_2O and Rb scatter and were mobile. CaO plots on a moderately well defined trend but the intersection with the MgO axis is too low because of Ca loss from the more magnesian samples. Date from Arndt and Nesbitt (1984), Arndt (1986a), Barnes (1983) and Arndt (unpublished).

are seen only for Na. Whereas this element appears to have been relatively immobile in the Zvishavane flows, data from most other komatiites scatter widely in variation diagrams and show no relationship to olivine control lines (Fig. 8). Many komatiites appear to have lost a large proportion (50–80%) of their original Na.

Mobility of Mg

In general MgO seems to have been relatively immobile in Archean komatiites. This is shown by the tight trends in diagrams of MgO vs various immobile elements, and by the appropriate intercepts with MgO axes in variation diagrams. Several authors (e.g. de Wit et al., 1983, 1987; Echeverría, 1982) have suggested, however, that high MgO contents in some komatiites may be due to Mg gain during interaction with sea water or hydrothermal fluids. De Wit et al. (1987) supported their interpretations with several arguments. They showed a positive correlation between MgO and H_2O, which they suggested might be a result of MgO uptake during hydration. In fact, this relationship probably arises from differences in the H_2O contents of the various secondary minerals that crystallize during alteration of komatiites of different compositions: serpentine and chlorite, which replace olivine in Mg-rich samples, have $H_2O > 10\%$, whereas amphibole and sodic plagioclase, which replace pyroxene and glass in Mg-poor samples, have $H_2O < 2\%$. The H_2O content correlates with the original olivine content, which in turn depends on MgO content, and a positive correlation between H_2O and MgO is entirely to be expected.

De Wit et al. (1987) plotted Mg/Fe ratios of olivine grains and host komatiites. In their compilation, only samples from Gorgona Island and Alexo have the relationship between olivine and host magma compositions predicted from Mg–Fe partitioning in mafic–ultramafic liquids (Roedder and Emslie, 1970; Bickle, 1982): in samples from Barberton and Munro Township, the komatiites appear to have anomalously high Mg/Fe ratios, a factor that the authors attributed to Mg gain during alteration. This argument conflicts with the disposition of whole-rock compositions of Barberton komatiites in variation diagrams. Figure 10 shows that the Barberton data plot on olivine control lines. If all these samples had gained a similar amount of MgO, the trend of the altered samples would have intercepted the x-axes at MgO values higher than those measured in olivines. Had some samples gained more MgO than others, the data would either scatter, or the trend would rotate. The only way of producing the correct MgO intercept in a suite of Mg-metasomatized samples is to have some form of "intelligent" MgO gain (to borrow a term from the Pb isotope geochemists): the amount of Mg gained must correlate systematically with the original Mg content, with less magnesian samples gaining more Mg through alteration than more magnesian samples. This type of behavior, which is illustrated in Fig. 11, seems most unlikely.

Further convincing evidence that komatiites retain their original MgO contents can be obtained by reconstructing major element compositions using modal analyses and estimated compositions of the igneous components. For example, the

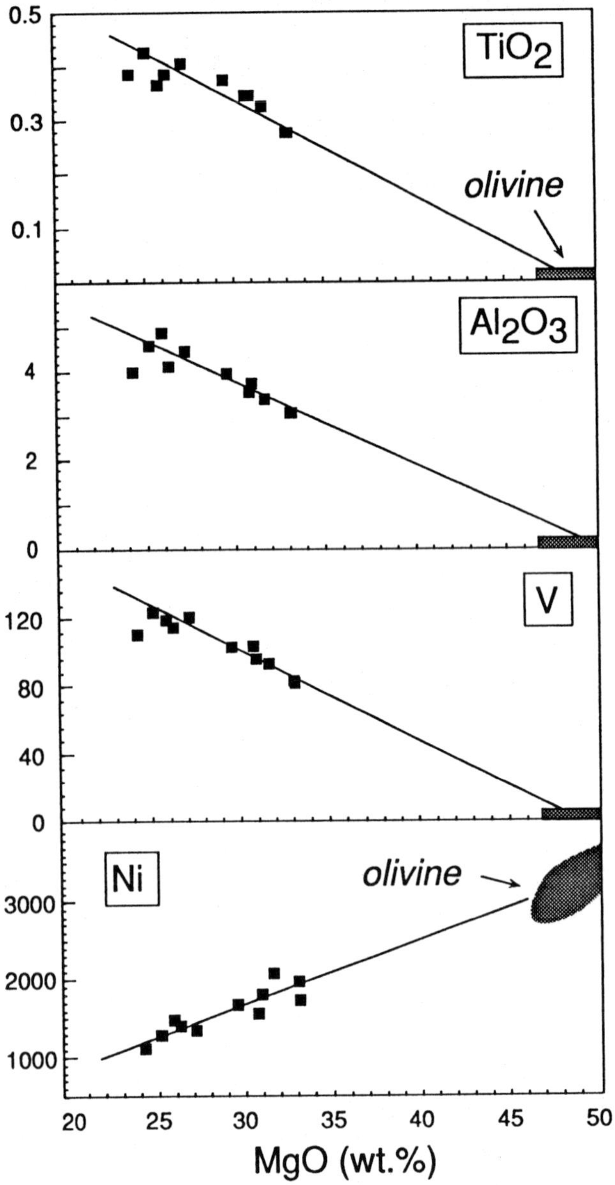

Fig. 10: Variation diagram for samples of aphyric komatiite from Barberton showing that they, too, plot on olivine control lines and apparently have not gained Mg. Data from Smith and Erlank (1982).

spinifex-textured komatiite M663 from Alexo contains about 55 modal percent olivine (Arndt, 1986a). The olivine, which has the composition Fo_{88-92}, and contains about 46–50% MgO (Arndt, 1986a), would have contributed about 0.55 $\times$ 48 = ~26% MgO to the rock composition, and the clinopyroxene and glass in the

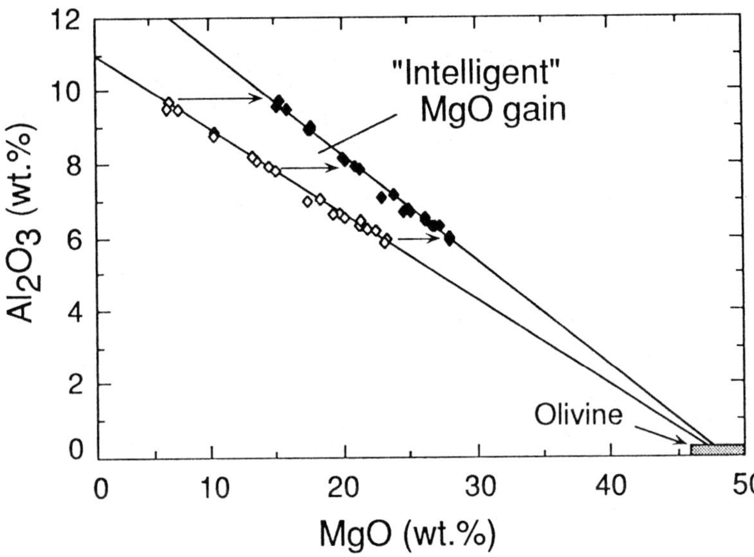

Fig. 11. Variation diagram illustrating the effects of MgO gain during alteration. The open symbols represent the measured compositions of samples from Munro Township (Arndt et al. 1977; Arndt and Nesbitt, 1982, 1984); the closed symbols represent samples to which MgO has been added. The data trend defined by the altered samples will intersect the olivine composition only if the amount of MgO gain is inversely proportional to MgO content, and even in this case the correct intercept will only be accidental.

matrix another ~2%, to bring the total to ~28% MgO. This value is very similar to the MgO content in the whole rock analysis, which is 28.4% MgO. A similar procedure also works for more altered samples, such as the Crixas komatiites (Arndt et al., 1989), and can be applied to any sample in which the modal composition can be measured, including the Barberton samples.

The apparent discrepancy between olivine compositions and the Mg-Fe ratios of komatiites, as highlighted by de Wit et al. (1987), probably results from several factors: (a) accumulation of olivine in many of samples; (b) a failure to analyze the most magnesian olivines in certain samples; (c) a decrease of Mg/Fe of some olivine during metamorphism.

On the other hand, it cannot be denied that individual samples in some komatiite suites have gained or lost a small amount of Mg. In the Alexo flow, for example, a sample from the flow-top breccia and another from the olivine cumulate layer plot below the trends defined by all other samples in most variation diagrams Arndt (1986a): these samples are very highly altered and contain no primary minerals, and both appear to have lost 1–2% MgO. Samples from the basal komatiite flow at Kambalda, Western Australia, plot on a reasonably well defined line with an appropriate MgO intercept, but in each diagram certain samples lie a little above the line and others lie below the line (Arndt and Lesher, 1992). This behavior may result from minor Mg mobility: samples above the lines have gained 1–2% MgO and those below have lost a similar amount.

Other types of mobile element behavior

Some studies of komatiite flows have demonstrated mobile element behavior more sporadic and less systematic than that discussed above. To illustrate this, a few examples will be presented.

(a) In the central part of the Alexo komatiite flow and in parts of a neighboring basaltic flow (Barnes, 1983; Arndt, 1986a; Lahaye et al., 1993), alkali elements, Sr and SiO_2 are strongly depleted and CaO is enriched. This type of behavior appears associated with a peculiar type of hydrothermal alteration that led to the formation of secondary calcic pyroxene and grossular (Arndt, 1977; Lahaye et al., 1993).

(b) In volcanic rocks from the Kambalda region, Australia, mainly in komatiitic basalts but also in some komatiites, the formation of ocelli has resulted in drastic changes in the abundances of many elements (Arndt and Jenner, 1986). Ocelli are small spherical to elliptical patches of relatively felsic material that lie in a more mafic matrix. They are strongly enriched in SiO_2, Na_2O, Sr and HREE, and are depleted in MgO and FeO relative to the matrix. Although it is not clear whether the ocelli are primary structures (produced by liquid immiscibility?) or form during later alteration, it is clear that their formation was associated with profound changes in the compositions of the lavas (Arndt and Jenner, 1986).

(c) In other komatiites, alteration has resulted in wholesale replacement of the original components. In many regions the influx of CO_2-rich fluids results in partial to complete carbonatization which is accompanied by drastic changes in the chemistry of the samples (e.g. Pyke, 1975; Tourpin et al., 1991). In southern Africa, some komatiites have been almost entirely replaced by quartz and sericite producing rocks, which in many cases retain their spinifex textures, with up to 90% SiO_2 (Duchac and Hanor, 1987).

The effects of alteration on the chemical and isotopic compositions of komatiites were highlighted by studies by Arndt et al. (1989) in the Crixas greenstone belt of Brazil and by Gruau et al. (1991) and Tourpin et al. (1991) in the greenstone belts of Finland. These studies demonstrated that even in rocks in which volcanic textures are preserved, the passage of fluids (probably CO_2-rich) can cause complete mobilization of most major and trace elements, including the reputably immobile Al and Ti, REE and high-field-strength elements. In the Finnish komatiites this alteration took place around 1800 Ma, some 900 Ma after the time of emplacement of the rocks, and had the effect of erasing all isotopic memory of the original magmatic compositions of the rocks (Gruau et al., 1991).

Crustal contamination

Another complicating factor is the tendency of komatiites to assimilate crustal rocks. Nisbet (1982), Huppert et al. (1984) and Huppert and Sparks (1985) pointed out that a low-viscosity liquid like komatiite flows turbulently. Heat is transmitted efficiently from the very hot magma to the wall or floor rocks, causing them to melt and to become assimilated by the komatiite. Evidence for this process is found at Kambalda in the form of channels attributed to thermal erosion, distinctive chemical and isotopic compositions of komatiites and basalts consistent with crustal contamination (Chauvel et al., 1985; Arndt and Jenner, 1986; Lesher and Arndt, 1994), and, most convincingly, the presence of zircon xenocrysts in the contaminated komatiitic basalts (Compston et al., 1986). Detailed studies in other regions have also produced evidence of thermal erosion by komatiites (Davis and Lesher, 1993), and of contamination of komatiites, leading to the question of whether any komatiite preserves its original magmatic composition. This question was addressed by Jochum et al. (1990) in their detailed study of the trace-element compositions of a large suite of komatiites from all over the world. Jochum et al. showed that certain element ratios, such as Th/Nb or La/Nb, are very sensitive measures of contamination, and that when used in combination with Nd or Pb isotopic data, they can effectively discriminate between contaminated and uncontaminated rocks. Using this technique, it is possible to select suites of komatiites essentially unaffected by the contamination process.

Conclusion: can we identify non-contaminated, unaltered komatiites?

The cases mentioned in the preceding sections include some examples of komatiites that are so strongly affected by alteration or contamination that they preserve next to nothing of their original magmatic compositions. They well illustrate the problems of interpreting the chemical and isotopic compositions of Archean volcanic rocks. However, they should not be used as a reason to write off all chemical data from komatiites. By carefully screening the data, the crust-contaminated komatiites can be eliminated and a distinction can be made between mobile and immobile elements in each komatiite suite. Such screening is usually only possible if the original volcanic structures and textures are preserved. The ability to select suites of samples from individual flows is invaluable because it allows the olivine-control-line tool to be used with confidence. This cannot be done with highly metamorphosed rocks such as amphibolites, serpentinites and talc schists from high-grade polymetamorphic terranes, and any chemical data from such suites must be treated to extreme caution. However, when screening is carried out, it is found that certain komatiites from almost every greenstone belt (though not from supracrustal belts in high-grade terranes), retain enough of their magmatic chemistry to characterize their initial chemical and isotopic compositions. By selecting suites such as those from Munro and Alexo in the Abitibi belt, Zvishevane in Zimbabwe, the Cape Smith belt in Canada, and with certain prudence the Barberton komatiites of South Africa and the Kambalda komatiites from Australia, a record of the chemical and isotopic evolution of komatiites and their mantle sources can be constructed.

CHEMICAL TYPES

The petrologically most important element in komatiites is aluminium, and variations CaO/Al_2O_3 (a somewhat unreliable parameter because of the mobility of Ca), or Al_2O_3/TiO_2 (better, because these elements are relatively immobile), form a integral part of all komatiite classifications. Most komatiites from the type area in the Barberton Greenstone Belt are characterized by high CaO/Al_2O_3 (>1) or low Al_2O_3/TiO_2 (10–15), and this characteristic formed part of Viljoen and Viljoen's (1969b) original definition of komatiite. Correlated with relatively low Al_2O_3/TiO_2 is a depletion in HREE (Figs. 12 and 13), which can be expressed as high Gd/Yb. Komatiites from most other regions have Al_2O_3/TiO_2 ratios around 20, a value that is close to the chondritic value and distinctly higher that of the rocks from the Barberton greenstone belt. The chondritic Al_2O_3/TiO_2 is accompanied by chondritic CaO/Al_2O_3 and flat HREE. The variations in these ratios forms the basis of a chemical subdivision of komatiites into either two or three groups. Nesbitt et al. (1979) divided komatiites into two groups: *Al-depleted komatiites* with low Al_2O_3/TiO_2 and depleted HREE; and *Al-undepleted komatiites* with chondritic Al_2O_3/TiO_2 and flat HREE. Jahn et al. (1982) recognized three types (illustrated in Fig. 7.14): *Group 1*, which correspond to the Al-undepleted komatiites; *Group 2*, the Al-depleted komatiites; and *Group 3*, a complement of the

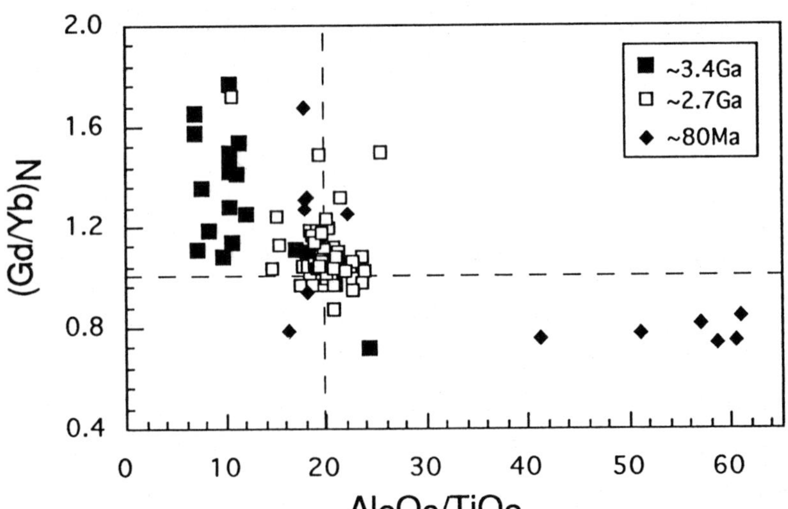

Fig. 12. Diagram showing that the older komatiites from Barberton and Pilbara have low Al/Ti and high Gd/Yb; late Archean komatiites have approximately chondritic ratios (as indicated by the dashed lines); and young komatiites from Gorgona Island have high Al/Ti and low Gd/Yb. The form of the diagram is from Jahn et al. (1982) but the data are newly plotted and come from a large number of sources. $(Gd/Yb)_N$ is the value normalized using the primitive mantle values of Hofmann (1988).

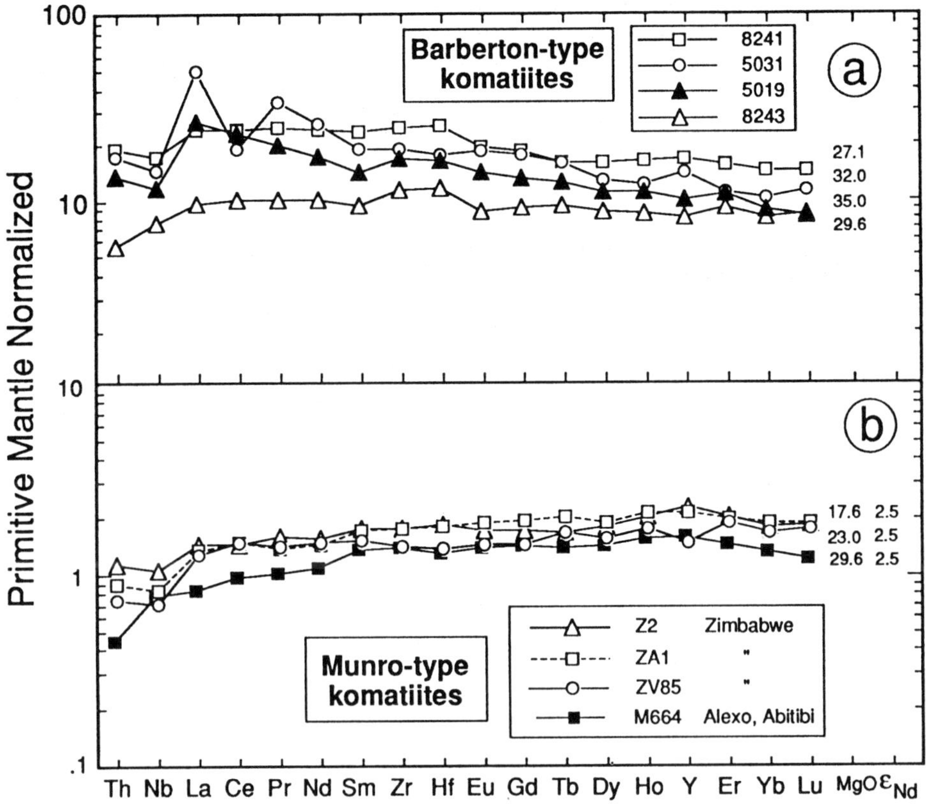

Fig. 13. Mantle-normalized trace element data for (a) komatiites from Barberton, South Africa (3.4 Ga); and (b) komatiites from Zimbabwe and Canada (2.7 Ga).

Al-depleted komatiites, with high Al_2O_3/TiO_2 and relatively enriched HREE (low Gd/Yb).

I have always found these terms unsatisfactory. "Al-undepleted" is an extremely awkward term. Some Al-depleted komatiites owe their low Al_2O_3/TiO_2 to high TiO_2 rather than low Al_2O_3. And finally I have never been able to remember which of Jahn et al. (1982) groups is which. In this chapter, I have therefore decided to follow a procedure adopted by several other authors and will call those komatiites with low Al_2O_3/TiO_2, and depleted HREE *Barberton-type* komatiites, and those with chondritic Al_2O_3/TiO_2 and HREE *Munro-type* komatiites.

It should be pointed out at this stage that there are several factors that complicate the simple correlations between CaO/Al_2O_3, Al_2O_3/TiO_2 and HREE. One is the mobility of Ca during hydrothermal alteration and metamorphism, which leads to spuriously high and low CaO/Al_2O_3 values and makes the ratio an unreliable measure of magma composition. A second factor is relative enrichment of moderately incompatible elements such as Ti and Gd by processes independent from those that cause the principal variation in Al_2O_3/TiO_2 or Gd/Yb. Certain komatiites from the Aldan Shield in USSR have high Gd/Yb, low Al_2O_3/TiO_2 but chondritic CaO/Al_2O_3 (Puchtel et al., 1993). There is no reason to believe that Ca have been lost systematically from these rocks, and the effect is attributed to low-degree melting or to enrichment of the source or magma with material that contains high Ti, LREE and other elements with similar geochemical characteristics.

Komatiite type seems to correlate with age (Jahn et al., 1982; Nesbitt et al., 1982; Gruau et al., 1990; Jahn 1990). As is shown in Fig. 12, Barberton-type komatiites predominate in ~3.4 Ga old terrains such as Barberton in South Africa and Pilbara in Australia, and Munro-type komatiites, although present in both these regions, are relatively rare. Conversely, in ~2.7 Ga old and younger terrains, Munro-type komatiites are the norm and Barberton-type komatiites are known from only a few localities (Newton Township, Ontario — Cattell and Arndt, 1987); Crixas, Brazil — Arndt et al., 1989). The Al-enriched types (Group III of Jahn et al. (1982)) are relatively rare but seem more abundant in older regions.

THE ORIGIN OF KOMATIITE MAGMA

It is now generally accepted that komatiites form by partial melting deep in the mantle, from sources that ascend from still greater depths. There are two principal lines of evidence. The first is the high temperature inferred for komatiite lavas on the basis of experimental studies. The eruption temperature of a komatiite with ~30 wt. MgO can estimated as around 1600°C using the relationship $T = 1000 + MgO \times 20$ from Nisbet (1982). As Jarvis and Campbell (1983) pointed out, and as has been confirmed by subsequent studies (Miller et al., 1991; Nisbet, 1993), it is most unlikely that magmas with such high temperatures arose from ambient mantle. More probably their source was anomalously hot, as in the central conduit of a mantle plume (Campbell et al., 1989). The relationship between the path followed by an ascending komatiite magma and the liquidus and solidus of mantle peridotite is shown in Fig. 14. Simple analysis of this diagram shows that in the most favorable case — that of a magma that separates completely from its source and rises to the surface with no further interaction with wall rocks — a komatiite magma with 30 wt. MgO and an eruption temperature of 1600°C should have segregated from its source at a depth of over 200 km, and this source would have started to melt at some depth greater than 400 km. The source would have been 300–400°C hotter than ambient convecting mantle, and, as argued first by Jarvis and Campbell (1983), it may have come from a deeper thermal boundary layer, perhaps at 670 km, or more probably at the core–mantle boundary.

An important factor in this discussion is the density of komatiite magma. Several important experimental studies (Rigden et al., 1984; Agee and Walker,

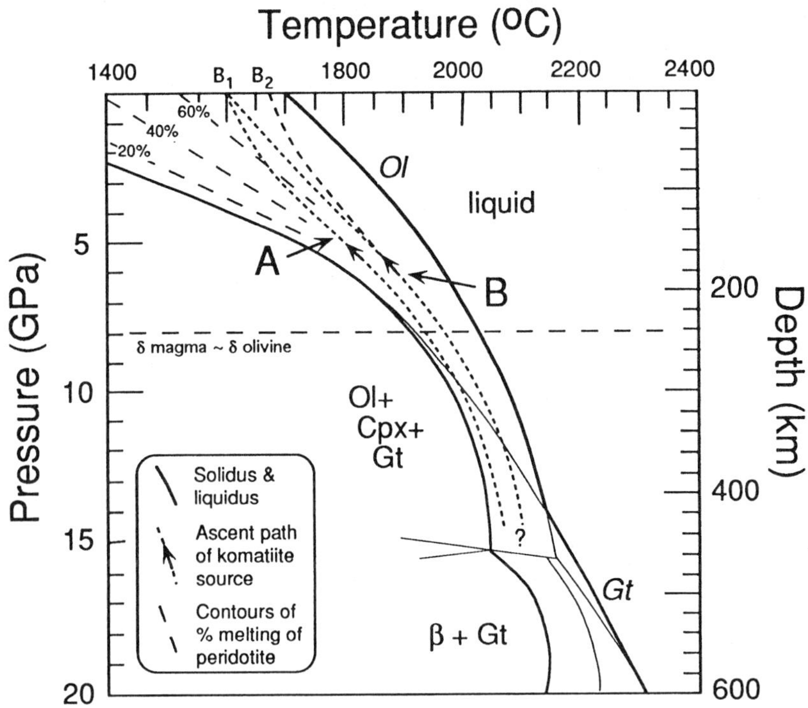

Fig. 14: Schematic diagram illustrating the manner in which komatiite might form and ascend to the surface. Two possible ascent paths are shown. The first (A) corresponds to a magma formed by about 20% partial melting at 500 km depth and follows an adiabatic path to the surface. When this magma reaches the surface, it has a temperature of 1600°C but is some 300°C above its liquidus, which can be estimated from the position of the 20% melting contour (from Hirose and Kushiro, 1993). To produce a magma that erupts at a liquidus temperature requires a higher degree of partial melting at depth and loss of heat to the surroundings during ascent, as shown by the path B. Phase relations for mantle peridotite are based on the experimental data of Herzberg et al. (1990 and unpublished) and the ascent paths are adapted from the study of Nisbet et al. (1993).

1988; Miller et al., 1991) have demonstrated that at pressures greater than about 8 GPa, the density of komatiite may exceed that of mantle minerals such as olivine and pyroxene. In the example shown in Fig. 14 there would be little tendency for liquid to segregate until the mantle source reaches the limit of neutral buoyancy, shown by the line labeled δ magma ~ δ olivine. During this passage the level of partial melting increases progressively, to reach 30% or more. This figure is consistent with the high levels of partial melting predicted for komatiite on the basis of modeling using major and trace elements.

The second argument for deep melting is the major and trace element patterns in certain komatiites, which point to the segregation of high-pressure phases. The depletion of Al and HREE in Barberton-type komatiites is best explained by the fractionation of majorite garnet, a phase that is on the liquidus of komatiite only at pressures greater than about 10–12 GPa (Green, 1975, 1981; Herzberg, 1983, 1992; Herzberg and Ohtani, 1988; Wei et al., 1990).

Detailed analysis of Fig. 14. reveals additional complications. Magma A, which forms by segregation of a 20% partial melt at 240 km, would reach the surface with a temperature of about 1600°C. This temperature is about 250°C above the liquidus temperature of the magma, which can be estimated from the 20% melting contour in Fig. 14: ~1300°C at 0 GPa. This magma is a superheated picrite, not a komatiite. True komatiites form at higher degrees of melting or at greater depths. Path B–B_1 is that of a magma that forms by about 60% partial melting but interacts with wall rocks or with shallower-level melts as it ascends from the site of initial magma segregation to the surface. This magma also reaches the surface with a temperature of 1600°C. Path B–B_2 represents a magma that segregates around 200 km then follows an adiabatic gradient. It reaches the surface with a temperature around 1680°C, some 80°C above its liquidus of about 1600°C. This magma is a superheated komatiite. The probable path of the most magnesian komatiites probably lies between the two.

In an earlier section, it was explained how spinifex and other textures peculiar to komatiites can only be explained if these magmas underwent an earlier period of superheating. The treatment of the magma ascent paths provides a basis for this theory. According to Huppert et al. (1985), komatiite magmas, once segregated from their mantle source, would ascend rapidly through the overlying mantle. They traverse the lithosphere in fractures, their ascent driven by buoyancy difference between magma and wall rocks. For a low-viscosity liquid like komatiite, ascent velocities probably are of the order of meters per second. Even at this rate, a komatiite would take several days to move from its source to the surface, and for much of this passage it would be superheated. Lofgren (1983) has shown that heating for several hours at temperatures only 50–100°C above the liquidus is sufficient to destroy the crystal embryos that facilitate homogeneous nucleation. The same process acting on komatiite is probably sufficient to strongly influence the subsequent crystallization pattern of the komatiite, and to produce spinifex textures and other features characteristic of these ultramafic lavas.

KOMATIITES AS MANTLE WITNESSES

The final point I wish to discuss is the extent to which the composition of a komatiite can be taken to represent that of the Archean mantle. In past years many attempts have been made to estimate the isotopic and trace-element composition of Archean depleted mantle, and these estimates have fueled debate on the nature of the earliest continental crust and the rate at which it grew. In the estimation of depleted mantle compositions, a large proportion of the data are from komatiites, and the assumption is made that the source of these rocks is equivalent to modern depleted upper mantle. The problems in screening the komatiite data to eliminate samples affected by alteration or crust contamination have been mentioned above. While these problems are not insurmountable (at least for rocks younger than about 3.5 Ga), a more fundamental flaw with the approach remains. If komatiites form by deep melting of the conduits of mantle plumes, their sources will be quite separate from the Archean upper mantle, and in a sense analogous to those of modern oceanic island basalts. Just as the compositions of modern OIB cannot be used to estimate the composition of present-day depleted upper mantle, the compositions of komatiites should not be used to estimate the composition of the Archean counterpart. Archean basalts probably represent a better source of information, but many of these may also be related to plume sources. Until Archean mid-ocean ridge basalt is positively identified, the exact composition of Archean upper mantle will remain undefined.

SUMMARY

(1) Spinifex is a texture unique to komatiites. It appears unrelated to any specific aspect of the chemical or mineralogical composition of the lavas (picritic rocks with compositions generally similar to komatiites crystallize without forming the texture). Nor can it be attributed to a special feature of the eruptive environment (in sequences of interlayered komatiites and tholeiites, only the komatiites have spinifex). It is suggested that komatiites develop the texture at least in part because they became superheated following separation from their mantle source.

(2) The prominent layering of spinifex-textured flows develops through gravitative settling of olivine phenocrysts. Convection probably is subdued during most of the cooling history of typical komatiite flows, but may have been vigorous during or soon after the emplacement of the most magnesian magmas.

(3) Although all komatiites are hydrothermally altered and metamorphosed, the intensity of this alteration and the effect it has on chemical compositions varies considerably. Certain elements such as the alkalis, Si, Ca, and Eu exhibit mobile behavior in almost all komatiite suites. Other elements such as Al and Ti, the other high-field-strength elements and most REE are generally immobile, even though in restricted and specific cases even these elements have been perturbed. There is firm evidence that most komatiite samples have compositions similar to those of the original magmas, except for the addition of H_2O, and no indication that metasomatism has significantly increased their MgO contents.

(4) Komatiites form by partial melting deep in the mantle, from sources that ascend from still greater depths. Most komatiites probably formed in mantle plumes that started to melt before they passed the transition from lower to upper mantle. After segregation from their source, the komatiite magmas followed a path that was close to adiabatic and steeper than the liquidus. This caused superheating

of the magma which is a factor leading to the crystallization of spinifex texture after the lavas erupt on the surface. The sources of Barberton-type (Al-depleted) komatiites were slightly hotter than those of Munro-type (Al-undepleted) komatiites. The extent of partial melting for Barberton-type lavas was relatively large as they traversed the region where majorite-garnet was a near-liquidus phase. The fractionation of this mineral was the cause of their distinctive chemical compositions.

(5) Because komatiites probably come from a deep-seated plume source and form by melting at great depths, they should not be used to estimate the composition of Archean depleted upper mantle.

ACKNOWLEDGMENTS

I thank Claude Herzberg for a copy of his unpublished manuscript of peridotite melting, and Mike Lesher, Mike Cheadle, Kent Condie and Claude Herzberg for constructive reviews of this manuscript.

REFERENCES

Agee, C.B. and Walker, D., 1988. Static compression and olivine flotation in ultrabasic silicate liquid. J. Geophys. Res., 93: 3437–3449.

Aitken, B.G. and Echeverría, L.M., 1984. Petrology and geochemistry of komatiites and tholeiites from Gorgona Island, Colombia. Contrib. Mineral. Petrol., 86: 94–105.

Arndt, N.T., 1976. Melting relations of ultramafic lavas (komatiites) at 1 atm and high pressure. Carnegie Inst. Wash. YB, 75: 555–562.

Arndt, N.T., 1977. Mineralogical and chemical variations in two thick, layered komatiitic lava flows. Carnegie Inst. Wash. YB, 76: 494–502.

Arndt, N.T., 1982. Proterozoic spinifex-textured basalts of Gilmour Island, Hudson Bay. Geol. Surv. Can., 83-1A: 137–142.

Arndt, N.T., 1984. Magma mixing in komatiitic lavas from Munro Township, Ontario. In: A. Kröner, G.N. Hanson and A.M. Goodwin (Eds.), Archaean Geochemistry. Springer-Verlag, Berlin, pp. 99–115.

Arndt, N.T., 1986a. Differentiation of komatiite flows. J. Petrol., 27: 279–301.

Arndt, N.T., 1986b. Spinifex and swirling olivines in a komatiite lava lake, Munro Township, Canada. Precambrian Res., 34: 139–155.

Arndt, N.T. and Jenner, G.A., 1986. Crustally contaminated komatiites and basalts from Kambalda, Western Australia. Chem. Geol., 56: 229–255.

Arndt, N.T. and Lesher, C.M., 1992. Fractionation of REE by olivine and the origin of Kambalda komatiites, Western Australia. Geochim. Cosmochim. Acta, 56: 4191–4204.

Arndt, N.T., Naldrett, A.J. and Pyke, D.R., 1977. Komatiitic and iron-rich tholeiitic lavas of Munro Township, northeast Ontario. J. Petrol., 18: 319–369.

Arndt, N.T. and Nesbitt, R.W., 1982. Geochemistry of Munro Township basalts. In: N.T. Arndt and E.G. Nisbet (Eds.), Komatiites. George Allen and Unwin, London, pp. 309–330.

Arndt, N.T., 1984. Magma mixing in komatiitic lavas from Munro Township, Ontario. In: A. Kröner, G.N. Hanson and A.M. Goodwin (Eds.), Archaean Geochemistry. Springer-Verlag, Berlin, pp. 99–115.

Arndt, N.T. and E.G. Nisbet, 1982. What is a komatiite? In: N.T. Arndt and E.G. Nisbet (Eds.), Komatiites. George Allen and Unwin, London, pp. 19–28.

Arndt, N.T., Teixeira, N.A. and White, W.M., 1989. Bizarre geochemistry of komatiites from the Crixás greenstone belt, Brazil. Contrib. Mineral. Petrol., 101: 187–197.

Baragar, W.R.A. and Lamontagne, C.G., 1980. The Circum-Ungava Belt in eastern Hudson Bay: the geology of the Sleeper Islands and parts of the Ottawa and Belcher Islands. Current Research, Part A, Geol. Surv. Canada, Paper 80-1a: 89–94.

Barnes, R.G., Lewis, J.C. and Gee, R.D., 1973. Archean ultramafic lavas from Mount Clifford. Geol. Survey of Western Australia Annual Report, pp. 59–70.

Barnes, S.-J., 1983. A comparative study of olivine and clinopyroxene spinifex flows from Alexo, Abitibi greenstone belt, Canada. Contrib. Mineral. Petrol., 83: 293–308.

Barnes, S.J., Hill, R.E.T. and Gole, M.J., 1988. The Perseverance ultramafic complex, Western Australia: the product of a komatiite lava river. J. Petrol., 29: 305–331.

Bédard, J.H.J., 1987. The development of compositional and textural layering in Archean komatiites and in Proterozoic komatiitic basalts from Cape Smith, Quebec, Canada. In: I. Parsons (Ed.), Origins of Igneous Layering. NATO ASI Series C, 196: 399–418.

Beswick, A.E., 1982. Some geochemical aspects of alteration and genetic relations in komatiitic suites. In: N.T. Arndt and E.G. Nisbet (Eds.), Komatiites. George Allen and Unwin, London, pp. 281–308.

Bickle, M.J., Arndt, N.T., Nisbet, E.G., Orpen, J.L., Martin, A., Keays, R.R. and Renner, R., 1993. Geochemistry of the igneous rocks of the Belingwe greenstone belt: alteration, contamination and petrogenesis. In: M.J. Bickle and E.G. Nisbet (Eds.), The Geology of the Belingwe Greenstone Belt, Zimbabwe. Balkema, Rotterdam, pp. 175–214.

Campbell, I.H., Griffiths, R.W. and Hill, R.I., 1989. Melting in an Archaean mantle plume: heads it's basalts, tails it's komatiites. Nature, 339: 697–699.

Cattell, A. and Arndt, N.T., 1987. Low- and high-alumina komatiites from a Late Archean sequence, Newton Township, Ontario. Contrib. Mineral Petrol., 97: 218–227.

Chauvel, C., Dupré, B. and Jenner, G.A., 1985. The Sm-Nd age of Kambalda volcanics is 500 Ma too old! Earth Planet Sci. Lett., 74: 315–324.

Collerson, K.D., Jesseau, C.W. and Bridgwater, D., 1976. Contrasting types of bladed olivine in ultramafic rocks from the Archean of Labrador. Can. J. Earth Sci., 13: 442–450.

Compston, W., Williams, I.S., Campbell, I.H. and Gresham, J.J., 1986. Zircon xenocrysts from the Kambalda volcanics: age constraints and direct evidence for older continental crust below the Kambalda–Norseman greenstones. Earth Planet. Sci. Lett., 76: 299–311.

Davis, P. and Lesher, C.M., 1993. Thermal erosion. Geol. Soc. Canada. Ann. Meeting.

de Wit, M.J., Hart, R. and Pyle, D., 1983. Mg-metasomatism of the oceanic crust; Implications for the formation of ultramafic rock types. EOS, 64: 333.

de Wit, M.J., Hart, R.A. and Hart, R.J., 1987. The Jamestown ophiolite complex, Barberton mountain belt: a section through 3.5 Ga oceanic crust. J. Afr. Earth Sci., 6: 681–730.

Donaldson, C.H., 1979. An experimental investigation of the delay in nucleation of olivine in mafic lavas. Contrib. Mineral. Petrol., 69: 21–32.

Donaldson, C.H., 1982. Spinifex-textured komatiites: A review of textures, mineral compositions, and layering. In: N.T. Arndt and E.G. Nisbet (Eds.), Komatiites. George Allen and Unwin, London, pp. 211–244.

Duchac, K. and Hanor, J.S., 1987. Origin and timing of the metasomatic silicification of an early

Archaean komatiite sequence, Barberton Mountain Land, South Africa. Precambrian Res., 37: 125–146.

Echeverría, L.M., 1980. Tertiary or Mesozoic komatiites from Gorgona Island, Colombia; field relations and geochemistry. Contrib. Mineral. Petrol., 73: 253–266.

Echeverría, L.M., 1982. Komatiites from Gorgona Island, Colombia. In: N.T. Arndt and E.G. Nisbet (Eds.), Komatiites. George Allen and Unwin, London, pp. 199–210.

Green, D.H., 1975. Genesis of Archaean peridotitic magmas and constraints on Archaean geothermal gradients and tectonics. Geology, 3: 15–18.

Green, D.H., 1981. Petrogenesis of Archean ultramafic magmas and implications for Archaean tectonics. In: A. Kröner (Ed.), Precambrian Plate Tectonics. Elsevier, Amsterdam, pp. 469–490.

Groves, D.I., Korkiakkoski, E.A., McNaughton, N.J., Lesher, C.M. and Cowden, A., 1986. Thermal erosion by komatiites at Kambalda, Western Australia and the genesis of nickel ores. Nature, 319: 136–138.

Gruau, G., Chauvel, C., Arndt, N.T. and Cornichet, J., 1990. Al-depletion in komatiites and garnet fractionation in the early Archean mantle: Hf isotopic constraints. Geochim. Cosmochim. Acta, 54: 3095–3101.

Gruau, G., Tourpin, S., Fourcade, S. and Blais, S., 1991. Loss of isotopic (Nd, O) and chemical (REE) memories during metamorphism: new evidence from eastern Finland. Contrib. Mineral. Petrol., 112: 68–82.

Herzberg, C., 1992. Depth and degree of melting of komatiite. J. Geophys. Res., 97: 4521–4540.

Herzberg, C., Gasparik, T. and Sawamoto, H., 1990. Origin of mantle peridotite: constraints form melting experiments to 16.5 GPa. J. Geophys. Res., 95: 15779–15803.

Herzberg, C. and Ohtani, E., 1988. Origin of komatiite at high pressures. Earth Planet. Sci. Lett., 88: 321–329.

Herzberg, C.T., 1983. Solidus and liquidus temperatures and mineralogies for anhydrous garnet–lherzolite to 15 GPa. Phys. Earth Planet. Int., 32: 193–202.

Hirose, K. and Kushiro, I., 1993. Partial melting of dry peridotites at high pressures: Determination of compositions of melts segregated from peridotite using aggregates of diamond. Earth Planet. Sci. Lett., 114: 477–489.

Hofmann, A.W., 1988. Chemical differentiation of the Earth: the relationship between mantle, continental crust, and oceanic crust. Earth Planet. Sci. Lett., 90: 297–314.

Huppert, H.E. and Sparks, R.S.J., 1985. Komatiites I: Eruption and flow. J. Petrol., 26: 694–725.

Huppert, J.E., Sparks, R.S.J., Turner, J.S. and Arndt, N.T., 1984. Emplacement and cooling of komatiite lavas. Nature, 309: 19–22.

Jahn, B.-M., 1990. Early Precambrian basic rocks of China. In: R.P. Hall and D.J. Hughes (Eds.), Early Precambrian Basic Magmatism. Blackie, Glasgow, pp. 294–316.

Jahn, B.M., Gruau, G. and Glickson, A.Y., 1982. Komatiites of the Onverwacht Group, South Africa: REE chemistry, Sm-Nd age and mantle evolution. Contrib. Mineral. Petrol., 80: 25–40.

Jarvis, G.T. and Campbell, I.H., 1983. Archean komatiites and geotherms: Solution to an apparent contradiction. Geophys. Res. Lett., 10: 1133–1136.

Jochum, K.P., Arndt, N.T. and Hofmann, A.W., 1990. Nb-Th-La in komatiites and basalts: constraints on komatiite petrogenesis and mantle evolution. Earth Planet. Sci. Lett., 107: 272–289.

Lahaye, Y., Arndt, N.T., Gruau, G. and Fourcade, S., 1993. Preservation of Archean mantle signature during alteration: Alexo komatiite, Canada. Terra Abstr., 5: 37.

Lajoie, J. and Gélinas, L., 1978. Emplacement of Archaean peridotitic komatiites in Lamotte Township, Quebec. J. Can. Earth Sci., 15: 672–677.

Lesher, C.M., 1983. Localization and genesis of komatiite-associated Fe-Ni-Cu sulphide mineralization at Kambalda, Western Australia. PhD thesis Univ. Western Australia (unpublished).

Lesher, C.M., 1989. Komatiite-associated nickel sulfide deposits. In: J.A. Whitney and A.J. Naldrett (Eds.), Ore Deposition associated with Magmas. Reviews in Economic Geology, El Paso, pp. 45–96.

Lesher, C.M. and Arndt, N.T., 1994. Geochemistry, petrogenesis, and volcanic evolution of contaminated komatiites at Kambalda, Western Australia. Lithos., in press.

Lesher, C.M. and Groves, D.I., 1986. Controls on the formation of komatiite-associated nickel–copper sulfide deposits. In: G.H.Friedrich et al. (Eds.), Geology and Metallogeny of Copper Deposits., Springer-Verlag, Berlin, 43–62.

Lewis, J.D. and Williams, J.R., 1973. The petrology of an ultramafic lava near Murphy Well, Eastern Goldfields, Western Australia. Geol. Surv. West. Austr. Ann. Rep. 1972, pp. 60–68.

Lofgren, G.E., 1980. Experimental studies on the dynamic crystallization of silicate melts. In: R.B. Hargraves (Ed.), Physics of Magmatic Processes. Princeton University Press, Princeton, pp. 487–552.

Lofgren, G.E., 1983. Effect of heterogeneous nucleation on basaltic textures: a dynamic crystallization study. J. Petrol., 24: 229–255.

McDonough, W.F. and Ireland, T.R., 1993. Intraplate origin of komatiites inferred from trace elements in glass inclusions. Nature, 365: 432–434.

Miller, G.H., Stolper, E.M. and Ahrens, T.J., 1991. The equation of state of molten komatiite, 2: application to komatiite petrogenesis and the Hadean mantle. J. Geophys. Res., 96: 11849–11864.

Muir, J.E. and Comba, C.D.A., 1979. The Dundonald deposit: an example of volcanic-type nickel-sulfide mineralization. Can. Mineral., 17: 351–360.

Nesbitt, R.W., 1971. Skeletal crystal forms in the ultramafic rocks of the Yilgarn Block, Western Australia: Evidence for an Archaean ultramafic liquid. Geol. Soc. Australia, 3: 331–347.

Nesbitt, R.W., Jahn, B.M. and Purvis, A.C., 1982. Komatiites: an early Precambrian phenomenon. J. Volcanol. Geotherm. Res., 14: 31–45.

Nesbitt, R.W., Sun, S.S. and Purvis, A.C., 1979. Komatiites: geochemistry and genesis. Can. Mineral., 17: 165–186.

Nisbet, E.G., 1982. The tectonic setting and petrogenesis of komatiites. In: N.T. Arndt and E.G. Nisbet (Eds.), Komatiites. George Allen and Unwin, London, pp. 501–520.

Nisbet, E.G., Cheadle, M.J., Arndt, N.T., and Bickle, M.J., 1993. Constraining the potential temperature of the Archaean mantle: a review of the evidence from komatiites. Lithos, 30: 291–307.

Nisbet, E.G., Arndt, N.T., Bickle, M.J., Cameron, W.E., Chauvel, C., Cheadle, M., Hegner, E., Kyser, T.K., Martin, A., Renner, R. and Roedder, E., 1987. Uniquely fresh 2.7 Ga komatiites from the Belingwe greenstone belt, Zimbabwe. Geology, 15: 1147–1150.

Nisbet, E.G., Bickle, M.J. and Martin, A., 1977. The mafic and ultramafic lavas of the Belingwe greenstone belt, Rhodesia. J. Petrol., 18: 521–566.

Oliver, R.L., Nesbitt, R.W., Hansen, D.M. and Franzen, N., 1972. Metamorphic olivine in ultramafic rocks from Western Australia. Contrib. Mineral. Petrol., 36: 335–342.

Pyke, D.R., 1975. On the relationship of gold mineralization and ultramafic volcanic rocks in the Timmins area: Ontario, Canada. Ontario Div. Mines Misc. Paper. 63, 23 pp.

Pyke, D.R., Naldrett, A.J. and Eckstrand, O.R., 1973. Archean ultramafic flows in Munro Township, Ontario. Geol. Soc. Am. Bull., 84: 955–978.

Puchtel, I.S., Zhuravlev, D.Z., Samsonov, A.V. and Arndt, N.T., 1993. Petrology and geochemistry of metamorphosed komatiites and basalts from the Tungurcha greenstone belt, Aldan Shield. Precambrian Res., 62: 399–418.

Renner, R., 1989. Cooling and Crystallization of Komatiite Flows from Zimbabwe. Ph.D. thesis.

Univ. Cambridge, 162 pp. (unpubl.).
Renner, R., Nisbet, E.G., Cheadle, M.J., Arndt, N.T., Bickle, M.J. and Cameron, W.E., 1993. Komatiite flows from the Reliance Formation, Belingwe Belt, Zimbabwe: I — petrography and mineralogy. J. Petrol. (in press)
Rigden, S.B., Ahrens, T.J. and Stolper, E.M., 1984. Densities of liquid silicates at high pressures. Science, 226: 1071–1074.
Roeder, P.L. and Emslie, R.F., 1970. Olivine–liquid equilibrium. Contrib. Mineral. Petrol., 29: 275–282.
Smith, H.S. and Erlank, A.J., 1982. Geochemistry and petrogenesis of komatiites from the Barberton greenstone belt, South Africa. In: N.T. Arndt and E.G. Nisbet (Eds.), Komatiites. George Allen & Unwin, London, pp. 347–398.
Smith, H.S., Erlank, A.J. and Duncan, A.R., 1980. Geochemistry of some ultramafic komatiite lava flows from the Barberton Mountain Land, South Africa. Precambrian Res., 11: 399–415.
Thompson, 1983. Book Review: Komatiites. J. Petrol., 24: 319–320.
Tourpin, S., Gruau, G., Blais, S. and Fourcade, S., 1991. Resetting of REE and Nd and Sr isotopes during carbonatization of a komatiite flow from Finland. Chem. Geol., 90: 15–29.
Turner, J.S., Huppert, H.E. and Sparks, R.S.J., 1986. Komatiites II: Experimental and theoretical investigations of post–emplacement cooling and crystallization. J. Petrol.. 27: 397–437.
Usselman, T.M., Hodge, D.S., Naldrett, A.J. and Campbell, I.H., 1979. Physical constraints on the characteristics of nickel-sulfide ore in ultramafic lavas. Can. Mineral., 17: 361–372.
Viljoen, M.J. and Viljoen, R.P., 1969a. Archaean vulcanicity and continental evolution in the Barberton region, Transvaal. In: T. N. Clifford and I. Gass (Eds.), African Magmatism and Tectonics. Oliver and Boyd, Edinburgh, pp. 27–39.
Viljoen, M.J. and Viljoen, R.P., 1969b. The geology and geochemistry of the lower ultramafic unit of the Onverwacht Group and a proposed new class of igneous rocks. Spec. Publ. Geol. Soc. South Africa, 2: 55–85.
Viljoen, R.P. and Viljoen, M.J., 1982. Komatiites — An historical review. In: N.T. Arndt and E.G. Nisbet (Eds.), Komatiites. George Allen and Unwin, London, pp. 5–18.
Wei, J.F., R.G. Tronnes and Scarfe, C. M., 1990. Phase relations of alumina-undepleted and alumina-depleted komatiites at pressures of 4–12 GPa. J. Geophys. Res., 95: 15817–15828.

Chapter 2

ARCHEAN VOLCANIC PATTERNS

P.C. THURSTON

INTRODUCTION

This chapter summarizes what is known about Archean volcanic regimes and their relationship to younger volcanic systems. Most Archean volcanic rocks are concentrated within "greenstone belts", long linear belts of relatively low grade supracrustal rocks dominated by volcanic rocks which are surrounded and rarely underlain by granitoid rocks. These belts are the "granite–greenstone" subprovinces of Card and Ciesielski (1986) which typify Archean cratons the world over (Goodwin, 1991).

The Archean has been characterized as an active regime with abundant, largely bimodal volcanics (Thurston et al., 1985), sediments representing alluvial–fluvial and resedimented associations with few sedimentary platforms (Ojakangas, 1985), early extensional tectonic regimes succeeded by tight isoclinal folding and diapiric emplacement of granitoids (e.g. Schwerdtner et al., 1978). Greenstone belts were generally thought to represent a single tectonic environment such as, amalgamated island arcs (Langford and Morin, 1976; Condie, 1986), back arc basins (Tarney et al., 1976; Condie, 1986), or collapsed continental rifts (Goodwin, 1981; Henderson, 1981). The diversity of Proterozoic orogens included the development of epicratonic basins (Windley, 1984) with shallow water platformal clastic and chemical sediments and largely basaltic volcanic rocks. Proterozoic volcanic rocks were seen largely as island arc and intra-plate volcanics (Taylor, 1987) and it was viewed as the era of the onset of plate tectonics (e.g. Hoffman, 1973).

As early as the seminal papers of Talbot (1973) and Langford and Morin (1976) there were advocates for plate tectonics in the Archean. However, development of the most convincing evidence had to await the development of: (1) recognition and understanding of large shear zones occasioned by the gold exploration boom of the 1980s (e.g. Robert and Brown, 1986); (2) a comprehensive high precision U-Pb zircon geochronologic data base for the Superior Province (Corfu and Davis, 1992); (3) trace element geochemistry of volcanic rocks and its application to paleotectonic interpretation; and (4) geothermobarometry. With this new database as a foundation, the Superior Province, the world's largest Archean craton, has been interpreted (Card, 1990; Williams et al., 1992) as a product of plate tectonic processes. Such a conclusion is critical to understanding Archean volcanism,

(A)

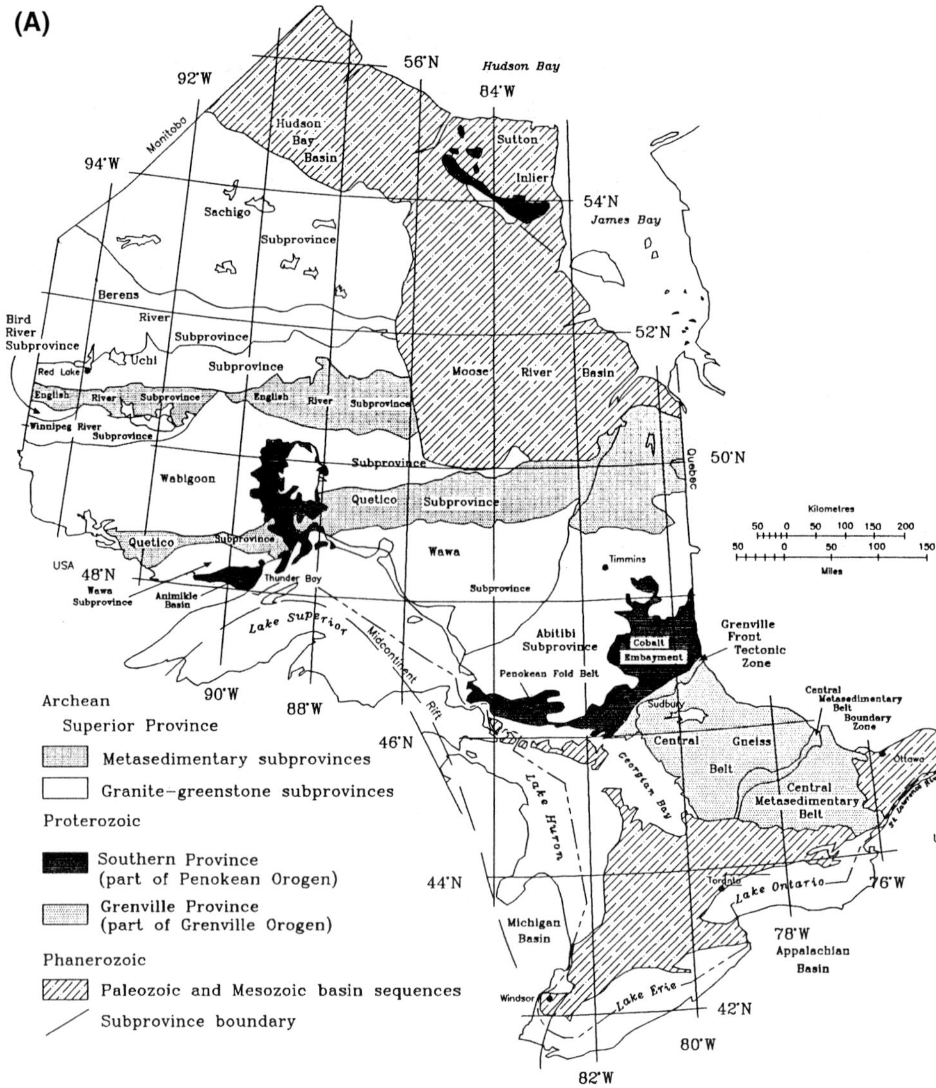

Fig. 1. Subdivisions of the Superior Province. (A) Subdivisions after Card and Ciesielski (1986) establishing the distribution of sedimentary and granite–greenstone subprovinces.

because it means that the best understood example of Archean volcanism may not differ substantially from younger orogens and their volcanic systems.

Card and Ciesielski (1986) subdivided the Superior Province into largely fault-bounded high metamorphic rank sedimentary and lower metamorphic rank granite–greenstone subprovinces generally tens of kilometers wide by hundreds of kilometers long (Fig. 1).

(B)

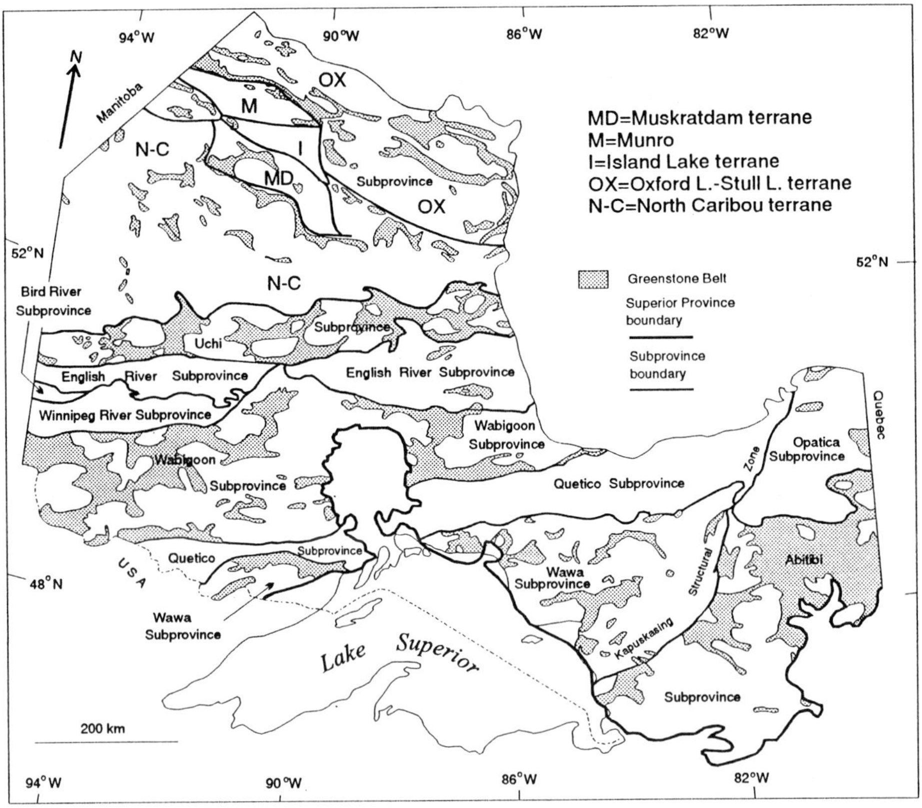

Geological map of the western part of the Superior Province showing major tectonic subdivisions and greenstone belts.

Fig. 1 (*continued*). (B) Subdivisions of the Superior Province after Williams et al. (1992), reflecting revised subdivisions of the previously poorly known northwestern part of the Province.

Geochronologic data (e.g. Corfu and Ayres, 1991) showed that individual greenstone belts only a few tens of kilometers in length could contain several disparate units separated by thrust faults (Ayres and Corfu, 1991) spanning almost 300 My. Such age ranges and the wide spectrum of rock types within greenstone belts suggested belts were subdivisible. Williams et al. (1992) and the recently released tectonic map of Ontario (Ontario Geological Survey, 1992) used the concept of lithotectonic assemblage, developed for terrane analysis in the Cordillera (Tipper et al., 1981; Gabrielse and Yorath, 1989), to subdivide greenstone belts in a large portion of the Superior Province. Tectonic assemblages are defined as "packages of stratified volcanic and/or sedimentary rock units built during a discrete interval of time in a common depositional or volcanic setting. Assemblages may be bounded by structural discontinuities, unconformities or intrusions" (Tipper et al., 1981; Gabrielse and Yorath, 1989).

(A)

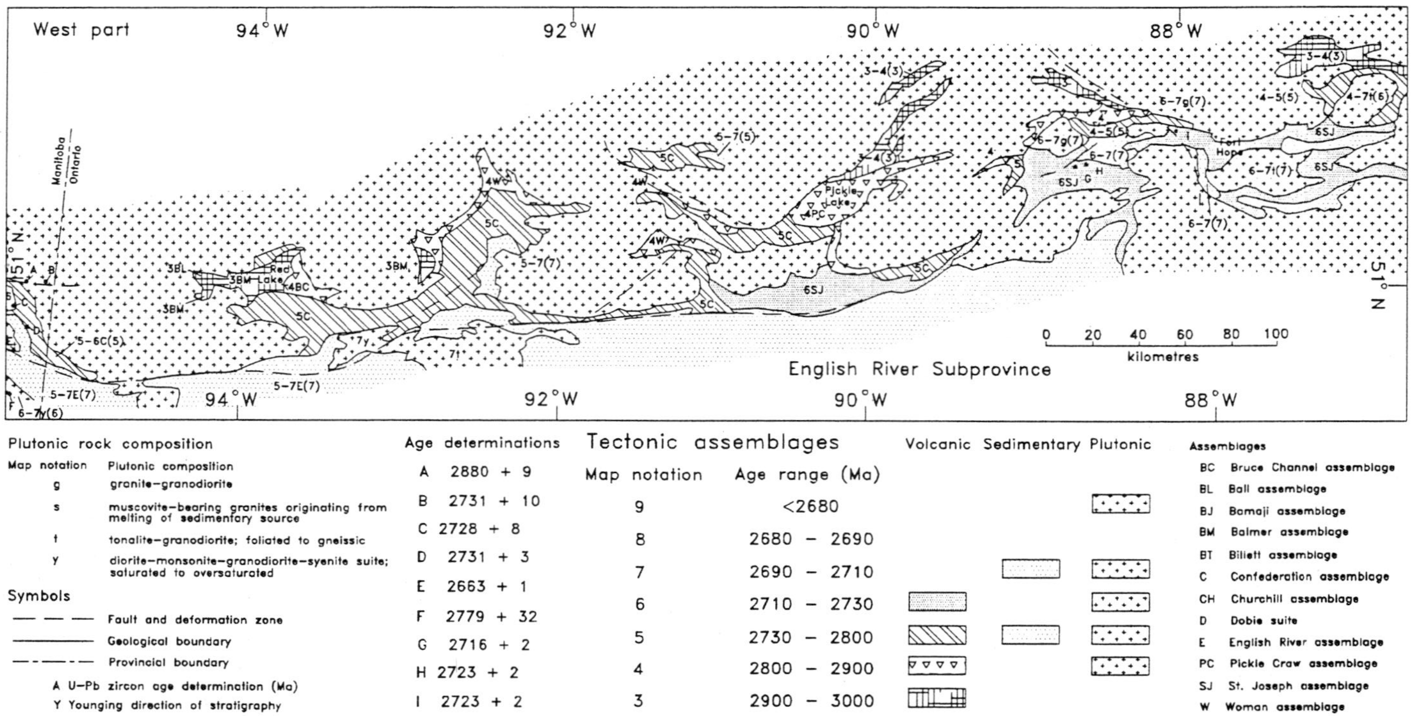

Fig. 2. The varying scale of assemblages within the Superior Province. (A) Assemblages within the Uchi subprovince after Stott and Corfu (1991) demonstrate the large scale of some assemblages.

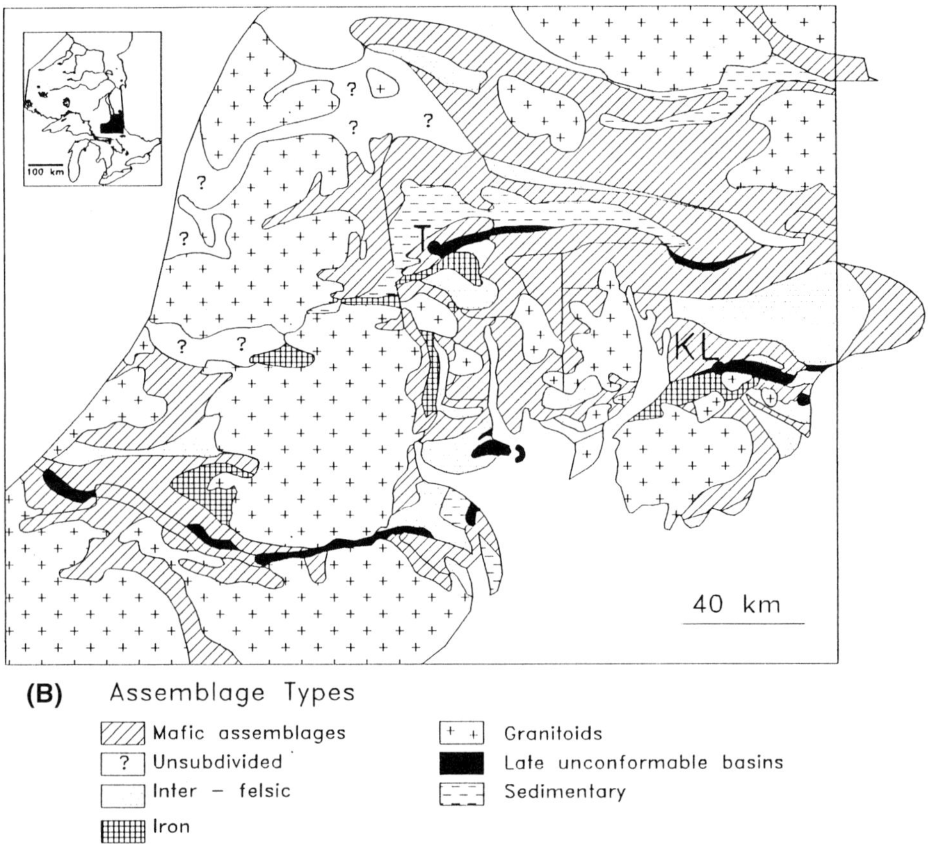

Fig. 2 (*continued*). (B) Assemblages of the western part of the Abitibi subprovince after Jackson and Fyon (1991) demonstrate the smaller scale of assemblages along the southern margin of the Superior Province.

The assemblage concept is applicable to several scales in Archean greenstone belts with adequate geochronologic and structural data: single assemblages hundreds of kilometers long exist in the Uchi Subprovince (Stott and Corfu, 1991) whereas the areally subordinate Ontario part of the Abitibi subprovince has been subdivided into 71 assemblages (Jackson and Fyon, 1991; Jackson et al., 1993) (Fig. 2). Using the assemblage approach, and an expanded geochronologic database (Corfu and Davis, 1992) subdivisions of the Superior Province changed in a few short years (Fig. 1). Analysis of several hundred dated assemblages in the central part of the Superior Province reveals an age span of <20 Ma within single assemblages and as much as 300 Ma between assemblages within a single greenstone belt (Williams et al., 1992).

The assemblage types recognized in Superior Province greenstone belts are listed in Table 1. Detailed examination of the stratigraphy, primary structures,

TABLE 1

Assemblage type	Rock types	Primary structures and biota	Isotopic data	References
Quartz arenite	quartz arenite, carbonate, shale, iron formation (oxide, silicate), tholeiite, komatiite, minor wacke	herring bone, trough, and hummocky cross-strat, stromatolites in sediments pillows and grade air fall tuff	detrital zircons in 2.9–2.85 Ca range, no isotopic inheritance	Thurston and Chivers (1990), Stevenson and Patchett (1990), Cortis et al. (1989)
Re-sedimented quartz arenite	quartz arenite, wacke, conglomente, tholeiite	scours, sheet flows, trough, hummocky cross-beds, pillow	detrital zircons in 3–2.9 Ca range, noise topic inheritance	Cortis et al. (1989), Thurston (1990)
Carbonate	carbonate, wacke, siltstone, tholeiite, komatiite, banded chert	stromatolites, karst sinks, accretionary lapilli radiating gypsum rosettes, flaser bedding, desiccation cracks	ages in 1.9–3.5 Ca range, no isotopic inheritance	Thurston and Chivers (1990), Wilks (1986)
Iron type	flows and pyroclastics of various magmatic affinities overlain by oxide and sulphide iron formation	parallel laminated "beds" of iron formation — primary structures in volcanic rocks variable	assemblages older than most others in a similar tectonic region	Jackson and Fyon (1991), Jackson et al. (1993)
Mafic assemblages	(1) komatiite–tholeiite ± iron formation; (2) komatiite–tholeiite, felsic volcanics, iron formation; (4) tholeiites ± minor komatiites	pillowed flows, massive flows with thin pillowed tops, distal facies planar bedded sulfidic argillites, minor clastic sediments, shallow dipping shield volcanic edifice	variably enriched/depleted mantle, rarely showing evidence of contamination, most assemblages show no isotopic inheritance	Thurston et al. (1993), Jackson et al. (1993), Barley et al. (1989)
Felsic assemblages	andesite to rhyolite flows and pyroclastic rocks	pillows, massive flows, rubbly flows, domes, tuff to breccias, variably distributed some in collapsed structures, abundant pumice, grading of beds, some indication of welding in some areas, vertical succession of primary structures indicates ash flow processes dominate in felsic units	some units show inheritance, most do not	Thurston and Chivers (1990), Corfu (1993)

Unconformable basins	alkaline to calc alkaline flows and pyroclastics; conglomerates, wackes, silts	cross-beds, scours, channels in sedimentary units, rubbly flows, bedded tuffs in volcanic rocks, fanglomorates	develop late in history of subprovinces, volcanic rocks from enriched mantle	Varne (1985), Jackson et al. (1993)
Continental style volcanism	andesite–rhyote complexes with related coarse to find clastic sediment fans	cross-beds, sheet flows in sediments; rubbly subaerial flows in volcanic rocks with minor ash-flows	nil	Giles and Hallberg (1982), Thurston et al. (1993)
Ophiolites	serpentinite, dunite → gabbro, basaltic flows, sulphidic argillite, ultramafic sediment	pillowed flows, thin-bedded sulphidic argillite, rare sheeted dikes, serpentinite melange	no inheritance	Thurston et al. (1993), Helmstaedt and Scott (1992)
Sedimentary	coarse to find clastic rocks, minor pillowed flows	traction deposits with turbidite style primary structures ranging from alluvial to abyssal environment	younger than major volcanism	Jackson et al. (1993), Thurston et al. (1993), Devaney and Williams (1989)

geochemistry and isotopic characteristics permit comparison to modern analogues based on previous compilations of analogues (Thurston and Chivers, 1990; Williams et al., 1992; Jackson et al., 1993). The following discussion of the various assemblage types emphasizes physical volcanology but dwells, where appropriate, upon other aspects in order to clarify their origin and role in Archean processes. This paper emphasizes Superior Province examples because of the author's experience and the extensive database of geochronology (Corfu and Davis, 1992), geochemistry and isotopic research. The comparisons made with Superior Province data can be extended to other cratons as is made clear in the sections on various assemblage types and the concluding discussion.

GREENSTONE BELT ASSEMBLAGE TYPES

Platform assemblages

Platform greenstone belt assemblages are of four types: quartz arenite-bearing, carbonate bearing assemblages (Thurston and Chivers, 1990), re-sedimented quartz arenite bearing assemblages (Thurston, 1990), and the "iron" assemblages of Jackson et al. (1993). The quartz arenite and carbonate-bearing sequences occur in several shields (Thurston and Chivers, 1990 and references therein). Where undisturbed, they lie unconformably on older greenstones or granitoids. Both types are characterized by primary structures and biota indicative of shallow water deposition (Thurston and Chivers, 1990) for the lower parts of the assemblages. The upper parts of these assemblages consist of komatiite and tholeiite flows with limited evidence of fractionated derivatives including minor subaqueously deposited felsic airfall and ash-flow tuffs (Thurston et al., 1987). Using the primary structure criteria of Dimroth et al. (1978), most flows represent a relatively proximal submarine facies. The limited thickness and lateral extent of individual flows and lack of ponded coarse-textured flow centres suggest that eruption rates were not extremely high (cf. Greeley, 1982; Coffin and Eldholm, 1991). Though lack of vesicularity is not a reliable indicator of depth of eruption (Bryan and Moore, 1977), the lack of units with abundant vesicles suggests that eruption depths for volcanics in the upper parts of these assemblages were substantial.

A re-sedimented quartz arenite assemblage (Cortis, 1991) has been described on the margin of the North Caribou terrane in the central part of the Superior Province. Analysis of the primary structures, stratigraphy, and geochronology indicates this assemblage formed by cannibalization of platformal quartz arenites and re-deposition in a submarine fan environment on a cratonic margin.

The "Iron" assemblages of the Abitibi subprovince (Fig. 1) includes assemblages containing volcanic rocks of a variety of magmatic affiliations, with the common element being the presence of sulphide or oxide facies iron formation (Jackson et al., 1993). This assemblage type is considered distinct because the variety of volcanic rocks (komatiites, tholeiites, calcalkaline) precludes classification on this basis and the amount of iron formation is greater than in other assemblage types. Within the restricted age range of the Abitibi subprovince (Corfu and Davis, 1992) the iron type assemblages tend to be older than other assemblage types. Iron assemblages are distributed adjacent to large batholiths and in structural culminations (Jackson et al., 1993). Contacts with other assemblages tend to be depositional contacts with variable degrees of later modification. Jackson et al. (1993) interpret the iron assemblages as an older amalgamation of assemblages forming structural/stratigraphic basement, which has been capped by pelagic, chemically precipitated iron formation. The younger Minas Supergroup of the Quadrilatero Ferrifera in the Brazilian shield may be another example of this assemblage type, lying unconformably above older greenstones (Marshak et al., 1993).

Genetic constraints

The provenance of the quartz in the quartz arenite-bearing and re-sedimented quartz arenite assemblages can be a powerful petrogenetic constraint. Quartz-rich sedimentary rocks could originate as felsic volcanic debris (Ojakangas, 1985; Ayres, 1983), by production of silcrete through extensive tropical weathering of a variety of source materials, or as granitoid debris. Over 80% of the coarse clastic debris in Sachigo subprovince quartz arenite-bearing assemblages is tonalitic, supporting the latter origin (Cortis et al., 1989). Cortis (1991) summarized geochemical evidence for rapid weathering of granitoid source rocks to yield quartz-rich sediments. Sandstone petrography discrimination diagrams (after Dickinson and Suczek, 1979) show that the Sachigo data concentrate within the continental block setting (Cortis et al., 1989) (Fig. 3). In the Superior Province, quartz and carbonate-bearing platform assemblages are concentrated in the Sachigo subprovince, and tend to lie in the age range 3.0–2.85 Ga (Thurston and Chivers, 1990; Williams et al., 1992). These data led Thurston and Chivers (1990) to suggest that platform assemblages developed relatively early in the evolution of the Superior Province. Scattered evidence from other shields — Yilgarn (Gee, 1982), Slave (Schau, 1977), India (Srinivasan and Ojakangas, 1986), Baltic (Kroner et al., 1981), Zimbabwe (Bickle et al., 1975), and perhaps the Aldan shield (Kazansky and Moralev, 1981) — suggest similar patterns.

Lu-Hf model ages (Stevenson and Patchett (1990) from some Superior Province quartz arenite-bearing platform assemblages indicate that the sialic substrate is not much older than the units containing the detrital grains. Given the structural position and geochronological data summarized above, the platform assemblages are seen as epicratonic sequences developed on older granitoid and greenstone units (Thurston and Chivers, 1990; Thurston et al., 1991; Jackson et al., 1993). Initial development was in shallow water based upon primary structure data (Table 1). Tholeiitic/komatiitic volcanic rocks of the quartz arenite and carbonate-bearing platform assemblages display primary structures indicative of deep water

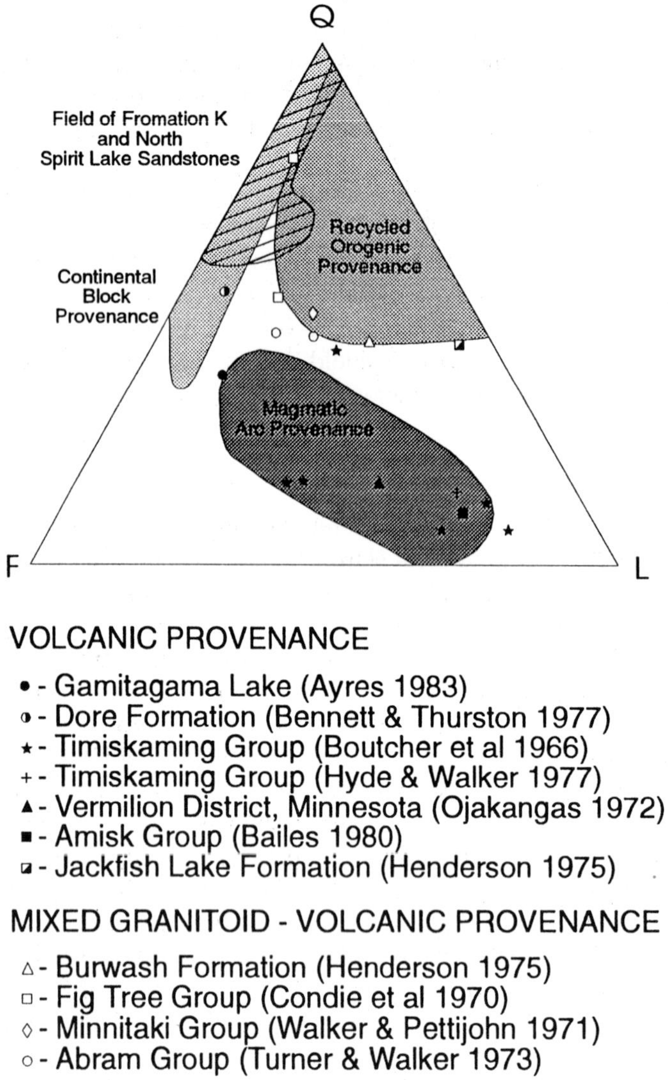

Fig. 3. Quartz–feldspar–lithic fragment ternary diagram (after Dickinson and Suczek, 1979) of quartz-rich metasediments from the North Caribou terrane (after Cortis, 1990).

deposition, such as the presence of sulphidic argillite and silicate facies iron formation, pillows and scarcity of vesicles and hyaloclastite units. These criteria suggest the water depth was certainly >1000 meters. The volcanic rocks are relatively enriched in Zr and Y similar to modern flood basalts (Thurston, 1990a) as well as being enriched in LIL elements and Ti as compared to modern oceanic and arc komatiites. The interpreted platform vs submarine fan depositional environments between the quartz arenite-bearing and the re-sedimented quartz arenite assemblages may relate to timing of rifting. Undisturbed platforms with upper volcanic units appear to have rifted late whereas the resedimented association rifted relatively early based upon the lower stratigraphic position of the volcanic units (Thurston, 1990).

The quartzite–carbonate association emblematic of cratonic stability is unusual in the Archean (Ojakangas, 1985) and more common in Paleoproterozoic stable platforms. The age and distribution of platform assemblages, particularly in the Superior Province, as well as other lines of evidence, were used to suggest the presence of a 2.8 Ga orogenic event as well as the 2.7 Ga "Kenoran" orogeny terminating the Archean in the Superior Province (Williams et al., 1992).

Mafic assemblages

Large parts of greenstone belts in several shields are typified by assemblages comprising tholeiitic and komatiitic volcanic rocks (Kaapvaal (de Wit et al., 1992), Yilgarn (Ahmat et al., 1990), Pilbara (Hickman et al., 1990), Zimbabwe (Bickle et al., 1975), Baltic (Goodwin, 1991; Kozhevnikov, 1992), Superior (Williams et al., 1992; Jackson et al., 1993)). The relative proportions of komatiite to tholeiite are quite variable as well as the extent of development of intermediate to felsic volcanic derivatives and iron formation.

Typical mafic assemblages with poorly developed felsic derivatives in the Abitibi subprovince consist of high magnesium basalt (HMB) tholeiites, and siliceous high magnesium basalt (SHMB (terminology of Redman and Keays, 1985) capped by ocellar, variolitic or plagioclase phyric SHMB, chemical sediments or felsic tuffs (Thurston, 1990). Interflow sedimentary units tend to be thin airfall tuffs or sulphidic argillites. Individual flows tend to reflect the architecture of rapid eruption: they are relatively thick (10s of metres) and exhibit thin pillowed upper parts, a thick massive to gabbroic interior, and a lateral extent measured in hundreds of metres. The larger scale stratigraphic unit of mafic volcanic assemblages commonly has a lateral extent of up to several tens of kilometers and is interpreted as a shield volcano with very low slope (cf. Dimroth et al., 1985). Thurston and Chivers (1990) equated this type of volcanism with mafic plain volcanism of Greeley (1982) which is characterized by rapid eruption rates. This assemblage type includes units with abundant high-Fe tholeiites which tend to develop shield volcanoes with more pronounced slopes than the more magnesian tholeiitic sequences described above (Fig. 4) (Dimroth et al., 1985).

Assemblages of this type containing andesitic to rhyolitic rocks are rare in general but are well developed in the Abitibi subprovince in Ontario (Jackson et al., 1993). Several geochemical studies show the evolved units (FII and FIII rhyolites of Lesher et al., 1986) can be modelled by extensive fractionation of mafic tholeiitic parent magma (e.g. Hart, 1984; Barrie et al., 19991). Rb-Sr, Sm-Nd and Lu-Hf studies (Barrie and Shirey, 1991; Corfu and Noble, 1992)

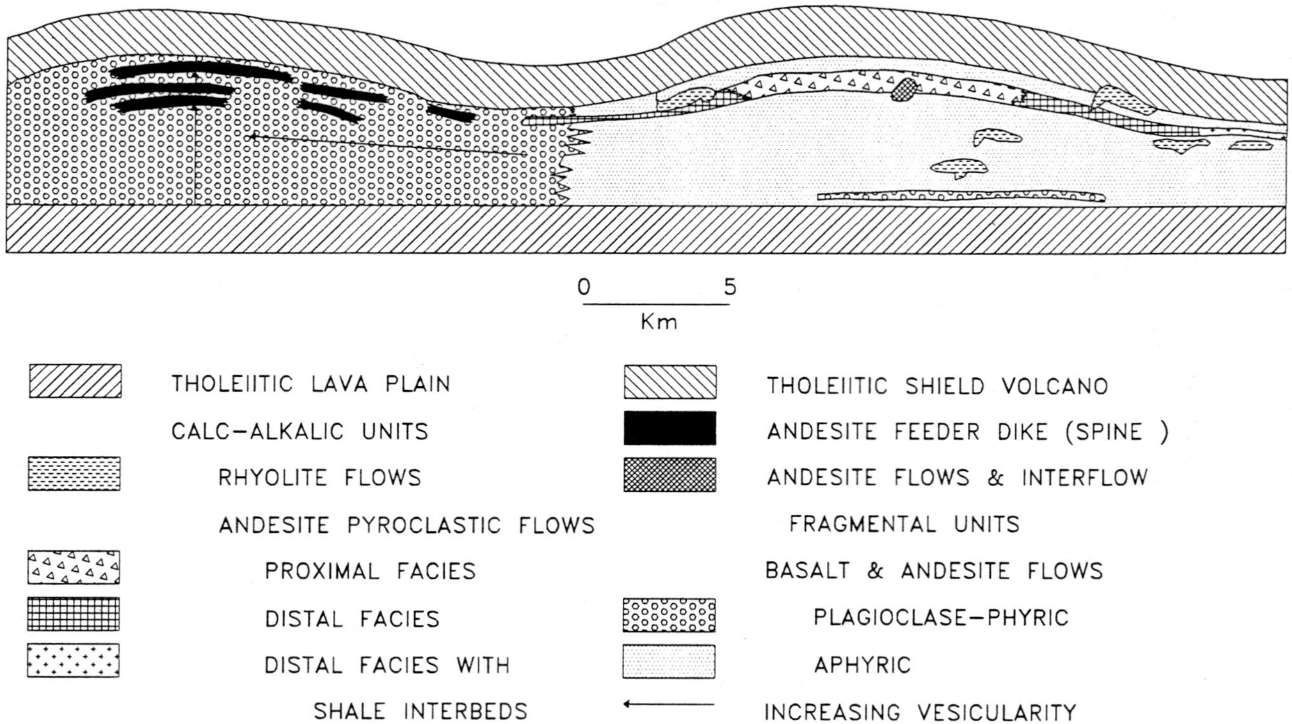

Fig. 4. Generalized diagram showing the relationship between mafic plain and Hawaiian type eruptions in the Abitibi subprovince after Dimroth et al. (1982).

indicate that the mafic assemblage type is derived from depleted mantle with no contribution from older chemically evolved sources.

Mafic volcanic sequences, largely tholeiites with minor komatiites, occur within the Quetico (Soo, 1988) and Pontiac (Morin et al., 1993) sedimentary subprovinces of the Superior Province. These sequences contain rare iron formation intercalations, thick flows, vesicles and hyaloclastite, all suggesting deep water deposition and relatively low rates of eruption.

Genetic constraints

Primary structures, flow morphology, and stratigraphy indicate most mafic assemblages were produced by subaqueous volcanism with large shield volcanoes and relatively deep water sedimentation with little terrigenous input (Thurston, 1990a). Jackson et al. (1993) compare the geochemistry of Abitibi subprovince mafic assemblages to modern analogues: (a) mafic assemblage basalts are similar to mid-ocean ridge and back-arc spreading centre basalts and (b) felsic metavolcanic units of mafic assemblages are similar to felsic rocks of back-arc basins and "anomalous" ocean ridge environments such as Iceland (Fig. 6). These authors explain the variations in proportions of tholeiites, komatiites and felsic derivatives by the interaction of two processes: (a) modelling of the earth's heat budget (MacKenzie and Bickle, 1988) predicts a hotter Archean mantle, and hence more abundant eruption of komatiitic lavas relative to tholeiites at that time; (b) increased crustal thickness typical of "anomalous" ocean ridge environments (Wood, 1978; Macdonald et al., 1990; Nicholson et al., 1991) yields mainly tholeiitic basalts at the main spreading axis. The "off-axis" environment is characterized by crust three times as thick and lesser heat flow with derivative magmas including tholeiitic, intermediate, and felsic types. This may be explained by melting of the base of thick mafic crust (Sigmarsson et al., 1991), or fractionation within crustal magma chambers (Macdonald et al., 1990). Thus, increasing degrees of mantle partial melting will yield more komatiitic magmas, whereas increasing crustal thickness will yield increasing amounts of fractionation.

Mafic assemblages were erupted from shield volcanoes in a subaqueous environment. There is no isotopic evidence for eruption in an ensialic regime (see above), save for some examples in the Yilgarn Block (e.g. Arndt and Jenner, 1986). Comparison of the geochemistry of mafic assemblages in granite–greenstone and sedimentary subprovinces (e.g. the Abitibi and the Quetico subprovince) shows a close fit with the geochemistry of oceanic ridges and back arc basins (Ludden, 1989; Barley et al., 1989; Thurston 1990). The abundance of both komatiites and tholeiites, the rapid eruption rates indicated by flow architecture (Thurston and Chivers, 1990; Thurston, 1990) and the Co, Cr, and Ni-rich nature of the rocks suggest the mafic assemblages may well represent oceanic plateaus, such as the plume-generated Large Igneous Provinces of Coffin and Eldholm (1991) (as suggested by Storey et al., 1991; Desrochers et al., 1993). Felsic derivatives in mafic assemblages are comparable to the felsic rocks of back arc basins and "anomalous mid-ocean ridge settings.

In terms of physical volcanology, the slower eruption rates that produce thinner flows and steeper slopes to the volcanic edifice as mafic assemblages evolve (Dimroth et al., 1985) may represent back arc or rifted arc sequences. This is based on the observation that the "mafic plain" sequences of Dimroth (1985) contain few felsic derivatives, whereas the assemblages with felsic derivatives are geochemically comparable to extensional settings (Jackson and Fyon, 1991).

Intermediate to felsic volcanic assemblages

Intermediate to felsic assemblages occur in all shields, but have been most extensively studied in the Superior Province, in large part due to their spatial association with volcanogenic massive sulphide mineralization (e.g. Lesher et al., 1986). These assemblages represent bimodal volcanism with basalt or andesite and rhyolite. The mafic to intermediate volcanic rocks are pillowed flows representative of subaqueous shield volcanism (Thurston et al., 1986). A shallowing upward aspect to these sequences is demonstrated by the increase in amygdules, hyaloclastites with or without cross-stratification (Dimroth et al., 1985), and locally intense chloritization and silicification with development of autobrecciation and devitrification related to hydrovolcanic activity (e.g. Gibson et al., 1983). The upper, felsic part of these assemblages are classified into two types based upon the caldera cycle of Smith and Bailey (1968):

(1) *Non-resurgent sequences*: These sequences display relatively minor volumes of siliceous flows and pyroclastic rocks capping a mafic shield volcano. The felsic volcanic rocks are dominated by relatively distal air-fall tuffs with smaller amounts of coarse breccias, typically mass flow deposits, and occasional fine grained turbiditic clastic sedimentary rocks deposited in a subaqueous environment (e.g. Johns, 1985). Some examples include a large scale ash-flow deposit (e.g. Thurston, 1980) as the last major volcanic event. These sequences lacked a sufficient supply of mafic input into subjacent felsic magma chambers to support resurgent volcanism (Cathles et al., 1983).

(2) *Resurgent sequences*: In this type of sequence, a basal basaltic to andesitic shield volcano is capped by an upper felsic volcanic sequence dominated by an extensive large ash-flow deposit (Thurston et al., 1985; Ayres, 1983), commonly consisting of multiple flow units succeeded by a caldera collapse structure with areally restricted products of resurgent volcanism. These resurgent units commonly include quartz/feldspar phyric endogenous domes with related flows, talus deposits, products of phreatomagmatic eruption, and small-scale ash-flow deposits. The resurgent parts of these sequences commonly display some evidence of compositional zonation: (a) they become more mafic upward on a scale of hundreds of metres, (b) at constant silica there is enrichment in heavy REE upward (e.g. Thurston, 1986), and (c) individual ash-flow depositional units show gradation in the composition of essential lithic fragments (Thurston, 1981). Typically, within a single depositional unit, essential lithic fragments and pumice will vary

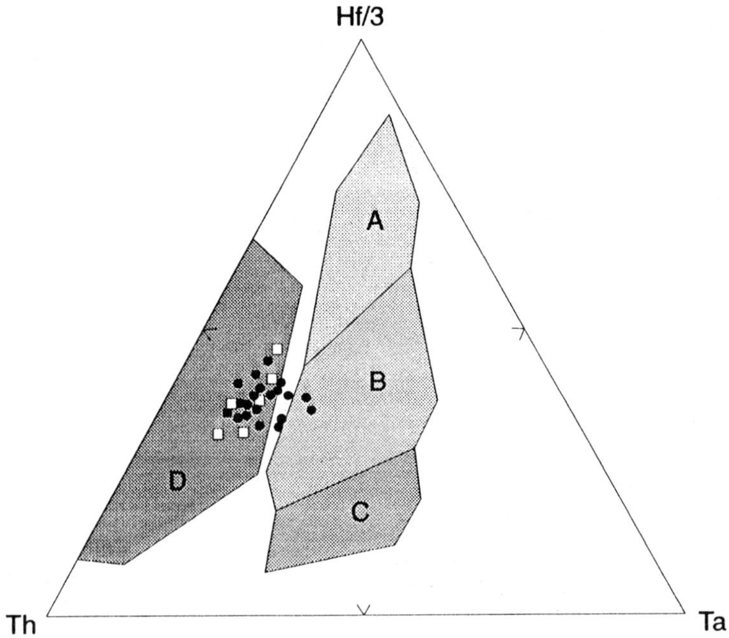

Fig. 5. Th-Hf-Ta diagram for upper Noranda Subgroup andesites, dacites and rhyolites of the Blake River assemblage. Data from: squares, Capdevila et al. (1982); dots, Ujjike and Goodwin (1987). Compositional fields from Wood et al. (1979) are: A: N-type MORB, B: E-type MORB; C: within plate basalts; D: magmas at destructive plate margins.

from rhyolite at the base to andesite toward the top. This gradation is a direct indication of compositional zonation in the magma chamber from which the ash-flows erupted.

The bulk composition of felsic volcanic assemblages is generally calcalkaline with trace element systematics similar to both continental and island arc systems, complete with Ti, Nb and Ta anomalies on primitive mantle normalized element concentration graphs (e.g. Sylvester et al., 1987; Capdevila et al., 1982). Examples from the Superior Province include the continental magmatic arcs in the Wawa Subprovince (Sylvester et al., 1987) and the 2.8 Ga arc which formed along the south margin of the North Caribou terrane (the former north part of the Uchi subprovince (Stott and Corfu, 1991) (Fig. 1). The Blake River assemblage in the Abitibi subprovince (Capdevila et al., 1982) is another well documented example. Similar sequences occur in other shields including examples in the Yilgarn (Hallberg et al., 1976), the Baltic Shield (Gaal, 1986), the Slave Province (Kusky, 1990), and the Zimbabwean craton where 2.7 Ga upper Bulawayan greenstones (east of Harare) include substantial rhyolitic flows (N. Baglow, pers. comm. 1987).

Volumetrically, these assemblages contain a higher proportion of felsic volcanic rocks than other assemblage types. Most sequences do however include rocks of andesitic to basaltic bulk composition. Trace element distributions are characterized by light REE enrichment and fractionated REE patterns (e.g. Capdevila et al., 1982; Smith, 1980; Hallberg et al., 1976). The Hf-Ta-Th relations are typical of arc volcanics using the compositional fields of Wood et al. (1979) (Fig. 5). Some sequences show isotopic inheritance indicative of passage through older sialic crust such as the Skead assemblage in the Abitibi subprovince (e.g. Corfu, 1993). There is some evidence these intermediate to felsic assemblages evolve from basal mafic units (Primitive island arcs?) consisting of thick, laterally extensive massive mafic flows with thin pillowed tops typical of the mafic plain setting (Dimroth et al., 1985) to shield volcanoes with thinner pillowed basaltic to andesitic flows and steeper dips to the edifice as a function of increasing magma viscosity as Fe tholeiites become prominent in the stratigraphy. The upper felsic part of these assemblages such as in the Noranda area are dominated by felsic ash-flow volcanism (Ayres, 1983; Thurston et al., 1985).

Genetic constraints

The intermediate to felsic assemblages of greenstone belts have been compared by many authors to island and continental magmatic arcs on the basis of geochemistry, volcanic style, and sedimentation patterns (Capdevila et al., 1982; Thurston and Ayres, 1985; Williams et al., 1992). Island arcs and continental arcs have been successfully distinguished in Archean settings based upon geochemical studies. For example, an island arc origin has been postulated for the Blake River assemblage (Capdevila et al., 1982) whereas a continental magmatic arc setting has been proposed for part of the Wawa subprovince (Sylvester et al., 1987). It is interesting to note that Archean continental magmatic arcs appear to lack volcanogenic massive sulphide mineralization (cf. Fyon et al., 1992) implying a less vigourous heat flow than is the case for Archean island arc assemblages.

Late Unconformable basins

Late Unconformable basins are greenstone belt assemblages characterized by alluvial–fluvial sediments and alkaline (shoshonitic) to calcalkaline volcanic rocks which unconformably overlie older greenstones. These assemblages occur along strike-slip shear zones marking major sequence boundaries in greenstone belts or at subprovince boundaries (Williams et al., 1992). Examples of these assemblages are found in the Pilbara (Krapez and Barley, 1987), Yilgarn (Ahmat et al., 1990), Baltic (Jatulian of Kulikov and Golubev, 1984), and Superior cratons (Williams et al., 1992) (Fig. 6). The sedimentary sequences display an orderly arrangement of depositional facies (Hyde, 1980; Mueller and Donaldson, 1992) suggesting the basin outlines are more or less as presently preserved. Structural analysis indicates these sequences were deposited after nappe style early deformation of the older greenstones, unroofing of undeformed granitic rocks and prior to the late deformation involving late movement on shear zones (Williams et al., 1992; Krapez and Barley, 1987).

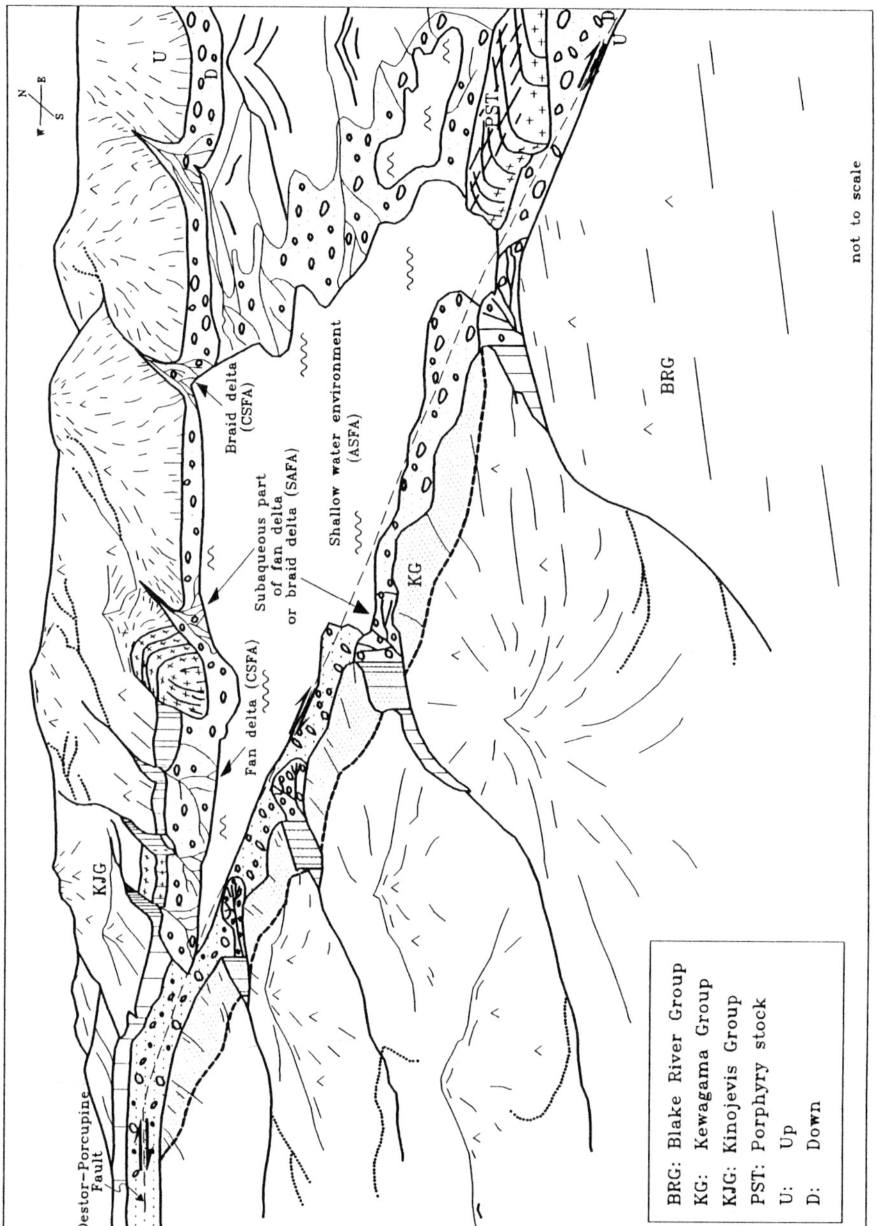

Fig. 6. Generalized view of the architecture of Late Unconformable basin assemblages. Note the fault bounded character of the basin and the discontinuous nature of units across and along strike. Volcanic strata are largely relatively proximal pyroclastic units and flows. (After Mueller and Donaldson, 1992).

Genetic constraints

Volcanic rocks in Late Unconformable basins are calcalkaline to shoshonitic rocks enriched in LIL elements and depleted in Ti, Ta and Nb (Brooks et al., 1982; Wyman and Kerrich, 1988) similar to volcanics in mature arcs. The origin of these sequences is controversial. Alternatives include pull-apart basins (e.g. Krapez and Barley, 1987; Thurston and Chivers, 1990; Mueller and Donaldson, 1992), piggy-back basins (Kusky, 1991; Turner, 1992) or simply late rift-related flysch basins (Corfu et al., 1991).

Continental style volcanism

Archean greenstone belts include subaerially deposited sedimentary–volcanic assemblages characterized by andesite–rhyolite volcanic piles with related alluvial fan sedimentary units, mud flows, and cross-bedded quartz arenites. Examples include the Marda Complex (Hallberg et al., 1976) and the Welcome Well complex (Giles and Hallberg, 1982) (Fig. 7) in the Yilgarn Block and an andesite–quartz arenite association in the Baltic craton at Hisovaara (Kozhevnikov et al., 1992). Geochemical data on the Marda complex suggests a magmatic arc origin (Hallberg et al., 1976).

Another possible example is the relatively undeformed 2.7 Ga Hamersley Basin of shallow marine to continental mafic volcano–sedimentary rocks overlain by extensive shallow-water iron formation, argillite, and sandstone (reviewed by Hickman et al., 1990). This Hamersley sequence overlies the older greenstones of the Pilbara Block and represents a rift–fill sequence that developed after cratonization of the underlying ~3–3.5 Ga Pilbara greenstones.

Continental style assemblages may represent: (1) subaerial equivalents of felsic associations described above; (2) assemblages transitional with the turbidite assemblages described below; or (3) rift-related sequence comparable to the 2.7 Ga Hamersley Basin in Australia. The principal difficulty in identifying the tectonic setting of the continental-style assemblages is the uncertain contact relations between examples of this assemblage type and surrounding assemblages.

Archean ophiolites

Ophiolites are fragments of oceanic crust consisting from base to top, of ultramafic rocks, gabbro, sheeted dikes, and pillow lavas, which have been emplaced onto continents. They vary from a 6- km pillow–dike–gabbro sequence in Oman (Hopson et al., 1981) to a complete sequence only 3 km thick in the Josephine ophiolite (Harper, 1984), to the Alpine–Apennine ophiolites which lack sheeted dikes and consist of serpentinite and minor gabbro overlain by pillows and breccias (Abbate et al., 1980). The differences in rock types are related to spreading rates at ocean ridges, with fast spreading yielding voluminous volcanism, nearly continuous shield volcanoes, less faulting, and commonly a narrow

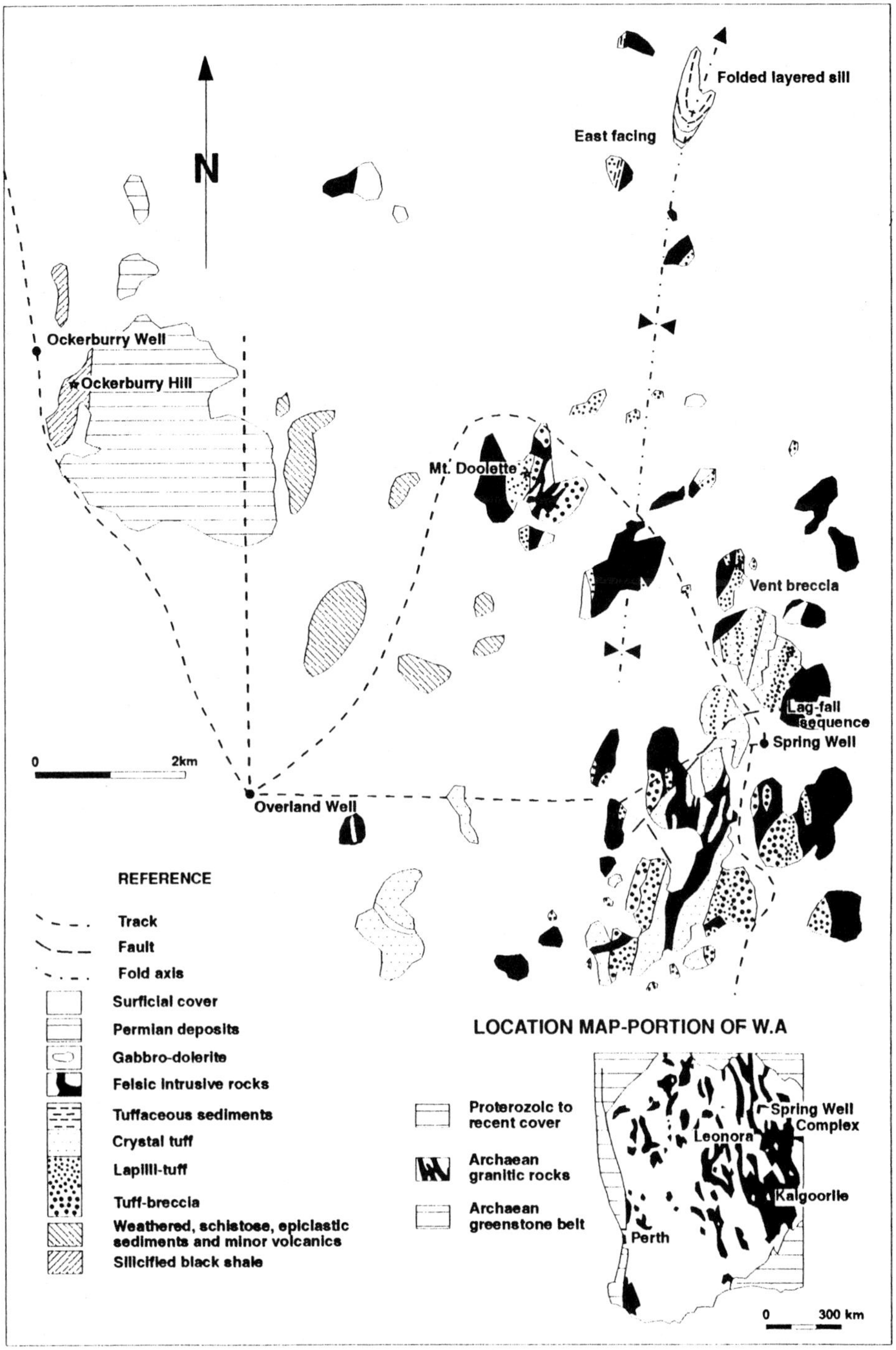

Fig. 7. A view of the Spring Well complex, a late continental assemblage in the Yilgarn Block (after Hallberg et al., 1980).

magma chamber (Solomon and Toomey, 1992). Slow spreading is characterized by discontinuous volcanoes, short-lived magma chambers, and in general extension is more prominently taken up by faulting, resulting in exposure of ultramafic oceanic basement.

Moores (1982) recognized two ophiolite types: (a) Tethyan type obducted onto passive margins, and (2) Cordilleran type forming the basement of accreted terranes. Ophiolites have chemical affinities ranging from MORB (Pearce et al., 1984) to island arc "suprasubduction zone ophiolites, many of which display more than one geochemical affinity (e.g. Troodos-Robinson et al., 1983). Coleman (1984) concludes that suprasubduction zone ophiolites form mainly in back-arc basins and MORB ophiolites form in small ocean basins.

Archean ophiolites are rare (McCall, 1981), but difficulties with their recognition is in large part due to deformation and metamorphism. A major distinguishing feature of ophiolites is the sheeted dike complex, a possible example of which has been recognized in the 2.7 Ga Slave Province (Helmstaedt et al., 1986) (Fig. 8). Other possible examples include parts of the Barberton greenstone belt (de Wit et al., 1987), and a dismembered mafic unit in the South pass area of the Wyoming craton (Harper, 1985).

Ophiolites also occur along sutures including the young suture zones extending from the Himalayas to the Alps. Linear "suture zones" (Brown and Coleman, 1972) occur in Proterozoic greenstone belts of the Arabian shield between arc assemblages (Fig. 9). Similar suture zones occur in the Proterozoic Svecokarelides of Finland (Gaal, 1982), and between continental rise sediments and the Great Bear magmatic zone of the Wopmay Orogen (Hoffman, 1980). Pod-like ultramafic

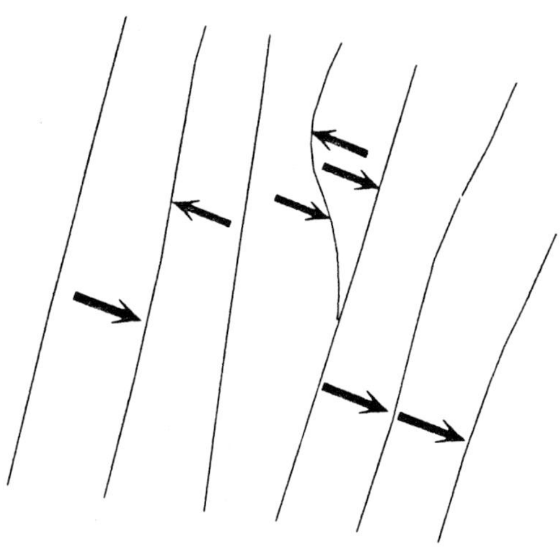

Fig. 8. Sketch of the general form of dike polarity within the ophiolitic complex of the Kam Group in the Archean Yellowknife greenstone belt of the Slave Province.

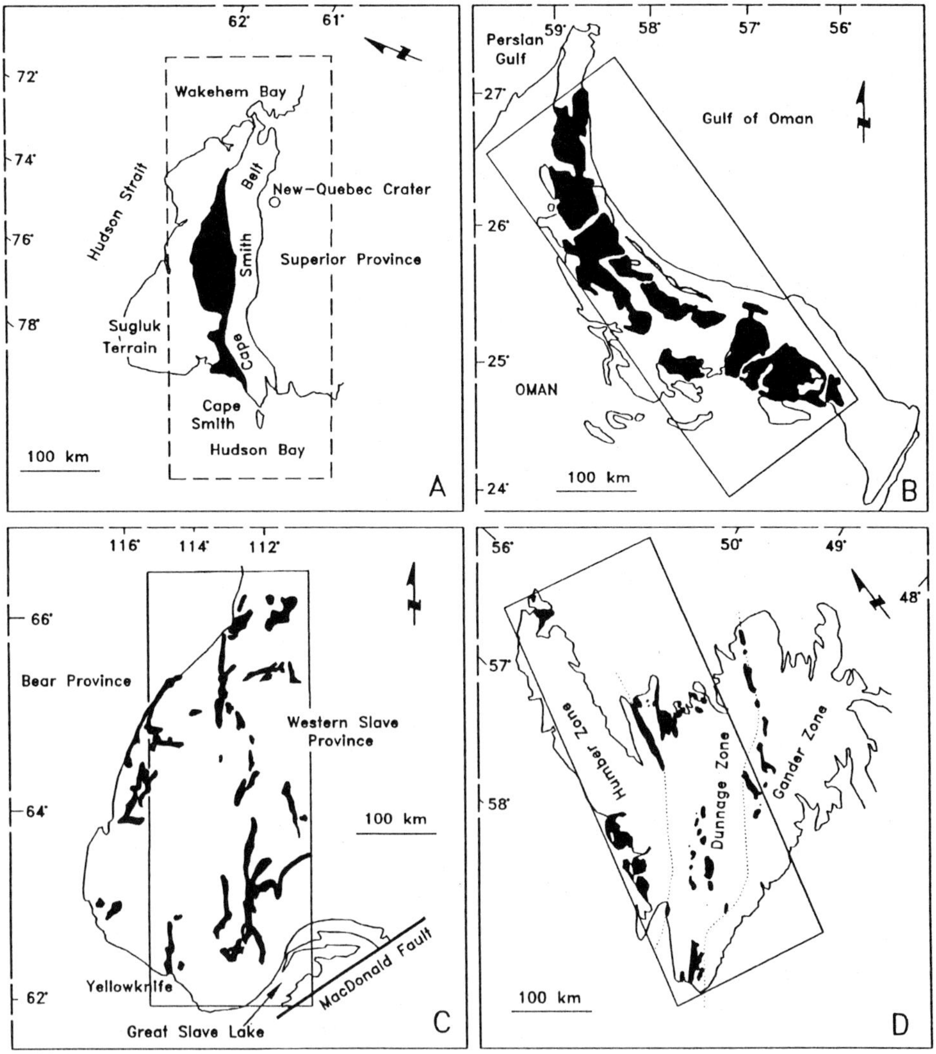

Fig. 9. Map patterns and size of ophiolitic sutures in Precambrian terranes compared to greenstone terrains. A: Cape Smith belt, which includes 1.8 Ga ophiolites, B: Samail ophiolite, Oman; C: greenstone belts of the western Slave Province, Canada; D: Newfoundland. The pattern of the Slave Province greenstone belts compares well with those of the Arabian shield and the Dunnage Zone of Newfoundland (after Helmstaedt and Scott, 1992).

bodies occur along major shear zones bounding subprovinces in the Superior Province (Williams, 1991). The composition and location of these pods along major tectonic boundaries in Archean shields, for example between sedimentary and granite–greenstone subprovinces of the Superior Province suggest the pods may be ophiolitic fragments.

The apparent scarcity of Archean ophiolites may be due to a thicker oceanic crust perhaps similar to modern oceanic plateaux (Sleep and Windley, 1982; Moores, 1986; Burke, 1988; Hoffman and Ranalli, 1988), produced by more common eruption of komatiitic liquids from a hotter Archean mantle (Bickle, 1978; Mackenzie and Bickle, 1988). Possibly Archean oceanic crust may have been too thick to be obducted or crustal delamination may have occurred within the Archean oceanic crust rather than at the base as is now the case. The continuing search for Archean ophiolites must bear in mind that many modern examples (Alpine–Apennine type) may not contain sheeted dikes and display a variety of geochemical affinities. As well, the mechanical difficulties imposed by thicker Archean oceanic crust may decrease the preservation potential of Archean ophiolites.

Sedimentary assemblages

Sedimentary assemblages comprising fine to coarse clastic sediments, largely traction deposits, are important in parts of most Archean greenstone belts (Ojakangas, 1985 and references therein). The rocks range from distal turbidites to coarse proximal conglomerates (Hyde, 1978; Wood, 1980; Ojakangas, 1985), forming distinct assemblages which are not interbedded with the upper parts of volcanic assemblages. Examples from many shield areas appear similar, based upon map relationships in, for example, the Slave Province (Kusky, 1990), the Baltic shield (Kozhevnikov et al., 1992; the Yilgarn (Ahmat et al., 1990) and the Pilbara Block (Krapez, 1993).

Two Superior Province examples serve to constrain the nature and probable tectonic setting of sedimentary assemblages. Along the interface between the 2.7 Ga Wabigoon subprovince greenstones and the Quetico subprovince metasediments to the south, Devaney and Williams (1989) described three fault-bounded belts consisting of basal oceanic volcanic rocks (Soo, 1988) overlain by clastic sediments (Fig. 1). From north to south, the sedimentary units in the three bands represent alluvial–fluvial, submarine fan and abyssal environments (Fig. 10). The shear zone kinematics indicate south-side down. Percival and Williams (1989) suggested these facies and structural relationships indicate the Quetico sedimentary subprovince represents a prograding accretionary wedge on the edge of an older arc sequence to the north. In the Abitibi subprovince Jackson et al. used geologic and geochronologic relationships to postulate a similar origin for their so-called "turbidite" assemblages. In the Abitibi subprovince, Archean volcanism ceased at about ~2700 Ma and Late Unconformable Basin assemblages developed between 2685–2675 Ma. The sedimentary assemblages do not contain interbeds of the earlier mafic or intermediate to felsic assemblages, or clasts of the alkaline volcanic rocks from Late Unconformable basin assemblages. Detrital zircon data (Davis, 1991) from "turbidite" or sedimentary assemblages indicate ages as young

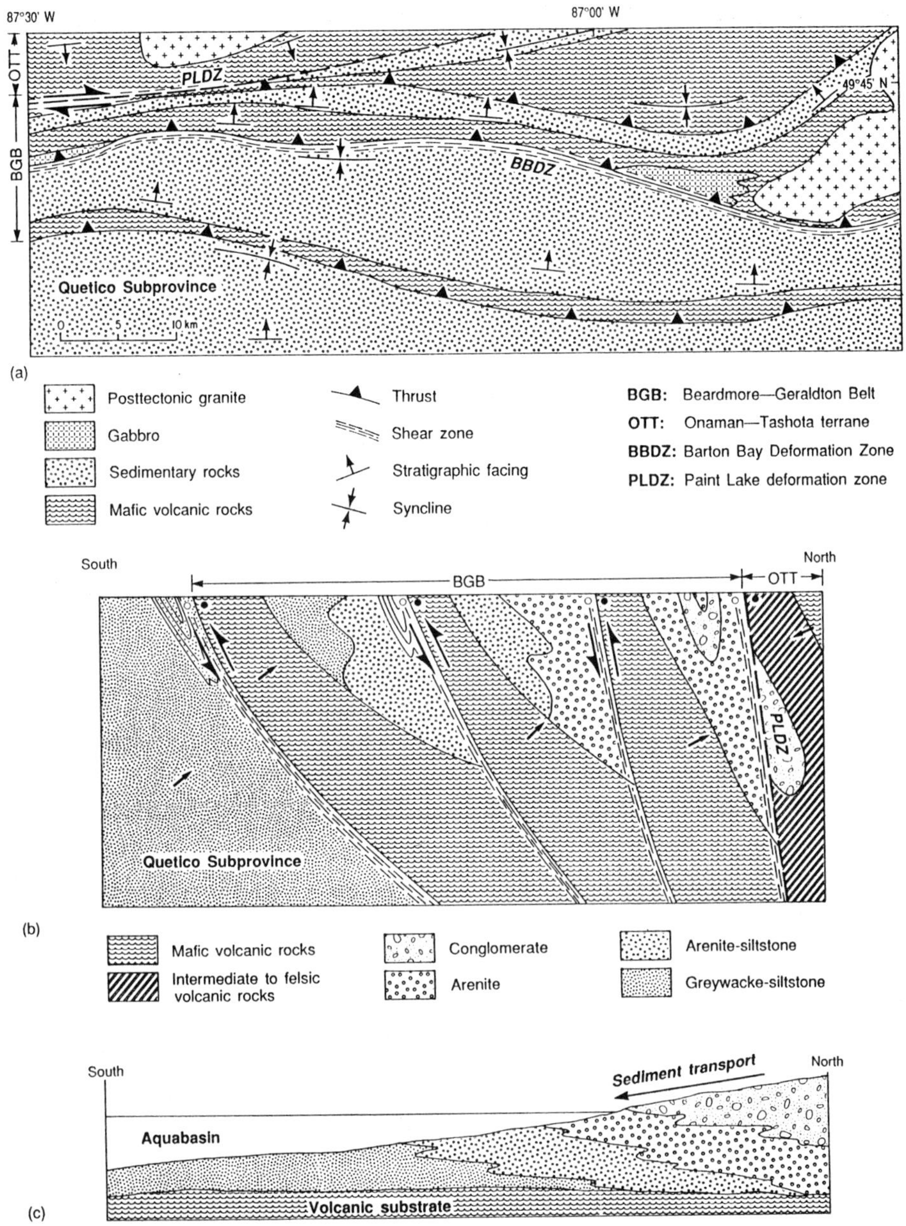

Fig. 10. Relationships along the interface between the Wabigoon and the Quetico Subprovince in the Beardmore Geraldton area (after Devaney and Williams, 1989).

as ~2686 Ma but not younger. Alluvial fluvial sediments representing Late Unconformable basin assemblages contain zircons as young as 2679 Ma (Corfu et al., 1991). In the Timmins area the Three Nations Late Unconformable assemblage (~2675 Ma) overlies folded 2700 Ma Krist assemblage felsic pyroclastic rocks.

The above examples show that the sedimentary assemblages post-date the main period of mafic and intermediate to felsic assemblage volcanism and the onset of early regional deformation in the Superior Province. These sedimentary assemblages mark a transition from submarine volcanism to subaerial alluvial–fluvial sedimentation and the first appearance of detrital zircons as old as 2.9 Ga (Gariepy et al., 1984). The development of sedimentary assemblages in both areas marks the development of later regional structures which are related to a super-terrane collisional process (Dimroth et al., 1983; Williams et al., 1992).

RELATIONS BETWEEN ASSEMBLAGES

Studies in the Superior Province (Williams et al., 1992) indicate that granite–greenstone subprovinces can be subdivided into assemblages varying in scale from a few kilometers in length to several hundred kilometers as indicated above. To understand the evolution of Archean greenstone belts as collages of assemblages, it is important to fit into their developmental chronology the major deformational events seen from a Superior Province perspective where accretionary prisms (Percival and Williams, 1989) are a first order feature. Contrasting the deformational regime in sedimentary subprovinces with that in granite–greenstone subprovinces has allowed a more complete understanding of the deformation processes and their relationship to tectonic models.

Williams et al. (1992) have erected a generalized chronology of greenstone belt deformation events which is similar to that observed in other Archean cratons (Kusky and Vearncombe, 1994). The early deformation in Superior Province greenstones consists largely of recumbent folding and thin-skinned thrusting which may be related to the juxtaposition of assemblages to form greenstone belts (Fig. 11). These early folds were then rotated into a subvertical orientation by later upright folding or doming by granitoid batholiths. The later upright folding was penecontemporaneous with and preceded development of large subprovince-bounding strike-slip shear zones. The early deformation is seen only within the granite–greenstone subprovinces whereas the later deformation affects the adjoining sedimentary subprovinces. Therefore Williams et al. (1992) have interpreted the later deformation to be due to docking of sedimentary subprovinces against granite–greenstone subprovinces. This interpretation is reinforced by the consistent dextral kinematics of the subprovince bounding shear zones across the Superior Province, explained as a product of transpressive deformation (Stott, 1986; Card, 1990).

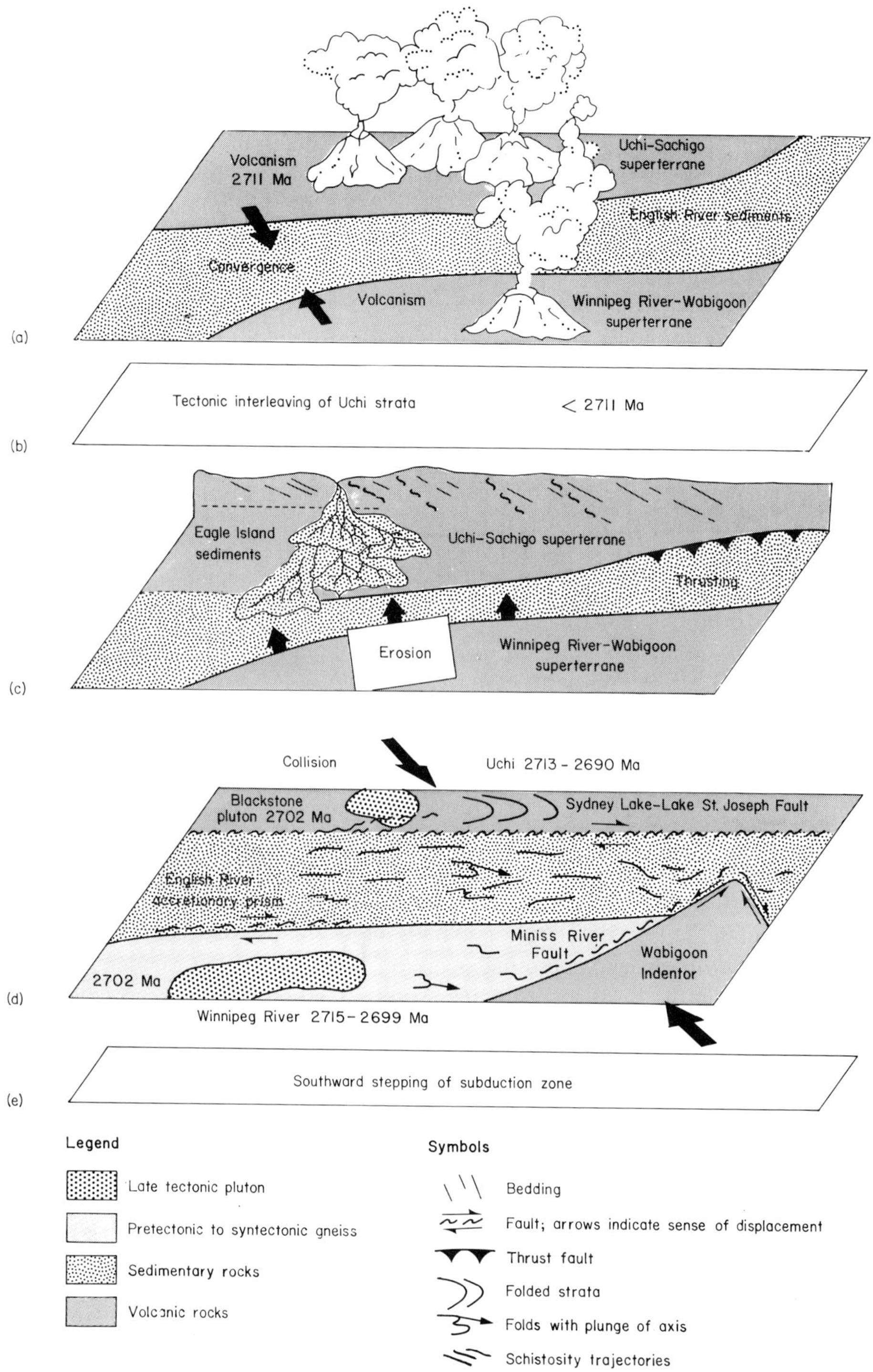

Fig. 11. Cartoon after Stott and Corfu (1991) illustrating the early deformation of Uchi subprovince and confinement of later deformation to the margins of the Uchi and the entire English River subprovince.

DISCUSSION

Archean greenstone belts are tectonic collages representing many paleoenvironments and a time span of up to 300 My in the Superior Province. Greenstone assemblages representing discrete paleoenvironments are bounded by depositional, or more commonly, tectonic contacts. Most shields contain Archean analogues to epicratonic platforms of several types, oceanic plateaus (submarine mafic plains), ophiolites, continental and island arcs, continental style magmatism, and several types of successor basins including pull-apart basins and piggyback basins. The diversity of ages and tectonic environments present within single greenstone belts (e.g. Williams et al., 1992) suggest that future research will show individual assemblages will display distinct isotopic signatures as indicated in the work of Corfu and Stott (1993).

The structural chronology of greenstone belts worldwide is remarkably similar (Kusky and Vearncombe, 1994). This chronology of early thrust-based deformation (e.g. Ayres and Corfu, 1991) is followed by shouldering aside of supracrustal rocks and steepening of earlier folds concomitant with granitoid intrusion (Williams et al., 1992) followed by late deformation concentrated along major strike slip zones forming major tectonic boundaries. This late deformation has been compared to transpressional deformation in modern orogens (Stott, 1986; Hudleston et al., 1988; Williams et al., 1992).

What does the variability in types of assemblages and paleoenvironments making up Archean greenstone belts say about tectonic processes? The diversity of ages and paleoenvironments in greenstone blet assemblages argues for accretionary tectonics on the scale of the greenstone belt. On the craton scale within the Superior Province, there is a systematic southward younging in the age of volcanism, granitic plutonism and the age of activity on subprovince-bounding shear zones (Card, 1990; Williams et al., 1992). These points coupled with the confinement of the later deformation to margins of granite–greenstone subprovinces and the sedimentary subprovinces are all points in favour of plate tectonic processes within the Superior Province. Similar tectonic schemes are advocated for the Slave (Kusky, 1990), Yilgarn (Ahmat et al., 1990), Pilbara (Hickman et al., 1990), Baltic (Gaal, 1986), and Aldan (Dook, 1989) shields.

Spatial and secular patterns

Geochronologic evidence summarized by Thurston and Chivers (1990) for the Superior Province as a whole suggests epicratonic/platformal assemblages were formed relatively early in the history of the craton. Epicratonic basins tend to develop at zones of convergence and mantle downwelling (Peltier et al., 1992); therefore, facies analysis of platform assemblages over broad areas may in future reveal the architecture of early orogenic belts within specific cratons.

Thurston and Chivers (1990) noted that Unconformable basins developed relatively late in the history of a given greenstone belt. The subsequent identification of a plethora of assemblage types leaves us with the notion that platformal assemblages tend to develop early and Unconformable Basins relatively late in greenstone belt chronology.

Inferences for Archean processes

Knowledge of volcanic processes can be used to constrain what we know about the Archean earth. Ash-flow volcanism with evidence for compositionally zoned magma chambers is dominant in Archean arc sequences (Ayres, 1983; Thurston et al., 1985). There is some evidence that Archean ash flows yield ejecta volumes similar to modern systems varying from 10^2 to 10^6 km^3 (Thurston et al., 1985; Ayres and Thurston, 1986). Development of compositional zonation in modern ash-flow magma chambers requires keeping the magma chambers sufficiently hot for the zonation to develop (Smith, 1979). This in turn requires continued input of hot, mantle-derived mafic melt into the base of the magma chamber. Shaw (1985) drew attention to the fact that in such a chamber, input of excessive basalt would result in basaltic volcanism and whereas a scant supply of magma would force the system to freeze in place yielding compositionally zoned felsic plutons. Therefore, the existence of abundant Archean ash-flow volcanism implies that the rate of mafic magma supply in Archean arc assemblages may have been roughly similar to that in modern systems otherwise, the same consequences would have obtained: basaltic volcanism or felsic plutons.

Since ash-flow systems lose a maximum of 10% of their volume in an eruption, with the remainder staying in a domical magma chamber (Smith, 1979), geometrical considerations show that the crustal thickness necessary to contain such a chamber is at least 30–40 km. Given the volumetrically large Archean ash-flows (Thurston et al., 1985), we can then assume that Archean crustal thickness in arc systems was similar to that of modern analogues.

If volcanic processes on continents and island arc systems of Archean age were similar to modern analogues as suggested here, and if the Archean mantle was warmer than the present mantle (Bickle, 1978; Mackenzie and Bickle, 1988), then as suggested by Thurston et al. (1993) heat loss in the Archean earth must have been concentrated in the oceans as indicated below.

Heat loss in Archean oceans could have been manifested through faster spreading rates, greater ridge lengths, or the development of large oceanic plateaus. The geochemistry and physical volcanology of Archean mafic assemblages may point to possible choices amongst these alternatives.

If greater ridge lengths were present in the Archean they would result in development of smaller plates. However, Sleep (1992) concluded on the basis of analysis of the duration of strike-slip faulting in the Superior Province, that in the Late Archean, plates were comparable in size to those involved in Cordilleran orogenesis. Hence, greater ridge lengths may not represent an alternative in the Late Archean.

It has been recently recognized that there are periodic igneous events resulting in large outpourings of basaltic and picritic magma and the development of so-called "Large Igneous Provinces" (LIP) (Coffin and Eldholm, 1991) in continental and oceanic settings. On the continents they result in such features as the Deccan Traps and the Columbia River basalts, whereas in oceanic settings they result in production of oceanic plateaus such as the Rockall plateau. The presence of primitive melts such as picrites and komatiites, the volume of the magmatic systems, and their high eruption rates manifest in volcanic architecture suggest an origin for these magmas in deep mantle-derived plumes. These Archean oceanic plateaus represent an Archean manifestation of Coffin and Eldholm's (1991) Large Igneous Provinces hypothesis in an oceanic setting, fitting well with the assertion by many (e.g. Jarvis and Campbell, 1983) that komatiite/tholeiite sequences represent Archean plume magmatism. Several authors (Storey et al., 1991; Desrochers et al., 1993; Chown et al., 1992) have compared Archean mafic assemblages to LIPs based on the occurrence of komatiite-like rocks in modern LIPs. The concentration of komatiite–tholeiite mafic assemblages in the southern volcanic zone of the Abitibi subprovince could lead to the speculative suggestion that the late Archean was a time of intense continent growth. This growth took place in part by aggregation of the products of a plume event or events which had generated the many mafic assemblages in that area in the interval 2.72–2.70 Ga. The LIPs and their komatiitic rocks may not be unique to the Archean (cf. Taylor, 1987; Gibbs, 1987) but the prominence of mafic assemblages in Archean greenstone belts renders them easily distinguishable from many Proterozoic greenstones.

Therefore it is concluded that Archean heat loss was accommodated in part through faster spreading (Thurston et al., 1993) and more rapid production of thicker oceanic crust (Hoffman and Ranalli (1988), but most importantly through a great abundance of Large Igneous Provinces, mainly submarine plateaus (Thurston et al., 1993). Possible continental analogues include sequences such as the Kambalda komatiites in Western Australia (Arndt and Jenner, 1986).

Uniqueness of the Archean

Komatiites have long been maintained to be unique to the Archean (e.g. Storey et al., 1991). However, recent work has demonstrated their presence in the oceanic volcanic units of Proterozoic age (e.g. Baragar and Scoates, 1987) and to a minor extent in Phanerozoic oceanic plateaus (Storey et al., 1991). Archean greenstone belts appear to lack pyroxene and plagioclase phyric andesites prominent in younger orogens (Gill, 1981; Thurston and Fryer, 1983) except possibly in scarce late continental assemblages (e.g. Hallberg et al., 1976).

The unique aspects of Archean volcanism may be used to constrain the types of plumbing or fractionation processes in the Archean. For example, the FeO, Ti, and Cr content of Archean basaltic liquids is higher than those for younger basalts (Basaltic Volcanism Study Project, 1981). These characteristics are usually explained as the result of igneous fractionation in basaltic systems. Alternatively they may be explained as the products of dryer, hotter plume-related Archean magmatic systems (Jarvis and Campbell, 1983). These dryer, hotter Archean basaltic liquids may not have cooled below the pyroxene and plagioclase liquids to yield orogenic andesites typical of younger orogens.

The abundance of ash-flows in Archean arcs (Ayres, 1983; Thurston et al., 1985) and the similar rate of supply of basaltic magma to Archean and Phanerozoic arcs (Thurston et al., 1993) suggests that plumbing in Archean arcs differs little from today. The mafic assemblages (particularly those with abundant komatiites) are considered oceanic plateaus (Storey et al., 1991; Chown et al., 1992; Desrochers et al., 1993) based upon geochemical, stratigraphic, and volcanological similarities.

Provincial-scale structures or depositional environments more prevalent in the Archean relative to younger orogens include: (1) widespread occurrence of epicratonic platforms rich in Fe and Mn; (2) mafic assemblages containing tholeiites and komatiites are more common in Archean relative to Proterozoic greenstone belts (e.g. Gibbs, 1987); and (3) the unique architecture of komatiite sequences (Hill et al., 1990) with thin laterally extensive flows at the margins and olivine-rich crystal mush units complete with large-scale thermal erosion beneath the komatiite sequences (Huppert and Sparks, 1985).

The epicratonic platforms may be due to the lesser average depth of Archean oceans relatively free of continents (Windley, 1984), representing an ideal condition for development of widespread shallow water platformal environments. Continuing seawater-dominated alteration of volcanic rocks could explain the abundance of altered siliclastic sedimentary rocks on the Pilbara (Dunlop and Buick, 1981) and the Kaapvaal cratons (de Wit et al., 1987).

In general, however, the discovery of the diversity of assemblage types within greenstone belts renders Archean greenstone belts similar to younger accretionary orogens or greenstone belts. Detailed differences will exist from belt to belt in the relative proportions of assemblage types. For example, Jackson et al. (1993) characterize the Abitibi subprovince as produced by microplate interaction. Such a process has given rise to vastly different assemblage distributions in time and space relative to the Uchi subprovince (Corfu and Stott, 1991) with individual assemblages hundreds of kilometers in length.

Implied in the secular variation from early platform assemblages to late unconformable basins is the notion that, at a craton scale, the secular variation in assemblage types reflects epicratonic deposition followed by rifting of earlier Archean orogens and then by an ocean opening and closing cycle. Re-examination of the Superior Province reveals that the architecture of this craton is not a simple matter of arc–arc collisions (Langford and Morin, 1976) but a much more complex scenario.

Implied in the larger scale possible secular variation of assemblage types as well is the idea that the shield to shield regularity in the age of epicratonic sequences (Thurston and Chivers, 1990) may be a function of persistence of cyclical continental aggregation and break-up cycles seen in the Proterozoic (Hoffman, 1988) that persisted into the Archean.

Archean greenstones as a component of crustal evolution

This chapter has shown that Archean greenstone belts comprise a variety of assemblage types, with a certain regularity of secular variation most convincingly displayed in the Superior Province, and possibly present in other Archean cratons. The pattern progresses from early epicratonic assemblages through abundant mafic assemblages indicative of the LIP environment with limited relatively late development of arc sequences and late unconformable basins (Thurston and Chivers, 1990; Thurston 1990a; Jackson et al., 1993). The pattern persists into Proterozoic greenstone belts with a general trend from oceanic assemblages with komatiites and tholeiites (Pharaoh et al., 1987) in the Early Proterozoic, to few recognized komatiites in Late Proterozoic belts. Belts with oceanic assemblages are concentrated along the edges of shields including the West African (Boher et al., 1992), the Baltic, and Laurentian shields (Pharaoh et al., 1987). Assemblages with calcalkaline geochemistry emblematic of subduction appear in the later stages of development of both Archean (e.g. Jackson et al., 1993) and Proterozoic greenstone belts (Lewry et al., 1987). Within-plate geochemical signatures are present in Proterozoic (Wyborn et al., 1987) and Archean (Sylvester et al., 1987) greenstone belts. It is speculated that the relative abundance of the various assemblage types may be related to cycles of continental aggregation (Hoffman, 1987) from the Mid-Archean to the beginning of the Phanerozoic. Any apparent scarcity of komatiite-bearing assemblages in the Phanerozoic may relate to the state of documentation of the modern oceanic plateaus more than any other factor.

SUMMARY

This chapter reviewed the physical volcanology and sedimentology of the lithotectonic assemblage types making up Archean greenstone belts. The various assemblage types and their modern analogues are:

(1) Platforms including quartz arenite-bearing, and carbonate-bearing types. The geochemistry of the associated volcanic units is comparable to flood basalts. Iron assemblages represent pelagic sedimentation of iron formation intercalated with volcanic rocks of various magmatic affinities.

(2) Mafic assemblages include varying proportions of tholeiite, komatiite,

felsic volcanics and iron formation. Physical volcanological and geochemical evidence suggests this assemblage type represents oceanic volcanism.

(3) Intermediate to felsic assemblages are bimodal sequences with physical volcanological and geochemical attributes of arc magmatism. The felsic component of these assemblages is dominated by ash-flow volcanism.

(4) Late unconformable basins contain characteristic alluvial–fluvial sediments and alkaline to calcalkaline volcanics distributed along major shear systems. They are compared to pull-apart basins, piggy-back basins or wrench basins.

(5) Continental volcanic sequences comprising subaerial volcanics and aprons of clastic sediments are considered on geochemical and volcanological evidence to represent magmatic arc activity.

(6) Ophilitic assemblages may exist in the Archean given the variation in stratigraphy and geochemistry of modern ophiolites. Examples of possible Archean ophiolitic material include intrusions caught up in subprovince bounding shear zones and possible ophiolitic volcanics in the Slave and Wyoming cratons.

(7) Sedimentary assemblages in Archean greenstone belts post-date early greenstone volcanism and deformation. Their sedimentology, the volcanology of minor volcanic intercalations, structural and geochronological relations suggest they are accretionary prisms.

(8) The secular variation of assemblage types on several shields progresses from early platforms and iron assemblages to late development of unconformable basins and sedimentary assemblages.

(9) The similarity of ash-flow volcanism in intermediate to felsic assemblages to modern arc systems suggests similar rates of basaltic magma supply and crustal thickness. this relationship implies that most Archean heat loss was through the oceans.

(10) Archean spreading rates and plate sizes were comparable to modern systems. Therefore, the required heat loss likely involves an Archean analogue to Large Igneous Provinces.

REFERENCES

Abbate, E., Bortolotti, V. and Principl, G., 1980. Apennine ophiolites: a peculiar oceanic crust. Ofioliti, 5: 59–96.

Ahmat, A.L. Griffin, T.J., McGoldrick, P.J., Morris, P.A., Swager, C.P. Witt, W.K., Wyche, S. and Hunter, W.M., 1990. Kalgoorlie Granite–Greenstone Terrain. In: S.E. Ho, J.E. Glover, J.S. Myers and J.R. Muhling (Eds.), Third International Archaean Symposium Excursion Guidebook. Univ. of Western Australia, Perth, W.A., pp. 203–304.

Arndt, N.T., and Jenner, G.A., 1986. Crustally contaminated komatiites and basalts from Kambalda, Western Australia. Chem. Geol., 56: 229–255.

Arndt, N.T. and Nesbitt, R.W., 1982. Geochemistry of Munro Township basalts. In: N.T. Arndt and E.G. Nesbit (Eds.), Komatiites. George Allen and Unwin, London, pp. 309–329.

Ayres, L.D., 1983. Bimodal volcanism in Archean greenstone belts exemplified by greywacke composition, Lake Superior Park, Ontario. Can. J. Earth Sci., 20: 1168–1194.

Ayres, L.D. and Corfu, F., 1991. Stacking of disparate volcanic and sedimentary units by thrusting in the Archean Favourable lake greenstone belt, central Canada. Precambrian Res., 50: 221–238.

Ayres, L.D. and Thurston, P.C., 1985. Archean supracrustal sequences in the Canadian Shield: an overview. In: L.D. Ayres, P.C. Thurston, K.D. Card, and W. Weber (Eds.), Evolution of Archean Supracrustal Sequences. Geol. Assoc. Can. Spec. Paper 28: 343–380.

Baragar, W.R.A. and Scoates, R.F.J., 1987. Volcanic geochemistry of the northern segments of the Circum-Superior Belt of the Canadian Shield. In: R.C. Pharaoh, R.D. Beckinsale and D. Rickard (Eds.), Geochemistry and Mineralization of Proterozoic Volcanic Suites. Geol. Soc. London, Spec. Publ. 33: 113–131.

Barley, M.E., Eisenlohr, B.N., Groves, D.I., Perring, C.S. and Vearncombe, J.R., 1989. Late Archean convergent margin tectonics and gold mineralization: a new look at the Norseman–Wiluna belt, Western Australia. Geology, 17: 826–829.

Barrie, C.T. and Shirey, S.B., 1991. Nd and Sr-isotope systematics for the Kamiskotia–Montcalm area: implications for the formation of late Archean crust in the western Abitibi Subprovince, Canada. Precambrian Res., 46: 58–76.

Barrie, C.T., Gorton, M.P., Naldrett, A.J., and Hart, T.R., 1991. Geochemical constraints on the petrogenesis of the Kamiskotia gabbroic complex and related basalts, Western Abitibi subprovince, Ontario, Canada. Precambrian Res., 50: 173–199.

Basaltic Volcanism Study Project 1981.

Bickle, M.J., 1978. Heat loss from the Earth: a constraint on Archaean tectonics from the relation between geothermal gradients and the rate of plate production. Earth Planet. Sci. Lett., 40: 301–315.

Bickle, M.J., Martin, A. and Nisbet, E.G., 1975. Basaltic and peridotitic komatiites and stromatolites above a basal unconformity in the Belingwe greenstone belt, Rhodesia. Earth Planet. Sci. Lett., 27: 155–162.

Brooks, C., Ludden, J.N., Pigeon, Y. and Hubregtse, J.J.M.W., 1982. Volcanism of shoshonite to high K andesite affinity in an Archean arc environment, Oxford Lake Manitoba. Can. J. Earth Sci., 19: 55–67.

Brown, G.F. and Coleman, R.G., 1972. The tectonic framework of the Arabian Peninsula. 24th International Geological Congress, 3: 300–304.

Bryan, W.B. and Moore, J.G., 1977. Compositional variations of young basalts in the Mid-Atlantic Ridge rift valley near lat. 36°49′ N. Geol. Soc. Am. Bull., 88: 556–570.

Burke, K., 1988. Tectonic evolution of the Caribbean. Annu. Rev. Earth Planet. Sci., 16: 201–230.

Capdevila, R., Goodwin, A.M., Ujike, O. and Gorton, M.P., 1982. Trace element geochemistry of Abitibi volcanic rocks and crustal growth in southwestern Abitibi belt. Can. Geol., 10: 418–422.

Card, K.D., 1990.The Superior Province of the Canadian Shield: a product of Archean plate convergence. Precambrian Res., 48: 99–156.

Card, K.D. and Ciesielski A., 1986. Subdivisions of the Superior Province of the Canadian Shield. Geosci. Can., 13: 5–13.

Cattell, A. and Arndt, N., 1987. Low- and high-alumina komatiites from a Late Archean sequence, Newton Township, Ontario. Contrib. Mineral. Petrol., 97: 218–227.

Cathles, L.M., Guber, A.L., Lenagh, T.C. and Dudas, F., 1983. Kuroko-type massive sulfide deposits of Japan: products of an aborted island-arc rift. In: Kuroko and Related Volcanogenic Massive Sulfide Deposits. Economic Geology Monograph, 5, pp. 96–114.

Chown, E.H., Daigneault, R., Mueller, W. and Mortensen, J., 1992 Tectonic evolution of the Northern Volcanic Zone, Abitibi belt, Quebec. Can. J. Earth Sci., 29: 2211–2225.

Coffin, M.F. and Eldholm, O. (Eds.), 1991. Large Igneous Provinces. JOI/USSAC Workshop

Report. The University of Texas at Austin Institute for Geophysics Technical Report No. 114.

Condie, K.C., 1986. Geochemistry and tectonic setting of early Proterozoic supracrustal rocks in the southwestern United States. J. Geol., 94: 845–864.

Corfu, F., 1993. The evolution of the Southern Abitibi Greenstone Belt in light of precise U-Pb geochronology. Econ. Geol., 88: 1323–1340.

Corfu, F. and Ayres, L.D., 1991. U-Pb geochronology in the Archean Favourable lake greenstone belt, northwestern Ontario, Canada: four magmatic events spanning 290 Ma. Precambrian Res., 50: 210–220.

Corfu, F. and Davis, D.W., 1992. A U-Pb geochronological framework for the Western Superior Province, Ontario. In: Geology of Ontario, Ontario Geological Survey Special Volume 4, Part 2, pp. 1335–1348.

Corfu, F., Jackson, S.L. and Sutcliffe, R.H., 1991. U-Pb ages and tectonic significance of late Archean alkalic magmatism and nonmarine sedimentation: Timiskaming Group, southern Abitibi belt, Ontario. Can. J. Earth Sci., 28: 489–503.

Corfu, F. and Noble, S.R., 1992. Genesis of the southern Abitibi greenstone belt, Superior Province, Canada: evidence from zircon Hf-isotope analyses using a single filament technique. Geochim. Cosmochim. Acta, 56: 2081–2097.

Corfu, F. and Stott, G.M., 1993. Age and petrogenesis of two Late Archean Magmatic Suites, Northwestern Superior Province, Canada: Zircon U-Pb and Lu-Hf isotopic relations. J. Petrol., 34: 817–838.

Cortis A.L., Ayres, L.D. and Thurston, P.C., 1989. A quartz-rich Archean fan delta, Sandy Lake Ontario. Geological Association of Canada/Mineralogical Association of Canada, Program with Abstracts 14: A36.

Cortis, A.L., 1991. Geology, Provenance and Depositional environment of the Keewaywin Formation, Sandy lake Greenstone belt, Northwestern Ontario. Unpublished M.Sc. thesis, Univ. of Manitoba, 265 p.

Davis, D.W., 1991. Age constraints on deposition and provenance of Archean sediments in the southern Abitii and Pontiac subprovinces from U-Pb analyses of detrital zircons. Geological Association of Canada/Mineralogical Association of Canada Joint Annual Meeting, Program with Abstracts, p. A29.

Defant, M.J., Maury, R.C., Ripley, E.M., Feigenson, M.D. and Jacques, D., 1991. An example of island-arc petrogenesis: geochemistry and petrology of the southern Luzon arc, Philippines. J. Petrol., 32: 445–500.

de Wit, M.J., Hart, R.A. and Hart, R.J., 1987. The Jamestown Ophiolite Complex, Barberton Mountain belt: a section through 3.5 Ga oceanic crust. J. African Earth Sci., 6: 681–730.

de Wit, M.J., Roering, C., Hart, R.J., Armstrong R.A., de Ronde, C.E.J., Green R.W.E., Tredoux, M., Peberdy, E. and Hart, R.A., 1992. Formation of an Archaean continent. Nature, 357: 553–562.

Desrochers, J.-P., Hubert, C., Ludden, J.N. and Pilote, P., 1993. Accretion of Archean oceanic plateau fragments in the Abitibi greenstone belt, Canada. Geology, 21: 451–454.

Devaney, J.R. and Williams, H.R., 1989. Evolution of an Archean subprovince boundary: A sedimentological and structural study of part of the Wabigoon–Quetico boundary in northern Ontario. Can. J. Earth Sci., 26: 1013–1026.

Dickinson W.R. and Suczek, C.A., 1979. Plate tectonics and sandstone compositions. Am. Ass. Petrol. Geol. Bull., 63: 2164–2182.

Dimroth, E., Cousineau, P., Leduc, M. and Sanschagrin, Y., 1978. Structure and organization of Archean subaqueous basalt flows, Rouyn-Noranda area, Quebec, Canada. Can. J. Earth Sci., 15: 902–918.

Dimroth, E., Imreh, L., Cousineau, P., Leduc, M. and Sanschagrin, Y., 1985. Paleogeographic analysis of mafic submarine flows and its use in the exploration for massive sulphide deposits. In: L.D. Ayres, P.C. Thurston, K.D. Card and W. Weber (Eds.), Evolution of Archean Supracrustal Sequences. Geol. Assoc. Can., Spec. Paper, 28, pp. 203–222.

Dimroth, E., Imreh, L., Goulet, N. and Rocheleau, M., 1983. Evolution of the south-central segment of the Archean Abitibi belt. Quebec. Part III, Plutonic and metamorphic evolution and geotectonic model. Can. J. Earth Sci., 20: 1374–1388.

Fowler, A.D. and Jensen, L.S., 1989. Quantitative trace element modelling of the crystallization history of the Kinojevis and Blake River groups, Abitibi greenstone belt, Ontario. Can. J. Earth Sci., 26: 1356–1367

Frey, F.A., Garcia, M.O., Wise, W.S., Kennedy, A., Gurriet, P. and Albarede, F., 1991. The evolution of Mauna Kea volcano, Hawaii: petrogenesis of tholeiitic and alkalic basalts. J. Geophys. Res., 96: 14,347–14375.

Fyon, J.A., Breaks, F.W., Heather, K.B., Jackson, S.L., Muir, T.L., Stott, G.M. and Thurston, P.C., 1992. Metallogeny of Metallic Mineral Deposits in the Superior Province of Ontario. Geology of Ontario, Ontario Geological Survey, Special Volume 4, Part 2, pp. 1091–1174.

Gaal, G., 1982. Proterozoic tectonic evolution and late Svecokarelian plate deformation of the central Baltic shield; Geol. Rundschau, 71: 158–170.

Gaal, G., 1986. 2200 Million years of crustal evolution: the Baltic shield. Geol. Soc. Finl. Bull, 58: 149–168.

Gabrielse, H. and Yorath, C.J., 1989. DNAG #4, The Cordilleran orogen in Canada. Geoscience Canada, 16: 67–83.

Gariepy, C., Allegre, C.J. and Lajoie, J., 1984. U-Pb systematics in single zircons from the Pontiac sediments, Abitibi greenstone belt. Can. J. Earth Sci., 21: 1296–1304.

Gee, R.D., 1982. Southern Cross 1:250,000 map explanatory notes. Geological Survey of Western Australia, 22 pp.

Gibbs, A.K., 1987 Proterozoic volcanic rocks of the northern Guiana shield, South America, In: T.C. Pharaoh, R.D. Beckinsale, and D. Rickard (Eds.), Geochemistry and Mineralization of Proterozoic volcanic Suites. Geological Society of London, Special Publication 33, pp. 275–288.

Gibson, H.L., Watkinson, D.H. and Comba, C.D.A., 1983. Silicification: hydrothermal alteration in an Archean geothermal system within the Amulet Rhyolite Formation, Noranda, Quebec. Econ. Geol., 78: 954–971.

Giles, C.W. and Hallberg, J.A., 1982. The Genesis of the Archaean Welcome Well Volcanic Complex, Western Australia. Contrib. Mineral. Petrol., 80: 307–318.

Gill, J.B., 1981. Orogenic Andesites and Plate Tectonics, Springer-Verlag, New York, 390 pp.

Goodwin A.M., 1981. Archean plates and greenstone belts. In: A. Kroner (Ed.), Precambrian Plate Tectonics. Elsevier, Amsterdam, pp. 105–135.

Goodwin, A.M., 1991. Precambrian Geology, The Dynamic Evolution of the Continental Crust. Academic Press, New York, 666 pp.

Greeley, R., 1982. The Snake River plain, Idaho: representative of a new category of volcanism. J. Geophys Res. 87: 2705–2712.

Hallberg, J.A., Johnston, C. and Bye, S.M., 1976. The Archaean Marda Igneous Complex, Western Australia. Precambrian Res., 3: 111–136.

Harper, G.D., 1984. The Josephine Ophiolite, Northwestern California. Geol. Soc. Am. Bull., 95: 1009–1026

Harper, G.D., 1985. Dismembered Archean Ophiolite, wind River Mountains, Wyoming (USA). Ofioliti 10: 297–306.

Hart, T.R., 1984. The Geochemistry and Petrogenesis of a Metavolcanic and Intrusive Sequence in the Kamiskotia Area: Timmins, Ontario. M.Sc. thesis, University of Toronto, 179 pp. (unpublished).

Hawkesworth, C.J., O'Nions, R.K., Pankhurst, R.J., Hamilton, P.J., and Evensen, N.M., 1977. A geochemical study of island-arc and back-arc tholeiites from the Scotia Sea. Earth Planet. Sci. Lett., 86: 253–262.

Helmstaedt, H., Padgham, W.A. and Brophy, J., 1986. Multiple dikes in the Kam Group, Yellowknife Greenstone Belt: Evidence for Archean sea-floor spreading? Geology, 14: 562–566.

Helmstaedt, H.H. and Scott, D.J., 1992. The Proterozoic ophiolite problem; In: K.D. Condie (Ed.), Proterozoic Crustal Evolution, Elsevier, Amsterdam, pp. 55–95.

Henderson, J.B., 1981. Archaean basin evolution in the Slave Province, Canada. In: A. Kroner (Ed.), Precambrian Plate Tectonics. Elsevier, Amsterdam. pp. 213–236.

Hickman, A.H., Thorne, A.M., Trendall, A.F., Barley, M.E., Blake, T.S., Barnes, S.J., Horwitz, R.C., Collins, W.J., Teyssier, C., Krapez, B. and Simonson, B.M., 1990. Excursion No. 5: Pilbara and Hamersley Basin. In: S.E. Ho, J.E. Glover, J.S. Myers and J.R. Muhling (Eds.), Third International Archaean Symposium, Perth, 1990. Excursion Guidebook. Geology Department, University of Western Australia, Publication No. 21.

Hill, R.E.T., Barnes, S.J., Gole, M.J. and Dowling S.E., 1990. The physical volcanology of komatiites in the Norseman–Wiluna Belt, pp. 362–397. In: Third International Archaean Symposium, Perth, 1990, Excursion Guidebook, Geology Department, University of Western Australia, Publication No. 21.

Hoffman, P.F., 1973. Evolution of an early Proterozoic continental margin: the Coronation Geosyncline and associated aulocogens of the northwestern Canadian Shield. Philosoph. Trans. Roy. Soc. London A273: 547–581.

Hoffman, P.F., 1980. Wopmay Orogen: a Wilson-cycle of early Proterozoic age in the northwest of the Canadian Shield. In Geological Association of Canada Special Paper 20: 523–549.

Hoffman, P.F., 1988. United plates of America, the birth of a craton; Early Proterozoic assembly and growth of Laurentia. Annu. Rev. Earth Planet. Sci., 16: 543–603.

Hoffman, P.F. and Ranalli, G., 1988. Oceanic flake tectonics. Geophys. Res. Lett.,. 15: 1077–1080.

Hopson, C.A., Coleman, R.G., Gregory, R.T., Pallister, J.S. and Bailey, E.H., 1981. Geological section through the Samail ophiolite and associated rocks along a Muscat–Ibra transect, southeastern Oman Mountains. J. Geophys. Res., 86: 2527–2544.

Hudleston, P.J., Schultz-Ela, D. and Southwick, D.L., 1988. Transpression in an Archean greenstone belt, northern Minnesota; Can. J. Earth Sci., 25: 1060–1068.

Huppert H.E., and Sparks, R.S.J., 1985. Komatiites I: eruption and flow. J. Petrol., 26: 694–725.

Hyde, R.S., 1978. Sedimentology, Volcanology, Stratigraph, and Tectonic Setting of the Archean Timiskaming Group, Abitibi Greenstone Belt, Northeastern Ontario, Canada. Ph.D. Thesis. McMaster University, Hamilton, Ont., 422 pp.

Hyde, R.S., 1980. Sedimentary facies in the Archean Temiskaming Group, and their tectonic implications, Abitibi greenstone belt, northeastern Ontario, Canada. Precambrian Res., 12: 141–160.

Ikeda, Y. and Yuasa, M., 1989. Volcanism in nascent back-arc basins behind the Shichto ridge and adjacent areas in Izu–Ogasawara arc, northwest pacific: evidence for mixing between E-type MORB and island arc magmas at the initiation of back-arc rifting. Contrib. Mineral. Petrol., 104: 377–393.

Jackson S.L. and Fyon A.J., 1991. Western Abitibi Subprovince. In: Geology of Ontario. Ontario Geological Survey Special volume 4, Part 1, pp. 405–482.

Jackson, S.L., 1993. The Precambrian geology of Pacaud and Catharine townships and parts of adjacent townships, district of Timiskaming. Ontario Geological Survey, Open File Report.

Jackson, S.L., Fyon, J.A. and Corfu, F., 1993. Review of Archean supracrustal assemblages of the southern Abitibi greenstone belt in Ontario, Canada: products of microplate interactions within a large-scale plate-tectonic setting. Precambrian Res., 65: xxx–xxx.

Jarvis, G.T. and Campbell, I.H. 1983. Archean komatiites and geotherms: solution to an apparent contradiction. Geophys. Res. Lett., 10: 1133–1136.

Johns, G.W.,, 1985. A volcanic facies interpretation of the Berry River Formation, In: G.P. Beakhouse (Ed.), Institute on Lake Superior Geology, 31st Meeting, Field Trip Guidebook, pp. 105–156.

Johnson, R.W., Jaques, A.L., Langmuir, C.H., Perfit, M.R., Staudigel, H., Dunkley, P.N., Chappell, B.W., Taylor, S.R. and Baekisapa, M., 1987. Ridge subduction and forearc volcanism: petrology and geochemistry of rocks dredged from the western Solomon arc and Woodlark basin. In: B. Taylor and N.F. Nixons (Eds.), Marine Geology, Geophysics, and Geochemistry of the Woodlark Basin–Solomon Islands. Circum Pacific Council for Energy and Mineral Resources. Earth Sci. Ser., 7: 155–226.

Kazansky, V.I. and Moralev, V.M., 1981. Archean geology and metallogeny of the Aldan Shield, USSR, In: J.E. Glover and D.I. Groves (Eds.), Archaean Geology Geological Society of Australia, Special Publication 7, pp. 111–120.

Klein, E.M., Langmuir, C.H. and Staudigel, H., 1991. Geochemistry of basalts from the southeast Indian ridge, 115E–138E. J. Geophys. Res., 96: 2089–2017.

Kozhevnikov, V.N., Slabunov, A.Y. and Systra, Y.Y. 1992. Guidebook of the geological excursion on the Archaean of Northern Karelia. Karelian Research Centre, Russian Academy of Science, Institute of Geology, 64 pp.

Krapez, B. and Barley, M.B., 1987. Archaean strike-slip faulting and related ensialic basins: evidence from the Pilbara block, Australia. Geol. Mag., 124: 555–567.

Kroner, A., Puustinen, K. and Hickman, M., 1981. Geochronology and an Archean tonalitic gneiss dome in northern Finland and its relation with an unusual overlying volcanic conglomerate and komatiitic greenstone. Contrib. Mineral. Petrol., 76: 33–41.

Kulikov, V.S. and Golubev, A.I., 1984. Geological Field Trips in Karelia, Guidebook. Inst. of Geology Academy of Sciences of the USSR Karelian Branch, 106 pp.

Kusky, T.M., 1990. Accretion of the Archean Slave province. Geology, 17: 63–67.

Kusky, T.M., 1991. Structural development of an Archean orogen, Western Point Lake, Northwest Territories. Tectonics, 10: 820–841.

Kusky, T.M. and Vearncombe, J.R., 1994. Structure of Archean greenstone belts. In: M.J. de Wit and L. Ashwal (Eds.), Tectonic Evolution of Greenstone Belts. Oxford Monograph on Geology and Geophysics, Oxford University Press.

Langford F.F. and Morin J.A. 1976.The development of the Superior province of northwestern Ontario by merging island arcs. Am. J. Sci., 276: 1023–1034.

Lesher, C.M., Goodwin, A.M., Campbell, I.H. and Gorton, M.P. 1986. Trace-element geochemistry of ore-associated and barren, felsic metavolcanic rocks in the Superior Province, Canada. Can. J. Earth Sci., 23: 222–237.

Lewry, J.F., Macdonald, R., Livesey, C., Meyer, M., Van Schmus, R. and Bidkfor, M.E., 1987. U-Pb geochronology of accreted terranes in the Trans-Hudson Orogen, Northern Saskatchewan, Canada. In: T.C. Pharaoh, R.D. Beckinsale and D. Rickard (Eds.), Geochemistry and Mineralization of Proterozoic Volcanic Suites. Geological Society Special Publication No. 33, pp. 147–166.

Ludden, J., 1989. Geochemical constraints on the origin of late Archaean magmas: examples from the Superior Province of Canada. Geol. Ass. Canada/Mineral Ass. Canada, Prog. Abstr., 14, p. A8.

Macdonald, R., McGarvie, D.W., Pinkerton, H., Smith, R.L. and Palacz, Z.A., 1990. Petrogenetic evolution of the Torfajokull volcanic complex, Iceland. I. Relationship between the magma types. J. Petrology, 31: 183–202.

MacKenzie, D. and Bickle, M.J., 1988. The volume and composition of melt generated by extension of the lithosphere. J. Petrology, 29: 625–679.

Marshak, S., Alkmim, F.F., Jordt-Evangelista, H. and Brueckner, H.K., 1993. Aspects of the Precambrian Tectonic Evolution of the Southern Sao Francisco Craton and its Eastern margin, Brazil. Geological Society of America, Abstracts with Program, Boston Meeting, p. A298.

McCall, G.J.H., 1981. Progress in research into the early history of the Earth: A review, 1970–1980. In: J.E. Glover and D.I. Groves (Eds.), Archaean Geology, Second International Archaean Symposium, Perth 1980. Geological Society of Australia Special Publication 7, pp. 3–18.

Moores, E.M., 1982. Origin and emplacement of ophiolites. Rev. Geophys. Space Physics, 20: 735–760.

Moores, E.M., 1986. The Proterozoic ophiolite problem, continental emergence, and the Venus connection. Science, 234: 65–68.

Morin, D., Jebrak, M., Bardoux, M. and Goulet, N., 1993. Pontiac metavolcanic rocks within the Cadillac tectonic zone, McWatters, Abitibi Belt, Quebec. Can. J. Earth Sci., 30: 1521–1530.

Mueller, W. and Donaldson, J.A. 1992. Development of sedimentary basins in the Archean Abitibi belt, Canada: an overview. Can. J. Earth Sci., 29: 2249–2265.

Nicholson, H., Condomines, M., Fitton, J.G., Fallick, A.E., Gronvold, K. and Rogers, G., 1991. Geochemical and isotopic evidence for crustal assimilation beneath Krafla, Iceland. J. Petrology., 32: 1005–1020.

Ojakangas, R.W., 1985. Review of Archean clastic sedimentation, canadian shield: major felsic volcanic contributions to turbidite and alluvial fan–fluvial facies associations. In: L.D. Ayres, P.C. Thurston, K.D. Card and W. Weber (Eds.), Evolution of Archean Supracrustal Sequences. Geological Association of Canada Special Paper 28, pp. 23–47.

Ontario Geological Survey 1992. Tectonic assemblages of Ontario. Ontario Geological Survey Maps, 2575–2578.

Pearce, J.A., Lippard, S.J. and Roberts, S., 1984. Characteristics and tectonic significance of suprasubduction zone ophiolites, In: B.P. Kokellar and M.F. Howell (Eds.), Marginal Basin Geology, Geological Society of London Special Publication 16, pp. 77–94.

Peltier, W.R., Forte, A.M., Mitrovica, J.X. and Dziewonski, A.M., 1992. Earth's gravitational field: seismic tomography resolves the enigma of the Laurentian anomaly. Geophys. Res. Lett., 19: 1555–1558.

Percival. J.A. and Williams, H.R., 1989. The late Archean Quetico accretionary complex, Superior Province, Canada. Geology, 17: 23–25.

Pharaoh, T.C., Warren, A. and Walsh, N.J., 1987. Early Proterozoic metavolcanic suites of the Northernmost part of the Baltic Shield, In: T.C. Pharaoh, R.D. Beckinsale and D. Rickard (Eds.), Geochemistry and Mineralization of Proterozoic Volcanic suites. Geological Society, Special Publication 33, pp. 41–58.

Redman, B.A. and Keays, R.R., 1985. Archaean basic volcanism in the Eastern Goldfields Province, Western Australia. Precambrian Res., 30: 113–152.

Robert F. and Brown A.C., 1986. Archean gold-bearing quartz veins at the Sigma mine, Abitibi greenstone belt, Quebec. Part I. Geologic relations and formation of the vein system. Econ. Geol., 81: 578–592.

Robinson, P.T., Melson, W.G., O'Hearn, T. and Schmincke, H.-U., 1983. Volcanic glass compositions of the Troodos ophiolite, Cyprus.

Schau, M., 1977. Komatiites and quartzites in the Archean Prince Albert Group. In: W.R.A. Baragar,

J.M. Coleman and J.M. Hall (Eds.), Volcanic Regimes in Canada. Geological Association of Canada, Special Paper 16, pp. 341–354.

Schwerdtner, W.M., Sutcliffe, R.H. and Trueng, B., 1978. Patterns of total strain in the crustal region of immature diapirs. Can. J. Earth Sci., 12: 1437–1447.

Shaw, H.R., 1985. Links between magma-tectonic rate balances, plutonism and volcanism. J. Geophys. Res., 90: 11275–11288.

Sigmarsson, O., Hemond, C., Condomines, M., and Fourcade, S., 1991. Origin of silicic magma in Iceland revealed by Th isotopes. Geology, 19: 621–624.

Sleep, N.H., 1992. Archean Plate Tectonics: what can be learned from continental geology. Can. J. Earth Sci., 29: 2072–2086.

Sleep, N.H. and Windley, B.F., 1982. Archean plate tectonics: Constraints and inferences. J. Geology, 90: 363–379.

Smith, I.E.M., 1980. Geochemical evolution of the Blake River Group, Abitibi greenstone belt. Superior Province. Can. J. Earth Sci., 17: 1292–1299.

Smith, R.L., 1979. Ash-flow magmatism. In: C.E. Chapin and W.E. Elston (Eds.), Ash-flow Tuffs. Geological Society of America, Special Paper 180, pp. 1–27.

Smith, R.L. and Bailey, R.A., 1968. Resurgent cauldrons. In: R.R. Coats, R.L. Hay and C.A. Anderson (Eds.), Studies in Volcanology. Geological Society of America Memoir 116, pp. 613–662.

Solomon, S.C. and Toomey, D., 1992. The structure of mid-ocean ridges. Annu. Rev. Earth Planet. Sci., 20: 329–364.

Soo, K.Y., 1988. A geochemical and petrological study of volcanic rocks in the Beardmore–Geraldton Archaean greenstone belt, Northwestern Ontario. Unpublished M.Sc thesis, Brock University, St. Catherines, Ont. 149 pp.

Srinivasan, R., and Ojakangas, R.W., 1986. Sedimentology of quartz pebble conglomerates and quartzites of the Archean Bababudan Group, Dharwar Craton, south India, evidence for early crustal stability. J. Geology, 94: 199–214.

Stern, R.J., Lin, P., Morris, J.D.,Jackson, M.C., Fryer, P., Bloomer, S.H. and Ito, E., 1990. Enriched back-arc basin basalts from the northern Mariana Trough: implications for the magmatic evolution of back-arc basins. Earth Planet. Sci. Lett., 100: 210–225.

Stevenson, R.K. and Patchett, P.J., 1990. Implications for the evolution of continental crust from Hf isotope systematics of Archean detrital zircons. Geochim. Cosmochim. Acta, 54: 1683–1697.

Storey, M., Mahoney, J.J., Kroenke, L.W. and Saunders, A.D., 1991. Are oceanic plateaus sites of komatiite formation? Geology, 19: 376–379.

Stott, G.M., 1986. A Structural Analysis of the Central Part of the Archean Shebandowan Greenstone Belt and a Crescent-shaped Granitoid Pluton, Northwestern Ontario. Ph.D. Thesis, University of Toronto, Toronto, Ontario, 285 pp. (unpublished).

Stott, G.M. and Corfu, F., 1991. Uchi subprovince. In: The Geology of Ontario, Ontario Geological Survey, Special Volume 4, Part 1, pp. 144–236.

Sylvester, P.J., Attoh, K. and Schulz, K.J., 1987. Tectonic study of late Archean bimodal volcanism in the Michipicoten (Wawa) Greenstone Belt, Ontario. Can. J. Earth Sci., 24: 1120–1134.

Talbot, C.J., 1973. A plate tectonic model for the Archaean crust. Philosoph. Trans. Roy. Soc. London, A273: 413–427.

Tarney J., Dalziel, I.W.D. and de Wit, M.J., 1976. Marginal basin 'rocas verdes' complex from Southern Chile: a model for Archaean greenstone belt formation. In: B.F. Windley (Ed.), The Early History of the Earth. Wiley-Interscience London, pp. 131–136.

Taylor, S.R., 1987. Geochemical and petrological significance of the Archaean–Proterozoic boundary, In: T.C. Pharaoh, R.D. Beckinsale and D. Rickard (Eds.), Geochemistry and Mineralization of Proterozoic Volcanic Suites. Geological Society of London, Special Publication, 33, pp. 3–8.

Taylor, S.R. and Hallberg, J.A., 1977. Rare-earth elements in the Marda calc-alkaline suite: an Archean geochemical analogue of Andean type volcanism. Geochim. Cosmochim. Acta, 41: 1125–1129.

Thurston, P.C. 1986. Volcanic cyclicity in mineral exploration; the caldera cycle and zoned magma chambers. Ont. Geol. Survey Misc. Paper 129, pp. 104–123.

Thurston, P.C., Byerly, G.R. and Harper, G.D., 1993. Volcanic aspects of Archean Greenstone belts. In: M. de Wit and L. Ashwal (Eds.), Tectonic Evolution of Greenstone Belts. Oxford University Press.

Thurston, P.C., 1980. Subaerial volcanism in the Archean Uchi–Confederation volcanic belt. Precambrian Res., 12: 79–98.

Thurston, P.C., 1981. Economic evaluation of Archaean felsic volcanic rocks using REE geochemistry. In: J.E. Glover and D.I. Groves (Eds.), Archaean Geology, Special Publication No. 7, Geological Society of Australia, pp. 439–450

Thurston, P.C., 1981. The volcanology and trace element geochemistry of cyclical volcanism in the Archean Confederation lake area, Northwestern Ontario. Ph.D. thesis, University of Western Ontario, London, 553 pp.

Thurston, P.C., 1990. The Superior Province emphasizing greenstone belts. In: S.E. Ho, F. Robert and D.I. Groves (Compilers), Gold and Base-metal Mineralization in the Abitibi Subprovince, Canada, with Emphasis on the Quebec Segment. Geology Department, University of Western Australia, Publication 24.

Thurston, P.C., 1990a. Early Precambrian basic rocks of the Canadian Shield. In: R.P. Hall and D.J. Hughes (Eds.), Early Precambrian Basic Magmatism. Blackie and Son Limited, London, pp. 221–247.

Thurston, P.C., Ayres, L.D., Dimroth, E., Easton, R.M. and Johns, G.W., 1986. Archean volcanology — progress to date. In: International Volcanology Congress Abstract, p. 358.

Thurston, P.C., Ayres, L.D., Edwards, G.R., Gelinas, L., Ludden, J.N. and Verpaelst, P., 1985. Archean bimodal volcanism. In: L.D. Ayres, P.C. Thurston, K.D. Card and W. Weber (Eds.), Evolution of Archean Supracrustal Sequences. Geological Association of Canada Special Paper 28, pp. 7–22.

Thurston, P.C., Cortis, A.L. and Chivers, K.M., 1987. A reconnaissance re-evaluation of a number of northwestern greenstone belts: evidence for an early Archean sialic crust. In R.B. Barlow, M.E. Cherry, A.C. Colvine, B.O. Dressler and O.L. White (Eds.), Summary of Field Work and Other Activities 1987. Ontario Geological Survey, Miscellaneous Paper 137, pp. 4–24.

Thurston, P.C., Osmani, I.A. and Stone, D. 1991. Northwestern Superior Province: Review and terrane analysis. In: Geology of Ontario, Ontario Geological Survey, Special Volume 4, Part 1, pp. 81–144.

Thurston, P.C. and Fryer, B.J., 1983. The geochemistry of repetitive cyclical volcanism from basalt through rhyolite in the Uchi-Confederation greenstone belt, Canada. Contrib. Mineral. Petrol., 83: 204–226.

Tipper, H.W. Woodsworth, G.J. and Gabrielse, H., 1981. Tectonic assemblage map of the Canadian Cordillera. Geological Survey of Canada, Map 1505A, Scale 1:2,000,000.

Turner, J.P., 1992. Evolving alluvial stratigraphy and thrust fault development in the West Jaca piggyback basin, Spanish Pyrenees. Geological Society of London, 149, pp. 51–63.

Ujike, O., 1985. Geochemistry of Archean alkalic volcanic rocks from the Crystal Lake area, east of Kirkland Lake, Ontario, Canada. Earth Planet. Sci. Lett., 73: 333–344.

Varne, R., 1985. Ancient subcontinental mantle: a source for K-rich orogenic volcanics. Geology, 13: 405–408.

Williams, H.R. Stott, G.M. and Thurston, P.C. 1992. Tectonic Evolution of Ontario: Summary and Synthesis, Part 1: Revolution in the Superior Province. In: Geology of Ontario, Ontario Geological Survey, Special Vol. 4, Part 2. pp. 1253–1294.

Williams, H.R., 1991. Quetico subprovince. In: Geology of Ontario, Ontario Geological Survey, Special Vol. 4, part 1, pp. 383–403.

Windley, B.F., 1984. The Archean–Proterozoic boundary. Tectonophysics, 105: 43–53.

Wood, D.A., 1978. major and trace element variations in the Tertiary lavas of eastern Iceland and their significance with respect to the Iceland geochemical anomaly. J. Petrology, 19: 393–436.

Wood, D.A., Joron, J.L. and Treuil, M., 1979. A re-appraisal of the use of trace elements to classify and discriminate between magma series erupted in different tectonic settings. Earth Planet. Sci. Lett., 45: 326–336.

Wood, J., 1980. Epiclastic sedimentation and stratigraphy in the North Spirit Lake and Rainy Lake areas: a comparison. Precambrian Res., 12: 227–255.

Wyborn, L.A.I., Page, R.W. and Parker, A.J., 1987. Geochemical and geochronological signatures in Australian Proterozoic igneous rocks. In: T.C. Pharaoh, R.D. Beckinsale and D. Rickard (Eds.), Geochemistry and Mineralization of Proterozoic Volcanic Suites. Geological Society, Special Publication No. 33: 377–394.

Wyman D. and Kerrich, R., 1988. Alkaline magmatism, major structures, and gold deposits; Implication for greenstone belt gold metallogeny. Economic Geol., 83: 454–461.

Chapter 3

GREENSTONES THROUGH TIME

KENT C. CONDIE

INTRODUCTION

Although prior to 20 years ago, greenstones were thought to represent an Archean phenomenon, it is now clear that greenstones have formed throughout geologic time (Condie, 1989; 1990a; Taira et al., 1992). It seems to be equally clear that all greenstones do not represent the same tectonic setting, nor do the proportions of preserved greenstones of a given age and tectonic setting necessarily reflect the original proportions of that tectonic setting (Condie, 1990a; 1993). To better understand the tectonic settings of greenstones and their role in the history of continents and ocean basins, it is of interest to compare greenstones through time. Some studies have compared the geochemistry of greenstone volcanics of different ages with emphasis on basalts, and have recognized secular changes, especially at the Archean/Proterozoic (A/P) boundary (Condie, 1989; 1990a; Arndt, 1991). Others have focused on granitoids associated with greenstones, and have reported secular geochemical changes, again with the most prominent changes at the A/P boundary (Martin, 1986; 1987). Many investigators have compared Archean greenstones with young arc volcanics, pointing out similarities and differences and their possible meaning in terms of tectonic setting (Thurston and Chivers, 1990; Taira et al., 1992; Williams et al., 1992). Although a large data base exists in the scientific literature, few investigators have compared Archean, Proterozoic, and Phanerozoic greenstones in terms of lithologic and stratigraphic features.

In this chapter, lithologic, stratigraphic, and geochemical parameters will be compared in greenstones through time in an attempt to better understand the evolution of tectonic settings from the Archean to the present. Although this is an awesome task, we now have enough stratigraphic, lithologic, geochemical, and precise U/Pb zircon geochronological data to begin such comparisons.

The first problem encountered in comparing Archean and post-Archean greenstones is that of deciding on what constitutes a "greenstone". This is a nontrivial question in that the tighter the restraints on greenstone definition, the greater the number of secular differences that may be recognized. At one extreme, it is possible to define a greenstone as any supracrustal sequence in which volcanic rocks occur, while at the other extreme, one may formulate specific lithologic, stratigraphic, and geochemical constraints for a greenstone. From lithologic and geochemical data, it is clear that several different tectonic settings are represented in successions that have been referred to as greenstone belts. Even among Archean greenstones there is a great deal of diversity in lithologic packages and geochemistry (Condie, 1981; 1989; Nisbet, 1987; Thurston, 1990; Thurston and Chivers, 1990; Williams et al., 1992). One way to compare greenstones with time is to consider only submarine successions in which volcanic rocks are important constituents (Condie, 1989; 1990a; 1993). This eliminates cratonic basin successions, continental flood basalts, and most continental rift successions that have been grouped with greenstones by some investigators. This study includes only greenstones in which the combined submarine volcanic (including contemporary sills) and volcaniclastic sediment (including graywackes) component exceeds 50%, and the remaining rocks are chiefly chert or carbonate. Arkose, conglomerate, quartz arenite, shale, and related sediments are less than 15% in the greenstones included. This definition also excludes the Archean platform type greenstones (Thurston and Chivers, 1990), which may contain significant amounts of quartz arenite and fluvial sediments, and in some cases, iron formation. In addition, successions that contain basalts or other volcanics significantly enriched in LILE (large ion lithophile elements), such as basalts characteristic of continental rifts, are also excluded, as are ophiolites since they have not been confirmed to exist in the Archean (Condie, 1989; Helmstaedt and Scott, 1992). Thus from a modern perspective, the greenstones considered in this chapter are those that may have formed as island arcs, continental-margin arcs, submarine plateaus, oceanic islands, and in some instances in the Archean, as oceanic crust.

A second problem relevant to greenstone definition relates to terranes. Although tectonostratigraphic terranes are now widely recognized in Phanerozoic continental crust, it has only been recently that terranes have been described in Precambrian crust (Kusky, 1989; Myers, 1992; Park, 1991; Condie, 1992; Desrochers et al., 1993; Nutman et al., 1989; 1993; Williams et al., 1992). Some stratigraphic successions in Archean greenstone belts are now known to include sections from two or more terranes. A major question that one faces in unraveling the history of oceanic terranes is that of how many greenstone assemblages constitute a given terrane. For instance, in Wrangellia, a superterrane accreted to western North America during the Cretaceous, single stratigraphic sections have several major unconformities and faults that break the sections (Plafker et al., 1989). Do we consider each segment of a section between faults and unconformities as a separate assemblage or do we group the entire succession into one assemblage? In this study, I will follow the usage of Williams et al. (1992) and use the term tectonic assemblage to refer to a package of supracrustal rocks deposited during a discrete time interval, and bounded by structural discontinuities, unconformities, or intrusions. Thus, each segment of a broken stratigraphic section is considered as a distinct assemblage. The term terrane will follow the usage of Jones et al. (1983) and refer to a fault-bounded segment of crust with a geologic history distinct from adjacent terranes. Terranes may contain from one to many tectonic assemblages, and may accrete into superterranes before finally colliding with, and becoming part of a continent. In this respect, most greenstone belts can be considered as terranes (or superterranes) comprising one or more greenstone assemblages (Williams et al., 1992; Desrochers et al., 1993; Kimura et al., 1993). For instance, the Archean Abitibi greenstone belt is a terrane composed of several to many assemblages (Lafleche et al., 1992; Williams et al., 1992; Desrochers et al., 1993) and the Wawa subprovince in western Ontario (Williams et al., 1992) is a superterrane comprising several terranes.

GREENSTONE TECTONIC ASSEMBLAGES AND TERRANES

Stratigraphic assemblages have been measured and described in numerous greenstone belts, with most of the detailed studies, curiously, coming from Archean greenstones. Published ages, thicknesses, and lithologic proportions in Archean, Proterozoic, and Phanerozoic greenstone tectonic assemblages are tabulated from the literature in Tables 1–3. Maximum ages are inferred from the oldest stratigraphically dated volcanics in a section, and minimum ages are estimated from either the youngest dated volcanic unit or the age of intrusive syntectonic granitoids. While in Phanerozoic successions, ages are constrained chiefly by fossil assemblages, Precambrian greenstones ages are based on U/Pb zircon ages.

Measured stratigraphic thicknesses of greenstone assemblages range from <3 to >20 km. However, these should not be considered as original thicknesses in that unconformities, faults, and intrusions may have eliminated portions of sections, and deformation may have duplicated sections. Rather, the thicknesses given in Tables 1–3 are "preserved thicknesses", and in some cases they may be maximum values, while in others they are minimum values. Lithologic proportions in each greenstone assemblage are estimated from published stratigraphic sections referenced in the tables, and tabulated data probably have errors of up to 25%.

A plot of the difference between maximum and minimum greenstone assemblage age, herein referred to greenstone duration, versus preserved thickness shows that most Precambrian assemblages accumulated in <50 My, while Phanerozoic assemblages typically record durations of up to 150 My (Fig. 1). Archean greenstones show a larger variability in measured stratigraphic thickness than post-Archean greenstones, and all sections >12 km thick are Archean in age. Detailed field work in the western Superior Province, which probably has the best geochronologic control of any Archean granite–greenstone province, has shown that many of the thick greenstone sections described in the earlier literature are either tectonically repeated or mixtures of more than one tectonic assemblage, and that most of these assemblages are <10 km thick (Williams et al., 1992; P.C. Thurston, pers. comm., 1993) (Fig. 1). Similar tectonic duplication has been described in the Kalgoorlie greenstone belt in the Yilgarn Province of Western Australia (Martyn, 1987). The fact that some Archean greenstone assemblages,

TABLE 1

Age, thickness, and lithologic proportions (in percent) in Phanerozoic greenstone assemblages

Ref.	Max age (Ma)	Min age (Ma)	Thickness (km)	Basalt	Felsic Volc	And	Gw	Chert-BIF	Carb	Flow	Frag
1	34	9	6.1	5.0	7.5	0	87.5	tr	0	40	60
2	75	60	4.2	8.3	12.5	0	75.0	4.2	0	39	61
3	260	248	1.5	14.5	7.5	3.5	72.5	0	2.0	75	25
4	200	100	4.3	18.8	0	0	41.4	36	3.8	80	20
5	240	140	9.75	71.6	0	0	10	4.2	14.2	83	17
6	245	125	4.0	15.0	6.2	9.0	58.5	3.0	8.3	50	50
7	97	17	7.0	9.7	0	3.7	83.1	3.5	0	72	28
8	125	60	7.1	14.2	0	10.0	75.2	2.6	0	90	10
9	130	70	7.0	6.1	37.2	23.6	31.2	1.9	0	53	47
10	300	185	6.8	28.9	9.9	13.1	4.6	1.2	42.3	40	60
11	225	175	3.5	15.4	54.2	9.6	12.5	2.2	6.1	9.2	90.8
12	275	190	3.2	8.2	33.6	4.5	44.6	1.8	7.3	88	12
13	280	230	1.8	30.2	0	6.1	53.3	1.6	8.8	70	30
14	300	165	4.3	13.3	12.5	0	60.5	1.4	12.3	51.5	48.5
15	440	395	2.15	11.6	0	0	32.6	2.4	53.4	50	50
16	440	365	3.75	58.1	12.5	5.3	19.8	4.3	0	65	35
17a	230	190	5.3	57.0	0	0	9.1	9.4	24.5	80	20
17b	180	150	2.3	0	0	0	82.4	1.6	15.0		
18	300	255	3.3	21.6	0	27.0	35.2	5.4	10.8	50	50
19	200	110	4.2	14.7	9.6	7.7	54.2	3.5	10.3	59	41
20	85	70	5.8	20.3	0	20.5	44.4	2.8	12.0	50	50
21	60	45	4.5	11.7	0	0	85.8	2.5	0	48	52
22	480	430	6.7	22.4	19.4	4.1	46.0	7.1	1.0	54	46
23	470	430	6.5	32.3	13.9	34.9	11.5	4.3	3.1	29.6	70.4
24	440	430	6.3	79.3	0	7.2	5.6	7.9	0	58.2	41.8
25	460	425	6.2	68.6	8.1	0	16.1	3.2	4.0	52	48

Felsic Volc = felsic volcanics; And = andesite; Koma = komatiite; Gw = graywackes and related volcaniclastic sediments; BIF = banded Fe formation; Carb = carbonates; Frag = fragmental volcanics; tr = trace. Reference key: 1. Muro Group, Japan (Kumon et al., 1988); 2. Ryujin Fm., Japan (Kumon et al., 1988); 3. Takitimu Gp., NZ (Houghton and Landis, 1989); 4. Tamba Gp., Japan (Imoto, 1984); 5. Karmutsen Fm., BC (Barker et al., 1989); 6. Alexander terrane I, BC (Admiralty subterrane) (McClelland and Gehrels, 1990); 7. Alexander terrane II, n. Kupreanof Isl., BC (McClelland et al., 1992); 8. Alexander terrane III, Duncan Canal, BC (McClelland et al., 1992); 9. Alexander terrane IV, Cape Fanshaw, BC (McClelland et al., 1992); 10. Stikine terrane, BC (Brown et al., 1991); 11. Blue Mtns, OR (Charvet et al., 1990); 12. Pine Forest, CA (Wyld, 1990); 13. N. Sierra Nevada, CA (Wyld, 1990); 14. Eastern Klamaths, CA (Wyld, 1990); 15. Yreka terrane (Horseshoe Gulch), CA (Potter et al., 1990); 16. Yreka terrane (Gregg Ranch), CA (Potter et al., 1990); 17. Wrangellia, Wrangell Mtns., AK (Plafker et al., 1989); 18. Southern Wrangellia, BC (Plafker et al., 1989); 19. Peninsular terrane, AK (Plafker et al., 1989); 20. Chugach terrane, AK (Plafker et al., 1989); 21. Prince William terrane, AK (Plafker et al., 1989); 22. Tumut Region, E Australia (Stuart-Smith et al., 1992); 23. Cutwell Gp., Nfld (Kean and Strong, 1975); 24. Moretons Harbour Gp, Nfld (Strong and Payne, 1973); 25. Terrane 5, Scandinavia (Stephens and Gee, 1989).

TABLE 2

Age, thickness, and lithologic proportions (in percent) in Proterozoic greenstone assemblages.

Ref	Max age (Ma)	Min age (Ma)	Thickness (km)	Basalt	Felsic Volc	And	Gw	Chert-BIF	Carb	Flow	Frag
1	1270	1200	11.4	42.2	10.0	22.6	14.4	0	10.8	74.3	25.7
2	620	415	6.2	54.2	11.8	22.9	9.9	1.2	0	59.7	40.3
3	2100	2075	8.7	33.0	17.8	24.1	23.9	1.2	0	36	64
4	2100	2080	10.0	56.1	3.1	31.7	5.4	2.7	1.0	67.5	32.5
5	1650		7.0	34.8	10.5	10.0	20.3	11.4	13.0	55	45
6	2100	2037	5.0	54.0	21.5	6.2	14.7	3.6	0	77	23
7	1000	708	7.3	36.6	25.2	0	37.0	1.2	0	59	41
8	1900	1885	9.0	28.8	10.6	12.5	47.1	1.0	0	49	51
9	1900	1870	12.0	3.0	73.2	11.3	11.5	0	1.0	17	83
10	1750	1732	7.5	42.3	15.2	29.7	10.3	2.5	0	40	60
11	1750	1730	5.8	38.0	14.3	38.9	6.8	2.0	0	52.7	47.3
12	1750	1728	7.9	10.6	39.1	19.2	28.7	2.4	0	20	80
13	1720	1720	10.4	15.2	31.4	19.6	31.9	1.4	0.5	29.5	70.5
14	1900		2.0	29.8	18.7	23.9	25.6	2.0	0	53.8	46.2
15	1900	1880	7.2	22.0	6.8	15.5	54.2	1.5	0	65	35
16	1900	1870	7.5	37.2	38.8	10.6	11.4	2.0	0	50	50
17	2100		9.2	36.7	14.8	22.2	24.8	1.5	0	71.1	28.9
18	2100		8.5	39.0	15.6	22.8	20.8	1.8	0	89	11
19	2100		6.5	41.1	16.9	21.7	19.0	1.3	0	85	15
20	1730	1710	6.3	28.8	30.1	10.2	30.6	0.3	0	42	58
21	1900	1865	2.6	57.2	11.0	21.0	8.8	2.0	0	63	37
22	1900	1882	9.9	13.0	12.0	62.8	11.2	1.0	0	59	41
23	1740	1720	2.0	51.0	31.3	5.4	20.0	1.2	0	40	60
24	1780	1762	3.6	46.0	32.5	8.2	21.1	1.0	0.6	14	86
25	600	400	24.0	7.2	33.6	16.6	41.4	1.2	0	7.1	92.9
26	1900	1850	6.0	59.5	0	18.2	21.0	1.3	0	80	20
27	1900	1865	7.2	54.7	10.2	34.1	0	1.0	0	78	22
28	1770	1740	5.0	29.0	30.0	3.0	35.0	2.0	0	58	42
29	1770	1735	7.0	53.2	28.6	5.2	11.1	1.2	0	35	65
30	1720	1700	5.0	44.0	15.0	1.0	39.5	1.5	0	50	50
31	1700	1675	7.5	21.0	16.0	10.0	53.5	1.6	0	35	65
32	1950	1920	10.8	47.2	9.5	9.0	0	2.1	0	71.5	28.5
33	1900	1875	9.7	20.0	13.5	22.5	0	1.2	0	35.7	64.3
34	1900	1872	9.5	24.8	7.5	15.7	0	1.0	0	51.6	48.3

See Table 1 for abbreviations. 1. Bishop Corners, Hastings area, Canada (Moore, 1977); 2. E-Cent Egypt, Pan-African (Stern, 1981); 3. Nangodi greenstone, NE Ghana (Attoh, 1982);4. Rio Itapicuru greenstone, Brazil (Davison et al., 1988); 5. Ljusnarsberg area, Sweden (Parr and Richard, 1987); 6. Mazaruni greenstone, Guiana (Renner and Gibbs, 1987); 7. Telemark area, Norway (Brewer and Atkin, 1987); 8. Viljakkala, Tampere, Finland (Kahkonen, 1989); 9. Sloan River greenstone (Hoffman and Cecile, 1974); 10. Big Bug Gp, Arizona (Vance, 1989); 11. Green Gulch volcanics, Arizona (Vance, 1989); 12. Ash Creek Gp, Arizona (Vance, 1989); 13. Red Rock, Alder, East Verde Gps, Arizona (Karlstrom, et al., 1990); 14. Lynn Lake greenstone, Sask. (Gilbert et al., 1980); 15. Nasijarvi, Finland (Kahkonen, 1989); 16. Camsell

River Gp, NWT, Canada (Hoffman and McGlynn, 1977); 17. Issineru greenstone, Guiana (Gibbs et al., 1982); 18. Purum greenstone, Guiana (Gibbs, A.K., unpub. data, 1982); 19. Kaburi-Omai Gui greenstone, Guiana (Gibbs et al., 1982); 20. Cochetopa succession, Colorado (Knoper and Condie, 1988); 21. W Amisk Lake greenstone, Manitoba (Fox, 1976); 22. Grythytle, Sweden (Vivallo and Rickard, 1984); 23. Salida greenstone, Colorado (Boardman, 1986); 24. Green Mtn Fm, Wyoming (Condie and Shadel, 1984); 25. Carolina terrane, N Carolina (Harris and Glover, 1988); 26. Doublet terrane, Saskatchewan (Wardle et al., 1991); 27. Amisk greenstone, Flin Flon, Manitoba (Syme, 1991); 28. Dubois greenstone, Colorado (Knoper and Condie, 1988); 29. Taos, New Mexico (Condie and McCrink, 1982); 30. Pecos, New Mexico (Roberston and Condie, 1989); 31. Pedernal Hills, New Mexico (McKee, 1988); 32. Pielevesi-Kiuruvesi, Finland (Park, 1991); 33. Bergslagen, Sweden (Park, 1991); 34. Skellefte, Sweden (Skiold, 1988; Park, 1991).

TABLE 3

Age, thickness, and lithologic proportions (in percent) in Archean greenstone assemblages

Ref	Max age (Ma)	Min age (Ma)	Thickness (km)	Basalt	Koma	Felsic Volc	And	Gw	Chert-BIF	Carb	Flow	Frag
1	2690	2638	15.2	71.0	0	14.0	3.0	11.0	1.0	0	9	19.0
2	2733	2696	8.0	43.0	11.0	18.0	15.0	10.0	3.0	0	70.3	29.7
3	2725	2711	10.0	75.2	0	5.0	12.0	5.2	2.6	0	84	16
4	2747	2700	9.0	90.7	3.9	1.7	2.2	1.0	0.5	0	85	15
5	2840	2740	12.3	48.2	0.7	15.3	24.2	9.7	1.9	0	58	42
6	2700		6.8	56.4	10.0	3.0	5.0	19.0	3.5	3.1	93	7.0
7	2720	2710	3.0	51.1	0.4	7.7	32.7	7.1	1.0	07	6.2	23.8
8	2730	2682	8.5	50.5	0.3	6.8	14.6	24.6	3.2	0	85.5	14.5
9	3450	3435	15.2	61.9	22.3	2.3	0	5.8	5.9	0.8	84.6	15.4
10	3810	3790	0.32	56.6	8.7	1.0	0	17.0	6.2	10.5	88.5	11.5
11	2700	2665	8.0	48.8	8.0	4.5	15.4	18.3	1.3	3.7	75.2	24.8
12	2730	2690	16.6	60.5	3.0	10.5	8.6	15.0	1.9	0.5	72.2	27.8
13	2500	2530	10.9	38.5	8.6	24.0	12.9	13.9	1.0	1.1	63.2	36.8
14	3400		8.0	65.0	24.9	1.0	0	8.0	1.1	0	90	10
15	3200		5.0	9.0	1.5	10.0	3.0	71.0	2.5	3.0	35	65
16	3000		7.5	65.5	1.0	0	0	22.5	10.0	1.0	80	20
17	3000	2950	10.5	56.6	9.4	11.7	0	14.0	8.3	0	79	21
18	3500	3450	14.6	76.5	4.8	6.5	6.2	3.7	2.2	0.1	69	31
19	2700	2675	8.1	50.3	12.6	14.3	3.9	12.3	6.6	0	77.6	22.4
20	2700	2675	17.5	63.4	7.1	3.6	24.3	0	1.6	0	71.4	28.6
21	2690	2660	19.4	51.4	12.0	4.7	0	28.6	3.3	0	88.9	11.1
22	2700	2670	15.1	43.0	7.8	6.3	21.4	17.7	2.8	1.0	85.8	14.2
23	2700		19.0	50.6	10.9	14.7	5.0	9.7	7.6	1.5	86.5	13.5
24	2700	2672	5.6	55.2	0	9.0	14.5	18.9	2.4	0	29.3	70.7
25	2700	2675	7.4	72.8	8.2	5.3	0	2.7	2.0	0	75.8	24.2
26	2900		7.0	18.6	25.7	31.4	18.6	0	5.7	0	56.7	43.3
27	3300	3250	7.5	15.8	2.9	11.5	0	54.7	15.1	0	74.4	25.6
28	3000	2935	8.0	39.0	0	6.3	4.1	47.5	3.1	0	70.4	29.6
29	2800	2755	5.2	40.9	16.2	18.9	13.5	9.5	1.0	0	81.1	18.9

For abbreviations see Table 1. 1. Yellowknife, NWT, Canada (Baragar, 1972); 2. Shebandowan, Ontario (Wilson and Morrice, 1977); 3. Kakagi Lake, Ontario (Wilson and Morrice, 1977); 4. Abitibi, Duparquet, Ontario (Goodwin, 1977); 5. Uchi Lake, Ontario (Goodwin, 1967); 6. Rio das Velhas,

Brazil (Ladeira and Roeser, 1983); 7. Lake of the Woods, Ontario (Goodwin, 1977); 8. Sioux Lookout, Ontario (Blackburn et al., 1985); 9. Onverwacht Gp, Barberton, S Africa (Lowe et al., 1985); 10. Isua, SW Greenland (Nutman et al., 1984); 11. Belingwe (younger section), Zimbabwe (Nisbet, 1987); 12. Ely greenstone, Minnesota (Schultz, 1980); 13. Qingyuan, China (Zhai et al., 1983); 14. Holenarsipur, India (Pichamuthu and Srinivasan, 1983); 15. Javanahalli, India (Pichamuthu and Srinivasan, 1983); 16. Bababudan, India (Pichamuthu and Srinivasan, 1983); 17. Luke Creek, W Australia (Watkins and Hickman, 1992); 18. Warrawoona Megasequence, W Australia (Barley, 1993); 19. Lawlers-Mt. Keith, W Australia (Naldrett and Turner, 1977); 20. Scotia, W. Australia (Wilson and Morrice, 1977); 21. Kalgoorlie (Coolgardie), W. Australia (Swager et al., 1992); 22. Que Que, Zimbabwe (Harrison, 1970); 23. Tati, Botswana (Key et al., 1976); 24. South Pass, Wyoming (Bayley et al., 1973); 25. Kalgoorlie (Oro Banda), W. Australia (Swager et al., 1992); 26. Belingwe (older section), Zimbabwe (Nisbet, 1987); 27. Gorge Creek Megasequence, W Australia (Krapez, 1993); 28. Roebourne Megasequence, W Australia (Krapez, 1993); 29. Mt Farmer, W. Australia (Watkins and Hickman, 1992).

however, may be >10 km thick as suggested by the original estimates, is attested to by progressive geochemical changes with stratigraphic height in thick basaltic sections such as occur in the Warrawoona Megasequence in Western Australia and the Blake River Group in the Abitibi belt (Goodwin, 1977; Glikson and Hickman, 1981a,b; Thurston et al., 1985).

From the data in Tables 1–3, it is possible to calculate a preservation rate for greenstone assemblages by dividing preserved thicknesses by duration times. Results indicate that Archean greenstones, including the anomalously thick examples, have preservation rates chiefly >0.1 km/My, whereas post-Archean assemblages

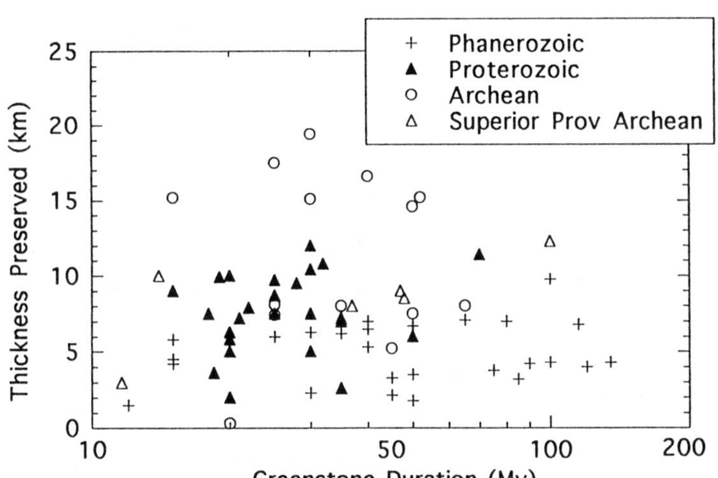

Fig. 1. Preserved thickness versus duration of greenstone tectonic assemblages of various ages. Data and references given in Tables 1–3.

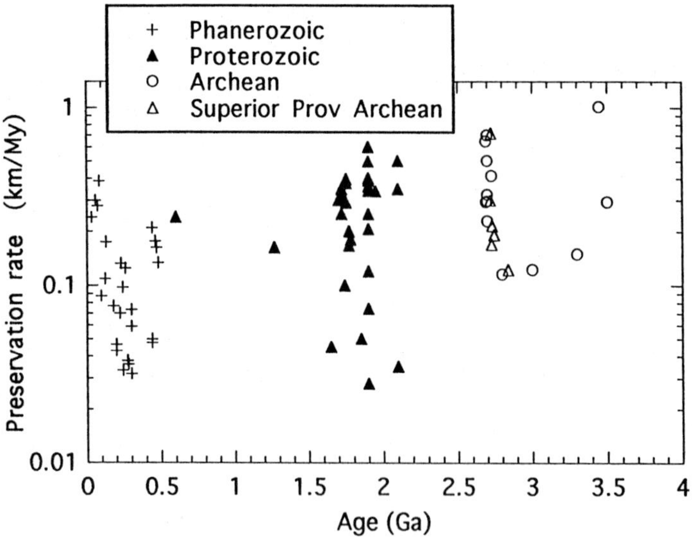

Fig. 2. Preservation rate versus age of greenstone tectonic assemblages. Data and references given in Tables 1–3. Preservation rate = preserved thickness divided by greenstone duration.

extend to much lower preservation rates (0.03 km/My) (Fig. 2). The mean Archean preservation rate of about 0.35 km/My is approximately twice the mean post-Archean rate. The results in Figs. 1 and 2 could mean that Archean greenstone eruption rates were greater than post-Archean rates, or alternatively, that greater proportions of Archean greenstones are preserved than of post-Archean greenstones. Greenstone preservation rate is a function of many factors including original eruption rate, the amount of protective underplating by low-density felsic magmas, the intensity and frequency of uplift and erosion, and how much section is removed by faulting. For the Archean when heat generation in the mantle was 3–4 times the present rate (Bickle, 1986), the results are consistent with greater eruption rates of oceanic volcanics. Widespread felsic underplating with buoyant TTG magmas in the Archean also may have increased the chances of survival of Archean greenstones.

From zircon chronologies of Archean and Proterozoic greenstone terranes, it is possible to compare their deformational and intrusion histories to Phanerozoic counterparts. However, this is a difficult task in that most greenstone belts have undergone complex, multiple deformations and granitoid intrusion. In many cases two ages can be obtained from greenstones. Eruption age (greenstone formation age) can be estimated either from U/Pb zircon ages of felsic volcanics in greenstone assemblages or from fossil distributions in associated carbonates (Phanerozoic only). The timing of greenstone–continent collisions can be approximated from the last major deformation of a greenstone, and is herein estimated from zircon ages of syntectonic granitoids intrusive into greenstones. Final suturing of greenstones to continents, however, may occur 10–30 My later than collision as recorded by discordant, stitching plutons.

Histograms of greenstone eruption and collision ages are given in Fig. 3. Perhaps the most striking feature of the histograms is the lack of greenstones with either eruption or collision ages in the time intervals 2450–2200 and 1650–1350 Ma. From an extensive literature survey, not one single high-quality greenstone zircon age was reported in these age windows. Prominent peaks in both greenstone eruption and collision age occur at 2700 and 1900 Ma. Eruption ages also peak at about 1300 Ma and collision ages at about 1100 Ma. No consistent pattern of ages, however, emerges from greenstones <1000 Ma nor >2700 Ma. The three peaks in greenstone eruption age are similar to peaks in the age of juvenile granitoids that have long been recognized (Moorbath, 1977; McCulloch and Wasserburg, 1978; Nelson and DePaolo, 1985; Condie, 1990b), suggesting a common explanation for both age distributions.

Although many Archean greenstones differ from younger greenstones by the presence of komatiite (Nisbet and Walker, 1982; Nisbet, 1987) (Tables 1–3), the proportion of basalt also is greater in Archean greenstones (Goodwin, 1977; Condie, 1981; 1989). The lithologic proportions summarized in Tables 1–3 and plotted in Fig. 4 clearly show that the volume of basalt + komatiite exceeds that of intermediate + felsic volcanics in most Archean greenstones. Those Archean examples with >80% basalt + komatiite are the mafic plains of Thurston and Chivers (1990), such as the Blake River Group in the Abitibi belt and the Onverwacht Group in the Barberton belt. These may represent submarine plateaus (Kusky and Kidd, 1992), although perhaps some of them are remnants of Archean oceanic crust. The Archean greenstones with 40–80% basalt + komatiite (komatiite is minor in these) are more like young oceanic arc systems. On the whole, the Proterozoic greenstones for which data are available have smaller proportions of basalt (an almost no komatiite) and range from 20–50% of felsic volcanics + andesite (Fig. 4), and in this respect are like continental-margin arc systems. Phanerozoic greenstones are more variable than Precambrian greenstones, and appear to represent a greater diversity of oceanic tectonic settings.

The data in Fig. 5 suggest a broad inverse correlation between the abundances of graywacke and basalt + komatiite in greenstones, with most of the Phanerozoic greenstones characterized by a high proportion of graywacke and correspondingly smaller amounts of basalt. In general, the amount of variation within an age group increases in the order Archean, Proterozoic, and Phanerozoic. If the sampling is representative in each age group, these results suggest that oceanic tectonic settings were less diverse in the Archean than afterwards. Although there is a great deal of overlap in the proportion of flows to fragmental volcanics in greenstones (Tables 1–3), the flow/fragmental ratio is generally higher in Archean greenstones (with a mean value of 4.4 ± 3.0) than in post-Archean greenstones (1.7 ± 1.7 and 2.0 ± 2.2 for Proterozoic and Phanerozoic greenstones, respectively) (Fig. 6).

(a)

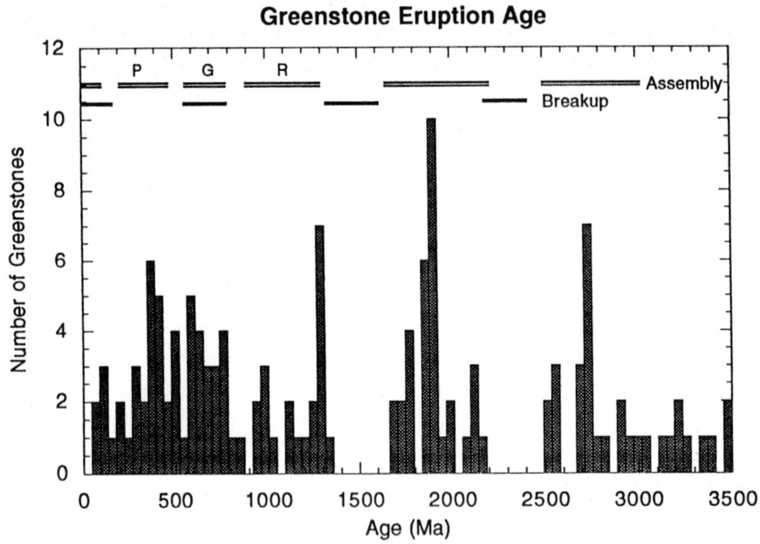

(b)

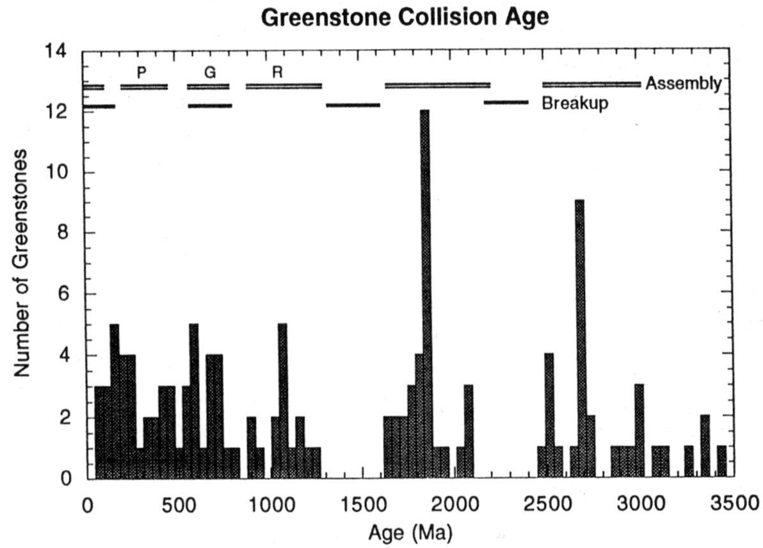

Fig. 3. Histograms of greenstone eruption (a) and continental collision ages (b) through time. Also shown are times of growth and breakup of supercontinents. Major references: Sinha and Bartholomew (1984); Bickford and Boardman (1984); Corfu and Grunsky (1987); Skiold (1988); Karlstrom and Bowring (1988); Kroner et al. (1988); Kumon et al. (1988); Barker et al. (1988); Kusky (1989); Robertson and Condie (1989); Premo and Van Schmus (1989); Stephens and Gee (1989); Pflafker et al (1989); Kahkonen et al. (1989); Sims et al. (1989); Kamo and Davis, 1991; Park (1991); Samson and Patchett (1991); Watkins and Hickman, 1992; Chown et al. (1992); Boher et al. (1992); Swager et al., 1992; Gower et al. (1992); Williams et al. (1992); Turek et al. (1992); Peucat et al. (1993); Krapez, (1993); Bevier et al. (1993); Kimura et al. (1993); Hayasaka (1993); van Staal and Colman-Sadd (1993); P.C. Thurston, pers. comm. (1993).

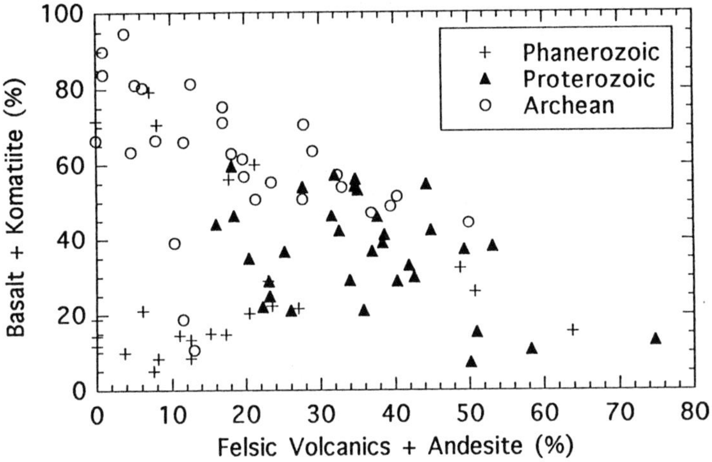

Fig. 4. Plot of percent of basalt + komatiite versus felsic volcanics + andesite in greenstone assemblages. Data and references given in Tables 1–3.

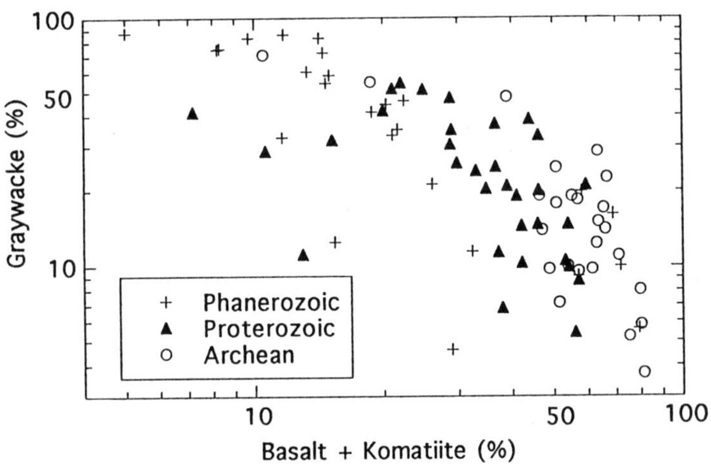

Fig. 5. Plot of percent of graywacke (including related volcaniclastic sediments) versus basalt + komatiite in greenstone assemblages. Data and references given in Tables 1–3.

Also, there appears to be an absence of Archean greenstone assemblages with flow/fragmental ratios <1. The consistently high flow/fragmental ratio in Archean greenstones suggests they were erupted chiefly as submarine flows in deep water, whereas a greater proportion of post-Archean greenstone volcanics were erupted in shallow water.

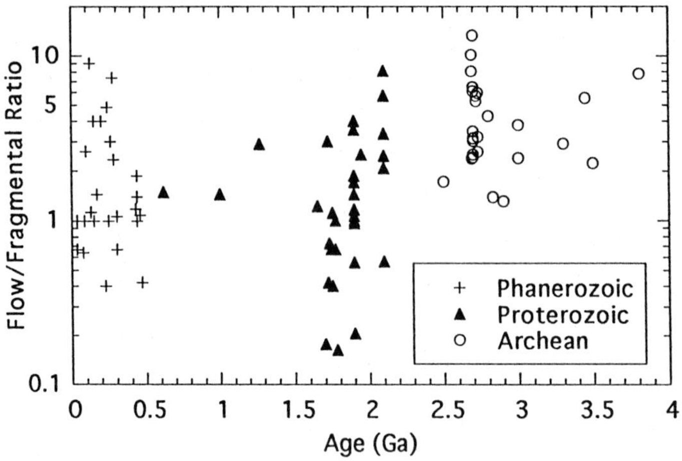

Fig. 6. Ratio of flows to fragmental volcanics versus age in greenstone assemblages. Data and references given in Tables 1–3.

The only significant difference between Proterozoic and Phanerozoic greenstones is a greater proportion of felsic and intermediate volcanics at the expense of volcaniclastic sediments and graywackes in Proterozoic greenstones (Tables 1–3). It is possible these differences represent differences in tectonic setting, since most of the Phanerozoic greenstone data come from Cordilleran terranes in western North America and most of the Proterozoic data come from Early Proterozoic greenstones of central and southern Laurentia. Most of the Phanerozoic Cordilleran terranes are oceanic terranes (island arcs, oceanic crust, submarine plateaus, etc.), whereas most the Laurentian Proterozoic terranes appear to represent continental-margin arcs (Condie, 1992). Thus, the infrequency of felsic and intermediate volcanics in the Cordilleran greenstones is not surprising, nor is the common presence of these rocks in the Laurentian Proterozoic greenstones. To see if these or other differences are characteristic of the Proterozoic/Phanerozoic transition, in general, must await more detailed descriptions of greenstones of both ages.

GREENSTONE GEOCHEMISTRY

General features

Previous studies have shown that selective preservation of greenstones formed in some tectonic settings may lead to disproportionate amounts of these greenstones in the geologic record (Condie, 1990a). Thus, in identifying geochemical changes in greenstones with time, caution must be used to compare greenstones from similar tectonic settings as dictated by their lithologic and stratigraphic features. Employing the greenstone definition given in the Introduction, several significant changes in composition of greenstone basalts with time have been recognized. For instance, basalts with island-arc geochemical affinities seem to have dominated in the Archean and those with calc-alkaline affinities, in the post-Archean (Condie, 1989; 1990a). Basalts with MORB geochemical characteristics are rare in greenstones of all ages and most Archean basalts, regardless of lithologic assemblage, are enriched in Ni and related transition metals compared to post-Archean basalts (Arndt, 1991; Arndt et al., 1993). We have found that important geochemical differences in basalts with time can be monitored with two element ratios (Th/Ta and La/Yb) and by their Ni contents.

Th/Ta–La/Yb relationships

Th/Ta and La/Yb ratios in basalts are particularly sensitive to mantle source composition and to mixing processes, and they are changed very little by different degrees of fractional crystallization or partial melting (Condie, 1989; 1990a; Jochum et al., 1991). They also appear to be relatively insensitive to alteration and metamorphism. MORB and submarine plateau basalts (SPB) are similar, both having relatively low ratios (Th/Ta <2; La/Yb chiefly <5) that reflect depleted mantle sources (Fig. 7a). Some MORB, such as those from the North Atlantic have La/Yb values >5 suggesting contamination with enriched mantle sources. In striking contrast to MORB-SPB, both continental-margin arc (CAB) and island arc basalts (IAB) show enriched mantle sources with Th/Ta >5 and La/Yb, although more variable, generally >2 (Fig. 7b). Some arc basalts have Th/Ta ratios even higher than average upper continental crust (>10). If Fiji is representative of oceanic arcs, both Th/Ta and La/Yb ratios increase in basalts as the arc evolved, as shown by the change in these ratios from the Early to the Mature stage (Gill, 1987) (Fig. 7b). Presumably, this reflects an increase in the amount of LILE enrichment in the mantle wedge with time. Enriched and HIMU (sources with high U/Pb ratios) mantle sources (Zindler and Hart, 1986) do not appear to have contributed to the production of most arc-related basalts. Although continental flood (CFB) and rift basalts (CRB) overlap arc-related basalts in Th/Ta–La/Yb space (Fig. 7c,d), none shows Th/Ta ratios >10 and most CRB have Th/Ta ratios <5. Some groups, such as the Ethiopia rift basalts, seem to have a major contribution from enriched or HIMU mantle source components. Others, such as the Siberian traps, may define mantle mixing curves with enriched or HIMU sources as one end member and depleted mantle (DM) as the other (Fig. 7c). Two or more populations appear to occur in the Deccan traps, with at least one suggestive of mixing of enriched or HIMU end members with depleted mantle.

Major processes responsible for the distribution of Th/Ta and La/Yb ratios in modern basalts fall into three broad categories:

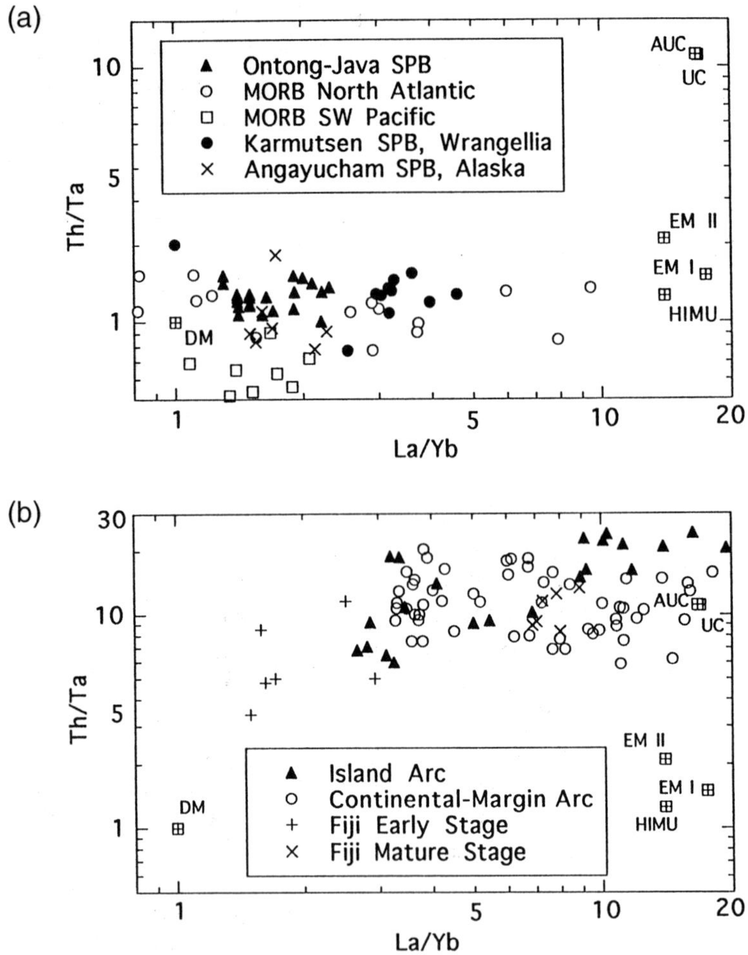

Fig. 7. Th/Ta versus La/Yb plots of various young basalts and greenstone basalts. Mantle and crustal end members: AUC, Archean upper continental crust; UC, upper continental crust; DM, depleted mantle; EM I and EM II, enriched mantle; HIMU, high-mu mantle (Zindler and Hart, 1986; Weaver, 1991; Condie, 1993). (a) MORB and submarine plateau basalts (SMP) (Ontong-Java, Karmutsen, Angayucham). References: Mahoney et al. (1992), Tarney et al. (1979), Floyd (1985), Barker et al. (1988; 1989). (b) Subduction-related basalts. Compiled from many sources; Fiji data from Gill (1987).

1. Metasomatism in which soluble LILE (such as Th and La) are transmitted to the mantle wedge by fluids rising from descending slabs, producing a subduction geochemical component (SGC). This SGC is then acquired by basalts produced in mantle wedges.
2. Contamination of basalts by continental crust, defining a mixing curve between depleted mantle (DM) and upper continental crust (UC and AUC in Fig. 7).

(c)

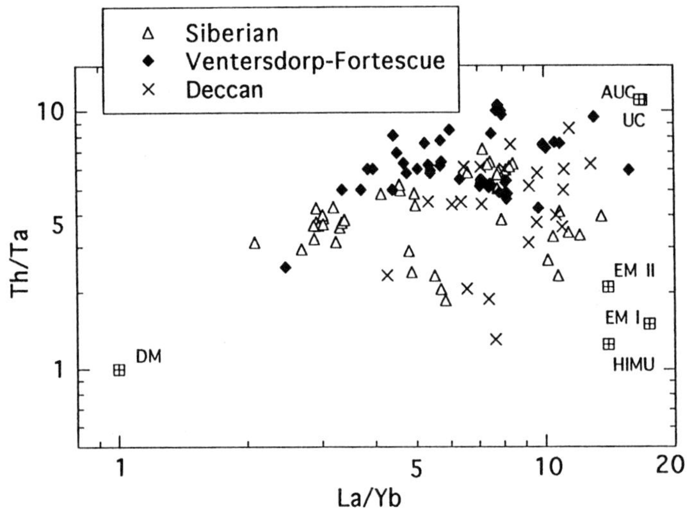

(d)

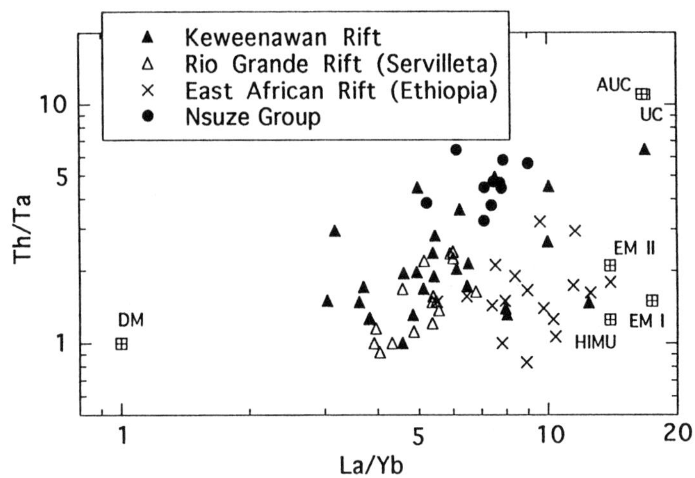

Fig. 7. (*continued*). (c) Continental flood basalts. References: Crow and Condie (1988), Lightfoot et al. (1991), Lightfoot and Hawkesworth (1988), Nelson et al. (1992). (d) Continental rift basalts. References: BVTP (1981), Massey (1983), Dungan et al. (1986), Hart et al. (1989).

3. Mixing of depleted with enriched mantle sources. This produces basalts that lie close to mixing curves between enriched or HIMU sources and depleted mantle. Such mixed sources may occur in the subcontinental lithosphere or in mantle plumes.

Th/Ta–La/Yb data from Archean and Proterozoic greenstone basalts fall within or near the broad mixing trajectories between depleted mantle (DM) and average

(e)

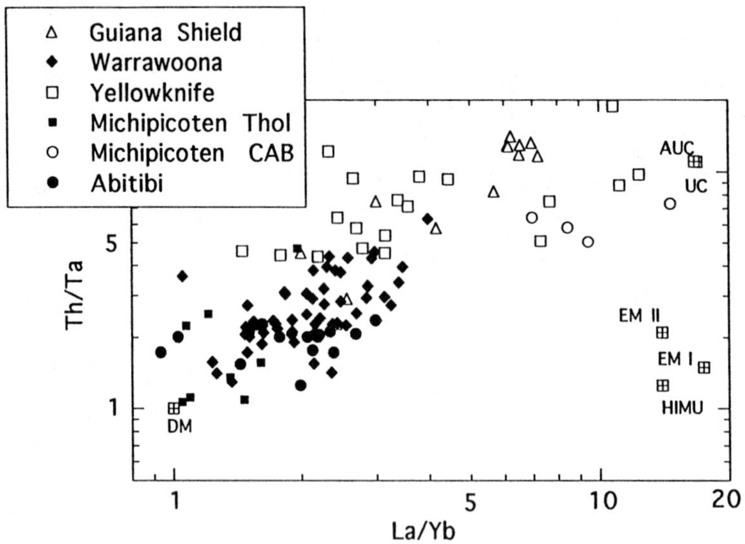

(f)

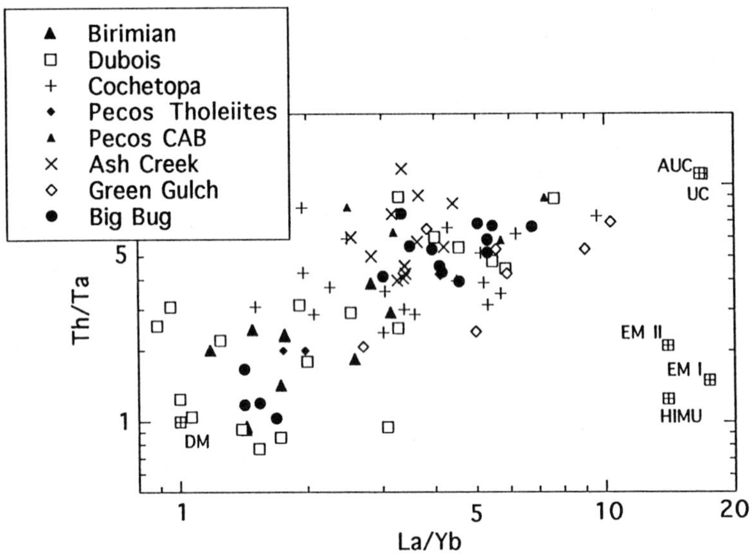

Fig. 7. (*continued*). (e) Archean greenstone basalts. References: Gibbs (1982), A.M. Goodwin, unpub. data (1992), Sylvester et al. (1987), Lafleche et al. (1992); Condie et al., unpub. data (1993). Solid symbols are mafic plain basalts and open symbols are basalts with arc affinities. (f) Proterozoic greenstone basalts. References: Sylvester and Attoh (1992), Knoper and Condie (1988), Robertson and Condie (1989), Vance (1989).

upper continental crust (UC) (Fig. 7e,f). Few, if any, show significant contributions from enriched or HIMU mantle sources. If mantle isochrons of 2.0–1.8 Ga accurately reflect the ages of these sources (Hart, 1984; Zindler and Hart, 1986), it is not surprising that pre-2.0 Ga basalts show no contributions of such sources. Many Archean and some post-Archean greenstones have thick basal successions of submarine basalt (± komatiite), and have been referred to as mafic plains (Picard and Piboule, 1986; Thurston and Chivers, 1990). Most basalts from Precambrian mafic plains have Th/Ta and La/Yb distributions similar to those of modern basalts derived from depleted mantle sources. Examples are tholeiites from the Warrawoona, Abitibi, and Michipicoten greenstone belts (Fig. 7f), where the Th/Ta and La/Yb ratios partially overlap those of young submarine plateau basalts, such as Ontong-Java, Karmutsen and Angayucham basalts (Fig. 7a). Basalts from some Archean greenstones such as Yellowknife in Canada, and basalts from Archean mafic to felsic cycles, such as occur in Guiana, Abitibi, and Michipicoten greenstones (calc-alkaline basalts), have elevated Th/Ta and La/Yb ratios, not unlike modern arc-related basalts. These results indicate that basalts with geochemical affinities to young basalts from both submarine plateaus and arcs occur in Archean greenstone belts, and in many instances, in the same greenstone belt in different tectonic assemblages. Although arc-like successions have long been recognized in Archean greenstones, only recently has lithologic and structural evidence been described that supports the existence of submarine plateaus in these greenstones (Kusky and Kidd, 1992; Desrochers et al., 1993), as suggested by the geochemical data described above.

Results from detailed stratigraphic sampling of the Warrawoona Megasequence in the Pilbara craton in Western Australia show systematic upward decreases in Th/Ta ratios (Fig. 8). Also decreasing upwards in each unit are the La/Yb, Th/Nb, Ti/Zr, Zr/Y, and Ti/V ratios. Similar changes in composition with stratigraphic height are also reported in other Archean greenstones (Capdevila et al., 1982; Thurston et al., 1985; Sylvester et al., 1987; Lafleche et al., 1992). These geochemical trends, which also have been identified in continental flood basalts from the Deccan Plateau in India and from Southwest Madagascar (Lightfoot and Hawkesworth, 1988; Mahoney et al., 1991), suggest the LILE contribution was greatest during the early stages of basalt eruption. Several explanations have been proposed for such geochemical variation. Thurston et al. (1985) have suggested that this trend in the Blake River Group of the Abitibi belt is due to successive melts derived from more depleted mantle. Alternately, a LILE-enriched lithosphere could have contributed more to the earliest basalts and as it stretched, in response perhaps to a rising mantle plume, the relatively hot plume-derived magmas could become more abundant with time (Carlson, 1991). Contamination of komatiite and basalt by continental crust also is capable of producing magmas enriched in LILE (Arndt and Jenner, 1986; Barley, 1986; Gruau et al., 1987). If crustal contamination is responsible for the stratigraphic geochemical trends, it could be that conduit walls became coated with chilled basalt and komatiite,

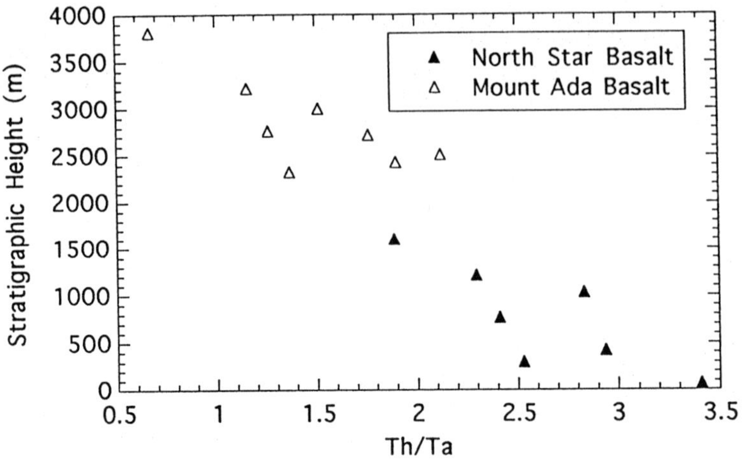

Fig. 8. Stratigraphic variation of the Th/Ta ratio in the McPhee-Reward section of the Early Archean Warrawoona Megasequence, Western Australia. Location of section and other geochemical results given in Glikson and Hickman (1981 a,b). Data from unpublished results of the author.

isolating later magmas from the wall rocks, thus decreasing the amount of contamination with time (Nisbet, 1987).

Consistent with crustal contamination in the production of the Archean Warrawoona basalts in Western Australia is the fact that these basalts fall on a broad mixing curve between felsic and komatiitic (or high-Mg basalt) end members on the Th/Ta–La/Yb plot (as represented by DM and AUC on Fig. 7e). Also, in these basalts the Th/Ta ratio increases with decreasing epsilon-Nd values (Gruau et al., 1987). Oxygen isotope distributions are also compatible with crustal contamination. The North Star basalts have high $\delta^{18}O$ values (8–10 per mil; unpub. data, A. Campbell, 1994) corresponding to the high Th/Ta ratios, and the Mount Ada basalts have low $\delta^{18}O$ values (5.5–6 per mil) corresponding to low Th/Ta ratios. An increasing database of Nd isotopic results from Archean greenstone basalts seems to limit the amount of upper continental crustal contamination to <20% (Shirey and Hanson, 1986; Galer and Goldstein, 1990; Arndt et al., 1993). Epsilon-Nd values for many LILE-enriched Archean greenstones fall near the depleted mantle curve of DePaolo (1981) allowing <10% crustal contamination, and major element contents of these rocks are consistent with this conclusion.

Proterozoic greenstone basalts show similar Th/Ta–La/Yb distributions to Archean examples (Fig. 7f). However, the proportion of mafic plain basalts that are derived from depleted sources (Th/Ta and La/Yb each <3) appears to be greater in Archean than in Proterozoic greenstones. Most Proterozoic greenstone basalts share geochemical affinities, including Th/Ta and La/Yb ratios, with modern arc-related basalts. Also, systematic changes in chemical composition of Proterozoic and Phanerozoic greenstone basalts with stratigraphic height have not been recognized. This feature may be due the fact that post-Archean greenstone successions seem to be more fragmented with unconformities and faults. Only the Birimian basalts in West Africa may contain a significant proportion of basalts erupted as submarine mafic plains (submarine plateaus) (Sylvester and Attoh, 1992). However, the relatively low Th/Ta and La/Yb ratios of some basalts from the Dubois greenstone in central Colorado, the Pecos greenstone in northern New Mexico, and part of the Big Bug Group (the Bluebell basalts) in central Arizona (Condie, 1986a; Knoper and Condie, 1988; Vance, 1989; Condie, 1992) (Fig. 7f) suggest these greenstones also contain remnants of submarine mafic plains.

Ni and Mg number relationships

Numerous studies have now documented that at the same Mg number, Archean basalts are enriched in Ni, Fe and Cr and depleted in Al_2O_3 compared to most post-Archean basalts (Condie, 1984; Arndt, 1991; Arndt et al., 1993). However, until recently, the chemical compositions of modern submarine plateau basalts were not adequately represented in the population of basalts to which the Archean database was compared. Modern submarine and arc basalts seem to fall into one of three populations on a Ni versus Mg number plot (Fig. 9a):

1. The subduction trend, which includes most arc-related basalts. This appears to be a fractional crystallization trend in which Ni decreases rapidly with falling Mg number (a Mg number of 45 corresponds to <10 ppm Ni).
2. The MORB trend, in which Ni decreases less rapidly with Mg number (a Mg number of 45 corresponds to about 50 ppm Ni).
3. The submarine plateau trend in which Ni decreases very little with Mg number. At a Mg number of 45, submarine plateau basalts typically have Ni contents of 80–100 ppm. The higher Ni contents in this case appear to reflect larger degrees of melting in a mantle plume.

Trends 2 and 3 are also recognized in continental flood and rift basalts of various ages (Fig. 9b). In Archean flood basalts, both trends may occur within the same volcanic field, as for instance in basalts from the Fortescue Group (Fig. 9b). The high-Ni trend of submarine plateau basalts may be characteristic of mantle plumes, where the temperature is higher than ambient mantle, and thus the Ni content of derivative basaltic magma is greater (Arndt, 1991). The MORB and arc Ni-Mg number trends are probably produced by varying degrees of melting of depleted mantle and fractional crystallization of the derivative basalts.

Archean greenstone basalts, whether they come from mafic plain successions or mafic to felsic cycles, all show relatively high Ni contents compared to post-Archean basalts at comparable Mg numbers (Fig. 9c). As proposed by Arndt (1991), this probably reflects higher mantle temperatures and corresponding higher degrees of melting in the Archean mantle, regardless of tectonic setting. Campbell et al. (1990) suggested that the close association of komatiites and

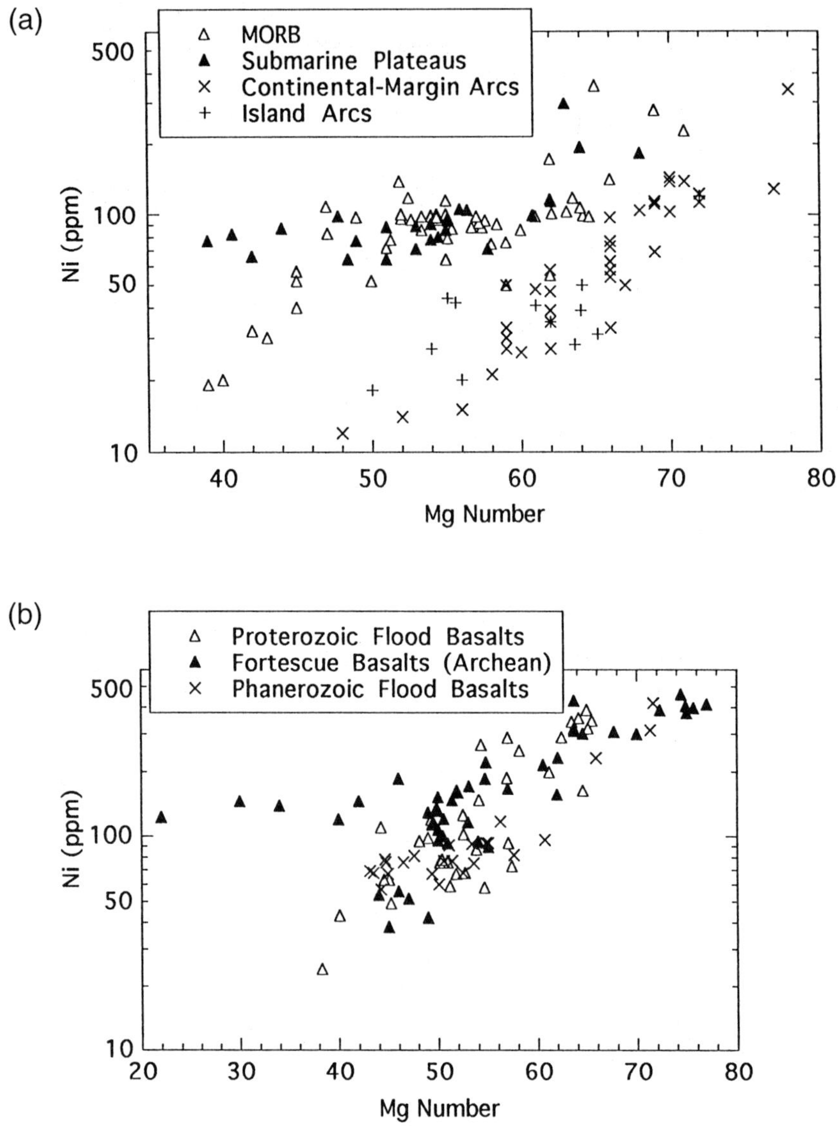

Fig. 9. Ni versus Mg number plots of various young basalts and greenstone basalts. See Fig. 7 for references. (a) MORB, submarine plateau basalts, and subduction-related basalts. (b) Flood basalts.

tholeiites in Archean greenstones could be explained by a mantle plume source in which the komatiites came from the hot plume tail and the basalts from the cooler head. The occurrence of young basaltic komatiites in submarine plateaus also supports a plume source (Storey et al., 1991). In the case of Archean mafic plain basalts, a mantle plume source is especially attractive (Arndt et al., 1993), whereas

(c)

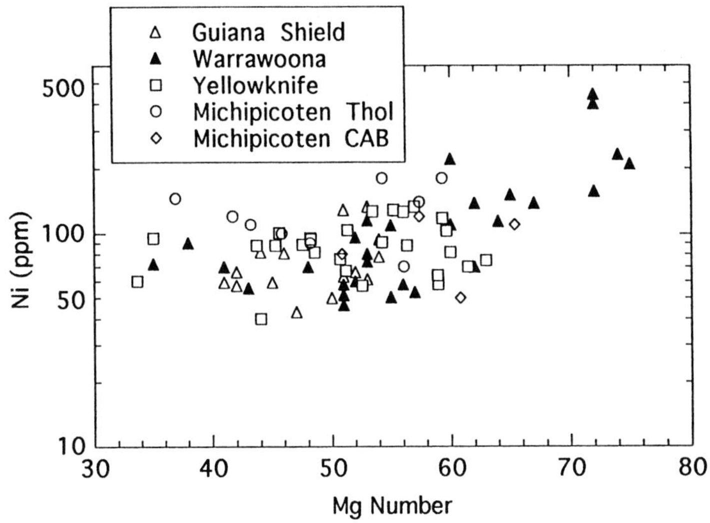

(d)

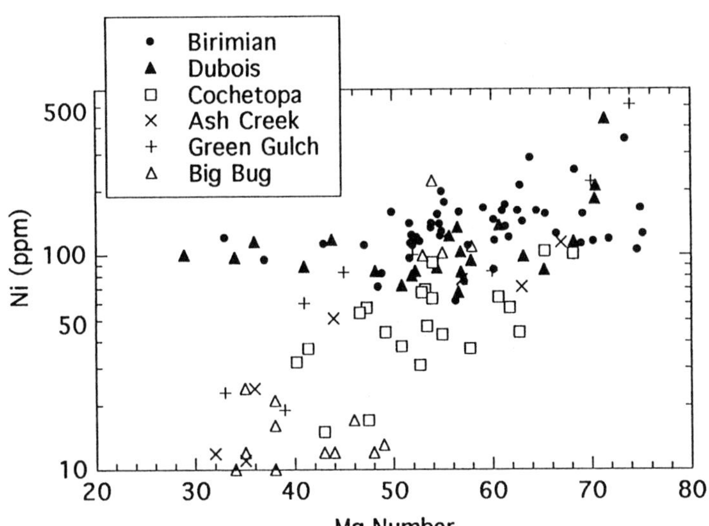

Fig. 9. (*continued*). (c) Proterozoic greenstone basalts. (d) Archean greenstone basalts.

the calc-alkaline basalts from greenstone mafic to felsic cycles may come from subduction zones with elevated mantle wedge temperatures.

Unlike Archean greenstone basalts, most analyzed Proterozoic greenstone basalts show steep Ni-Mg number trends, similar to arc-related basalts (Fig. 9d). A few, such as the Dubois and Birimian basalts, exhibit a high Ni trend, consistent with their containing at least portions of Proterozoic submarine plateaus.

DISCUSSION

Archean greenstone peculiarities

Although many Archean greenstones evolved with time from submarine mafic plains to shallow or subaerial stratovolcanic centers of intermediate to felsic composition (Thurston and Chivers, 1990; Watkins and Hickman, 1992), the mafic plain component generally dominates in these types of greenstones. A similar evolution has been reported from only a few post-Archean greenstones, such as the Birimian in West Africa (Sylvester and Attoh, 1992). From the comparisons of greenstones through time, it is clear that submarine mafic plains are more common in Archean greenstones than they are in post-Archean greenstones. At least two explanations need to be considered to explain this observation: (1) the rate of preservation of mafic plains was greater in the Archean than afterwards, and/or (2) the frequency of mantle plumes (that produced the mafic plains) was greater in the Archean. With our present database it is not possible to choose between these alternatives, and there are results consistent with both possibilities. Archean greenstones, in general, may owe their preservation to underplating with TTG (tonalite–trondjhemite–granodiorite) (Condie, 1986b). Large volumes of TTG were intruded into greenstones in the Late Archean as the volume of preserved continental crust increased. The chief source of TTG may have been the roots of thickening mafic plains, which were partially melted under hydrous conditions. As the TTG magmas buoyantly moved upwards, they may have trapped greenstones, preventing them from being recycled into the mantle. Post-Archean granitoid production, by whatever means, may not have resulted in such widespread underplating of greenstones.

More frequent production of mantle plumes in the Archean is consistent with higher heat production in the mantle at this time (Campbell and Griffiths, 1992). If plumes were more abundant in the Archean mantle and mafic plains (submarine plateaus) are the products of plume magmatism, they should be more abundant in Archean greenstones than in younger greenstones. The increased proportion of mafic plains in Archean greenstones could thus, reflect a hotter Archean mantle. Recent studies of the relative buoyancies of oceanic terranes suggest still another way that Archean submarine plateaus may have been preferentially preserved. Submarine plateaus with less than about 20 km of mafic oceanic crust (capping 100 km of lithosphere) should be subductable, whereas thicker plateaus with ≥30 km of mafic crust should not (Cloos, 1993). If larger volumes of magma were produced in Archean mantle plumes than afterwards (due to higher mantle temperatures), thick submarine plateaus may have been common, and when these collided with growing continents they became part of the continent. Post-Archean submarine plateaus may have been on the average much thinner, due to smaller amounts of magma produced in plumes, and thus more frequently subducted into the mantle.

Because unambiguous ophiolites have not as yet been described from the Archean, the question still exists as to whether they are really absent, or if they are deformed to such a degree that important components (such as sheeted dikes) are difficult to recognize (de Wit et al., 1987; Helmstaedt and Scott, 1992). If indeed the higher heat production in the Archean mantle resulted in thicker oceanic crust (Sleep and Windley, 1982), the mechanism of producing oceanic crust may have differed from that which produces ophiolites. Although some investigators have suggested that Archean mafic plains are remnants of Archean oceanic crust, the coexistence of komatiite and basalt and basalt geochemistry (as discussed previously) do not favor this interpretation. Perhaps the apparent absence of Archean ophiolites is a problem of selective preservation. It could be that in the Archean only submarine plateaus and arcs were underplated with TTG, thus enhancing their preservability, while oceanic crust was largely or entirely recycled into the mantle.

Unlike many Late Archean greenstones whose total lifespans were <50 My (Williams et al., 1992), Early Archean greenstones (>3.5 Ga) appear to have had long, complex tectonic histories. Isua in SW Greenland, Barberton in Southern Africa, and Pilbara in Western Australia were all subjected to multiple deformation and plutonism over a time span of up to one billion years (de Wit et al., 1992; Nutman et al., 1993; Krapez, 1993). Even individual successions within these greenstone belts, such as the Onverwacht Group in the Barberton suite and the Roebourne Megasequence in the Pilbara craton (Lowe et al, 1985; Krapez, 1993), have prolonged tectonic histories of up to 100 My. Unlike Late Archean terranes, which accreted into cratons almost as they formed, Early Archean terranes appear to have bounced around like bumper cars. Why didn't they accrete into supercontinents? Perhaps there were too few continents, and the probability of collision was low.

Greenstones and supercontinents

If as the existing data suggest there is an absence of greenstones with ages of 2450–2200 and 1650–1350 Ma, several alternative explanations need to be considered:

1. plate tectonics stopped during these times and greenstones did not form;
2. platform sediments selectively cover greenstones of these ages;
3. greenstones of these ages have not as yet been sampled and dated; or
4. greenstones of these ages were recycled into the mantle, and thus, are not preserved in the geologic record.

It seems extremely unlikely in a steadily cooling mantle in which heat is lost dominantly by plate formation at ocean ridges, that plate tectonics should stop and restart again several times. Rapid variations in mantle temperature are strongly inhibited by silicate rheology, and thus sharp changes in the cooling history of the mantle are unlikely (Gurnis and Davies, 1986; Davies, 1992). Hence, the first model is not considered a viable explanation. Although recent dating of basement rocks in western Canada suggest that 2.3–2.0 Ga granitoids may underlie part of this region (Ross et al., 1991), it would seem highly implausible that young platform sediments should somehow selectively cover greenstones (and related juvenile felsic rocks) formed during both greenstone age gaps on all continents. Hence, this explanation is also unlikely. Inadequate regional sampling can lead to apparent time gaps in the distribution of isotopic ages, as evidenced most recently by the discovery of extensive juvenile crust, including greenstones, of about 2.1 Ga age in West Africa and in the Guiana Shield in South America (Boher et al., 1992). Before this discovery, few if any juvenile crust ages of 2.1 Ga had been recognized on the continents. In the last few years, however, our geographic database of U/Pb zircon ages has increased remarkably, thus decreasing the probability of completely missing rocks of a given age. This is not say that greenstones that formed in the two apparent age gaps will not be discovered in the future, but that the total volume of greenstones preserved in the continents with these ages is probably very small.

Gurnis and Davies (1985, 1986) showed that crustal age distributions may reflect selective recycling of young crust into the mantle, and that preferential recycling of Phanerozoic crust can lead to an apparent peak in crustal growth at 2 to 3 Ga. This should occur because young crust is more abundant in young orogenic belts, which are elevated and preferentially eroded and the derivative sediments are carried to the sea floor where they can be subducted. Could such a process lead to time gaps in which greenstones are absent? I think not. Although recycling of young crust may change the distribution of crustal ages, it is difficult to conceive of any way it could selectively eliminate greenstones of a given age on all continents. Another problem with this explanation is the assumption that young collisional mountains are composed dominantly Phanerozoic rocks. The Nd model ages of suspended loads of major rivers (T_{DM} = 1.5–1.7 Ga; Goldstein and Jacobsen, 1988a,b) are only slightly younger than the average Nd T_{DM} model age of the continental crust of 1.8 Ga (Jacobsen, 1988). Thus, it would seem that it is largely old crust and not young crust that is being carried to the oceans to become available for subduction. The proportion of old to new crust carried to oceans by rivers during the Precambrian may have been even greater, since many Precambrian collisional orogens appear to contain large amounts of reworked older crust (Windley, 1992). One of the best known examples is the Grenville belt, where there is considerable isotopic evidence for a large component of reworked older crust (Condie, 1990b).

The fact that we see oceanic greenstones in the geologic record is due to their incorporation into relatively buoyant continental crust, which protected them from subduction. These greenstones may collide and accrete to continents or they may be underplated with felsic magmas, such as the extensive TTG that underplates Archean greenstones. Could there have been periods of time during which collisions of greenstones with continents were less frequent during which most greenstones were recycled into the mantle? A possible time of increased greenstone subduction is during fragmentation of supercontinents. When supercontinents are fragmented over mantle upwellings (Hoffman, 1989), subduction zones should shift to regions over mantle downwellings in oceanic areas. Opening ocean basins between continental fragments may have few if any convergent margins, and hence neither oceanic nor continental-margin arc systems would form in these areas. Furthermore, continental collisions would be infrequent and greenstones may not be preserved. Oceanic islands, plateaus, and arcs should move to areas over mantle downwellings increasing their chances of subduction. Cloos (1993) has shown that oceanic terranes (arcs, plateaus, islands, oceanic crust) should be subducted if they are less than 15–20 km thick, and it would be only the very thick submarine plateaus or oceanic arcs that survive for >20 My that may avoid subduction. These should later be accreted to continents. Thus, if any thick submarine plateaus formed during times of continental fragmentation, they should show up as terranes (or as parts of terranes) accreted to continents at later times.

The supercontinent model is attractive to explain the 2450–2200 and 165 1350 Ma greenstone age gaps in that these two gaps correspond to times probable supercontinent fragmentation (Hoffman, 1989; Barley and Grov 1992), as shown in Fig. 3. This is perhaps not too surprising in that supercontin aggregation phases are defined in part from the ages of juvenile continental cru of which greenstones are important components. If the correlation of supercor nent breakup and missing greenstones is real, in the future greenstone age distributions could be one of our most precise methods to date supercontinent assembly and dispersal. Could supercontinent fragmentation lead the complete absence of greenstones in the 2450–2200 and 1650–1350 Ma time windows? This seems improbable in that not all greenstones of these ages should have been subducted. The same question applies to felsic crust, which should be less susceptible to subduction if it is over 15–20 km thick (Cloos, 1993). Juvenile felsic magmas can be produced in two ways: partial melting of the mafic roots of greenstones (arcs, submarine plateaus, etc.), or fractional crystallization of mafic magmas produced in mantle wedges above subduction zones. If most mafic crust is subducted before it attains thicknesses of 20 km, the first mechanism of felsic magma production cannot be important for most oceanic terranes. What about continental-margin arcs, however? Surely, some should survive around fragments of dispersing supercontinents. As components of the two oldest supercontinents (ca. 2.7 and 1.9 Ga) become better defined, we can identify geographic regions in which to concentrate our search for surviving remnants of such continental-margin arc greenstones.

If indeed, the rate at which continental crust has been extracted from the mantle has decreased with time as dictated by a cooling earth (Armstrong, 1991), how is a smooth continental growth curve maintained at times of supercontinent breakup? Collision and accretion of oceanic terranes to continents is an important mechanism of continental growth. Clearly, if the subduction rate of oceanic terranes increases, recycling them before they collide with continents, some other mechanism must increase in importance to maintain a steadily falling continental growth curve. One possibility is an increase in the rate of mafic underplating of continents. As a supercontinent disperses over a mantle upwelling, increased mantle plume activity may lead to an enhanced rate of crustal underplating from mafic magmas derived from plumes.

It would appear that the period of time from 2.7 to 1.0 Ga is the only time period that shows well-defined episodisity in the ages of greenstones and juvenile felsic granitoids. Although the database of zircon ages >2.7 Ga is still small, there is no convincing evidence for a greenstone age gap prior to this time (Fig. 3), suggesting that supercontinents did not form prior to the Late Archean. The same pattern seems to characterize the last 1.0 Gy, where we have an excellent isotopic age database. During this period, greenstones seemed to have formed and collided with continents at a rather continuous rate, unrelated to the assembly and fragmentation of supercontinents. Perhaps this is due to a significant overlap in time between assembly and breakup stages of post-1.0-Ga supercontinents. For instance, most results indicate that the Late Proterozoic supercontinent Rodinia was fragmenting as Gondwana was forming at 800–600 Ma (Fig. 3) (Hoffman, 1991; Unrug, 1992). Pangea began to breakup about 180 Ma, not long after completion when SE Asia collided with China (200–195 Ma) (Scotese, 1991). While still in the dispersal phase, terrane collisions began about 100 Ma in the Cordillera of North America and in northeastern Asia, and India collided with Tibet about 50 Ma. One test for the supercontinent time overlap idea is the geographic location of post-1.0-Ga greenstones, as they should have accreted to the growing supercontinent, not the dispersing one. This seems to be the case for 900–600 Ma greenstones, which were accreted to Gondwana along the Avalonian–Cadomian belt and along the Pan-African belt in NE Africa and Arabia (Murphy and Nance, 1991). The widespread greenstone terrane collisions in the Late Cretaceous and Early Tertiary around the margins of the North Pacific basin could record a change from the maximum dispersal phase of Pangea to the first assembly phase of a new supercontinent, and thus also be consistent with the proposed model.

From the age distributions of greenstones and juvenile crust in general, it would appear that earth history can be divided into three segments: (1) >2.7 Ga, when greenstones formed and collided continuously, although probably not forming a supercontinent; (2) 2.7–1.0 Ga, where a clear episodisity is apparent in greenstone eruption and collision ages, and with two greenstone age gaps corresponding to supercontinent fragmentation; and (3) <1.0 Ga, when again greenstones appear to have formed and collided rather continuously, perhaps in response to significant overlap in the times of dispersal and assembly of supercontinents. Why the pattern of supercontinent evolution should have changed 1.0 Ga is unknown, but it would seem to be related to a change in mantle convection patterns. Prior to 1.0 Ga, upwelling and downwelling zones may have occurred in relatively few and rather well defined regions of the mantle, whereas after this time these zones may have been less stable and tended to become mixed both in time and space. Such changes could have resulted in a rather clean break between the dispersal and assembly phases of supercontinents before 1.0 Ga, whereas after this time it led to considerable overlap in these phases.

SUMMARY

If our current database of greenstone ages is representative, neither eruption nor collision ages of oceanic greenstones are represented in the time intervals 2450–2200 and 1650–1350 Ma. Prominent peaks in both age distributions occur at 2.7 and 1.9 Ga, similar to those observed in juvenile felsic granitoids. The two greenstone age gaps correlate with times of probable supercontinent fragmentation, and may reflect less frequent greenstone-continent collisions at these times, resulting in more greenstones being recycled into the mantle by subduction. Collision and accretion of greenstone terranes to continents is an important mechanism of continental growth. To avoid a decrease in continental growth rate during times of increased greenstone recycling into the mantle, as dictated by a gradually cooling mantle, requires an increasing rate of some other continental growth mechanism. One possibility is an increase in the rate of mafic underplating of continents, in response to increased mantle plume activity during supercontinent dispersal.

From the age distributions of greenstones and juvenile crust in general, earth history can be divided into three segments: (1) >2.7 Ga, when greenstones formed and collided continuously, although probably not forming a supercontinent; (2) 2.7–1.0 Ga, where a clear episodisity is apparent in greenstone eruption and collision ages, with two greenstone age gaps caused by extensive greenstone recycling during supercontinent fragmentation; and (3) <1.0 Ga, when again greenstones appear to have formed and collided rather continuously, perhaps in response to significant overlap in the times of dispersal and assembly of supercontinents. If this three-fold time division is correct, it implies that prior to 1.0 Ga upwelling and downwelling zones may have occurred in relatively few and rather well defined regions of the mantle, whereas after this time these zones were less stable and tended to become mixed both in time and space.

Basalt geochemistry suggests that both submarine plateaus and arc systems were present by the Early Archean. Most basalts from Archean mafic plains have Th/Ta and La/Yb distributions similar to those of modern submarine plateau basalts derived from depleted mantle sources. A greater proportion of submarine mafic plains in Archean greenstones probably reflects: (1) an enhanced preservation rate of Archean greenstones due to magma underplating with TTG, and/or (2) a greater frequency of plumes in the Archean caused by higher mantle temperatures. Archean greenstone basalts all show relatively high Ni, Fe and Co contents and low Al_2O_3 compared to post-Archean basalts at comparable Mg numbers. This observation is generally thought to reflect higher degrees of melting in the Archean mantle in response to higher mantle temperatures.

So where is Archean greenstone research going in the future? One area of continuing interest is whether any Archean oceanic crust is preserved, and if so how do we identify it? If it comes from a depleted mantle reservoir like MORB and it is relatively unaltered, it should have MORB-like geochemical and isotopic characteristics. Some of the Archean mafic plain assemblages, especially those without or with a few komatiites, should be carefully studied with this possibility in mind. Another question of interest is that of how did the apparently old, thick mantle lithosphere form beneath some Archean crust? Is it the residue of early mantle plumes from which komatiites have been extracted? Perhaps geochemical studies of Archean mantle xenoliths will help resolve this problem. Finally, isotopic geochemists are fervently looking for an Early Archean enriched reservoir that must have formed to complement the depleted reservoir from which most Archean igneous rocks appear to have been derived (Galer and Goldstein, 1991). Could some Archean greenstone assemblages represent fragments of an Early Archean "enriched" mafic crust formed some 4 Ga and mostly recycled into the mantle? It would seem that at least a few minor fragments of such crust would have escaped recycling, and if so may be hidden in Archean greenstone terranes like the Warrawoona Megasequence in Western Australia or the Acasta gneisses in northern Canada. It would seem worthwhile to continue to look for such enriched crust in the oldest preserved greenstones.

REFERENCES

Armstrong, R.L., 1991. The persistent myth of crustal growth. Austral. J. Earth Sci., 38: 613–630.

Arndt, N.T., 1991. High Ni in Archean tholeiites. Tectonophys., 187: 411–420.

Arndt, N.T. and Jenner, G.A., 1986. Crustally contaminated komatiites and basalts from Kambalda, Western Australia. Chem. Geol., 86-229–255.

Arndt, N.T., Albarede, F., and Nisbet, E.G., 1994. Mafic and ultramafic magmatism. In: M.J. de Wit and L.D. Ashwal (Eds.), Tectonic Evolution of Greenstone Belts. Cambridge Univ. Press.

Attoh, K., 1982. Structure, gravity models, and stratigraphy of an Early Proterozoic volcanic-sedimentary belt in NE Ghana. Precambrian Res., 18: 275–290.

Ayres, L.D. and Corfu, F., 1991. Stacking of disparate volcanic and sedimentary units by thrusting in the Archean Favourable Lake greenstone belt, central Canada. Precambrian Res., 50: 221–238.

Baragar, W.R.A., 1972. Some physical and chemical aspects of Precambrian volcanic belts of the Canadian shield. Can. Dept. Energy, Mines, Resources, Publ. Earth Physics Branch, 42(3): 129–140.

Barker, F., Jones, D.L, Budahn, J.R. and Coney, P.J., 1988. Ocean plateau-seamount origin of basaltic rocks, Angayucham terrane, central Alaska. J. Geol., 96: 368–374.

Barker, F., Brown, A.S., Budahn, J.R., and Plafker, G., 1989. Back-arc with frontal arc component origin of Triassic Karmutsen basalt, BC, Canada. Chem. Geol., 75: 81–102.

Barley, M.E., 1986. Incompatible-element enrichment in Archean basalts: a consequence of contamination by older sialic crust rather than mantle heterogeneity. Geology, 14: 947–950.

Barley, M.E., 1993. Volcanic, sedimentary and tectonostratigraphic environments of the 3.46 Warrawoona megasequence: a review. Precambrian Res., 60: 47–67.

Barley, M.E. and Groves, D.I., 1992. Supercontent cycles and the distribution of metal deposits through time. Geology, 20: 291–294.

Bayley, R.W., Proctor, P.D., and Condie, K.C., 1973. Geology of the South Pass area, Fremont Country, Wyoming. U.S. Geol. Survey Prof. Pap., 793, 39 pp.

Bevier, M.L., Barr, S.M., White, C.E., and Macdonald, A.S., 1993. U/Pb geochronologic constraints on the volcanic evolution of the Mira terrane, SE Cape Breton Island, Nova Scotia. Can. J. Earth Sci., 30: 1–10.

Bickford, M.E. and Boardman, S.J., 1984. A Proterozoic volcano-plutonic terrane, Gunnison and Salida area, Colorado. J. Geol., 92: 657–666.

Bickle, M.J., 1986. Implications of melting for stabilization of the lithosphere and heat loss in the Archean. Earth Planet. Sci. Lett., 80: 314–324.

Blackburn, C.E., Bond, W.D., Breaks, F.W., Davis, D.W., Edwards, G.R., Poulsen, K.H., Trowell, N.F., and Wood, J., 1985. Evolution of Archean volcanic-sedimentary sequences of the western Wabigoon subprovince and its margins: a review. Geol. Assoc. Canada, Spec. Pap. 28: 89–116.

Boardman, S.J., 1986. Early Proterozoic bimodal volcanic rocks in central Colorado, USA, Part 1. Precambrian Res., 34: 1–36.

Boher, M., Abouchami, W., Michard, A., Albarede, F., and Arndt, N.T., 1992. Crustal growth in West Africa a 2.1 Ga. J. Geophys. Res., 97: 345–369.

Brewer, T.S. and Atkin, B.P., 1987. Geochemical and tectonic evolution of the Proterozoic Telemark supracrustals, southern Sweden. Geol. Soc. London, Spec. Publ., No. 33: 471–488.

Brown, D.A., Logan, J.M., Gunning, M.H., Orchard, M.J., and Bamber, W.E., 1991. Stratigraphic evolution of the Paleozoic Stikine assemblage in the Stikine and Iskut Rivers area, NW British Columbia. Can. J. Earth Sci., 28: 958–972.

BVTP, 1981. Basaltic Volcanism on the Terrestrial Planets. Pergamon Press, New York, 1286 pp.

Campbell, I.H. and Griffiths, R.W., 1992. The changing nature of mantle hotspots through time: implications for the chemical evolution of the mantle. J. Geol., 92: 497–523.

Campbell, I.H., Hill, R.E.T., and Griffiths, R.W., 1990. Melting in an Archean mantle plume: heads its basalts, tails its komatiites. Nature, 339: 697–699.

Capdevila, R.A.M., Goodwin, A.M., Ujike, O. and Gorton, M.P., 1982. Trace element geochemistry of Archean volcanic rocks and crustal growth in SW Abitibi belt, Canada. Geology, 10: 418–422.

Carlson, R.W., 1991. Physical and chemical evidence on the cause and source characteristics of flood basalt volcanism. Austral. Jour. Earth Sciences, 38: 525–544.

Charvet, J., Lapierre, H., Rouer, O., Coulon, C., Campos, C., Martin, P. and Lecuyer, C., 1990. Tectonomagmatic evolution of Paleozoic and Early Mesozoic rocks in the eastern Klamath Mountains, California, and the Blue Mountains, eastern Oregon-western Idaho. Geol. Soc. America, Spec. Pap., 255: 255–276.

Chown, E.H., Daigneault, R., and Mueller, W., 1992. Tectonic evolution of the northern volcanic zone, Abitibi belt, Quebec. Can. J. Earth Sci., 29: 2211–2225.

Cloos, M., 1993. Lithospheric buoyancy and collisional orogenesis: subduction of oceanic plateaus, continental margins, island arcs, spreading ridges, and seamounts. Geol. Soc. Am. Bull., 105: 715–737.

Condie, K.C., 1981. Archean Greenstone Belts. Elsevier, Amsterdam, 434 pp.

Condie, K.C., 1984. Secular variation in the composition of basalts: an index to mantle evolution. J. Petrol., 26: 545–563.
Condie, K.C., 1986a. Geochemistry and tectonic setting of Early Proterozoic supracrustal rocks in the Southwestern United States. J. Geol., 94: 845–864.
Condie, K.C., 1986b. Origin and early growth rate of continents. Precambrian Res., 32: 261–278.
Condie, K.C., 1989. Geochemical changes in basalts and andesites across the Archean–Proterozoic boundary: identification and significance. Lithos, 23: 1–18.
Condie, K.C., 1990a. Geochemical characteristics of Precambrian basaltic greenstones. In: R.P. Hall and D.J. Hughes (Eds.), Early Precambrian Basic Magmatism. Blackie Publ., Glasgow, UK, pp. 40–55.
Condie, K.C., 1990b. Growth and accretion of continental crust: inferences based on Laurentia. Chem. Geol., 83: 183–194.
Condie, K.C., 1992. Proterozoic terranes and continental accretion Southwestern North America. In: K.C. Condie (Ed.), Proterozoic Crustal Evolution. Elsevier, Amsterdam, pp. 447–480.
Condie, K.C., 1993. Chemical composition and evolution of the upper continental crust: contrasting results from surface samples and shales. Chem. Geol., 104: 1–37.
Condie, K.C. and McCrink, T.P., 1982. Geochemistry of Proterozoic volcanic and granitic rocks from the Gold Hill–Wheeler Peak area, northern New Mexico. Precambrian Res., 19: 141–166.
Condie, K.C. and Shadel, C.A., 1984. An Early Proterozoic volcanic arc succession in SE Wyoming. Can. J. Earth Sci., 21: 415–427.
Condie, K.C., Liu, J., Glikson, A.Y., Hickman, A.H. and Davy, R., 1993. Archean basalts from the Pilbara craton, Western Australia: geochemical evidence for two mantle sources. Can. J. Earth. Sci., in press.
Corfu, F. and Grunsky, E.C., 1987. Igneous and tectonic evolution of the Batchawana greenstone belt, Superior Province: a U/Pb zircon and titanite study. J. Geol., 95: 87–105.
Corfu, F. and Wood, J., 1986. U/Pb zircon ages in supracrustal and plutonic rocks; North Spirit Lake area, NW Ontario. Can. J. Earth Sci., 23: 967–977.
Corfu, F. and Stott, G.M., 1986. U/Pb ages for late magmatism and regional deformation in the Shebandowan belt, Superior Province, Canada. Can. J. Earth Sci., 23: 1075–1082.
Crow, C. and Condie, K.C., 1988. Geochemistry and origin of Late Archean volcanics from the Ventersdorp Supergroup, South Africa. Precambrian Res., 42: 19–37.
Davies, G.F., 1992. On the emergence of plate tectonics. Geology, 20: 963–966.
Davis, D.W. and Edwards, G.R., 1982. Zircon U/Pb ages from the Kakagi Lake area, Wabigoon subprovince, NW Ontario. Can. J. Earth Sci., 19: 1235–1245.
Davison, J., Teixeira, J.B.G., Silva, M.G., Neto, M.B.R., and Matos, F.M.V., 1988. The Rio Itapicuru greenstone belt, Bahia, Brazil: structure and stratigraphic outline. Precambrian Res., 42: 1–17.
DePaolo, D.J., 1981. Neodymium isotopes in the Colorado Front Range and crust–mantle evolution in the Proterozoic. Nature, 291: 193–196.
Desrochers, J-P., Hubert, C., Ludden, J.N., and Pilote, P., 1993. Accretion of Archean oceanic plateau fragments in the Abitibi greenstone belt, Canada. Geology, 21: 451–454.
de Wit, M.J., Hart, R.A. and Hart, R.J., 1987. The Jamestown ophiolite complex, Barberton Mountain belt: a section through 3.5 Ga oceanic crust. J. African Earth Sci., 5: 681–730.
de Wit, M.J., Roering, C., Hart, R.J., Armstrong, R.A., Ronde, C.E.J., Green, R.W.E., Tredoux, M., Peberdy, E., and Hart, R.A., 1992. Formation of an Archean continent. Nature, 357: 553–562.
Dungan, M.A., Lindstrom, M.M., McMillan, N.J., Moorbath, S., Hoefs, J. and Haskin, L.A., 1986. Open system magmatic evolution of the Taos Plateau volcanic field, Northern New Mexico. J. Geophys. Res., 91: 5999–6028.
Floyd, P.A., 1985. Petrology and geochemistry of oceanic intraplate sheet-flow basalts, Nauru Basin,

DSDP Leg 89. Init. Repts. DSDP Leg 89, Washington, DC, pp. 471–497.
Fox, J.S., 1976. Some comments of the volcanic stratigraphy and economic potential of the West Amist Lake area, Saskatchewan. Geol. Div., Sask. Res. Council, Circ. 9.
Galer, S.J.G. and Goldstein, S.L., 1990. Early mantle differentiation and its thermal consequences. Geochim. Cosmochim. Acta, 55: 227–239.
Gibbs, A.K., 1982. Petrology of the Lower Proterozoic greenstone belts of the Guiana shield. Unpubl. ms.
Gilbert, H.P., Syme, E.C., and Zwanzig, H.V., 1980. Geology of the metavolcanic and volcaniclastic metasedimentary rocks in the Lynn Lake area. Manitoba Dept. Energy Mines, Geol. Pap. GP80-1.
Gill J.B., 1987. Early geochemical evolution of an oceanic island arc and backarc: Fiji and the South Fiji Basin. J. Geol., 95: 589–615.
Glikson, A.Y. and Hickman, A.H., 1981a. Geochemistry of Archean volcanic successions, eastern Pilbara block, Western Australia. Austral. Bur. Mines Min. Resources, Record 1981/36.
Glikson, A.Y. and Hickman, A.H., 1981b. Geochemical stratigraphy and petrogenesis of Archean basic–ultrabasic volcanic units, eastern Pilbara block, Western Australia. Spec. Publ. Geol. Soc. Australia, 7:287–300.
Goldstein, S.J. and Jacobsen, S.B., 1988a. Nd and Sr isotopic systematics of river water suspended material: implications for crustal evolution. Earth Planet. Sci. Lett., 87: 249–265.
Goldstein, S.J. and Jacobsen, S.B., 1988b. Rare earth elements in river waters. Earth Planet. Sci. Lett., 89: 35–47.
Goodwin, A.M., 1967. Volcanic studies in the Birch–Uchi Lakes area of Ontario. Ontario Dept. Mines, Misc. Pap. 6.
Goodwin, A.M., 1977. Archean volcanism in Superior Province, Canadian shield. In: W.R. Baragar, L.C. Coleman and J.M. Hall (Eds.), Volcanic Regimes in Canada. Geol. Assoc. Canada, Spec. Pap., 16: 205–242.
Gower, C.F., Scharer, U., and Heaman, L.M., 1992. The Labradorian orogeny in the Grenville Province, eastern Labrador, Canada. Can. J. Earth Sci., 29: 1944–1957.
Gruau, G., Jahn, B.M., Glikson, A.Y., Davy, R., Hickman, A.H., and Chauvel, C., 1987. Age of the Archean Talga-Talga Subgroup, Pilbara block, Western Australia, and early evolution of the mantle: new Sm-Nd isotope evidence. Earth Planet. Sci. Lett., 85: 105–116.
Gurnis, M. and Davies, G.F., 1985. Simple parametric models of crustal growth. J. Geodynamics, 3: 105–135.
Gurnis, M. and Davies, G.F., 1986. Apparent episodic crustal growth arising from a smoothly evolving mantle. Geology, 14: 396–399.
Harris, C.W. and Glover, L., 1988. The regional extent of the ca. 600 Ma Virgilian deformation: implications for stratigraphic correlation in the Carolina terrane. Geol. Soc. Am. Bull., 100: 200–217.
Harrison, N.M., 1970. The geology of the country around Que Que, Zimbabwe. Rhodesian Geol. Surv., Bull., 67.
Hart, S., 1984. A large-scale isotope anomaly in the Southern Hemisphere mantle. Nature, 309: 753–757.
Hart, W.K., Woldegabriel, G., Walter, R.C., and Mertzman, S.A., 1989. Basaltic volcanism in Ethiopia: constraints on continental rifting and mantle interactions. J. Geophys. Res., 94: 7731–7748.
Hayasaka, Y., 1992. Maizuru terrane. Unpub. ms., 13 pp.
Helmstaedt, H.H. and Scott, D.J., 1992. The Proterozoic ophiolite problem. In: K.C. Condie (Ed.), Proterozoic Crustal Evolution. Elsevier, Amsterdam, p. 55–96.

Hoffman, P.F., 1989. Speculations on Laurentia's first gigayear. Geology, 17: 135–138.

Hoffman, P.F., 1991. Did the breakout of Laurentia turn Gondwanaland inside-out? Science, 252: 1409–1412.

Hoffman, P.F. and Cecile, M.P., 1974. Volcanism and plutonism, Sloan River map area, Great Bear Lake, Dist. Mackenzie. Geol. Surv. Canada, Pap. 74–1, Pt. A.

Hoffman, P.F. and McGlynn, J.C., 1977. Great Bear batholith: a volcano-plutonic depression. Geol. Assoc. Canada, Spec. Pap., 16: 169–192.

Houghton, B.F. and Landis, C.A., 1989. Sedimentation and volcanism in a Permian arc-related basin, southern New Zealand. Bull. Volcanol., 51: 433–450.

Hynes, A., 1990. Two-stage rifting of Pangea by two different mechanisms. Geology, 18: 323–326.

Imoto, N., 1984. Late Paleozoic and Mesozoic cherts in the Tamba belt, SW Japan. Bull. Kyoto Univ. Education, Ser. B, No. 65: 15–40.

Jacobsen, S.B., 1988. Isotopic constraints on crustal growth and recycling. Earth Planet. Sci. Lett., 90: 315–329.

Jochum, K.P., Arndt, N.T., and Hofmann, A.W., 1991. Nb-Th-La in komatiites and basalts: constraints on komatiite petrogenesis and mantle evolution. Earth Planet. Sci. Lett., 107: 272–289.

Jones, D.L., Howell, D.G., Coney, P.J., and Monger, J.W.H., 1983. Recognition, character and analysis of tectonostratigraphic terranes in western North America. In: M. Hashimoto and S. Nyeda (Eds.), Advances in Earth and Planetary Sciences. Terra Scient. Publ. Co., Tokyo, pp. 21–35.

Kahkonen, Y., Huhma, H. and Aro., K. 1989. U/Pb and Rb/Sr whole rock isotope studies of Early Proterozoic volcanic and plutonic rocks near Tampere, Southern Finland. Precambrian Res., 45: 27–43.

Kamo, S. and Davis, D., 1991. A review of geochronology from the Barberton Mountain Land. In: Two Cratons and an Orogen, Excursion Guidebook, IGCP Project 280, Dept. Geology, Univ. Witwatersrand, Johannesburg, S.A., pp. 59–68.

Karlstrom, K. and Bowring, S.A., 1988. Early Proterozoic assembly of tectonostratigraphic terranes in SW North America. J. Geol., 96: 561–576.

Karlstrom, K., Doe, M.F., Wessels, R.L., Bowring, S.A., Dann, J.C., and Williams, M.L., 1990. Juxtaposition of Proterozoic crustal blocks: 1.65–1.60 Ga Mazatzal orogeny. Geol. Soc. Am., Cordilleran Sect., Guidebook 1990, pp. 114–123.

Kean, B.F. and Strong, D.F., 1975. Geochemical evolution of an Ordovician island arc of the central Newfoundland Appalachians. Amer. J. Sci., 275: 97–118.

Key, R.M., Litherland, M. and Hepworth, J.V., 1976. The evolution of the Arechean crust of NE Botswana. Precambrian Res., 3: 375–413.

Kimura, G., Ludden, J.N., Desrochers, J.P., and Hori, R., 1993. A model of ocean–crust accretion for the Superior province, Canada. Lithos, 30: 337–355.

Knoper, M.W. and Condie, K.C., 1988. Geochemistry and petrogenesis of Early Proterozoic amphibolites, west-central Colorado. Chem. Geol., 67: 209–225.

Krapez, B., 1993. Sequence stratigraphy of the Archean supracrustal belts of the Pilbara block, Western Australia. Precambrian Res., 60: 1–45.

Krogh, T.E. and Turek, A., 1982. Precise U/Pb zircon ages from the Gamitagama greenstone belt, southern Superior Province. Can. J. Earth Sci., 19: 859–867.

Kroner, A., Compston, W., Zhang, G., Guo, A-L., and Todt, W., 1988. Age and tectonic setting of Late Archean greenstone-gneiss terrain in Henan Province, China, as revealed by single-grain zircon dating. Geology, 16: 211–215.

Kumon, F. and others, 1988. Shimanto belt in the Kii Peninsula, SW Japan. Modern Geol., 12: 71–96.

Kusky, T.M., 1989. Accretion of the Archean Slave Province. Geology, 17: 63–67.

Kusky, T.M. and Kidd, W.S.F., 1992. Remnants of Archean oceanic plateau, Belingwe greenstone belt. Geology, 20: 43–46.

Ladeira, E.A. and Roeser, M.P., 1983. Petrography of the Rio das Velhas greenstone belt, Minas Gerais, Brazil. Zbl. Geol. Palaont., 1: 430–445.

Lafleche, M.R., Dupuy, C., and Bougault H., 1992. Geochemistry and petrogenesis of Archean mafic volcanic rocks of the southern Abitibi belt, Quebec. Precambrian Res., 57: 207–241.

Lightfoot, P.C. and Hawkesworth, C., 1988. Origin of Deccan trap lavas: evidence from combined trace element and Sr, Nd, and Pb isotope studies. Earth Planet. Sci. Lett., 91: 89–104.

Lightfoot, P.C., Sutcliffe, R.H., and Doherty, W., 1991. Crustal contamination identified in Keweenawan Osler Group tholeiites, Ontario: a trace element perspective. J. Geol., 99: 739–760.

Lowe, D.R., Byerly, G.R., Ransom, B.L., and Nocita, B.R., 1985. Stratigraphic and sedimentological evidence bearing on structural repetition in Early Archean rocks of the Barberton greenstone belt, South Africa. Precambrian Res., 27: 165–186.

Mahoney, J., Nicollet, C. and Dupuy, C., 1991. Madagascar basalts: tracking oceanic and continental sources. Earth Planet. Sci. Lett., 104: 350–363.

Mahoney, J., Storey, M., Duncan, R.A., Spencer, K.J., and Pringle, M., 1992. Geochemistry and geochronology of Leg 130 basement lavas: Nature and origin of the Ontong Java Plateau. Proc. ODP, Sci. Research, 130.

Martin, H., 1986. Effect of steeper Archean geothermal gradient on geochemistry of subduction zone magmas. Geology, 14: 753–756.

Martin, H., 1987. Petrogenesis of Archean trondhjemites, tonalites, and granodiorites from eastern Finland: major and trace element geochemistry. J. Petrol., 28: 921–953.

Martyn, J.E., 1987. Evidence for structural repetition in the greenstones of the Kalgoorlie District, Western Australia. Precamb. Res., 37: 1–18.

Massey, N.W.D., 1983. Magma genesis in a Late Proterozoic proto-oceanic rift. Precambrian Res., 21: 81–100.

McClelland, W.C. and Gehrels, G.E., 1990. Geology of the Duncan Canal shear zone: evidence for Early to Middle Jurassic deformation of the Alexander terrane, SE Alaska. Geol. Soc. Am. Bull., 102: 1378–1392.

McClelland, W.C., Gehrels, G.E., and Saleeby, J.B., 1992. Upper Jurassic–Lower Cretaceous basinal strata along the Cordilleran margin: implications for the accretionary history of the Alexander–Wrangellia–Peninsular terrane. Tectonics, 11: 823–835.

McCulloch, M.T. and Wasserburg, G.J., 1978. Sm-Nd and Rb-Sr chronology of continental crust formation. Science, 200: 1003–1011.

McKee, C.G., 1988. Geochemistry and tectonic setting of some Proterozoic rocks in the Pedernal Hills and Manzano Mountains, NM. M.S. Thesis, NM Instit. Mining Tech., Socorro, NM, 349 pp.

Moorbath, S., 1977. Ages, isotopes and evolution of the Precambrian continental crust. Chem. Geol., 20: 151–187.

Moore, J.M., Jr., 1977. Orogenic volcanism in the Proterozoic of Canada. In: W.R. Baragar, L.C. Coleman and J.M. Hall (Eds.), Volcanic Regimes in Canada. Geol. Assoc. Canada, Spec. Pap., 16: 127–148.

Murphy, J.B. and Nance, R.D., 1991. Supercontinent model for the contrasting character of Late Proterozoic orogenic belts. Geology, 19: 469–472.

Myers, J.S., 1992. Tectonic evolution of the Yilgarn craton, Western Australia. In: J.E. Glover and S.E. Ho (Eds.), The Archean: Terrains, Processes and Metallogeny. Geology Department, Univ. Western Australia, Publ. No. 22: 265–274.

Naldrett, A.J. and Turner, A.R., 1977. The geology and petrogenesis of a greenstone belt and related nickel sulfide minralization at Yakabindie, Western Australia. Precambrian Res., 5: 43–103.

Nelson, B.K. and DePaolo, D.J., 1985. Rapid production of continental crust 1.7 to 1.9 by ago: Nd isotopic evidence from the basement of the North American mid-continent. Geol. Soc. Am. Bull., j96: 746–754.

Nelson, D.R., Trendall, A.F., de Laeter, J.R., Grobler, N.J., and Fletcher, I.R., 1992. A comparative study of the geochemical and isotopic systematics of Late Archean flood basalts from the Pilbara and Kaapvaal cratons. Precambrian Res., 54: 231–256.

Nisbet, E.G., 1987. The Young Earth: An Introduction of Archean Geology. Allen & Unwin, Boston, 402 pp.

Nisbet, E.G. and Walker, D., 1982. Komatiites and the structure of the Archean mantle. Earth Planet. Sci. Lett., 60: 105–113.

Nutman, A.P., Allaart, J.H., Bridgwater, D., Rosing, M., and Dimroth, E., 1984. Stratigraphic and geochemical evidence for the depositional environment of the Early Archean Isua supracrustal belt, SW Greenland. Precambrian Res., 25: 365–396.

Nutman, A.P., Friend, C.R.L., Baadsgaard, H. and McGregor, V.R., 1989. Evolution and assembly of Archean gneiss terranes in the Godthabsfjord region, SW Greenland: Structural, metamorphic and isotopic evidence. Tectonics, 8: 573–589.

Nutman, A.P., Friend, C.R., Kinny, P.D., and McGregor, V.R., 1993. Anatomy of an Early Archean gneiss complex: 3900 to 3600 Ma crustal evolution in southern West Greenland. Geology, 21: 415–418.

Park, A.F., 1991. Continental growth by accretion: a tectonostratigraphic terrane analysis of the evolution of the western and central Baltic Shield, 2.5 to 1.75 Ga. Geol. Soc. America Bull., 103: 522–537.

Parr, J. and Rickard, D., 1987. Early Proterozoic subaerial volcanism and its relationship to Broken Hill type mineralization in central Sweden. Geol. Soc. London, Spec. Publ., No. 33: 81–94.

Peucat, J.J., Mahabaleswar, B., and Jayananda, M., 1993. Age of younger tonalitic magmatism and granulitic metamorphism in the South Indian transition zone; comparison with older Peninsular gneisses from the Gorur-Hassan area. J. Metamorphic Geol., 11: 879–888.

Picard, C. and Piboule, M., 1986. Petrologie des roches volcaniques du sillon de roches vertes archeennes de Matagami–Chibougamau a l'ouest de Chapais. Can. J. Earth Sci., 23: 561–578.

Pichamuthu, C.S. and Srinivasan, R., 1983. A billion year history of the Dharwar craton. In: S.M. Naqvi and J.J.W. Rogers (Eds.), Precambrian of South India. Geol. Soc. India, Bangalore, pp. 121–142.

Plafker, G., Nokleberg, W.J., and Lull, J.S., 1989. Bedrock geology and tectonic evolution of the Wrangellia, Peninsular, and Chugach terranes along the Trans-Alaska curstal transact in the Chugach Mountains and southern Copper River Basin, Alaska. J. Geophys. Res., 94: 4255–4295.

Potter, A.W. et al., 1990. Early Paleozoic stratigraphy, paleogeographic and biogeographic relations of the eastern Klamath belt, northern California. Geol. Soc. Am., Spec. Pap. 255: 57–74.

Premo, W. and Van Schmus, W.R., 1989. Zircon geochronology of Precambrian rocks in southeastern Wyoming and northern Colorado. Geol. Soc. Am., Spec. Pap., 235: 13–32.

Renner, R. and Gibbs, A.K., 1987. Geochemistry and petrology of the metavolcanic rocks of the Early Proterozoic Mazaruni greenstone belt, Northern Guiana. Geol. Soc. Lond., Spec. Publ., 33: 289–309.

Robertson, J.M. and Condie, K.C., 1989. Geology and geochemistry of Early Proterozoic volcanic and subvolcanic rocks of the Pecos greenstone belt, Sangre de Cristo Mountains, New Mexico. Geol. Soc. Am., Spec. Pap., 235: 119–146.

Rosen, O.M., Condie, K.C., Natapov, L.M., and Nozhkin, A.D., 1994. Archean and Early Proteroxoic evolution of the Siberian craton: a preliminary assessment. In: K. Condie (Ed.), Archean Crustal Evolution, Chap. 10, pp. xxx–xxx.

Ross, G.M., Parrish, R.R., Villeneuve, M.E., and Bowring, S.A., 1991. Geophysics and geochronology of the crystalline basement of the Alberta basin, western Canada. Can. J. Earth Sci., 28: 512–522.

Samson, S.D. and Patchett, P.J., 1991. The Canadian Cordillera as a modern analog of Proterozoic crustal growth. Austral. J. Earth Sci., 38: 595–611.

Scotese, C.R., 1991. Jurassic and Cretaceous plate tectonic reconstructions. Paleogeogr. Paleoclimatol. Paleoecol., 87: 493–501.

Schultz, K.J., 1980. The magmatic evolution of the Vermilion greenstone belt, NE Minnesota. Precambrian Res., 11: 215–246.

Shirey, S.B. and Hanson, G.N., 1986. Mantle heterogeneity and crustal recycling in Archean granite-greenstone belts: evidence from Nd isotopes and trace elements in the Rainy Lake area, Superior Province, Ontario, Canada. Geochim. Cosmochim. Acta, 50: 2631–2651.

Skiold, T., 1988. Implications of new U/Pb zircon chronology to Early Proterozoic crustal accretion in northern Sweden. Precambrian Res., 38: 147–164.

Sims, P.K., Van Schmus, W.R., Schulz, K.J., and Peterman, Z.E., 1989. Tectonostratigraphic evolution of the Early Proterozoic Wisconsin magmatic terranes of the Penokean orogen. Can. J. Earth Sci., 26: 2145–2158.

Sinha, A.K. and Bartholomew, M.J., 1984. Evolution of the Grenville terrane in the central Virginia Appalachians. Geol. Soc. Am. Spec. Pap. 194: 175–186.

Sleep, N.H., and Windley, B.F., 1982. Archean plate tectonics: constraints and inferences. J. Geol., 90: 363–379.

Stephens, M.B. and Gee, D.G., 1989. Terranes and polyphase accretionary history in the Scandinavian Caledonides. Geol. Soc. Am. Spec. Pap., 230: 17–30.

Stern, R.J., 1981. Petrogenesis and tectonic setting of Late Precambrian ensimatic volcanic rocks, central eastern desert of Egypt. Precambrian Res., 16: 195–230.

Storey, M., Mahoney, J.J., Kroenke, L.W., and Saunders, A.D., 1991. Are oceanic plateaus sites of komatiite formation? Geology, 19: 376–379.

Strong, D.F. and Payne, J.G., 1973. Early Paleozoic volcanism and metamorphism of the Moretons Harbour–Twillingate area, Newfoundland. Can. J. Earth Sci., 10: 1363–1378.

Stuart-Smith, P.G., Hill, R.I., Rickard, M.J., and Etheridge, M.A., 1992. The stratigraphy and deformation history of the Tumut region: implications for the development of the Lachlan Fold Belt. Tectonophysics, 214: 211–237.

Swager, C.P., Witt, W.K., Griffin, T.J., Ahmat, A.L., Hunter, W.M., McGoldrick, P.J., and Wyche, S., 1992. Late Archean granite-greenstones of the Kalgoorlie terrane, Yilgarn craton, Western Australia. In: J.E. Glover and S.E. Ho (Eds.), The Archean: Terrains, Processes and Metallogeny. Geology Department, Univ. Western Australia, Publ. No. 22: 107–122.

Sylvester, P.J. and Attoh, K., 1992. Lithostratigraphy and composition of 2.1 Ga greenstone belts of the West African craton and their bearing on crustal evolution and the Archean/Proterozoic boundary. J. Geol., 100: 377–393.

Sylvester, P.J., Attoh, K., and Schulz, K.J., 1987. Tectonic setting of late Archean bimodal volcanism in the Michipicoten greenstone belt, Ontario. Can. J. Earth Sci., 24: 1120–1134.

Syme, E.C., 1991. Stratigraphy and geochemistry of the Lynn Lake and Flin Flon metavolcanic belts, Manitoba. Geol. Assoc. Canada, Spec. Pap. 37: 143–161.

Taira, A., Pickering, K.T., Windley, B.F., and Soh, W., 1992. Accretion of Japanese island arcs and implications for the origin of Archean greenstone belts. Tectonics, 11: 1224–1244.

Tarney, J., Wood, D.A., Varet, J., Saunders, A.D., and Cann, J.R., 1979. Nature of mantle heterogeneity in the North Atlantic: Evidence from Leg 49 basalts. American Geophysical Union, Maurice Ewing Series, 2: 285–301.

Taylor, S.R. and McLennan, S.M., 1985. The Continental Crust: Its Composition and Evolution. Blackwell, Oxford, UK, 312 pp.

Thurston, P.C., 1990. Early Precambrian basic rocks of the Canadian Shield. In: R.P. Hall and E.J. Hughes (Eds.), Early Precambrian Basic Magmatism. Blackie & Sons, London, pp. 221–247.

Thurston, P.C., Ayres, L.D., Edwards, G.R., Gelinas, L., Ludden, J. N., and Verpaelst, P., 1985. Archean bimodal volcanism. Geol. Assoc. Canada, Special Pap. No. 28, pp. 7–22.

Thurston, P.C. and Chivers, K.M., 1990. Secular variation in greenstone sequence development emphasizing Superior Province, Canada. Precambrian Res., 46: 21–58.

Turek, A., Sage, R.P., and Van Schmus, W.R., 1992. Advances in the U/Pb zircon geochronology of the Michipicoten greenstone belt, Superior Province, Ontario. Can. J. Earth Sci., 29: 1154–1165.

Unrug, R., 1992. The supercontinent cycle and Gondwanaland assembly: Component cratons and the timing of suturing events. J. Geodynamics, 16: 215–240.

Vance, R.K., 1989. Geochemistry and tectonic setting of the Yavapai Supergroup, West-Central Arizona. PhD Dissert., New Mexico Inst. Mining & Tech., Socorro, NM, 461 pp.

van Staal, C.R. and Colman-Sadd, S.P., 1993. The central mobile belt of the northern Appalachians: a Paleozoic analog of an Archean granite-greenstone terrain. In: M.J. de Wit and L.D. Ashwal (Eds.), Tectonic Evolution of Greenstone Belts. Cambridge Univ. Press.

Vivallo, W. and Rickard, D., 1984. Early Proterozoic ensialic spreading-subsidence: evidence from the Garpenberg enclave, central Sweden. Precambrian Res., 26: 203–222.

Wardle, R.J., Ryan, B., and Nunn, G.A.G., 1991. Labrador segment of the Trans-Hudson orogen: crustal development through oblique convergence and collision. Geol. Assoc. Canada, Spec. Pap. 37: 353–369.

Watkins, K.P. and Hickman, A.H., 1992. Geology of the Murchison Province granite-greenstone terrain, Western Australia. In: S.E. Ho, J.E. Glover, J.S. Myers and J.R. Muhling (Editors), Third International Archean Symposium, Perth, 1990. Geology Dept., Univ. Western Australia, Publ. No. 21: 147–171.

Weaver, B.L., 1991. The origin of ocean island basalt end-member compositions: trace element and isotopic constraints. Earth Planet. Sci. Lett., 104: 381–397.

Williams, H.R., Stott, G.M., Thurston, P.C., Stucliffe, R.H., Bennett, G., Easton, R.M., and Armstrong, D.K., 1992. Tectonic evolution of Ontario: summary and synthesis. Ontario Geol. Surv., Spec. Vol. No. 4: 1255–1324.

Wilson, H.D.B. and Morrice, M.G., 1977. The volcanic sequence in Archean shields. In: W.R. Baragar, L.C. Coleman and J.M. Hall (Eds.), Volcanic Regimes in Canada. Geol. Assoc. Canada, Spec. Pap., 16: 355–374.

Windley, B.F., 1992. Proterozoic collisional and accretionary orogens. In: K.C. Condie (Ed.), Proterozoic Crustal Evolution. Elsevier, Amsterdam, pp. 419–446.

Wyld, S.J., 1990. Paleozoic and Mesozoic rocks of the Pine Forest Range, NW Nevada, and their relation to volcanic arc assemblages of the western U.S. Cordillera, in Paleozoic and Early Mesozoic paleogeographic relations. Geol. Soc. America, Spec. Pap. 255: 219–238.

Zhai, M., Yang, R., Lu, W., and Zhou, J., 1983. The Qingyuan Archean granite–greenstone belt and its geochemistry: Discussion. Unpub. ms.

Zindler, A. and Hart, S., 1986. Chemical geodynamics. Ann. Rev. Earth Planet. Sci., 14: 493–571.

Chapter 4

ARCHEAN GREENSTONE-RELATED SEDIMENTARY ROCKS

DONALD R. LOWE

INTRODUCTION

Archean sedimentary rocks provide the primary record of surface conditions and crustal evolution on the early earth. Although there is a sparse development of Proterozoic- and Phanerozoic-style craton cover and craton margin sedimentary rocks older than 2.5 Ga, the principal present-day repositories of the Archean sedimentary record are greenstone belts. This record begins with the 3.9 Gy-old banded iron-formation, felsic metatuffs and metavolcanic breccias, cherts, and calc-silicate rocks of the Isua supracrustal suite in West Greenland (Fig. 1); includes the well-preserved greenstone sequences of the 3.55–3.2 Gy-old Barberton Greenstone Belt, South Africa and Swaziland, and of the 3.5–3.2-Gy-old eastern Pilbara Block, Western Australia (Fig. 2); and culminates with a host of 3.0–2.5 Gy-old, Late Archean belts (Fig. 3), present on virtually all continents but best studied in the Superior and Slave Provinces of Canada (Card, 1990) and the Yilgarn Block, Western Australia (Gee et al., 1981; Pidgeon and Wilde, 1990). While the Archean sedimentary record is dominated by largely simatic greenstone-belt supracrustal sequences, the latest Archean saw the formation of enormous, buoyant blocks of continental crust that dominated both tectonic and sedimentary systems throughout post-Archean time.

Interpreting supracrustal rocks in Archean greenstone belts is complicated by a number of problems that tend to obliterate, blur, and bias the geologic record. In addition, there is strong evidence that there may be significant differences in make-up and tectonic settings of pre-3.0 Ga and post-3.0 Ga greenstone sequences (Lowe, 1980, 1982; Groves and Batt, 1984). The following discussion will review the principal types of sedimentary rocks and sedimentary associations in Archean greenstone belts, some of the problems complicating the interpretation of Archean greenstone-belt sedimentary sequences, similarities and differences between pre- and post-3.0 Ga greenstone belts, and the main implications of greenstone-belt sedimentary rocks toward interpreting the characteristics and evolution of the early earth.

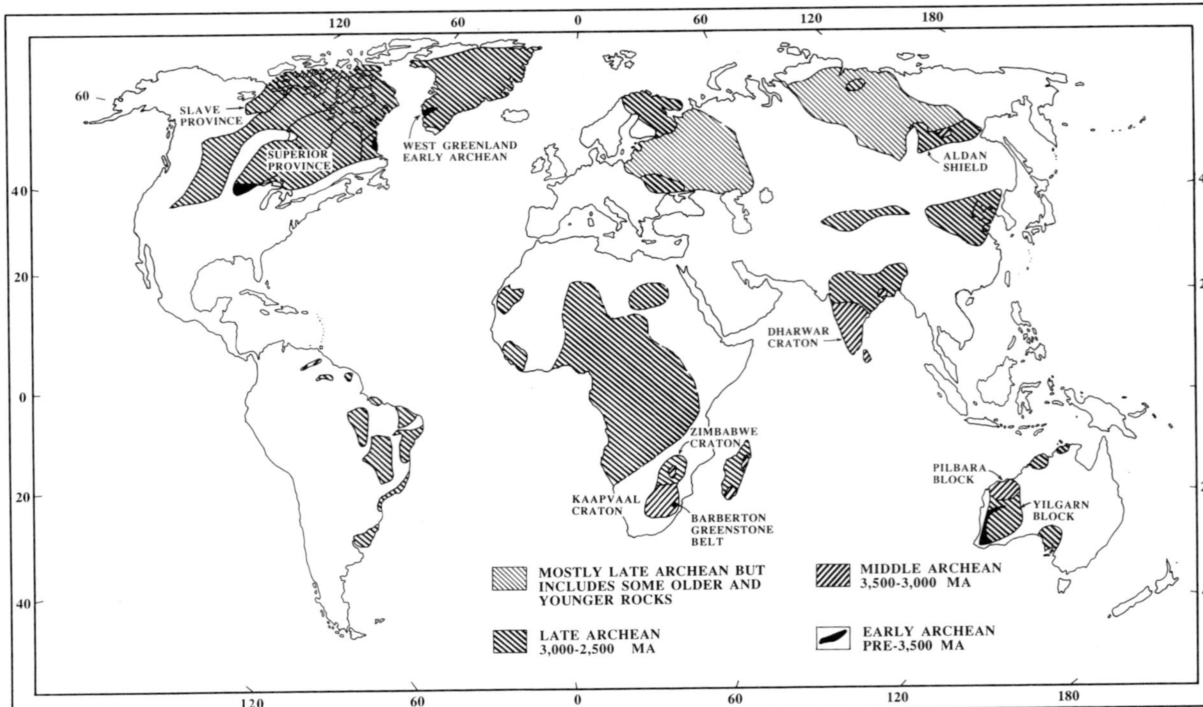

Fig. 1. General distribution of Archean basement blocks (from Lowe, 1992a) and location of some of the major exposures Archean rocks mentioned in the text.

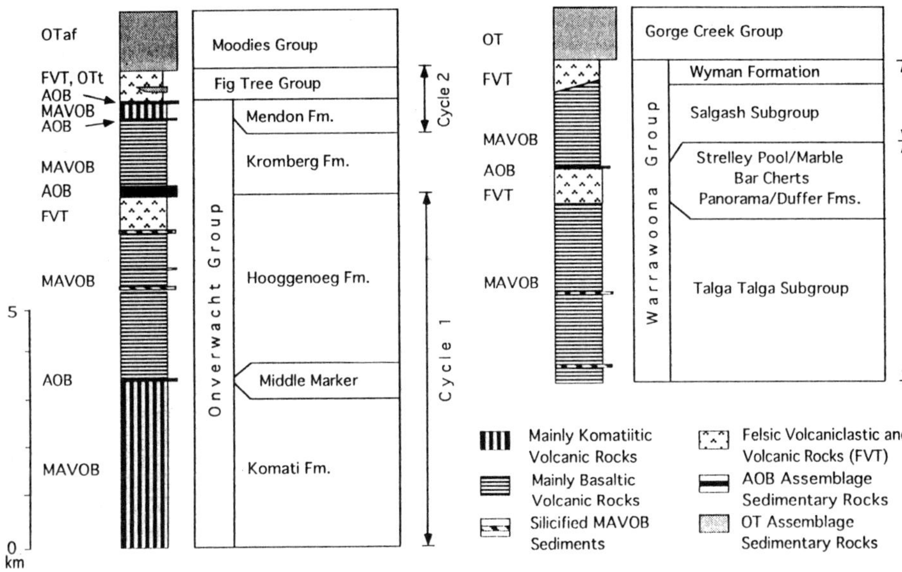

Fig. 2. Generalized stratigraphic columns of the pre-3.0 Ga Barberton Greenstone Belt, South Africa (left), and eastern Pilbara Block, Western Australia (right), showing the distribution of major lithofacies associations.

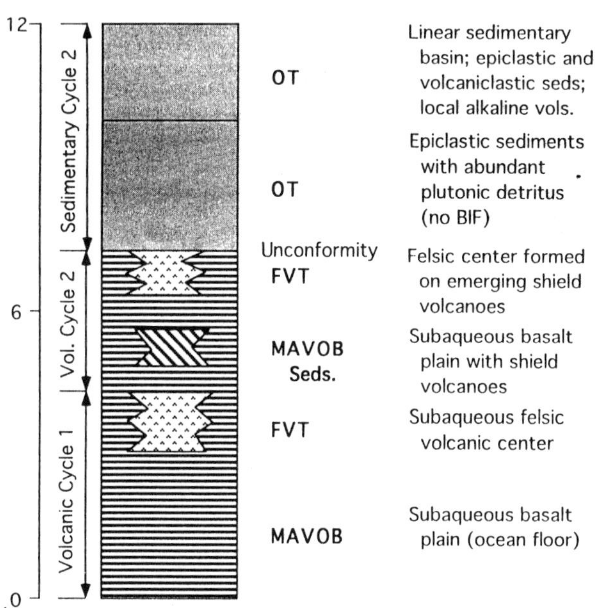

Fig. 3. Generalized stratigraphic column of rocks in the Northern Volcanic Zone, Abitibi Greenstone Belt, Superior Province, Quebec, Canada (from Chown et al., 1992) showing the distribution of major lithofacies associations. Interpreted tectonic and depositional settings on right.

GEOLOGIC SETTINGS OF ARCHEAN SEDIMENTARY ROCKS

Archean sedimentary rocks are preserved in three principal types of geologic terranes: (1) high-grade gneiss terranes, (2) greenstone belts, and (3) craton-margin and craton-cover sequences (Lowe and Ernst, 1992; Lowe, 1992a). Within high-grade terranes, such as the Limpopo Belt of southern Africa (van Biljon and Legg, 1983), the gneiss and granulite terrane of southern India (Naqvi and Rogers, 1983), and the West Greenland gneiss terrane (Bridgwater et al., 1976), all sedimentary rocks have been severely metamorphosed and are typically represented by varieties of felsic aluminous schists, quartz–mica schists, saccharoidal metaquartzites and jaspers, and calc-silicate rocks. Although the trace-element geochemistry of these metasediments can yield important information about their depositional protoliths, most details useful in interpreting their sedimentology and the nature of early exogenic systems have been obliterated or at least severely modified by metamorphism.

Major Archean craton-margin and craton-cover sequences are known only from rocks younger than 3.0 Ga on the Kaapvaal Craton, southern Africa, and Pilbara Craton, Western Australia, and resemble most younger cratonic sequences (Eriksson and Fedo, 1994). They include immature craton-derived, largely terrestrial orogenic, post-orogenic, and rift-related terrigenous and volcaniclastic sequences, such as the Witwatersrand and Pongola Supergroups of South Africa (Tankard et al., 1982) and the Fortescue Supergroup of Western Australia (Hickman, 1983; Blake, 1993), as well as mature, quartzite–carbonate–shale associations deposited under shallow-marine shelfal conditions, such as the Transvaal Supergroup of South Africa (Tankard et al., 1982). These generally well-preserved and only slightly metamorphosed units have provided important details about the Archean atmosphere, biosphere, and lithosphere.

By far the greatest volume of Archean sedimentary rocks is preserved within greenstone belts. The heterogeneous basement complexes of every modern continent include large tracts of Archean crust composed largely of plutonic rocks in which are embedded greenstone belts representing remnants of the pre-cratonic and syn-cratonization stages of craton evolution. These greenstone-belt supracrustal sequences vary greatly in makeup. Most consist largely of mafic volcanic rocks containing subordinate interbedded ultramafic volcanic rocks, felsic volcanic and volcaniclastic units, and sedimentary layers (Figs. 2 and 3). In a few areas, such as some greenstone belts of the Dharwar Craton of India (Naqvi and Rogers, 1983) and the Slave Province of northern Canada (Padgham, 1985), preserved Archean greenstone sequences are dominated by immature clastic sedimentary rocks.

The abundance of greenstone belts in the Archean and their apparent paucity in the Proterozoic, and the sudden appearance of voluminous craton-cover sequences in the Proterozoic and their rarity in the Archean have been interpreted to indicate that a fundamental change in the style of lithospheric tectonics and crustal evolution occurred at the Archean–Proterozoic boundary, 2.5 Ga (e.g. Taylor and

McLennan, 1985). Greenstone belts themselves have been regarded by some as uniquely Archean tectonostratigraphic domains. However, it is now generally recognized that metamorphosed supracrustal remnants in continental crust of all ages represent analogous pre-cratonic metavolcanic and metasedimentary sequences and that the abundance of Archean greenstone belts reflects the abundance of preserved Late Archean crustal blocks: Late Archean crust comprises at least 50% of the total preserved Precambrian crust today (Lowe, 1992a). Proterozoic and Phanerozoic crustal blocks include similar volcanic and sedimentary successions, often somewhat more metamorphosed than their Archean analogs, and generally, though not always, lacking komatiites. The abundance of greenstone belts of any given age reflects the abundance of preserved continental crust of that age not the actual areal extent of greenstone-type settings at that time.

PRINCIPAL TYPES OF SEDIMENTARY ROCKS IN ARCHEAN GREENSTONE BELTS

The general characteristics of sedimentary rocks in Archean greenstone belts have been summarized by Pettijohn (1943, 1970, 1972), Lowe (1980, 1982), Condie (1981), Groves and Batt (1984), and Ojakangas (1985) and the lithologies and stratigraphies of individual belts and blocks are presented in numerous local reports. Greenstone-belt sedimentary rocks can be discussed in term of four principal primary lithologic and genetic sedimentary rock suites (Lowe, 1980, 1982): (1) pyroclastic and autoclastic deposits, (2) terrigenous epiclastic sedimentary rocks, (3) orthochemical deposits, and (4) biogenic deposits.

Pyroclastic and autoclastic deposits

Most greenstone-belt supracrustal sequences are from 5 to 10 km thick and dominated by volcanic flow rocks (Figs. 2 and 3). Tholeiitic basalts predominate, but Mg-rich volcanic rocks, from high Mg basalts to komatiites, are widespread and are locally nearly as abundant as basalts, as in the Barberton Greenstone Belt. Andesitic volcanic rocks are poorly represented or absent in pre-3.0 Gy-old greenstone belts but comprise thick successions in some Late Archean belts, such as those of the Superior Province, Canada (Baragar and Goodwin, 1968; Condie, 1982; Jensen, 1985) and Yilgarn Block, Western Australia (Giles, 1982; Hallberg et al., 1976). Felsic volcanic and volcaniclastic units containing flows of dacitic to rhyolitic magmas characterize greenstone sequences throughout the Archean. Fragmental volcanic rocks, including pyroclastic and autoclastic deposits and their locally reworked equivalents, are widespread but quantitatively minor components of all of these volcanic sequences. Epiclastic sediments formed by the weathering and erosion of lithified volcanic rocks will be discussed elsewhere.

Mafic and ultramafic pyroclastic and autoclastic deposits

Basaltic greenstone-belt sequences are commonly pillowed and most were erupted in subaqueous, commonly deep-water settings as extensive lava plains, shield volcanoes, or volcanic platforms or plateaus (e.g. Dimroth et al., 1985b; Chown et al., 1992). Because mafic pyroclastic volcanism and phreatomagmatic explosions are unlikely at water depths below 150 to 500 m (Colgate and Sigurgeirsson, 1973; Sigurdsson, 1982a,b), most greenstone-belt mafic volcanic sequences contain a small proportion of pyroclastic and autoclastic detritus (Dimroth et al., 1985a). Thin interflow layers of altered basaltic tuff and reworked tuff occur within some mafic flow sequences, probably representing distal airborne debris from distant subaerial eruptions. Shallow-water basaltic sequences show the development of pillow, flow-top, and hyaloclastic breccias and lapilli and ash tuffs (Viljoen and Viljoen, 1969a; Dimroth et al., 1985a; Mueller, 1991). Thick, commonly lenticular units of basaltic lapillistone (Fig. 4) and tuff represent cinder cones, tuff rings, and similar accumulations of coarse mafic fragmental volcaniclastic debris erupted from shallow subaqueous or subaerial vents. One such unit of mafic lapillistone in the Kromberg Formation in the Barberton Greenstone Belt locally reaches 1000 m thick and shows massive fall-deposited and current-worked, cross-stratified (Fig. 4) facies (Viljoen and Viljoen, 1969b; Ransom, 1987).

Fig. 4. Current-deposited mafic lapillistone in the Kromberg Formation, Barberton Greenstone Belt, South Africa. This fragmental unit, which locally reaches 1000 m thick, is interbedded with and grades laterally into basaltic volcanic rocks and includes both sediment-flow and current-deposited facies (Ransom, 1987). The photo shows a layer with large-scale cross-stratification between layers with upper flow regime plane bedding.

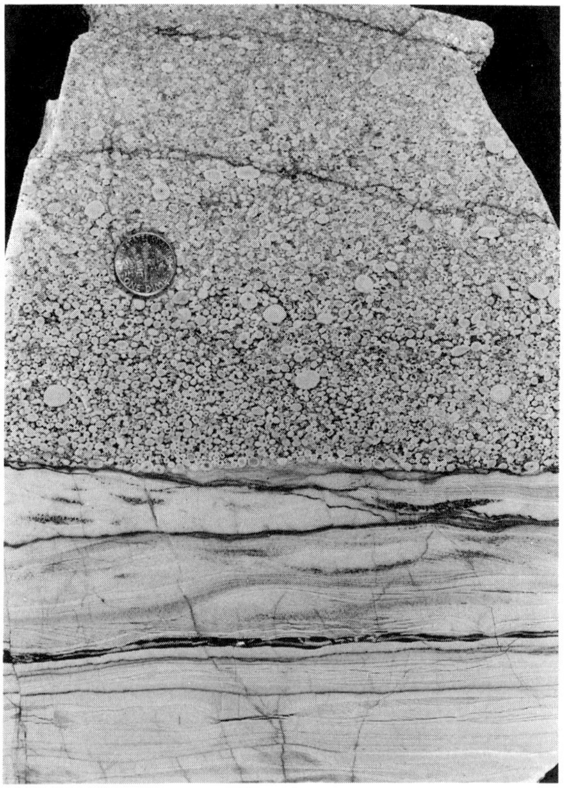

Fig. 5. Silicified komatiitic (basaltic komatiitic) pyroclastic units of the Msauli Chert (Reimer, 1975; Stanistreet et al., 1980; Heinrichs, 1984; Lowe, in review, b) in the Mendon Formation, Barberton Greenstone Belt. A layer of fall-deposited accretionary lapilli, many with nuclei, is underlain by cross-laminated ash showing well-preserved ripples and mixed with small amounts of carbonaceous detritus (black particles).

Because of the low volatile content and low viscosities of ultramafic liquids, komatiitic eruptions might be expected to produce less pyroclastic detritus than even basaltic eruptions. However, in the 3.55–3.2 Gy-old Barberton Greenstone Belt and in the 3.5–3.2 Gy-old eastern Pilbara Block, rather common thin layers of fine-grained pale greenish to gray, impure chert (Fig. 5) have recently been shown to have formed by the silicification of komatiitic tuffs (Lowe, in review, a). Although composed mainly of microcrystalline mosaics of sericite, chlorite, and quartz, these layers are everywhere interbedded with or cap sequences of peridotitic or basaltic komatiites, contain high Ni and Cr contents, and group with komatiites in Zr-Al-Ti diagrams. They commonly contain accretionary lapilli (Figs. 5 and 6) composed of dust and blocky ash grains (Fig. 6) that probably formed through thermal fragmentation (Heinrichs, 1984). In the Barberton Green-

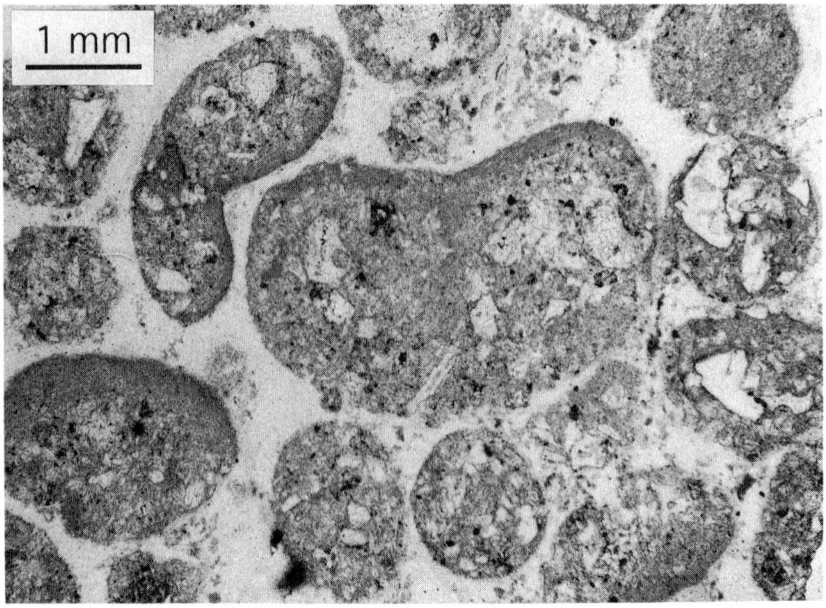

Fig. 6. Photomicrograph of silicified komatiitic accretionary lapilli, Msauli Chert, Mendon Formation, Barberton Greenstone Belt. Although primary textures are well preserved, all of the pyroclastic detritus has been altered to a micromosaic of quartz, phyllosilicate, and oxide grains. The lapilli were composed of dust- and blocky angular ash-sized vitric grains produced by thermal fragmentation (Heinrichs, 1984). Some lapilli appear to be compound particles formed by welding of individual lapilli. Concentric layering is largely absent, although some grains show a partial outer coating of fine dust (dark layer). Plane light.

stone Belt (Lowe and Knauth, 1978; Stanistreet et al., 1981; Reimer, 1975, 1983; Heinrichs, 1984), many individual layers of accretionary lapilli and ash less than a meter thick can be traced regionally, commonly over distances of 10s of km. These komatiitic tuff layers appear to reflect massive shallow-water hydroclastic explosions associated with the terminal stages of komatiitic volcanic eruptions.

Andesitic pyroclastic and autoclastic deposits

Intermediate calc-alkaline volcanic successions are components of many Late Archean greenstone belts. Like modern intermediate volcanic successions, thick Archean andesitic sequences were generally erupted to form high-relief stratovolcanoes, and coarse autobreccias and lahar and debris-flow deposits representing resedimented autobreccias are common (e.g. Jolly and Hallberg, 1990).

Felsic pyroclastic and autoclastic deposits

The bulk of pyroclastic and autoclastic deposits in Archean greenstone belts is composed of felsic, mostly dacitic volcanic debris. Examples include 3.44–3.45 Gy-old dacitic breccias, conglomerates, and sandstones near the top of the Hooggenoeg Formation in the Barberton Greenstone Belt (Viljoen and Viljoen, 1969a; Lowe and Knauth, 1977); felsic flows, breccias, tuffs, conglomerates, and sandstones of the 3.43–3.46 Gy-old Duffer and Panorama Formations in the Warrawoona Group in the eastern Pilbara greenstone belts (Barley et al., 1979, 1984; DiMarco and Lowe, 1989a,b; Barley, 1993), and numerous felsic units in Late Archean belts in the Superior (Sangster, 1972; Dimroth, 1977; Dimroth and Demarcke, 1978; Tasse et al., 1978; Dimroth and Rocheleau, 1979; Thurston, 1980; Blackburn et al., 1985; Thurston et al., 1985; Car and Ayres, 1991; Mueller and White, 1992; and many others) and Slave (Lambert, 1988; Padgham, 1980) Provinces. Felsic volcanic units include thick, grossly lenticular proximal facies centered around an eruptive vent or vent complex (Fig. 7) and widespread, sheet-like distal deposits. Proximal facies are commonly composed largely of massive to crudely stratified volcanic autobreccias (Fig. 7a) interbedded with felsic flow rock, lapilli tuff and lapillistone, and the coarse current-worked or resedimented equivalents of autobreccias (Fig. 7b) and tuffs (Fig. 8a). Eruptive centers, where subaerial, locally include ash-flow tuffs (Dimroth and Demarcke, 1978; Thurston, 1980;) and were commonly flanked by coarse-grained alluvial systems and fan-deltas composed largely of reworked tuff and autoclastic materials. These proximal units generally grade distally into reworked, subaqueously-deposited conglomerate, sandstone, and siltstone (Fig. 8b); terrigenous debris derived by erosion of the vent complexes; and layers of subaqueously deposited tuff (Fig. 9), carbonaceous mudstone, and banded iron-formation. Subaqueous felsic eruptions form thick piles of coarse autobreccia and a variety of hydroclastic deposits, and their resedimented turbiditic and debris flow equivalents (e.g. Mueller and White, 1992). Both pyroclastic and epiclastic units are likely to be poorly developed or absent around deep-water felsic vents.

Alteration of fragmental volcanic deposits

The identification of fine-grained mafic and komatiitic pyroclastic units is commonly hampered by metamorphism and metasomatism. Low-grade greenschist-facies metamorphism is widespread and characterized by actinolite–chlorite–epidote–sericite–quartz–carbonate mineral assemblages (e.g. Barley, 1984; Mueller, 1991). Carbonation is locally intense (Veizer et al., 1989a), especially in more porous mafic and ultramafic lapillistones, and disseminated secondary iron-rich carbonate is common within many flow units. In both the Warrawoona Group, eastern Pilbara Block, and the Onverwacht Group, Barberton Belt, the author has seen units of impure, brownish-weathering, gray, iron-rich carbonate up to 10 m thick formed by the alteration of mafic to komatiitic lapillistones. Early metasomatism has removed soluble Ca, Mg, and Fe from and introduced K and silica into many komatiitic tuffs and komatiitic flows underlying cherts in the Barberton Belt and eastern Pilbara Block (Lowe and Byerly, 1986; Duchac and Hanor, 1987; Hanor and Duchac, 1990). The resulting rocks include a variety of impure K-rich cherts that have been locally misinterpreted as felsic tuffs.

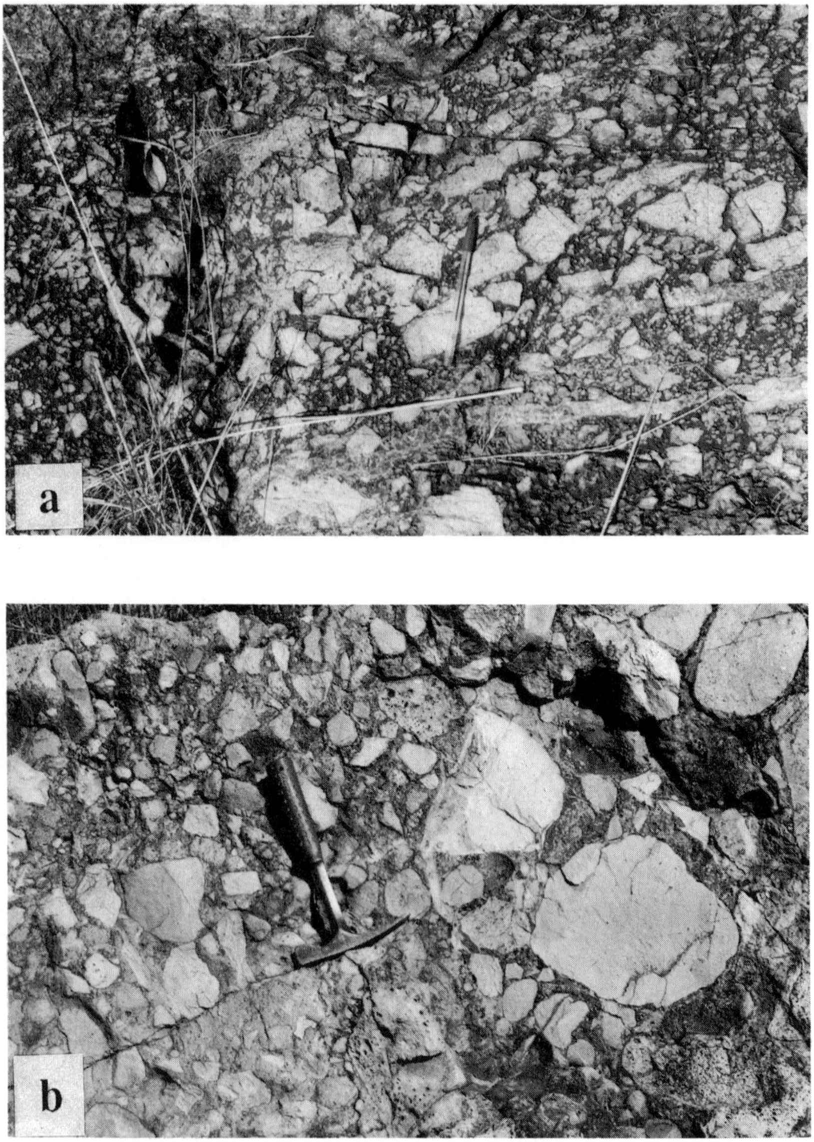

Fig. 7. Coarse felsic autoclastic units, upper Hooggenoeg Formation, Barberton Greenstone Belt. (a) Breccia composed of monolithologic dacitic clasts in matrix of carbonated dacitic ash. This unit, deposited by subaqueous debris flows, is part of a thick sequence of resedimented autoclastic and pyroclastic debris deposited adjacent to an eruptive center. (b) Conglomerate formed by reworking of autoclastic debris in alluvial and/or shallow-marine environments before resedimentation by subaqueous debris flows.

Fig. 8. Resedimented sand-sized pyroclastic and autoclastic debris, upper Hooggenoeg Formation, Barberton Greenstone Belt. (a) Cross-stratified ash deposited in shallow-water. (b) Large flame structure formed by loading of basal massive coarse-grained ash of one sedimentation unit into fine-grained, flat-laminated and cross-laminated ash at top of underlying sedimentation unit. Both units are thick-bedded turbidites deposited by submarine resedimentation of felsic debris.

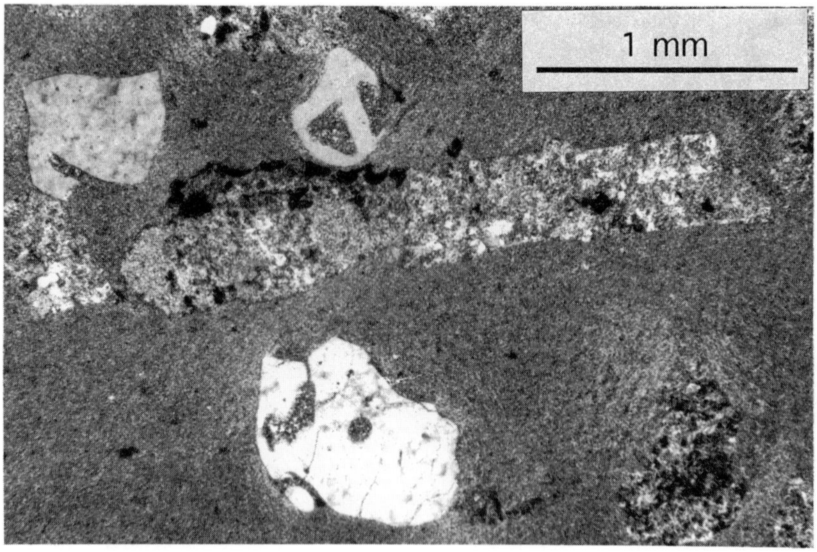

Fig. 9. Photomicrograph of fall-deposited dacitic ash, upper Hooggenoeg formation, Barberton Greenstone Belt. Phenocrysts of quartz (white) and plagioclase (recrystallized lath-shaped grain), and recrystallized lithic grains (irregular) in a matrix of recrystallized ash and dust. Recrystallized materials are now micromosaics of quartz, sericite, and oxide grains.

Because of the reactiveness of glassy and finely crystalline felsic volcaniclastic detritus, felsic pyroclastic and autoclastic units in greenstone belts are also commonly extensively altered (Fig. 9). Most have been recrystallized under low-grade metamorphic and/or metasomatic conditions to micromosaics of intergrown phyllosilicate minerals, silica, and oxides, commonly dominated by sericite and quartz (DiMarco et al., 1989; Cullers et al., 1993). Where metasomatism is extensive, the alteration mineral assemblages generally do not reflect the primary compositions of the rocks.

In spite of widespread alternation, in the absence of penetrative shearing, primary textures and structures are commonly well preserved in volcaniclastic units within Archean greenstone sequences.

Terrigenous epiclastic sedimentary rocks

Greenstone belts also include thick units of conglomerate, sandstone, and mudstone derived by erosion of lithified rocks within or around the basin of sedimentation. Three principal end-member types of terrigenous sediments can be distinguished: (1) synvolcanic, pre-deformation sediments derived by erosion of the greenstone-belt volcanic sequence itself; (2) late volcanic and syndeformation sediments derived by erosion of uplifted greenstone-belt volcanic rocks and deeper-level synvolcanic and syntectonic plutonic rocks exposed during deformation, and (3) sediments derived from older rocks around or beneath the basin of deposition. Many units are mixtures of these three end-member sediment types.

Synvolcanic epiclastic deposits

Because of their low viscosities and low volatile contents, mafic and ultramafic volcanic rocks within greenstone belts were erupted to form relatively flat submarine platforms and lava plains (Lowe and Knauth, 1977; Barley et al., 1979; Dimroth et al., 1985b; Chown et al., 1992; and many others). Although these lava sequences were subject to little synvolcanic erosion, local uplifts during later stages of volcanism locally yielded terrigenous sediments enriched in mafic and even ultramafic components (e.g. Dunlop and Buick, 1981; Feng et al., 1993). Felsic complexes, however, commonly formed steep-sided, high-standing volcanoes that, where subaerial, yielded large quantities of coarse, immature epiclastic sediment by erosion. Volcaniclastic conglomerates and sandstones deposited as alluvial, fan-delta, and turbiditic sediment aprons around felsic volcanic centers are present in felsic volcanic units in the Superior Province, in the upper 100–700 m of the Hooggenoeg Formation in the Barberton Greenstone Belt (Viljoen and Viljoen, 1969b; Lowe and Knauth, 1977), and in the Duffer and Panorama Formations in the eastern Pilbara Block (DiMarco and Lowe, 1989a,b). Where composed essentially entirely of felsic debris, these units are commonly difficult to differentiate from reworked pyroclastic and autoclastic deposits discussed above.

Syndeformational epiclastic deposits

Tectonism accompanying volcanism or late-stage deformation widely exposed deeper levels of greenstone belts as well as embedded plutons, principally of the tonalite–trondhjemite–granodiorite (TTG) suite, to erosion. Sediments derived from such uplifts are compositionally varied, depending on the make-up of the source terrane and the intensity of local weathering. Examples from pre-3.0 Gy-old greenstones include rocks of the synvolcanic Fig Tree Group and post-volcanic, pre- to synorogenic Moodies Group in the Barberton Greenstone Belt (Condie et al., 1970; Jackson et al., 1987; Heubeck, 1993; Heubeck and Lowe, 1992) and the Gorge Creek Group in the eastern Pilbara Block (Eriksson, 1980b, 1982). Thick heterogeneous clastic metasedimentary sequences are widespread in the Superior Province (Pettijohn, 1943, 1970, 1972; Ojakangas, 1985), including units within the greenstone successions (Turner and Walker, 1973; Teal and Walker, 1977; Hyde, 1980; Ojakangas, 1985; Mueller et al., 1991; Jackson et al., 1994) as well as discrete metasedimentary belts between granite–greenstone terranes, such as the Pontiac (Camire et al., 1993; Feng et al., 1993), Quetico (Wood, 1980; Sawyer, 1986) and English River Subprovinces (Card, 1990). Examples of turbiditic and debris-flow terrigenous units from the Abitibi Green-

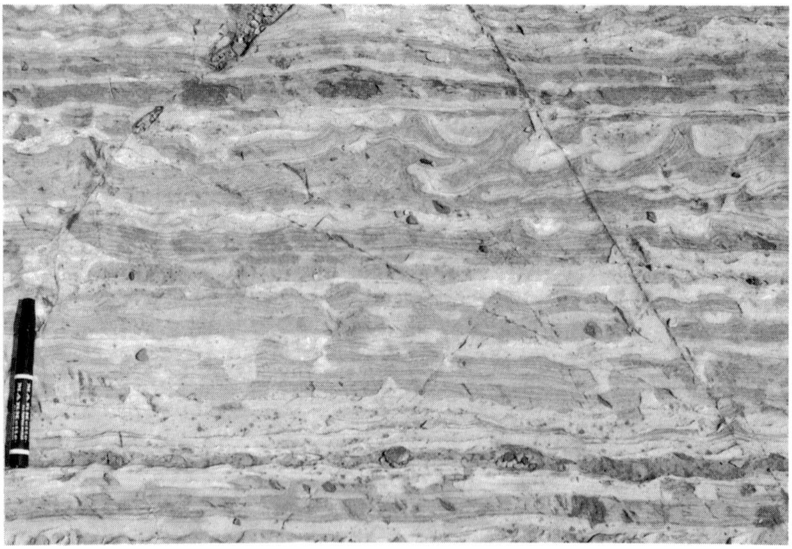

Fig. 10. Fine-grained, thin-bedded terrigenous turbidites interbedded with basaltic rocks in the southern Abitibi Greenstone Belt, Ontario, Canada.

stone Belt in the Superior Province are shown in Figs. 10, 11, and 12. Thick Late Archean greenstone-belt epiclastic sequences also comprise the bulk of the Slave Province, Canada (Henderson, 1975a), large parts of the northern Dharwar Craton, India (Naqvi and Rogers, 1983), and portions of many greenstone belts in the Yilgarn Block, Western Australia (Glikson, 1971; Marston, 1978; Hallberg, 1985).

Syndeformational epiclastic conglomerates are polymictic and contain clasts of underlying greenstone-belt volcanic rocks, greenstone-belt sedimentary rocks, and TTG and, locally, more differentiated plutonic rocks. The composition of the conglomerates varies widely as a function of the character of the source rocks and the degree of weathering. Where weathering has been severe, the conglomerates are dominated by resistant rock types, especially cherts. In the Barberton Greenstone Belt, many Fig Tree and Moodies Group conglomerates are composed almost exclusively of chert clasts eroded from thin silicified layers within the Onverwacht volcanic sequence and clasts of resistant, silicified felsic volcanic rock (Fig. 13a). More voluminous but less resistant mafic and komatiitic volcanic rocks in the source areas have been completely reduced by weathering. Where weathering was slight, conglomerates can contain a high proportion of mafic and even komatiitic clasts (e.g. Mueller et al., 1991). In the Moodies Group, where the basal conglomerate unconformably overlies deformed mafic and komatiitic volcanic rocks of the Onverwacht Group, the lowest conglomerate locally contains clasts of basalt and talcose ultramafic rock. Such labile constituents disappear a few meters above the basal contact.

Fig. 11. Terrigenous turbidites composed of massive coarse-grained basal T_a divisions layer overlain by flat- and cross-laminated T_b and T_c divisions, southern Abitibi Greenstone Belt, Ontario.

Sandstones in greenstone-belt epiclastic units are commonly graywackes or quartz-poor arenites composed largely of lithic debris and plagioclase-dominant feldspars derived by erosion of greenstone volcanic and sedimentary suites. Plutonic sources are occasionally represented by granitoid clasts in conglomerates, but more commonly by monocrystalline quartz or feldspar grains in associated sandstones. Where weathering has been effective and plutonic rocks have contributed a large proportion of the detritus, the resulting sandstones can be quartose, such as the Moodies Group in the Barberton Greenstone Belt (Fig. 13b).

Mudstones within greenstone belts commonly show high Ni and Cr contents, indicating partial derivation from greenstone-belt mafic and ultramafic rocks (e.g. Danchin, 1967; Taylor and McLennan, 1985). Because the mafic volcanic rocks are easily weathered under normal surface conditions, their weathering products are preferentially concentrated in the mudrocks. In contrast, quartzo-feldspathic

Fig. 12. Heterogeneous terrigenous debris-flow deposit interbedded with turbidites in Figs. 10 and 11, southern Abitibi Greenstone Belt, Ontario. This and interbedded turbidites and debris-flow deposits include spinifex-bearing ultramafic, mafic, and felsic clasts.

detritus from TTG plutons and felsic volcanic detritus, especially quartz, tends to concentrate in sandstones. Lithic debris from silicified volcanic and sedimentary rocks is common in greenstone-belt sandstones but dominates many conglomeratic units.

Epiclastic deposits representing pre-greenstone-belt sources

Mineral grains and lithic debris eroded from older extra-basinal or sub-basinal rocks are common components of greenstone belts. Some extra-basinal debris is present as xenocrystic components within the volcanic and volcaniclastic sequences (e.g. Compston et al., 1986; Kröner et al., 1992), but much occurs within epiclastic units (e.g. Kröner and Compston, 1988). Detrital zircons and geochemical signatures of older Late Archean rocks are present in both the Pontiac and Quetico Subprovince sedimentary units of the Superior Province (Davis et al., 1990; Feng et al., 1993). Thick sedimentary sequences composed largely of debris derived by the erosion of older extra-basinal crust and rock units are rare, but have been reported, including quartz-rich sandstones deposited upon and adjacent to 2.9 Gy-old and older basement in the western Slave Province (Easton, 1985; Padgham and Fyson, 1992), quartzites in the Chittradurga Greenstone Belt in India (Naqvi and Rogers, 1983), and local quartz-rich sandstones in the western Superior Province (Thurston and Chivers, 1990). All of these sequences include clean, cross-stratified, quartz-rich sandstones deposited under shallow-water, shelfal

Fig. 13. Terrigenous epiclastic deposits in the Barberton Greenstone Belt. (a) Chert-clast conglomerate in the Fig Tree Group. Most clasts are composed of massive black, white, banded, or laminated chert. Some clasts of dacitic volcanic rock are also present. (b) Current-deposited sandstone of the Moodies Group. The sandstone shows bi-directional cross-stratification and was probably deposited in a shallow-water, intertidal setting.

conditions and overlain by greenstone-belt volcanic and immature sedimentary sequences.

The interpretation of older components within greenstone belts must be approached with caution. Commonly, xenocrystic and detrital zircons, granitoid plutonic clasts, or quartz-rich sediments are taken as evidence for the presence of underlying or nearby pre-existing continental crust (e.g. Compston et al., 1986; Thurston and Chivers, 1990; Kröner and Compston, 1988). However, because all greenstone belts contain zircon-bearing felsic volcanic components and are intruded by synvolcanic and early post-volcanic zircon-bearing TTG plutonic rocks, magmatic or subaerial erosion of older volcanic and plutonic terranes can yield continent-like detrital materials that were not derived from thickened, evolved, stable continental blocks. In the Barberton belt, 3445 Ma zircons from Hooggenoeg-age felsic and/or TTG units are widely reworked into younger 3260–3225 My-old Fig Tree felsic pyroclastic and epiclastic units (Kröner and Compston, 1988; Kröner et al., 1992) and into post-3225 Ma Moodies sandstones (Kröner and

Compston, 1988), although the eastern Kaapvaal Craton did not form as a stable continental block until after 3,150 Ma.

Orthochemical deposits

The Archean oceans, like their modern counterparts, were supplied with dissolved materials by submarine hydrothermal systems, interaction between ocean water and the sea floor, and rivers and streams draining land areas. However, because Archean land areas were small and primarily non-continental in character and/or because of the extent and intensity of thermal activity on the Archean ocean floors, the geochemistry of the Archean oceans was dominated by the global oceanic crust (Veizer et al., 1982, 1989a,b).

Also in contrast to the modern oceans, the principal means of dissolved-sediment removal from the Archean oceans was inorganic chemical (orthochemical) precipitation. In the absence or paucity of silica secreting organisms, large-scale bio-mediated carbonate deposition, and similar biologically-controlled sedimentation processes, a steady-state Archean ocean would have balanced the input of dissolved solids by their removal through orthochemical sedimentation and rock-water reactions on and below the sea floor. The exotic compositions of some Archean sedimentary associations, especially the abundance of cherty sedimentary units, reflect the dominance of orthochemical over biological processes in removing some constituents, such as silica, from Archean sea water.

Perhaps the most widespread orthochemical sediment in Archean greenstone sequences was silica, now represented by layers of chert composed largely of granular microcrystalline quartz. Chert in greenstone belts has formed both by the diagenetic and metasomatic replacement of a wide variety of primary non-silica sediment types, including tuff, mudstone, and sandstone (Lowe and Knauth, 1977; Barley et al., 1979), as well as by the recrystallization of what were probably primary siliceous sedimentary layers. Silica is especially conspicuous as relatively pure chert bands, generally from a few mm to 10 cm in thickness, interlayered with a variety of other sediment types to form characteristically banded rocks (Figs. 14 and 15). Such layers or bands include jasper bands in banded iron-formation (BIF) (Fig. 14a), where they alternate with iron-rich layers; white chert bands in banded ferruginous cherts, where white chert layers alternate with impure carbonaceous, tuffaceous, argillaceous, and sideritic layers (Fig. 14b); and white chert bands alternating with black carbonaceous bands and, less commonly, jasper bands, in banded black-and-white cherts (Figs. 15a and 15b).

Probably the best studied Archean orthochemical and perhaps in part biomediated chemical sedimentary rock is oxide-facies BIF (Fig. 14a and James, 1954). Oxide-facies BIF is widely developed in subaqueous deposits both closely associated with volcanic centers (e.g. Mueller and White, 1992) and in areas far removed spatially and temporally from volcanic activity, although the ultimate source of the iron and silica was probably submarine hydrothermal systems. It

Fig. 14. Orthochemical deposits. (a) Banded iron-formation, Fig Tree Group, Barberton Greenstone Belt. Lighter bands are jasper, darker bands are weathered hematite. (b) Banded ferriginous chert, Fig Tree Group, Barberton Greenstone Belt. White bands are relatively pure or slightly calcareous chert. Darker bands are composed of intergrown clay, recrystallized ash, carbonaceous matter, and carbonate, mainly siderite where fresh. Both BIF and banded ferruginous cherts were deposited in quiet, deep-water settings.

Fig. 15. Orthochemical deposits. (a) Banded black-and-white chert at the base of the Kromberg Formation, Barberton Greenstone Belt. Black bands are composed of carbonaceous chert and white bands of relatively pure chert. (b) Banded black chert (dark gray), jasper (medium gray), and white chert (white) in the Marble Bar Chert, Warrawoona Group, eastern Pilbara Block. Pen for scale (arrow).

occurs in both the volcanic and sedimentary parts of greenstone-belts and most reflects deposition under fully subaqueous and generally deep-water conditions. A number of investigators have suggested that the Archean oceans were stratified, with separate and poorly mixed deep- and shallow-water masses (Klein and Beukes, 1989; Lowe, 1994; and many others) and that oxide-facies BIF was deposited on the sea floor below sites of mixing of deep, iron-rich, oxygen-poor and shallow, iron-poor, oxygen-bearing water masses. Such mixing zones may have developed in areas of dynamic upwelling or in thermal plumes above submarine volcanic vents. Much of the oxygen required for iron oxidation and precipitation may have been contributed by organisms living within the near-surface mixing zones (Cloud, 1968, 1976; Schidlowski, 1976).

Shallow-water Archean successions generally lack iron-rich sedimentary rocks but contained a variety of orthochemical units when deposited, including local bedded barite, sparse carbonate, and evaporites. Barite (Fig. 16) occurs as an early sediment in the 3.26 to 3.23 Ga-old Fig Tree Group in the Barberton Greenstone Belt (Heinrichs and Reimer, 1977; Reimer, 1990) and has apparently formed by the early replacement of gypsum in the Pilbara belts (Groves et al., 1981; Buick and Dunlop, 1990). Primary sedimentary limestone and dolomite are difficult to

Fig. 16. Layers of coarsely crystalline barite, possibly formed by the replacement of gypsum (Buick and Dunlop, 1990), North Pole Chert, Warrawoona Group, eastern Pilbara Block.

identify with certainty in the pre-3.0 Ga-old greenstone belts because of the abundance of secondary replacement carbonate. It occurs locally in the Fig Tree Group, where it precipitated in an agitated, wave-active zone along the margin of a submerged shallow-water fan-delta platform (Lowe and Nocita, in review). Some orthochemical carbonate may have been a component of evaporitic sequences in the Warrawoona Group in the Pilbara (Groves et al., 1981; Lowe, 1983; Buick and Dunlop, 1990) and in the Barberton Greenstone Belt (Worrell, 1985; Lowe and Worrell, in review). Thick stromatolitic limestone and dolomite units occur widely in Late Archean greenstone belts (Henderson, 1975b; Martin et al., 1980; Wilks and Nisbet, 1985).

Siderite is a widespread minor component of many Archean greenstone-belts. It was generally deposited in quiet, deep-water settings along with silica, fine-grained volcanic ash, clays, and carbonaceous matter during periods of volcanic quiescence and low clastic influx (Lowe, in review, a). These sediments have widely been diagenetically altered to banded ferruginous rocks that consist of relatively pure chert bands up to 10 cm thick alternating with layers that are mixtures of the other depositional components (Fig. 14b). They may represent condensed sections that were deposited over extended periods of time characterized by extremely low sedimentation rates. The mixing of fine-grained orthochemical (siderite and silica), biogenic (carbonaceous matter), and epiclastic (clay) components may also indicate that these layers represent background sediments that rained slowly out of the Archean oceans in deep-water settings on a day-to-day basis (Lowe, in review, a).

There is considerable evidence that gypsum may have been a widely developed primary, shallow-water, evaporitic sediment in older, pre-3.0 Ga-old greenstone belts (Fig. 17). Barite pseudomorphs after evaporitic gypsum (Groves et al., 1981; Buick and Dunlop, 1990) and a variety of silicified evaporitic gypsum facies (Lowe, 1983) have been reported from the Warrawoona Group and layers of silicified evaporitic gypsum have been described from the upper part of the Onverwacht Group in the Barberton Greenstone Belt (Worrell, 1985; Lowe and Worrell, in review).

Biogenic deposits

Although a wide variety of Archean sedimentary rocks, including iron-formation, may have been influenced by biological activities, the main sediment types that owe their existence primarily to organic processes are stromatolitic carbonates and carbonaceous matter in cherts, carbonates, and shales. Stromatolitic carbonates and other bio-mediated carbonate units are unknown from the older, pre-3.0 Ga-old greenstones, but occur widely in Late Archean greenstone sequences in Canada (Henderson, 1975b; Hofmann et al., 1985), Zimbabwe (Martin et al., 1980), and elsewhere (Walter, 1983). These units were deposited on localized shallow-water platforms formed along the margins of older crustal

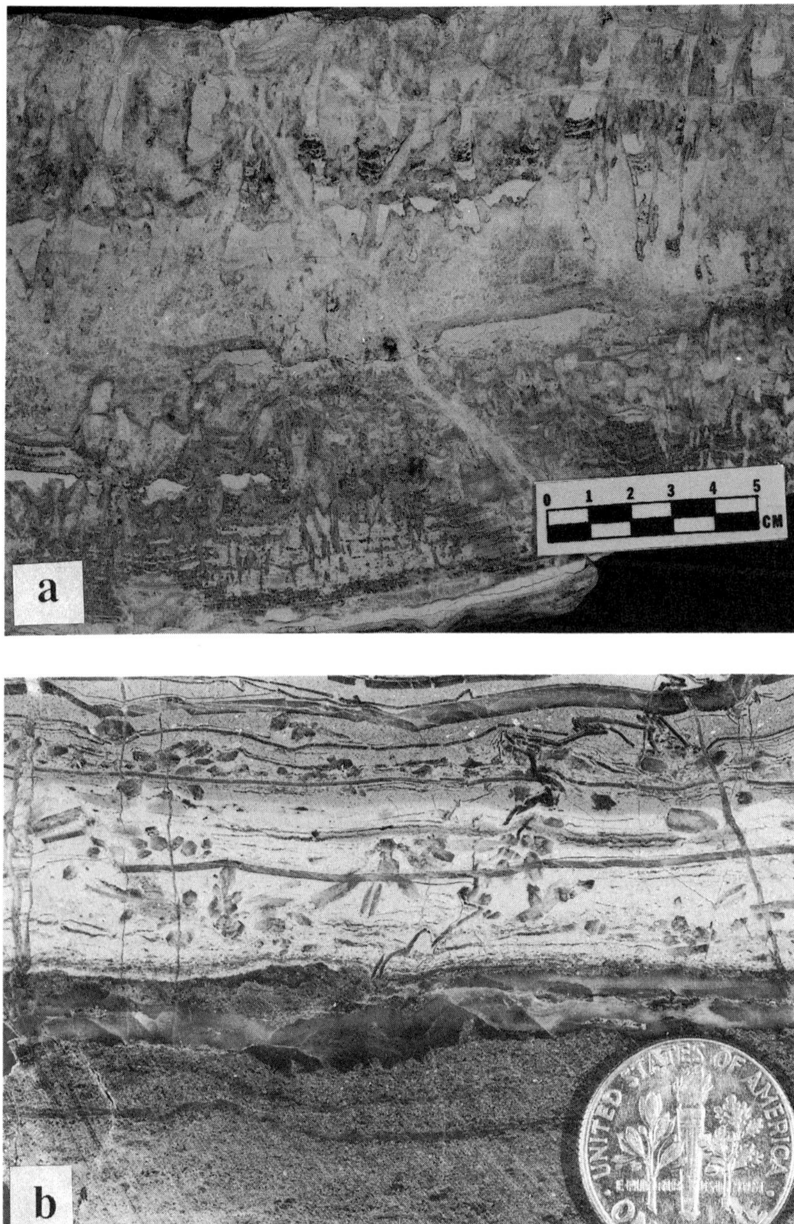

Fig. 17. Orthochemical deposits, Barberton Greenstone Belt. (a) Layers of upward-radiating silicified crystals, possibly representing primary gypsum (Worrell, 1985), that grew within fine sediment, possibly carbonate. The sediment was deposited in a small coastal brine pond, lower Kromberg Formation. (b) Silicified, intrastratal, pseudohexagonal, authigenic crystals, possibly representing gypsum, from the middle part of the Hooggenoeg Formation. Surrounding sediment is fine, silicified komatiitic ash.

blocks (Wilks and Nisbet, 1985), on the eroded tops of subaerial felsic volcanic complexes (Henderson, 1975b; Hofmann et al., 1985), or within locally developed shallow-water parts of the volcanic or sedimentary successions (Martin et al., 1980).

Carbonaceous sedimentary rocks are widespread in both shallow- and deep-water Archean successions (Walsh, 1989, 1992). Carbonaceous cherts containing mat-like carbonaceous laminations (Fig. 18a), floccule-like composite carbonaceous particles (Fig. 18b), current-transported carbonaceous matter (Fig. 19a), and fine comminuted carbonaceous matter (Fig. 19b) are abundant in shallow-water, pre-3.0-Ga old greenstone sequences in Australia and South Africa (Lowe and Knauth, 1977; Barley et al., 1979; Lowe, 1983; Walsh, 1989). These rocks reflect the widespread presence of primitive prokaryotic bacterial communities in Archean shallow-water environments. Carbonaceous sedimentary rocks deposited under quiet, deep-water conditions include carbonaceous shales and banded ferruginous cherts. The carbonaceous matter in deep-water units is mainly finely comminuted detritus, probably representing organisms living in the overlying water column.

ARCHEAN GREENSTONE-BELT SEDIMENTARY ASSOCIATIONS

The vertical and lateral variability of most Archean greenstone successions; the dominance of volcanic processes; the unusual lithologies, such as komatiites and banded iron-formation, which are absent or rare in Phanerozoic sequences; extensive diagenetic modification; and widespread faulting and metamorphism make it difficult to recognize associations of Archean sedimentary rocks that might represent or characterize individual large-scale tectonic or depositional systems. Jackson et al. (1994) recognized five main supracrustal assemblages in the southern Abitibi Greenstone Belt in Ontario: (1) THOLKOM-type assemblages dominated by volcanic rocks of tholeiitic or komatiitic affinity; (2) INTFEL-type assemblages dominated by intermediate to felsic rocks, largely of calc-alkalic affinity; (3) IRON-type assemblages containing substantial deposits of iron formation; (4) TURB-type assemblages dominated by turbiditic sedimentary deposits; and (5) ALUFLU-type assemblages made up largely of sedimentary rocks deposited in alluvial and fluvial environments.

Based largely of studies of sedimentary rocks in pre-3.0 Gy-old greenstone belts with more cursory examination of younger belts, the present author would suggest that the following six lithofacies associations appear to characterize sedimentary/supracrustal rocks in most Archean greenstone belts (Table 1):

(1) Mafic Anorogenic Volcaniclastic–Orthochemical–Biogenic Association (MAVOB),

(2) Anorogenic Orthochemical–Biogenic Association (AOB)

(3) Felsic/Intermediate Volcaniclastic–Terrigenous Association (FVT),

(4) Orogenic Terrigenous Association (OT).

(6) Anorogenic Polycyclic Terrigenous Association (APT)

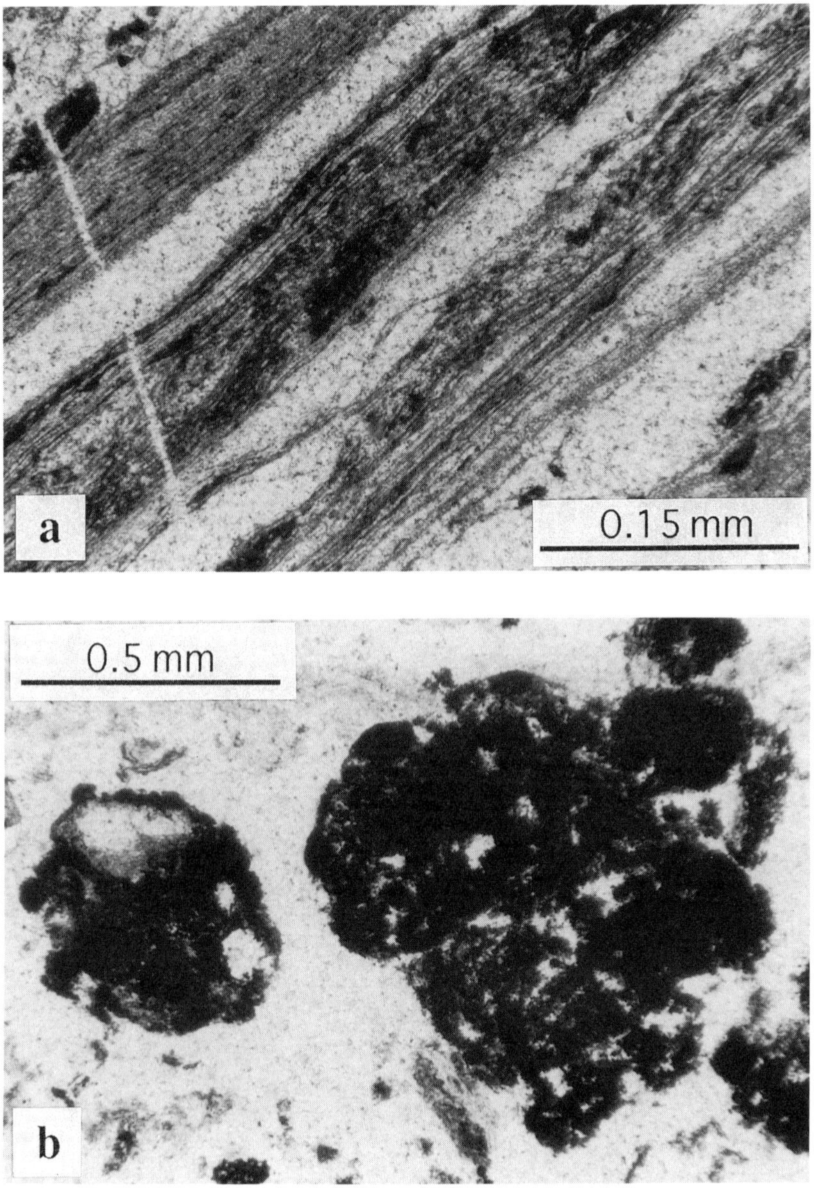

Fig. 18. Photomicrographs of biogenic deposits, Onverwacht Group, Barberton Greenstone Belt. (a) Fine carbonaceous laminations, probably representing bacterial mats. (b) Large floccule-like composite carbonaceous particles, common in shallow-water deposits in the Barberton Belt, set in a chert matrix.

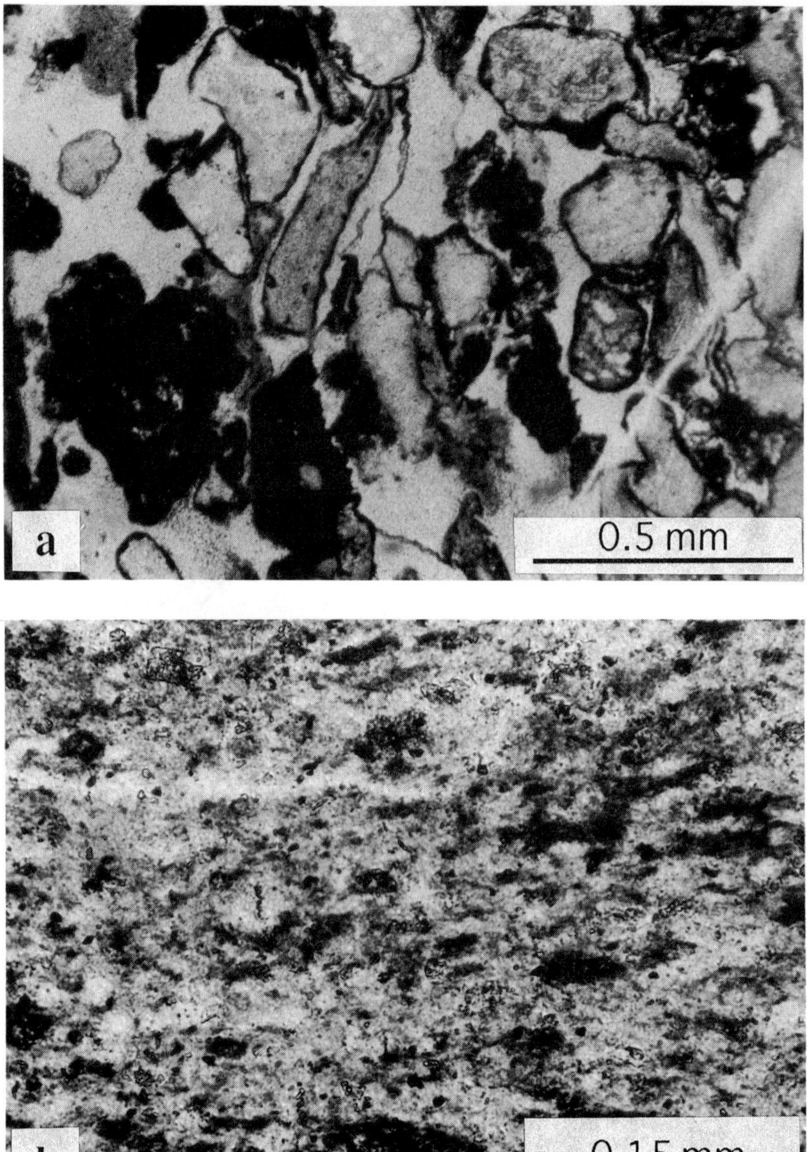

Fig. 19. Photomicrographs of biogenic deposits, Onverwacht Group, Barberton Greenstone Belt. (a) Rounded detrital carbonaceous grains mixed with sand-sized chert grains. (b) Carbonaceous chert composed of finely comminuted carbonaceous matter in chert matrix.

TABLE 1

Principal lithofacies associations of Archean greenstone belts and their interpretations

Lithofacies association	Composition	Interpretation
Mafic anorogenic volcaniclastic–orthochemical–biogenic association (**MAVOB**)	Mafic/Ultramafic vol. rx. Mafic tuff Black and banded cherts Evaporites, BIF, BFC*	Spreading centers Oceanic islands Oceanic plateaus
Anorogenic orthochemical–biogenic association (**AOB**)	Black and banded cherts BIF, BFC	Ocean floor Inactive subaqueous Volcanic islands and oceanic plateaus
Felsic volcaniclastic Terrigenous volcaniclastic–association (**FVT**)	Felsic vol., rx., Tuffs Felsic autoclastic seds. Felsic epiclastic seds. BIF, Carbonaceous mudst.	Convergent-junction magmatic arcs
Orogenic terrigenous assoc. (**OT**)	Cgl., Sst., Mdst. Local BIF	
Turbiditic assoc. (**OTf**)		Convergent-junction Trench, forearc, and back-arc basins
Alluvial–fluvial assoc. (**OTaf**)		Convergent-junction Foreland basins; Late orogenic basins
Anorogenic polycyclic terrigenous association (**APT**)	Qtzose sandst. Shale Carbonates Evaporites, BIF	Older-block margins Older-block cover

*__Banded ferruginous chert__. Generally formed by silicification of a primary deep-water sediment composed of admixed silica, carbonaceous matter, siderite, clay, and, locally, fine ash.

Mafic Anorogenic Volcaniclastic–Orthochemical–Biogenic Association (MAVOB)

Both pre-3.0 and post-3.0 Gy-old Archean greenstone belts consist largely of mafic and ultramafic volcanic rocks comprising sections that commonly reach 5 to 10 km in thickness. In preserved pre-3.0 Ga-old greenstone belts in South Africa and Western Australia (Fig. 2), these simatic sequences were deposited mainly under shallow-water, anorogenic "platformal" conditions (Lowe, 1980, 1982;

Groves and Batt, 1984). In Late Archean belts, similar basaltic and komatiitic sequences were deposited as deep-water lava plains, also mainly under anorogenic conditions (Goodwin, 1982; Dimroth et al., 1982, 1985b; Chown et al., 1992). Examples of the MAVOB association in pre-3.0 Ga belts include the Komati Formation, possibly the lower part of the overlying Hooggenoeg Formation, and the Mendon Formation in the Barberton Greenstone Belt (Fig. 2) and probably most of the Warrawoona Group (Fig. 2) in the eastern Pilbara Block, Western Australia (Barley et al., 1979; Hickman, 1983). Late Archean examples include thick basaltic and komatiitic sequences throughout the Superior Province (Card, 1990; Chown et al., 1992; Jackson et al., 1994) and in the Yilgarn Block (Gee et al., 1981; Hallberg, 1985; Barley et al., 1989). This association corresponds closely to the THOLKOM-type supracrustal assemblage of Jackson et al. (1994).

Autoclastic, pyroclastic, and epiclastic units in the MAVOB association locally reach over 1000 m thick. Interflow orthochemical and biogenic sedimentary layers are mostly thin, 1 to 5 m thick, but some reach over 50 m thick. In the pre-3.0 Gy-old, mainly shallow-water Archean belts, the primary sediments included beds of current-deposited, shallow-water mafic to komatiitic lapillistone (Fig. 4), accretionary lapilli (Fig. 5), and tuff (Lowe and Knauth, 1977; Barley et al., 1979; Lowe, 1980; 1982; in review, a). Large- and small-scale cross-bedding (Figs. 4 and 5), mud flasers, reactivation surfaces, and other evidence for wave- and current-activity is widespread. Although some of the graded layers of accretionary lapilli have been interpreted as turbidites (Stanistreet et al., 1981; Heinrichs, 1984), most are composed solely of pyroclastic debris, each is texturally and compositionally distinctive, and most occur interbedded with sandstones and mudstones deposited under shallow subtidal to intertidal conditions (Lowe, in review, b). These accretionary lapilli and ash layers were probably deposited as pyroclastic fall units. Shallow-water biogenic units include black carbonaceous cherts and black-and-white banded cherts (Fig. 15a) made up of silicified flat bacterial mats (Fig. 18a), granular carbonaceous particles (Fig. 18b), and current-deposited carbonaceous grains (Fig. 19a; Walsh, 1989). Orthochemical units include silicified evaporites (Figs. 16 and 17), mainly gypsum and carbonate, deposited under shallow-water, locally restricted, hypersaline conditions (Lowe, 1983; Buick and Dunlop, 1990). Virtually absent are iron-rich sediments, including oxide-facies BIF and sideritic units (Lowe, 1980, 1982).

Because sedimentation occurred mainly under anorogenic conditions in areas of rapid mafic and ultramafic volcanism, there were few tectonic uplifts and little volcanic relief. As a result, there are few terrigenous sedimentary rocks in the MAVOB association in the older greenstone belts, although thin epiclastic units derived from mafic and komatiitic volcanic rocks occur locally (e.g. Dunlop and Buick, 1981).

These pre-3.0 Gy-old platformal sequences formed broad, low-relief mafic volcanic edifices. Their low relief reflects the low viscosities of komatiitic and basaltic liquids. The eruption rates were commonly sufficiently high that thousands of meters of volcanic rock accumulated without weathering, erosion, or submarine alteration of the flow tops and with little or no deposition of sediment on the exposed flow surfaces. The platforms were broad, exceeding 100 km across in the eastern Pilbara Block (Lowe, 1983), and probably formed as large, anorogenic oceanic volcanoes with surfaces near sea level throughout most of their active, constructional stage of development.

A deep-water MAVOB facies is only locally developed in the pre-3.0 Gy-old greenstones but is widespread in Late Archean belts. In the older belts, it is composed mainly of thin layers of banded ferruginous, commonly sideritic or hematitic chert (e.g. Fig. 14b); locally, jasper; massive to banded carbonaceous chert composed mainly of finely divided carbonaceous material in a microquartz matrix (Fig. 19b); and tuffaceous layers interbedded with volcanic rocks and showing evidence of deposition under quiet, deep-water conditions (Lowe, 1980; 1982; in review, a). Such units are developed locally around the margins of the shallow-water volcanic platforms and, in a few areas in the Barberton Greenstone Belt, shallow-water sedimentary units can be traced laterally into deeper water facies. The deep-water association is also interbedded sporadically within largely shallow-water MAVOB sections where the rate of platform subsidence exceeded the rate of eruption and volcanic aggradation.

In the younger, post-3.0 Ga-old greenstone belts, there are few if any reported occurrences of the shallow-water, platformal facies in the MAVOB association, although some basaltic shield volcanoes aggraded into shallow water (e.g. Mueller, 1991). Deep-water MAVOB facies are dominated by pillowed tholeiitic and komatiitic volcanic rocks containing thin interbedded units of iron-formation, ferruginous chert, carbonaceous mudstone, and, locally, coarser epiclastic units (Jackson et al., 1994). Deposition occurred mainly in deep-water substantially removed from subaerial uplifts.

Anorogenic Orthochemical–Biogenic Association (AOB)

The uppermost volcanic rocks of the largely volcanic Onverwacht Group in the Barberton Greenstone Belt are capped regionally by 10–100 m of black, black-and-white banded, and banded ferruginous chert that was deposited following mafic volcanism and preceding the Fig Tree orogenesis and felsic volcanism. This chert unit, locally termed the Zwartkoppie Bar but assigned to the Mendon Formation (Fig. 2) by the present author (Lowe, 1992b; Lowe and Byerly, in review), hosts most of the gold deposits of the northern Barberton Greenstone Belt. It marks a regional interval of anorogenic, non-volcanogenic, generally quiet, deep-water sedimentation that began with the cessation of volcanism and continued during subsequent cooling and subsidence of the volcanic edifice to the start of Fig Tree orogenesis and felsic volcanism. This and similar thin regional fine-grained units in other belts probably mark condensed sections formed through prolonged, slow sedimentation of fine hemipelagic, orthochemical, and distal airborne volcaniclastic materials (Lowe, in review, a and c). AOB-type cherts may occur within the volcanic sequence, but, in areas of complex structure or without regional mapping, are difficult to distinguish from cherts in the MAVOB association, which generally mark local, relatively short pauses in volcanic activity rather than regional, long-term hiatuses. The Middle Marker, regionally overlying komatiites of the 3,500-m-thick Komati Formation and underlying basalts of the Hooggenoeg Formation in the southern Barberton Greenstone Belt (Fig. 2 and Viljoen and Viljoen, 1969a,b), may represent a shallow-water facies (Lanier and Lowe, 1982) of the AOB association.

Similar thin sedimentary layers may be present in Late Archean belts but have not been specifically recognized or described.

Felsic/Intermediate Volcaniclastic–Terrigenous Association (FVT)

Felsic volcanic and volcaniclastic units are characteristic components of Archean greenstone-belt sequences. The high-standing, commonly subaerial stratovolcanoes and volcanic complexes, widely associated uplifts of underlying portions of the greenstone-belt sequences, and locally exposed subvolcanic plutons provided major sources of compositionally varied debris to the local and regional sedimentary systems. The FVT association includes rocks deposited sufficiently close to the volcanic centers that they consist predominantly of volcanic or juvenile pyroclastic and autoclastic rocks. Felsic volcanic units include those erupted under relatively anorogenic conditions and composed of oligomictic felsic volcanic and volcaniclastic materials as well as others containing debris derived from uplifts of underlying portions of the greenstone belts, subvolcanic plutons, and even older pre-greenstone-belt rocks. The latter reflect concurrent volcanism and deformation. The Duffer and Panorama Formations in the eastern Pilbara Block, the felsic volcanic unit at the top of the Hooggenoeg Formation in the Barberton Belt, and many of the felsic units in the Abitibi Belt (e.g. Dimroth and Rocheleau, 1979; Car and Ayres, 1991; Mueller and White, 1992) are composed of juvenile felsic–clast conglomerates, breccias, tuffs, and sandstones (Figs. 7 and 8). The southern facies of the Fig Tree Group in the Barberton Greenstone Belt contains dacitic tuffs (Fig. 9) interbedded polymictic chert- and felsic–volcanic–clast conglomerates (Fig. 13a), reflecting eruptive units deposited concurrently with debris eroded from both the felsic volcanic centers and uplifts of older portions of the greenstone sequence (Nocita and Lowe, 1990).

More distal sequences in both associations include fine tuffs (Fig. 9), resedimented felsic sandstone and siltstone, and minor interbedded or capping units of carbonaceous mudstone. Oxide- and sulfide-facies iron-formation (Fig. 14a) are common accessory sediment types in deep-water settings.

Orogenic Terrigenous Associations (OTt and OTaf)

Most greenstone belts include thick orogenic terrigenous clastic units containing few or no volcanic, autoclastic, or pyroclastic rocks. Two principal associations can be distinguished: (1) thick sequences of largely turbiditic strata, such as the northern facies of the Fig Tree Group in the Barberton Belt (Condie et al., 1970; Eriksson, 1980a), the Gorge Creek Group in the eastern Pilbara Block (Eriksson, 1980b), metasedimentary units in the Pontiac and related metasedimentary subprovinces of the Superior Province (e.g. Feng et al., 1993), and similar units of the TURB-type assemblage in the Abitibi Belt (Jackson et al., 1994) and (2) sequences composed mainly of alluvial and fluvial deposits, such as the Moodies Group in the Barberton Belt (Eriksson, 1977, 1978, 1979; Heubeck, 1993) and the Duparquet Formation (Mueller et al., 1991), Timiskaming Group (Hyde, 1980), and similar units of the ALUFLU-type association of Jackson et al. (1994) in the Abitibi Belt. These are here termed OTt and OTaf associations, respectively.

The OTt association is dominated by lithic sandstones or graywackes characterized by low to moderate quartz percentages and showing graded bedding, sole marks, and internal structures indicating deposition by turbidity currents or related sediment flows (Figs. 10, 11, and 12). Debris flows are common within this facies (Ojakangas, 1985). The sections tend to be thick, commonly thousands of meters; highly deformed; and extensively effected by metamorphism or low-grade metasomatism. Petrographic and mineralogical studies (e.g. Condie et al., 1970; Feng et al., 1993) commonly indicate compositionally varied source areas, often including felsic volcanic rocks, parts of the TTG intrusive suite, minor mafic and ultramafic components, and, in some sequences, older, pre-greenstone rocks.

The OTaf association tends to include more feldspathic and, in some areas, quartzose sandstones, reflecting deeper unroofing of the subvolcanic plutonic complexes. Locally developed volcanic units and associated plutons in the OTaf tend to be alkalic in composition (BenOthman et al., 1990; Jackson et al., 1994). The sandstones are characterized abundant cross-lamination and cross-stratification, scour, lenticular units, and lithofacies characteristic of alluvial, fluvial, fan-delta, and, locally, shallow-marine (Fig. 13b) deposits.

Many investigators have noted the apparent bimodality of greenstone-belt epiclastic sequences, with abundant deep-water and alluvial units and few shallow-water deposits (e.g. Turner and Walker, 1973; Teal and Walker, 1977; Young, 1978; Dimroth and Rocheleau, 1979; Hyde, 1980; Blackburn et al., 1985; Nocita and Lowe, 1990; Jackson et al., 1994), although lateral equivalence of these contrasting facies is often difficult to demonstrate. In many areas, these two contrasting facies may be diachronous (Ojakangas, 1985; Jackson et al., 1994) or separated by faults that have juxtaposed the deposits of separate basins (Mueller et al., 1991). Many epiclastic units probably represent fan-deltas that included interfingering subaerial, shallow-water fan-delta platform, and deeper-water turbiditic fan-delta-front facies (Nocita and Lowe, 1990). Where diachronous, largely turbiditic basins appear to predominate during the volcanic and early orogenic evolution of greenstone belts and alluvial sequences during their late orogenic and post-orogenic evolution (Card, 1990; Mueller et al., 1991; Jackson et al., 1994). The former basins were probably developed mainly during accretionary, collisional stages of crustal evolution and the latter commonly during late transpressional or strike-slip stages (Card, 1990; Mueller et al., 1991).

Subsequent greenstone-belt evolution generally included late stage deformation, potassic plutonism, uplift, and erosion. Later-deposited sediments are usually cratonic in character, including craton cover or cratonic rift sequences, such as the Pongola Supergroup of southern Africa (Tankard et al., 1982).

Anorogenic Polycyclic Terrigenous Association (APT)

Although mature, quartzose, shallow-water sandstones and carbonate units are not common in Archean greenstones, they are developed locally, mainly in the younger, post-3.0 Ga belts. Examples include quartzite-rich sandstones at the base of the greenstone succession in the Chitradurga greenstone belt, India (Naqvi, 1986); in some Late Archean belts in the Superior Province, Canada (Thurston and Chivers, 1990); in the western Slave Province (Easton, 1985; Padgham and Fyson, 1992), in the Steep Rock Group of the southern Superior Province (Wilks and Nisbet (1985, 1988), and in the Buhwa Greenstone Belt, Zimbabwe (Eriksson and Fedo, 1994). These rocks reflect the presence of older, pre-3.0-Ga-old continental blocks during formation of the younger greenstone belts. However, extensive and deep weathering of coeval TTG sources may have also produced first-cycle, compositionally mature quartzose sandstones, such as the Moodies Group in the Barberton belt, that are difficult to distinguish from those derived from older continental basement blocks without careful radiometric dating of detrital zircons.

DISCUSSION

Long-term evolution of Precambrian depositional and tectonic systems

Phanerozoic sedimentation has been dominated by a number of well-defined, plate-tectonic-controlled petrologic suites and basin types (Dickinson, 1984; Ingersoll, 1988). In terms of preserved sediment volumes, the most important Phanerozoic depositional settings are continental rifts, passive continental margins, and foredeeps and foreland basins. These major basin types reflect the main stages of the Phanerozoic Wilson cycle: continental rifting, intraplate quiescence with passive continental margins, and continental convergence and suturing. Virtually all sediments deposited in these settings show derivation from continental blocks. The same tectonic settings — spreading, intraplate, and convergent — exist in fully oceanic regimes, but the sediments deposited along mid-ocean spreading centers and on intraplate oceanic abyssal plains form a relatively thin cover on oceanic crust that is ultimately fated to disappear through subduction. Somewhat more voluminous and preservable sequences accumulate along oceanic convergent margins because of the high-standing volcanic vents, but the overall Phanerozoic record of oceanic arc settings is substantially less than that of continental arc settings.

In contrast, Archean greenstone sequences only locally contain major sedimentary components derived from continental sources. They are instead dominated by volcanic and sedimentary components of oceanic or at least non-cratonic affinities.

The early Proterozoic geologic record is, in many respects, transitional between that of the Archean, dominated by greenstone belts, and that of the late Proterozoic and Phanerozoic, dominated by cratonic sedimentary sequences (Lowe, 1992a). The earliest major Proterozoic supracrustal sequences, deposited between about 2100 Ma and 1700 Ma, include both Archean-type greenstone and Phanerozoic-type craton-related successions. Thick continent-sourced quartzite–shale–carbonate continental rift, continental margin, and craton cover sequences, such as the Coronation Supergroup of northern Canada and the Labrador Trough sequence of eastern Canada (Hoffman, 1988, 1989), reflect the existence of large Late Archean cratonic blocks while juvenile volcanogenic greenstone successions, represented by Birimian rocks of western Africa (Wright et al., 1985; Taylor et al., 1992), metasedimentary and metavolcanic remnants in the early Proterozoic basement of the southwestern United States (Bickford, 1988), and the Svecofennian Province of Scandinavia (Gorbatschev and Gaal, 1987; Pharoah and Brewer, 1990) continued to accumulate in others areas and ultimately evolved into new, Proterozoic blocks of continental crust.

Contrasts between preserved rock sequences over geologic time lead to the inference that the Archean world included few large continental blocks until 2.7–2.6 Ga and that the Archean tectonic cycle was dominated by the interaction of largely oceanic plates and small micro- or quasi-continental blocks (Lowe and Ernst, 1992; Lowe, 1992a). If so, then the major Phanerozoic depositional settings would not have existed on the Archean earth.

What, then, were the main Archean basin types?

If our techniques of interpreting Phanerozoic sedimentary and volcanic sequences can be applied as well to the Archean, the main Archean greenstone-belt lithofacies associations must bear some relationship to the principal Archean basin types and depositional/tectonic settings in which greenstone-belt sequences were deposited. The present author would suggest the relationships outlined below.

Depositional settings of Archean lithofacies associations

Archean greenstone belts are diverse in their makeup and probably represent a number of different tectonic, magmatic, and depositional settings. The question, "What was the tectonic setting of Archean greenstone belts?", is no more mean-

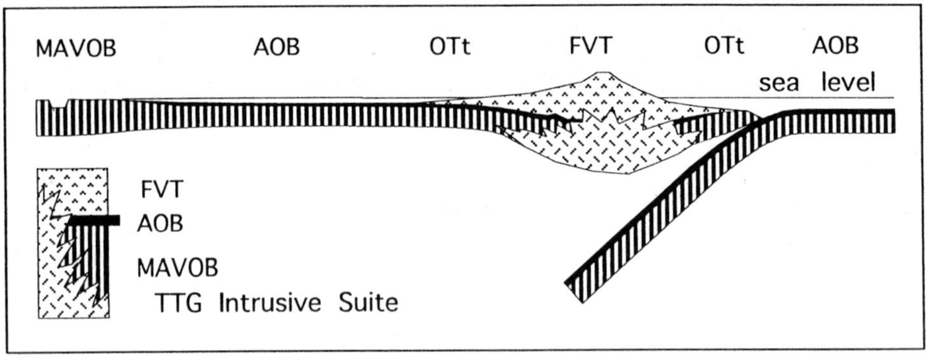

Fig. 20. Tectonic interpretations of the major Archean greenstone-belt lithofacies associations during early stages of greenstone-belt evolution. Oceanic plate forms at spreading center or over hot spot, where MAVOB association is deposited under conditions ranging from shallow- to relatively deep-water. As individual sites move away from the locus of volcanism and into an intraplate setting, it cools and subsides, and volcanism-dominated sedimentation is replaced by the slow deposition of orthochemical and biogenic sediments (AOB association). Crust eventually reaches convergent plate junction where, through polarity shift, it becomes situated above subduction zone where it is overplated and underplated by low-density felsic magmas. Sediments deposited in trench, forearc, and back-arc settings are mainly arc-derived turbiditic clastic units of the OTt association. The generalized stratigraphic column on the left is common to many individual greenstone-belt cycles in both older and younger greenstone belts.

ingful than, "What is the tectonic setting of Cenozoic volcanic and sedimentary sequences?". Suggestions that all greenstone belts or all Cenozoic sedimentary sequences can be described in terms of a single tectonic or depositional setting greatly oversimplify their complex evolutionary histories. Figures 20 and 21 illustrate the generalized depositional settings of the main Archean lithofacies associations as discussed below.

MAVOB association

The mafic anorogenic association accumulated in areas characterized voluminous mafic to ultramafic volcanism and low rates of pyroclastic and terrigenous sediment influx. In the early Archean, the resulting volcanic edifices commonly stood as high, flat, shallow-water, platforms surrounded by deeper-water. The high eruption rates, common lack of detrital quartz and feldspar, and absence of orogenesis suggest that these rocks represent deposition on large, simatic, anorogenic oceanic volcanic islands (Lowe and Knauth, 1977; Barley et al., 1979; Lowe, 1980, 1982). In contrast, Late Archean MAVOB associations accumulated as largely subaqueous lava plains. In terms of modern analogs, MAVOB volcanic rocks and eruptive styles are most similar to those of oceanic spreading centers or isolated oceanic islands associated with mantle plumes or "hot spots". The Mendon Formation in the Barberton Greenstone Belt shows evidence of having

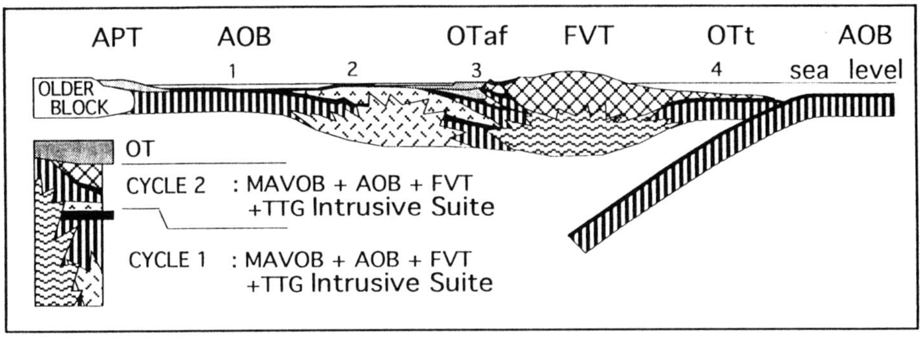

Fig. 21. Tectonic interpretations of the major Archean greenstone-belt lithofacies associations during the later stages of greenstone-belt evolution. Arc-assemblage 1 (cycle 1) developed in Fig. 20 is succeeded by formation of a second major magmatic complex along the margin of the older complex and the deposition of greenstone-belt cycle 2. Back-arc thrusting is associated with deposition of lithofacies OTaf (3) while OTt accumulates in trench and forearc settings (4). Note that AOB association can form in both quite, deep-water, sea-floor settings (1) and shallow-water settings on inactive volcanic platforms (2).

accumulated during rifting, probably along a spreading axis (Lowe and Byerly, in review; Lowe, in review, c).

A number of investigators have also suggested formation of the mafic and ultramafic Archean greenstone units through back-arc spreading (Tarney et al., 1976; Jackson et al., 1994; and many others). The predominantly deep-water setting of Late Archean belts and the cyclic interbedding of mafic lava plain and felsic calc-alkaline units is consistent with such an interpretation. Some predominantly basaltic MAVOB sequences, such as the Hooggenoeg Formation in the Barberton Belt, might also characterize early stages of subduction-related volcanism.

AOB association

The AOB association forms thin, fine-grained, clastic-poor, largely non-volcanogenic, mainly deep-water sedimentary units capping thick MAVOB successions and underlying FVT or OT associations within pre-3.0 Gy-old greenstone belts. These units mark intervals, possibly quite long, of volcanic and orogenic quiescence during which the volcanic surfaces subsided and were maintained under deep-water conditions before the beginning of orogenesis and arc magmatism. These conditions most closely resemble those characterizing intraplate settings on modern oceanic plates. The eruption of the mafic volcanic rocks along mid-ocean spreading centers, in regions of back-arc spreading, or over oceanic hot spots, is followed by movement of the plate and/or volcanic edifice away from the site of volcanism, cooling and subsidence, and deep-water deposition of a mantle of fine pelagic detritus. The probable absence of Archean silica- and carbonate-secreting planktonic organisms would have greatly reduced intraplate sedimentation rates, compared with open-ocean rates today, and restricted sedimentation to orthochemical deposits, evidently mainly silica and siderite, and minor hemipelagic carbonaceous, tuffaceous, and argillaceous materials. If the oceans were stratified (Klein and Beukes, 1989), with little or only slow mixing of shallow- and deep-water masses, and continents provided few nutrients through weathering, erosion, and runoff, ocean surface waters away from zones of active dynamic upwelling and mixing may have been nutrient-poor biological deserts, providing very little net organic debris to the ocean floor (Lowe, 1994).

FVT and OT associations

The felsic volcanic and comagmatic TTG plutonic units in Archean granite–greenstone terranes are generally regarded as the extrusive and intrusive phases, respectively, of magmatism along subduction-related magmatic arcs (Bickle et al., 1983; Abbott and Hoffman, 1984; Arkani-Hamed and Joly, 1989; and many others). Thick portions of the associated greenstone-belt basaltic sequences may also reflect magmatism in arc settings. These rocks indicate that the mafic–ultramafic spreading center, oceanic island, and/or back-arc spreading center assemblages (MAVOB association) and overlying oceanic intraplate sediments (AOB association) commonly came to be situated above subduction zones during later stages of their evolution. Calc-alkaline felsic volcanism was widely accompanied by deformation, probably reflecting both intra-arc deformation and back-arc fold-and-thrust activity, leading to the formation of mixed volcaniclastic and epiclastic orogenic detrital suites. The OTt association has been related to sedimentation within trenches and forearc settings that became parts of growing accretionary complexes. The sediments represent uplifted, more mature and deeply eroded magmatic arcs as well as older rocks where arcs developed as continental or Andean type systems. Similar sedimentary suites might be expected within back-arc basins associated with back-arc fold-and-thrust belts.

The latest stages of greenstone-belt evolution involved formation of foreland basins (Jackson et al., 1987; Heubeck, 1993) and strike-slip or oblique-slip, pull-apart, and successor basins (Mueller et al., 1991). Sedimentation was dominated by alluvial, fluvial, and shallow-water sediments of the OTaf association.

APT association

Local mature sedimentary units within or marginal to Late Archean greenstone belts probably reflect erosion of either uplifted intra-belt TTG complexes, in which case they actually represent the FVT or OT association, or of older basement blocks, in which case they may represent micro-continental rift and marginal sequences.

Archean tectonics and sedimentation

Although there is still considerable controversy regarding the nature of Archean tectonics (e.g. Hamilton, 1993), available evidence has led most workers to infer the existence of an active plate-tectonic geodynamic system (Burke et al., 1976; Windley, 1976; Hoffman, 1989; Barley and Groves, 1990; Card, 1990; Card and King, 1992; de Wit et al., 1992; Kröner and Layer, 1992; Lowe and Ernst, 1992; Sleep, 1992; and many others). Widespread calc-alkaline TTG plutonism and felsic volcanism having geochemical and petrologic similarities to modern igneous suites of subduction-related magmatic arcs (e.g. Bickle et al., 1983; Abbott and Hoffman, 1984), the abundance of tholeiitic and komatiitic volcanism resembling that on modern tectonically quiescent oceanic islands, plateaus, and spreading centers (e.g. Nisbet, 1982; Kusky and Kidd, 1992), and the dominance of horizontal shortening in late greenstone-belt tectonics (e.g. Bickle et al., 1980; de Wit, 1982; Sleep, 1992; Lowe, in review, c) all suggest the existence of an Archean regime dominated by horizontal plate movements, plate boundaries, and magmatism resembling those on the modern earth. Differences between Archean magmatic suites, including the abundance of komatiites and the more bimodal character of arc-related volcanic rocks, probably reflect differences in the thermal regimes of the Archean and modern earth (Nisbet, 1982; Abbott and Hoffman, 1984; Bickle, 1986; Martin, 1986; Drummond and Defant, 1990; Sleep, 1992).

Unless preserved Archean supracrustal sequences selectively preserve non-cratonic depositional and tectonic settings, in contrast to their Proterozoic and Phanerozoic analogs which are biased toward cratonic settings and sediments, the pre-2.7-Ga Archean world also differed from the Proterozoic and Phanerozoic in having been dominated by a geodynamic system in which continental blocks played a minor role (Lowe and Ernst, 1992; Lowe, 1992a). Although continental blocks clearly existed in early Archean time, possibly before 3.5 Ga and certainly by 3.2–3.0 Ga, geologic and sedimentological evidence indicates that Archean greenstone systems, especially before 3.0 Ga, show only local evidence for the influence of continental blocks (Lowe, 1992a). In contrast, most post-3.0 Ga Archean greenstones locally contain mature quartzitic sedimentary units and craton-derived detritus and show development against and overlap onto older cratonic basement, reflecting the presence and influence but not yet dominance of large, older crustal blocks.

Based on preserved greenstone sequences, the pre-3.0-Ga Archean tectonic cycle began with anorogenic komatiitic and basaltic magmatism along spreading centers and/or over mantle plumes. Most 3.0–2.5 Gy-old belts, as well as Proterozoic and Phanerozoic greenstone enclaves in continental blocks, probably included volcanic sequences representing similar tectonic settings (e.g. Pharoah and Brewer, 1990), but these assemblages have commonly been obliterated by tectonism and magmatism, and preserved Late Archean and younger greenstone belts are dominated by subduction-related volcanic and sedimentary rocks.

In some belts, such as the central and northern Barberton Greenstone Belt, early mafic and ultramafic volcanism was followed by magmatic and tectonic stasis in a deep-water, probably intraplate setting. Intraplate stasis was followed by basaltic and felsic magmatism along convergent plate boundaries. In other sequences, evidence for a prolonged period of residence in an intraplate setting is lacking. Such belts or blocks may have evolved rapidly from divergent to convergent settings as a consequence of abrupt plate reorganization or preserved rocks may record only the convergent-margin stage of development. The tectonic cycle within individual belts culminated with regional orogeny and shortening reflecting the welding of the belts within accretionary complexes that were intruded, metamorphosed, and thickened to form new continental blocks.

Preserved and well-studied greenstone belts suggest that there are major differences in the sedimentary suites and inferred depositional environments of pre-3.0 and post-3.0 Ga greenstone belts (Lowe, 1980, 1982; Groves and Batt, 1984). Sedimentological studies suggest that MAVOB associations in the older belts formed broad, flat, shallow-water volcanic platforms or plateaus whereas in the younger belts, MAVOB rocks accumulated as subaqueous lava plains and shield volcanoes. There are also more andesitic and possibly less komatitic volcanic rocks in most younger belts and important differences in mineralization (Groves and Batt, 1984). It appears that pre-3.0 Ga belts are dominated by mid-ocean spreading center, hot spot, or oceanic plateau MAVOB associations whereas post-3.0 Ga belts are dominated by MAVOB and FVT sequences formed through back-arc spreading and arc volcanism and sedimentation.

The inferred Archean tectonic cycle included three main stages: (1) initial formation of a mafic crustal layer along oceanic or perhaps back-arc spreading centers and/or over hot spots, (2) cooling and subsidence of the mafic layer in intraplate settings, and (3) crustal consumption along convergent plate boundaries. Preserved greenstone belts record a variation in this cycle in which the formed and cooled mafic oceanic crust was thickened and partially stabilized through subduction-related TTG plutonism (underplating) and felsic volcanism (overplating) along the convergent plate boundary instead of being recycled into the mantle through subduction. This type of crustal accretion and thickening may have occurred preferentially along continental or Andean-type convergent margins, although in the Archean the continental blocks involved would have been microcontinents. The Archean tectonic cycle was the analog of the present-day continent-dominated Wilson cycle but in a world without continents.

SUMMARY

Sedimentary rocks in Archean greenstone belts represent four principal petrologic suites: (1) volcaniclastic and pyroclastic rocks, (2) terrigenous sedimentary units, (3) orthochemical deposits, and (4) biogenic sedimentary units. They can also grouped into five main lithofacies associations reflecting the principal tectonic and depositional settings on the Archean earth: (1) a mafic anorogenic volcaniclastic–orthochemical biogenic association (MAVOB), (2) anorogenic orthochemical–biogenic association (AOB), (3) felsic volcaniclastic and terrigenous association (FVT), (4) orogenic terrigenous associations (OTt and OTaf)), and (5) anorogenic polycyclic terrigenous association (APT).

Early Archean greenstone belts preserve the record of a world in which continental blocks had little influence on sedimentation or lithospheric dynamics. Most are compound terranes formed in a range of tectonic and depositional settings reflecting the evolutionary stages of the Archean continental crust. They include thick successions of komatiitic and basaltic rocks representing oceanic islands or spreading centers (MAVOB association) capped by thin cherty units representing stratigraphically attenuated but long-lived oceanic intraplate depositional systems (AOB). The upper parts of Archean greenstone sequences reflect orogenesis and accretion along convergent plate boundaries (FVT and OT). By the Late Archean, the presence of older microcontinental blocks resulted in the local development of mature continental margin sequences (APT) during the early rifting stages of greenstone-belt evolution. Archean APT-association sequences are thin and poorly developed compared to their Phanerozoic analogs, reflecting the small size of Archean microcontinental blocks. There is some suggestion that pre-3.0 Gy-old greenstones are dominated by rocks formed during oceanic spreading and/or hot spot volcanism whereas Late Archean belts appear to be dominated by the culminating stage of crustal evolution involving magmatism, deformation, and accretion along convergent plate margins.

There are many remaining unanswered or poorly answered questions regarding the nature and interpretation of sedimentary rocks in Archean greenstone belts. Our present knowledge is based on a limited number of well-studied terranes; many large shield areas in Asia, South America, Africa, and Antarctica remain poorly known. Only two well-preserved, relatively unsheared pre-3.0 Ga greenstone sequences are known, the Kaapvaal and eastern Pilbara cratons, and studied greenstone belts in both show remarkably similar ages and lithofacies. Are these belts representative of tectonic and depositional systems on the early Archean earth or only of a narrow range of settings that were selectively preserved from a time when virtually all crust was recycled? Archean sedimentary deposits provide the only direct record of Archean climate, ocean, and atmospheric evolution, but much of their potential to yield clues regarding surficial conditions on the early earth remains untapped. In part, this reflects difficulties in distinguishing primary from diagenetic and metamorphic components and structures, especially in largely orthochemical deposits, such as the wide variety of Archean banded cherty sedimentary rocks. In addition, Archean clastic sequences are difficult to analyze using modern petrographic and petrofacies techniques, in part because of widespread, low-grade metamorphism and metasomatism and in part because of the exotic nature of many of the Archean detrital components, especially chert debris, which can represent virtually any type of silicified volcanic or sedimentary rock, including komatiites and basalts. Suggested differences between early and late Archean greenstone belts, although based on a reasonably large number of observations, remain to be systematically documented and interpreted. It is not clear whether such differences, if real, reflect differences in kind or merely in degree between early and late Archean depositional and tectonic systems. Finally, although there is a growing consensus that plate tectonics provides a rational basis for interpreting Archean magmatism, sedimentation, tectonics, and crustal evolution, there are important differences between the Archean and Phanerozoic geologic records that may not simply reflect variations on the plate tectonic theme related to higher heat loss on the early earth (Hamilton, 1993). The geologic record has proven to be a relatively problematic basis for recognizing processes that are fundamentally different from those with which we are familiar. The evolution of the plate tectonics paradigm itself provides an excellent example of how difficult it has been to interpret unfamiliar processes based solely on the geologic record. Were Archean crustal development controlled by a fundamentally different geodynamic regime that yielded rock types and structural styles that are grossly similar to those of the Phanerozoic world, we might fail to recognize the nature and uniqueness of that regime. Studies of the sedimentology and sedimentary petrology of Archean greenstone belts must continue to focus on identifying the nature of the early geodynamic system and processes leading to the formation and evolution of the continental crust.

ACKNOWLEDGEMENTS

This research was supported by NASA grants NCA 2-332 and NCC 2-721 from the Exobiology Program, NASA grant NAG 9-344 from the Planetary Materials and Geochemistry Program, and grant EAR 8904830 from the National Science Foundation. The author is grateful to the many persons who have aided in the course of field studies of Archean greenstone belts, especially Dr. Gary Byerly (LSU), who has been a close friend and field partner, and the Anglo American and Anglo Vaal mining companies of South Africa. Ken Eriksson, John Grotzinger, Wulf Mueller, and an anonymous reviewer read the manuscript and provided many helpful comments.

REFERENCES

Abbott, D.H. and Hoffman, S.E., 1984. Archaean plate tectonics revisited, 1. Heat flow, spreading rate, and the age of subducting oceanic lithosphere and their effects on the origin and evolution of continents. Tectonics, 3: 429–448.

Arkani-Hamed, J. and Joly, W.T., 1989. Generation of Archean tonalites. Geology, 17: 307–310.

Baragar, W.R.A. and Goodwin, A.M., 1968. Andesites and Archean volcanism in the Canadian Shield. In: A.R. McBirney (Ed.), Proceedings of the Andesite Conference. Oregon Dept. Geol. Min. Indust. Bull., 65: 121–142.

Barley, M.E., 1984. Volcanism and hydrothermal alteration, Warrawoona Group, East Pilbara. In: J.R. Muhling, D.I. Groves, and T.S. Blake (Eds.), Archaean and Proterozoic Basins of the

Pilbara, Western Australia: Evolution and Mineralization Potential. Geol. Dept. Ext. Serv. Univ. W. Aust. Spec. Publ., 9: 23–36.

Barley, M.E., 1993. Volcanic, sedimentary and tectonostratigraphic environments of the ~3.46 Ga Warrawoona Megasequence: a review. Precambrian Res., 60: 47–67.

Barley, M.E. and Groves, D.I., 1990. Deciphering the tectonic evolution of Archaean greenstone belts: the importance of contrasting histories to the distribution of mineralization in the Yilgarn Craton, Western Australia: Precambrian Res., 46: 3–20.

Barley, M.E., Dunlop, J.S.R., Glover, J.E., and Groves, D.I., 1979. Sedimentary evidence for an Archean shallow-water volcanic-sedimentary facies, eastern Pilbara Block, Western Australia. Earth Planet. Sci. Lett., 43: 74–84.

Barley, M.E., Eisenlohr, B.N., Groves, D.I., Perring, C.S., and Vearncombe, J.R., 1989. Late Archean convergent margin tectonics and gold mineralization: A new look at the Norseman–Wiluna belt, Western Australia. Geol., 17: 826–829.

Barley, M.E., Sylvester, G.C., Groves, D.I., Borley, G.D., and Rogers, N., 1984. Archean calc-alkaline volcanism in the Pilbara Block, Western Australia. Precambrian Res., 24: 285–319.

BenOthman, D., Arndt, N.T., White, W.M., and Jochum, K.P., 1990. Geochemistry and age of Timiskaming alkali volcanics and the Otto syenite stock, Abitibi, Ontario. Can. J. Earth Sci., 27: 1304–1311.

Bickford, M.E., 1988. The accretion of Proterozoic crust in Colorado: Igneous, sedimentary, deformational, and metamorphic history. In: W.G. Ernst (Ed.), Metamorphic and Crustal Evolution of the Western United States. Prentice-Hall, Englewood Cliffs, New Jersey, pp. 411–430.

Bickle, M.J., 1986. Implications of melting for stabilization of the lithosphere and heat loss in the Archaean. Earth Planet. Sci. Lett., 80: 314–324.

Bickle, M.J., Bettenay, L.F., Boulter, C.A., Groves, D.I., and Morant, P., 1980. Horizontal tectonic interaction of an Archaean gneiss belt and greenstones, Pilbara Block, Western Australia. Geology, 8: 525–529.

Bickle, M.J., Bettenay, L.F., Barley, M.E., Groves, D.I., Chapman, H.J., Campbell, I., and de Laeter, J.R., 1983. A 3,500 Ma plutonic and volcanic calc-alkaline province in the Pilbara Archaean. Contr. Mineral. Petrol., 84: 25–35.

Blackburn, C.E., Bond, W.D., Breaks, F.W., Davis, D.W., Edwards, G.R., Poulsen, K.H., Trowell, N.F., and Wood, J., 1985. Evolution of Archean volcanic-sedimentary sequences of the western Wabigoon Subprovince and its margins: a review. In: L.D. Ayres, P.C. Thurston, K.D. Card, and W. Weber (Eds.), Evolution of Archean Supracrustal Sequences. Geol. Assoc. Can. Spec. Pap., 28: 89–116.

Blake, T.S., 1993. Late Archaean crustal extension, sedimentary basin formation, flood basalt volcanism and continental rifting: the Nullagine and Mount Jope Supersequences, Western Australia. Precambrian Res., 60: 185–241.

Bridgwater, D., Keto, L., McGregor, V.R., and Myers, J.S., 1976. Archaean gneiss complex of Greenland. In: A. Escher, and W.S. Watt (Eds.), Geology of Greenland. Geol. Surv. Greenland, pp. 18–75.

Buick, R. and Dunlop, J.S.R., 1990. Evaporitic sediments of Early Archaean age from the Warrawoona Group, North Pole, Western Australia. Sedimentology, 37: 247–277.

Burke, K., Dewey, J.F., and Kidd, W.S.F., 1976. Dominance of horizontal movements, arc and microcontinental collisions during the later permobile regime. In: B.F. Windley (Ed.), The Early History of the Earth. John Wiley & Sons, London, pp. 113–130.

Camire, G.E., Lafleche, M.R., and Ludden, J.N., 1993. Archaean metasedimentary rocks from the northwestern Pontiac Subprovince of the Canadian Shield: chemical characterization, weathering and modelling of the source areas. Precambrian Res., 62: 285–305.

Car, D. and Ayres, L.D., 1991. A thick debris flow sequence, Lake of the Woods greenstone terrane, central Canada: resedimented products of Archean vulcanian, plinian and dome-building eruptions. Precambrian Res., 50: 239–260.

Card, K.D., 1990. A review of the Superior Province of the Canadian Shield, a product of Archean accretion. Precambrian Res., 48: 99–156.

Card, K.D. and King, J.E., 1992. The tectonic evolution of the Superior and Slave provinces of the Canadian Shield: introduction. Can. J. Earth Sci., 29: 2059–2065.

Chown, E.H., Daigneault, R., Mueller, W., and Mortensen, J.K., 1992. Tectonic evolution of the Northern Volcanic Zone, Abitibi belt, Quebec. Can. J. Earth Sci., 29: 2211–2225.

Cloud, P., 1968. Atmospheric and hydrospheric evolution on the primitive earth. Science, 160: 729–736.

Cloud, P., 1976. Major features of crustal evolution. Geol. Soc. South Africa Annexure, 79: 1–33.

Colgate, S.A. and Sigurgeirsson, T., 1973. Dynamic mixing of water and lava. Nature, 244: 552–555.

Compston, W., Williams, I.S., Campbell, I.H., and Gresham, J.J., 1986. Zircon xenocrysts from the Kambalda volcanics: age constraints and direct evidence for older continental crust below the Kambalda-Norseman greenstones. Earth Planet. Sci. Lett., 76: 299–311.

Condie, K.C., 1981. Archean Greenstone Belts. Elsevier, Amsterdam, 434 pp.

Condie, K.C., 1982. Archaean andesites. In: R.S. Thorpe (Ed.), Andesites: orogenic andesites and related rocks. John Wiley, New York, pp. 575–590.

Condie, K.C., Macke, J.E., and Reimer, T.O., 1970. Petrology and geochemistry of early Precambrian graywackes from the Fig Tree Group, South Africa. Geol. Soc. Am. Bull., 81: 2759–2776.

Cullers, R.L., DiMarco, M.J., Lowe, D.R., and Stone, J. 1993. Geochemistry of a silicified, felsic volcaniclastic suite from the early Archaean Panorama Formation, Pilbara Block, Western Australia: an evaluation of depositional and post-depositional processes with special emphasis on the rare-earth elements. Precambrian Res., 60: 99–116.

Danchin, R.V., 1967. Chromium and nickel in the Fig Tree Shale from South Africa. Science, 158: 261–262.

Davis, D.W., Pezzutto, F., and Ojakangas, R.W., 1990. The age and provenance of metasedimentary rocks in the Quetico subprovince, Ontario, from single zircon analyses: Implications for Archean sedimentation and tectonics in the Superior Province. Earth Planet. Sci. Lett., 99: 195–205.

de Wit, M.J., 1982. Gliding and overthrust nappe tectonics in the Barberton Greenstone Belt. J. Struct. Geol., 4: 117–136.

de Wit, M.J., Roering, C., Hart, R.J., Armstrong, R.A., de Ronde, C.E.J., Green, R.W.E., Tredoux, M., Peberdy, E., and Hart, R.A., 1992. Formation of an Archaean continent. Nature, 357: 553–562.

Dickinson, W.R., 1984. Interpreting provenance relations from detrital modes of sandstones. In: G.G. Zuffa (Ed.), Provenance of Arenites. D. Reidel, Dordrecht, pp. 333–361.

DiMarco, M.J., and Lowe, D.R., 1989a. Stratigraphy and sedimentology of an early Archean felsic volcanic sequence, Eastern Pilbara Block, Western Australia, with special reference to the Duffer Formation and implications for crustal evolution. Precambrian Res., 44: 147–169.

DiMarco, M.J. and Lowe, D.R., 1989b. Shallow-water volcaniclastic deposition in the Early Archean Panorama Formation, Warrawoona Group, eastern Pilbara Block, Western Australia. Sediment. Geol., 64: 43–63.

DiMarco, M.J., Ferrell, R.E., Jr., and Lowe, D.R., 1989. Polytypes of 2:1 dioctahedral micas in silicified volcaniclastic sandstones, Warrawoona Group, Pilbara Block, Western Australia. Am. J. Sci., 289: 649–660.

Dimroth, E., 1977. Archean subaqueous autoclastic volcanic rocks, Rouyn-Noranda area, Quebec: Classification, diagenesis and interpretation. Geol. Surv. Can. Pap., 77-1A: 513–522.

Dimroth, E. and Demarcke, J., 1978. Petrography and mechanisms of eruption of the Archean Dalembert Tuff, Rouyn-Noranda, Quebec, Canada. Can. J. Earth Sci., 15: 1712–1723.

Dimroth, E. and Rocheleau, M., 1979. Volcanology and sedimentology of Rouyn-Noranda area, Quebec: Geol. Assoc. Can. Guidebook Field Trip A-1, Universite Laval, Quebec, 193 pp.

Dimroth, E., Imreh, L., Rocheleau, M., and Goulet, N., 1982. Evolution of the south-central part of the Archean Abitibi belt, Quebec, Part I: stratigraphic and paleogeographic model. Can. J. Earth Sci., 19: 1729–1758.

Dimroth, E., Imreh, L., Cousineau, P., Meduc, M., and Sanschagrin, Y., 1985a. Paleogeographic analysis of mafic submarine flows and its use in the exploration for massive sulphide deposits. In: L.D. Ayres, P.C. Thurston, K.D. Card, and W. Weber (Eds.), Evolution of Archean Supracrustal Sequences. Geol. Assoc. Can. Spec. Pap., 28: 203–222.

Dimroth, E., Rocheleau, M., Mueller, W., Archer, P., Brisson, H., Fortin, G., Jutras, M., Lefebvre, C., Piche, M., Pilote, P., and Simoneau, P., 1985b. Paleogeographic and paleotectonic response to magmatic processes: a case history from the Archean sequence in the Chibougamau area, Quebec. Geol. Rundschau, 74: 11–32.

Drummond, M.S. and Defant, M.J., 1990. A model for trondhjemite–tonalite– dacite gensis and crustal growth via slab melting: Archaean to modern comparisons. J. Geophys. Res., 95: 21503–21521.

Duchac, K.C. and Hanor, J.S., 1987. Origin and timing of the metasomatic silicification of an early Archean komatiite sequence, Barberton Mountain Land, South Africa. Precambrian Res., 37: 125–146.

Dunlop, J.S.R. and Buick, R., 1981. Archaean epiclastic sediments derived from mafic volcanics, North Pole, Pilbara Block, Western Australia. Spec. Publ. Geol. Soc. Aust., 7: 225–233.

Easton, R.M., 1985, The nature and significance of pre-Yellowknife Supergroup rocks in the Point Lake area, Slave structural province, Canada: In: L.D. Ayres, P.C. Thurston, K.D. Card, and W. Weber (Eds.), Evolution of Archean Supracrustal Sequences. Geol. Assoc. Can. Spec. Pap., 28: 153–168.

Eriksson, K.A., 1977. Tidal deposits from the Archaean Moodies Group, Barberton Mountain Land, South Africa. Sediment. Geol., 18: 257–281.

Eriksson, K.A., 1978. Alluvial and destructive beach facies from the Archaean Moodies Group, Barberton Mountain Land, South Africa and Swaziland. In: A.D. Miall (Ed.), Fluvial Sedimentology. Can. Soc. Petrol. Geol. Mem., 5: 287–311.

Eriksson, K.A., 1979. Marginal marine depositional processes from the Archaean Moodies Group, Barberton Mountain Land, South Africa: evidence and significance. Precambrian Res., 8: 153–182.

Eriksson, K.A., 1980a, Hydrodynamic and paleogeographic interpretation of turbidite deposits from the Archean Fig Tree Group of the Barberton Mountain Land, South Africa. Geol. Soc. Am. Bull., 91: 21–26.

Eriksson, K.A., 1980b, Archaean platform-to-trough sedimentation, east Pilbara Block, Australia. Spec. Publs. Geol. Soc. Aust., 7: 235–244.

Eriksson, K.A., 1982. Geometry and internal characteristics of Archaean submarine channel deposits, Pilbara Block, Western Australia. J. Sed. Petrology, 52: 383–393.

Eriksson, K.A. and Fedo, C.M., 1994. Archean synrift and stable-shelf sedimentary successions. In: K.C. Condie (Ed.), Archean Crustal Evolution. Elsevier, Amsterdam, Chap. 5, pp. 171–204.

Feng, R., Kerrich, R., and Maas, R., 1993. Geochemical, oxygen, and neodymium isotope compositions of metasediments from the Abitibi greenstone belt and Pontiac Subprovince, Canada:

Evidence for ancient crust and Archean terrane juxtaposition. Geochim. Cosmochim. Acta, 57: 641–658.

Gee, R.D., Baxter, J.L., Wilde, S.A., and Williams, I.R., 1981. Crustal development in the Archaean Yilgarn Block, Western Australia. Geol. Soc. Australia Spec. Publ., 7: 43–56.

Giles, C.W., 1982. The geology and geochemistry of the Archaean Spring Well Felsic Volcanic Complex, Western Australia. J. Geol. Soc. Australia, 29: 205–220.

Glikson, A.Y., 1971. Archaean geosynclinal sedimentation near Kalgoorlie, Western Australia. Spec. Publ. Geol. Soc. Aust., 3: 443–460.

Goodwin, A.M., 1982. Archean volcanoes in southwestern Abitibi Belt, Ontario and Quebec: form, composition, and development. Can. J. Earth Sci., 19: 1140–1155.

Gorbatschev, R. and Gaal, G., 1987. The Precambrian history of the Baltic Shield. In: A. Kroner (Ed.), Proterozoic Lithospheric Evolution. American Geophys. Union Ser., 17: 149–159.

Groves, D.I., and Batt, W.D., 1984. Spatial and temporal variations of Archean metallogenic associations in terms of evolution of granitoid–greenstone terrains with particular emphasis on the Western Australian Shield. In: A. Kröner, G.N. Hansen, and A.M. Goodwin (Eds.), Archaean Geochemistry. Springer-Verlag, Berlin, pp. 73–98.

Groves, D.I., Dunlop, J.S.R., and Buick, R., 1981. An early habitat of life. Sci. Am., 245: 64–73.

Hallberg, J.A., 1985. Geology and mineral deposits of the Leonora–Laverton area: Northeastern Yilgarn Block. Perth, Australia, Hesperian Press, 140 pp.

Hallberg, J.A., Carter, D.N., and West, K.N., 1976. Archaean volcanism and sedimentation near Meekatharra, Western Australia. Precambrian Res., 3: 577–595.

Hamilton, W.B., 1993. Evolution of Archean mantle and crust. In: J.C. Reed Jr., M.E. Bickford, R.S. Houston, P.K. Link, D.W. Rankin, P.K. Sims, and W.R. Van Schmus (Eds.), Precambrian: Conterminous U.S. Geol. Soc. Am., The Geology of North America, C-2: 597–614.

Hanor, J.S. and Duchac, K.C., 1990. Isovolumetric silicification of Early Archean komatiites: geochemical mass balances and constraints on origin. J. Geol., 98: 863–877.

Heinrichs, T.K., 1984. The Umsoli Chert, turbidite testament for a major phreatoplinian event at the Onverwacht/Fig Tree transition (Swaziland Supergroup, Archaean, South Africa). Precambrian Res., 24: 237–283.

Heinrichs, T.K. and Reimer, T.O., 1977. A sedimentary barite deposit from the Archean Fig Tree Group of the Barberton Mountain Land (South Africa). Econ. Geol., 72: 1426–1441.

Henderson, J.B., 1975a, Sedimentology of the Archean Yellowknife Supergroup at Yellowknife, District of Mackenzie. Geol. Surv. Can. Bull., 246, 62 pp.

Henderson, J.B., 1975b. Archean stromatolites in the northern Slave Province, Northwest Territories, Canada. Can. J. Earth Sci., 12: 1619–1630.

Heubeck, C., 1993. Geology of the Archean Moodies Group, central Barberton Greenstone Belt, South Africa. Stanford University, Ph.D. thesis, 450 pp.

Heubeck, C. and Lowe, D.R., 1992. Petrographic evolution and provenance of the Archean Moodies Basin, Barberton Greenstone Belt, South Africa. Geol. Soc. Am., Absts. Prog., 24: 179.

Hickman, A.H., 1983. Geology of the Pilbara Block, Western Australia. Geol. Surv. W. Australia. Bull., 127, 268 pp.

Hoffman, P.F., 1988. United plates of America, the birth of a craton: Early Proterozoic assembly and growth of Laurentia. Ann. Rev. Earth Planet. Sci., 16: 543–603.

Hoffman, P.F., 1989. Precambrian geology and tectonic history of North America. In: A.W. Bally and A.R. Palmer (Eds.), The Geology of North America — An Overview. Geol. Soc. Am., The Geology of North America, A: 447–512.

Hofmann, H.J., Thurston, P.C., and Wallace, H., 1985. Archean stromatolites from Uchi Greenstone Belt, northwestern Ontario. In: L.D. Ayres, P.C. Thurston, K.D. Card, and W. Weber (Eds.),

Evolution of Archean Supracrustal Sequences. Geol. Assoc. Can. Spec. Pap., 28: 125–132.

Hyde, R.S., 1980. Sedimentary facies is the Archean Timiskaming Group and their tectonic implications, Abitibi Greenstone Belt, northeastern Ontario, Canada. Precambrian Res., 12: 161–195.

Ingersoll, R.V., 1988. Tectonics of sedimentary basins. Geol. Soc. Am. Bull., 100: 1704–1719.

Jackson, M.P.A., Eriksson, K.A., and Harris, C.W., 1987 Early Archean foredeep sedimentation related to crustal shortening: a reinterpretation of the Barberton sequence, southern Africa. Tectonophysics, 136: 197–221.

Jackson, S.L., Fyon, J.A., and Corfu, F., 1994. Review of Archean supracrustal assemblages of the southern Abitibi greenstone belt in Ontario, Canada: products of microplate interaction within a large-scale plate-tectonic setting. Precambrian Res., 65: 183–205.

James, H.L., 1954. Sedimentary facies of iron-formation. Econ. Geol., 49: 235–293.

Jensen, L.S., 1985. Stratigraphy and petrogenesis of Archean metavolcanic sequences, southwestern Abitibi Subprovince, Ontario. In: L.D. Ayres, P.C. Thurston, K.D. Card, and W. Weber (Eds.), Evolution of Archean Supracrustal Sequences. Geol. Assoc. Canada Spec. Pap., 28: 66–88.

Jolly, W.T. and Hallberg, J.A., 1990. Role of crustal contamination in heterogeneous Archean volcanics from the Leonora region, Western Australia. Precambrian Res., 48: 75–98.

Klein, C. and Beukes, N.J., 1989. Geochemistry and sedimentology of a facies transition from limestone to iron-formation deposition in the early Proterozoic Transvaal Supergroup, South Africa. Econ. Geol., 84: 1733–1774.

Kröner, A. and Compston, W., 1988. Ion microprobe ages of zircons from early Archaean granite pebbles and graywacke, Barberton Greenstone Belt, southern Africa. Precambrian Res., 38: 367–380.

Kröner, A. and Layer, P.W., 1992. Crust formation and plate motion in the early Archean. Science, 256: 1405–1411.

Kröner, A., Byerly, G.R., and Lowe, D.R., 1992. Chronology of early Archaean granite–greenstone evolution in the Barberton Mountain Land, South Africa, based on precise dating by single zircon evaporation. Earth Planet. Sci. Lett., 103: 41–54.

Kusky, T.M. and Kidd, W.S.F., 1992. Remnants of an Archean oceanic plateau, Belingwe greenstone belt, Zimbabwe. Geology, 20: 43–46.

Lambert, M.B., 1988. Cameron River and Beaulieu River volcanic belts of the Archean Yellowknife Supergroup, District of Mackenzie, Northwest Territories. Geol. Surv. Can. Bull., 382, 145 pp.

Lanier, W. P. and Lowe, D. R., 1982. Sedimentology of the Middle Marker (3.4 Ga), Onverwacht Group, Transvaal, South Africa. Precambrian Res., 18: 237–260.

Lowe, D.R., 1980. Archean sedimentation. Ann. Rev. Earth Planet. Sci., 145–167.

Lowe, D.R., 1982. Comparative sedimentology of the principal volcanic sequences of Archean greenstone belts in South Africa, Western Australia, and Canada: implications for crustal evolution. Precambrian Res., 17: 1–29.

Lowe, D.R., 1983. Restricted shallow-water sedimentation of 3.4 Byr-old stromatolitic and evaporitic strata of the Strelley Pool Chert, Pilbara Block, Western Australia. Precambrian Res., 19: 239–283.

Lowe, D.R., 1985. Sedimentary environment as a control on the formation and preservation of Archean volcanogenic massive sulphide deposits. In: L.D. Ayres, P.C. Thurston, K.D. Card, and W. Weber (Eds.), Evolution of Archean Supracrustal Sequences. Geol. Assoc. Canada Spec. Pap., 28: 193– 202.

Lowe, D.R., 1992a. Major events in the geological development of the Precambrian earth. In: J.W. Schopf and C, Klein (Eds.), The Proterozoic Biosphere. Cambridge University Press, Cambridge, pp. 67–75.

Lowe, D.R., 1992b. Geology of the Barberton Greenstone Belt: an overview. In: L.D. Ashwal (Ed.), Two Cratons and an Orogen. IGCP Proj. 280, Dept. Geol., Univ. Witwatersrand, Johannesburg, pp. 47–58.

Lowe, D.R., 1994. Early environments: constraints and opportunities for early evolution. In: S. Bengtson (Ed.), Early Life on Earth. Nobel Symposium 84, Columbia Univ. Press, New York, pp. 24–35.

Lowe, D.R., in review, a. Petrology and sedimentology of cherts in the Swaziland Supergroup, Barberton Greenstone Belt, South Africa. Geol. Soc. Am.

Lowe, D.R., in review, b. Shallow-water sedimentation of the Msauli Chert. Geol. Soc. Am.

Lowe, D.R., in review, c. Geologic evolution of the Barberton Greenstone Belt and vicinity. Geol. Soc. Am.

Lowe, D.R. and Byerly, G.R., in review. Stratigraphy of the west-central part of the Barberton Greenstone Belt, South Africa. Geol. Soc. Am.

Lowe, D.R., and Ernst, W.G., 1992. The Archean geologic record. In: J.W. Schopf and C. Klein (Eds.), The Proterozoic Biosphere. Cambridge University Press, Cambridge, pp. 13–19.

Lowe, D.R. and Byerly, G.R., 1986. Archean flow-top alteration zones formed initially in a low-temperature sulphate-rich environment. Nature, 324: 245–248.

Lowe, D.R. and Knauth, L.P., 1977. Sedimentology of the Onverwacht Group (3.4 billion years), Transvaal, South Africa, and its bearing on the characteristics and evolution of the early earth. J. Geol., 85: 699–723

Lowe, D.R., and Knauth, L.P., 1978. The oldest marine carbonate ooids reinterpreted as volcanic accretionary lapilli, Onverwacht Group, South Africa. J. Sed. Petrology, 48: 708–722.

Lowe, D.R. and Nocita, B.R., in review. Stratigraphy and sedimentology of the Mapepe Formation: sedimentation in an evolving foreland basin. Geol. Soc. Am.

Lowe, D.R. and Worrell, G.F., in review. Sedimentology and mineralogy of silicified evaporites in the basal Kromberg Formation. Geol. Soc. Am.

Marston, R.J., 1978. The geochemistry of Archaean clastic metasediments in relation to crustal evolution, northeastern Yilgarn Block, Western Australia. Precambrian Res., 6: 157–175.

Martin, A., Nisbet, E.G., and Bickle, M.J., 1980. Archean stromatolites of the Belingwe Greenstone Belt, Zimbabwe (Rhodesia). Precambrian Res., 13: 337–362.

Martin, H., 1986. Effect of steeper Archaean geothermal gradient on geochemistry of subduction-zone magmas. Geology, 14: 753–756.

Mueller, W., 1991. Volcanism and related slope to shallow-marine volcaniclastic sedimentation: an Archean example near Chibougamau, Quebec, Canada. Precambrian Res., 49: 1–22.

Mueller, W. and White, J.D.L., 1992. Felsic fire-fountaining beneath Archean seas: pyroclastic deposits of the 2730 Ma Hunter Mine Group, Quebec, Canada. J. Volcanol. Geotherm. Res., 54, 117–134.

Mueller, W., Donaldson, J.A., Dufresne, D., and Rocheleau, M., 1991. The Duparquet Formation: sedimentation in a late Archean successor basin, Abitibi greenstone belt, Quebec, Canada. Can. J. Earth Sci., 28: 1394–1406.

Naqvi, S.M., 1986. Geochemical characters and tectonic evolution of the Chitradurga Schist Belt: An Archaean suture (?) of the Dharwar Craton, India. In: M.J. de Wit and L.D. Ashwal (Eds.), Workshop on Tectonic Evolution of Greenstone Belts. Lunar and Planetary Institute Technical Rept., 86-10: 160–161.

Naqvi, S.M. and Rogers, J.J.W., 1983. Precambrian of Southern India. Geol. Soc. India Mem., 4: 575 pp.

Nisbet, E.G., 1982. The tectonic setting and petrogenesis of komatiites. In: N.T. Arndt and E.G. Nisbet (Eds.), Komatiites. Allen and Unwin, London, pp. 501–526.

Nocita, B.W. and Lowe, D.R., 1990. Fan-delta sequence in the Archean Fig Tree Group, Barberton Greenstone Belt, South Africa. Precambrian Res., 48: 375–393

Ojakangas, R.W., 1985. Review of Archean clastic sedimentation, Canadian Shield: major felsic volcanic contributions to turbidite and alluvial fan-fluvial facies associations. In: L.D. Ayres, P.C. Thurston, K.D. Card, and W. Weber (Eds.), Evolution of Archean Supracrustal Sequences. Geol. Assoc. Can. Spec. Pap., 28: 23–48.

Padgham, W.A., 1980. An Archean ignimbrite at Yellowknife and its relationship to the Kam Formation basalts. Precambrian Res. 12: 99–113.

Padgham, W.A., 1985. Observations and speculations on supracrustal successions in the Slave structural province. In: L.D. Ayres, P.C. Thurston, K.D. Card, and W. Weber (Eds.), Evolution of Archean Supracrustal Sequences. Geol. Assoc. Can. Spec. Pap., 28: 133–151.

Padgham, W.A. and Fyson, W.K., 1992. The Slave Province: a diatinct Archean craton. Can. J. Earth Sci., 29: 2072–2086.

Pettijohn, F.J., 1943. Archean sedimentation. Geol. Soc. Am. Bull., 54: 925–972.

Pettijohn, F.J., 1970. The Canadian Shield — a status report, 1970. In: A.J. Baer (Ed.), Symposium on Basins and Geosynclines of the Canadian Shield. Geol. Soc. Can. Pap., 70-40: 239–265.

Pettijohn, F.J., 1972. The Archean of the Canadian Shield: A resume. In: B.R. Doe and D.K. Smith (Eds.), Studies in Mineralogy and Precambrian Geology. Geol. Soc. Am. Mem., 135: 131–149.

Pidgeon, R.T. and Wilde, S.A., 1990. The distribution of 3.0 Ga and 2.7 Ga volcanic episodes in the Yilgarn Block of Western Australia. Precambrian Res., 48: 309–325.

Pharaoh, T.C. and Brewer, T.S., 1990. Spatial and temporal diversity of early Proterozoic volcanic sequences — comparisons between the Baltic and Laurentian shields. Precambrian Res., 47: 169–189.

Ransom, B.L., 1987. The paleoenvironmental, magmatic, and geologic history of the 3,500 Myr Kromberg Formation, west limb of the Onverwacht Anticline, Barberton Greenstone Belt, South Africa. Louisiana State Univ., MS thesis, 103 pp.

Reimer, T.O., 1975. Paleogeographic significance of the oldest known oolite pebbles in the Archaean Swaziland Supergroup (South Africa). Sed. Geol., 14: 123–133.

Reimer, T.O., 1983. Accretionary lapilli and other spheroidal rocks from the Swaziland Supergroup of the Barberton Mountain Land, South Africa. In: T. Peryt (Ed.), Coated Grains. Springer-Verlag, Berlin, 619–634.

Reimer, T.O., 1990. Archaean baryte deposits of southern Africa. J. Geol. Soc. India, 35: 131–150.

Sangster, D.F., 1972. Precambrian volcanogenic massive sulphide deposits in Canada: a review. Geol. Surv. Can. Pap. 72–22, 44 p.

Sawyer, E.W., 1986. The influence of source rock type, chemical weathering and sorting on the geochemistry of clastic sediments from the Quetico metasedimentary belt, Superior Province, Canada. Chem. Geol., 55: 77–95.

Schidlowski, M., 1976. Archaean atmosphere and evolution of the terrestrial oxygen budget. In: B.F. Windley (Ed.), The Early History of the Earth. John Wiley and Sons, London, pp. 525–535.

Sigurdsson, H., 1982a. Volcanogenic sediments in island arcs. In: L.D. Ayres (Ed.), Pyroclastic Volcanism and Deposits of Cenozoic Intermediate to Felsic Volcanic Islands with Implications for Precambrian Greenstone-Belt Volcanoes. Geol. Assoc. Can. Short Course Lecture Notes, 2: 221–293.

Sigurdsson, H., 1982b. Subaqueous volcanogenic sediments in ocean basins. In: L.D. Ayres (Ed.), Pyroclastic Volcanism and Deposits of Cenozoic Intermediate to Felsic Volcanic Islands with Implications for Precambrian Greenstone-Belt Volcanoes. Geol. Assoc. Can. Short Course Lecture Notes, 2: 294–342.

Sleep, N.H., 1992. Archean plate tectonics: what can be learned from continental geology? Can. J.

Earth Sci., 29: 2066–2071.

Stanistreet, I.G., de Wit, M.J., and Fripp, R.E.P., 1981. Do graded units of accretionary spheroids in the Barberton greenstone belt indicate Archaean deep water environment? Nature. 293: 280–284.

Tankard, A.J., Jackson, M.P.A., Eriksson, K.A., Hobday, D.K., Hunter, D.R., and Minter, W.E.L., 1982. Crustal Evolution of Southern Africa. Springer- Verlag, New York, 523 pp.

Tarney, J., Dalziel, I.W.D., and de Wit, M.J., 1976. Marginal basin 'Rocas Verdes' Complex from S. Chile: A model for Archaean greenstone belt formation. In: B.F. Windley (Ed.), The Early History of the Earth. John Wiley & Sons, London, pp. 131–146.

Tasse, N., Lajoie, J., and Dimroth, E., 1978. The anatomy and interpretation of an Archean volcaniclastic sequence, Noranda region, Quebec. Can. J. Earth Sci., 15: 874–888.

Taylor, P.N., Moorbath, S., Leube, A., and Hirdes, W., 1992. Early Proterozoic crustal evolution in the Birimian of Ghana: constraints from geochronology and isotope geochemistry. Precambrian Res., 56: 97–111.

Taylor, S.R. and McLennan, S.M., 1985. The Continental Crust: Its Composition and Evolution. Blackwell Scientific Publications, Oxford, 312 p.

Teal, P.R. and Walker, R.G., 1977. Stratigraphy and sedimentology of the Archean Manitou Group, northwestern Ontario. Geol. Surv. Can. Pap., 77-1A: 181– 184.

Thurston, P.C., 1980. Subaerial volcanism in the Archean Uchi-Confederation volcanic belt: Precambrian Res., 12: 79–98.

Thurston, P.C. and Chivers, K.M., 1990. Secular variation in greenstone sequence development emphasizing Superior Province, Canada. Precambrian Res., 46: 21–58.

Thurston, P.C., Ayres, L.D., Edwards, G.R., Gelinas, L., Ludden, J.N., and Verpaelst, P., 1985. Archean bimodal volcanism. In: L.D. Ayres, P.C. Thurston, K.D. Card, and W. Weber (Ed.), Evolution of Archean Supracrustal Sequences. Geol. Assoc. Can. Spec. Pap., 28: 7–21.

Turner, C.C. and Walker, R.G., 1973. Sedimentology, stratigraphy, and crustal evolution of the Archean greenstone belt near Sioux Lookout, Ontario. Can. J. Earth Sci., 10: 817–845.

van Biljon. W.J. and Legg, J.H., 1983, The Limpopo Belt. Spec. Publ. Geol. Soc. S. Africa, 8, 203 p.

Veizer, J., Compston, W., Hoefs, J., and Nielsen, H., 1982. Mantle buffering of the early oceans. Naturwissenschaften, 69: 173–180

Veizer, J., Hoefs, J., Ridler, R.H., Jensen, L.S., and Lowe, D.R., 1989a. Geochemistry of Precambrian carbonates: I. Archean hydrothermal systems. Geochim. Cosmochim. Acta, 53: 845–857

Veizer, J., Hoefs, J., Lowe, D.R., and Thurston, P.C., 1989b. Geochemistry of Precambrian carbonates: II. Archean greenstone belts and Archean sea water. Geochim. Cosmochim. Acta, 53: 859–871.

Viljoen, M.J. and Viljoen, R.P., 1969a. The geology and geochemistry of the lower ultramafic unit of the Onverwacht Group and a proposed new class of igneous rocks. Geol. Soc. S. Africa Spec. Publ., 2: 55–86.

Viljoen, R.P. and Viljoen, M.J., 1969b. The geological and geochemical significance of the upper formations of the Onverwacht Group. Geol. Soc. S. Africa Spec. Publ., 2: 113–152.

Walsh, M.M., 1989. Carbonaceous cherts of the Swaziland Supergroup, Barberton Mountain Land, South Africa. Louisiana State University, PhD thesis, Baton Rouge, Louisiana, 199 pp.

Walsh, M.M., 1992. Microfossils and possible microfossils from the Early Archean Onverwacht Group, Barberton Mountain Land, South Africa: Precambrian Res., 54: 271–293.

Walter, M.R., 1983. Archean stromatolites: Evidence of the earth's earliest benthos. In: J.W. Schopf (Ed.), Earth's Earliest Biosphere. Princeton Univ. Press, Princeton, NJ., 187–213.

Wilks, M.E. and Nisbet, E.G., 1985. Archaean stromatolites from the Steep Rock Group, northwestern Ontario, Canada. Can. J. Earth Sci., 22: 792–799.

Wilks, M.E. and Nisbet, E.G., 1988. Stratigraphy of the Steep Rock Lake Group, northwest Ontario: A major Archean unconformity and Archean stromatolites. Can. J. Earth Sci., 25: 370–391.

Windley, B.F., 1976. New tectonic models for the evolution of Archaean continents and oceans. In: B.F. Windley (Ed.), The Early History of the Earth. John Wiley & Sons, London, pp. 105–112.

Wood, J., 1980. Epiclastic sedimentation and stratigraphy in the North Spirit Lake and Rainy Lake areas: a comparison. Precambrian Res., 12: 227–255.

Worrell, G.F., 1985. Sedimentology and mineralogy of silicified evaporites in the basal Kromberg Formation, South Africa. Louisiana State University, MS thesis, Baton Rouge, Louisiana, 152 pp.

Wright, J.B., Hastings, D.A., Jones, W.B., and Williams, H.R., 1985. Geology and Mineral Resources of West Africa. Allen and Unwin, Boston, 188 p.

Young, G.M., 1978. Some aspects of the evolution of the Archean crust. Geosci. Canada, 5: 140–149.

Chapter 5

ARCHEAN SYNRIFT AND STABLE-SHELF SEDIMENTARY SUCCESSIONS

KENNETH A. ERIKSSON and CHRISTOPHER M. FEDO

INTRODUCTION

Archean sedimentary rocks, defined as exceeding 2.5 billion years in age, are preserved mainly in greenstone belts that are widely considered to record "permobile" tectonic settings (e.g., Windley, 1984). However, the Archean rock record also contains sedimentary rocks that were deposited following early cratonization. These are developed widely on and along the margins of extensive cratons and also are preserved locally in high-grade terranes as well as in greenstone belts. Such sedimentary successions are predominantly of synrift and stable-shelf affinity. The former include conglomerate, sandstone and mudstone associated with mafic and/or felsic volcanic rocks. In contrast, stable-shelf successions are dominated by quartz arenite, stromatolitic carbonate, mudstone and iron-formation. We exclude foreland-basin successions, such as the upper Witwatersrand Supergroup, because these indicate tectonic reactivation between periods of cratonization.

This paper documents examples of 3.1–2.9 Ga and 2.7–2.5 Ga synrift and stable-shelf sedimentary successions from the Kaapvaal and Zimbabwe Cratons and Limpopo Province of southern Africa, the Pilbara Craton of Western Australia, the Superior Province of Canada, and the Indian Craton (Fig. 1). Reference is also made to examples from the Yilgarn Craton of Western Australia, the Wyoming Province of the USA, and western Greenland. The compiled data are used to argue that the proportion of synrift and stable-shelf sedimentary rocks increased through the Archean Era in response to growth of continents from 4.0 to 2.5 Ga.

EVIDENCE FROM THE 3.2–2.9 GA RECORD

Synrift volcano-sedimentary and stable-shelf sedimentary successions that developed between 3.2 and 2.9 Ga are preserved on the Kaapvaal Craton as the Dominion Group, the Nsuze Group of the Pongola Supergroup, and the coeval lower Witwatersrand Supergroup and Mozaan Group of the Pongola Supergroup (Fig. 2). Additional examples of these associations are discussed from the Central Zone of the high-grade Limpopo Province of southern Africa, and from greenstone belts in Zimbabwe (Buhwa), India (Bababudan) and Canada (Steep Rock).

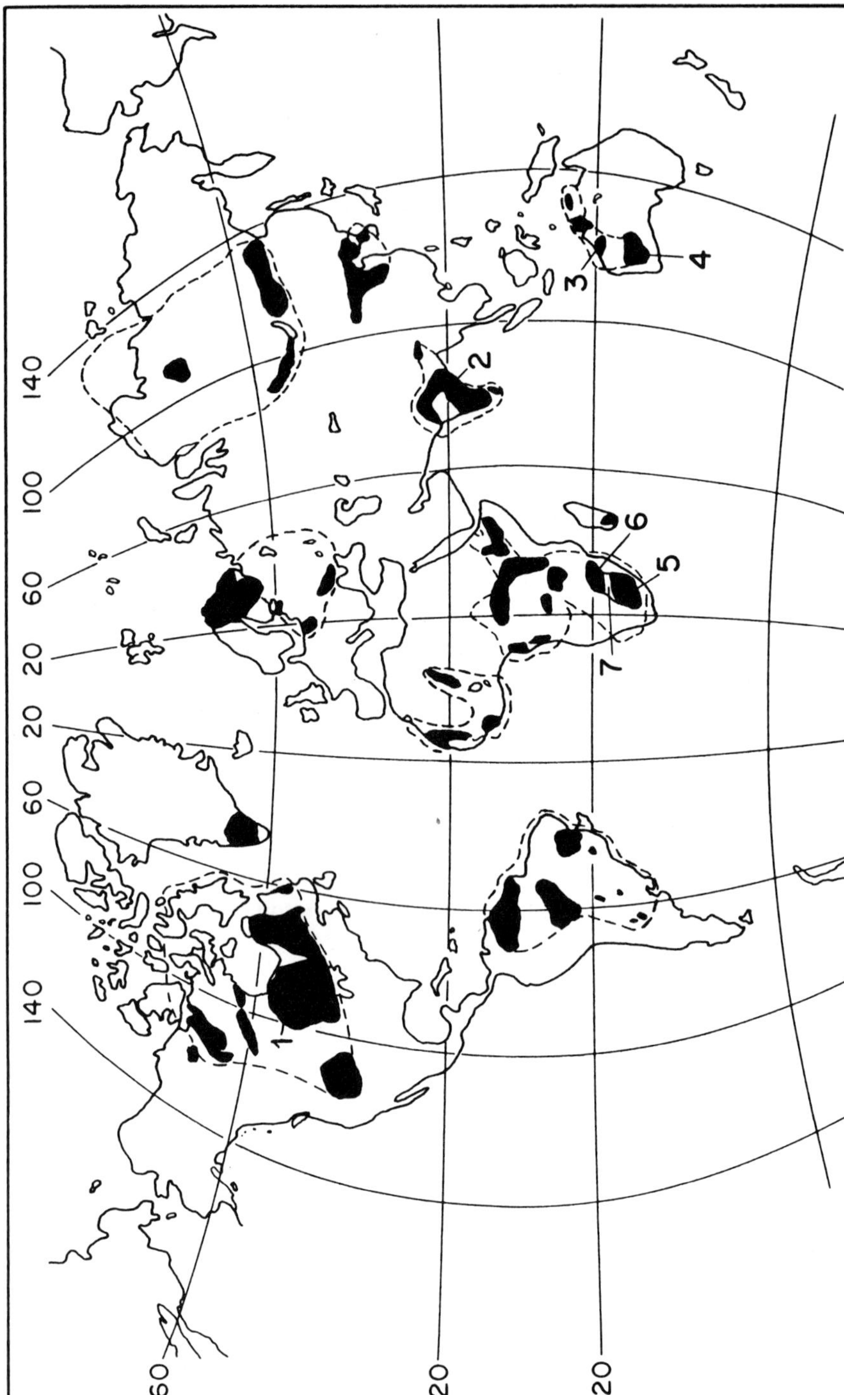

Fig. 1. Distribution of Archean crustal provinces (from Condie, 1981). Dashed lines demarcate regions that probably consist of Archean crust. Crustal provinces referred to in text are: 1. Superior; 2. Indian; 3. Pilbara; 4. Yilgarn; 5. Kaapvaal; 6. Zimbabwe; 7. Limpopo.

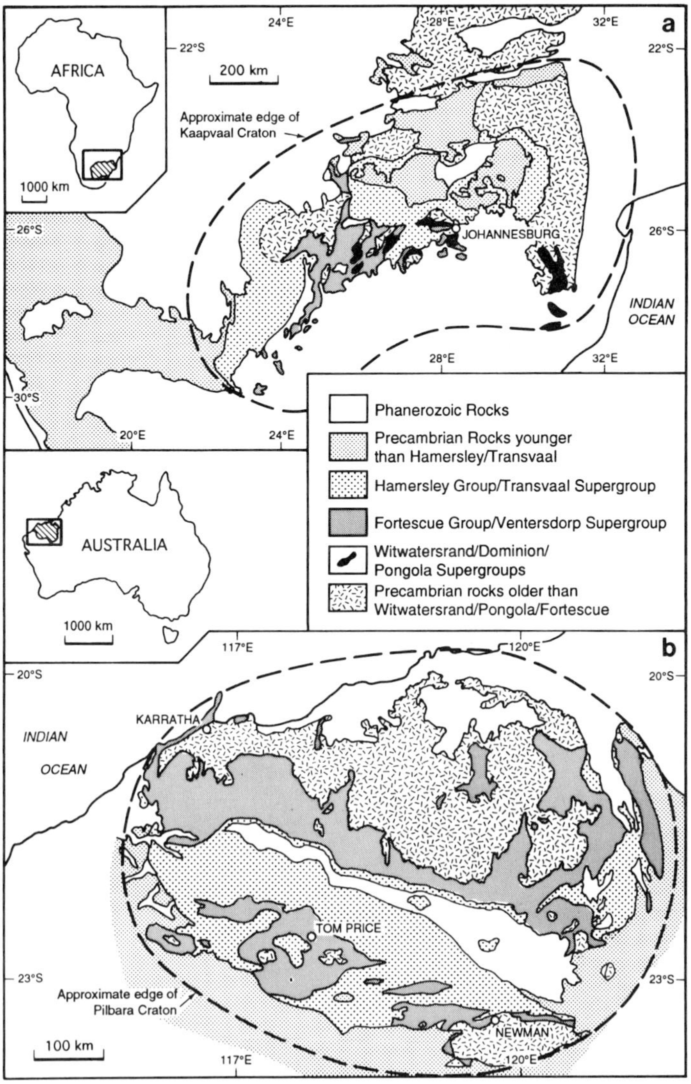

Fig. 2. Simplified geological maps of: (a) Kaapvaal Craton, and (b) Pilbara Craton (modified from Nelson et al., 1992).

Dominion and Nsuze Groups

Sedimentary rocks in the Dominion and Nsuze Groups are associated with mafic to felsic volcanic rocks (Fig. 3). Both groups overlie ca. 3100 Ma basement and display abrupt thickness variations, from 0.5 to 2 km for the Dominion Group and from 2 to 6 km for the Nsuze Group (Button, 1981; Bowen et al., 1986). Age

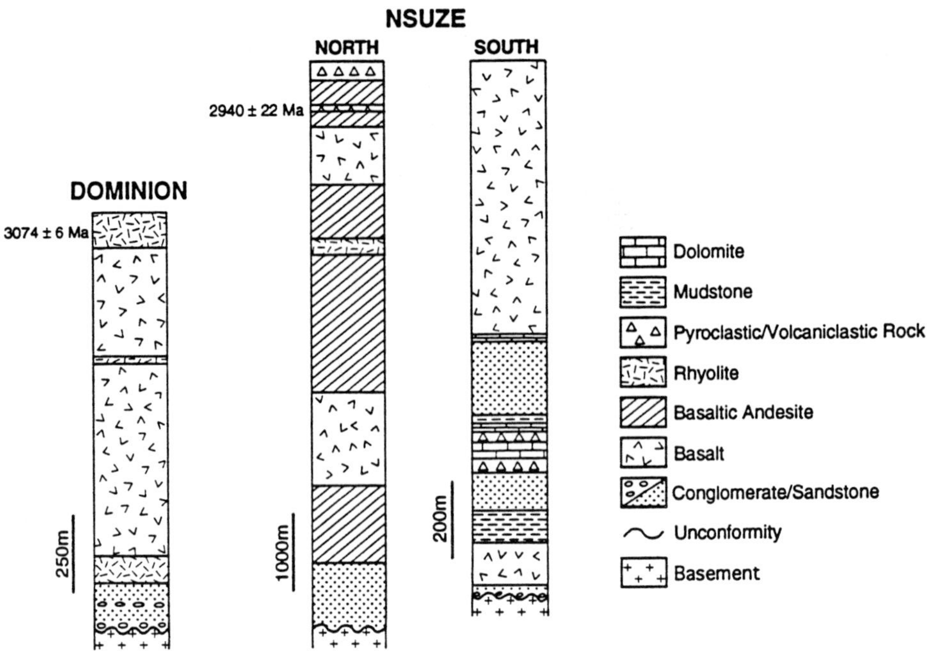

Fig. 3. Selected and generalized lithostratigraphic columns of the Dominion and Nsuze Groups (from Armstrong et al., 1982; Bowen et al., 1986; Beukes and Lowe, 1989).

constraints suggest that the Dominion and Nsuze Groups are not correlative. A quartz feldspar porphyry from near the top of the Dominion Group is dated at 3074 ± 6 Ma (Armstrong et al., 1991), whereas felsic volcanic rocks at the top of the Nsuze Group are dated at 2940 ± 22 Ma (Hegner et al., 1984) and 2984 ± 2.6 Ma (Hegner et al., 1993). Older ages from the Nsuze Group comparable to those obtained for the Dominion Group are considered to be due to crustal contamination (Hegner et al., 1984). Irrespective of whether the two groups correlate in part or not at all, both accumulated at around 3000 Ma.

Sedimentary rocks comprise the basal Dominion Group and also are present as thin interbeds within the overlying volcanic intervals (Fig. 3). The basal sedimentary unit overlies a paleosol developed on granite (Grandstaff et al., 1986) and consists predominantly of cross-bedded sandstone with minor conglomerate and mudstone. Deposition took place in braided rivers on a gentle, southwesterly directed paleoslope (Tankard et al., 1982). Pebble concentrations at the base of the sedimentary interval and a few tens of meters higher in the stratigraphy accumulated in paleovalleys and host placer uraninite and pyrite mineralization.

In the northern outcrop belt of the Nsuze Group, sedimentary rocks are confined to a basal, ca. 800 m-thick succession that overlies a 2–10 m paleosol developed on granite (Fig. 3; Armstrong et al., 1982). The predominant sedimentary rock types are quartz wackes and feldspathic arenites. Locally, conglomerates overlie irregular erosional surfaces and form the base of 1–2 m thick, upward-fining cycles. Trough cross beds are the most common sedimentary structure; these indicate that flow was mainly toward the southeast. The above criteria are interpreted by Watchorn and Armstrong (1981) to indicate a distal braided-river setting. Lateral continuity of the sedimentary succession is attributed to rapid lateral migration of channels in a tectonically active environment. Lenticular sedimentary units intercalated within the overlying volcanic succession also are interpreted as braided-alluvial deposits.

Within inliers south of the main outcrop belt of the Pongola Supergroup (Fig. 2), a ca. 600 m thick succession of sedimentary rocks is associated with amygdaloidal basalts (Fig. 3), which are pillowed in places. A basal conglomerate–feldspathic sandstone–quartz arenite interval is related to transgression of shelf sandstones across a distal braidplain. Mudstones above the lower volcanic unit contain upward-fining, channel-fill deposits of possible tidal origin (Matthews, 1967). The overlying sandstone–dolomite interval (Fig. 3) contains quartz arenite, mudstone, stromatolitic carbonate and reworked tuffs. Upward-fining, sandstone–mudstone cycles are interpreted as progradational, tidal-flat deposits (Von Brunn and Mason, 1977). Stromatolitic carbonates also are interpreted as tidal facies (Beukes and Lowe, 1989). Conical stromatolites associated with cross-bedded carbonate sandstones accreted in the deeper parts, whereas composite columnar stromatolites up to 1.6 m high formed along margins of tidal channels. Flat algal laminites and small domical stromatolites developed on adjacent tidal flats. Beukes and Lowe (1989) considered that the stromatolites were produced by filamentous, oxygen-producing cyanobacteria.

Presence of bimodal volcanic rocks and immature, basement-derived conglomerates and sandstones, and abrupt variations in thickness support a synrift setting for the Dominion and Nsuze groups (Bickle and Eriksson, 1982; Burke et al., 1985). The Dominion Group and northern outcrops of the Nsuze Group are entirely continental. In contrast, southern outcrops of the Nsuze Group record a probable tidal influence suggesting that the rift basin opened to an ocean to the south.

Lower Witwatersrand Supergroup and Mozaan Group

The lower Witwatersrand Supergroup unconformably overlies paleosols developed on 3.1 Ga basement (Button and Tyler, 1981) or Dominion Group, whereas the Mozaan Group is unconformable on the Nsuze Group. Age constraints indicate that these two stratigraphic successions are correlative. Detrital zircons from the lower Witwatersrand Supergroup have minimum ages of 2970 Ma (U-Pb, Barton et al., 1989; Robb et al., 1990) whereas a volcanic interval from near the top of this succession is dated at 2914 ± 8 Ma (U-Pb zircon, Armstrong et al., 1991). The

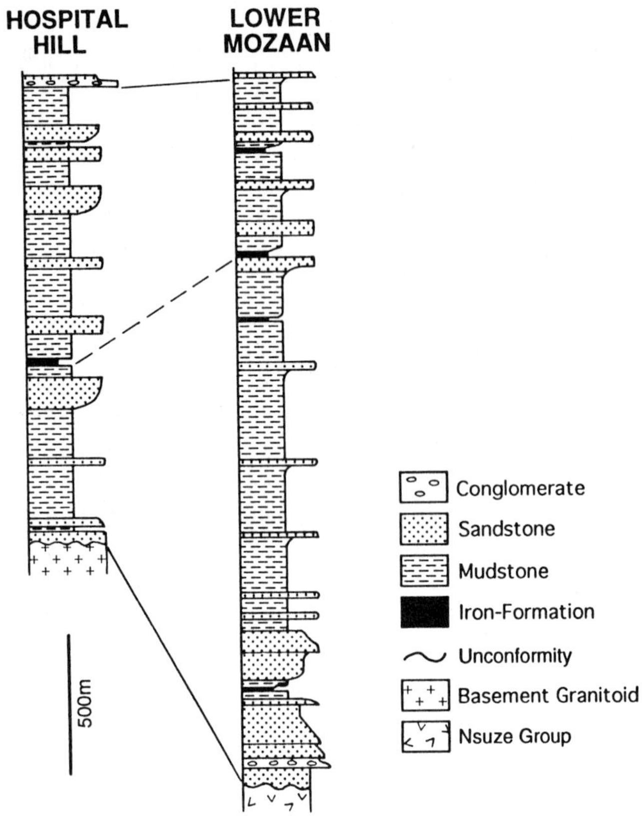

Fig. 4. Generalized lithostratigraphic columns of the Hospital Hill Subgroup of the Witwatersrand Supergroup and lower Mozaan Group of the Pongola Supergroup (from Beukes and Cairncross, 1991).

age of the Mozaan Group is constrained by the 2940 and 2984 Ma ages on the uppermost layers in the Nsuze Group and the ca. 2870 Ma age of the intrusive Usushwana Complex (Sm-Nd, Hegner et al., 1984; U-Pb, Hegner et al., 1993). Thicknesses of these two stratigraphic successions also are comparable. The lower Witwatersrand Supergroup is 3500–4500 m thick whereas the Mozaan Group is 5000 m thick. Figure 4 shows lithologic sections through the lower parts of these two successions (Tankard et al., 1982; Beukes and Cairncross, 1991).

Both the lower Witwatersrand Supergroup and Mozaan Group are dominated by quartz arenite and ferruginous mudstone with subordinate iron-formation conglomerate and quartz wacke (Beukes and Cairncross, 1991). Conglomerates and wackes are of braided-alluvial origin, and locally contain placer mineralization (Saager et al., 1986). Cross-bed measurements indicate a southeasterly to northeasterly paleoslope for the Mozaan Group and a southerly paleoslope for the

Fig. 5. Herringbone cross bedding from the Brixton Formation of the lowermost Witwatersrand Supergroup. Lens cap for scale.

lower Witwatersrand Supergroup. Iron-formation, mudstone and quartz arenite are arranged in coarsening-upward intervals that record progradation of proximal marine shelf sands and distal shelf muds over starved distal basin environments represented by iron-formation. The latter are interpreted as Archean analogues of modern-day pelagic sediments (Eriksson, 1983a). Abundant hummocky cross stratification within the proximal to distal facies transition indicates that the shelf was influenced mainly by storms (Beukes and Cairncross, 1991). Tidal overprint is recorded locally by herringbone cross bedding (Fig. 5), and modified ripples including flat-topped double-crested and ladderback varieties (Von Brunn and Hobday, 1976; Eriksson et al., 1981). Water-level marks and dewatering pits on some of the ripples (Fig. 6) indicate emergence of tidal flats.

Compositional and textural maturity of arenites in the lower Witwatersrand Supergroup and Mozaan Group indicates that sedimentation took place in a slowly subsiding basin on the Kaapvaal Craton. Close temporal and spatial association with the older synrift Dominion and Nsuze Groups suggests that development and evolution of this stable-shelf basin can be attributed to thermal subsidence associated with cooling of the previously thinned lithosphere (Bickle and Eriksson, 1982; Stanistreet and McCarthy, 1991), rather than in a foreland basin related to crustal shortening as proposed by Burke et al. (1986).

Fig. 6. Megaripples with small ripples superimposed in troughs from the Orange Grove Quartzite of the lowermost Witwatersrand Supergroup. Also note water-level marks and dewatering pits on sides of megaripples. Similar ripples are developed on modern tidal flats. Scale is 20 cm long.

Beitbridge Complex: Central Zone, Limpopo Province

Within the Central Zone of the Limpopo Province (Fig. 7), the Sand River Gneisses, with an age of 3.3–3.2 Ga (U-Pb zircon, Retief et al., 1990), are recognized as basement to younger cover rocks represented by the Beitbridge Complex (Fig. 8; Fripp, 1983). The basement-cover relationship is based on field evidence which indicates that the oldest recognized structural fabrics are confined to the Sand River Gneisses (Fripp, 1983). Four lithological associations comprise the Beitbridge Complex, namely, quartzite–pyroxenetic amphibolite (Fig. 9); biotite–garnet–cordierite–sillimanite gneiss; magnetite quartzite, calc-silicate gneiss and marble, and grey gneiss. Detrital zircons (Fig. 10) in the quartzites range in age from 3.8–3.2 Ga (Brandl and Barton, 1991) providing a *maximum* depositional age of 3.2 Ga for the Beitbridge Complex. Intrusive into the cover rocks are the Messina Layered Intrusion and Bulai granitoid (Fig. 8). The latter is dated at ca. 2650 Ma (Pb-Pb, Barton et al., 1979) and provides a minimum age for the Beitbridge Complex.

In the absence of primary structures such as cross stratification, and because of great structural complexity, a stratigraphic succession cannot be established for

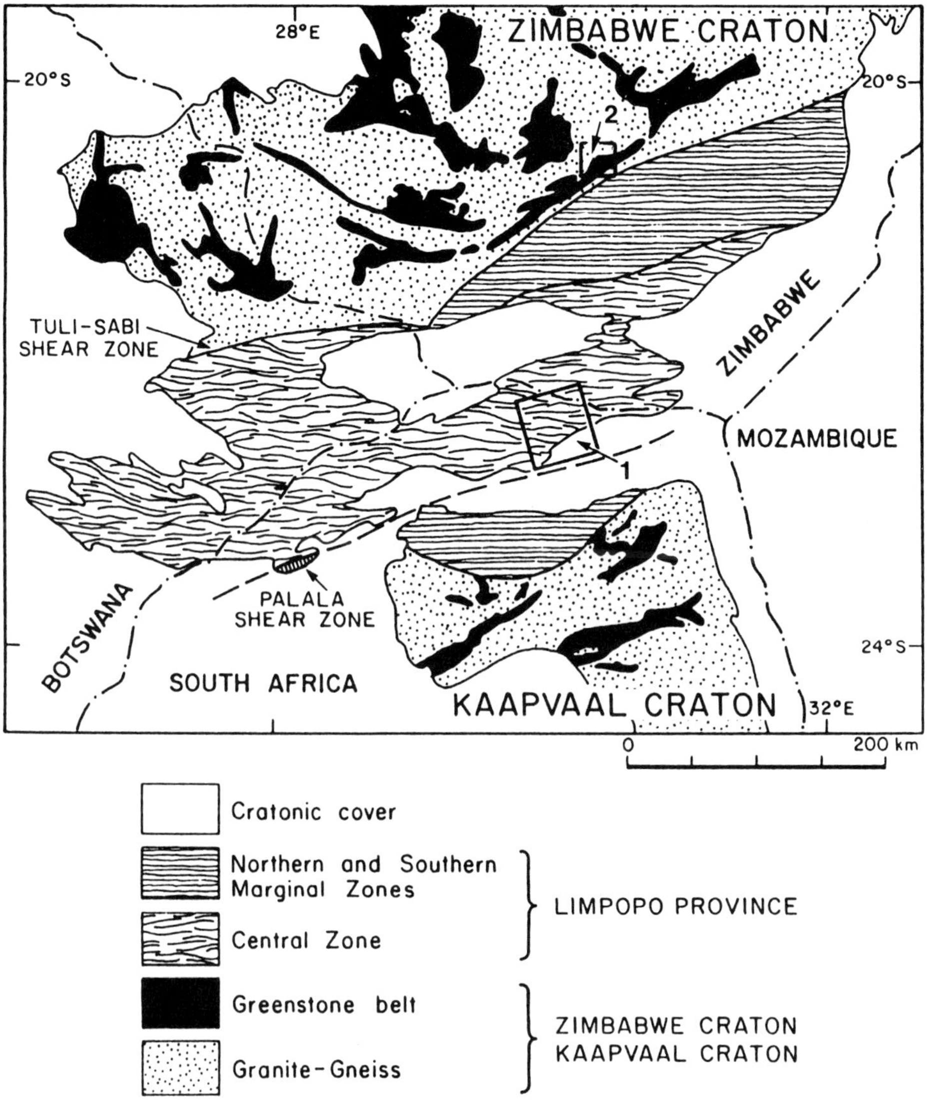

Fig. 7. Generalized map of the Limpopo Province and adjoining Zimbabwe and Kaapvaal Cratons (from Tankard et al., 1982). Inset 1 shows location of Sand River Gneiss and Beitbridge Complex discussed in text. Inset 2 shows location of that portion of Buhwa Greenstone Belt discussed in text.

the Beitbridge Complex. However, it is possible to interpret protoliths for the cover rocks (Eriksson et al., 1988). Quartzites are presently up to 100 m thick in single occurrences and consist of up to 99.5% quartz. Metapelite partings are rare. The presence of detrital zircons with rounded cores, as well as other heavy

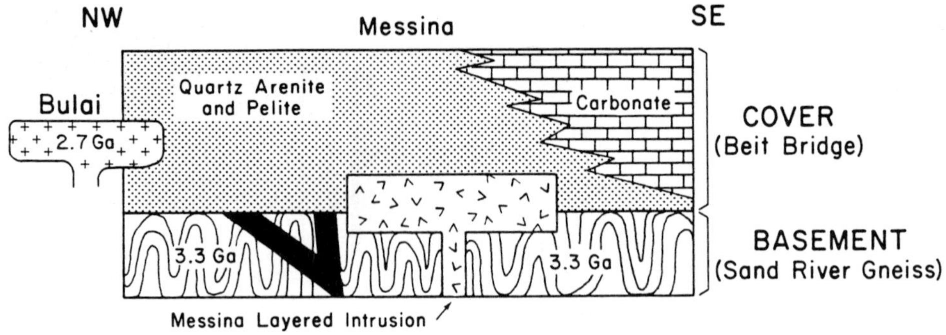

Fig. 8. Idealized stratigraphic cross section illustrating temporal relationships of basement, cover and intrusive events in the Central Zone of the Limpopo Province.

Fig. 9. Quartzite–amphibolite lithological association from the Beitbridge Complex, Limpopo Province. Protoliths are interpreted as detrital quartz arenite and mafic igneous (probably intrusive), respectively.

minerals such as rutile and tourmaline favors a detrital origin for the quartzites. Thus, they are interpreted as original quartz arenites rather than cherts. Associated pyroxenitic amphibolites are geochemically indistinguishable from basaltic

Fig. 10. Zircons with rounded cores and angular, metamorphic overgrowths from the Beitbridge Complex, Limpopo Province. Cores range in age from 3.8–3.2 Ga whereas overgrowths are ca. 1.9 Ga old (Brandl and Barton, 1991). Scale bar is 0.2 mm long.

tholeiites of intrusive or extrusive origin (Fripp, 1983). The greater abundance of amphibolite in proximity to the Messina Layered Intrusion than elsewhere in the central zone led Eriksson et al. (1988) to argue for an intrusive origin for the protolith.

The protoliths of two of the other lithological associations comprising Beitbridge Complex are more easily defined. Marble in the calc-silicate gneiss–marble association attains thicknesses, for single outcrops, of at least 100 m and, is developed in a belt parallel and to the south of quartzite occurrences. Marbles displaying abundant compositional layering are visualized as original sedimentary carbonate rocks. Calc-silicate gneiss is considered to have originally been calcareous pelite. Associated, but subordinate magnetite quartzite may represent either metamorphosed iron-formation or iron-rich quartzite. Most workers agree that biotite–garnet–cordierite–sillimanite gneisses represent metapelites and the major element geochemistry of those in the Beitbridge Complex closely resembles average "geosynclinal" shales (Brandl, 1983; Fripp, 1983). Taylor et al. (1986) and Boryta and Condie (1990) argue that the mineralogy, and major, trace and rare earth element (REE) geochemistry supports a pelitic origin for these paragneisses. The protolith of the fourth lithologic association, namely the gray gneisses is highly problematical. On the basis of their geochemistry, Fripp (1983) considers that these gneisses may have been felsic to intermediate volcanic and/or terrigenous sedimentary rocks, whereas Brandl (1983) favors an arkosic protolith.

On the basis of the above discussion, it is possible that the Beitbridge Complex contained few, if any, volcanic rocks and an estimate of lithologic proportions of primary surficial deposits, assuming a sedimentary origin for the gray gneisses, is quartz arenite — 17%; limestone and calcareous mudstone — 17%; mudstone — 50%; arkose/feldspathic wacke — 16%.

In many respects, the Beitbridge Complex is similar lithologically to the widely-developed, cratonic-shelf, quartz arenite–carbonate association present in younger, unmetamorphosed sequences including the late-Archean, Chuniespoort–Ghaap Group of the Transvaal Supergroup (Fig. 2) that is discussed later. The central Appalachian, Cambro-Ordovician passive-margin sequence appears to be a particularly appropriate analogue; it is notably rich in mudstone because of its near shelf-edge position in contrast to time-equivalent inboard platform sequences to the west which consist primarily of quartz arenite and carbonate (Rankin et al., 1989; Simpson and Eriksson, 1990). However, the data base for the Beitbridge Complex is inadequate to draw a direct analogy with a passive margin and the more general term *stable shelf* thus is preferred.

Buhwa Greenstone Belt, Zimbabwe

The Buhwa greenstone belt in Zimbabwe (Fig. 7) consists of a sedimentary-rock dominated succession that is intruded by a tonalitic batholith. An unconformity with older gneissic basement, termed the Tokwe segment (Wilson, 1990), is not exposed in the area, but is inferred from relative age relationships and what absolute dates are available. Dodson et al. (1988) dated detrital zircons, extracted from the thick quartzite unit; these yielded concordant U-Pb ages ranging from 3.1 Ga to 3.8 Ga. The tonalite that intrudes the greenstone belt has not yet been dated isotopically, but is correlated with the ca. 2.9 Ga Chingezi and Mashaba tonalites (Hawkesworth et al., 1979; Taylor et al., 1991) approximately 30 km north, near the Belingwe Greenstone Belt. Field relationships and available dates thus suggest a depositional age for the succession of less than than 3.1 Ga, but greater than 2.9 Ga.

The succession at Buhwa is divided into eastern and western associations (Fedo and Eriksson, 1993, 1994). The eastern association consists of intercalations of recrystallized chert, greenstone, and less abundant iron-formation; the sedimentary units probably represent deposition in a pelagic setting (cf. Eriksson, 1983a). The western association consists of a thick siliciclastic-dominated succession that, in general, fines up section (Fig. 11). The base of the succession consists of a ca. 1 km-thick orthoquartzite that is divisible into three units: a lower and upper quartzite separated by a thick section of interbedded quartzite and phyllite. Quartzites consist of clean quartz sand and rare flakes of fuchsite and contain a variety of wave-produced sedimentary structures, including trough cross-bedding and

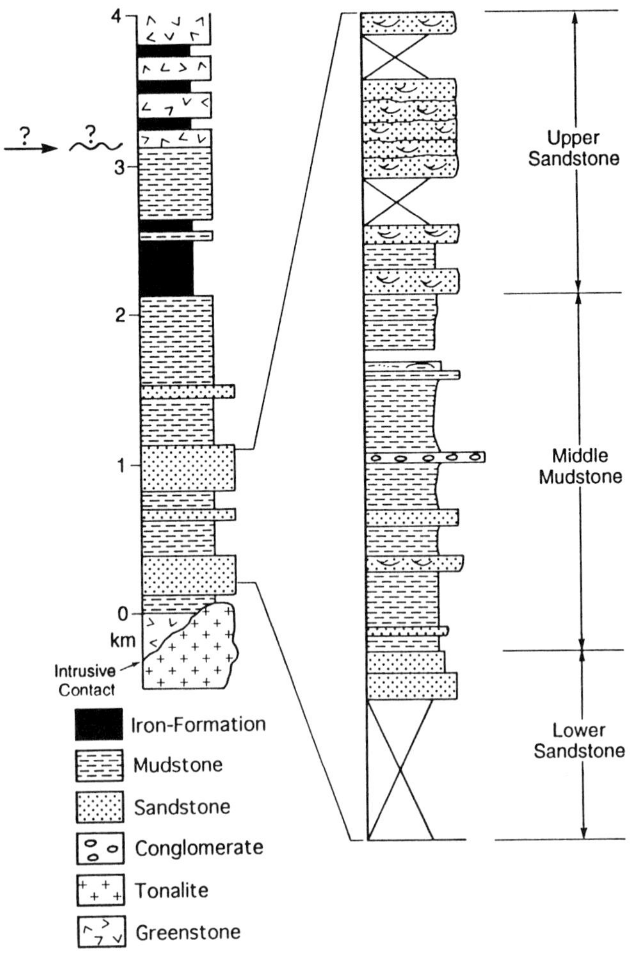

Fig. 11. Lithostratigraphic column of cover rocks in the Buhwa Greenstone Belt, Zimbabwe. Nature of contact between main sedimentary succession and overlying greenstone-iron formation interval is uncertain. It may be an unconformity or a fault.

symmetrical ripple marks, which suggest deposition on a shallow shelf. A ca. 1 km thick section that consists mostly of phyllite overlies the quartzite. The lower 500–600 m are very poorly exposed. The upper part consists of flat- and wavy-laminated alternations of mudstone (now phyllite) and siltstone or very fine-grained sandstone with rare conglomerate. A thick succession of iron-formation, which hosts the iron-ore deposits at Buhwa, conformably overlies the phyllites and consists of millimeter- to centimeter-scale alternations of recrystallized white or red chert and hematite. An abundance of fine-grained rocks overlying the quartzite suggests that low-energy conditions prevailed as the quartzite shelf was submerged.

Vertical and lateral facies trends permit interpretation of the Buhwa cover succession as a transition from stable-shelf to deeper-water sedimentation (Fedo and Eriksson, 1994). The upward-fining siliciclastic succession capped by orthochemical sedimentary rocks in the western association indicates a systematic deepening and is reminiscent of a Phanerozoic transgressive shelf deposit. The eastern association is more difficult to assess in terms of stratigraphic relationships because of the lack of any facing criteria, lithologic homogeneity, and structural complexity. Nevertheless, the consistent association of metachert, iron-formation, and mafic-to-ultramafic volcanic rocks points to below wave-base, more basinward, conditions for the succession.

Bababudan Group, Dharwar Craton, India

The Bababudan Group in southern India consists predominantly of siliciclastic sedimentary and mafic volcanic rocks unconformably overlying basement gneisses. A paleosol is developed at the contact. Basement Peninsular Gneiss is dated at 3358 ± 66 Ma (Rb-Sr whole rock, Beckinsale et al., 1980) whereas mafic volcanics are 3020 ± 230 Ma old (Sm-Nd, Drury et al., 1983). Younger granites are dated at 3175 ± 45 Ma and 3190 ± 100 Ma (Pb-Pb, Taylor et al., 1984). Thus, the Bababudan Group is probably 3200–3000 Ma in age (Srinivasan and Ojakangas, 1986).

Sedimentary rocks are dominated by cross-bedded sandstone and quartz-pebble conglomerate containing placer uraninite, gold and pyrite. A braided-alluvial origin is inferred for these facies (Srinivasan and Ojakangas, 1986). Paleoflow was to the south, southeast and east. Associated mafic volcanic rocks also are of subaerial origin. Subordinate pelite and iron-formation (Fig. 12) are interpreted as shallow-marine deposits. The relatively mature clastic sedimentary rocks were derived by intense weathering of granite-gneiss and deposited on a stable platform. Associated mafic volcanic rocks are considered to reflect episodic rifting of the stable platform (Srinivasan and Ojakangas, 1986).

Steep Rock Group, Superior Province

Strata of the Steep Rock Group rest unconformably on the ca. 3.0 Ga Marmion complex (Joliffe, 1955; Wilks and Nisbet, 1988). Depositional age of the Steep Rock Group is not well constrained but is believed to be ca. 2.9 Ga. Northeast of the Steep Rock Group is the Lumby Lake greenstone succession, which has yielded dates around 3000 Ma (U-Pb zircon, Davis and Jackson, 1988). Post-tectonic granites dated at 2870 Ma intrude probable correlatives of the Steep Rock Group (U-Pb zircon, reported in Thurston and Chivers, 1990).

The Steep Rock Group overlies a paleosol developed on the Marmion complex. The basal Wagita Formation is a variably thick (0–150 m) mixture of siliciclastic debris that includes conglomerate, pebbly arenite, and pelite deposited in an

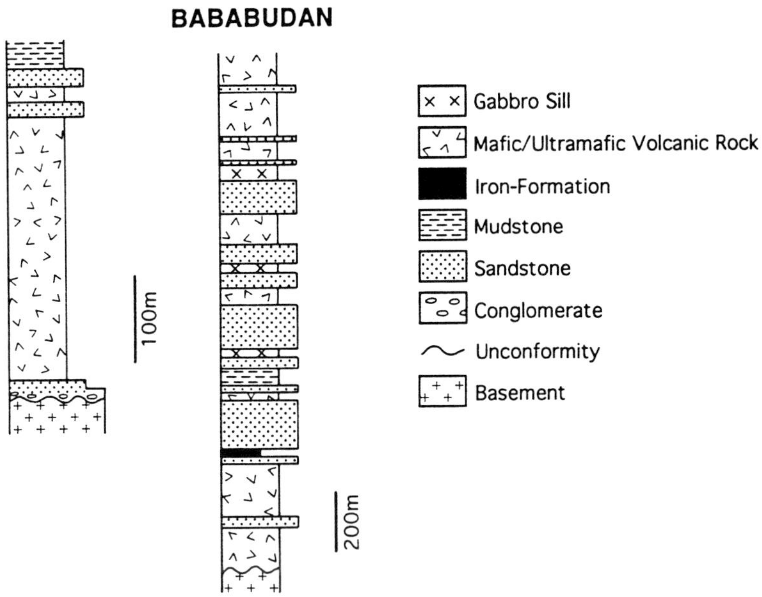

Fig. 12. Lithostratigraphic columns of the Bababudan Group in the Dharwar Craton of India (after Srinivasan and Ojakangas, 1986).

alluvial setting (Fig. 13; Wilks and Nisbet, 1988). Clast analysis reveals that all debris in the Wagita Formation was derived from the underlying Marmion complex (Wilks and Nisbet, 1988). Overlying the Wagita Formation is the 0–500 m-thick Mosher Carbonate, which contains a significant occurrence of Archean stromatolites. While the carbonate unit is extensively faulted, the observed thickness variation may be in part due to a major karstic unconformity, beneath which the entire carbonate section has been locally removed (Grotzinger, 1989). Wilks and Nisbet (1985) subdivided the Mosher Carbonate based on stromatolite morphology. Laminar stromatolites lie at the base and are overlain by domal to columnar stromatolites. Giant columnar forms with up to 1 m of synoptic relief follow and the succession is capped by domal stromatolites. A shallow subtidal to intertidal setting similar to Hamelin Pool, Shark Bay, Australia, was suggested by Nisbet and Wilks (1989) and Wilks and Nisbet (1985, 1988) as the depositional environment for the Mosher Carbonate. An ore zone consisting mostly of goethite and hematite, which probably represents weathered iron-formation overlies the carbonate (Fig. 13) and is in turn overlain by a thick, dominantly volcanic pile that includes pyroclastic rock at the base tectonically overlain by mafic volcanics and komatiites (Grotzinger, 1989).

The progression from alluvial sediments of variable thickness into a thick carbonate succession suggests that lower Steep Rock Group strata developed in response to rifting followed by transgression and development of a stable shelf (Wilks and Nisbet, 1985; Grotzinger, 1989). Whether the carbonates represent a

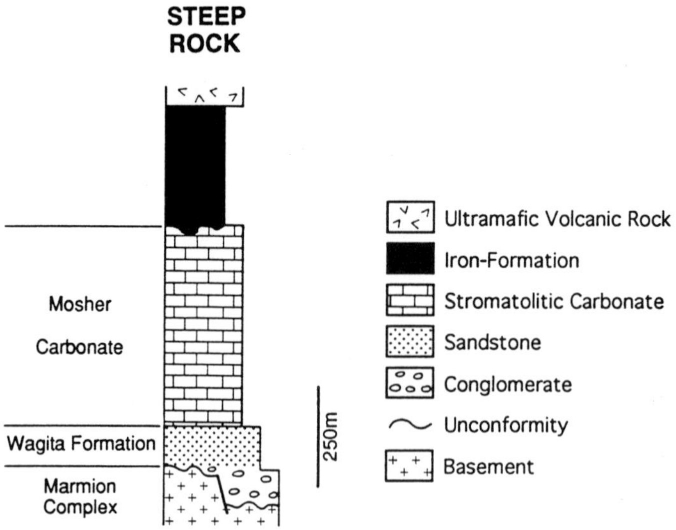

Fig. 13. Lithostratigraphy of the Steep Rock Group in the Superior Province Canada (adapted from Wilks and Nisbet, 1985).

shallowing- or deepening-upward succession is still in question (discussed by Grotzinger, 1989). Nevertheless, a major sea-level fall is recorded by the karstic unconformity at the top of the carbonate succession and is thought to have been caused by either rift processes (Wilks, 1986) or development of a peripheral bulge during subsequent crustal shortening (Grotzinger, 1989).

Other examples

Other examples of cratonic sedimentary associations are preserved locally in many parts of the world. Conglomerate, arenite, pelite and iron-formation in the high-grade, Narryer and Jack Hills areas of the western Yilgarn Craton have maximum ages of 3.1 Ga (Maas and McCulloch, 1991). Detrital zircons include a population with ages between 4.10 and 4.27 Ga (Froude et al., 1983; Compston and Pidgeon, 1986). The siliciclastic sediments in the Narryer region are interpreted as braided-alluvial deposits (Eriksson et al., 1988). Other examples include 3.3–2.8 Ga quartz arenites (now quartzites) containing detrital zircons up to 3.96 Ga old, and iron-formations in the Wyoming Province (Mueller et al., 1992), the greater than 3.05 Ga Malene supracrustals including quartzite, metapelite and marble in Greenland (Chadwick, 1990) and 3.2–3.1 Ga conglomerate, arenite, iron-formation containing stromatolites, and pelite of the Iron Ore Group in the Singhburn Craton of eastern India (Naqvi and Rogers, 1987). In addition to the Steep Rock succession, numerous other local, ca. 2.9 Ga occurrences of conglomerate, quartz arenite, pelite and iron-formation ± stromatolitic carbonate are known from the Superior Province (Thurston and Chivers, 1990).

EVIDENCE FROM THE 2.7–2.5 GA RECORD

Synrift volcano-sedimentary and stable-shelf sedimentary successions that developed between 2.7 and 2.5 Ga are represented by the greenstone-associated Manjeri Formation in Zimbabwe (Fig. 14) and by the Ventersdorp Supergroup and the Chuniespoort–Ghaap Groups of the Transvaal Supergroup on the Kaapvaal Craton, and by the Fortescue and Hamersley Groups on the Pilbara Craton (Fig. 1).

Manjeri Formation, Ngezi Group

The Manjeri Formation is the basal unit of the Ngezi Group in the Belingwe Greenstone Belt (Fig. 14). In the past, the Ngezi Group has been referred to as the "upper greenstones" or the "Upper Bulawayan" (e.g., Stagman, 1978). Manjeri Formation strata unconformably overlie ca. 2.9 Ga greenstones in the western and southeastern part of the Belingwe belt, and paleosols developed on 3.5 Ga granitic

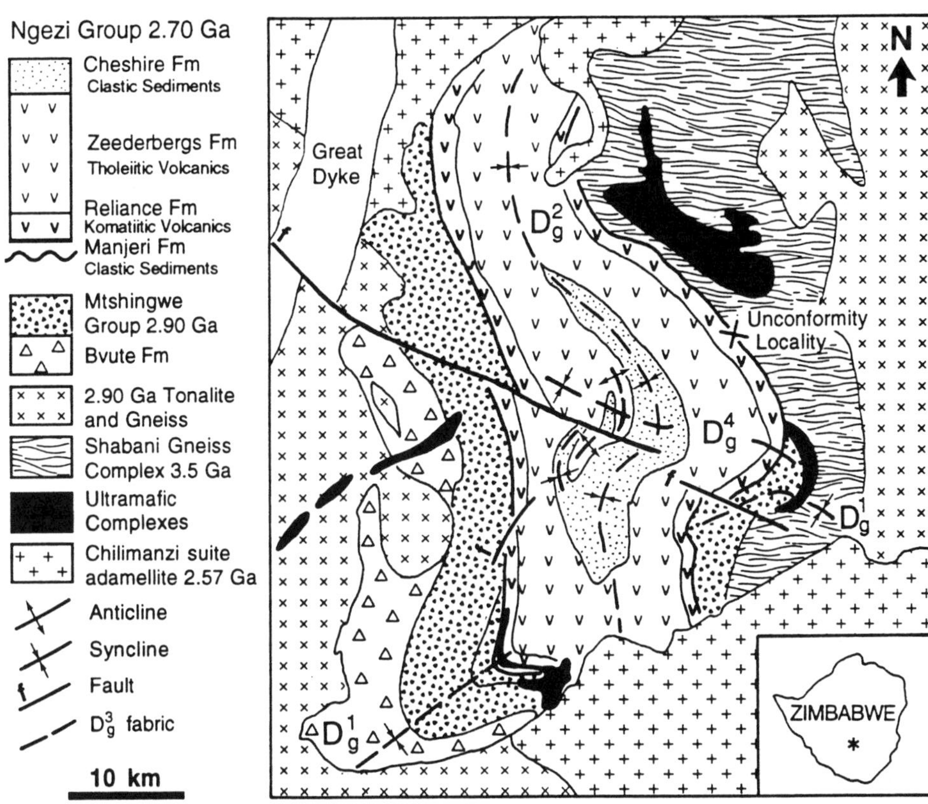

Fig. 14. Geological map of the Belingwe Greenstone Belt, Zimbabwe. Note the position of the Manjeri Formation at the base of the Ngezi Group.

Fig. 15. Nonconformity between 3.5 Ga gneisses and the ca. 2.7 Ga Manjeri Formation in the Belingwe Greenstone Belt, Zimbabwe. Backpack on contact for scale.

basement in the east (Fig. 15), where the best exposures of the unconformity are present (Bickle et al., 1975). Mafic and ultramafic lavas of the overlying Reliance Formation yield parallel Pb-Pb isochrons with ages of 2692 ± 9 Ma and 2675 ± 173 Ma (Chauvel et al., 1993). Most workers generally accept 2.7 Ga as the age of the Manjeri Formation. The conformable nature of the Manjeri and Reliance Formations has been recently questioned by Kusky and Kidd (1992) who envisage a tectonic contact; however, field and laboratory analyses by Blenkinsop et al. (1993) have reconfirmed the conformable nature of the contact.

Thickness of the Manjeri Formation varies from 0 to 250 m, although typically it ranges from 50 to 100 m (Martin et al., 1993). The formation consists predomi-

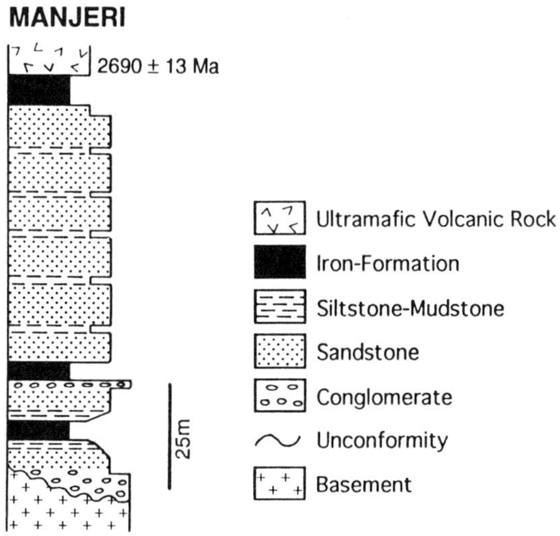

Fig. 16. Lithostratigraphic column of the Manjeri Formation at the unconformity locality in the Belingwe Greenstone Belt, Zimbabwe (see Fig. 10 for location of column).

nantly of siliciclastic detritus and sulfide- and oxide-facies iron-formation. Locally, stromatolite-bearing limestone is developed. At the type locality of the Manjeri Formation (Fig. 14), basal strata overlie a paleosol developed on granite. Facies consist of intercalated quartz-pebble conglomerate and trough cross-stratified sandstones of probable alluvial origin, fine- to medium-grained sandstone that displays ripple cross-lamination, plane bedding, and hummocky cross-stratification all of shallow-shelf origin, less common mudstone, and oxide-facies iron-formation (Fig. 16). Paleocurrent information is sparse, although unimodal and bimodal bipolar patterns (Nisbet et al., 1993) are consistent with alluvial environments being reworked in a shallow, tidally influenced setting. Storm overprint is recorded by the hummocky cross stratification. This typically shallow-water assemblage is overlain by multiple, stacked cycles of granular and coarse sandstone grading to fine sandstone, and capped by mudstone (Fig. 16). These meter-scale cycles are interpreted collectively as a thick turbidite succession related to drowning of the underlying shelf sediments (Bickle et al., 1975; Grotzinger, 1989).

The collection of locally mature sandstones of open-marine shelf affinity laterally associated with stromatolitic carbonates at the base of the Manjeri Formation is consistent with deposition on a transgressive, slowly subsiding, stable shelf or platform as was suggested by Grotzinger (1989). Abrupt upward-deepening is indicated by the turbidite, or "drowning," succession. The cause of increased subsidence is conjectural, but is almost certainly tectonically driven. Kusky and Kidd (1992) suggested that thrust loading as a result of obduction of an oceanic plateau (Reliance Formation and overlying Zeederburgs Formation) may have been the driving mechanism. However, Blenkinsop et al. (1993) have demonstrated the autochthoneity of the volcanic succession and have alternatively suggested that the subsidence could be extension-related.

Ventersdorp Supergroup and Fortescue Group

In both the Ventersdorp and Fortescue successions, sedimentary rocks are associated with basalts and subordinate felsic volcanic rocks. Thickness variations are extreme, from less than 500 to greater than 5000 m (Winter, 1976; Blake and Groves, 1987). Age constraints suggest that these stratigraphic intervals developed at roughly the same time; the Ventersdorp Supergroup accumulated around 2700 Ma (U-Pb zircon, Armstrong et al., 1991), whereas the Fortescue Group developed between 2756 and 2684 Ma (U-Pb zircon, Arndt et al., 1991).

Both the Ventersdorp Supergroup and Fortescue Group are comprised of three regional unconformity bounded stratigraphic subdivisions (Fig. 17). Lowermost subdivisions consist predominantly of subaerial flood basalts of probable pre-rift origin; conglomerates (Fig. 18) and sandstones of braided alluvial origin are developed locally at the base of the Fortescue Group. Basal volcanic intervals are unconformably overlain by subdivisions dominated by sedimentary rocks that developed in synrift basins. Unconformably overlying, uppermost subdivisions consist of mixed sedimentary and mafic volcanic rocks that record regional subsidence (Winter, 1976; Blake and Groves, 1987).

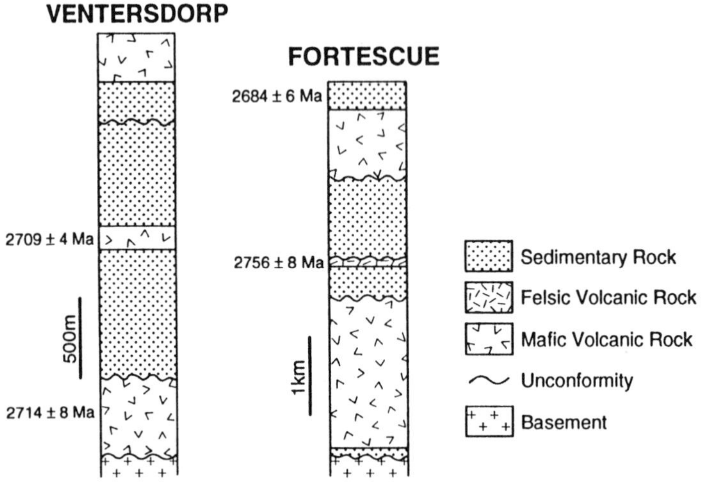

Fig. 17. Selected and generalized lithostratigraphic columns showing the three-fold unconformity bounded subdivisions of the Ventersdorp Supergroup on the Kaapvaal Craton and Fortescue Group on the Pilbara Craton (adapted from Buck, 1980; Blake and Groves, 1987). Both successions display considerable thickness variation not shown here.

Fig. 18. Conglomerate containing granitoid and greenstone clasts from the base of the Fortescue Group.

Extensive alluvial-fan deposits are present in the middle subdivision of both successions. In the Ventersdorp Supergroup, these consist of wedge-shaped coarse scree and debris-flow conglomerates reaching thicknesses approaching 2000 m thick (Winter, 1976, 1990; Buck, 1980). Clasts consist of lower Ventersdorp, Witwatersrand and basement rocks derived from adjacent highlands. Alluvial-fan facies interfinger with lacustrine mudstone, marl, stromatolitic chert and limestone, and bedded chert interpreted as an alkaline-lake deposit (Buck, 1980; Karpeta, 1993). Middle Fortescue Group debris-flow deposits contain chert clasts ranging in diameter from a few centimeters to five meters. Near Nullagine in the Pilbara Craton, a number of overlapping debris-flow lenses are recognizable; these grade upward into horizontally stratified pebbly sandstones of probable sheetflood origin (Eriksson, 1983b). Alluvial-fan facies grade basinward into lakes that were also fed by longitudinal braided-river systems (Barley et al., 1992).

The nature of lithospheric extension that promoted accumulation of thick basaltic volcanic and coarse alluvial-fan successions, is controversial. Flood basalts dominate the lowermost and uppermost subdivisions. In contrast, Buck (1980) favored a system of horsts and grabens for the middle Ventersdorp Supergroup, whereas Karpeta (1993) argued for alternate-polarity half grabens akin to the East African Rift (cf. Rosendahl, 1987). Fortescue rift basins are considered to record WNW–ESE extension along a major, shallow-dipping detachment fault (Barley et al., 1992).

Chuniespoort–Ghaap and Hamersley Groups

These stratigraphic intervals represent the latest Archean stable-shelf sedimentary successions preserved. Tuffs from toward the top of the Campbellrand Subgroup (Fig. 19) yield U-Pb zircon ages of 2552 ± 11 Ma (Barton et al., 1994) and 2520 ± 2/3 (J. Grotzinger, pers. comm., 1993). Outcrops and suboutcrops of the Chuniespoort-Ghaap Group extend over approximately 350 000 km^2 (Fig. 20) but these stratigraphic intervals probably accumulated across much of the Kaapvaal Craton and around its southwestern, western and possibly northern margins (Fig. 20; Beukes, 1987). Thicknesses of the Campbellrand–Ghaap Group are upward of 1.6 km (Fig. 19). Granitic paleosols are widely developed beneath the basal siliciclastic unit that underlies the Chuniespoort Group (Fig. 19; Button and Tyler, 1981). The Hamersley Group is preserved only along the eastern flank of the Pilbara Craton and in the main basin on the southern margin of the craton where it attains a thickness of approximately 2 km (Figs. 2, 21; Simonson et al., 1993). Tuff beds from near the base and near the top of the Hamersley Group are dated at 2603 ± 7 Ma and 2470 ± 4 Ma, respectively (Fig. 21; Trendall et al., 1990; Hassler, 1993).

The Chuniespoort–Ghaap and Hamersley Groups comprise stable-shelf and basinal facies. Within the Malmani Dolomite and Campbellrand Subgroup (Fig. 19) recognizable depositional environments are platform, platform edge, slope and basin (Fig. 20). Stratigraphic evolution of these dominantly carbonate successions can be compared with established passive-margin models (Read, 1982). An early homoclinal ramp evolved into a rimmed shelf followed by a drowned platform (Beukes, 1987). The homoclinal ramp margin is represented by the basal siliciclastic unit in which platformal quartz arenites of tidal-shelf and tidal- flat origin pass southward into platform-margin, oolitic carbonates that grade into basinal mudstones (Fig. 19). Basinal mudstone followed by limestone with ferruginous dolomite were deposited during drowning of the platform. A temporally more persistent rimmed shelf evolved from the drowned ramp. Platform facies consist of tidal-flat and shallow-subtidal, recrystallized stromatolitic dolomite with chert, and deeper subtidal, fine-grained, iron-poor stromatolitic dolomite (Fig. 19). Tidal-flat associations are dominated by flat algal laminites displaying evidence for desiccation and microdigitate stromatolites. Large stromatolitic domes (Fig. 22), with associated beach rosettes and wave rippled sands developed further down the paleoslope. Deeper subtidal stromatolitic mounds (Fig. 23) are from 5 to 20 m wide and up to 40 m long. Interlayering of tidal flat-shallow subtidal and deeper subtidal stratigraphic units can be related to periodic transgressions resulting in partial or, in one case, complete drowning of the platform (Fig. 19). Platform-edge facies are confined to the Campbellrand Subgroup and

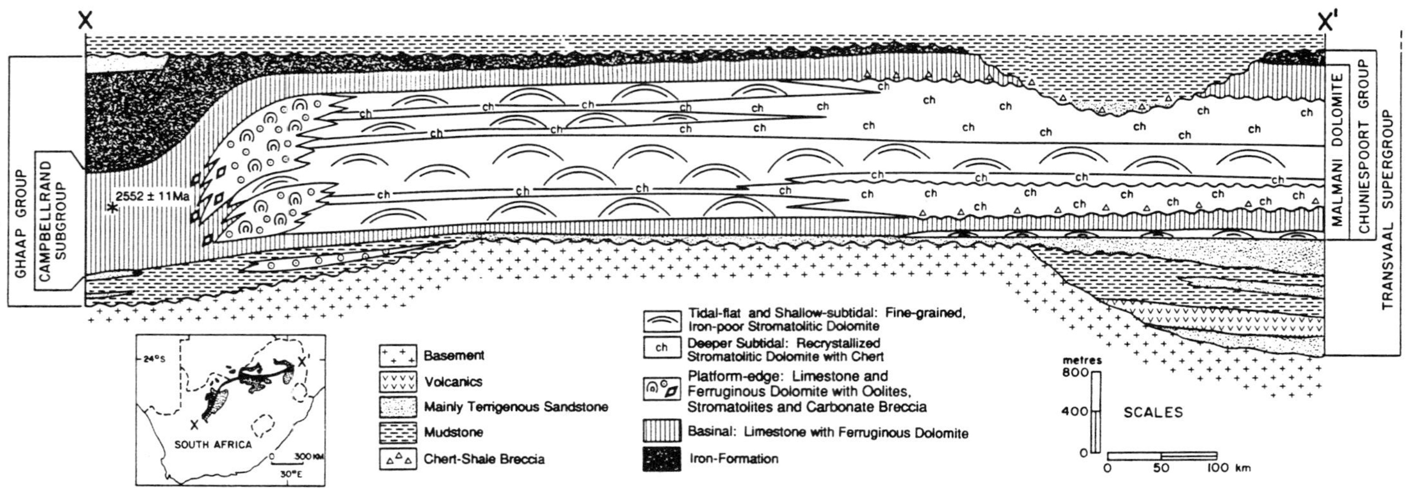

Fig. 19. Generalized lithostratigraphic cross section of Ghaap and Chuniespoort Groups of the Transvaal Supergroup, also showing relationship to underlying and overlying stratigraphic units (adapted from Eriksson et al., 1976; Beukes, 1982). Line of section chosen because slope and basinal facies are preserved only in the southwest. Inset shows location of section.

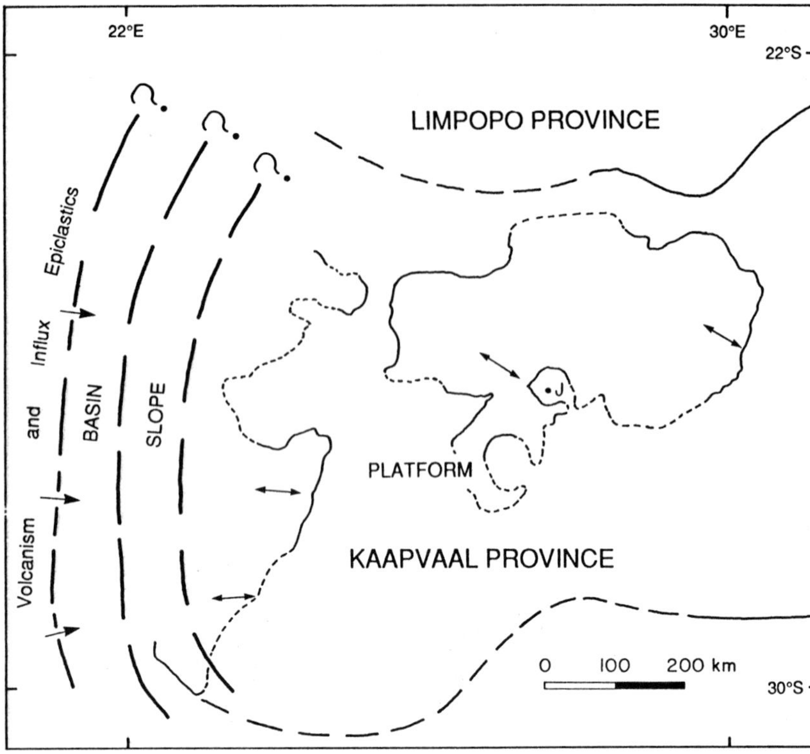

Fig. 20. Paleogeographic map for the Campbellrand and Malmani Subgroups on the Kaapvaal Province (from Beukes, 1987). Outlined areas are present outcrop/suboutcrop belts but the two subgroups probably were deposited across the entire Kaapvaal Province.

consist of linear belts of limestone and ferruginous dolomite containing oolites, oncolites and columnar stromatolites. Platform-edge stabilization related to binding by cyanobacteria produced relatively steep slopes fronting the rimmed shelf. Breccias and slump folds characterize the slope. Limestone and ferruginous dolomite lacking stromatolites, carbonaceous mudstone and chert horizons or proto iron-formation (Button, 1976) comprise the basinal facies (Fig. 19). Carbonates are mainly fine-grained and of probable pelagic origin; carbonate turbidites are locally developed. Final drowning of the platform resulted in blanketing by basinal limestones and ferruginous dolomites. Absence of stromatolites in basinal facies implies deposition below the photic zone. Overlying iron-formations in the Chuniespoort and Ghaap Groups largely mimic the depositional environments of the carbonates although stabilized platform margins are not developed. Instead, the margin had more of a ramp morphology in which basinal, banded iron-formations graded into platformal, clastic-textured (granular), iron-formations (Beukes and Klein, 1990).

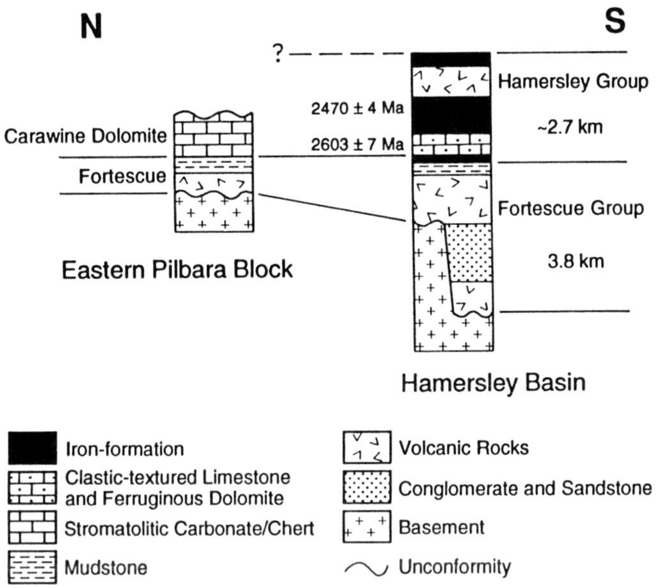

Fig. 21. Lithostratigraphic columns for the "platformal" Carawine Dolomite and coeval "basinal" Hamersley Group in the Pilbara Block, also showing relationship to Fortescue Group (adapted from Simonsin et al., 1993).

In contrast to the Chuniespoort and Ghaap Groups that are dominated by platform facies, the Hamersley Group displays preferential preservation of basinal facies (Eriksson,1983b). Platform sedimentation probably took place across much of the Pilbara Craton (Fig. 2) but these facies are preserved only along the eastern margin of the craton as the Carawine Dolomite (Fig. 21; Simonson et al., 1993). This stratigraphic interval is dominated by platformal facies including tidal-flat and shallow-subtidal recrystallized dolomite and chert and deeper subtidal chert-free, fine-grained dolomite with large-scale stromatolitic mounds. Also present are possible platform-edge and slope facies containing flat-pebble conglomerates, soft-sediment roll-up structures and breccias (Simonson et al., 1993). Basinal facies are confined to the main Hamersley Basin to the south and include limestone and ferruginous dolomite containing turbidites derived from the platform to the north, mudstone, banded iron-formation and pyroclastic and volcaniclastic beds (Simonson et al., 1993). Stratigraphic evidence suggests that the Hamersley Group and Carawine Dolomite are at least in part coeval indicating that platform carbonates and basinal banded iron-formations were accumulating simultaneously (Simonson et al., 1993).

Fig. 22. Tidal flat-shallow subtidal stromatolitic domes from the Chuniespoort Group. Pen for scale.

DISCUSSION AND BROADER IMPLICATIONS

The preserved Archean stratigraphic record reflects an increase in synrift stable-shelf sedimentary successions from 3.2 to 2.5 Ga. Is this a true record of the previous extent of these sedimentary rocks or is it a function of tectonic recycling? Veizer and Jansen (1985) have argued that of all tectonic settings, platformal sedimentary rocks have the least potential for tectonic recycling. Thus, the preserved rock record probably is a close approximation of the extent of synrift and stable-shelf sedimentary rocks formed during the Archean.

The oldest preserved continental crust is represented by 3.96 gneisses in the Slave Province of Canada (Bowring et al., 1989), whereas the oldest preserved cover successions probably are the ca. 3.4 Ga Mkhondo and Mahamba sequences in Swaziland (Hunter and Wilson, 1988). It is possible that other, older cover successions have been recycled. More extensive sedimentary successions of rift and stable-shelf affinity developed during the time period 3.1-2.9 Ga either on small continental nuclei, or on cratons with dimensions of thousands of square kilometers. Examples of the former are present in the Superior and Wyoming Provinces and in the Indian and Yilgarn Cratons. In contrast, the apparently coeval, stable-shelf sediments of the lower Witwatersrand Supergroup, Mozaan

Fig. 23. Deeper subtidal, large stromatolitic mounds from the Chuniespoort Group.

Group, Beitbridge Complex and Buhwa Greenstone Belt developed on an extensive craton or, more likely, on several smaller cratons that subsequently were accreted. These small cratons included the southern half of the Kaapvaal Craton, the Central Zone of the Limpopo Province and the Tokwe Segment in southern Zimbabwe. Available rare-earth-element and trace-element geochemistry on pelites indicate that differentiated granites comprised at least part of this craton or these cratons in southern Africa (Taylor et al., 1986; Wronkiewicz and Condie, 1987, 1989; Boryta and Condie, 1990). The ca. 2.7 Ga Manjeri Formation developed following initial stabilization of the Zimbabwe Craton and is considered as a precursor of the extensive stable-shelf facies of the lower Transvaal Supergroup and Hamersley Group that post-date widespread cratonization on a global scale by 2.6 Ga. Temporal increase in the volume of synrift and stable-shelf sedimentary rocks support crustal growth models that invoke growth of cratons through the Archean (e.g., Veizer and Jansen, 1979; McLennan and Taylor, 1982). This analysis refutes models that infer rapid growth of continents in the Early Archean followed by tectonic recycling (e.g., Fyfe, 1978; Armstrong, 1981).

Prior to 3.2 Ga, continents probably comprised less than five percent of the Earth's crust and, as a consequence, sedimentation and volcanism took place mainly in oceanic and island-arc settings (Lowe, 1992). Progressive growth of stable continents favored expansion of shallow epicontinental seas inhabited by stromatolite-producing cyanobacteria. Thus, it is likely that oxygen productivity increased from 3.2 Ga onward (Cloud, 1988). The common association of iron-formation with stable-shelf sedimentary facies can be related to this increase in oxygen productivity. Prior to 3.2 Ga, oceans are likely to have become progressively enriched in Fe^{2+} as well as Mn^{2+} resulting from submarine hydrothermal activity (Derry and Jacobsen, 1990). These metals were precipitated on stable shelves above the pycnocline in response to upwelling of reduced bottom waters (Button et al., 1982; Veizer, 1983).

The common succession of synrift sedimentary and volcanic rocks overlain by post-rift, stable-shelf sedimentary rocks in the examples discussed suggests an evolutionary relationship between the two. Crustal extension to accommodate the Nsuze and Fortescue Groups was followed by regional subsidence to produce the Mozaan Group and Hamersley Group basins. Regional subsidence can be attributed to cooling of the previously thinned and heated lithosphere (Bickle and Eriksson, 1982). In the case of the Dominion Group–lower Witwatersrand Supergroup and Ventersdorp Group–Ghaap/Chuniespoort Group, this relationship is less clear because a significant hiatus exists between the synrift and post-rift successions. Either the Dominion Group and lower Witwatersrand Supergroup are genetically unrelated or age constraints are inadequate. The undated uppermost Ventersdorp Supergroup may be genetically related to the Ghaap/Chuniespoort Group (Winter, 1976). In this regard, it is interesting to note that the Chuniespoort–Ghaap Groups were previously assigned to the Proterozoic rather than the late-Archean Era. Of the other examples discussed, the Manjeri Formation and Steep Rock Group also may contain synrift and post-rift components (Eriksson et al., in press).

CONCLUSIONS

1. Archean synrift and stable-shelf sedimentary successions dating to 3.2 Ga, are preserved on cratons as well as in high-grade terranes and greenstone belts.
2. The sedimentary facies discussed here developed in synrift basins in association with bimodal volcanic rocks and in post-rift, stable-shelf basins. Facies include conglomerate, feldspathic and quartz arenite, mudstone, stromatolitic carbonate and iron-formation that were deposited in alluvial, lacustrine and shallow-marine environments.
3. Occurrence of Archean synrift and stable-shelf sedimentary rocks implies that the early Earth was not entirely "permobile" but contained stable continents at least by 3.2 Ga and probably much earlier.
4. Paleosols that are commonly present beneath the cratonic sedimentary successions provide further evidence for crustal stability during the Archean era.
5. The preserved stratigraphic records suggest an increase in the volume of synrift and stable-shelf sedimentary successions from 3.2 to 2.5 Ga in response to growth of stable continents during the late Archean.

ACKNOWLEDGMENTS

Research in the Buhwa and Belingwe Greenstone Belts is supported by National Science Foundation Grant EAR-9104876 to KAE. We have benefitted greatly from discussions with Tom Blenkinsop, Scott McLennan, Ross Taylor, and Jim Wilson. The manuscript has benefitted from reviews by K.C. Condie, D.R. Lowe, and H. de la R. Winter. We thank Sharon Chiang for drafting the figures.

REFERENCES

Armstrong, N.V., Hunter, D.R. and Wilson, A.H., 1982. Stratigraphy and petrology of the Archaean Nsuze Group, northern Natal and southeastern Transvaal, South Africa. Precambrian Res., 19: 75–107.

Armstrong, R.A., Compston, W., Retief, E.A., Williams, I.S. and Welke, H.J., 1991. Zircon ion microprobe studies on the age and evolution of the Witwatersrand triad. Precambrian Res., 53: 243–266.

Armstrong, R.L., 1981. Radiogenic isotopes: The case for crustal recycling on a near steady-state no-continental-growth Earth. Phil. Trans. Roy. Soc. Lond., A301: 43–472.

Arndt, N.T., Nelson, D.R., Compston, W., Trendall, A.F. and Thorne, A.M., 1991. The age of the Fortescue Group, Hamersley Basin, Western Australia: From ion microprobe zircon U-Pb results. Austr. J. Earth Sci., 38: 261–281.

Barley, M.E., Blake, T.S. and Groves, D.I., 1992. The Mount Bruce Megasequence Set and eastern Yilgarn Craton: Examples of late Archaean to early Proterozoic divergent and convergent craton margins and controls on mineralization. Precambrian Res., 58: 55–70.

Barton, E.S., Compston, W., Williams, I.S., Bristow, J.W., Hallbauer, D.K. and Smith, C.B., 1989. Provenance ages for the Witwatersrand Supergroup and the Ventersdorp Contact Reef: Constraints from ion microprobe U-Pb ages of detrital zircons. Econ. Geol., 84: 2012–2019.

Barton, E.S., Alterman, W., Williams, I.S., and Smith, C.B., 1994. U-Pb zircon age for a tuff in the Campbell Group, Griqualand West sequence, South Africa: implications for early Proterozoic sedimentation rates. Geology, 22: 343–346.

Barton, J.M., Ryan, B., Fripp, R.E.P. and Horrocks, P., 1979. Effects of metamorphism on the Rb-Sr and U-Pb systematics of the Singelele and Bulai gneisses, Limpopo Mobile Belt, southern Africa. Geol. Soc. S. Afr. Trans., 82: 259–269.

Beckinsale, R.D., Drury, S.A. and Holt, R.W., 1980. 3360 Myr old gneisses from the South India Craton. Nature, 283: 469–470.

Beukes, N.J., 1987. Facies relations, depositional environments and diagenesis in a major early Proterozoic stromatolitic carbonate platform to basinal sequence, Campbellrand Subgroup, Transvaal Supergroup, Southern Africa. Sed. Geol., 54: 1–46.

Beukes, N.J. and Cairncross, B., 1991. A lithostratigraphic–sedimentological reference profile for the Late Archaean Mozaan Group, Pongola Sequence: Application to sequence stratigraphy and correlation with the Witwatersrand Supergroup. S. Afr. J. Geol., 94: 44–69.

Beukes, N.J. and Klein, C., 1990. Geochemistry and sedimentology of a facies transition from microbanded to granular iron-formation in the early Proterozoic Transvaal Supergroup, South Africa. Precambrian Res., 47: 99–139.

Beukes, N.J. and Lowe, D.R., 1989. Environmental control on diverse stromatolite morphologies in the 3000 Myr Pongola Supergroup, South Africa. Sedimentology, 36: 383–397.

Bickle, M.J. and Eriksson, K.A., 1982. Evolution and subsidence of early Precambrian sedimentary basins. Phil. Trans. Roy. Soc. Lond., A305: 225–247.

Bickle, M.J., Martin, A. and Nisbet, E.G., 1975. Basaltic and peridotitic komatiites, stromatolites and a basal unconformity in the Belingwe greenstone belt, Rhodesia. Earth Planet. Sci. Lett., 27: 155–162.

Blake, T.S. and Groves, D.I., 1987. Continental rifting and the Archean–Proterozoic transition. Geology, 15: 229–232.

Blenkinsop, T.G., Fedo, C.M., Bickle, M.J., Eriksson, K.A., Martin, A., Nisbet, E.G. and Wilson, J.F., 1993. Ensialic origin for the Ngezi Group, Belingwe Greenstone Belt, Zimbabwe. Geology, 21: 1135–1138.

Boryta, M. and Condie, K.C., 1990. Geochemistry and origin of the Archaean Beit Bridge complex, Limpopo Belt, South Africa. J. Geol. Soc. Lond., 147: 229–239.

Bowen, T.B., Marsh, J.S., Bowen, M.P. and Eales, H.V., 1986. Volcanic rocks of the Witwatersrand Triad, South Africa. 1: Description, classification and geochemical stratigraphy. Precambrian Res., 31: 297–324.

Bowring, S.A., Williams, I.S. and Compston, W., 1989. 3.96 Ga gneisses from the Slave Province, Northwest Territories, Canada. Geology, 17: 971–975.

Brandl, G., 1983. Geology and geochemistry of various supracrustal rocks of the Beit Bridge Complex east of Messina. Geol. Soc. South Afr. Spec. Publ., 8: 103–112.

Brandl, G. and Barton, J.M., 1991. Pre-Limpopo geology of the Central Zone. In: Ashwal, L.D. (Ed.), Two Cratons and an Orogen-Excursion Guidebook and Review Articles for a Field Workshop through Selected Archaean Terranes of Swaziland, South Africa and Zimbabwe. IGCP Project 280, Dept. of Geology, Univ. Witwatersrand, Johannesburg, pp. 235–249.

Buck, S.G., 1980. Stromatolite and ooid deposits within the fluvial and lacustrine sediments of the Precambrian Ventersdorp Supergroup of South Africa. Precambrian Res., 12: 311–330.

Burke, K., Kidd, W.S.F. and Kusky, T.M., 1985. The Pongola structure of Southeastern Africa: The World's oldest preserved rift? J. Geodynamics, 2: 35–49.

Burke, K., Kidd, W.S.F. and Kusky, T.M., 1986. Archean foreland basin tectonics in the Witwatersrand, South Africa. Tectonics, 5: 439–456.

Button, A., 1976. Iron-formations as an end member in carbonate sedimentary cycles in the Transvaal Supergroup, South Africa. Econ. Geol., 71: 193–201.

Button, A., 1981. The cratonic environment. In: Hunter, D.R. (Ed.), Precambrian of the Southern Hemisphere. Elsevier, Amsterdam, pp. 501–510.

Button, A., Brock, T.D., Cook, P.J., Eugster, H.P., Goodwin, A.M., James, H.L., Margulis, L., Nealson, K.H., Nriagu, J.O., Trendall, A.F. and Walter, M.R., 1982. Sedimentary iron deposits, evaporites and phosphorites. In: Holland, H.D. and Schidlowski, M. (Eds.), Mineral Deposits and Evolution of the Biosphere. Dahlem Konferenzen-Dahlem Workshop Report 16, Berlin, Springer, pp. 259–273.

Button, A. and Tyler, N., 1981. The character and economic significance of Precambrian paleoweathering and erosion surfaces in southern Africa. Econ. Geol., 75: 686–709.

Chadwick, B., 1990. The stratigraphy of a sheet of supracrustal rocks within orthogneisses and its bearing on Late Archaean structure in southern West Greenland. J. Geol. Soc. Lond., 147: 639–652.

Chauvel, C., Dupre, B. and Arndt, N.T., 1993. Pb and Nd isotopic correlation in Belingwe komatiites and basalts. In: Bickle, M.J. and Nisbet, E.G. (Eds.), The geology of the Belingwe greenstone belt: A study of the evolution of Archaean continental crust. Geol. Soc. Zimbabwe Spec. Publ., 2: 167–174.

Cloud, P.E., 1988. Oasis in Space: Earth History from the Beginning: W.W. Norton and Co., New York, 508 p.

Compston, W. and Pidgeon, R.T., 1986. Jack Hills, evidence of more very old detrital zircons in Western Australia. Nature, 321: 766–769.

Condie, K.C., 1981. Archean Greenstone Belts. Developments in Precambrian Geology 3, Elsevier, Amsterdam, 434 pp.

Davis, D.W., and Jackson, M.C., 1988. Geochronology of the Lumby Lake greenstone belt: a 3 Ga complex within the Wabigoon subprovince, northwest Ontario. Geol. Soc. Am. Bull., 100: 818–824.

Derry, L.A. and Jacobsen, S.B., 1990. The chemical evolution of Precambrian seawater: evidence from REE's in banded iron formations. Geochim. Cosmochim. Acta, 54: 2965–2977.

Dodson, M.H., Compston, W., Williams, I.S. and Wilson, J.F., 1988. A search for ancient zircons in Zimbabwean sediments. J. Geol. Soc. Lond., 145: 977–983.

Drury, S.A., Holt, R.W., Van Calsteren, P.C. and Beckinsale, R.D., 1983. Sm-Nd and Rb-Sr ages for Archean rocks from western Karnataka, South India. Geol. Soc. Ind. J., 24: 454–459.

Eriksson, K.A., 1983a. Archaean iron-formations: Environments of deposition and controls on formation. J. Geol. Soc. Austr., 30: 473–482.

Eriksson, K.A., 1983b. Archaean and early Proterozoic sedimentation styles in the Kaapvaal Province, South Africa and Pilbara Block, Australia. Proceedings International Symposium on Archaean and Early Proterozoic Geological Evolution. Brazilian Geol. Soc., Rev. Brasil. Geosci., 12: 121–131.

Eriksson, K.A., Kidd, W.S.F. and Krapez, B., 1988. Basin analysis in regionally metamorphosed and deformed Early Archean terrains: Examples from southern Africa and Western Australia. In: Kleinspehn, K.L. and Paola, C. (Eds.), New Perspectives in Basin Analysis. Springer-Verlag, New York, pp. 371–404.

Eriksson, K.A., Krapez, B. and Fralick, P.W., Sedimentological aspects of greenstone belts. In: deWit, M.J. and Ashwal, L. (Eds.), Tectonic Evolution of Greenstone Belts. Oxford University Press, in press.

Eriksson, K.A., Truswell, J.F. and Button, A., 1976. Palaeoenvironmental and geochemical models from a lower Proterozoic succession in South Africa. In: Walter, M.R. (Ed.), Stromatolites, Developments in Sedimentology, Elsevier, Amsterdam, pp. 635–643.

Eriksson, K.A., Turner, B.R. and Vos, R.G., 1981. Evidence for tidal processes from the lower part of the Witwatersrand Supergroup. Sed. Geol., 29: 309–325.

Fedo, C.M. and Eriksson, K.A., 1993. Chronological evolution of the Archean (~3.0 Ga) Buhwa Greenstone Belt. 16th International Colloquium on African Geology, Ezulweni, Swaziland, pp. 125–126.

Fedo, C.M., and Eriksson, K.A., 1994. Geologic setting and ideas concerning the origin of the iron-ore deposits at Buhwa, Zimbabwe. In: Blenkinsop, T.G., and Tromp, P.L. (Eds.), Sub-Saharan Economic Geology 1993. A.A. Balkema, Rotterdam.

Fripp, R.E.P., 1983. The Precambrian geology of the area around the Sand River near Messina, Central Zone, Limpopo Mobile Belt. Geol. Soc. S. Afr. Spec. Publ., 8: 89–102.

Froude, D.O., Ireland, T.R., Kinny, P.D., Williams, I.S., Compston, W., Williams, I.R. and Myers, J.S., 1983. Ion microprobe identification of 4,100–4,200 Myr-old terrestrial zircons. Nature, 304: 616–618.

Fyfe, W.S., 1978. The evolution of the Earth's crust: Modern plate tectonics to ancient hot spot tectonics? Chem. Geol., 23: 89–114.

Grandstaff, D.E., Edelman, M.J., Foster, R.W., Zbinden, E., and Kimberley, M.M., 1986. Chemistry and mineralogy of Precambrian paleosols at the base of the Dominion and Pongola Groups

(Transvaal, South Africa). Precambrian Res., 32: 97–131.

Grotzinger, J.P., 1989. Facies and evolution of Precambrian carbonate depositional systems: Emergence of the modern platform archetype. In: Crevello, P.D., Wilson, J.L., Sarg, J.F., and Read, J.F. (Eds.), Controls on carbonate platform and basin development. Soc. Econ. Paleontol. Mineralog. Spec. Publ., 44: 79–106.

Hassler, S.W., 1993. Depositional history of the Main Tuff Interval of the Wittenoom Formation, late Archean-early Proterozoic Hamersley Group, Western Australia. Precambrian Res., 60: 337–359.

Hawkesworth, C.J., Bickle, M.J., Gledhill, A.R., Wilson, J.F. and Orpen, J.L., 1979. A 2.9 b.y. event in the Rhodesian Archaean. Earth Planet. Sci. Lett., 43: 285–297.

Hegner, V.E., Kröner, A. and Hofmann, A.W., 1984. Age and isotope geochemistry of the Archean Pongola and Usushwana suites, southern Africa: A case for crustal contamination of the mantle derived magma. Earth Planet. Sci. Lett., 70: 267–279.

Hegner, V.E., Kröner, A. and Hofmann, A.W., 1993. Trace element and isotopic constraints on the origin of the Archaean Pongola and Usushwana igneous suites in Swaziland. 16th International Colloquium on African Geology, Ezulweni, Swaziland, pp. 147–149.

Hunter, D.R. and Wilson, A.H., 1988. A continuous record of Archean evolution from 3.5 Ga to 2.6 Ga in Swaziland and northern Natal. S. Afr. J. Geol., 91: 57–74.

Joliffe, A.W., 1955. Geology and iron ores at Steep Rock Lake. Econ. Geol., 50: 373–398.

Karpeta, W.P., 1993. Volcanism and sedimentation in part of a late Archean rift: The Hartbeesfontein basin, Transvaal, South Africa. Basin Res., 5: 1–19.

Kusky, T.M. and Kidd, W.S.F., 1992. Remnants of an Archaen oceanic plateau, Belingwe greenstone belt, Zimbabwe. Geology, 20: 43–46.

Lowe, D.R., 1992. Major events in the geologic development of the Precambrian Earth. In: Schopf, J.W., and Klein, C. (Eds.), The Proterozoic Biosphere. Cambridge University Press, pp. 67–75.

Maas, R. and McCulloch, M.T., 1991. The provenance of Archean clastic metasediments in the Narryer Gneiss Complex, Western Australia: Trace element geochemistry, Nd isotopes, and U-Pb ages for detrital zircons. Geochim. Cosmochim. Acta, 55: 1915–1932.

Martin, A., Nisbet, E.G., Bickle, M.J. and Orpen, J.L., 1993. Rock units and stratigraphy of the Belingwe greenstone belt: The complexity of the tectonic setting. In: Bickle, M.J. and Nisbet, E.G. (Eds.), The Geology of the Belingwe Greenstone Belt, Zimbabwe: A Study of the Evolution of Archaean Continental Crust. A.A. Balkema, Rotterdam, Geol. Soc. Zimbabwe Spec. Publ., 2: 13–37.

Matthews, P.E., 1967. The pre-Karroo formations of the White Umfolozi inlier, northern Natal. Geol. Soc. S. Afr. Trans., 70: 39–63.

McLennan, S.M. and Taylor, S.R., 1982. Geochemical constraints on the growth of the continental crust. J. Geol., 90: 342–361.

Mueller, P.A., Wooden, J.L. and Nutman, A.P., 1992. 3.96 Ga zircons from an Archean quartzite, Beartooth Mountains, Montana. Geology, 20: 327–330.

Naqvi, S.M. and Rogers, J.J.W., 1987. The Precambrian Geology of India. Clarendon Press, New York, 223 p.

Nelson, D.R. Trendall, A.F., de Laeter, J.R., Grobler, N.J. and Fletcher, I.R., 1992. A comparative study of the geochemical and isotopic systematics of late Archaean flood basalts from the Pilbara and Kaapvaal Cratons. Precambrian Res., 54: 231–256.

Nisbet, E.G., Martin, A, Bickle, M.J.and Orpen, J.L., 1993. The Ngezi Group: Komatiites,basalts, and stromatolites on continental crust. In: Bickle, M.J. and Nisbet, E.G. (Eds.), The Geology of the Belingwe Greenstone Belt: A Study of the Evolution of Archaean Continental Crust. Geol. Soc. Zimbabwe Spec. Publ., 2: 121–165.

Nisbet, E.G. and Wilks, M.E., 1989. Archaean stromatolite reef at Steep Rock Lake, Atikokan, northwestern Ontario. In: Geldsetzer, H.H.J., James, N.P. and Tebbutt, G.E. (Eds.), Reefs, Canada and adjacent areas. Can. Soc. Petrol. Geolog. Memoir, 13: 89–92.

Rankin, D.W., Drake, A.A., Jr., Glover, L., III, Goldsmith, R., Hall, L.M., Murray, D.P., Ratcliffe, N.M., Read, J.F., Secor, D.T., Jr., and Stanley, R.S., 1989. Pre-orogenic terranes In: Hatcher, R.D., Jr., Thomas, W.A., and Viele, G.W. (Eds.), The Appalachian-Ouachita Orogen in the United States: Boulder, Colorado. Geological Society America, The Geology of North America, Vol. F-2, pp. 7–100.

Read, J.F., 1982. Carbonate platforms of passive (extensional) continental margins: Types, characteristics and evolution. Tectonophysics, 85: 195–212.

Retief, E.A., Compston, W., Armstrong, R.A. and Williams, I.S., 1990. Characteristics and preliminary U-Pb ages from Limpopo Belt lithologies. In: Van Reenen, D.D. and Roering, C.D. (Eds.), The Limpopo Belt: A Field Workshop on Granulites and Deep Crustal Tectonics. Rand Afrik. Univ. Publ., 289 pp.

Robb, L.J., Davis, D.W. and Kamo, S.L., 1990. U-Pb ages on single detrital zircon grains from the Witwatersrand Basin, South Africa: Constraints on the age of sedimentation and on the evolution of granites adjacent to the basin. J. Geol., 98: 311–328.

Rosendahl, B.R., 1987. Architecture of continental rifts with special reference to East Africa. Annu. Rev. Earth Planet. Sci., 15: 445–503.

Saager, R., Stupp, H.D., Utter, T. and Matthey, H.O., 1986. Geological and mineralogical notes on placer occurrences in some conglomerates of the Pongola Sequence. In: Anhaeusser, C.R. and Maske, S. (Eds.), Mineral Deposits of Southern Africa, 1, pp. 473–487, Geological Society South Africa, Johannesburg.

Simonson, B.W., Schubel, K. A. and Hassler, S. W., 1993. Carbonate sedimentology of the early Precambrian Hamersley Group of Western Australia. Precambrian Res. 60: 287–335.

Simpson, E.L. and Eriksson, K.A., 1990. Early Cambrian progradational and transgressive sedimentation patterns in Virginia: an example of the early history of a passive margin. J. Sed. Petrol., 60: 84–100.

Srinivasan, R. and Ojakangas, R.W., 1986. Sedimentology of quartz-pebble conglomerates and quartzites of the Archean Bababudan Group, Dharwar Craton, South India: Evidence for early crustal stability. J. Geol., 94: 199–214.

Stagman, J.G., 1978. An outline of the geology of Rhodesia. Geol. Soc. Rhodesia Bull., 50, 126 pp.

Stanistreet, I.G. and McCarthy, T.S., 1991. Changing tectono-sedimentary scenarios relevant to the development of the Late Archaean Witwatersrand Basin. J. Afr. Earth Sci., 13: 65–81.

Tankard, A.J., Jackson, M.P.A., Eriksson, K.A., Hobday, D.K., Hunter, D.R. and Minter, W.E.L., 1982. Crustal Evolution of Southern Africa: 3.8 Billion Years of Earth History. Springer-Verlag, New York, 523 pp.

Taylor, P.N., Chadwick, B., Moorbath, S., Ramakrishnan, M. and Viswanatha, M.N., 1984. Petrography, chemistry, and isotopic ages of Peninsular Gneiss, Dharwar and volcanic rocks, and the Chitradurga Granite with special reference to the late Archean evolution of the Karnataka Craton, southern India. Precambrian Res., 23: 349–375.

Taylor, P.N., Kramers, J.D. Moorbath, S., Wilson, J.F., Orpen, J.L. and Martin, A., 1991. Pb/Pb, Sm-Nd and Rb-Sr geochronology in the Archean Craton of Zimbabwe. Chem. Geol., 87: 175–196.

Taylor, S.R., Rudnick, R.L., McLennan, S.M. and Eriksson, K.A., 1986. Rare earth element patterns in Archean high-grade metasediments and their tectonic significance. Geochim. Cosmochim. Acta, 50: 2267–2279.

Thurston, P.C. and Chivers, K.M., 1990. Secular variation in greenstone sequence development emphasizing Superior Province, Canada. Precambrian Res., 46: 21–58.

Trendall, A.F., Compston, W., Williams, I.S., Armstrong, R.A., Arndt, N.T., McNaughton, N.J., Nelson, D.R., Barley, M.E., Beukes, N.J., deLaeter, J.R., Retief, E.A. and Thorne, A. M., 1990. Precise zircon U-Pb chronological comparison of the volcano-sedimentary sequences of the Kaapvaal and Pilbara Cratons between about 3.1 and 2.4 Ga. Third International Archaean Symposium, Perth, pp. 81–83.

Veizer, J., 1983. Geologic evolution of the Archean-Early Proterozoic Earth. In: Schopf, J.W. (Ed.), Earth's Earliest Biosphere: Its Origin and Evolution. Princeton University Press, Princeton, pp. 240–259.

Veizer, J. and Jansen, S.L., 1979. Basement and sedimentary recycling and continental evolution. J. Geol., 87: 341–370.

Veizer, J. and Jansen, S.L., 1985. Basement and sedimentary recycling. 2. Time dimension to global tectonics. J. Geol., 93: 625–643.

Von Brunn, V. and Hobday, D.K., 1976. Early Precambrian tidal sedimentation in the Pongola Supergroup of South Africa. J. Sed. Petrol., 46: 670–679.

Von Brunn, V. and Mason, T.R., 1977. Siliciclastic-carbonate tidal deposits from the 3000 m.y. Pongola Supergroup, South Africa. Sed. Geol., 18: 245–255.

Watchorn, M.B. and Armstrong, N.V., 1981. Contemporaneous sedimentation and volcanism at the base of the early Precambrian Nsuze Group, South Africa. Geol. Soc. S. Afr. Trans., 83: 231–238.

Wilks, M.E., 1986. The geology of the Steep Rock Group, N.W. Ontario: A major Archaean unconformity and Archaean stromatolites (M.S. Thesis). University of Saskatchewan, 206 pp.

Wilks, M.E. and Nisbet, E.G., 1985. Archaean stromatolites from the Steep Rock Group, northwestern Ontario, Canada. Can. J. Earth Sci., 22: 792–799.

Wilks, M.E. and Nisbet, E.G., 1988. Stratigraphy of the Steep Rock Group, northwest Ontario: A major Archean unconformity and Archean Stromatolites. Can. J. Earth Sci., 25: 370–391.

Wilson, J.F., 1990. A craton and its cracks: Some of the behavior of the Zimbabwe block from the late Archaean to the Mesozoic in response to horizontal movements, and the significance of some of its mafic dyke fracture patterns. J. Afr. Earth Sci., 10: 483–501.

Windley, B.F., 1984. The Evolving Continents. John Wiley and Sons, 399 pp.

Winter, H. de la R., 1976. A lithostratigraphic classification of the Ventersdorp Succession. Geol. Soc. S. Afr. Trans., 79: 31–48.

Winter, H. de la R., 1990. Discussion on "evolution of the late Archean volcano-sedimentary basins of the Platberg Group near Welkom, Orange Free State". S. Afr. J. Geol., 93: 869–875.

Wronkiewicz, D.J. and Condie, K.C., 1987. Geochemistry of Archean shales from the Witwatersrand Supergroup, South Africa: Source-area weathering and provenance. Geochim. Cosmochim. Acta, 51: 2401–2416.

Wronkiewicz, D.J, and Condie, K.C., 1989. Geochemistry and provenance of sediments from the Pongola Supergroup, South Africa: Evidence for a 3.0-Ga-old continental craton. Geochim. Cosmochim. Acta, 53: 1537–1549.

Chapter 6

THE ARCHEAN GREY GNEISSES AND THE GENESIS OF CONTINENTAL CRUST

H. MARTIN

INTRODUCTION

The solar system formed in an already old Universe, about 10 Ga after the Big Bang. The earth accreted from dust and gas in the inner part of the solar nebula. Probably due to intense solar activity, volatile elements were rejected to the periphery of the nebula where giant planets (Jupiter, Saturn, Uranus, Neptune) formed, whereas terrestrial planets (Mercury, Venus, Earth and Mars) accreted in the relatively gas-poor region. Whatever the earth accretion process was (homogeneous vs. heterogeneous), it resulted in a shelled structure. All of the layers did not differentiate at the same time or by the same mechanism. It is now commonly admitted that the core-mantle differentiation occurred very early, before 4.5 Ga or even during the accretion event itself (Ringwood, 1970, 1983, 1992; Clark et al., 1972; Vidal and Dosso, 1978; Wetherill, 1978; Salomon et al., 1981; Wänke, 1981; Taylor, 1992, 1993; Hofmann, 1993). Both accretion and core formation released more than 85% of the total earth energy (Flasar and Birch, 1973; Kaula, 1979; Wänke, 1981). This rapid energy release induced the melting of the outer part of the mantle, thus creating a magma ocean. Crystallization and fractionation of garnet in the magma ocean resulted in an early mantle stratification (Hofmeister, 1983; Agee and Walker, 1988; Anderson, 1989; Ringwood, 1992). In contrast to the Moon, there is no remnant or geochemical evidence for an anorthositic crust (Taylor, 1975), and if it existed, it was totally destroyed and recycled by meteoritic bombardment (between 3.8 and 4.0 Ga) and by mantle convection (Ringwood, 1992). If core-mantle differentiation operated very rapidly during the first 0.1 Ga of the earth history, the outer shells (crust, ocean, atmosphere) differentiated and evolved more slowly. For instance the continental crust formed later through progressive and/or episodic processes, which are still operating today.

The genesis and differentiation of the continental crust started early in the history of the earth. Some detrital zircons from Mount Narryer and Jack Hills in Australia were dated at 4.18 Ga (Froude et al., 1983; Compston and Pidgeon, 1986), and one sample registered a $^{207}Pb/^{206}Pb$ age as old as 4.276 ± 0.006 Ga, which corresponds to a minimum estimate of the true age of zircon crystallization. These data demonstrate the existence of an early and yet complex continental crust of unknown size. On the other hand, the oldest known rocks are gneiss remnants from the Acasta Gneisses in Canada, whose zircons were dated at 3.962 ± 0.003 Ga (Bowring et al., 1990) and the more widely-exposed Amîtsoq gneisses in Greenland, that yield a 3.822 ± 0.005 Ga zircon age (Kinny, 1986). The scarcity of zircons and the lack of preserved rocks older than 3.82 Ga indicate that although crust existed since at least 4.3 Ga, it corresponded to small volumes or/and was destroyed prior to 3.82 Ga. The destruction of the crust could have been caused by heavy meteorite bombardment 4.0 to 3.8 Ga ago, which resulted in disruption and recycling of the crust into the mantle. It could also be a consequence of the small size of the crustal components, which were reincorporated into the mantle by the strong and efficient early Archean mantle convection. After 3.8 Ga, the continental crust reached a critical size and acquired a buoyant behaviour, preserving it from large scale recycling into the mantle. The continental crust grew continuously from that time until today, with periods of accelerated growing rates (i.e. super-events of Moorbath, 1976).

Depending on the model (Fig. 1): (1) most of the continents formed prior to 4.0 Ga (Fyfe, 1978; Armstrong, 1981; Reymer and Schubert, 1984) or (2) the rate of crustal growth from the mantle increased from 3.8 to 2.5 Ga and progressively decreased until now (Veizer and Jansen, 1979; McLennan and Taylor, 1982; Allègre, 1985; Taylor and McLennan, 1985, Condie, 1990). However, the majority of the crustal growth models suggests that about 75% of the earth's crust was generated prior to 2.5 Ga.

Paradoxically, this long period (from 3.8 to 2.5 Ga) of large-scale continental crust genesis and differentiation was not a major focus of research until the end of the sixties and the beginning of the seventies (i.e. Condie, 1967; Anhausser et al., 1969; Viljoen and Viljoen, 1969a, 1969b; Goldich et al., 1970; Goodwin and

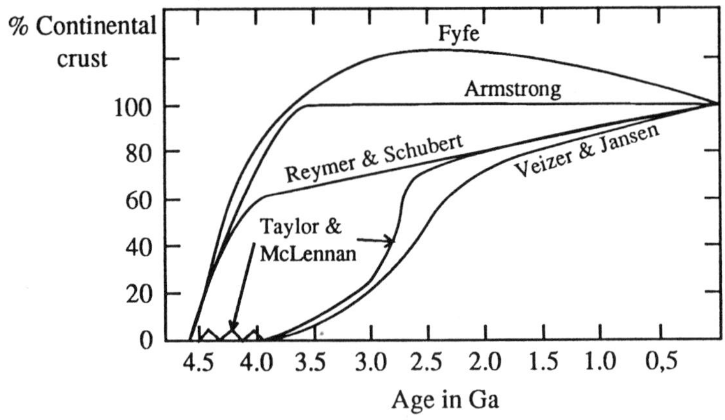

Fig. 1. Models of crustal growth (from Taylor and McLennan, 1985). Even if they widely differ one from the other, all indicate that about 75% of the continental crust was generated before 2.5 Ga. Data are from: Fyfe, (1978); Veizer and Jansen, (1979); Armstrong, (1981); Reymer and Schubert, (1984); Taylor and McLennan, (1985).

Ridler, 1970; Hunter, 1970; Ermanovics, 1971; Glikson, 1971; White et al., 1971; Windley and Bridgwater, 1971; Arth and Hanson, 1972; McGregor, 1973; Stowe, 1973; Wilson, 1973; Pitchamuthu, 1974). The development of extensive studies during the two last decades is mainly due to advances in trace element geochemistry and modelling, isotope geochemistry and geochronology, and experimental petrology. One of the main results of these developments was to demonstrate that Archean continental crust is different from modern continental crust in both composition and rock relative proportions. This indicates that the crustal sources and/or the petrogenetic mechanisms of crustal growth were different during the first half of the earth history.

The purpose of this paper is: (1) to outline the main characteristics of the Archean continental crust, (2) to address the problem of its petrogenesis using petrological and geochemical approaches, (3) to discuss experimental petrology data, and (4) to compare Archean crustal petrogenetic mechanisms with their modern equivalents and to discuss the possible geodynamic environments of crustal formation.

FIELD DATA AND PETROLOGY

Archean terrains are widespread all over the world (Fig. 2). According to Condie (1981), they can be divided into three associations: (1) granite-greenstone terrains metamorphosed from greenschist to high-grade amphibolite facies (North America, Australia, North Europe, Southern Africa, India, South America); (2) high-grade terrains metamorphosed from middle amphibolite to upper granulite facies (Greenland, Labrador, Scotland, China, Russia, North and Central Africa); and (3) cratonic basin sediments affected by very low-grade metamorphism (Kaapvaal, Sargur in India). Most of the differences between the three types consist in metamorphic grade and level of erosion. Within the first two of these associations, three lithologic components are recognized (Windley and Bridgwater, 1971): (1) a granite–gneiss basement, (2) supracrustal belts, and (3) late-stage granitic intrusives generated by recycling of the granite–gneiss basement. This chapter will only discuss the granite–gneiss basement, which is made up of grey gneisses ($\geq$ 95%) and subordinate high-Mg granodiorites (sanukitoids). These high-Mg granodiorites are exposed mainly in the late Archean terrains (from 2.8 to 2.5 Ga), where they represent less than 5% of the whole basement (Jahn et al., 1988; Stern and Hanson, 1991; Jayananda et al., 1992). Our interest will focus on the grey gneisses.

Archean granite–gneiss basement is exposed as huge areas of medium to fine grained grey gneisses. Their field aspect is highly variable: homogeneous strongly foliated gneisses, regularly banded gneisses, agmatitic gneisses, highly chaotic to nebulitic migmatites, etc. In many places, the grey gneisses are the basement on which the volcanic and sedimentary greenstone belts are emplaced. Generally, the

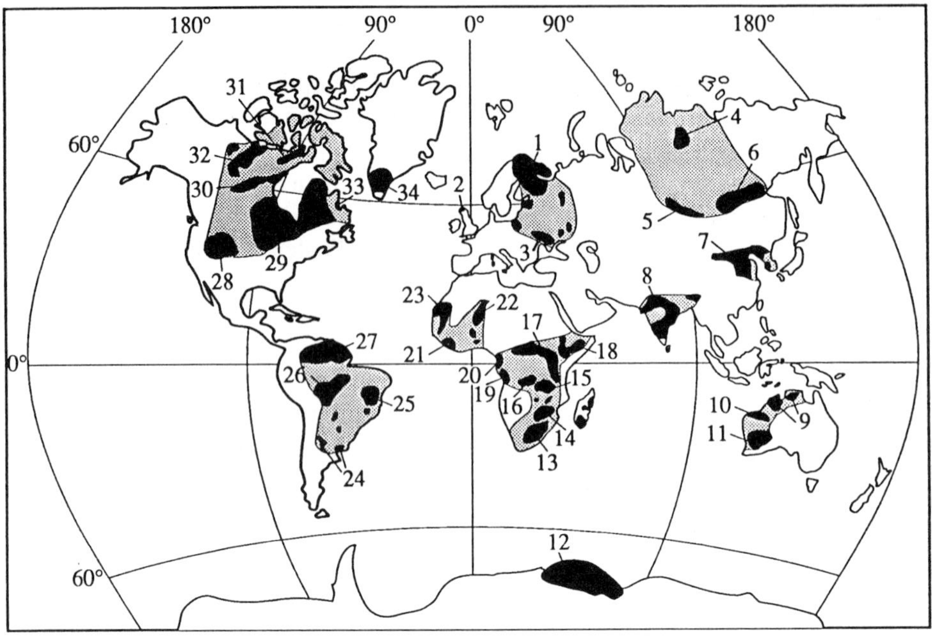

Fig. 2. Distribution of Archean provinces (after Condie, 1981; and Goodwin, 1991). The exposed Archean terranes are in black, and the areas underlain by Archean rocks are in grey. (1) Baltic Shield; (2) Scottish Shield; (3) Ukrainian Shield; (4) Anabar Shield; (5) Baikal, Sayan and Yienisei fold belts; (6) Aldan Shield; (7) Sino-Korean, Tarim and Yangtze cratons; (8) Indian Shield; (9) Litchfield, Rul Jungle and Nanambu Complexes; (10) Pilbara block; (11) Yilgarn block; (12) Napier Complex; (13) Kaapvaal Craton; (14) Zimbabwe Craton; (15) Zambian Block; (16) Kasaï Craton; (17) Central Africa Craton; (18) Ethiopian Block; (19) Chaillu Craton; (20) Cameroon N'tem Complex; (21) Man Shield; (22) Tuareg Shield; (23) Reguibat Shield; (24) Rio de la Plata and Luis Alves Massifs; (25) São Francisco Craton; (26) Guapore Craton; (27) Guiana Shield; (28) Wyoming Province; (29) Superior Province; (30) Kaminak Group; (31) Committe Bay Block; (32) Slave Province; (33) Labrador Shield; (34) Greenland Shield.

greenstone belts have an elongated shape and are roughly parallel and concordant with the gneissic foliation. In several places, the grey gneiss–greenstone belt contact is tectonic and/or obliterated by the intrusion of late-stage granites.

Typical microgranular enclaves (Didier, 1973) are very rare in grey gneisses, but the abundance of xenoliths varies widely from one area to another. In many places the intense deformation of the rocks does not allow a distinction to be made between stretched xenoliths and disrupted and stretched dykes or supracrustal layers (Martin, 1985).

Archean grey gneisses are quartz and sodic plagioclase rich and K-feldspar poor, with a microcline/plagioclase ratio of about 0.07, and the mafic mineral content is generally lower than 15% (Table 1). The most frequent mineral association is oligoclase + quartz + biotite, but oligoclase + quartz + biotite + microcline,

TABLE 1

Average modal compositions of Archean grey gneisses

	Finland	Labrador	India	Zimbabwe	Brazil	Ontario	Average
	n=26	n=11	n=9	n=4	n=18	n=21	n=89
Quartz	29.1	20.1	34.9	26.3	24.70	30.2	27.82
Na-plagioclase	46.5	65.2	50.9	59.5	61.30	59.5	55.90
K-feldspar	1.8	6.0	5.7	3.3	4.33	3	3.58
Biotite	16.1	8.7	3.6	7.9	7.28	4.7	9.08
Hornblende	1.9			2.1			0.65
Epidote	2.9		4.9		1.33	0.8	1.80
Sphene	0.4				0.08		0.13
Apatite	0.1				0.03		0.04
Zircon	0.2				0.03		0.06
Mgt + Ilm	0.3			0.9	0.39	0.1	0.23
Accessories	0.7				0.53	1.7	0.71
Total	100	100	100	100	100	100	100

Data are from: Finland (Hyppönen, 1983; Martin, 1985; Luukkonen, 1988), Labrador (Barton, 1975), India (Sengupta et al., 1991), Zimbabwe (Condie and Allen, 1980), Brazil (personal data), Ontario (Hanson et al., 1971). n = number of samples.

or oligoclase + quartz + biotite + green hornblende are also found. Accessory minerals are epidote (pistacite, zoisite and clinozoisite), allanite, sphene, zircon, apatite, ilmenite, magnetite and infrequent pyrite. Secondary minerals, when present, include muscovite, chlorite and carbonates.

In the Q-A-P modal diagram from Streckeisen (1975), most Archean grey gneisses plot in the tonalitic field (Fig. 3A) except for a few points that fall in the granodioritic domain. When compared with the classical magmatic series as defined by Lameyre and Bowden (1982), the grey gneisses follow a typical low-K_2O calc-alkaline trend (Fig. 3B). This evolution is similar to that reported for a Proterozoic trondhjemitic suite from Southwest Finland (Fig. 3C; Hietanen, 1943; Arth et al., 1978). Nevertheless one important difference must be noted: the least differentiated members (mela-gabbros and mela-tonalites) are unknown in the Archean grey gneiss suites. Other petrographic characters of the grey gneisses, such as the presence of hornblende, a deficiency of muscovite and other alumina silicates, the presence of both magnetite and ilmenite, the relative abundance of allanite and sphene, and occasional pyrite indicate that these rocks belong to the I-type granitoids of Chappell and White (1974) or to the M-type of Didier et al. (1982).

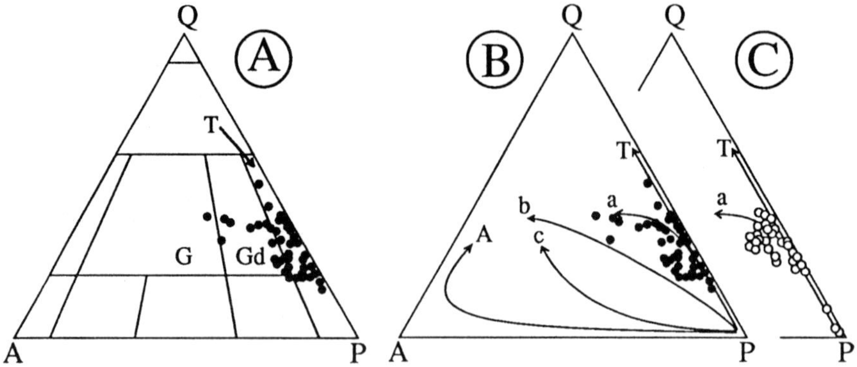

Fig. 3. Modal Q-A-P (Quartz-Alkali-feldspar-Plagioclase) triangles. Filled circles correspond to Archean grey gneisses whose average modal compositions are reported in Table 1. A: Archean grey gneisses plotted within the granitic rock classification diagram of Streckeisen (1975): T = tonalite; Gd = granodiorite; G = granite. B: Archean grey gneisses belong to the low-K calc-alkaline (trondhjemitic) series as defined by Lameyre and Bowden (1982). T = tholeiitic series; A = alkaline series; calc-alkaline series subdivided into, a = low-K, b = intermediate-K, and c = high-K. C: Data from the typical Early Proterozoic trondhjemitic suite from Southwest Finland (Hietanen, 1943; Arth et al., 1978). Unlike this suite, Archean grey gneisses do not possess the less differentiated members that plot near the P pole.

In most Archean grey gneisses, epidotes and sphene are always associated with biotite and together form small masses elongated in the foliation. Collerson and Bridgwater (1979) interpreted this as reflecting metasomatic processes: potassium-rich fluids destabilize hornblende in a zoisite + clinozoisite + biotite assemblage. Martin (1985) suggested that the potassium necessary for biotite formation is provided by K-feldspar through a reaction similar to that proposed by Strens (1965): 4 K-feldspar + 2 hornblende + 7 water $\rightarrow$ 2 clinozoisite + 4 biotite + 4 quartz. It has been recently shown that the epidote + biotite + oligoclase + quartz association can be a magmatic near-solidus assemblage at water pressures higher than 8 kbar (Zen and Hammarstrom, 1984; Van der Laan and Wyllie, 1992).

GEOCHEMICAL CHARACTERISTICS

Major elements

Table 2 gives average compositions of Archean grey gneisses. These rocks are silica and alumina rich but have low $Fe_2O_3^*$ + MgO + MnO + TiO_2 contents (< 5%). Modal composition indicates that they belong to a K_2O-poor calc-alkaline suite, and this is corroborated by the low K_2O/Na_2O ratios, which are generally lower than 0.5 (average = 0.36). Table 2 also compares the average major-element compositions of Archean grey gneisses with the main geochemical features of

TABLE 2

Average major-element composition of Archean grey gneisses compared with an average composition compiled by Condie (1981) and with a standard trondhjemite proposed by Barker (1979)

	Weight % oxide			
	Average n=355	Std. dev.	Average Condie (1981)	Trondhjemite Barker (1979)
SiO_2	69.79	4.90	69.66	68–75
Al_2O_3	15.56	1.20	15.86	>15 High-Al
Fe_2O_3*	3.12	1.50	3.16	
MnO	0.05	0.03	0.04	
MgO	1.18	0.70	1.14	
CaO	3.19	1.00	3.38	1.5–4.4
Na_2O	4.88	0.75	4.70	4.0–5.5
K_2O	1.76	0.70	1.59	<2.0
TiO_2	0.34	0.16	0.35	
P_2O_5	0.13	0.10	0.11	
Fe_2O_3*+MgO	4.30		4.30	<3.65
Fe_2O_3*/MgO	2.64		2.77	2.2–3.3
K_2O/Na_2O	0.36		0.34	<0.5
A/CNK	0.99		1.02	
Mg#	0.43		0.42	

A/CNK = mol $Al_2O_3/(CaO+Na_2O+K_2O)$; Mg# = mol MgO/(MgO+FeO*).

trondhjemites as defined by Barker (1979). Archean grey gneisses match the typical trondhjemitic composition except for two significant features: (1) their Fe_2O_3* + MgO is of about 4.3% in comparison to 3.65% in Barker's average trondhjemite (however the Fe_2O_3*/MgO ratios remain similar), and (2) CaO is about 3.2%, a value intermediate between typical trondhjemites (1.5–3.0%) and calcic trondhjemites (4.4–4.5%). Based on Al_2O_3 content, Barker and Arth (1976) subdivide trondhjemites into high- and low-Al_2O_3 groups; most Archean grey gneisses have Al_2O_3 > 15% at SiO_2 = 70% and consequently belong to the high-Al_2O_3 group. On the other hand, other major-element data (A/CNK < 1.1; K_2O/Na_2O, low; Na > 3.2; normative corundum <1%) support the conclusions drawn from petrography, which indicate that grey gneisses are I-type (Chappell and White, 1974) or M-type (Didier et al., 1982) granitoids.

O'Connor (1965) proposed a normative An-Ab-Or triangular classification for plutonic rocks containing more than 10% modal or normative quartz (Fig. 4). Most grey gneisses plot in the tonalite and trondhjemite fields, with a smaller number falling in the granodiorite field. The clustering of points around the triple junction

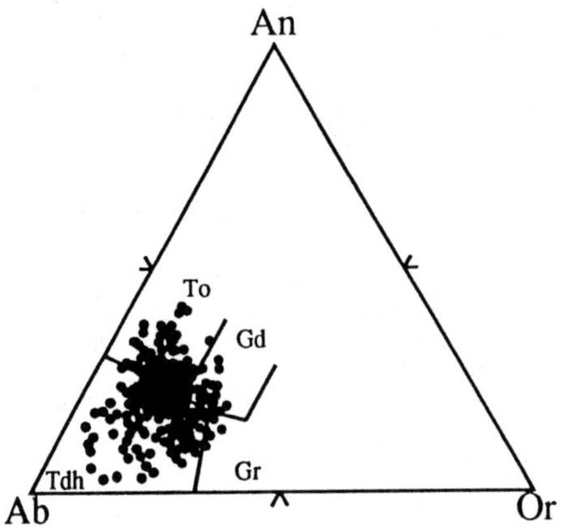

Fig. 4. Normative An-Ab-Or diagram of Archean grey gneisses (O'Connor, 1965). The fields are those drawn by Barker (1979) in order to correspond to the IUGS petrographic terminology. To = tonalite; Tdh = trondhjemite; Gd = granodiorite; Gr = granite.

of the three compositional fields is typical of juvenile Archean continental crust irrespective of geographic location and age (Barker, 1979; Glikson, 1979; Condie, 1981, 1993). This leads to the commonly given name of TTG (Tonalite, Trondhjemite, Granodiorite) association (Jahn et al., 1981; Martin et al., 1983a).

In the A-F-M diagram (Fig. 5A), Archean TTG plot in the calc-alkaline domain, on the most fractionated extremity of the trend drawn by the Proterozoic trondhjemitic suite from Southwest Finland (Barker and Arth, 1976). Most of the data plot close to the A apex and do not define a complete calc-alkaline trend but only represent the more differentiated members. Barker and Arth (1976) proposed two triangular diagrams in order to discriminate between the trondhjemitic and the classical calc-alkaline suites. The former shows Na-enrichment in the course of differentiation, whereas the later evolves through K-enrichment. In the Q-Ab-Or normative diagram (Fig. 5B), TTG are rather scattered, but most data plot near the trondhjemitic trend, without any affinity to the classical calc-alkaline trend. The K-Na-Ca triangle (Fig. 6A) provides additional information. The TTG widely scatter along and around the trondhjemitic trend and never display the K-enrichment typical of classical calc-alkaline suites. As previously pointed out from field data and also apparent in both Q-A-P and A-F-M triangles, Archean TTG suites do not possess the less differentiated members (mela-gabbros, mela-diorites, and mela-tonalites) typical of Southwest Finland suite. This suggests that they were generated through distinct petrogenetic mechanisms. In the K-Na-Ca diagram (Fig. 6), Archean TTG scatter around the trondhjemitic line and do not define any specific trend. For this reason in the K-Na-Ca triangle Archean TTG are characterized by a field of trondhjemitic affinity rather than by a single linear trend (Fig. 6B).

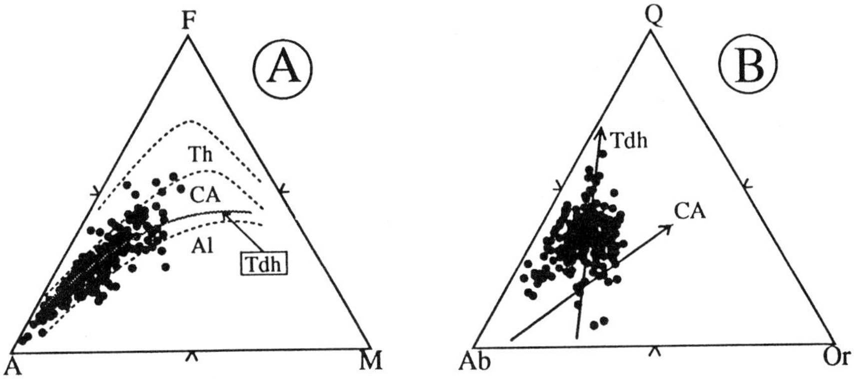

Fig. 5. A: A-F-M (A = $Na_2O + K_2O$; F = $FeO + 0.9*Fe_2O_3$; M = MgO) diagram evidencing the calc-alkaline character of TTG magmas. The fields are from Kuno (1968). Th = tholeiitic; CA = calc-alkaline; Al = alkaline. The Tdh line corresponds to the differentiation trend of the standard trondhjemitic suite from Southwest Finland (Barker and Arth, 1976). Note that the Archean TTG do not possess the less differentiated end-members. B: Normative Q-Ab-Or triangle (Barker and Arth, 1976) showing that TTG coincide with the trondhjemitic trend (Tdh) rather than with a classical calc-alkaline one (CA).

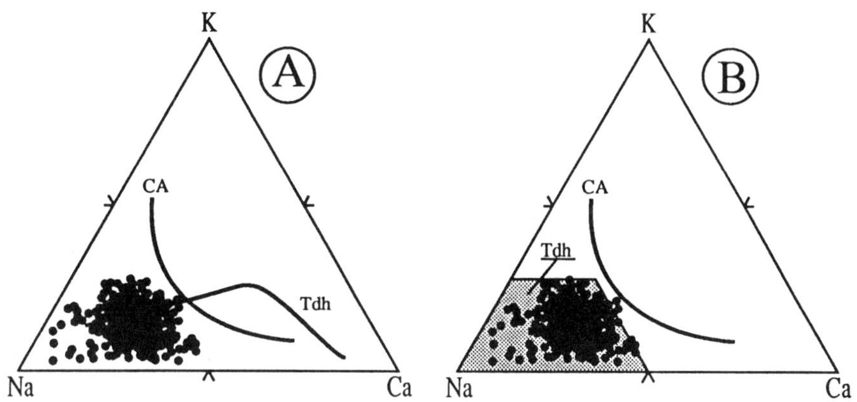

Fig. 6. K-Na-Ca diagrams. In diagram (A) the classical calc-alkaline (CA) and the trondhjemitic (Tdh) trends are distinguished (Barker and Arth, 1976). TTG plot and scatter along the trondhjemitic trend near the Na pole and far from the classical calc-alkaline line. Nevertheless, TTG do not define a real trend of differentiation, but rather plot in a field of trondhjemitic affinity (Tdh grey field in diagram B).

Trace elements

The main trace-element geochemical features of Archean TTG are given in Table 3 and summarized in a primordial-mantle normalized diagram (Fig. 7). TTG

TABLE 3

Average trace element composition of Archean TTG

	ppm		
	Average (*n*=355)	Std. dev.	Average Condie (1993)
Rb	55	30	65
Ba	690	500	660
Th	6.9	6	8
U	1.6	1.3	2
Ta	0.71	0.5	0.7
Nb	6.4	4	7.5
Sr	454	200	435
Hf	4.5	2	4
Zr	152	110	160
Cr	29	30	22
Y	7.5	4	13
Sc	4.7	2	5
Ni	14	10	13
V	35	23	37
La	32	20	30
Ce	56	32	56
Nd	21.4	3	22
Sm	3.3	1.5	3.4
Eu	0.92	0.4	1
Gd	2.2	1.1	2.98
Tb	0.31	0.1	0.45
Dy	1.16	0.7	
Er	0.59	0.3	
Yb	0.55	0.3	1
Lu	0.12	0.1	0.17
Rb/Sr	0.12		0.15
$(La/Yb)_N$	38.4		19.8
Eu/Eu*	0.99		0.95

$Eu/Eu^* = Eu_N/((Sm_N+Gd_N)/2)$.

are Sr rich (454 ppm) and generally display low Rb/Sr ratios (0.05 < Rb/Sr < 1.0, with an average of 0.12). These values are highly dependant on metamorphic grade, with some granulitic rocks depleted in Rb (as well as in U and Th) (Tarney et al., 1979; Weaver and Tarney, 1981; Tarney et al., 1982). As with most of the other crustal rocks, Archean TTG have prominent negative Nb-Ta and Ti anomalies (e.g., Tarney et al., 1982; Condie, 1989, 1993; Jahn, 1994).

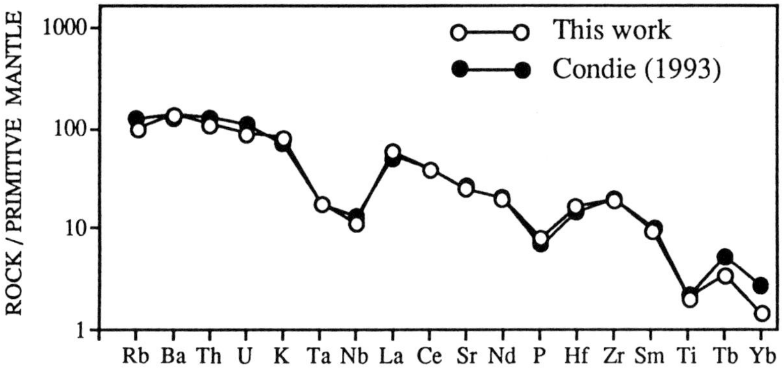

Fig. 7. Primitive mantle normalized average compositions of Archean TTG (data from Table 3). Their trace element signature includes low contents in HREE (Yb) and negative anomalies in Ta, Nb, P and Ti. Primordial mantle values are from Taylor and McLennan (1985).

A phosphorus negative anomaly is also common and could reflect apatite fractionation during differentiation.

Transition-element contents are typically low (Ni = 14 ppm; Cr = 29 ppm; V = 35 ppm). One of the more characteristic features of Archean TTG is their rare earth element (REE) patterns (Fig. 8), which are strongly fractionated and generally display similar degrees of fractionation for both light (LREE) and heavy REE (HREE). The average $(La/Yb)_N$ is 38.4 but in some cases it can be higher than 150, Yb content is always low ($Yb_N = 2.6$), and REE patterns show a concave form at the HREE end. Generally, Archean TTG do not have significant negative or positive Eu anomalies, but in some cases, positive Eu anomalies have been reported (Martin, 1985, 1987a), and were interpreted as reflecting the fractionation of REE-rich accessory mineral phases (i.e. allanite).

The Sr-Nd isotopic compositions of Archean TTG are roughly close to mantle values at the time of their emplacement. For instance, $^{87}Sr/^{86}Sr$ initial ratios (I_{Sr}) of TTG typically range from 0.701 to 0.703, and measurements made on mafic volcanic rocks systematically show that the 2.7-Ga mantle had an I_{Sr} of about 0.7011 (Hart, 1977, Moorbath, 1977, O'Nions and Pankhurst, 1978; Peterman, 1979; Martin et al., 1983b). This could be interpreted as indicating either a short time interval between the extraction from the mantle and the crustal accretion, or the existence of mantle heterogeneity in the Early Archean. $\varepsilon_{Nd}(T)$ values range from +4 to –3 (i.e., Jahn, 1994), and possibly reflect Archean mantle heterogeneity, but the existence of positive $\varepsilon_{Nd}(T)$ in terrains as old as 3.9 Ga (Bowring et al., 1989; McCulloch and Bennett, 1993) favours an already depleted mantle at that time. Early mantle depletion is generally considered as an evidence for sialic crust formation prior to 3.9 Ga. This hypothesis is now supported by the discovery of very old zircons in both Australia (Froude et al., 1983; Compston and Pidgeon, 1986) and Canada (Bowring et al., 1989, 1990), which assert the existence of at least some continental crust since 4.3 Ga.

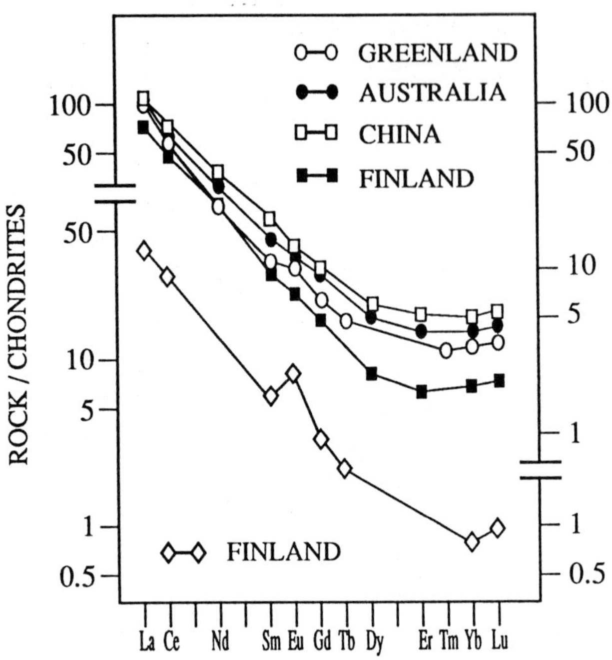

Fig. 8. Chondrite normalized REE patterns for Archean TTG. All are strongly fractionated with low HREE contents and a concave shape at the HREE extremity. Generally, these rocks do not have significant Eu anomalies. The only exceptions are the positive Eu anomalies associated with low-ΣREE samples. Data are from: Greenland (Compton, 1978); Australia, (Jahn et al., 1981); China, (Jahn et al., 1988); Finland (Martin, 1987a). Chondrite normalization values are from Masuda et al. (1973) divided by 1.2.

Comparison with sanukitoids

Over the past decade, magnesium-rich monzodioritic to granodioritic rocks have been described in a number of late Archean terrains from Canada (Shirey and Hanson, 1984, 1986; Sutcliffe, 1989; Stern and Hanson, 1991; Evans and Hanson, 1992), the USA (Mueller et al., 1983), China (Jahn et al., 1988), and India (Jayananda et al., 1995). Because of their geochemical similarity to Mg-rich andesites (sanukites) from Setouchi belt in Southwest Japan they were called the "sanukitoid suite" (Shirey and Hanson, 1984; Stern et al., 1989; Stern and Hanson, 1991). Sanukitoids comprise less than 5% of juvenile Archean continental crust and can be discriminated from TTG on the basis of the following criteria: (1) in contrast to Archean TTG, sanukitoids are always associated with low-SiO_2 dioritic, monzonitic and lamprophyric magmas; (2) in a K-Na-Ca triangle sanukitoid suites show K-enrichment and follow a classical calc-alkaline trend; (3) TTG have low Mg-numbers (< 0.45) and low Cr contents (< 50 ppm), whereas sanukitoids have high Mg-numbers (0.45–0.68) and high-Cr concentrations (50–150 ppm); (4) sanukitoids are always post-kinematic, whereas TTG are pre- to syn-kinematic; (5) contrary to TTG, sanukitoids can strongly interact with pre-existing TTG, inducing partial melting (Jayananda et al., 1992, 1995); (6) sanukitoids appear to be restricted to small late-Archean domains unlike TTG, which are geographically widespread throughout the whole Archean. In some places, as in the Closepet batholith in south India, high-Mg magmatism seems to be related to mantle plume activity beneath an already mature and thick sub-continental lithosphere (Martin et al., 1993; Peucat et al., 1993; Jayananda et al., 1995).

PETROGENESIS

In the past twenty years, several models have been proposed for the origin of trondhjemitic magmas. Figure 9 from Barker (1979) summarizes the more representative ones. Trondhjemitic suites are assumed to be generated either (1) by partial melting of gabbro, quartz eclogite or amphibolite, or (2) by fractional crystallization of a wet, basaltic or low-K andesitic magma. The chemistry of the source, as well as the modal composition of the residue or cumulate, intimately controls the ultimate character (high- or low-Al_2O_3) of the trondhjemitic suite.

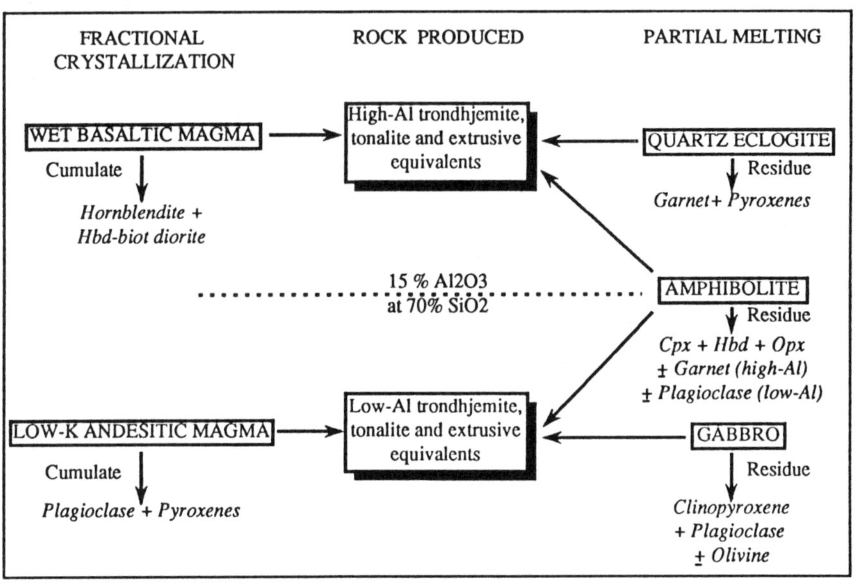

Fig. 9. Diagram summarizing the possible petrogenetic mechanisms for both high- and low-Al_2O_3 trondhjemites (from Barker, 1979).

Although Archean TTG belong exclusively to the high-Al_2O_3 trondhjemite group, at least five families of models have been proposed to explain their genesis and differentiation:

(1) Fractional crystallization of a wet basaltic magma leaving a cumulate made up mainly of hornblende ± plagioclase ± biotite (Arth et al., 1978; Arth, 1979; Barker, 1979; Maaløe, 1982; Smith et al., 1983). As pointed out by Arth et al. (1978), Meijer (1983) and Spulber and Rutherford (1983), more than 75% fractional crystallization is required to generate trondhjemitic liquids from a basaltic source. Such a process should produce enormous amounts of mafic and intermediate rocks, about 3-times the volume of the residual felsic magma. This mafic–intermediate-felsic magmatic association exists in some Proterozoic terranes such as the trondhjemitic suite from Southwest Finland (Arth et al., 1978), but it is unknown in Archean terranes, where TTG are not genetically linked with mafic or intermediate magmatism.

Hildreth and Moorbath (1988) proposed that some high-K dacites from the Andes were derived through fractional crystallization of a mafic magma produced by fusion of the mantle and contaminated by substantial amounts of continental crust. As already discussed by Defant and Drummond (1990), this process is highly improbable because of the large viscosity contrast between mafic and felsic components, which precludes large scale mixing and assimilation.

(2) Direct melting of the mantle (Moorbath, 1975; Peterman and Barker, 1976; Stern and Hanson, 1991). The generation of felsic magmas by partial melting of the mantle requires very low degrees of melting (< 5%). However, even at such low degrees of melting, theoretical calculations based on REE demonstrate that it is impossible to account for the Yb depletion, the concave shape of HREE, and the high La/Yb ratios typical of Archean TTG (Fig. 10; Martin et al., 1983a; Jahn et al., 1984; Martin, 1987a). The magmatic suites derived through partial melting of the mantle should display high Mg-numbers as well as high-Ni and Cr contents. Although these characteristics are commonly observed in sanukitoids, they are unknown in Archean TTG. On an other hand, experimental melting of a wet mantle source (metasomatized or not) (Green, 1976; Green and Ringwood, 1977; Mysen and Boettcher, 1975a,b; Wyllie, 1977) is only able to generate andesitic liquids. Nicholls and Ringwood (1973) and Graviou (1984) considered that subsequent fractional crystallization of olivine and/or spinel at depth can drive magma composition from andesitic to dacitic, but on no account can this process explain the trace-element behaviour in Archean TTG.

(3) Partial melting of Archean greywackes (Arth and Hanson, 1975). This mechanism considers that rapid (T < 0.05 Ga) erosion, sedimentation and burial of a basalt-dacite pile can generate greywackes, whose partial melting can produce the quartz-monzonites associated with dacites and quartz-diorites. Although this model explains the REE and some trace-element characteristics of Archean TTG, it cannot account for major- and other trace-element behaviour (see discussion in Martin, 1985).

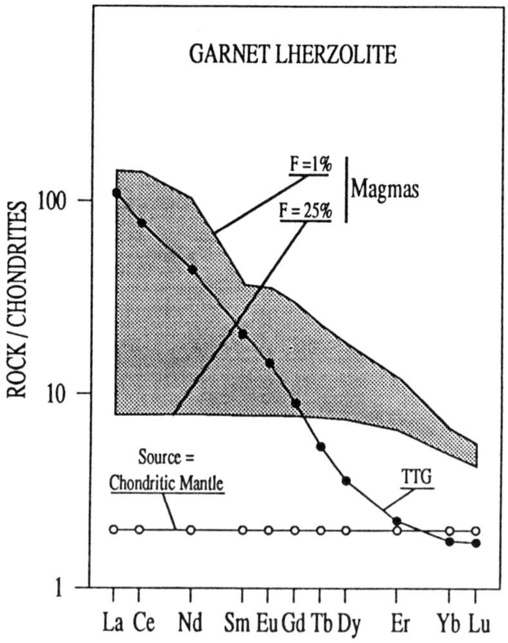

Fig. 10. Chondrite-normalized REE pattern for TTG compared to model magmas generated by partial melting of a mantle source (from Martin, 1987a). This diagram demonstrates that melting of garnet lherzolite cannot produce the fractionated REE patterns, the Yb depletion, or the concave shape of HREE, typical of Archean TTG, even for low degrees (F = 1%) of partial melting. Open circle = mantle source (2 × chondrites); filled circle = TTG; grey field = domain of REE patterns for liquids generated by 1 to 25% melting of the source.

(4) Partial melting of quartz-eclogite (Condie and Lo, 1971; Arth and Hanson, 1972, 1975; Hanson and Goldich, 1972; Glikson, 1976; Compton, 1978; Jahn et al., 1981; Gower et al., 1983; Rapp et al., 1991). The HREE depletion typical of Archean TTG is generally explained by the presence of garnet in the residue of partial melting, and this is why the eclogite model is very attractive. Nevertheless, the remaining problem is that high-pressure rocks like eclogites are unknown in Archean terranes.

(5) Partial melting of garnet amphibolite (Barker and Arth, 1976; Hunter et al., 1978; Barker, 1979; Tarney et al., 1979, 1982; Condie, 1981, 1986; Martin et al., 1983a; Sheraton and Black, 1983; Martin, 1986, 1987a, 1993; Johnson and Wyllie, 1988; Ellam and Hawkesworth, 1988; Arculus and Ruff, 1990; Nédélec et al., 1990; Poidevin, 1991; Winther and Newton, 1991; Rapp et al., 1991, Neymark et al., 1993; Bickle et al., 1993). In this model, partial melting of an Archean tholeiitic garnet amphibolite leaves a residue made up of hornblende + garnet + clinopyroxene + minor plagioclase. It is important to note that the fusion of a garnet-free amphibolite is unable to cause the HREE depletion typical of Archean

TTG. Consequently, it appears that both garnet and hornblende are absolutely essential residual phases during the genesis of TTG magmas. Melt generation may or may not be followed by small degrees of fractional crystallization.

The most recent research on the production of TTG magmas in Archean continental crust favour an origin by partial melting of garnet-bearing amphibolite or hornblende eclogite, a model which is supported by both geochemical and experimental data.

Geochemical data

General mechanism

One of the most likely petrogenetic processes, that accounts for all geochemical characteristics of Archean TTG is a three-stage model (Martin, 1985, 1987a), which is summarized as follows (Fig. 11):

(1) The first stage consists in partial melting of the mantle that generates large volumes of tholeiitic magma. The calculated chemical composition of these mafic magmas is significantly different from that of modern MORB in that they are LREE enriched with slightly fractionated REE patterns ($2 < (La/Yb)_N < 3$; $Yb_N \approx 10$). This composition is similar to the average composition of tholeiites from the Kuhmo–Suomussalmi belt in eastern Finland (Jahn et al., 1980), and intermediate between the enriched (TH2) and depleted (TH1) Archean tholeiites (Condie, 1981). In order to explain the positive $\varepsilon_{Nd}(T)$ observed in even the oldest known Archean mafic and felsic rocks (Jahn, 1994), several authors assume production of LREE-enriched mafic crust in the very Early Archean. Partial melting of this mafic crust generated the early continental crust and caused the mantle depletion (Chase and Patchett, 1988; Galer and Goldstein, 1991). During emplacement of

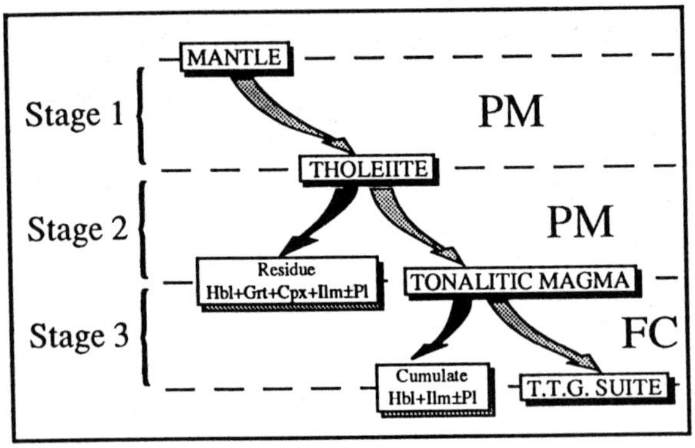

Fig. 11. Schematic diagram summarizing the different stages of petrogenesis of TTG. Hbl = hornblende; Grt = garnet; Pl = plagioclase; Cpx = clinopyroxene; Ilm = ilmenite; PM = partial melting; FC = fractional crystallization (from Martin, 1993).

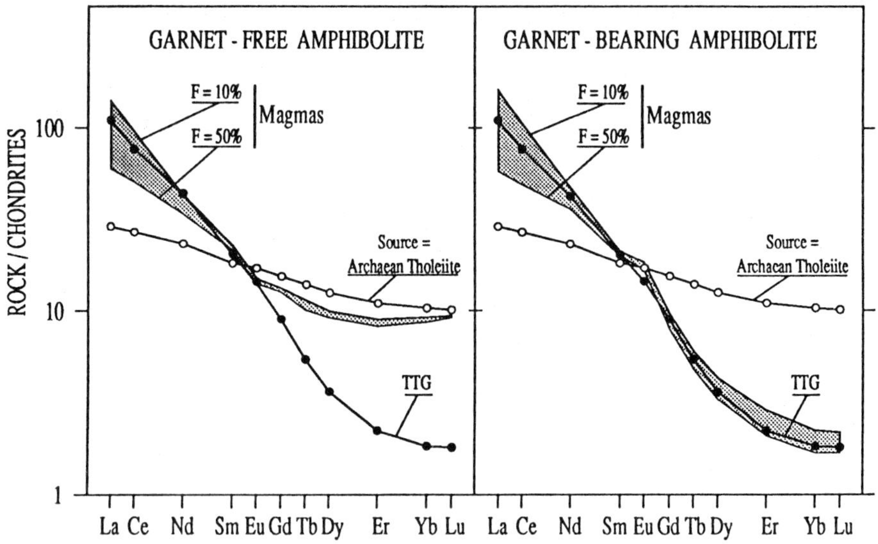

Fig. 12. Chondrite-normalized REE patterns for TTG compared with model magmas generated by partial melting of garnet-free and garnet-bearing (25%) amphibolites (from Martin, 1987a). Partial melting of a garnet-free amphibolite can generate LREE-enriched magmas but is unable to produce significant HREE depletion. However, melting of a garnet-bearing amphibolite perfectly fits TTG composition, thus indicating that both hornblende and garnet must be residual minerals during TTG genesis. Open circles = Archean tholeiite source; filled circles = TTG; grey field = domain of REE patterns for liquids generated by 10 to 50% melting of the source.

the tholeiites (in ridge-like systems, for instance) alteration may have accompanied interaction with sea water.

(2) The second stage is partial melting of Archean tholeiite transformed into garnet amphibolite or hornblende eclogite. The residue of partial melting is dominated by hornblende and garnet with subordinate amounts of clinopyroxene and plagioclase and the degree of partial melting remains low and rarely exceeds 30%. Both hornblende and garnet must be residual phases in order to account for trace-element and REE behaviour (Fig. 12). This important conclusion, also supported by experimental data, shows that melting operated at relatively moderate temperature and pressure (650–850°C and 10–20 kbar), before the amphibolite completely dehydrated. This process does not generate either positive or negative Eu anomalies. The Sr and Nd isotopic composition of TTG indicates that the melting of amphibolite or hornblende eclogite occurred rapidly (≤ 0.1 Ga) after their differentiation from the mantle, since mantle-like isotopic characteristics were preserved. It is during this stage that the important geochemical characteristics of TTG are acquired.

(3) The last stage consists in fractional crystallization. This mechanism is not reported in all Archean TTG, but is well documented in some places. Its role can be

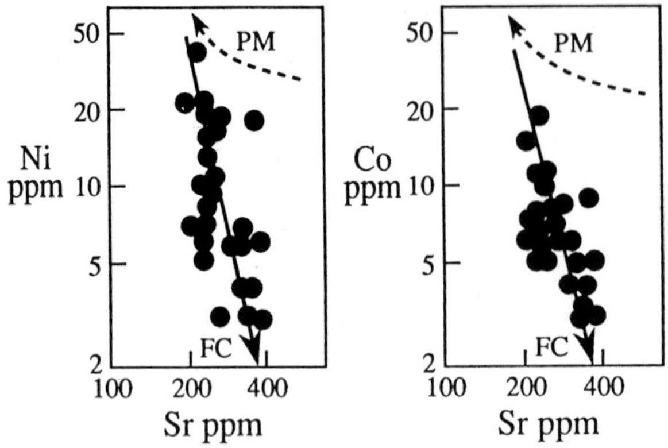

Fig. 13. Log of incompatible element (Sr) versus log of compatible elements (Ni, Co) demonstrating the role that fractional crystallization can play in the differentiation of TTG suites. In this type of diagram, partial melting (PM) draws a sub-horizontal trend (dotted lines), whereas fractional crystallization (FC) defines a sub-vertical trend (black line). Fractional crystallization is not reported in all TTG suites and it is always of small extend (less than 30%). The data (filled circles) are TTG from eastern Finland (Martin, 1985), and they all plot along a fractional crystallization trend.

demonstrated by field and petrological data or using a log(incompatible element) vs. log(compatible element) diagram (Fig. 13), which discriminates between fractional crystallization (sub-vertical trend) and partial melting (sub-horizontal trend) (Cocherie, 1986). The degree of fractional crystallization is always low and never exceeds 30%, a feature which is probably due to the high viscosity of felsic magmas, that prevents large-scale mineral separation. At this step of the process, the role of accessory minerals, such as allanite, apatite, zircon, etc., can be prominent and totally control some trace-element behaviour. For instance, the unusual Eu positive anomaly as reported in some TTG is acquired during this stage through allanite and/or apatite fractionation (Martin, 1987a).

Nb-Ta-Ti anomalies

Archean TTG have negative Nb-Ta-Ti anomalies, as is the case for many continental crustal rocks (Fig. 7), and the interpretation of these anomalies is still under debate. In modern island arcs they are assumed to be caused by the contamination of the magma or/and its source by crustal components. Jahn (1994) calculated that only 2% contamination of a primitive mantle by upper continental crust would generate a strong negative Nb-Ta anomaly without significantly changing the major-element composition. This interpretation may not be appropriate during the early history of the earth. In the Archean, the volume of continental crust was probably small and its recycling was negligible or even non-existent as indicated by the low I_{Sr} and the positive or slightly negative $\varepsilon_{Nd}(T)$ typical of Archean TTG.

It has also been suggested that Ti-bearing minerals, such as sphene, rutile, and ilmenite, which have high mineral/liquid partition coefficients (KD) for HFSE (high field strength elements) (Green and Parson, 1986; Ryerson and Watson, 1987) could have been residual phases during magma genesis, causing HFSE to have a compatible behaviour. Hofmann (1988) proposed that in modern subduction zones, partial melting of hydrated subducted oceanic crust generates magmas with sufficiently low Ti solubility to keep rutile stable in the residue. On an other hand, Kelemen et al. (1990) consider that the HFSE depletion in arc magmas reflects interactions between the magma and a depleted peridotite. Olivine, orthopyroxene and spinel have higher KD for HFSE than for other incompatible elements, thus inducing their relative depletion in the magma. If this process operated during TTG genesis, Nb-Ta-Ti negative anomalies could testify the slow ascent of the magma through a depleted mantle.

An alternative explanation was provided by Drummond and Defant (1990), who take into consideration the role played by amphibole. KD values for Nb-Ta-Ti between amphibole and intermediate and felsic liquids are high (Ti = 6; Nb = 4), (Pearce and Norry, 1979; Lemarchand et al., 1987). Since these KD values are greater than 1, the presence of hornblende among the residual or cumulative phases would produce HFSE depletion. This hypothesis is very attractive because: (1) hornblende is one of the main fractionating minerals during TTG magma genesis, behaving as a near-liquidus phase during both partial melting and fractional crystallization; and (2) the model does not necessitate an hypothetical crustal contamination, which may not have been important during generation of the earliest continental crust. It has also been suggested (Drummond and Defant, 1990) that garnet could significantly contribute to HFSE depletion, but this possibility cannot be discussed further because of the lack of reliable KD data for HFSE in garnet.

Experimental data

In the last twenty years, experimental melting has been undertaken in order to understand the petrogenesis of calc-alkaline magmatism. Both lherzolite and basalt have been studied as possible source rocks. Lherzolite experimental melting generates liquids whose composition is basaltic andesite and andesite. Silica-rich magmas (dacitic) have also been produced but the application to their genesis is still under debate. They are considered either as primary liquids (Kushiro et al., 1972; Mysen and Boettcher, 1975 a,b) or as products of subsequent fractional crystallization of olivine at depth (Green, 1973, 1976; Nicholls and Ringwood, 1973). Whatever the process is in nature, basaltic andesite and andesite liquids are commonly produced, yet this range of composition is very rare in Archean TTG suites. On the other hand, when experimental dacitic liquids are generated, they have both high Mg-numbers and high CaO contents ($\geq$ 0.6 and 6–10%, respectively), which are clearly distinct from TTG values (0.43 and 3.19% respectively;

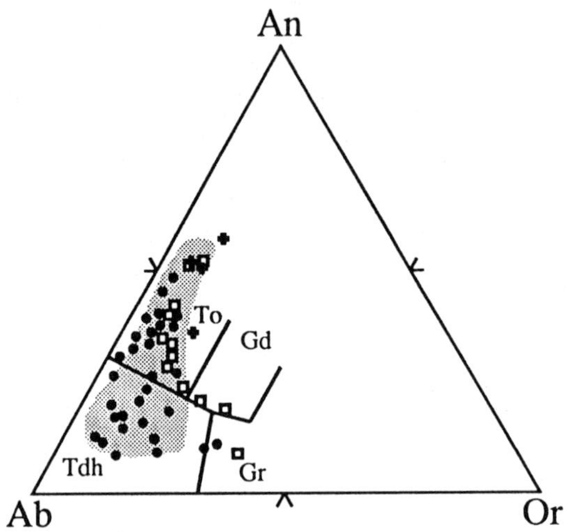

Fig. 14. Normative An-Ab-Or diagram of O'Connor (1965), and Barker (1979) showing the composition of liquids generated by experimental fusion of basalts and amphibolites. Open squares = Holloway and Burnham (1972) and Helz (1976); black crosses = Rushmer (1991); filled circles = Rapp et al. (1991); grey field = Winther and Newton (1991). To = tonalite; Tdh = trondhjemite; Gd = granodiorite; Gr = granite.

Table 2). Experimental petrology data demonstrate that TTG parental magmas cannot be generated from direct partial melting of a mantle source, which agree with geochemical modelling.

Also, several melting experiments were made on basaltic sources under various conditions. Holloway and Burnham (1972) melted a tholeiite at water-undersaturated conditions, and they obtained tonalitic liquids that do not show trondhjemitic affinity (Fig. 14). Similarly, Helz (1976) studied water-saturated melting of alkali-basalt and tholeiite at 5 kbar, here she also generated tonalitic and granodioritic liquids. As shown by Beard and Lofgren (1989, 1991), most of the experimental melting of basalt in water-saturated systems produces Al-, Ca-, K-rich and Fe-poor liquids when compared with Archean TTG. They conclude that TTG magma cannot be produced under water-saturated conditions, but rather by dehydration melting. Water content in altered basalts from subducted oceanic crust ranges up to 7% (Humphris and Thompson, 1978), where it is stored as OH^- in minerals and it is released when these minerals break-down. To constrain the production of TTG magma, dehydration melting experiments were performed at water-undersaturated conditions, either on amphibolites with no water added to the system or on natural unaltered basalts with less than 5% water added. Tonalitic and trondhjemitic liquids were produced by dehydration melting of low-K tholeiite at 1–7 kbar (Beard and Lofgren, 1991) and at 8 kbar (Rushmer, 1991). Melting begins at about 800–850°C and amphibole is stable until 950°C. Residual phases are plagioclase + amphibole + clinopyroxene + orthopyroxene + ilmenite or

TABLE 4

Composition of liquids generated by experimental dehydration melting of amphibolites compared with average TTG composition

	Weight % oxide				
	Experiments			TTG	
	1	2	3	4	5
SiO_2	69.75	68.85	69.76	69.79	4.9
Al_2O_3	16.89	16.62	15.59	15.56	1.2
Fe_2O_3*	2.96	3.57	3.8	3.12	1.5
MnO	0.09	0.06	0.03	0.05	0.03
MgO	1.26	0.72	0.71	1.18	0.7
CaO	3.93	3.16	3.16	3.19	1.0
Na_2O	4.2	5.29	4.5	4.88	0.75
K_2O	1.31	1.28	1.81	1.76	0.70
TiO_2	0.54	0.75	0.85	0.34	0.16
K_2O/Na_2O	0.31	0.24	0.40	0.36	
Mg#	0.46	0.29	0.27	0.43	

(1) Average from Rapp et al. (1991); (2) Sample #1 (1000°C, 16 kbar) from Rapp et al. (1991); (3) Average from Winther and Newton (1991); (4) Average and (5) standard deviation for Archean TTG (this work); Mg# = mol MgO/(MgO+FeO*).

plagioclase + clinopyroxene + orthopyroxene + ilmenite + magnetite. It is important to note that during these low-pressure experiments garnet does not appear in the residual assemblage.

Vapour-absent experiments on metabasalts were conducted by Rapp et al. (1991) for pressure ranging from 8 to 32 kbar, Winther and Newton (1991) on high-Al basalt and an average Archean tholeiite between 5 and 30 kbar, and Wolf and Wyllie (1993) on natural low-K calcic amphibolite. The melts generated for 10 to 40% fusion are high-Al_2O_3 tonalites and trondhjemites (Fig. 14) and equilibrated with residues made up of plagioclase + amphibole ± orthopyroxene ± ilmenite at low pressure (8 kbar), garnet + amphibole ± plagioclase ± clinopyroxene ± ilmenite at 16 kbar, and garnet + clinopyroxene ± rutile at higher pressure. Typically, an average K_2O/Na_2O ratio is of about 0.3 with $Na_2O = 4.9$ (Table 4). Rapp et al. (1991) and Wolf and Wyllie (1993) calculated REE patterns of liquids generated by dehydration melting at low pressure in the garnet-absent system, the REE patterns are flat or poorly fractionated without any HREE impoverishment (Fig. 15). At higher pressures, when garnet is a residual phase, REE are strongly fractionated ($30 \leq (La/Yb)_N \leq 50$) with HREE depletion. Wolf

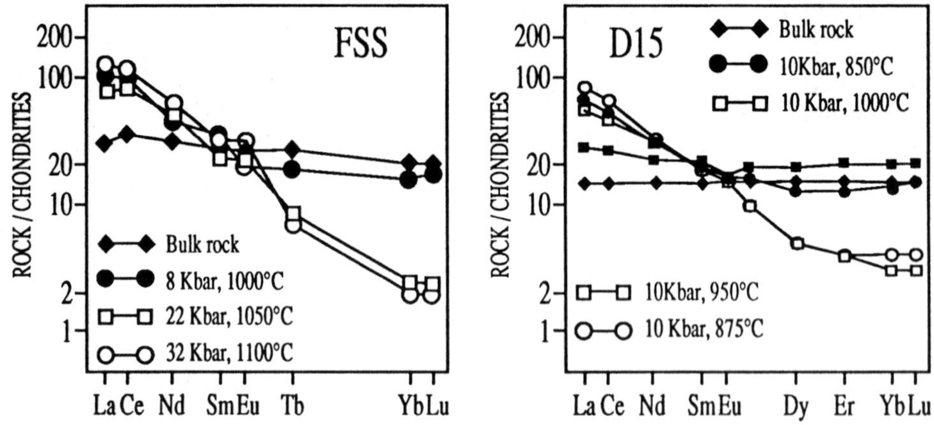

Fig. 15. Chondrite normalized REE patterns of liquids generated by experimental melting. FSS = tholeiite (Rapp et al. 1991); D15 = low-K calcic amphibolite (Wolf and Wyllie, 1993). At relatively low pressure, when garnet is not stable in the residue (filled symbols), HREE contents remain high, more or less similar to the parent rock and LREE display various degrees of enrichment. When garnet is stable in the residue (open symbols), TTG-like LREE-enriched and HREE-depleted patterns are obtained.

and Wyllie (1993) indicated that dehydration melting of amphibolite can begin at temperatures as low as 740°C (P = 10 kbar), but the common temperature range seems to be between 825 and 1000°C. Wolf and Wyllie (1991) considered that the most suitable conditions for melt extraction from amphibolitic residue are realized at 10 kbar in a temperature interval of 850–900°C. When hornblende dehydrates, the released water lowers the viscosity of the liquid, whereas at higher temperatures, the melt fraction increases and becomes water-undersaturated, and its viscosity increases (McKenzie, 1984) thus precluding an efficient segregation. Nevertheless, typical TTG compositions are obtained for pressures significantly higher than 10 kbar (≥ 16 kbar).

In summary, it appears that experimental data are consistent with geochemical conclusions and demonstrate that it is possible to generate Archean TTG magmas by partial melting of an altered tholeiite at intermediate to high pressures. In addition, experiments indicate that: (1) HREE depletion observed in TTG is obtained only when garnet is a residual phase; (2) hornblende and/or Fe-Ti oxides (rutile, ilmenite) are common residual minerals, thus accounting for Ti-Nb-Ta negative anomalies in Archean TTG; (3) dehydration melting is the most reliable process for TTG magma production thus precluding excess water in the system; and (4) the particular composition of the liquids is highly dependent on the source composition. Consequently, the relative homogeneity of Archean TTG could reflect the homogeneity of their source, as well as the constancy of melting conditions.

COMPARISON BETWEEN ARCHEAN AND MODERN GRANITOIDS

Since Poldervaart (1955), the discrepancy between Archean and modern granitoids in terms of rock composition and field association was widely documented. The timing of this fundamental change is thought to have occurred 2.5 Ga ago at the Archean–Proterozoic boundary. For instance, (1) high-MgO komatiitic lavas are widespread in the Archean but almost disappear after 2.5 Ga; (2) per-alkaline magmas as well as eclogites are unknown or very rare in Archean terranes, they appear in significant amounts only since the Early Proterozoic; (3) andesites are minor components in Archean greenstone belts but represent huge volumes in Phanerozoic arc systems. The average composition of the upper continental crust expressed in sedimentary records, also shows chemical breaks at the Archean–Proterozoic boundary, as illustrated by the Eu/Eu* and Th/Sc ratios (Taylor and McLennan, 1985; Condie, 1993).

TTG are by far the major constituents of Archean continental crust, where they represent new magmas added to the crust. For this reason, they will be subsequently referred as juvenile magmas. A major difficulty in addressing the question of change in composition and petrogenesis of juvenile continental crust with time, is the role of crustal recycling processes. As the volume of continents increased, the probability of their recycling (by anatectic or sedimentary processes) also increased. This mechanism, subordinate during the Archean, became prominent after 2.5 Ga. In order to compare similar and equivalent processes, the subsequent comparisons will preclude all magmas generated by reworking of older crustal materials, and focus only on juvenile magmas. The identification of juvenile igneous rocks is based on their mantle-like isotopic characteristics (Sr, Nd, Pb). For post-Archean period, only I-type (Chappell and White; 1974) or M-type (Didier et al., 1982) granitoids were selected, whereas A-type granitoids were rejected.

Variation of juvenile granitoid composition

One of the more obvious differences between Archean TTG and post-Archean juvenile granitoids is that most of the latter are associated with mafic to intermediate plutonic rocks and that they generally contain large amounts of microgranular enclaves (Didier, 1973). As mentioned previously, Archean juvenile granitoids are TTG in composition with low K_2O/Na_2O ratios (< 0.5; Table 5). In a K-Na-Ca plot they show Na-enrichment and plot in the trondhjemitic field (Fig. 6B, 16). Alternatively, post-Archean granitoids are typically granodioritic to granitic in composition with an average K_2O/Na_2O ratio of 0.92 (Table 5) and individual K_2O/Na_2O values that can be greater than 1 (Ronov and Yaroshensky, 1976). In the K-Na-Ca plot, they define a classical calc-alkaline trend, which is characterized by K-enrichment during differentiation (Fig. 16). Mg numbers are similar (≈ 0.4) despite Fe_2O_3*+MgO being higher in the post-2.5 Ga magmas.

TABLE 5

Average major-element composition of Archean TTG and Modern arc plutonic rocks. For modern granitoids only samples belonging to the I-type (Chappell and White; 1974) or to the M-type (Didier et al., 1982) granitoids were selected

	Weight % oxide				
	Archean TTG T>2.5 Ga		Modern granitoids T<0.2 Ga		
	Average (*n*=355)	Std. dev.	Average *n*=250	Std. dev.	Condie (1993)
SiO_2	69.79	4.90	68.10	6.20	67.2
Al_2O_3	15.56	1.20	15.07	1.60	15.5
Fe_2O_3*	3.12	1.50	4.36	2.00	4.22
MnO	0.05	0.03	0.09	0.10	—
MgO	1.18	0.70	1.55	1.00	1.70
CaO	3.19	1.00	3.06	0.64	3.50
Na_2O	4.88	0.75	3.68	0.49	3.50
K_2O	1.76	0.70	3.40	1.10	3.30
TiO_2	0.34	0.16	0.54	0.32	0.55
P_2O_5	0.13	0.10	0.15	0.08	0.12
Fe_2O_3*+MgO	4.30		5.91		5.92
Fe_2O_3*/MgO	2.64		2.81		2.48
K_2O/Na_2O	0.36		0.92		0.94
A/CNK	0.99		0.98		0.99
Mg#	0.43		0.41		0.44

A/CNK = mol $Al_2O_3/(CaO+Na_2O+K_2O)$; Mg# = mol MgO/(MgO+FeO*).

As with major elements, trace elements also show significant differences as summarized in Table 6 and in Fig. 17. When compared with TTG, modern granitoids are enriched in all incompatible elements (Rb, Ba, Th, U, K). For instance, Archean TTG are Rb-poor and Sr-rich such that the Rb/Sr ratio changes from 0.12 to 0.35 at the Archean–Proterozoic boundary. Both old and young granitoids display similar Nb, Ta, P and Ti negative anomalies (Fig. 17), whereas Eu exhibits significant negative anomalies in the more recent rocks (Eu/Eu* = 0.99 and 0.57 before and after 2.5 Ga respectively). Another significant difference is the strong depletion in Sc, Y, Yb, and HREE observed in Archean TTG (Table 6). Figure 18 is a $(La/Yb)_N$ vs. (Yb_N) graph showing a compilation of more than 355 Archean TTG and 400 post-Archean juvenile granitoids. When younger than 0.2 Ga, the rocks (250) were selected in an established subduction zone environ-

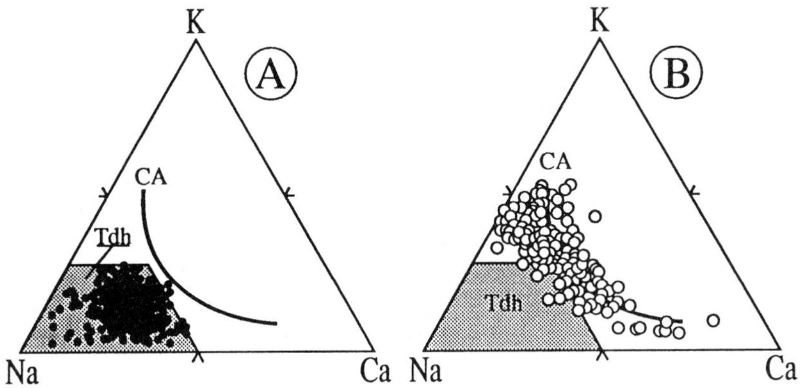

Fig. 16. K-Na-Ca plots for Archean TTG (A) and post-2.5 Ga juvenile granitoids (B). The TTG plot in the trondhjemitic field, whereas their modern equivalents show a K-enrichment and follow a classical calc-alkaline trend. (CA) = calc-alkaline trend; (Tdh) = trondhjemitic field.

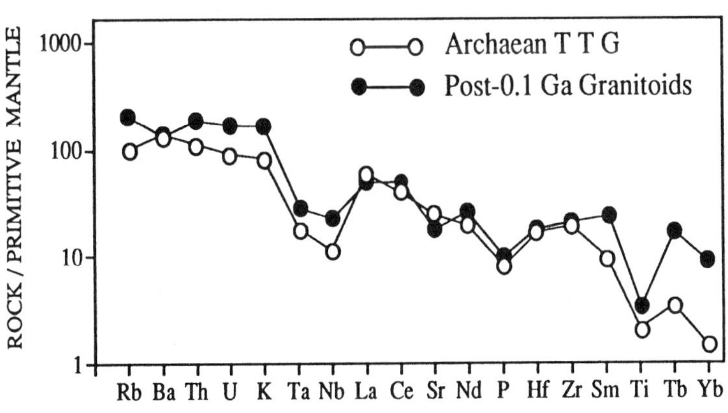

Fig. 17. Primitive-mantle normalized average compositions of Archean TTG (open circles) compared with modern juvenile granitoids (filled circles). Both present similar Ta, Nb, P and Ti negative anomalies, but the modern granitoids differ by their higher Rb, Th, U, K and HREE contents.

ment. In such a diagram, the $(La/Yb)_N$ ratio is representative of the mean degree of fractionation of the REE pattern, and each pattern is reduced to a single point in the $(La/Yb)_N$ vs. (Yb_N) space (Jahn et al., 1981; Martin et al., 1983a; Martin, 1986). Only two age groups are distinguished.

(1) Archean TTG (Fig. 18, inset A). REE patterns are strongly fractionated ($5 < (La/Yb)_N \leq 150$) with low Yb contents ($0.3 < Yb_N < 8.5$). The degree of fractionation is the same for LREE and HREE, there is no significant positive or negative Eu anomalies, and the HREE show a concave shape.

TABLE 6

Average trace-element composition of Archean TTG and Modern arc plutonic rocks

	ppm				
	Archean TTG T>2.5 Ga		Modern granitoids T<0.2 Ga		
	Average (n=355)	Std. dev.	Average n=250	Std. dev.	Condie (1993)
Rb	55	30	110	50	90
Ba	690	500	715	205	800
Th	6.9	6	11.8	6.5	10
U	1.6	1.3	3.0	1.0	2.5
Ta	0.71	0.5	1.1	0.3	0.65
Nb	6.4	4	12.1	5.0	8
Sr	454	200	316	150	350
Hf	4.5	2	4.7	2.0	4.2
Zr	152	110	171	53	140
Cr	29	30	23	15	33
Y	7.5	4	26	5	25
Sc	4.7	2	13.3	4.0	14.0
Ni	14	10	10.5	8.0	15.0
V	35	23	76	45	80
La	32	20	31	9	25
Ce	56	32	67	17	47
Nd	21.4	3	27	7	19.5
Sm	3.3	1.5	5.3	14	4.6
Eu	0.92	0.4	1.0	0.5	1.1
Gd	2.2	1.1	5.5	1	4.67
Tb	0.31	0.1	1.0		0.76
Dy	1.16	0.7	5.2	0.1	—
Er	0.59	0.3	3.0	1	—
Yb	0.55	0.3	3.2	0.5	1.7
Lu	0.12	0.1	0.5	0.1	0.28
Rb/Sr	0.12		0.35		0.26
$(La/Yb)_N$	38.4		6.4		9.7
Eu/Eu*	0.99		0.57		0.73

$Eu/Eu^* = Eu_N/((Sm_N+Gd_N)/2)$.

(2) Post-Archean juvenile granitoids (Fig. 18, inset B). These rocks have moderately fractionated REE patterns ($(La/Yb)_N \leq 20$) with relatively high Yb contents ($5 < Yb_N \leq 20$). Unlike TTG, the HREE are relatively flat and clearly less

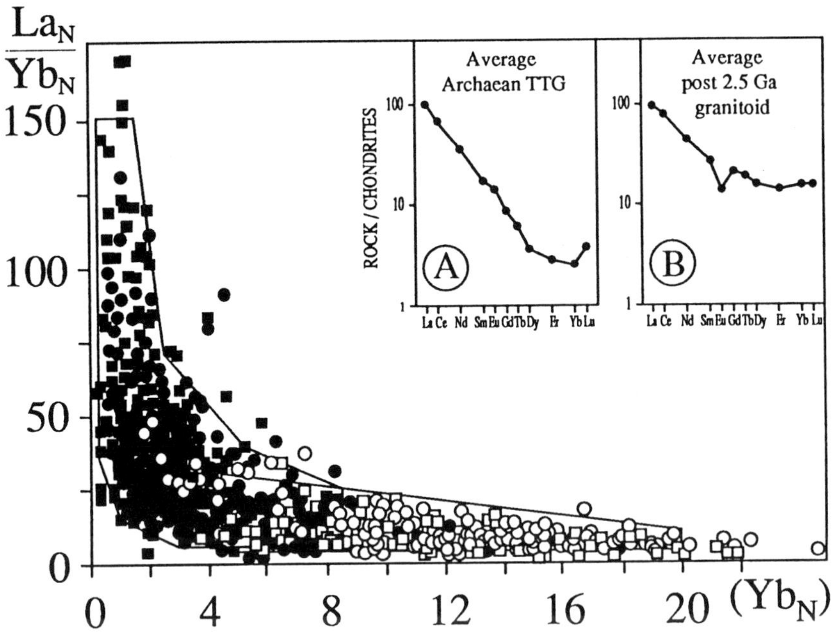

Fig. 18. Compilation of $(La/Yb)_N$ and Yb_N values for Archean TTG and post-Archean juvenile granitoids. Four rock groups are distinguished on the basis of their age of emplacement: T > 3.0 Ga (filled circles); 3.0 Ga < T < 2.5 Ga (filled squares); 2.5 < T < 0.2 Ga (open circles); T < 0.2 Ga (open squares). Only two chemical groups can be distinguish, they show a major break at the Archean-Proterozoic boundary: TTG emplaced before 2.5 Ga are characterized by low-Yb contents ($0.3 < Yb_N < 8.5$) and correlated strongly fractionated REE patterns ($5 < (La/Yb)_N < 150$); granitoids generated after 2.5 Ga have high-Yb concentrations ($4.5 < Yb_N < 20$) and moderately fractionated REE patterns ($(La/Yb)_N \leq 20$). The two insets represent average REE patterns for Archean TTG (A), and post-Archean granitoids (B), (Table 6).

fractionated than LREE. They also do not have a concave shape and Eu commonly exhibits a significant negative anomaly (Eu/Eu* = 0.57).

These are the more prominent features, but Fig. 18 requires additional scrutiny:

- In the $(La/Yb)_N$ vs. (Yb_N) plot, four rock groups were made as function of their age. No significant difference can be evidenced (1) between TTG emplaced before 3.0 Ga and between 3.0 and 2.5 Ga or (2) between juvenile granitoids generated in the 2.5–0.2 Ga range and after 0.2 Ga. This allows consideration of the Archean–Proterozoic boundary as a major break in the earth's history.
- The two groups of points overlap for $4.5 \leq Yb_N \leq 8.5$, such as this part of the diagram is not discriminant. REE behaviour modelling should explain this overlap.
- The $(La/Yb)_N$ and (Yb_N) values are inter-correlated and when insets A and B are compared, it clearly appears that the change in REE patterns relative to time is due mainly to changes in Yb rather than in La content (Table 6).

Ytterbium evolves from low concentrations in Archean TTG to high concentrations in post-2.5 Ga granitoids. As previously discussed, both geochemical modelling and experimental petrology demonstrate that the low Yb values (as well as other geochemical characteristics) observed in TTG are a direct consequence of the presence of hornblende + garnet in the residue of partial melting of the tholeiitic source. This constraint is of major importance because it fixes the thermodynamic conditions of amphibolite dehydration melting in the stability field of hornblende and garnet. Reciprocally, the high Yb concentrations in post-2.5 Ga juvenile granitoids can be interpreted as indicating the lack of hornblende and garnet in their source. Consequently, these differences reflect secular modifications in thermodynamic conditions of melting and/or in source composition (Martin, 1985, 1986; Defant and Drummond, 1990; Drummond and Defant, 1990).

Petrogenetic model

Field, experimental and theoretical research have established that plate tectonic-like mechanisms operated during the Archean, including creation and destruction of oceanic lithosphere as well as collision between rigid crustal blocks. This type of kinematic setting would be more or less equivalent to rifting, subduction and collision modern environments (Bickle, 1978; Condie, 1981; Arndt, 1983; Nisbet and Fowler, 1983; Abbott and Hoffman, 1984; Campbell and Jarvis, 1984; Martin, 1986, 1987 a, b, 1993; Nisbet, 1987; Ellam and Hawkesworth, 1988; Mueller and Wooden, 1988; Arkani-Hamed and Jolly, 1989; Jégouzo and Blais, 1991; McGregor et al., 1991; Nutman and Collerson, 1991; DeWit et al., 1992; Hale, 1992; Maruyama, 1992; Maruyama et al., 1992; McCulloch, 1992; Nutman et al. 1992; Treloar et al. 1992; Ludden et al., 1993; Jahn, 1994). On the other hand, the secular compositional change in juvenile continental crust (as well as in other earth components) indicates that if plate tectonics operated during the first half of the earth history, at least in detail, the processes were different.

Today, most of the new continental crust is generated in arc systems. Wyllie (1979, 1983) addressed the problem of the place where calc-alkaline magmas are generated in modern subduction zones as a function of heat distribution between the subducted lithosphere and the mantle wedge. Four possibilities are considered.

(1) Hot oceanic crust and cold mantle. Calc-alkaline magmas are produced by melting of the subducted slab, and they traverse the mantle wedge, that is too cold to undergo partial melting.

(2) Hot oceanic crust and hot mantle. Calc-alkaline magmas can be generated either by melting of the subducting oceanic slab (magmas which cross the mantle wedge are able to initiate its melting), or by melting of the mantle wedge. In the latter case, fluids released by dehydration of the subducted slab metasomatize the mantle wedge and induce its fusion.

(3) Cold oceanic crust and cold mantle. The subducted slab only dehydrates, it cannot melt, and the liberated fluids do not intersect the hydrous solidus of the

mantle wedge. The only possibility to generate magmas would be the melting of the base of the overlying continental crust.

(4) Cold oceanic crust and hot mantle. Here also the oceanic crust dehydrates, but the liberated fluids intersect the hydrous solidus of the mantle wedge, and cause its partial melting.

In summary, it appears that the place where calc-alkaline magmas are generated is controlled by a competition between dehydration and partial melting processes in the subducted slab. In fact, partial melting of the subducted crust can occur only when the latter is abnormally hot, a drastic condition that is rarely realized in modern subduction environments (Wyllie, 1983; Martin, 1987b; Defant and Drummond, 1990; Drummond and Defant, 1990). Today the most widespread scenario is the subduction of cold oceanic crust into hot mantle.

In present day arc environments, the oceanic crust is old when it begins to subduct. Its average age is of about 60 Ma (Bickle, 1978), but ranging up to 180 Ma. Parson and Sclater (1977) calculated that, for identical initial temperature and composition, an 80-Ma-old lithosphere is twice as cold as a 20-Ma-old one (Fig. 19). Consequently, in most modern subduction zones, the oceanic lithosphere behaves as a cold slab. When it penetrates into the mantle it exchanges energy and heat with the mantle, such that the latter cools down and the slab warms up. The fluids released by slab dehydration, migrate upward and also tend to cool the overlying mantle (Peacock, 1990, 1993). These processes should result in the progressive cooling of the mantle wedge, if hot material is not continuously brought to the wedge by mantle convection (Toksöv and Hsui, 1978; Anderson et al., 1980; Honda and Uyeda, 1983).

In a subduction zone, isotherms retrogress along the Benioff plane, and the geothermal gradient along the slab–mantle interface remains low (≤ 10°C/km;

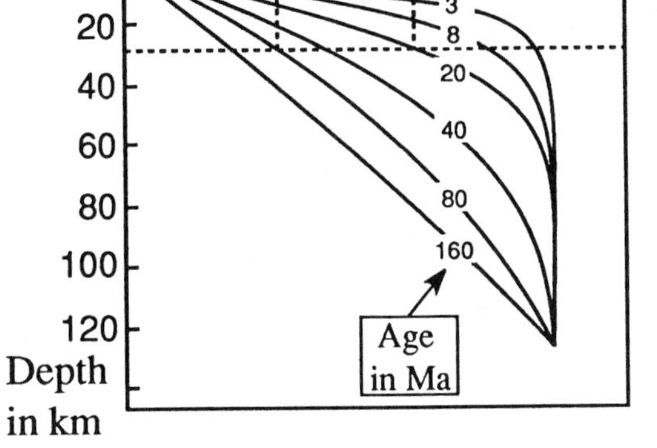

Fig. 19. Depth vs. temperature diagram (after Parson and Sclater, 1977) showing isotherms in subducted lithosphere as a function of its age. As example, dotted lines indicate that at 30 km depth, the temperature of a 20-Ma-old lithosphere is about twice that of a 80-Ma-old one.

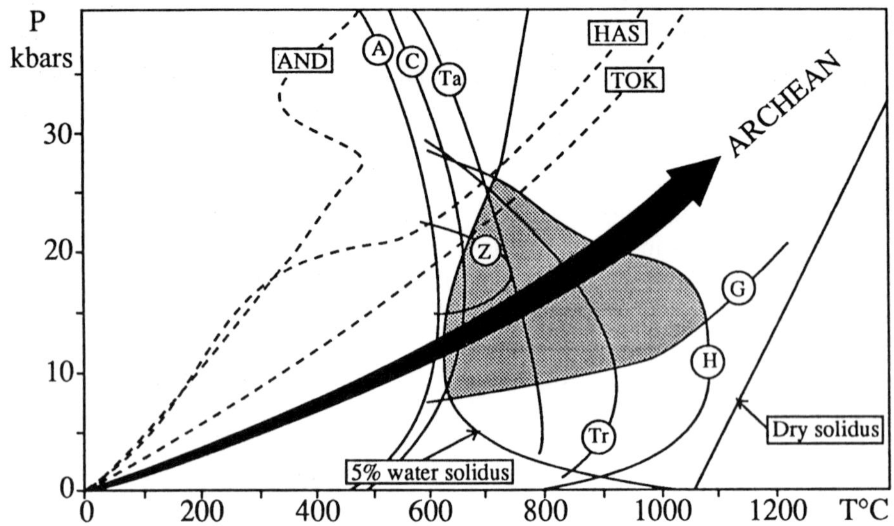

Fig. 20. P-T diagram displaying the dry and 5% hydrous solidus of tholeiite (Wyllie, 1971; Green, 1982). The main dehydration reactions of the oceanic lithosphere are also drawn: H = hornblende-out; A = anthophyllite-out; C = chlorite-out; Ta = talc-out; Tr = tremolite-out; Z = zoisite-out. The stability field of garnet is delimited by the G line. The grey field is the P-T domain where a magmatic liquid generated by partial melting of an hydrated tholeiite can coexist with a hornblende- and garnet-bearing residue. In modern subduction zones, the geothermal gradients along the Benioff plane are low (HAS = Hasebe et al., 1970; TOK = Toksöv et al., 1971); they must be even lower when calculations account for the endothermic character of dehydration reactions (AND = Anderson et al., 1978). These gradients are such that slab dehydration generally occurs before melting can begin. During the Archean, when geothermal gradients were higher (black arrow), melting began in the subducted slab before dehydration, in the hornblende + garnet stability field. (from Martin, 1993).

Fig. 20; Hasebe et al., 1970; Toksöv et al., 1971; Bird et al., 1975; Drummond and Defant, 1990). According to Toksöv et al. (1971) and Defant and Drummond (1990), a 10°C/km thermal gradient intersects the melt + garnet + hornblende stability field at relatively high pressures. However, as a slab subducts, it is metamorphosed and progressively dehydrates and Fig. 20 shows that, even with a 10°C/km thermal gradient, dehydration curves are intersected before reaching the solidus of hydrous tholeiite. At temperatures of about 750°C (temperature of the 5% hydrous–tholeiite solidus at 20 kbar, Wyllie, 1971; Green, 1982), the oceanic lithosphere should be almost completely dehydrated and melting of such a dry tholeiite cannot occur at low temperatures but requires a temperature higher than 1200°C at 20 kbar. However, the residence time of fluids in the slab is not negligible, and fluids could be retained into the slab at temperatures greater than the temperature of dehydration reactions (Philippot and Selverstone, 1991; Philippot, 1993). Consequently, as pointed out by Defant and Drummond (1990) melting of the slab in the garnet + hornblende stability field is possible, but only in a very narrow P-T window. This condition is realized when a very young and hot oceanic crust is subducted (Martin, 1986, 1987b; Drummond and Defant, 1990; Peacock, 1990, 1993).

Lower thermal gradients (i.e. Hasebe et al., 1970) cannot produce magmas equilibrated with a garnet + hornblende-bearing residue. On an other hand, Delany and Helgerson (1978) and Anderson et al. (1978) considered the exothermic character of dehydration reactions, and showed that they cool or inhibit the slow heating of a subducted slab. Figure 20 illustrates this relationship in the case of an old and cold subducted slab, and displays that the resulting calculated thermal gradient totally precludes partial melting of the oceanic lithosphere in or outside of the garnet + hornblende stability field. These very low thermal gradients were calculated in the case of a cold subducted slab, nevertheless, it remains that dehydration reactions are exothermic and contribute to cool the subducted slab and to lower the thermal gradients. Consequently, even if in some specific cases, thermal gradients as high as 10°C/km can be reached in modern subduction zones, thus permitting melting of the subducted slab, the most ordinary situation consists in lower thermal gradients that favour dehydration, fluid migration, and that do not allow slab melting.

The fluids liberated by dehydration of the subducted crust are mainly H_2O with subordinate amounts of CH_4 and CO_2, and their density is 0.3–0.5 times that of the surrounding rocks (Peacock, 1990, 1993). Hence, they will tend to migrate upwards and to diffuse from the slab to the mantle wedge. Depending on fluid/mineral partition coefficients, the fluid phase is selectively enriched in elements such as LILE and LREE (Nicholls, 1974; Mysen, 1979; Taylor and Nesbitt, 1988). It has also been proposed that this mechanism could be responsible for Nb-Ta-Ti anomalies in arc magmas (Tatsumi et al., 1986). For instance the fluid/mineral partition coefficients are by far lower for Nb (and Ta) than for Rb, Sr and REE (Tatsumi and Nakamura, 1986), leading to a LILE enrichment and correlated Nb-Ta impoverishment of the fluid. Partial melting of a peridotitic source metasomatized by this kind of fluid would produce Nb-Ta depleted magmas.

The fluids are not only LILE and LREE carriers, but they also hydrate the mantle wedge and lower its solidus temperature. Experimental data have shown that partial melting of an hydrous peridotite generates calc-alkaline magmas that can subsequently evolve through high-pressure olivine crystallization and/or crustal contamination and hybridization to form a typical basalt–andesite–dacite–rhyolite (BADR) suite (Kushiro et al., 1972; Green, 1973, 1976; Mysen and Boettcher, 1975 a,b; Mann, 1983; Meijer, 1983). In this context, the primordial source of calc-alkaline magmas is metasomatized mantle peridotite, whose partial melting leaves a residue of olivine + pyroxenes. Depending on the depth of melting, garnet or spinel can also play a subordinate role, but garnet + hornblende cannot be major residual phases. The bulk distribution coefficients (D) for HREE are always lower than 1 in these assemblages, so HREE have an incompatible behaviour resulting in the high HREE contents ($5 < Yb_N \leq 20$; Fig. 18) of the modern arc calc-alkaline magmas.

Geochemical modelling and experimental data demonstrate that TTG magmas can be generated by dehydration melting of an Archean tholeiite, only when garnet and hornblende coexist in the residue. In Fig. 20, this constraint can be met when P-T conditions suitable for TTG genesis are in the grey field. It was previously mentioned that present-day geothermal gradients along Benioff planes intersect this grey field only in few restricted situations.

As TTG production always occurs in the grey P-T domain in Fig. 20, this implies that geothermal gradients were higher during the first half of earth history. Gradients of 25 to 30°C/km along the Benioff plane (black arrow in Fig. 20) have been proposed for Archean subduction environments (Martin, 1986, 1987b). With such gradients, for pressures ranging from 8 to 18 kbar the relative positions of dehydration curves and 5% hydrous tholeiitic solidus are inverted. Except for anthophyllite → forsterite + talc + H_2O reaction, which slightly precedes the solidus temperature, a subducted slab can reach temperatures of 650–700°C before it dehydrates (Fig. 20). At these temperatures, water is still present so dehydration melting can take place in the subducted slab. Consequently, during the Archean and unlike modern arc systems, subducted oceanic crust was able to start dehydration melting at relatively low temperatures. Garnet and hornblende were the main residual phases and, because of their high KD for HREE (KD >> 1), they control the compatible behaviour of these elements accounting for their low concentrations in TTG magmas ($0.3 < Yb_N < 8.5$), (Table 6; Fig. 18; Martin, 1986, 1987b).

The model proposed here is summarized in Table 7. It accounts for secular changes in juvenile granitoid compositions by changing the place where calc-alkaline magmas are generated, the latter is controlled by earth heat production and distribution. Such a model needs further testing and evaluation since it implies higher heat production in the Archean and considers that plate tectonics operated at that time. It also must be confronted with alternative models that do not necessitate subduction (i.e. underplating from mantle plumes; Arndt and Goldstein, 1989; Kröner, 1991; Kröner and Layer, 1992).

Test of the proposed model

Geochemical test

In the subduction model, differences between Archean and modern juvenile granitoids mainly reflect differences in their source composition: garnet-bearing amphibolite to hornblende–eclogite before 2.5 Ga and mantle peridotite after 2.5 Ga. Theoretical magma compositions can be easily calculated using simple equations as proposed by Shaw (1970), and results are shown on a $(La/Yb)_N$ vs. (Yb_N) plot (Fig. 21). The Archean situation was computed using $(La/Yb)_N = 2.5$ and $Yb_N = 10$ for a tholeiitic source. Melting residues considered are (1) garnet free

TABLE 7

Summary of relationships between thermal regimes in the oceanic lithosphere, calc-alkaline magma sources and REE characteristics of derivative magmas

Age	Archean T>2.5 Ga	Post-Archean T<2.5 Ga
Age and temperature of the subducted slab	Young (T<25 Ma) and hot	Old (T>>25 Ma) and cold
Geothermal gradient along the Benioff Zone	High: from 25 to 30°C/km	Low: lower or equal to 10°C/km
Dehydration of the subducted slab	No: water is available for dehydration melting	Yes: the subducted slab is dry when it reaches its hydrous-solidus T°
Source of the magmas	Oceanic crust transformed into garnet-bearing amphibolite or hornblende eclogite	Metasomatized peridotite in the mantle wedge
Residue of melting	Garnet and hornblende ± Cpx	Olivine and pyroxenes with minor spinel or garnet
Magma composition	TTG	BADR
Yb_N	Low: ($0.3 < Yb_N < 8.5$)	High: ($4.5 < Yb_N \leq 20$)

amphibolite, (2) 10% and (3) 25% garnet-bearing amphibolite, and (4) eclogite. In all cases, La has incompatible and Yb compatible behaviour, but the degree of REE fractionation remains very low ($(La/Yb)_N < 20$) when the garnet free amphibolite residue is considered. In this case, the modelled magma compositions do not fall in the field of Archean TTG. However, in the other three models where garnet is a residual mineral, results predict high La/Yb ratios and low Yb contents in very good agreement with the values measured in Archean TTG.

Modelling of mantle anatexis is more difficult because of fluid metasomatism. Two possibilities were checked: an unenriched mantle lherzolite (M) with $(La/Yb)_N = 1$ and $Yb_N = 2$, and a LREE enriched, fluid metasomatized mantle lherzolite (mM) with $(La/Yb)_N = 8$ and $Yb_N = 3$, calculated using Mysen's (1979) equations. These values are similar to those proposed by several authors (Sun et al., 1979; Chauvel and Jahn, 1984; Graviou and Auvray, 1990). Whatever the mantle source is, olivine + orthopyroxene + clinopyroxene are residual phases together with subordinate amounts of garnet or spinel for low degrees of melting. Both La and Yb behave as incompatible elements, and as such the magmatic liquid is enriched in these two REE. Depending on the model, Yb_N can be higher than 20 but $(La/Yb)_N$ never exceeds 30. As discussed in a previous section, mantle partial melting generates mafic to intermediate magmas that can evolve to felsic compositions through fractional crystallization of olivine at depth. This latter possibility

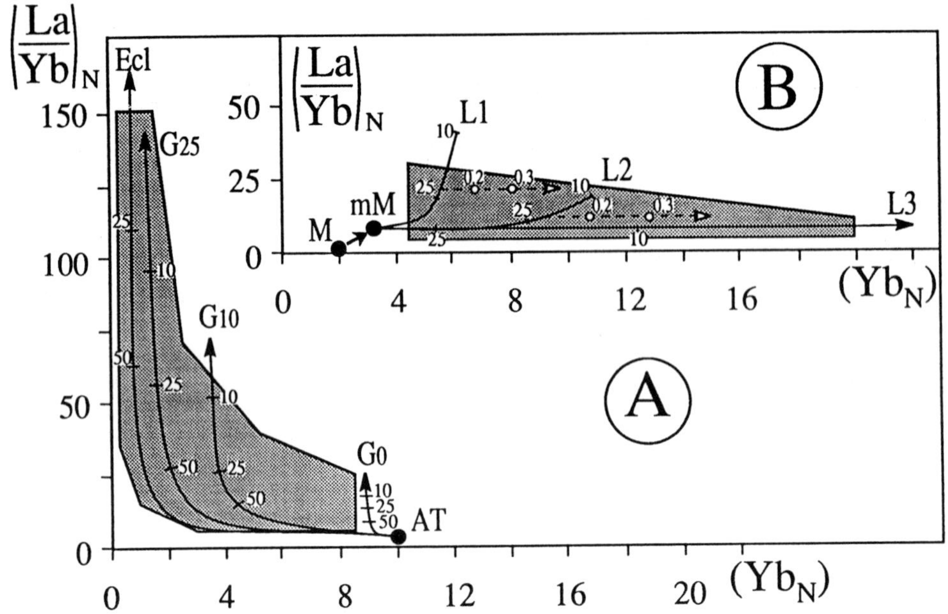

Fig. 21. $(La/Yb)_N$ versus Yb_N diagram for both Archean (A) and post-Archean (B) juvenile granitoids. The curves were calculated using the equations of Shaw (1970). The partial melting of a tholeiitic source was modelled in A, where the residue of melting is garnet-free amphibolite (G0), 10% or 25% garnet-bearing amphibolite (G10 and G25), or eclogite (Ecl). Inset B represents melting of a metasomatized (mM) mantle (M = chondritic mantle) whose composition is assumed to be 10% and 5% garnet lherzolite (L1 and L2, respectively) and 5% spinel lherzolite (L3). Dotted lines correspond to the fractional crystallization of olivine at depth in a magma generated by 20% melting of a garnet lherzolite. Numbers indicate the degree of both partial melting (10, 25, 50%) or fractional crystallization (0.2 = 20%, 0.3 = 30%). In both cases, there is good agreement between the analytical data (grey fields) and the theoretical calculations.

was taken into account in the modelling, but because of the very low partition coefficients of REE between olivine and magma (K_{DLa} = 0.0004 and K_{DYb} = 0.0015), the Yb content of the magmatic liquid very rapidly increases without significantly changing the La/Yb ratio (Fig. 21). In all cases, the theoretical modelling of lherzolite melting at mantle depths matches the REE distribution of post-2.5 Ga juvenile granitoids, but it is definitively unable to account for the low Yb and high La/Yb of Archean TTG.

Because of the distinct compositions of the mantle ($Yb_N = 3$) and of the tholeiite ($Yb_N = 10$), the computed melting curves (Fig. 21) cross-cut and overlap, thus predicting the existence of a compositional window where both mantle and tholeiite derived magmas can have identical REE compositions. This is consistent with the observed overlap between Archean and post-Archean fields (Fig. 18) for Yb_N values ranging from 4.5 to 8.5.

Modern analogue of Archean subduction

Table 7 indicates that the site of magma generation in subduction zones is correlated to the temperature of the subducted lithosphere, which in turn is controlled by the age of the slab when it begins to subduct (Fig. 19; Parson and Sclater, 1977; Abbott and Hoffman, 1984; Peacock, 1990, 1993). Hence, the hottest oceanic crust that can be subducted is an ocean ridge. In this case, Archean-like geothermal gradients could be locally attained and it should be possible to reach the hydrous-tholeiitic solidus before dehydration of the slab. Under these thermodynamic conditions, garnet and hornblende should be residual phases and the generated magma should display Yb depletion. Martin (1986, 1987b) proposed that this situation could provide a test for the Archean TTG petrogenetic model. The test consists of comparing geochemical characteristics (mainly REE) of calc-alkaline magmas emplaced in two adjacent regions corresponding to the subduction of both old and young lithosphere.

A well-documented example of ridge subduction exists in southern Chile, at the triple junction between the South-American, Nazca and Antarctic Plates. At the Taitao peninsula, the active Chile ridge is being subducted (Fig. 22; Lagabrielle

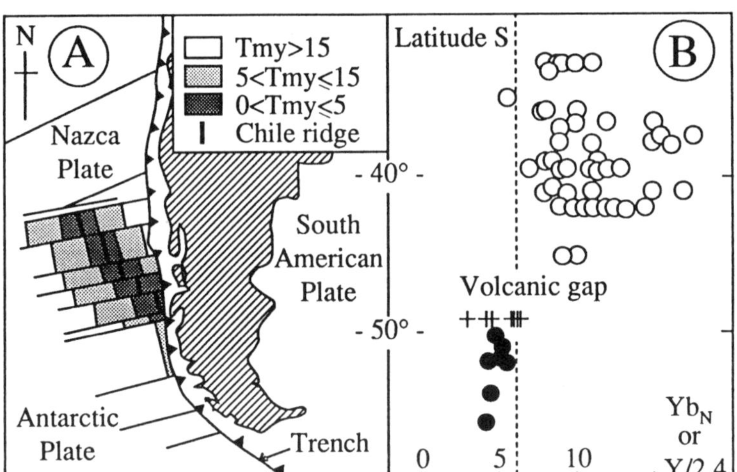

Fig. 22. A: A schematic map of South Chile shows the age of the oceanic lithosphere when it enters in the subduction zone adjacent to the South-American plate. North of latitude 44°S, the age of the Nazca plate is greater than 20 Ma (20–50 Ma), whereas south of 49°S the age of the Antarctic plate ranges from 0 to 20 Ma. B: (Yb_N or Y/2.4) vs. latitude plot. To the south, where subducted oceanic crust is young, the volcanic (filled circles) and plutonic (crosses) rocks display low Yb or Y/2.4 contents ($Yb_N < 5.5$) similar to Archean TTG and consistent with an origin by partial melting of the subducted slab in the hornblende + garnet P-T stability field. To the north, Yb_N contents of the volcanics (open circles) range from 8 to more than 20, indicating an origin by partial melting of the metasomatized mantle wedge. The two domains are separated by a volcanic gap. A vertical dashed line separates the two populations of rocks.

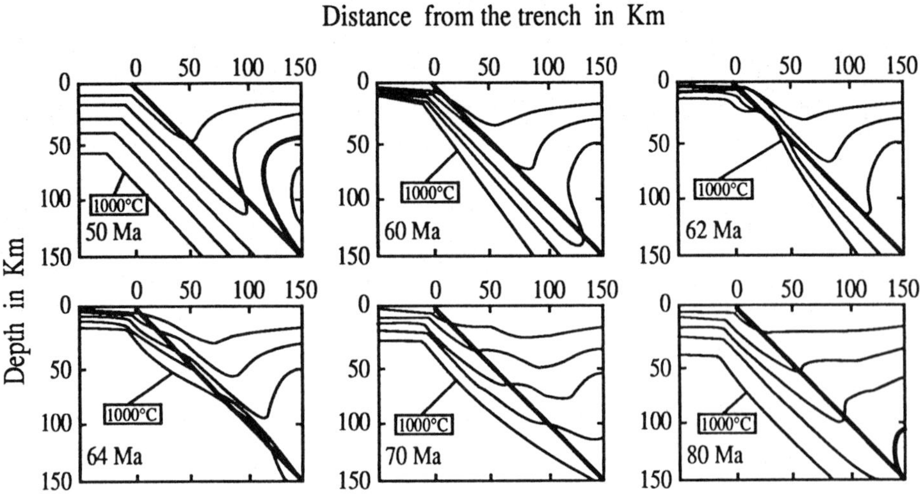

Fig. 23. Models of thermal evolution of a subduction zone when a ridge is subducted (after Delong et al., 1979). The black lines are the 200, 400, 600, 800 and 1000°C isotherms. The ridge enters in subduction at 60 Ma in a thermally stable system, where after 2 Ma, it induces a strong positive anomaly restricted in both time and space. In this situation, the 1000°C isotherm may intersect the Benioff plane allowing dehydration melting of the slab.

et al., 1994). Theoretical simulations made by Delong et al. (1979) demonstrated that when a ridge enters in an already thermally stabilized subduction zone, it creates a strong positive thermal anomaly. Figure 23 shows that at the passage of the ridge, isotherms rise such that the 1000°C isotherm intersects the slab-mantle interface. The thermal anomaly is restricted in both time (ca. 5 Ma) and space but, as it can easily exceed the 5% hydrous-tholeiitic solidus temperature, slab dehydration melting is allowed. Hence, the situation in south Chile can be summarized as follows: (1) north of 44°S the subducted slab is older than 20 Ma (50 Ma to the north of the Nazca plate) when it enters the subduction zone, (2) south of 49°S it is younger than 10 Ma. Only lavas from active volcanoes were considered.

North of 44°S the Yb_N contents in modern lavas range from 8 to 18 and some samples display values higher than 20 (Fig. 22). The REE patterns are typical of most modern subduction zones, where LREE are fractionated and HREE are rather flat. Such REE patterns are identical to those of post-2.5 Ga juvenile granitoids. Based on petrological and geochemical characteristics, these volcanics are considered to be derived from partial melting of the metasomatized mantle wedge (Lefèvre, 1979; Stern et al., 1984a,b; Lopez Escobar, 1984; Futa and Stern, 1988; Rogers and Hawkesworth, 1989). Consequently, north of 44°S, the subduction of and old and cold lithosphere is correlated with volcanics generated by melting of the mantle wedge.

Figure 22 shows that in the southern part of the Andean arc (south of 49°S), Yb_N contents of the lavas range between 4 and 5.5, values that are significantly lower than those observed further to the north. In the same diagram, Y/2.4 is shown (Y/2.4 can be considered as equivalent to Yb_N; Defant and Drummond, 1990) for a 4 Ma-old granodiorite from the Taitao Peninsula in Chile. The Y/2.4 ratios of this granodiorite range between 3.5 and 6.5, identical to those of associated present-day lavas, whose more detailed petrogenetic studies demonstrate that they originate from partial melting of the slab (Stern and Futa, 1982; Stern et al., 1984a,b). The degree of fractionation of LREE and HREE is similar, to those characteristic of Archean TTG.

The "modern analogue" test, applied to South Chile by Martin (1987b) was generalized by Defant and Drummond (1990). The results of their compilation (Fig. 24) can be summarized as follows. There is a world-wide correlation between the age of the lithosphere when it begins to subduct and the composition of associated calc-alkaline magmas: (1) when the subducted slab is younger than 30 Ma, TTD (tonalites, trondhjemites and dacites) magmas are produced, their Yb_N (and Y/2.4) contents are typically lower than 7.5 ($Y < 18$), consistent with an origin by partial melting of the slab; (2) when the subducted slab is older than 30 Ma, BADR (basalt, andesite, dacite, rhyolite) magmas are generated, and they exhibit Yb_N (and Y/2.4) contents that range from 6.5 to 25. This feature strongly favours a metasomatized peridotitic source without garnet and hornblende in the residue of melting.

In conclusion, there is a correlation between the age of the subducted lithosphere and the sites where calc-alkaline magmas are generated. Today, in exceptional environments, such as ridge subduction, very young and hot oceanic crust can be subducted, creating Archean-like geothermal gradients along the Benioff plane.

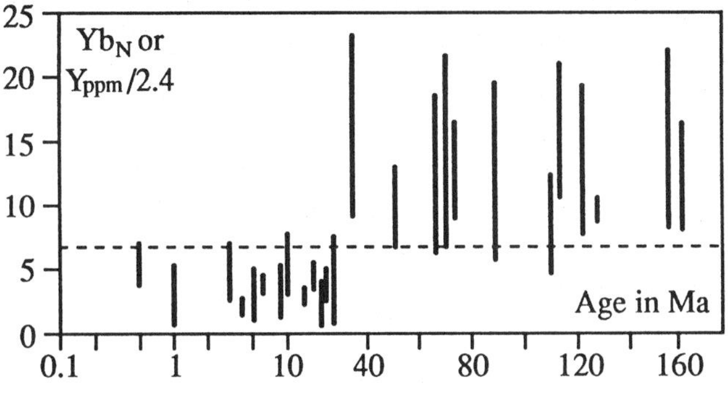

Fig. 24. Yb_N (or Y/2.4) vs. age of subducted lithosphere diagram (after Defant and Drummond, 1990). The horizontal scale is logarithmic when $T < 30$ Ma and arithmetic when $T > 30$ Ma. The black lines correspond to the compositional range of the arc magmatism. There is a world-wide correlation between the age of the subducted lithosphere and its geochemical characteristics.

Calc-alkaline magmas produced under these conditions have Archean-like geochemical signatures. This observation supports and reinforces the model of genesis of TTG by partial melting of young oceanic crust in a subduction environment.

DISCUSSION

The proposed petrogenetic model for Archean TTG implies that heat production was greater during the Archean, such that the thermal regime in subduction zones corresponded to the hot oceanic crust - hot mantle situation of Wyllie (1979, 1983). Heat distribution in subduction zones depends on both mantle wedge and descending lithosphere temperatures. Models of earth heat budget consider that radiogenic heat production was higher during the Archean and that it exponentially decreases since then. Abbott and Hoffman (1984) suggested that heat production during the Archean was roughly three times the present day value. Most estimations yield Archean-mantle temperatures 100 to 200°C higher than today (Jarvis and Campbell, 1983; Abbott and Hoffman, 1984; Campbell and Jarvis, 1984; Richter, 1985; Bickle, 1986, 1992; Nisbet et al, 1993). During the Archean, oceanic crust was composed of tholeiites and of unknown amount of komatiites. Depending on the authors, estimated eruption temperatures for these high-MgO komatiitic lavas range from 1525°C (26% MgO) to 1650°C (> 30% MgO) (Bickle, 1982, 1986, 1992; Arndt, 1983; Nisbet and Fowler, 1983; Nisbet, 1987; Nisbet et al., 1993). Whatever the exact komatiite-eruption temperature, it was significantly higher than the 1200–1350°C of basalt sources in present-day ridge systems (Forsyth, 1977; Bickle, 1978; Sleep and Windley, 1982; Richter, 1985). It was undoubtedly also higher than the exceptional Phanerozoic source temperature of 1400°C recorded by Gorgona Island komatiites (Echeverría, 1982). Therefore, it arises that both Archean mantle and oceanic lithospheres were hotter than today, allowing higher geothermal gradients in subduction zones. However, the P-T conditions recorded in crustal blocks older than 2.5 Ga do not show abnormally high geothermal gradients (Burke and Kidd, 1978; Wells, 1979; Condie, 1980; England and Bickle, 1984; Martin et al., 1984; Richter, 1985; Newton, 1990). This apparent contradiction between data and theoretical estimates can be explained when one considers that about 45% of the total earth's heat is lost in ridge and subduction zones (McKenzie, 1967; Sclater and Francheteau, 1970; Bickle, 1978). Hargraves (1986) calculated that heat loss by ridges is a function of the cubic root of the ridge length, and consequently, the higher Archean heat production could have been dissipated by longer ridge systems. Hence, two categories of Archean thermal zones can be defined: (1) large zones, roughly corresponding to continental lithosphere, where geothermal gradients have remained almost unchanged since about 4.0 Ga, and (2) smaller areas with higher Archean geothermal gradients, that correspond to ridge and subduction systems and act as outlets for the terrestrial heat.

As the earth volume is constant, a longer ridge system implies smaller plates (McKenzie and Weiss, 1975; Burke et al., 1976; Condie, 1980, 1981; Martin et al., 1984; Drummond and Defant, 1990). It has also been proposed that Archean spreading rates were greater, thus contributing to the efficiency of heat dissipation (Burke and Kidd, 1978; Bickle, 1978; Dewey and Windley, 1981; Sleep and Windley, 1982; Hargraves, 1986; Nisbet, 1987; Arculus and Ruff, 1990). Smaller plate dimensions as well as greater spreading rates, signify that Archean oceanic slabs were younger when they began to subduct. The average age of Archean subducted slabs has been estimated at 10–20 Ma, whereas presently it averages 60 Ma and can be older than 180 Ma (Bickle, 1978; Arndt, 1983; Nisbet and Fowler, 1983; Abbott and Hoffman, 1984; Abbott and Lyle, 1984; Nisbet, 1984, 1987; Drummond and Defant, 1990). As discussed previously, the temperature of subducted oceanic lithosphere is dependant on its initial temperature at the ridge (higher in the Archean) and its age when it begins to subduct (lower in the Archean). Consequently, before 2.5 Ga, the temperatures of both the mantle and subducted lithosphere were higher than today, thus accounting for greater geothermal gradients along Archean Benioff planes.

On the other hand, Peacock (1990) showed that the subduction of oceanic lithosphere corresponds to the introduction of a cool slab into a warm mantle (even when the subducted oceanic crust is young and hot), such that it rapidly cools the whole subduction zone precluding slab melting. He considers that P-T conditions required for slab melting cannot be realized during more than 50 Ma after subduction began. This age condition is exceptional today, but probably common during the Archean when oceanic plates subducted before reaching 20 Ma.

In this chapter, it has been assumed that plate tectonic-like mechanisms operated during the Archean, and because not all geologists agree with this premise, it needs to be discussed. Several authors addressed the existence of plate tectonics before 2.5 Ga on the basis of geochemical data. They compared Archean igneous rock chemical characteristics with those of modern magmas produced in well-known tectonic environments. This approach is questionable when old rocks (> 1.0 Ga) are considered, because even if plate tectonics operated at that time, both source composition and melting conditions could have been different, such that, in a given tectonic environment geochemical signatures could strongly differ. The occurrence of plate tectonics before 2.5 Ga must be demonstrated and constrained using theoretical models and structural evidences. Several calculations and models have shown that the physical conditions for plate tectonics existed for at least 4.0 Ga (Bickle, 1978; Arndt, 1983; Nisbet and Fowler, 1973;, Abbott and Hoffman, 1984; Campbell and Jarvis, 1984; etc.). As modern oceanic crust does not survive for more than 0.2 Ga, evidence of plate activity must be investigated in continental blocks. One of the more prominent and spectacular features of modern global tectonics is continental collision, which generates mountain belts and develops thrusting and horizontal structures. Collisional structures demonstrate that rigid plates existed, and that continental blocks moved relative to each other.

Plate motion on a constant volume planet can only be explained if crust is created in some places and destroyed in others. Recently, several detailed structural studies documented the existence of large-scale Archean thrusting in the Baltic shield (Jégouzo and Blais, 1991), Greenland (McGregor et al., 1991), South Africa (DeWit et al., 1992; Treloar et al., 1992), Canada (Ludden et al., 1993), and Australia, (Bickle et al. 1980). As an example, in Eastern Finland Jégouzo and Blais (1991) mapped two structural domains separated by a major thrust. Along the thrust plane are remnants of oceanic crust, including marine sediments, tholeiitic lavas, gabbros and ultramafic rocks. The structural position of the oceanic components is the same as that of modern ophiolites in alpine-type mountain belts. In addition, beneath the thrust, Jégouzo and Blais (1991) describe the development of local migmatisation and associated anatectic granites, that appear to have formed immediately after thrusting. This provides new and strong arguments to favour Archean plate tectonics. Other examples of rigid block collision or collage in the Archean are given by precise SHRIMP zircon age determinations (Nutman et al., 1992) which demonstrate the juxtaposition and amalgamation of small terranes between 2.8 and 2.7 Ga in West Greenland, and the paleomagnetic studies (Hale, 1992), that indicate Archean collage tectonics in Canada.

It has been suggested that Archean continental crust formed by melting of a tholeiitic source in a tectonic environment (underplating and/or above a mantle plume) that does not necessitate modern-type plate tectonics (Arndt and Goldstein, 1989; Kröner, 1991; Arndt, 1992; Kröner and Layer, 1992). Similar models were already proposed to account for greenstone belt formation and emplacement (Hunter, 1974; Condie and Hunter, 1976; Campbell et al. 1989; Hill, 1991, 1993; Hill et al., 1992). Nowadays, mantle plumes are known in many tectonic environments where they are related to mafic magmatism. In a few exceptional cases, such as in Iceland, small amounts of felsic magma are generated but they do not have TTG characteristics. On the other hand, late Archean K- and Mg-rich mafic to intermediate magmas produced in close association with lamprophyres and/or alkali basalts (e.g.; "Archean sanukitoids") could reflect mantle plume activity, as documented for instance in South India (Jayananda et al., 1995; Martin et al., 1992, 1993; Peucat et al., 1993). However, the chemical characteristics of these magmas are totally different from those of TTG. Consequently, it seems that if mantle plume activity can produce mafic to intermediate magmas in a restricted area and during a short period of time, it is unable to generate the enormous volumes of Archean TTG. Arndt (1992) proposed that Archean mantle plumes induced the melting of a very thick (40–60 km) Archean oceanic crust, but no direct or indirect evidence of such a thick mafic crust is preserved in Archean record. In addition, in order to melt the base of this mafic crust in the hornblende + garnet stability field, water must be added to the system since mantle plumes alone can easily supply heat, but cannot account for water enrichment. Alternatively, Atherton and Petford, (1993) consider that ignimbrites from Cordillera

Blanca (Peru) related to subduction of a 60-Ma-old oceanic crust could form by melting of young underplated basaltic crust, and in this environment water from slab-dehydration could be available for melting. This possibility was refuted for Archean TTG by Drummond and Defant (1990) who consider that relatively low-pressure melting of a basaltic source would leave calcic plagioclase in the residue, such that the derived magmas are Al_2O_3-, Sr- and Eu-depleted in contrast to TTG compositions. These authors also noticed the lack of identifiable positive gravity anomalies in Archean cratons (Newton, 1989) that should result from basalt underplating. Furthermore, Davies (1993) calculated that plumes could have operated through most of the earth history at about their present level of activity, but that they never were able to efficiently remove heat from earth's interior. He concluded that plumes could not have substituted for plate tectonics because plumes and plates are driven by different thermal boundary layers that operate independently.

One of the more striking evidences in favour of the Archean subduction model is that today, when Archean-like thermal regimes are created in subduction environments, TTG-like magmas are generated, whereas this kind of magmatism is totally unknown in association with plume systems. Moreover, in Catalina Island (California) subducted oceanic crust and associated sediments are exceptionally well preserved (Sorensen and Barton, 1987; Sorensen, 1988; Sorensen and Grossman, 1989; Bebout and Barton, 1993). The mafic remnants are transformed into garnet-bearing amphibolite or hornblende eclogite, and they record temperatures of about 650–750°C and pressures ranging from 9 to 11 kbar. Within the amphibolites, the authors describe migmatitic structures and veins which demonstrate that the amphibolites exceeded their solidus temperature and began to melt. These liquids have high-Al_2O_3 trondhjemitic compositions. This indicates that TTG magmas can be generated in subduction environments and by partial melting of the subducted oceanic crust.

One of the main parameters that controlled the evolution of juvenile granitoid compositions with time is the progressive decrease in heat production, which began immediately after the earth accretion. It resulted in a progressive modification of the thermal regimes that determine the sites where calc-alkaline magmas are generated. In South Chile, a geographical gap in active volcanism exactly corresponds to the place where calc-alkaline magma source changes from slab to mantle wedge. This change did not occur everywhere on the world at the same time. For instance, even today, in rare occasions, high-Al TTD magmas can be generated when a ridge is subducted. Similarly, the existence of relatively rare Archean andesites could indicate that mantle wedge exceptionally melted before 2.5 Ga. In spite of these few exceptions, the main transitional period in the site of TTG production roughly appears to be the Archean–Proterozoic boundary. This period is characterized by an almost complete lack of magmatic activity and crustal accretion between 2.5 and 2.3 Ga. This world-wide magmatic gap could be analogous to the volcanic gap observed in South Chile, which represents a transition zone. As suggested by Martin (1993) it can be tentatively proposed that the 2.5–2.3 Ga interval could be a period of magmatic inactivity due to the change from archaic to modern mechanisms in subduction zones. In addition, since 4.5 Ga earth cooling is an exponential progressive mechanism that resulted in a highly discontinuous magmatic record. When earth heat production exceeded a critical threshold (before 2.5 Ga), melting of the oceanic crust was favoured, and when heat production decreased and passed below the threshold, melting of a mantle lherzolite became the predominant mechanism.

SUMMARY

The main points of this chapter can be summarized as follows:

- Grey gneisses are very widespread in all Archean cratons, where they generally form the basement of the greenstone belts.
- The main mineralogical association is oligoclase + quartz + biotite (K-feldspar and hornblende are minor phases). In the modal classification of Lameyre and Bowden (1982), they define a low-K calc-alkaline (trondhjemitic) trend.
- In the normative An-Ab-Or classification diagram (O'Connor, 1965), Archean grey gneisses plot in the TTG (tonalite, trondhjemite, granodiorite) fields. In a K-Na-Ca triangle, they show trondhjemitic affinities, these features being typical of Archean granitoids.
- Both mineralogical and chemical compositions indicate that TTG belong to the I-type granitoids of Chappell and White (1974) or to the M-type of Didier et al. (1982).
- Trace element composition reveals Nb-Ta-Ti and P negative anomalies. REE patterns are strongly fractionated $(La/Yb)_N = 38.4$, with low HREE contents ($Yb_N = 2.6$) and no significant Eu anomaly ($Eu/Eu^* = 0.99$).
- TTG parental magma formed by partial melting of an Archean tholeiite transformed into garnet-bearing amphibolite or hornblende eclogite. Fractional crystallization can follow the melting stage, but the degree of fractional crystallization never exceeds 30%.
- Experimental melting of basalts and amphibolites result in tonalitic and trondhjemitic liquids. However, typical TTG are obtained only when garnet is a residual mineral phase and for pressures ≥ 16 kbar.
- Archean TTG differ from post-2.5 Ga calc-alkaline juvenile granitoids in both mineralogical and chemical composition. In the K-Na-Ca triangle post-Archean granitoids are granodiorites and granites; they define a classical calc-alkaline trend and do not have HREE depletion. This difference is assumed to reflect differences in both sources and petrogenesis. Today in modern subduction zones, geothermal gradients along the Benioff plane are low, such that the subducted slab dehydrates before it reaches its hydrous solidus temperature. Calc-alkaline magmas are generated by melting of the metasomatized

mantle wedge. During the Archean, geothermal gradients were higher and the hydrous solidus of the subducted crust was attained before it completely dehydrated. The source of TTG was the subducted slab.
- A modern analogue of Archean thermal regimes can be realized today when a very young and hot oceanic crust subducts. For instance, during ridge subduction, a thermal anomaly is created along the Benioff plane, it is restricted in both time and space, but it simulates Archean-like gradients. Modern andesites produced in this environment are totally different from typical arc andesites and show all the chemical characteristics of Archean TTG. This conclusion was generalized to all young subducted lithospheres all over the world by Defant and Drummond (1990).
- It has also been proposed that Archean TTG can be generated by melting of a garnet amphibolite in a plume environment. The more striking evidence that favours the subduction model is that today, when Archean-like thermal regimes are created in subduction environments, TTG-like magmas are generated, whereas this kind of magmatism is totally unknown in association with plume systems.
- Since the Archean, there has been a change in the site of continental crust genesis. The location of calc-alkaline magma source in subduction-zone environments has migrated through time from the subducted slab to the metasomatized mantle wedge. This is a direct consequence of the progressive cooling of the earth.

ACKNOWLEDGEMENTS

Thanks are due to Bernard Auvray, Fred Barker, Kent Condie and Mark Drummond, for constructive scientific comments and language corrections, which greatly improved the final version of the manuscript. Jacques Bourgois and Yves Lagabrielle kindly provided granodiorite samples from Taitao Peninsula. I am also grateful to Arlette Falaise for her efficient technical assistance.

REFERENCES

Abbott, D.H. and Hoffman, S.E., 1984. Archaean plate tectonics revised. I. Heat flow, spreading rate, and the age of subducting oceanic lithosphere and their effects on the origin and evolution of continents. Tectonics, 3: 429–448.

Abbott, D.H. and Lyle, M., 1984. Age of oceanic plates at subduction and volatile recycling. Geophys. Res. Lett., 11: 951–954.

Agee, C.R. and Walker, D., 1988. Mass balance and phase density constraints on early differentiation of chondritic mantle. Earth Planet. Sci. Lett., 90: 144–156.

Allègre, C.J., 1985. The evolving Earth system. Terra Cognita, 5: 5–14.

Anderson, D.L., 1989. Theory of the Earth. Blackwell Scientific Publications, 366 pp.

Anderson, R.N., Delong, S.E. and Schwarz, W.M., 1978. Thermal model for subduction with dehydration in the downgoing slab. J. Geol., 86: 731–739.

Anderson, R.N., Delong, S.E. and Schwarz, W.M., 1980. Dehydration, astenospheric convection and seismicity in subduction zones. J. Geol., 88: 445–451.

Anhausser, C.R., Mason, P., Viljoen, M.J. and Viljoen, R.P., 1969. A reappraisal of some aspects of Precambrian shields geology.Geol. Soc. Am. Bull., 80: 2175–2200.

Arculus, R.J. and Ruff, L.J., 1990. Genesis of continental crust: evidence from island arcs, granulites, and exospheric processes. In: Granulites and Crustal Evolution (D. Vielzeuf and Ph. Vidal, eds.). Kluwer Academic Publishers, pp. 7–23.

Arkani-Hamed, J. and Jolly, W.T., 1989. Generation of Archaean tonalites. Geology, 17: 307–310.

Armstrong, R.L., 1981. Radiogenic isotopes: the case for crustal recycling on a near-steady-state no-continental-growth Earth. Phil. Trans. R. Soc. Lond., A301: 443–472.

Arndt, N.T., 1983. Role of a thin, komatiite-rich oceanic crust in the Archaean plate-tectonic process. Geology, 11: 372–375.

Arndt, N.T., 1992. Rate and mechanism of continent growth in the Precambrian. In: Evolving Earth Symposium (S. Maruyama, ed.), Okazaki, pp. 38–41.

Arndt, N.T. and Goldstein, S.L., 1989. An open boundary between lower continental crust and mantle: its role in crust formation and crustal recycling. Tectonophysics, 161: 201–212.

Arth, J.G., 1979. Some trace elements in trondhjemites, their implication to magma genesis and paleotectonic setting. In: Trondhjemites, Dacites and Related Rocks (F. Barker, ed.), Elsevier, Amsterdam, pp. 123–132.

Arth, J.G. and Hanson, G.N., 1972. Quartz diorites derived by partial melting of eclogite or amphibolite at mantle depth. Contrib. Mineral. Petrol., 37: 164–174.

Arth, J.G. and Hanson, G.N., 1975. Geochemistry and origin of the Early Precambrian crust of north-eastern Minnesota. Geochim. Cosmochim. Acta, 39: 325–362.

Arth, J.G., Barker, F., Peterman, Z.E. and Frideman, I., 1978. Geochemistry of the gabbro–diorite–tonalite–trondhjemite suite of south-west Finland and its implications for the origin of tonalitic and trondhjemitic magmas. J. Petrol., 19: 289–316.

Atherton, M.P. and Petford, N., 1993. Generation of sodium-rich magmas from newly underplated basaltic crust. Nature, 362: 144–146.

Barker, F., 1979. Trondhjemites: definition, environment and hypothesis of origin. In: Trondhjemites, Dacites and Related Rocks (F. Barker, ed.), Elsevier, Amsterdam, pp. 1–12.

Barker, F. and Arth, J.G., 1976. Generation of trondhjemitic–tonalitic liquids and Archaean bimodal trondhjemite–basalt suites. Geology, 4: 596–600.

Barton, J.M. Jr., 1975. Rb-Sr isotopic characteristics and chemistry of the 3.6 b.y. Hebron gneiss, Labrador. Earth Planet. Sci. Lett., 27: 427–435.

Beard, J.S. and Lofgren, G.E., 1989. Effect of water on the composition of partial melts of greenstones and amphibolites. Science, 244: 195–197.

Beard, J.S. and Lofgren, G.E., 1991. Dehydration melting and water-saturated melting of basaltic and andesitic greenstones and amphibolites at 1, 3 and 6.9 kb. J. Petrol., 32: 465–501.

Bebout, G.E. and Barton, M.D. (1993. Metasomatism during subduction: products and possible paths in the Catalina schist, California. Chem. Geol., 108: 61–92.

Bickle, M.J., 1978. Heat loss from the Earth: constraint on Archaean tectonics from the relationships between geothermal gradients and the rate of plate production. Earth Planet. Sci. Lett., 40: 301–315.

Bickle, M.J., 1982. The magnesium contents of komatiitic liquids. In: Komatiites, (N.T. Arndt and E.G. Nisbet, eds.). Allen and Unwin, London, pp. 479–494.

Bickle, M.J., 1986. Implications of melting for stabilisation of lithosphere and heat loss in the Archaean. Earth Planet. Sci. Lett., 80: 314–324.

Bickle, M.J., 1992. Archaean magmatism. In: Evolving Earth Symposium (S. Maruyama, ed.). Okazaki, pp. 43–46.

Bickle, M.J., Battenay, L.F., Boulter, C.A., Groves, D.I. and Morant, P., 1980. Horizontal tectonic interaction of the Archaean gneiss belt and greenstones, Pilbara block. Contrib. Mineral. Petrol., 84: 25–35.

Bickle, M.J., Battenay, L.F., Chapman, H.J., Groves, D.I., McNaughton, N.J., Campbell, I.H. and deLaeter, J.R., 1993. Origin of the 3500–3300 Ma calc-alkaline rocks in the Pilbara Archaean: isotopic and geochemical constraints from the Shaw batholith. Precambrian Res., 60: 117–149.

Bird, P., Toksöv, M.N. and Sleep, N.H., 1975. Thermal and mechanical models of continent–continent convergence zones. J. Geophys. Res., 80: 4405–4416.

Bowring, S.A., Williams, I.S. and Compston, W., 1989. 3.96 Ga gneisses from the Slave province, N.W.T. Canada. Geology, 17: 971–975.

Bowring, S.A., Houst, T.B. and Isachsen, C.E., 1990. The Acasta gneisses: remnants of Earth's early crust. In: Origin of the Earth (H.E. Newsom and J.H. Jones, eds.). Oxford Univ. Press, Houston, pp. 319–343.

Burke, K. and Kidd, W.S.F., 1978. Were Archaean geothermal gradients much steeper than those of today? Nature, 272: 240–241.

Burke, K., Dewey, J.F. and Kidd, W.S.F., 1976. Dominance of horizontal movements, arc and microcontinental collisions during the later permobile regime. In: The Early History of the Earth (B.F. Windley, ed.). Wiley, London, pp. 113–129.

Campbell, I.H. and Jarvis, G.T., 1984. Mantle convection and early crustal evolution. Precambrian Res., 26: 15–56.

Campbell, I.H., Griffiths, R.W. and Hill, R.I., 1989. Melting in an Archaean mantle plume: heads it's basalts, tails it's komatiites. Nature, 339: 697–699.

Chappell, B.W. and White, A.J.R., 1974. Two contrasting granite types. Pacific Geol., 8, 173–184.

Chase, C.G. and Patchett, P.J., 1988. Stored mafic/ultramafic crust and Early Archaean mantle depletion. Earth Planet. Sci. Lett., 91: 66–72.

Chauvel, C. and Jahn, B.M., 1984. Nd-Sr isotope and REE geochemistry of alkali basalts from the Massif Central, France. Geochim. Cosmochim. Acta, 48: 93–110.

Clark, A.M., Turekian, K.,K. and Grossman, L., 1972. Early history of the Earth. In: The Nature of the Solid Earth (E.C. Robertson, ed.). pp. 3–18.

Cocherie, A., 1986. Systematic use of trace-element distribution patterns in log–log diagrams for plutonic suites. Geochim. Cosmochim. Acta, 50: 2517–2522.

Collerson, K.D. and Bridgwater, D., 1979. Metamorphic development of Early Archaean tonalitic and trondhjemitic gneisses: Saglek area, Labrador. In: Trondhjemites, Dacites and Related Rocks (F. Barker, ed.). Elsevier, Amsterdam, pp. 206–273.

Compston, W. and Pidgeon, R.T., 1986. Jack Hills, evidence of more very old detrital zircons in Western Australia. Nature, 321: 766–769.

Compton, P., 1978. Rare-earth evidence for the origin of the Nûk gneisses, Buksefjorden region, southern West Greenland. Contrib. Mineral. Petrol., 66: 283–294.

Condie, K.C., 1967. Geochemistry of Early Precambrian graywackes from Wyoming. Geochim. Cosmochim. Acta, 31: 2135–2149.

Condie, K.C., 1980. Origin and early development of the earth's crust. Precambrian Res., 11: 183–197.

Condie, K.C., 1981. Archaean Greenstone Belts. Elsevier, Amsterdam, pp. 434.

Condie, K.C., 1986. Origin and early growth rate of continents. Precambrian Res., 32: 261–278.

Condie, K.C., 1989. Plate Tectonics and Crustal Evolution. Pergamon, Oxford, 3rd edn., 476 pp.

Condie, K.C., 1990. Growth and accretion of continental crust: Inferences based on Laurentia. Chem. Geol., 83: 183–194.

Condie, K.C., 1993. Chemical composition and evolution of the upper continental crust: contrasting results from surface samples and shales. Chemical Geol., 104: 1–37.

Condie, K.C. and Lo, H.H., 1971. Trace element geochemistry of the Louis lake batholith of early Precambrian age, Wyoming. Geochim. Cosmochim. Acta, 35: 1099–1119

Condie, K.C. and Hunter, D.R., 1976. Trace element geochemistry of Archaean granitic rocks from Barberton region, South Africa. Earth Planet. Sci. Lett., 29: 389–400.

Condie, K.C. and Allen, P., 1980. Origin of the Archaean migmatites from the Gwenoro Dam area, Zimbabwe, Rhodesia. Contrib. Mineral. Petrol., 74: 35–43.

Davies, G.F., 1993. Conjectures on the thermal and tectonic evolution of the Earth. Lithos, 30: 281–289.

Defant, M.J. and Drummond, M.S., 1990. Derivation of some modern arc magmas by melting of young subducted lithosphere. Nature, 347: 662–665.

Delany, J.M. and Helgerson, H.C., 1978. Calculation of the thermodynamic consequences of dehydration in subducting oceanic crust to 100 Kbar and >800°C. Am. J. Sci., 278: 638–686.

Delong, S.E., Schwarz, W.M. and Anderson, R.N., 1979. Thermal effects of ridge subduction. Earth Planet. Sci. Lett., 44: 239–246.

Dewey, J.F. and Windley, B.F., 1981. Growth and differentiation of the continental crust. Phil. Trans. R. Soc. Lond., 301: 189–206.

DeWit, M.J., Roering, C., Hart, R.J., Armstrong, R.A., De Ronde, C.E.J., Green, R.W.E., Tredoux, M., Pederby, E. and Hart, R.A., 1992. Formation of Archaean continents. Nature, 357: 553–562.

Didier, J., 1973. Granites and Their Enclaves. The Bearing of Enclave on the Origin of Granites. Elsevier, Amsterdam. 393 pp.

Didier, J., Duthou, J.L. and Lameyre, J., 1982. Mantle and crustal granites: genetic classification of orogenic granites and the nature of their enclaves. J. Volcanol. Geotherm. Res., 14, 125–132.

Drummond, M.S. and Defant, M.J., 1990. A model for trondhjemite–tonalite–dacite genesis and crustal growth via slab melting: Archaean to modern comparisons. J. Geophys. Res., 95: 21503–21521.

Echeverría, L.M., 1982. Komatiites from Gorgona island, Columbia. In: Komatiites, (N.T. Arndt and E.G. Nisbet, eds.). Allen and Unwin, London, pp. 199–209.

Ellam, R.M. and Hawkesworth, C.J., 1988. Is average continental crust generated at subduction zones? Geology, 16: 314–317.

England, P.C. and Bickle, M.J., 1984. Continental thermal and tectonic regimes during the Archaean. J. Geol., 92: 353–367.

Ermanovics, I.F., 1971. Granites, granite-gneiss and tectonic variation of the Superior Province in southeastern Manitoba. Geol. Assoc. Can. Spec. Pap., 9: 77–81.

Evans, O.C. and Hanson, G.H., 1992. Most late Archaean tonalites, trondhjemites and granodiorites (TTG) in the SW Superior Province were derived from mantle melts, not by melting of basalts. AGU abstract, V22D-3: 330.

Flasar, F.M. and Birch F., 1973. Energetics of core formation: a correction. J. Geophys. Res. 78: 6101–6103.

Forsyth, D., 1977. The evolution of the upper mantle beneath mid-ocean ridges. Tectonophysics, 38: 89–95.

Froude, D.O., Ireland, T.R., Kinny, P.D., Williams, I.S., Compston, W., Williams, I.R. and Myers, J.S., 1983. Ion microprobe identification of 4000–4200 Myr-old terrestrial zircons. Nature, 304: 616–618.

Futa, K. and Stern, C.R., 1988. Sr and Nd isotopic and trace element compositions of quaternary volcanic centres of the southern Andes. Earth Planet. Sci. Lett., 88: 253–262.

Fyfe, W.S., 1978. Evolution of the Earth's crust: modern plate tectonics to ancient hot spot tectonics? Chem. Geol., 23: 89–114.

Galer, S.J.G. and Goldstein, S.L., 1991. Early mantle differentiation and its thermal consequences. Geochim. Cosmochim. Acta, 55: 227–239.
Glikson A.Y., 1971. Primitive Archaean element distribution patterns: chemical evidence and geotectonic significance. Earth Planet. Sci. Lett., 12: 309–320.
Glikson, A.Y., 1976. Stratigraphy and evolution of primary and secondary greenstones. Significance of data from shields of the southern hemisphere. In: The Early History of the Earth (B.F. Windley, ed.). Wiley-Interscience, New York, pp. 257–278.
Glikson, A.Y., 1979. Early Precambrian tonalite–trondhjemite sialic nuclei. Earth Sci. Rev., 15: 1–73.
Goldich, S.S., Hedge, C.E. and Stern, T.W., 1970. Age of the Morton and Montevideo gneisses and related rocks, southwestern Minnesota. Geol. Soc. Am. Bull., 81: 3671–3996.
Goodwin, A.M. (Ed.), 1991. Precambrian Geology. Academic Press, 666 pp.
Goodwin, A.M. and Ridler, R.H., 1970. The Abitibi orogenic belt. Geol. Surv. Can. Pap., 70–40: 1–24.
Gower, C.F., Crocket, J.H. and Kabir, A., 1983. Petrogenesis of Archaean granitoid plutons from the Kenora area, English River sub province, Northwest Ontario, Canada. Precambrian Res., 22: 245–270.
Graviou, P., 1984. Pétrogenèse des magmas calco-alcalins: exemple des granitoïdes cadomiens de la région trégorroise (Massif Armoricain). Unpublished thesis, Rennes (France), 236 pp.
Graviou, P. and Auvray, B., 1990. Late Precambrian M-type granitoid genesis in the Cadomian belt of NW France. In: The Cadomian Orogeny (R.S. D'Lemos, R.A. Strachan and C.G. Topley, eds.). Geol. Soc. Spec. Publ., 51: 231–244.
Green, D.H., 1973. Experimental melting studies on a model upper-mantle composition at high-pressure under water-saturated and water-under saturated conditions. Earth Planet. Sci. Lett., 19: 37–53.
Green, D.H., 1976. Experimental testing of equilibrium partial melting of peridotite under water-saturated, high pressure conditions. Can. Mineral., 14: 255–268.
Green, D.H., 1982. Anatexis of mafic crust and high pressure crystallisation of andesite. In: Andesites (R.S. Thorpe, ed.). J. Wiley, New York, pp. 465–487.
Green, D.H. and Ringwood, A.E., 1977. Genesis of the calc-alkaline igneous rock suites. Contrib. Mineral. Petrol., 18: 105–162.
Green, T.H., and Parson, N.J., 1986. Ti-rich accessory phase saturation in hydrous mafic–felsic compositions at high P,T. Chemical Geol., 54: 185–201.
Hale, C.J., 1992. Paleomagnetic evidence for accretionary collage tectonics along the southern margin of the Superior craton in the late Archaean. In: Evolving Earth Symposium (S. Maruyama, ed.). Okazaki, pp. 15–16.
Hanson, G.N. and Goldich, S.S., 1972. Early Precambrian rocks in the Saganaga Lake, Northern Light Lake area, Minnesota–Ontario; Part 2: Petrogenesis. In: Studies in Mineralogy and Precambrian geology (B.R. Doe and D.K. Smith, eds.). Geol. Soc. Am., 135: 179–192.
Hanson, G.N., Goldich, S.S., Arth, J.G. and Yardley, D.H., 1971. Age of the early Precambrian rocks of the Saganaga Lake–Northern Light Lake area, Ontario–Minnesota. Can. J. Earth Sci., 8: 1110–1124.
Hargraves, R.B., 1986. Faster spreading or greater ridge length in the Archaean. Geology, 14: 750–752.
Hart, S.R., 1977. The geochemistry and evolution of the early Precambrian mantle. Contrib. Mineral. Petrol., 61: 109–128.
Hasebe, K., Fujii, N. and Uyeda, S., 1970. Thermal processes under island arcs. Tectonophysics, 10: 335–355.

Helz, R.T., 1976. Phase relations in basalts in their melting range at P(H_2O) = 5 kb. Part II. Melt compositions. J. Petrol., 17: 139–193.

Hietanen, A., 1943. Uber das Grundgebirge des Kalantigebietes im Sudwestlichen Finnland. Finl. Comm. Geol. Bull., 103: 1–105.

Hildreth, W. and Moorbath, S., 1988. Crustal contributions to arc magmatism in the Andes of central Chile. Contrib. Mineral. Petrol., 98: 455–489.

Hill, R.I., 1991. Starting plumes and continental break-up. Earth Planet. Sci. Lett., 104: 398–416.

Hill, R.I., 1993. Mantle plumes and continental tectonics. Lithos, 30: 193–206.

Hill, R.I., Campbell, I.H., Davies, G.F. and Griffiths, R.W., 1992. Mantle plumes and continental tectonics. Science, 256: 186–193.

Hofmann, A.W., 1988. Chemical differentiation of the Earth: the relationships between mantle, continental crust, and oceanic crust. Earth Planet. Sci. Lett., 90: 297–314.

Hofmann, A.W., 1993. Chemical and isotopic differentiation of the Earth: a tale of two time scales. In: Evolving Earth Symposium (S. Maruyama, ed.). Okazaki, pp. 164–165.

Hofmeister, A.M., 1983. Effect of a headen terrestrial magma ocean on crust and mantle evolution. J. Geophys. Res., 88: 4963–4983.

Holloway, J.R. and Burnham, C.W., 1972. Melting relations of basalt with equilibrium water pressure less than total pressure. J. Petrol., 13: 1–29.

Honda, S. and Uyeda, S., 1983. Thermal processes in subduction zones: a review and preliminary approach on the origin of arc volcanism. In: Arc Volcanism (D. Shimozuru and I. Yokoyama, eds.). pp. 117–140.

Humphris, S.E. and Thompson, G., 1978. Hydrothermal alteration of oceanic basalts by sea water. Geochim. Cosmochim. Acta, 42: 107–125.

Hunter, D.R., 1970. The Ancient gneiss complex in Swaziland. Trans. Geol. Soc. S. Afr., 73: 107–150.

Hunter, D.R., 1974. Crustal development in Kaapvaal craton, I. The Archaean. Precambrian Res., 1: 259–294.

Hunter, D.R., Barker, F. and Millard, H.T., 1978. The geochemical nature of the Archaean Ancient Gneiss Complex and granodioritic suite, Swaziland: a preliminary study. Precambrian Res., 7: 105–127.

Hyppönen, V., 1983. Ontojoen, Hiisijärven ja Kuhmon kartta-alueiden kallioperä. Explanation to the maps No. 4411, 4412 and 4413. Geological Survey of Finland. pp. 60.

Jahn, B.M., 1994. Géochimie des granitoïdes archéens et de la croûte primitive. In: La Géochimie de la Terre (R. Hagemann, J. Jouzel, M. Treuil and L. Turpin, eds.) Masson-CEA, (in press).

Jahn, B.M., Auvray, B., Blais, S., Capdevila, R., Cornichet, J., Vidal, F. and Hameurt, J., 1980. Trace-element geochemistry and petrogenesis of Finnish greenstone belts. J. Petrol., 21: 201–244.

Jahn, B.M., Glikson, A.Y., Peucat, J.J. and Hickman, A.H., 1981. REE geochemistry and isotopic data of Archaean silicic volcanics and granitoids from the Pilbara Block, western Australia: implications for the early crustal evolution. Geochim. Cosmochim. Acta, 45: 1633–1652.

Jahn, B.M., Vidal, Ph. and Kroner, A., 1984. Multi-chronometric ages and origin of Archaean tonalitic gneisses in Finnish Lapland: A case for long crustal residence time. Contrib. Mineral. Petrol., 86: 398–408.

Jahn, B.M., Auvray, B., Shen, Q.H., Liu, D.Y., Zhang, Z.Q., Dong, Y.J., Ye, X.J., Zhang, Q.Z., Cornichet, J. and Macé, J., 1988. Archaean crustal evolution in China: the Taishan complex, and evidence for juvenile crustal addition from long-term depleted mantle. Precambrian Res., 38: 381–403.

Jarvis, G.T. and Campbell, I.H., 1983. Archaean komatiites and geotherms: solution to an apparent contradiction. Geophys. Res. Lett., 10: 1133–1136.

Jayananda, M., Martin, H. and Mahabaleswar, B., 1992. The mechanisms of recycling of the Archaean continental crust: example of the Closepet granite, south India. In: Third International Archaean Symposium (J.E. Glover and S.E. Ho, eds.). Perth, pp. 213–222.

Jayananda, M., Martin, H., Peucat, J.J. and Mahabaleswar, B., 1995. Late Archaean crust–mantle interactions: The Closepet batholith, South India, evidence from Sr-Nd isotopes, major- and trace-element geochemistry. Contrib. Mineral. Petrol. (in press).

Jégouzo, P. and Blais, S., 1991. Thrusting, crustal thickening and granite formation in the Archaean of eastern Finland. Terra Abstracts, 3 (1): 33.

Johnson, A.D. and Wyllie, P.J., 1988. Constraints on the origin of Archaean trondhjemites based on phase relationships of Nûk gneiss with H_2O at 15 kbars. Contrib. Mineral. Petrol., 100: 35–46.

Kaula W.M., 1979. Thermal evolution of the Earth and Moon growing by planetesimal impacts. J. Geophys. Res., 84: 9999–10008.

Kelemen, P.B., Johnson, K.T.M., Kinzler, R.J. and Irving, A.J., 1990. High-field-strength element depletion in arc basalts due to mantle–magma interaction. Nature, 345: 521–524.

Kinny, P.D., 1986. 3820 Ma zircons from a tonalitic Amîtsoq gneiss in Godthåb district of southern West Greenland. Earth Planet. Sci. Lett., 79: 337–347.

Kröner, A., 1991. Tectonic evolution in Archaean and Proterozoic. Tectonophysics, 187: 393–410.

Kröner, A. and Layer, P.W., 1992. Crust formation and plate motion in the Early Archaean. Science, 256: 1405–1411.

Kuno, H., 1968. Differentiation of basalt magmas. In: The Poldervaart Treatise on Rocks of Basaltic Composition (H.H. Hess and A. Poldervaart, eds.). Interscience, Vol. 2, pp. 623–688.

Kushiro, I., Yoder, H.S. and Nishikawa, M., 1972. Compositions of coexisting liquid and solid phases formed upon melting of natural garnet and spinel lherzolites at high pressures. A preliminary report. Earth Planet. Sci. Lett., 14: 19–25.

Lagabrielle, Y., Le Moigne, J., Maury, R., Cotten, J. and Bourgois, J., 1994. Volcanic record of subduction of an active spreading ridge, Taitao Peninsula, southern Chile. (in press).

Lameyre, J. and Bowden, P., 1982. Plutonic rock type series: discrimination of various granitoid series and related rocks. J. Volcanol. Geotherm. Res., 14: 169–186.

Lefèvre, C., 1979. Un exemple de volcanisme de marge active dans les Andes du Pérou, du miocène à l'actuel: zonation et pétrogenèse des andésites et shoshonites. Unpublished thesis, Montpellier (France), 555 pp.

Lemarchand, F., Villemant, B. and Calas, G., 1987. Trace element distribution coefficients in alkaline series. Geochim. Cosmochim. Acta, 51: 1071–1081.

Lopez-Escobar, L., 1984. Petrology and chemistry of volcanic rocks of the Southern Andes. In: Andean Magmatism, Chemical and Isotopic Constraints (R.S. Harmon and B.A. Barreiro, eds.). Shiva Geology Series, Nantwich, pp. 47–71.

Ludden, J., Hubert, C., Barnes, A., Mikereit, B. and Sawyer, E., 1993. A three dimensional perspective on the evolution of the Earth's largest Archaean crust. LITHOPROBE seismic reflection images in the southwestern Superior Province. Lithos, 30: 357–372.

Luukkonen, E.J., 1988. Moisiovaaran ja Ala-Vuokin kartta-alueen kallioperä. Explanation to the maps No. 4421, 4423 and 4441. Geological Survey of Finland. pp. 90.

Maaløe , S., 1982. Petrogenesis of Archaean tonalites. Geol. Rundsch., 71: 328–346.

Mann, A.C., 1983. Trace-element geochemistry of high alumina basalt–andesite–dacite– rhyodacite lavas of the main volcanic series of Santori volcano, Greece. Contrib. Mineral. Petrol., 84: 43–57.

Martin, H., 1985. Nature, origine et évolution d'un segment de croûte continentale archéenne: contraintes chimiques et isotopiques. Exemple de la Finlande orientale. Mem. Doc. Centre Arm. Et. Struct. Socles, Rennes, 1: pp. 392.

Martin, H., 1986. Effect of steeper Archaean geothermal gradient on geochemistry of subduction-zone magmas. Geology, 14: 753–756.

Martin, H., 1987a. Petrogenesis of Archaean trondhjemites, tonalites and granodiorites from eastern Finland: Major and trace element geochemistry. J. Petrol., 28: 921–953.

Martin, H., 1987b. Archaean and modern granitoids as indicators of changes in geodynamic processes. Rev. Bras. Geoc., 17: 360–365.

Martin, H., 1993. The mechanisms of petrogenesis of the Archaean continental crust, comparison with modern processes. Lithos, 30: 373–388.

Martin, H., Chauvel, C. and Jahn, B.M., 1983a. Major and trace element geochemistry and crustal evolution of Archaean granodioritic rocks from eastern Finland. Precambrian Res., 21: 159–180.

Martin, H., Chauvel, C., Jahn, B.M. et Vidal, Ph., 1983b. Rb-Sr and Sm-Nd isotopic ages and geochemistry of Archaean granodioritic rocks from eastern Finland. Precambrian Res., 20: 79–91.

Martin, H., Auvray, B., Blais, S., Capdevila, R., Hameurt, J., Jahn, B.M., Piquet, D., Quérré, G. and Vidal, Ph., 1984. Origin and geodynamic evolution of the Archaean crust of Eastern Finland. Bull. Geol. Soc. Finland, 56: 135–160.

Martin, H., Auvray, B., Jayananda, M. and Peucat, J.J., 1992. Interactions croûte-manteau: l'exemple du granite tardi-archéen de Closepet (Inde du Sud). In: Symposium on Dynamique et bilans de la Terre, Toulon: T 5.

Martin, H., Peucat, J.J., Auvray, B. and Jayananda, M., 1993. The Archaean "sanukitoid" magmatism: example of the Closepet granite (south India). In: Abstract of EUG VII, Strasbourg, A3: 38.

Maruyama, S., 1992. Similarity and dissimilarity between Archaean and Phanerozoic accretionary complex and its cause. In: Evolving Earth Symposium (S. Maruyama, ed.). Okazaki, pp. 25–29.

Maruyama, S., Masuda, T., Nohda, S. and Appel, P., 1992. The earliest records on oceanic and continental crust from 3.8 Ga accretionary complex, Isua, Greenland. In: abstracts of the 29 th I.G.C. Kyoto, 1239, 5.

Masuda, A., Nakamura, N. and Tanaka, T., 1973. Fine structures of mutually normalized rare-earth patterns of chondrites. Geochim. Cosmochim. Acta, 37: 239–244.

McCulloch, M.T., 1992. Partial melting versus dehydration of subducted slabs and the Pb paradox. In: Abstracts of the 29th I.G.C. Kyoto, 1638, 181.

McCulloch, M.T. and Bennet, V.C., 1993. Evolution of the early Earth: constraints from ^{143}Nd–^{142}Nd isotopic systematics. Lithos, 30: 237–255.

McGregor, V.R., 1973. The early Precambrian geology of the Godthåb district, West Greenland. Phil. Trans. R. Soc. London, A-273: 243–258.

McGregor, V.R., Friend, C.R.L. and Nutman, A.P., 1991. The late-Archaean mobile belt through Gotthåbsfjord region, southern west Greenland: a continent–continent collision-zone? Bull. Geol. Soc. Denmark, 39: 179–197.

McKenzie, D.P., 1967. Some remarks on the heat flow and gravity anomalies. J. Geophys. Res., 72: 6261.

McKenzie, D.P., 1984. The generation and compaction of partially molten rocks. J. Petrol., 25: 713–765.

McKenzie, D.P. and Weiss, N., 1975. Speculations on the thermal and tectonic history of the Earth. R. Astron. Soc. Geophys. J., 42: 131–174.

McLennan, S.M. and Taylor, S.R., 1982. Geochemical constraints on the growth of the continental crust. J. Geol., 90: 342–361.

Meijer, A., 1983. The origin of low-K rhyolites from the Marianna frontal arc. Contrib. Mineral. Petrol., 96: 212–224.

Moorbath, S., 1975. Evolution of Precambrian crust from strontium isotopic evidence. Nature, 254: 395–398.

Moorbath, S., 1976. Age and isotope constraints for the evolution of Archaean crust. In: The Early History of the Earth (B.F. Windley, ed.). Wiley, pp. 351–360.

Moorbath, S., 1977. Aspects of the geochronology of ancient rocks related to continental evolution. In: The Continental Crust and Its Mineral Deposits (D.W. Strangway, ed.). Geol. Ass. Can. Spec. Paper, 20: 89–115.

Mueller, P.A. and Wooden, J.L., 1988. Evidence for Archaean subduction and crustal recycling, Wyoming province. Geology, 16: 871–874.

Mueller, P.A., Wooden, J.L., Schulz, K. and Bowes, D.R., 1983. Incompatible element rich andesitic amphibolites from the Archaean of Montana and Wyoming: evidence for mantle metasomatism. Geology, 11: 203–206.

Mysen, B.O., 1979. Trace element partitioning between garnet peridotite minerals and water-rich vapour: experimental data from 5 to 30 kbars. Am. Mineral., 64: 274–287.

Mysen, B.O. and Boettcher, A.L., 1975a. Melting of a hydrous mantle I. Phase relations of natural peridotite at high pressures and temperatures with controlled activities of water, carbon dioxide and hydrogen. J. Petrol., 16: 520–548.

Mysen, B.O. and Boettcher, A.L., 1975b. Melting of a hydrous mantle II. Geochemistry of crystals and liquids formed by anatexis of mantle peridotite at high pressures and temperatures as a function of the controlled activities of water, carbon dioxide and hydrogen. J. Petrol., 16: 549–593.

Nédélec, A., Nsifa, E.N. and Martin, H., 1990. Major and trace element geochemistry of the Archaean Ntem plutonic complex (South Cameroon): petrogenesis and crustal evolution. Precambrian Res., 47: 35–50.

Newton, R.C., 1989. Metamorphic fluids in the deep crust. Annu. Rev. Earth Planet. Sci., 17: 385–412.

Newton, R.C., 1990. The late Archaean high-grade terrain of South India and the deep structure of the Dharwar craton. In: Exposed Cross-sections of the Continental Crust (M.H. Salisbury and D.M. Fountain, eds.). Kluwer, The Netherlands, pp. 305–326.

Neymak, L.A., Kovach, V.P., Nemchin, A.A., Morozova, I.M., Kotov, A.B., Vinogradov, D.P., Gorokhovsky B.M., Ovchinnikova, G.V., Bogomolova, L.M. and Smelov, A.P., 1993. Late Archaean intrusive complexes in the Olekma granite–greenstone terrain (eastern Siberia): geochemical and isotopic study. Precambrian Res., 62: 453–472.

Nicholls, I.A., 1974. Liquid in equilibrium with peridotitic mineral assemblages at high water pressure. Contrib. Mineral. Petrol., 45: 289–316.

Nicholls, I.A. and Ringwood, A.E., 1973. Production of silica saturated tholeiitic magmas in island arcs. Earth Planet. Sci. Lett., 16: 243–246.

Nisbet, E.G., 1984. The continental and oceanic crust and lithosphere in the Archaean: isostatic, thermal and tectonic models. Can. J. Earth Sci., 21: 1426–1441.

Nisbet, E.G., 1987. The Young Earth: An Introduction to Archaean Geology. Allen & Unwin, Boston, pp. 402.

Nisbet, E.G. and Fowler, C.M.R., 1983. Model for Archaean plate tectonics. Geology, 11: 376–379.

Nisbet, E.G., Cheadle, M.J., Arndt, N.T. and Bickle, M.J., 1993. Contraining the potential temperature of the Archaean mantle: A review of the evidence from komatiites. Lithos, 30 : 291–307.

Nutman, A.J. and Collerson, K.D., 1991. Very early Archaean crustal-accretion complexes preserved in the North Atlantic craton. Geology, 19: 791–794.

Nutman, A.J., Friend, C.R.L. and McGregor, V.R., 1992. Unravelling the architecture of Archaean high-grade gneiss complexes: evidence for terrane amalgamation. In: Evolving Earth Symposium (S. Maruyama, ed.), Okazaki, pp. 47–52.

O'Connor, J.T., 1965. A classification for quartz-rich igneous rocks based on feldspar ratios. U.S. Geol. Surv. prof. pap., 525-B: 79–84.

O'Nions, R.K. and Pankhurst, R.J., 1978. Early Archaean rocks and geochemical evolution of the Earth's crust. Earth Planet. Sci. Lett., 38: 211–236.

Parson, B.A. and Sclater, J.G., 1977. An analysis of the variation of ocean floor bathymetry and heat flow with ages. J. Geophys. Res., 82: 803–827.

Peacock, S.M., 1990. Fluid processes in subduction zones. Science, 248: 329–337.

Peacock, S.M., 1993. Large-scale hydration of the lithosphere above subducting slabs. Chem. Geol., 108: 43–59.

Pearce, J.A. and Norry, M.J., 1979. Petrogenetic implications of Ti, Zr, Y and Nb variations in the volcanic rocks. Contrib. Mineral. Petrol., 69: 33–47.

Peterman, Z.E., 1979. Strontium isotope geochemistry of late Archean to late Cretaceous tonalites and trondhjemites. In: Trondhjemites, Dacites and Related Rocks (F. Barker, ed.). Elsevier, Amsterdam, pp. 133–147.

Peterman, Z.E. and Barker, F., 1976. Rb-Sr whole-rock age of trondhjemites and related rocks of the south western Trondheim region, Norway. U.S. Geol. Surv. Open File Rept., 76-670: 1–17.

Peucat, J.J., Gruau, G., Martin, H., Auvray, B., Fourcade, S., Jayananda, M., Bouhallier, H. and Choukroune, P., 1993. A 2.5 Ga. mega-plume in south India? In: Abstract of EUG VII, Strasbourg, C13: 321.

Philippot, P., 1993. Fluid-melt-rock interaction in mafic eclogites and coesite-bearing metasediments: Constraints on volatile recycling during subduction. Chem. Geol., 108: 93–112.

Philippot, P. and Silverstone, J., 1991. Trace-element-rich brines in eclogite veins: implications for fluid composition and transport during subduction. Contrib. Mineral. Petrol., 106: 417–430.

Pitchamuthu, C.S., 1974. The Dharvar craton. J. Geol. Soc. India, 15: 339–346.

Poidevin, J.L., 1991. Les ceintures de roches vertes de la République de Centre Afrique (Mbomou, Bandas, Boufoyo, Bogoin), contribution à la connaissance du Précambrien du nord du craton du Congo. Unpublished thesis, Clermont-Ferrand (France), 440 pp.

Poldervaart, A., 1955. Chemistry of the Earth's crust. G.S.A. Spec. Pap., 62: 119.

Rapp, R.P., Watson, E.B. and Miller, C.F., 1991. Partial melting of amphibolite/eclogite and the origin of Archaean trondhjemites and tonalites. Precambrian Res., 51: 1–25.

Reymer, A. and Schubert, G., 1984. Phanerozoic addition rates of the continental crust and crustal growth. Tectonics, 3: 63–78.

Richter, F.M., 1985. Models for the Archaean thermal regime. Earth Planet. Sci. Lett., 73: 350–360.

Ringwood, A.E., 1970. Origin of the Moon: The precipitation hypothesis. Earth Planet. Sci. Lett., 8, 131–140.

Ringwood, A.E., 1983. Geochemical relationships between the Earth's core and mantle. Lunar Planet. Sci., XIV: 646–647.

Ringwood, A.E., 1992. Thermal and geochemical regimes of the primordial Earth. In: Evolving Earth Symposium (S. Maruyama, ed.). Okazaki, pp. 71–72.

Rogers, G. and Hawkesworth, C.J., 1989. A geochemical traverse across the North Chilean Andes: evidence for crust generation from the mantle wedge. Earth Planet. Sci. Lett., 91: 271–285.

Ronov, A.B. and Yaroshensky, A.A., 1976. A new model for the chemical structure of the Earth's crust. Geochim. Int., 13: 89–121.

Rushmer, T., 1991. Partial melting of two amphibolites: contrasting experimental results under fluid-absent conditions. Contrib. Mineral. Petrol., 107: 41–59.

Ryerson, F.J. and Watson, E.B., 1987. Rutile saturation in magmas: implications for Ti-Nb-Ta depletion in island-arc basalts. Earth Planet. Sci. Lett., 86: 225–239.

Salomon, S.C., Ahrens, T.J., Cassen, P.M., Hsui, A.T., Minear, J.W., Reynolds, R.T., Sleep, N.H.,

Strangway, D.W. and Turcotte, D.L., 1981. Thermal history of the terrestrial planets, Chapter 9. In: Basaltic Volcanism on the Terrestrial Planets. Pergamon, New York, pp. 1129–1234.

Sclater, J.G. and Francheteau, J., 1970. The implications of terrestrial heat flow observations on current tectonic and geochemical models of the crust and upper mantle of the earth. J. R. Astron. Soc., 26: 515.

Sengupta, S., Paul, D.K., de Laeter, J.R., McNaughton, N.J., Bandopadhyay, P.K. and de Smeth, J.B., 1991. Mid-Archaean evolution of the Eastern Indian craton: geochemical and isotopic evidence from the Bonai pluton. Precambrian Res., 49: 23–37.

Shaw, D.M., 1970. Trace element fractionation during anatexis. Geochim. Cosmochim. Acta, 34: 237–243.

Sheraton, J.W. and Black, L.P., 1983. Geochemistry of Precambrian gneisses: relevance for the evolution of the east Antarctic shield. Lithos, 16: 273–296.

Shirey, S.B. and Hanson, G.N., 1984. Mantle derived Archaean monzodiorites and trachyandesites. Nature, 310: 222–224.

Shirey, S.B. and Hanson, G.N., 1986. Mantle heterogeneity and crustal recycling in Archaean granite-greenstone belts: evidence from Nd isotopes and trace elements in the Rainy Lake province, Ontario, Canada. Geochim. Cosmochim. Acta, 50: 2631–2651.

Sleep, N.H. and Windley, B.F., 1982. Archaean plate tectonics: constraints and inferences. J. Geol., 90: 363–379.

Smith, T.E., Choudhry, A.G. and Huang, C.H., 1983. The geochemistry and petrogenesis of the Archaean Gamitagama lake igneous complex, Southern Superior Province. Precambrian Res., 22: 219–244.

Sorensen, S.S., 1988. Petrology of amphibolite-facies mafic and ultramafic rocks from Catalina schist, southern California: metamorphism and migmatisation in a subduction zone metamorphic setting. J. Metam. Geol., 6: 405–435.

Sorensen, S.S. and Barton, M.D., 1987. Metasomatism and partial melting in a subduction complex: Catalina schist, southern California. Geology, 15: 115–118.

Sorensen, S.S. and Grossman, J.N., 1989. Enrichment in trace elements in garnet amphibolites from a paleo-subduction zone: Catalina schist, southern California. Geochim. Cosmochim. Acta, 53: 3155–3177.

Spulber, S.D. and Rutherford, M.J., 1983. The origin of rhyolite and plagiogranite in oceanic crust: An experimental study. J. Petrol., 24: 1–25.

Stern, C.R. and Futa, K., 1982. An Andean andesite derived directly from subducted MORB or from LIL depleted subcontinental mantle. Trans. Am. Geophys. Union, 63: 1148.

Stern, C.R., Futa, K. and Muehlenbachs, K., 1984a. Isotopic and trace element data for orogenic andesites from the austral Andes. In: Andean Magmatism, Chemical and Isotopic Constraints (R.S. Harmon and B.A. Barreiro, eds.). Shiva Geology Series, Nantwich, pp. 1–46.

Stern, C.R, Futa, K., Muehlenbachs, K., Dobbs, M., Munoz, J., Godoy, E. and Charrier, R., 1984b. Sr, Nd, Pb, O isotope composition of late Cenozoic volcanics; northernmost SVZ (33–34°S). In: Andean Magmatism, Chemical and Isotopic Constraints (R.S. Harmon and B.A. Barreiro, eds.). Shiva Geology Series, Nantwich, pp. 96–105.

Stern, R.A. and Hanson, G.N., 1991. Archaean high-Mg granodiorite: a derivative of Light Rare Earth enriched monzodiorite of mantle origin. J. Petrology, 32: 201–238.

Stern, R.A., Nesbitt, R.W. and McCulloch, M.T., 1989. Geochemistry and petrogenesis of siliceous high magnesian basalts of the Archaean and early Proterozoic. In: Boninites and Related Rocks (A.R. Crawford, ed.). Unwin Hyman, pp. 148–173.

Stowe, C.W., 1973. The older tonalite gneiss complex in the Selukwe area, Rhodesia. Geol. Soc. S. Afr. Spec. Publ., 3: 85–96.

Streckeisen, A., 1975. To each plutonic rock its proper name. Earth Sci. Rev., 12: 1–33.
Strens, R.G.J., 1965. Stability and relations of the Al-Fe epidotes. Miner. Mag., 35, 464–475.
Sun, S.S., Nesbitt, R.W.N. and Sharaskin, A.Y., 1979. Geochemical characteristics of mid-ocean ridge basalts. Earth Planet. Sci. Lett., 44: 119–138.
Sutcliffe, R.H., 1989. Magma mixing in late Archaean tonalitic and mafic rocks of the Lac des Iles area, Western Superior Province. Precambrian Res., 44: 81–101.
Tarney, J., Weaver, B.L. and Drury, S.A., 1979. Geochemistry of Archaean trondhjemitic and tonalitic gneisses from Scotland and E. Greenland. In: Trondhjemites, Dacites and Related Rocks (F. Barker, ed.). Elsevier, Amsterdam, pp. 275–299.
Tarney, J., Weaver, B.L. and Winley, B.F., 1982. Geological and geochemical evolution of the Archaean continental crust. Rev. Bras. Geoc., 12: 53–59.
Tatsumi, Y. and Nakamura, N., 1986. Composition of aqueous fluid from serpentinite in the subducted lithosphere. Geochim. J., 20: 191–196.
Tatsumi, Y., Hamilton, D.L. and Nesbitt, R.W., 1986. Chemical characteristics of fluid phase from the subducted lithosphere: evidence from high-pressure experiments and natural rocks. J. Volcanol. Geotherm. Res., 29: 293–309.
Taylor, R.N. and Nesbitt, R.W., 1988. Light rare-earth enrichment of supra subduction-zone mantle: evidence from the Trodos ophiolite, Cyprus. Geology, 16: 448–451.
Taylor, R.S., 1975. Lunar Science: A Post Apollo View. Pergamon Press, 372 pp.
Taylor, R.S., 1992. The origin of the Earth. In: Understanding the Earth, a New Synthesis (G.C. Brown, C.J. Hawkesworth and R.C.L. Wilson, eds.). Cambridge University Press, pp. 25–43.
Taylor, R.S., 1993. Early accretional history of the Earth and the Moon forming event. Lithos, 30: 207–221.
Taylor, S.R. and McLennan, S.M., 1985. The Continental Crust: Its Composition and Evolution. Blackwell Scientific, Oxford, 312 pp.
Toksöv, M.N. and Hsui, A.T., 1978. Numerical studies of back-arc convection and the formation of marginal basins. Tectonophysics, 50: 177–196.
Toksöv, M.N., Minear, J.W. and Julian, B.R., 1971. Temperature field and geophysical effects of a downgoing slab. J. Geophys. Res., 76: 1113–1138.
Treloar, P.J., Coward, M.P. and Harris, N.B.W., 1992. Himalayan–Tibetan analogies for the evolution of the Zimbabwe craton and Limpopo belt. Precambrian Res., 55: 571–587.
Van der Laan, S. and Wyllie, P.J., 1992. Constraints on Archaean trondhjemite genesis from hydrous crystallisation experiments on Nûk gneiss at 10–17 Kbar. J. Geol., 100: 57–68.
Veizer, J. and Jansen, S.L., 1979. Basement and sedimentary recycling and continental evolution. J. Geol., 87: 341–370.
Vidal, Ph. and Dosso, L., 1978. Core formation: catastrophic or continuous? Sr and Pb isotope geochemistry constraints. Geophys. Res. Lett., 5: 169–172.
Viljoen, M.J. and Viljoen, R.P., 1969a. A proposed new classification of the granitic rocks of the Barberton region. Geol. Soc. S. Afr. Spec. Publ., 2: 153–188.
Viljoen, M.J. and Viljoen, R.P., 1969b. The chemical evolution of the granitic rocks of the Barberton region. Geol. Soc. S. Afr. Spec. Publ., 2: 189–220.
Wänke, H., 1981. Constitution of terrestrial planets. Phil. Trans. R. Soc. Lond., A303: 287–302.
Weaver, B.L. and Tarney, J., 1981. Lewisian gneiss geochemistry and Archaean crustal development models. Earth Planet. Sci. Lett., 55: 171–180.
Wells, P.R.A., 1979. Chemical and thermal evolution of the Archaean sialic crust, Southern West Greenland. J. Petrol., 20: 187–226.
Wetherill, G.W., 1978. Accumulation of the terrestrial planets. In: Protostars and Planets (T. Gehrels, ed.). University Arizona Press, Tuscon, Arizona, 565 pp.

White, A.J.R., Jakes, P. and Christie, D.M., 1971. Composition of greenstones and the hypothesis of sea-floor spreading in Archaean. Geol. Soc. Aust. Spec. Publ., 3: 47–56.

Wilson, J.F., 1973. Granites and gneiss around Mashaba, Rhodesia. Geol. Soc. S. Afr. Spec. Publ., 3: 79–84.

Windley, B.F. and Bridgwater, D., 1971. The evolution of Archaean low- and high-grade terrains. Geol. Soc. Aust. Spec. Publ., 3: 33–46.

Winther, T.K. and Newton, R.C., 1991. Experimental melting of an hydrous low-K tholeiite: evidence on the origin of Archaean cratons. Bull. geol. Soc. Denmark, 39.

Wolf, M.B. and Wyllie, P.J., 1991. Dehydration-melting of solid amphibolite at 10 Kbar: textural development, liquid interconnectivity and applications to the segregation of magmas. Mineral. Petrol., 44: 151–179.

Wolf, M.B. and Wyllie, P.J., 1993. Dehydration-melting of amphibolite at 10 Kbar: effect of temperature, time and texture. Contrib. Mineral. Petrol. (in press).

Wyllie, P.J., 1971. The role of water in magma genesis and initiation of diapiric uprise in the mantle. J. Geophys. Res., 76: 1328–1338.

Wyllie, P.J., 1977. Effects of H_2O and CO_2 on magma generation in the crust and mantle. J. Geol. Soc. London, 134: 215–234.

Wyllie, P.J., 1979. Magmas and volatile components. Am. Mineral., 64: 469–500.

Wyllie, P.J., 1983. Experimental and thermal constraints on the deep seated parentage of some granitoid magmas in subduction zones. In: Migmatites, Melting and Metamorphism (M.P. Atherton and C.D. Gribble, eds.). Shiva Geology Series, Nantwich, pp. 37–51.

Zen, E. and Hammarstrom, J.M., 1984. Magmatic epidote and its petrologic significance. Geology, 12: 515–518.

Chapter 7

ARCHEAN GRANITE PLUTONS

PAUL J. SYLVESTER

INTRODUCTION

Potassium-rich granite, contrary to some opinions expressed in the past (e.g., Burke and Kidd, 1978), is a widespread and voluminous constituent of Archean cratons. Some of the granite occurs as highly metamorphosed banded gneiss but the majority forms massive to moderately foliated, medium- to coarse-grained plutons that largely retain an igneous mineralogy. Most of the plutons have areal exposures of 5–500 km^2 but a few exceed 1000 km^2.

Condie (1993) estimated that granite plutons make up ~20% of the rock exposed in Archean shields, placing them far ahead of tholeiite (~10%) and second only to the tonalite–trondhjemite–granodiorite (TTG) suite (~50%) in abundance. It is somewhat surprising therefore that, until recently, there have been only a handful of detailed studies of Archean granite plutons and, with few exceptions (e.g., Condie, 1981; Ridley, 1992; Wyborn et al., 1992), almost no attempt to place what is known about these rocks in a general context. Happily this situation has improved; many new data are available and we are now in a position to begin addressing first-order questions such as: Were all Archean granites melted from the same sorts of source materials? How do Archean granites differ from their younger counterparts? What do Archean granites tell us about tectonic and thermal regimes present during formation of the early continental crust?

In this chapter, answers are sought to these and other questions. We first review what is known about the geologic setting and chemical composition of Archean granite plutons of eight major shield regions and then use these data to draw conclusions about their petrogenesis and role in crustal evolution. As has become common practice in studies of post-Archean granites, following the pioneering work of Chappell and coworkers in the Lachlan Fold Belt of southeastern Australia (Chappell and White, 1974, 1992; Collins et al., 1982), a three-fold classification is used to describe Archean granites, albeit a descriptive rather than genetic one. Thus, there are calc-alkaline, strongly peraluminous and alkaline granites. Strongly peraluminous granites are distinguished from calc-alkaline and alkaline granites by the presence of primary, igneous minerals that are more aluminous than biotite, such as muscovite, alumino-silicates, garnet and/or cordierite, as discussed by Miller (1985). Where there is reasonable doubt about a primary origin for muscovite, as is often the case (White et al., 1986), a secondary origin is assumed. Alkaline granites are distinguished from calc-alkaline granites by a plot of $(Al_2O_3 + CaO)/(FeO_t + Na_2O + K_2O)$ vs. $100(MgO + FeO_t + TiO_2)/SiO_2$, as shown by Sylvester (1989). In the Lachlan Fold Belt, calc-alkaline granite plutons are thought to have been derived by partial melting of igneous rocks that had not been subjected to weathering at the earth's surface and are described as "I-types", while strongly peraluminous granite plutons are thought to result from partial melting of sedimentary rocks and are "S-types". Alkaline granite plutons are thought to be either highly-fractionated I-type granites or partial melts of restitic rocks, the latter being described as "A-type" granite plutons.

The granite plutons considered in this chapter consist of rocks that would be classified as alkali feldspar granite, syenogranite, monzogranite or granodiorite using the modal quartz–alkali feldspar–plagioclase (Q-A-P) diagram of Streckeisen (1976). In the normative orthoclase–albite–anorthite (Or-Ab-An) diagram of Barker (1979), the plutons plot in the granite and granodiorite fields (Fig. 1). Some Archean TTG rocks likewise fall in the granodiorite fields of the Q-A-P and Or-Ab-An plots but most Archean TTGs would be classified as tonalites or trondhjemites using these diagrams.

Chemical compositions derived for each granite pluton of this study are averages of chemical analyses of individual, constituent samples, except in a few cases where only one sample has been analyzed. All samples used in the averages are those claimed by the original investigators of the plutons to be the least affected by secondary alteration. In rare examples of composite calc-alkaline–strongly peraluminous plutons, averages were made for only the more abundant type of granite. Unless stated otherwise, all isotopic ages of the plutons referred to below are those determined on zircon or monazite using the U-Pb method of dating.

The Sr, O, Nd and Pb isotopic systematics of Archean granite plutons are not discussed in this chapter. While the Sr isotopic system has been used quite successfully to constrain the nature of the source materials of Phanerozoic granite plutons, it has often been proven to be too disturbed by post-magmatic, volatile alteration to be of much value in discerning the origin of Archean granite plutons (Beakhouse et al., 1988). The O, Nd and Pb systems seem to be less disturbed and hold some promise in helping to distinguish the source characteristics of Archean granite plutons but, as yet, too few data are available from which to draw general conclusions.

GEOLOGIC SETTING OF ARCHEAN GRANITE PLUTONS

Pilbara Block, Western Australia

Detailed studies of the Pilbara Block have been concentrated in the southeastern part of the craton. In this region, numerous granodiorite-granite and granite plutons have intruded older TTG gneisses, TTG plutons, mafic–felsic volcanics

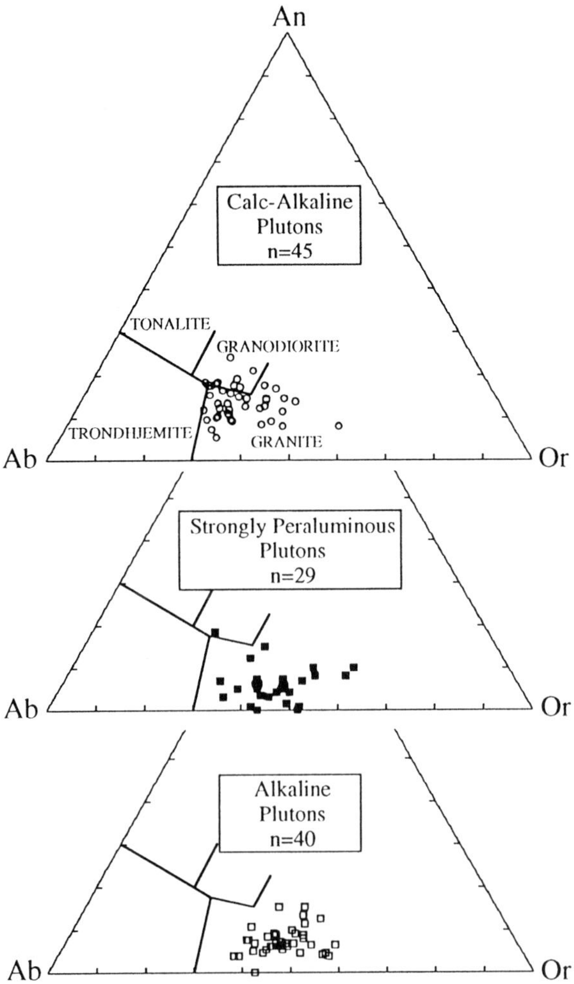

Fig. 1. Compositions of calc-alkaline, strongly peraluminous and alkaline Archean granite plutons plotted in the normative orthoclase (Or)-albite (Ab)-anorthite (An) diagram of Barker (1979). Data sources given in the text and/or Tables 1, 3 and 4.

and minor granites (Blockley, 1980; Hickman, 1983). The older rocks formed during magmatic events at ~3450 and 3325 Ma and were intercalated during horizontal deformation and kyanite–sillimanite-grade metamorphism at ~3300 Ma (Bickle et al., 1993).

There are reliable determinations of the ages of only four of the granitic plutons, all from the Shaw Batholith (Bickle et al., 1989, using the Pb-Pb whole rock isochron method of dating). The age determinations suggest two episodes of intrusion, one at ~2970 Ma and another at ~2850 Ma. The event at 2970 Ma preceded open upright folding and andalusite–sillimanite-grade metamorphism and formed calc-alkaline and strongly peraluminous granites, whereas the event at 2850 Ma followed folding and metamorphism and produced alkaline granites. Granites similar to those present in the Shaw Batholith have been described in the nearby Mount Edgar Batholith (Davy and Lewis, 1986) and Corunna Downs (Davy, 1988) Batholith, as well as in terrains located further south — the Kurrana Batholith and the Cooninia, Billinooka and Sylvania inliers (Williams, 1989; Tyler, 1991). Volcanism does not seem to have accompanied granite plutonism in the southeastern Pilbara.

Regional tectonic events occurring during emplacement of the granite plutons are poorly constrained. Tyler et al. (1992), however, postulated that shearing occurred along the northern margin of the Kurrana Batholith between 3000 and 2760 Ma. Krapez and Barley (1987) suggested that the Lalla Rookh Formation, 3000 m of coarse clastic sediments deposited in a pull-apart basin located just north of the Shaw Batholith, formed at about the same time.

Yilgarn Block, Western Australia

The Yilgarn Block differs from most other Archean cratons in that granodiorite-granite and granite plutons are much more voluminous than TTG gneisses and plutons. In the Norseman region, located in the southeastern part of the craton, major syn- to late-kinematic calc-alkaline granodiorite–granite plutons were emplaced at ~2685 and 2665 Ma, and were followed by minor post-kinematic alkaline granite plutons at ~2640 and 2600 Ma (Hill et al., 1992a). The plutons were preceded by mafic volcanism at ~2715 Ma, felsic volcanism at ~2940 Ma and, based on U-Pb ages of zircons inherited from the source regions of some of the granites, probably even older magmatic rocks (Hill et al., 1992b). Deformation associated with emplacement of the syn-kinematic plutons was synchronous with greenschist to amphibolite facies metamorphism and involved recumbent folding, thrusting, and oblique- and strike-slip shearing (Barley and Groves, 1990). The 2685 My-old granites were accompanied by felsic volcanism, whereas minor tonalite-trondhjemite plutonism was contemporaneous with formation of the 2665 My-old granites (Hill et al., 1992a).

Quite remarkably, ~2685, 2640 and 2600 My-old granite plutonism, very similar to that seen in the Norseman region, has been recognized 500–600 km to the northwest, in the Murchison Province (Watkins et al., 1991; Wiedenbeck and Watkins, 1993). Here, however, the granites were preceded by ~2760 My-old tonalitic plutons and felsic volcanic rocks, mafic volcanic rocks presumed to be somewhat older than 2760 My, and ~2920 My-old TTG and granite gneisses (Wiedenbeck and Watkins, 1993). Elsewhere in the Yilgarn Block, much less is known about the details of granite plutonism. Wilde and Pidgeon (1986), Pidgeon et al. (1990) and Hill et al. (1992a), however, have documented the presence of ~2665 and 2640 My-old granites in a region 300-500 west of Norseman.

Superior Province, Canada

Covering a huge expanse of two million km², the Superior Province exhibits a striking pattern of subparallel, broadly east-west trending, elongated subprovinces that are distinguished from one another by differences in lithology. Thus, in the southern part of the craton, which has been studied in the most detail, there are the Wawa-Abitibi, Wabigoon and Bird River subprovinces, which largely consist of volcanic and plutonic rocks; the Pontiac, Quetico and English River subprovinces, which are dominated by turbidite sedimentary rocks; and the Winnipeg River subprovince, more than 95% of which is plutonic rock (Card, 1990). The sedimentary subprovinces are unusual; in most other cratons, Archean sedimentary rocks are found mainly in association with greenstone belts (Williams, 1990).

For such a large region, most of the Archean rocks of the southern Superior Province formed in a surprisingly short time between ~2750 and 2645 Ma. The first half of this interval was dominated by mafic to felsic volcanism and TTG plutonism, the second by granite–granodiorite plutonism (Card, 1990; Sutcliffe et al., 1993). Major deformation and low-grade metamorphism occurred during the transition from the first type of magmatism to the second, at ~2700–2680 Ma, beginning with north-south shortening and ending with northwest dextral and northeast sinistral transcurrent faulting (Card, 1990).

Granite plutons of the southern Superior Province have calc-alkaline and strongly peraluminous compositions. The two types of plutons were intruded at about the same time and, in each of the Wawa-Abitibi (Arth and Hanson, 1975; Smith et al., 1985; Boily et al., 1990; Feng and Kerrich, 1992), Bird River (Cerny et al., 1987) and Quetico (Day and Weiblen, 1986; Percival, 1989) belts, side-by-side within a single subprovince. In other regions, however, there seems to be a strong relationship between the lithologic character of a subprovince and the compositions of the granite plutons present within it. Thus, calc-alkaline granite seems to form the predominant type of pluton in the volcanic–plutonic Wabigoon (Shirey and Hanson, 1985; Day, 1990) and plutonic Winnipeg River (Gower et al., 1983; Beakhouse and McNutt, 1991) subprovinces, whereas strongly peraluminous granite plutons seem to dominate in the sedimentary English River (Breaks et al., 1985) and Pontiac (Feng and Kerrich, 1992) subprovinces.

Slave Province, Canada

The Slave Province is commonly subdivided into western and eastern regions based largely on differences in the proportion of mafic to felsic volcanic rocks in greenstone belts: belts in the west are basalt-rich, whereas in the east, both basalt-rich and rhyodacite-rich belts occur (Padgham, 1985; Kusky, 1989). Felsic volcanic units and associated TTG plutons from both regions give isotopic ages that fall between ~2700 and 2665 My (Mortensen et al., 1988; van Breeman et al., 1989; Bevier and Gebert, 1991). In the Yellowknife area, in the south of the craton, ages as old as ~2715 My have been documented for the felsic volcanics, providing a minimum age for the initiation of mafic volcanism in the province (Isachsen et al., 1991). Pre-volcanic rocks, mainly TTG gneisses, seem to be restricted to the western Slave but syn- to post-volcanic, greywacke–mudstone turbidite sedimentary rocks are found province-wide. The latter rocks cover at least 4 times more area than the volcanic rocks (Padgham, 1985) and in some places form wide belts resembling those of the Superior Province, albeit not with as much lateral continuity.

Voluminous plutonism occurred throughout the Slave Province between ~2620 and 2580 Ma with peaks of activity at 2620, 2610–2605 and 2590–2580 Ma (Henderson et al., 1987; van Breeman et al., 1987, 1989; van Breeman and Henderson, 1988; Bevier and Gebert, 1991). The 2610–2605 My-old event was coeval with major isoclinal folding and low-pressure, high-temperature (andalusite-sillimanite) regional metamorphism; the 2590-2580 My-old event followed peak metamorphic conditions and was synchronous with NE- and NW-trending regional cross folding (King and Helmstaedt, 1989; King et al., 1990). Compositions of these plutons have not been well-documented but the 2620 and 2610–2605 My-old events seem to have involved mainly TTG magmatism, while the 2590–2580 My-old event seems to have consisted largely of strongly peraluminous granodiorite and granite magmatism (Drury, 1979; Frith and Fryer, 1985; Cerny and Meintzer, 1988; Kretz et al., 1989). Subordinate calc-alkaline granite plutons are spatially associated with the strongly peraluminous granites (Frith and Fryer, 1985) and may also have formed during the 2590–2580 My-old event.

Wyoming Province, USA

As in the Yilgarn Block, a large proportion of the Wyoming Province consists of granite–granodiorite plutons that are more voluminous than TTG gneisses and plutons. In the Wind River Range, located near the center of the province, protoliths of paragneiss-rich migmatite and felsic orthogneiss formed between ~3800 and 3300 Ma and were subjected to episodes of high-grade metamorphism at ~3200 and 2700 Ma (Aleinikoff et al., 1989). During final stages of the 2700 My-old event, which was accompanied by tight isoclinal folding, calc-alkaline granodiorite–granite plutons were emplaced (Koesterer et al., 1987). These were followed by many more calc-alkaline granodiorite–granite plutons at ~2630 Ma and strongly peraluminous granite plutons at ~2545 Ma, which together now make up ~60% of the Range (Stuckless, 1989). The 2630 My-old plutonic event has been recognized, along with ~2595 My-old strongly peraluminous granite plutonism, in the Granite Mountains, 100 km to the east (Stuckless and Meisch, 1981). In the Black Hills, located along the presumed eastern edge of the province, 450 km northeast of the Wind River Range, two tiny granite bodies are present: one formed at ~2550 Ma and is strongly peraluminous, the other is poorly-dated (but probably late Archean) and alkaline (Gosselin et al., 1990).

Not all of the Wyoming Province preserves a record of voluminous 2700–2550 My-old granite plutonism like that outlined above. In the Beartooth Mountains, 200 km north of the Wind River Range, calc-alkaline granite–granodiorite plutons formed at ~2780 and 2740 Ma, along with abundant tonalitic and trondhjemitic rocks, following ~2790 My-old andesitic volcanism and amphibolite facies metamorphism (Mueller et al., 1988). In the Owl Creek Mountains, 100 km northeast of the Wind River Range, an ~2730 My-old alkaline granite pluton is present (Stuckless et al., 1986).

Dharwar Craton, India

The Dharwar Craton appears to be a tilted block, essentially an exposed cross section of Archean crust, with paleopressures grading from 3 kbar in the north to 10 kbar in the south (Newton, 1990). Most work has been carried out in the western half of the craton where extensive, north-south trending sedimentary and volcanic belts that formed sometime between ~3130 and 2600 Ma (Taylor et al., 1984; Nutman et al., 1992) are surrounded by TTG gneisses and plutons formed largely at ~3300–3200 Ma and ~3000–2950 Ma (Taylor et al., 1984; Friend and Nutman, 1991). Granite plutons were intruded into the older rocks during upright folding (Naha et al., 1991) and transcurrent shearing (Jayananda and Mahabaleswar, 1991a) between ~2600 and 2510 Ma. The most spectacular example is the ~2513 My-old (Friend and Nutman, 1991), ~400 km long, ~20 km wide, multi-phase, calc-alkaline Closepet Granite, which cuts a north–south line through the center of the craton and hence is exposed at both middle and upper crustal levels (Allen et al., 1986; Newton, 1990; Jayananda and Mahabaleswar, 1991b). Also present in this region are two alkaline granite plutons (Taylor et al., 1984; Rogers, 1988), one emplaced ~2540 Ma (Pb-Pb whole rock–feldspar age; Meen et al., 1992), the other ~2600 Ma (Pb-Pb whole rock age; Taylor et al., 1984), and two poorly-dated strongly peraluminous granite plutons (Dhoundial et al., 1987).

Based on limited data, the eastern Dharwar Craton seems to be somewhat different than the western Dharwar Craton. Supracrustal belts in the east, although trending north-south as in the west, are narrower, smaller, more basalt-rich and greywacke-poor (Krogstad et al., 1989). They also tend to possess higher temperature mineral assemblages than do the western belts (Chadwick et al., 1992). Nonetheless, granite plutonism in the east seems to have occurred at the same time as in the west. Around the Kolar Schist Belt, in the southeast of the craton, an ~2530 My-old calc-alkaline granite and an ~2550 My-old alkaline granite were emplaced following ~2630–2610 My-old TTG plutonism and ~2700 My-old basaltic volcanism (Balakrishnan and Rajamani, 1987; Krogstad et al., 1989).

Kaapvaal Craton, Southern Africa

The Kaapvaal Craton preserves a large number of discrete episodes of major deformation and magmatism, each separated from the next by tens to hundreds of million years. Granite–granodiorite plutons are voluminous, perhaps even more so than TTG rocks (L.J. Robb, pers. comm.). In the Barberton Mountain Land, located along the eastern margin of the craton, two stages of TTG plutonism and associated thrust faulting at ~3445 and 3225 Ma were followed by the emplacement of mostly calc-alkaline granodiorite–granite batholiths at ~3105 Ma, during transcurrent shearing, and mostly alkaline granite plutons at ~3075, 2840 and 2720 Ma, during extension and oblique-slip faulting (Kamo and Davis, 1991; de Wit et al., 1992; Meyer et al., 1992; Robb et al., 1993). Although more poorly documented, similar events are recognized to the west, in the Witwatersrand Basin (Robb et al., 1991), and to the south, in Natal (Hunter, 1991). The Witwatersrand Basin, in addition, preserves major basaltic to rhyolitic volcanic rocks extruded ~3075 and 2715 Ma (Armstrong et al., 1991). In the northeastern part of the craton, syn- to post-kinematic granitic plutons associated with the Limpopo orogeny were intruded between ~2700 and 2660 Ma (Barton and van Reenen, 1992). It has been suggested that ~2785 My-old alkaline granites and rhyolites that occur in the northwestern part of the craton are also related to Limpopo events (Moore et al., 1993).

North Atlantic Craton, Southern West Greenland

The North Atlantic Craton, as exposed in southern West Greenland, largely consists of ~3870 to 2820 My-old, amphibolite and granulite facies, TTG and granite–granodiorite gneisses (McGregor et al., 1986; Nutman et al., 1993) intruded by the ~2800 My-old, calc-alkaline Ilivertalik granite complex (Myers, 1976; Pidgeon et al., 1976; Compton, 1978; Wells, 1979), the ~2660 My-old Qarusuk aplite and pegmatite dikes (McGregor et al., 1983) and the ~2530 My-old, calc-alkaline Qorqut granite complex (Baadsgaard, 1976; Brown et al., 1981; McGregor et al., 1986). The Ilivertalik granite was intruded following isoclinal folding and during or just before granulite facies metamorphism (Nutman et al., 1989); it is associated spatially and temporally with subordinate diorite and tonalite (Myers, 1976). Intrusion of the Qarusuk dikes was broadly contemporaneous with the formation of upright folds and steeply-dipping, north–northeast-trending shear zones (Brown et al., 1981; McGregor et al., 1983; Nutman et al., 1989). The Qorqut granite, which is not associated with co-magmatic mafic rocks, was emplaced after major deformation and forms the core of an antiform that plunges south–southwest and trends parallel to the strike of the shear zones associated with the Qarusuk dikes (Brown et al., 1981).

General characteristics

Perhaps the most striking characteristic of the geologic settings of Archean granite plutons is their near-synchronous emplacement across vast areas of individual cratons, largely following the formation of TTG gneisses, TTG plutons, and felsic volcanic rocks of greenstone belts. The granite plutons thus make up huge igneous provinces that clearly represent magmatic events different from those which produced the TTGs and felsic volcanic rocks. Almost all of the granite plutons were intruded during or just after episodes of transcurrent shearing that accompanied or followed closely on the heels of craton-wide folding and thrusting. There are cases, however, such as the 3075 My-old alkaline granites of the Kaapvaal Craton, and the 2970 My-old calc-alkaline and strongly peraluminous granites of the Pilbara Block, where granite plutonism may have occurred long after compressional deformation ceased.

Archean granite plutons consist mainly of granodiorite and true granite, with only subordinate amounts of tonalite. Calc-alkaline, strongly peraluminous and alkaline granite plutons are each quite abundant: of the total exposed area (21491 km^2) of well-analyzed plutons considered in this chapter, calc-alkaline and alkaline plutons each comprise 35%, while strongly peraluminous plutons comprise 30%. Within a single craton, alkaline plutons tend to be somewhat younger than calc-alkaline and strongly peraluminous plutons although there are examples (in the Dharwar Craton, for instance) where they are older. Early plutons are somewhat deformed and hence syn- to late-tectonic, whereas late plutons are undeformed and post-tectonic.

Basaltic volcanism preceded emplacement of the plutons by a short time in parts of some provinces, but not everywhere, and, in general, granite plutonism occurred in the absence of significant basaltic volcanism. It is therefore much more likely that the granites were derived by partial melting of quartzo-feldspathic igneous and sedimentary rocks in the crust than by the differentiation of basaltic magmas.

As students of Phanerozoic granites will know, the aforementioned characteristics of Archean granites are remarkably similar to those of the large, granite-dominated provinces formed at ~600 Ma in the Pan-African orogeny, ~400 Ma in the British Caledonides and the Lachlan Fold Belt, and ~300 Ma in the Hercynides of Europe (Pitcher, 1987). It is appropriate therefore to focus on these regions when comparing Archean granite plutons to their younger counterparts, as is done in the discussion that follows. Comparisons with modern continental-arc granitoids, such as in the Western Cordillera of North and South America, are less appropriate because in these regions, tonalite is more voluminous than true granite, a significant volume of mafic rock is present, and emplacement of granite occurred in rather narrow belts rather than over a broad areas (Pitcher, 1987).

A NORMALIZATION DIAGRAM FOR GRANITE PLUTONS

Before proceeding with an examination of the chemical compositions of Archean granites, it is instructive to briefly review what is known about the behavior of certain trace elements in granite melts by way of introducing a new normalization diagram for granite plutons. In comparing compositions of basalts with one another, it has proven useful to normalize their minor and trace element concentrations to those in their upper mantle source regions and plot the resulting enrichment factors with elements arranged in order of their bulk residue/basaltic melt concentration ratios or "partition coefficients" during melting (Thompson et al., 1984). Compositions of granite plutons can be plotted on the same diagrams, but with much less significance, because the likely source rocks of granites (quartzo-feldspathic igneous and sedimentary rocks) are not located in the mantle and, as shown below, the order of element partition coefficients (often referred to as the order of element compatibility) during crustal and mantle melting are very different from each other. Furthermore, granites are so far removed in composition from rocks of the mantle that granites of very different composition appear to be much the same on mantle-normalized diagrams.

In the following discussion, as an alternative to the commonly-used mantle-normalized multi-element diagrams, concentrations of minor and trace elements in granite plutons are normalized to concentrations of the same elements in the upper continental crust. Resulting enrichment factors are plotted with the elements arranged from left to right in order of increasing bulk solid/calc-alkaline granite melt partition coefficient. Actual values for the partition coefficients are not well-known but relative values are inferred to be inversely proportional to enrichment factors calculated for the average composition of "unfractionated felsic I-type" granite from the Lachlan Fold Belt, as given by Chappell and White (1992). This average, being based on 131 samples of granites that are thought neither to contain restitic crystals nor to have lost precipitated crystals, is probably our best estimate of the composition of typical calc-alkaline granite melt formed in large Phanerozoic granite–granodiorite provinces. The composition of the upper continental crust used in the normalization is that of Taylor and McLennan (1985). Normalization to the average upper continental crust, which is broadly tonalitic to granodioritic in composition, is more appropriate than normalization to an average composition of the total continental crust because the lower continental crust probably contains large volumes of mafic rocks (Rudnick and Presper, 1990), which are infertile sources of granite plutons (Winther and Newton, 1991; Chappell and White, 1992). Normalization to felsic rocks of the lower continental crust is another possibility but the average composition of such rocks is not nearly as well-known as that of the upper continental crust.

The resulting normalization diagram for unfractionated felsic I-type Lachlan granite is shown in Fig. 2. Also plotted are average compositions calculated by Chappell and White (1992) for "unfractionated felsic S-type" and "A-type" granites of the Lachlan Fold Belt. As for the I-type average, the S- and A-type averages are based on samples of granite that are thought by Chappell and White (1992) to represent compositions of melts. For the I-type granite, Fig. 2 shows that enrichment factors range from about 2 for Th, Rb and Y to between about 1.2 and 0.5 for La, Ce, Ba, Zr, Ti, Nb, Zn and P to about 0.4 for Sr and 0.25 for Cu. A similar pattern is seen for the S-type granite, the major difference being that the

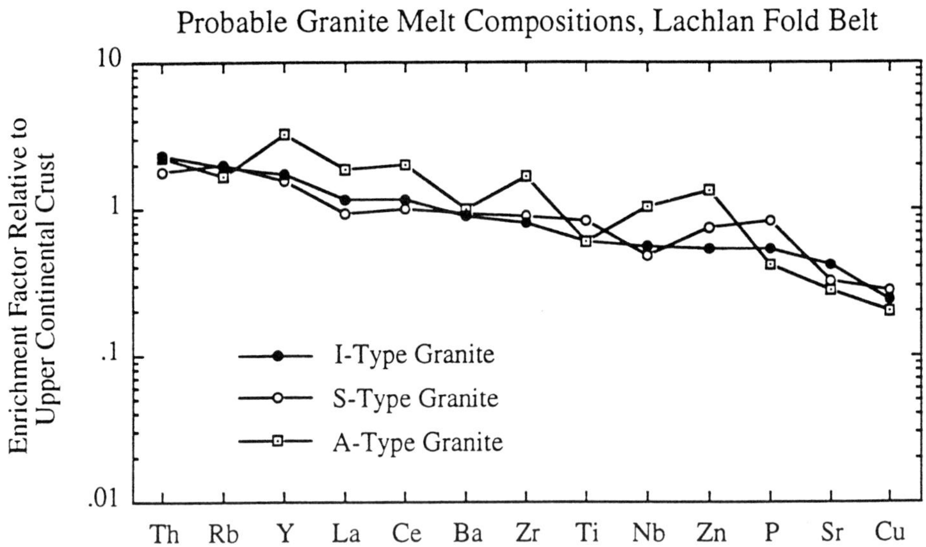

Fig. 2. Upper continental crust-normalized trace element diagram for average unfractionated I-type granite, average unfractionated S-type granite, and average A-type granite from the Lachlan Fold Belt of southeastern Australia (Chappell and White, 1992). Element concentrations used for the upper continental crust are (in ppm): Th = 10.7, Rb = 112, Y = 22, La = 30, Ce = 64, Ba = 550, Zr = 190, Ti = 3000, Nb = 25, Zn = 71, P = 740, Sr = 350 and Cu = 25 (Taylor and McLennan, 1985).

enrichment factor for Nb is much lower than that for Ti, as compared to in the I-type granite. The comparatively low Nb enrichment factor can be explained by the fact that Paleozoic greywacke, the probable source rock of S-type Lachlan granite (Chappell and White, 1992), has a much lower Nb content and Nb/Ti ratio (10 ppm and 0.002, respectively, according to Condie, 1993) than does the upper continental crust as a whole (25 ppm and 0.008, respectively, according to Taylor and McLennan, 1985).

The similarity of enrichment patterns for the I- and S-type granites in Fig. 2 suggests that during many of the different episodes of crustal melting that produced granite plutons, relative compatibilities of minor and trace elements were about the same, with Th, Rb and Y being strongly incompatible, Sr and Cu being strongly compatible, and La, Ce, Ba, Zr, Ti, Nb, Zn and P having intermediate compatibilities. The relative compatibilities of minor and trace elements seen in granites produced during crustal melting, however, are quite different from those seen in basalts produced during upper mantle melting. For instance, in many basalts, Y behaves as a strongly compatible element, while Nb is rather incompatible and Sr has an intermediate compatibility (Thompson et al., 1984). These differences can be attributed to the fact that, compared to the residues of basalt melts, the residues of granite melts are probably poorer in Y-bearing phases such as garnet, amphibole and clinopyroxene, and richer in Nb-bearing phases such as rutile, sphene and ilmenite, and Sr-bearing phases such as plagioclase and K-feldspar.

Despite the similarities of element partitioning shown by different kinds of granites, as emphasized above, the enrichment pattern of the A-type granite plotted in Fig. 2 suggests that element compatibilities were not the same during all kinds of crustal melting. In the A-types, Y, La, Ce, Zr, Nb and Zn seem to have behaved much more incompatibly than they did in the I- and S-type granites. As is discussed later in this chapter, the bulk composition and mineralogy of the source rocks of A-type granites are poorly known. Thus, the significance of the apparently distinct partitioning behavior of trace elements in A-type granites is unclear.

With the foregoing considerations in mind, we now turn our attention to describing the chemical compositions of the various kinds of Archean granite plutons and speculating on their significance in terms of possible differences in the compositions of the source rocks and melting regimes of granites through time.

CHEMICAL COMPOSITIONS OF CALC-ALKALINE GRANITE PLUTONS

Phanerozoic plutons

Table 1 gives the mean chemical composition of ten calc-alkaline granite plutons from large granite-dominated Phanerozoic provinces, as described above. There is a large literature discussing the origin of such granites with most workers concluding that they formed by partial melting of crustal igneous or meta-igneous rocks of intermediate composition. This idea has largely been confirmed by experimental partial melting studies in which tonalite sources produced melts with compositions similar to those of calc-alkaline granites (Skjerlie and Johnston, 1993). Analogous studies using basalt sources have instead produced melts of trondhjemitic composition (Winther and Newton, 1991). Calc-alkaline, strongly peraluminous or alkaline granites cannot be partial melts of mantle peridotite because, as discussed by Chappell and White (1992), their experimentally-determined liquidus mineralogy includes quartz and their oxygen isotopic compositions are high, which is not the case for mantle-derived melts.

Archean plutons

Compared to their Phanerozoic counterparts, Archean calc-alkaline granite plutons exhibit a wide range of compositions that can be divided into two subgroups. The mean composition of each subgroup is presented in Table 1. One subgroup, which for simplicity is referred to here as the "CA1-type" has higher mean concentrations of each of Y, TiO_2, FeO_t, MgO, CaO, P_2O_5, Sc, V, Zr, rare

TABLE 1

Mean chemical compositions of calc-alkaline granite plutons

	Phanerozoic	Archean	
		CA1-type	CA2-type
wt%			
SiO_2	68.86±0.79 (10)	70.00±0.51 (16)	71.88±0.38 (12)
TiO_2	0.48±0.04 (10)	0.40±0.03 (16)	0.23±0.01 (12)
Al_2O_3	14.9±0.21 (10)	14.63±0.20 (16)	14.69±0.16 (12)
FeO_t	3.14±0.31 (10)	2.72±0.18 (16)	1.65±0.08 (12)
MnO	0.07±0.01 (10)	0.05±0.00 (16)	0.03±0.00 (12)
MgO	1.30±0.24 (10)	0.84±0.11 (16)	0.48±0.03 (12)
CaO	2.69±0.28 (10)	2.28±0.14 (16)	1.69±0.10 (12)
Na_2O	3.62±0.16 (10)	3.89±0.12 (16)	4.45±0.18 (12)
K_2O	3.68±0.18 (10)	3.58±0.17 (16)	3.69±0.18 (12)
P_2O_5	0.16±0.02 (10)	0.17±0.02 (16)	0.08±0.01 (12)
LOI	0.96±0.12 (9)	0.78±0.07 (11)	0.81±0.11 (9)
ppm			
Li	32±6 (3)	42±19 (7)	22±7 (4)
Be	1.0±0 (2)	1.6±0.3 (2)	4.0±0.0 (1)
B			7±4 (2)
F	780±20 (2)	680±100 (6)	500±180 (4)
Sc	10±2 (7)	4.7±0.5 (9)	1.8±0.4 (3)
V	47±9 (9)	28±4 (11)	15±2 (8)
Cr	24±11 (8)	26±10 (15)	73±29 (8)
Co	10±2 (9)	13±4 (12)	14±5 (6)
Ni	13±6 (8)	12±2 (11)	12±4 (7)
Cu	7±2 (9)	19±7 (10)	13±3 (7)
Zn	50±3 (8)	59±3 (14)	41±5 (8)
Ga	17±1 (5)	18±2 (3)	17±1 (3)
Rb	131±8 (10)	117±8 (16)	125±11 (12)
Sr	243±29 (9)	479±73 (16)	455±77 (12)
Y	26±2 (10)	21±4 (16)	7.0±1.4 (12)
Zr	192±13 (9)	218±21 (16)	142±9 (12)
Nb	17±3 (9)	12±2 (15)	9±1 (11)
Sn	6±2 (3)	<4 (5)	<5 (4)
Cs		2.2±0.4 (9)	1.8±1.0 (3)
Ba	600±40 (10)	1300±150 (15)	1210±130 (12)
La	32±2 (9)	71±11 (14)	42±7 (10)
Ce	68±5 (10)	133±20 (14)	71±11 (11)
Nd	32±4 (6)	45±6 (11)	29±4 (6)
Sm	7.5±1.1 (3)	7.8±0.9 (13)	4.4±0.6 (8)
Eu	1.18±0.16 (3)	1.56±0.17 (13)	0.79±0.08 (8)
Gd	7.0±0.8 (2)	5.1±0.7 (6)	3.4±0.0 (1)

(continued)

TABLE 1 (*continuation*)

	Phanerozoic	Archean	
		CA1-type	CA2-type
Tb		0.73±0.09 (9)	0.41±0.09 (5)
Dy	5.4±0.3 (2)	3.2±0.6 (5)	2.2±0.0 (1)
Yb	2.0±0.3 (3)	1.55±0.24 (12)	0.51±0.11 (6)
Lu	0.29±0.12 (2)	0.24±0.03 (12)	0.09±0.03 (5)
Hf	5.6±0.0 (1)	5.8±0.7 (8)	4.6±0.8
Ta	1.5±0.0 (1)	0.93±0.11 (6)	0.29±0.12 (2)
Pb	18±2 (6)	23±2 (12)	28±5 (9)
Th	15±1 (9)	21±3 (14)	22±3 (10)
U	3.4±0.6 (7)	2.4±0.4 (11)	3.3±1.2 (7)
ratios			
Al_2O_3/TiO_2	33±3 (10)	39±2 (16)	66±2 (12)
CaO/Na_2O	0.79±0.12 (10)	0.59±0.04 (16)	0.38±0.02 (12)
Mg#	40±3 (10)	34±3 (16)	34±1 (12)
[La/Yb]n	14±2 (3)	38±6 (12)	80±24 (6)
Eu/Eu*	0.54±0.05 (3)	0.73±0.04 (13)	0.73±0.09 (8)
Rb/Y	5.4±0.7 (10)	7.1±0.8 (16)	25±6 (12)
Tb/Yb		0.55±0.05 (9)	1.08±0.26 (5)

Error on the mean is calculated as $\sigma_n\sqrt{(n-1)}$ where n, the number of plutons, is given in parentheses. FeO_t is total iron expressed as FeO. Mg# = 100 [molecular MgO/(MgO + FeO_t)]. [La/Yb]n is the chondrite-normalized La/Yb ratio. Eu/Eu* is the observed Eu concentration divided by the Eu concentration calculated assuming that the chondrite-normalized REE pattern had no Eu anomaly. Data sources: *Phaneroxoic plutons* — Buckenbowra (i = number of individual samples = 1) and Nelligen (i = 4) (Griffin et al., 1978); Eugowra (i = 1) (Wyborn et al., 1987); Glenbog (i = 1) (Chappell and White, 1992); Loch Doon (i = 5) (Tindle and Pearce, 1981; A. Tindle, pers. comm.); Benettuti (i = 4) (Zorpi et al., 1991); Bejar (i = 9) (Rittura et al., 1989); Trois Signeurs (i = 14) (Wickham, 1987); Sidarah (i = 5) and Al Jizl (i = 4) (unpublished analytical database from Jackson et al., 1984). *CA1-Type Archean plutons* — Long Lake Granodiorite (i = 10) (Wooden et al., 1982; Mueller et al., 1988); Louis Lake Granite (i = 5) and Granodiorite (i = 10) (Stuckless, 1989); Gamitagama (i = 8) (Smith et al., 1985); Yarlot Well (i = 6) and pluton 31 (i = 1) (Watkins and Hickman, 1990); Closepet (i = 24) (weighted average of the 4 main phases analyzed by Allen et al., 1986, and mapped by Jayananda and Mahabaleswar, 1991b); Edjudina pluton 39 (i = 10) (Cassidy et al., 1991); Eley (i = 9) and Mulgandinnah (i = 17) (Blockley, 1980; Bickle et al., 1989); Libery (i = 5) (Witt and Swager, 1989; Cassidy et al., 1991; Hill et al., 1992a; A. Thom, pers. comm.); Nelspruit-Hood (i = 3) (Glikson, 1976); Dalmein (i = 12) (Glikson, 1976; Robb, 1983; Robb et al., 1990); Marginal (i = 7) (Gower et al., 1983); Carbana Pool (i = 21) (Davy, 1988); Rest Island (i = 10) (Shirey and Hanson, 1986; Day, 1990). *CA2-Type Archean plutons* — Austin (i = 3) (Gower et al., 1983); Theatre Rocks (i = 1) (Hill et al., 1992a; A. Thom. pers. comm.); Baldhead River (i = 17) (Smith et al., 1985); Long Lake Granite (i = 21) (Wooden et al., 1982; Mueller et al., 1988); Quetico pink (i = 4) (Percival, 1989); Widgiemooltha pluton 69 (i = 1) (Cassidy et al., 1991); Garrison (i = 4) (Feng and Kerrich, 1992, including unpublished data base); Round Hill Bore foliated (i = 8) (Tyler et al., 1992; I. Tyler, pers. comm.); Kambha (i = 5) (Balakrishnam and Rajamani, 1987); Couttes Well (i = 3), Darn Hill (i = 3) and Moyagee (i = 8) (Watkins and Hickman, 1990).

earth elements (REEs) and Ta, lower mean concentrations of each of Na_2O and Cr, and lower mean Al_2O_3/TiO_2, Rb/Y and Tb/Yb ratios, than the other subgroup, referred to as the "CA2-type". Mean concentrations of each of SiO_2, Al_2O_3, K_2O, Co, Ni, Rb, Sr, Nb, Ba, Pb and Th, and mean molecular 100 $MgO/(MgO+FeO_t)$ ratios or "Mg numbers" are similar in the two types of plutons. In chondrite-normalized REE diagrams, as plotted in Fig. 3, the generally higher REE enrichment factors of individual CA1-type plutons are readily apparent. Nonetheless, it is also evident from the patterns in these plots that the two types of plutons share some REE characteristics, namely enrichment of the light REEs over the heavy REEs and either no Eu anomaly or just small negative Eu anomalies. Mean Eu/Eu* ratios (the fraction of Eu observed relative to that calculated assuming no anomaly) are 0.73 in both types of plutons (Table 1).

Calc-alkaline granite plutons are present in all eight cratons reviewed earlier in this chapter. In the Slave Province, Kaapvaal Craton and North Atlantic Craton, CA2-type plutons have yet to be documented, but in all other cratons, both CA1s and CA2s occur together. There are no consistent differences between CA1- and CA2-type plutons with respect to their distribution within cratons or the timing of their emplacement relative to deformation in those cratons. In some cases, such as the CA1 Long Lake Granodiorite and CA2 Long Lake Granite of the Wyoming Province (Mueller et al., 1988), the two kinds of plutons are in contact with each other but, in other cases, such as the CA1 Closepet Granite (Allen et al., 1986) and the CA2 Kambha Granite (Balakrishnan and Rajamani, 1987) of the Dharwar Craton, the two kinds are separated by great distances. Some CA1 and CA2 plutons within a single craton have essentially the same crystallization age; for example, in the Superior Province, this is the situation for the 2667 ± 2 My-old CA1 Gamitagama granites (Krogh and Turek, 1982) and the 2665 ± 2 My-old CA2 pink leucogranites of Quetico Park (Percival and Sullivan, 1988). In other cases, such as the ~2780 My-old Long Lake Granodiorite and the ~2740 My-old Long Lake Granite (Mueller et al., 1988), significant age differences exist. CA1-type plutons seem to be somewhat more voluminous than CA2-type plutons. Of the 28 plutons listed in Table 1, the mean area of the CA1s is 240 ± 100 km^2, whereas for CA2s, it is only 80 ± 40 km^2. Seven of the 16 CA1-type plutons, but only one of twelve CA2-type plutons, have areas greater than 100 km^2.

It is doubtful that CA1 and CA2 plutons formed from a common parent magma. If this were the case, one might expect the two kinds of plutons to be closely related in space and time within individual cratons but, as noted above, many CA1-type plutons seem to have formed independently of CA2-type plutons, and *vice versa*. Also, because there is no significant difference in the mean concentrations of each of Sr, Rb and Ba in the two types of plutons, it is not possible to relate them to a common magma through separation of different amounts of plagioclase, biotite or alkali feldspar, minerals that have very high partition coefficients for those elements (Arth, 1976). The fact that the two kinds of plutons have similar Mg numbers makes it difficult to relate them through separation of different amounts of mafic phases.

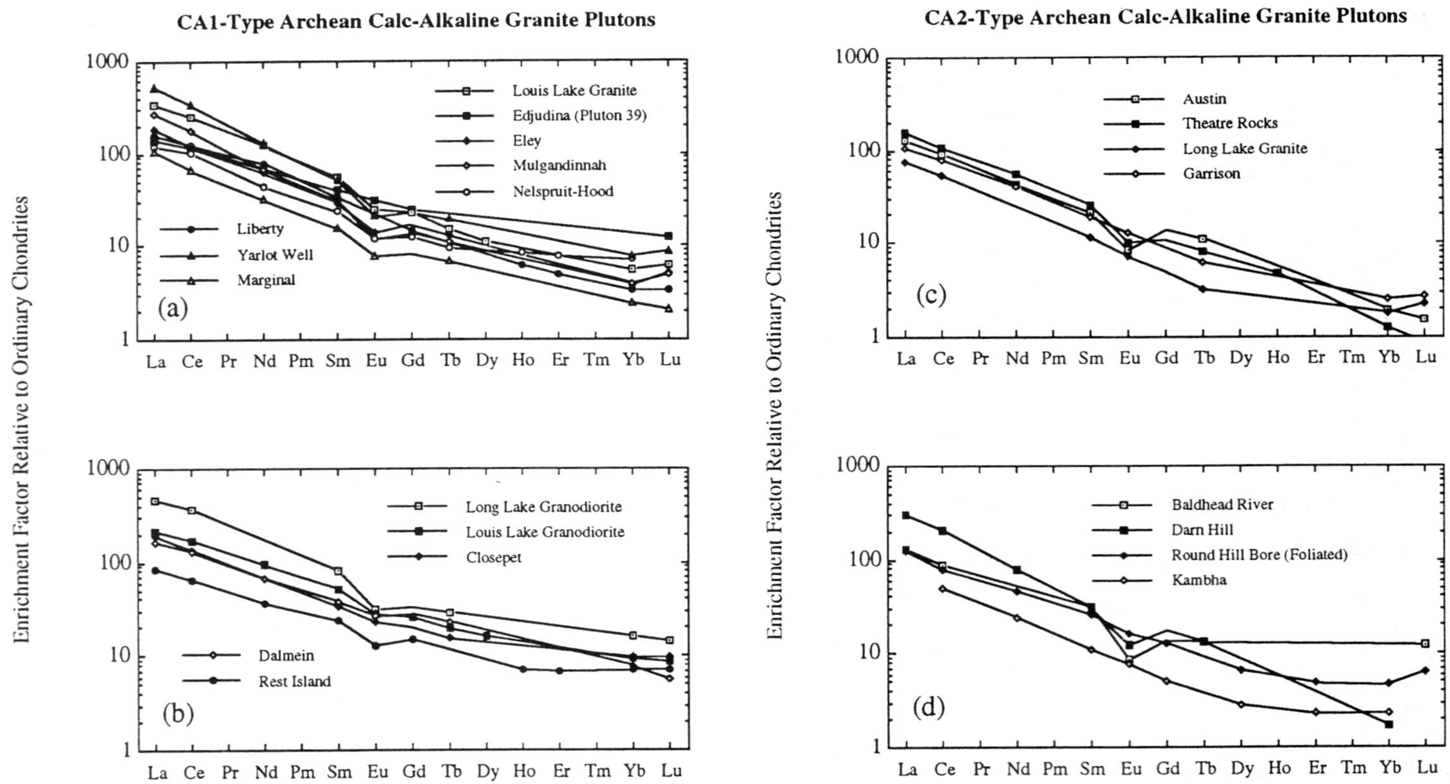

Fig. 3. Ordinary chondrite-normalized REE plots for (a, b) CA1-Type and (c, d) CA2-Type Archean calc-alkaline granite plutons. Sources of data for the plutons are listed in Table 1. REE concentrations for ordinary chondrites are those of Anders and Grevesse (1989) for C1 chondrites multiplied by 1.36 and are (in ppm): La = 0.319, Ce = 0.820, Pr = 0.121, Nd = 0.615, Sm = 0.200, Eu = 0.076, Gd = 0.267, Tb = 0.049, Dy = 0.330, Ho = 0.076, Er = 0.216, Tm = 0.033, Yb = 0.221, Lu = 0.033.

The foregoing difficulties suggest that, rather than forming from a common magma, the two types of plutons represent two fundamentally different kinds of magma produced during melting of the Archean continental crust. If this is true, how might have these two different kinds of magma formed? There seems to be three possible explanations: first, the two kinds of magma could be products of different degrees of partial melting of source rocks of the same composition; second, the magmas could have been melted from source rocks of different composition; and third, the magmas could have formed from the same source rocks located at different depths in the crust. None of these possibilities can be excluded from having had a role in the formation of the two types of calc-alkaline granite magmas, but only the third possibility can alone explain many of the differences in composition seen in those magmas. Let us see why this is the case by examining each of the three possibilities.

If CA1 and CA2 plutons formed from the same source rocks, but as a result of different degrees of partial melting, it would be expected that on a per source volume basis, small-volume plutons would be enriched in Th, Rb and Y, and depleted in P, Sr and Cu, relative to large-volume plutons. This is because, as noted above, Th, Rb and Y are strongly incompatible during partial melting of the crust, while P, Sr and Cu are strongly compatible. In fact, as can be seen in Table 1, there is little difference in the concentrations of each of Th, Rb, Sr and Cu in the two types of plutons, and *both* Y and P are enriched in the CA1s relative to the CA2s. Thus, differences other than in the degrees of partial melting must have been important in producing differences in composition seen in the plutons.

The argument that differences in the compositions of Archean calc-alkaline granite plutons simply reflect differences in the compositions of their source rocks is most properly evaluated by considering the composition of the Archean TTG suite. Because TTG rocks have intermediate compositions and make up the bulk of Archean cratons, they are the most likely sources of Archean calc-alkaline granite plutons. The Archean TTG suite, however, does not possess a spectrum of compositions that reflects that of Archean calc-alkaline granites. To illustrate, Fig. 4 is a plot of Al_2O_3/TiO_2 vs Rb/Y ratios for CA1 and CA2 plutons listed in Table 1 and several averages of large numbers of various suites of Archean TTG rocks taken from Tarney et al. (1979), Bickle et al. (1983;, 1993), Martin (1987), Feng and Kerrich (1992) and Condie (1993). The plutons show a wide range of Al_2O_3/TiO_2 and Rb/Y ratios with CA1s tending to have lower ratios than CA2s. In contrast, each of the averages of Archean TTGs (and, where available, standard deviations of those averages) plot within the lower part of the range, overlapping the ratios of the CA1-type plutons but not those of the CA2s. Some high-Al_2O_3/TiO_2, high-Rb/Y TTG rocks do exist (such as the Nuk trondhjemitic gneisses of Greenland, reported by McGregor, 1979) but, as suggested by the averages and standard deviations plotted in Fig. 4, such rocks do not seem to be a volumetrically significant component of the Archean TTG suite. Thus, differences in source composition were probably not the primary cause of differences in composition seen between the two kinds of plutons.

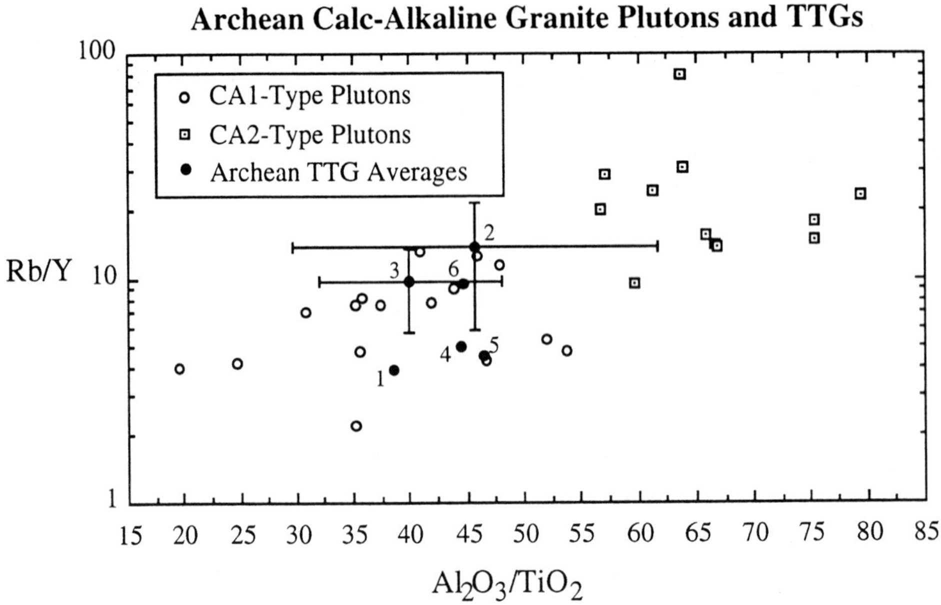

Fig. 4. Al_2O_3/TiO_2 vs Rb/Y plot for Archean calc-alkaline granite plutons and various averages of Archean TTG rocks. Sources of data for the plutons are listed in Table 1. Key to Archean TTG averages: 1 = average of mean synvolcanic TTG series, 34 samples, and mean syntectonic TGGM series, 51 samples, Abitibi southern volcanic zone, Canada (Feng and Kerrich, 1992); 2 = average of 48 grey gneisses, eastern Finland (Martin, 1987); 3 = average of 42 granitoid rocks of the Shaw Batholith, Western Australia (Bickle et al., 1983, 1993); 4 = worldwide TTG average (Condie, 1993); 5 = average of 62 grey gneisses, east and west Greenland (Tarney et al., 1979); and 6 = average of 83 grey gneisses, Scotland (Tarney et al., 1979). Standard deviations for the Archean TTG means are shown where available.

Probably the most important of all the variables involved in producing differences in composition among Archean calc-alkaline granite plutons was the depth at which melting took place. Recent experimental melting studies by Skjerlie and Johnston (1993) and Skjerlie et al. (1993) on a biotite-rich, hornblende-poor, Archean tonalite gneiss, similar to many rocks that compose the Archean TTG suite, indicate that under water-undersaturated conditions and temperatures of 900–950°C, biotite dehydration leads to the production of fluorine-poor granite melt similar in composition to many calc-alkaline granites. In experiments conducted at pressures of 6 kbar, as would be expected near the middle of the continental crust, an orthopyroxene-bearing, garnet- and hornblende-free residue formed in equilibrium with the melt, whereas at 10 kbar, corresponding to lower crustal depths, the residue contained garnet ± hornblende ± orthopyroxene. Other residual phases in the experimental charges included biotite, plagioclase, epidote, quartz, magnetite, ilmenite, apatite, zircon and alkali feldspar. Skjerlie et al.

(1993) showed that the volume of granite melt produced from the tonalite is large (>50%) if sedimentary rocks are interlayered with that tonalite, as might be common in the continental crust.

Because Y, heavy REEs, Sc, V and Ti are more strongly partitioned into garnet than into a coexisting granite melt (Sisson and Bacon, 1992) or orthopyroxene (Sisson, 1991), a granite melt separated from a garnet-bearing residue should have lower concentrations of Y, heavy REEs, Sc, V and Ti than do both its source rocks or a granite melt separated from a garnet-free, orthopyroxene-bearing residue. Conversely, because Yb is preferred over Tb in garnet and because the partition coefficient for Cr is smaller in garnet than in orthopyroxene (Sisson, 1991; Sisson and Bacon, 1992), the granite melt from the garnet-bearing residue will have a comparatively high Tb/Yb ratio and Cr concentration. These compositional relationships are precisely what is seen among Archean calc-alkaline granite plutons: CA2-type plutons have lower mean concentrations of Y, heavy REEs, Sc, V and Ti, and higher mean Tb/Yb ratios and Cr concentrations than do both CA1-type plutons (Table 1) and the average Archean TTG of Condie (1993). It is thus quite likely that melts which formed most CA2-type plutons equilibrated with garnet-bearing residues at lower crustal depths while melts that formed most CA1-type plutons equilibrated with garnet-free residues at mid-crustal depths. Strong geologic evidence that at least one CA1-type pluton originated in the middle crust comes from the crustal cross section exposed in the Dharwar Craton. There, the Closepet Granite is found at mid- and upper-crustal levels and does not appear to have migrated upward from the lower crust (Newton, 1990).

One consequence of CA1- and CA2-type plutons having been derived from different crustal depths may be the factor of three difference in size between them. If sedimentary rocks make up a larger fraction of the middle crust than the lower crust and, as noted above, the volume of granite melt produced from a tonalite interlayered with sedimentary rocks is much larger than that from a tonalite alone, then CA1 plutons derived from the middle crust would be expected to be larger than CA2 plutons derived from the lower crust, as is observed. Another consequence of different derivation depths for CA1 and CA2 plutons may have been that CA2 melts, having a greater distance available to rise in the crust, were able to separate from the residual minerals of their source regions more completely than were CA1 melts. Thus, the reason that CA1-type plutons have higher mean concentrations of Fe and Mg than do CA2-type plutons (Table 1) may be that most CA1s retained larger amounts of a major restitic mafic mineral such as orthopyroxene. Similarly, the comparatively high concentrations of Ca, P, Zr and light REEs in CA1-type plutons (Table 1) are consistent with them retaining a larger proportion of restitic plagioclase, epidote, apatite, zircon and allanite, phases rich in one or more of those elements. Although data for Ta are limited, the apparently low Nb/Ta ratios of CA1-type plutons (Table 1) suggest that they inherited large amounts of residual sphene or rutile, phases that strongly partition Ta over Nb (Green and Pearson, 1987).

To pursue these ideas, least-squares major-element–oxide mass–balance calculations have been used to determine the amounts of likely restitic phases that must be removed from CA1 and CA2 plutons in order to yield compositions similar to those of the 6- and 10-kbar experimental melts, respectively, of Skjerlie and Johnston (1993). Compositions for plagioclase, orthopyroxene and epidote used in the calculation are those reported by Skjerlie and Johnston (1993) and Skjerlie et al. (1993). Compositions used for phases not analyzed by the latter are those of near-stoichiometric minerals occurring in granitic host rocks, as tabulated by Deer et al. (1966).

Results of the calculation are shown in Table 2. Concentrations of major element oxides predicted for the melts are very similar to those observed, except for Na_2O, which is known to have a low measured concentration as a result of loss during analysis (Skjerlie and Johnston, 1993). The model melt compositions shown are achieved, for both CA1 and CA2 plutons, by a combination of orthopyroxene + plagioclase + epidote + apatite + sphene ± magnetite subtraction and alkali feldspar ± magnetite addition. Thus, in order to relate the melt compositions of Skjerlie and Johnston (1993) to the compositions of Archean granite plutons, fractional crystallization of alkali feldspar ± magnetite is required to have occurred in addition to restite phase inheritance. The more significant result, however, is that calculated proportions of each of the restite phases in CA1-type plutons exceed those of the same phases in CA2-type plutons, consistent with the notion that CA1s separated from their source rocks less completely than did CA2s.

Archean plutons compared to their Phanerozoic counterparts

Almost all calc-alkaline granite plutons from large granite-dominated Phanerozoic provinces, as described earlier, have chemical compositions that are more similar to those of CA1-type Archean plutons than to CA2-type Archean plutons. As can be seen in Table 1, ratios of oxides or elements that are low in the CA1-type plutons, such as Al_2O_3/TiO_2 and Rb/Y, are also low in the Phanerozoic plutons. Ratios that are high in the CA1s, such as CaO/Na_2O, are correspondingly high in the Phanerozoic plutons. Thus, if the CA1s formed by melting of the middle crust, rather than the lower crust, as suggested above, this may also have been the case for the large majority of calc-alkaline granite plutons of granite-dominated Phanerozoic provinces.

Although more similar to the CA1s than to CA2s, Phanerozoic calc-alkaline plutons have some compositional characteristics distinct from both kinds of Archean plutons. For instance, Table 1 shows that the Phanerozoic plutons have a higher mean Mg number than do either of their Archean counterparts. There are also several trace element differences. Figure 5 is an upper continental crust-normalized minor and trace element plot for nine of the ten Phanerozoic calc-alkaline granite plutons used to compile the mean composition of Table 1. Figure 6 is an analogous plot for Archean calc-alkaline granite plutons. Compared to both CA1 and CA2 Archean plutons, Phanerozoic plutons have comparatively flat patterns: except for Cu, which is strongly depleted in some plutons relative to the upper continental crust, element enrichment factors vary between ~0.4 and 2. In contrast, the patterns of

TABLE 2

Mass balance modelling of Archean calc-alkaline granite plutons

	CA1-type pluton		CA2-type pluton	
	Model melt	6-kbar, 925°C measured melt	Model melt	10-kbar, 950°C measured melt
wt%				
SiO_2	73.79	72.97	72.50	70.92
TiO_2	0.09	0.10	0.19	0.18
Al_2O_3	14.18	13.97	14.61	14.27
FeO_t	1.06	1.02	1.54	1.54
MgO	0.24	0.25	0.17	0.19
CaO	0.68	0.68	1.01	0.96
Na_2O	3.77	2.03	4.37	1.78
K_2O	4.73	4.68	4.44	4.33
P_2O_5	0.10	0.10	0.04	0.04
Melt–Mineral Mixing Proportions				
melt	0.865		0.986	
opx	–0.036		–0.022	
plagioclase	–0.098		–0.038	
K-feldspar	0.051		0.058	
magnetite	–0.004		0.008	
apatite	–0.002		–0.001	
sphene	–0.009		–0.001	
epidote	–0.037		–0.018	

FeO_t is total iron expressed as FeO. Measured experimental melt compositions for tonalite melting at 6-kbar, 925°C and 10-kbar, 950°C are from Skjerlie and Johnston (1993). Phase compositions used in the modelling are from Deer et al. (1966), except for orthopyroxene (=opx) and epidote, which are from Skjerlie and Johnston (1993), and plagioclase, which is from Skjerlie et al. (1993). Composition of orthopyroxene used in each model is that produced at the pressure and temperature of the corresponding equilibrium melt. For plagioclase, only the composition produced in the 10-kbar run is known and so that was used for modelling both the 6- and 10-kbar melts. For epidote, the composition present in the starting material was used because the compositions present in neither the 6- nor 10-kbar charges are known.

Archean plutons exhibit a much larger range, with element enrichment factors varying between ~0.08 and 6. Normalized ratios of elements of similar compatibility, Th/Rb, La/Y, Ba/Zr and Sr/P, are within ~40% of unity in most of the Phanerozoic plutons, but tend to be greater than ~1.5 in most of the Archean plutons.

As discussed above, both Archean and Phanerozoic calc-alkaline granite plutons were probably derived by partial melting of tonalitic rocks. It is likely therefore that some of the differences in composition between the Archean and Phanerozoic plutons are related to differences in the composition of those

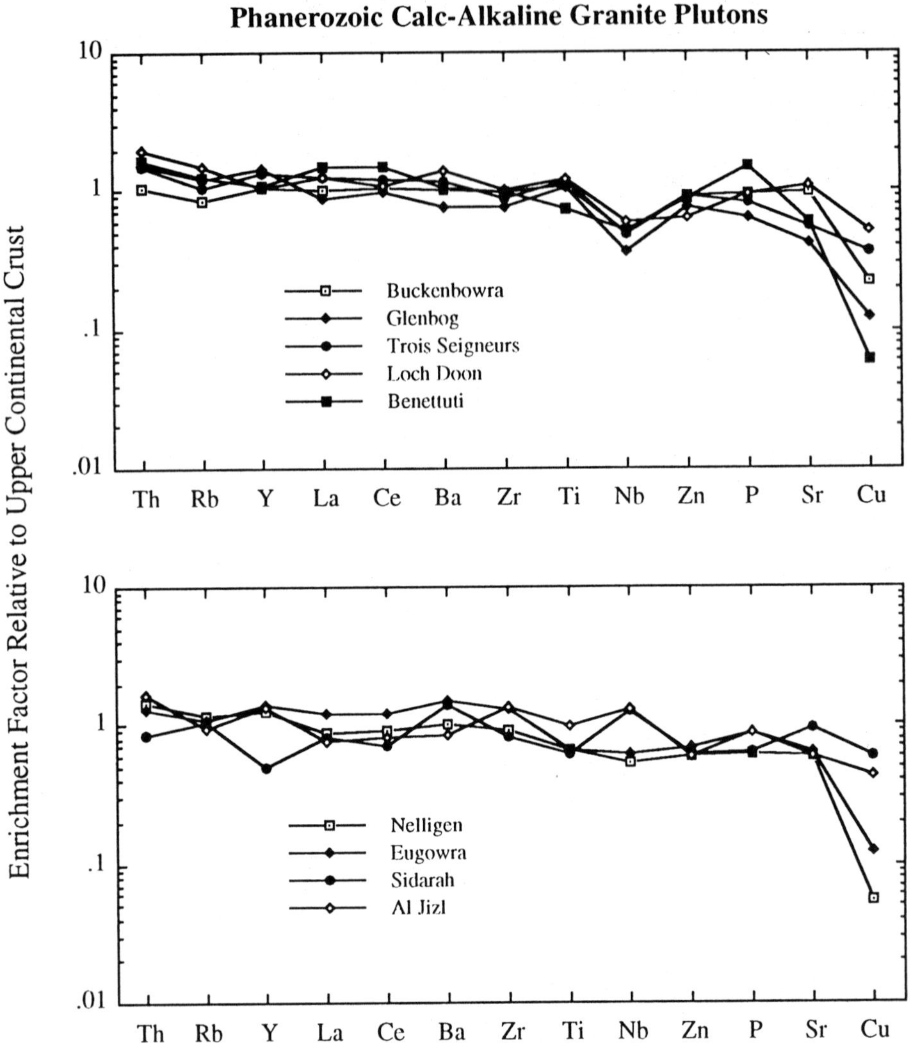

Fig. 5. Upper continental crust-normalized trace element diagram for Phanerozoic calc-alkaline granite plutons. Sources of data for the plutons are listed in Table 1. A pattern for the Bejar pluton is not shown because the data are incomplete. Composition of the upper continental crust is given in the caption to Fig. 2.

tonalites. For instance, the comparatively high La/Y and Sr/P ratios and low Mg numbers of the Archean plutons may be in large part inherited from the Archean TTG suite. Condie (1993) estimated that the mean Sr/P ratio of Archean TTG rocks is ~35% higher than that of Phanerozoic TTG rocks, while the mean La/Y ratio of the former is more than twice as high as the latter. Also, the mean Mg number of his Archean TTG is 36, whereas that for his Phanerozoic TTG is 44.

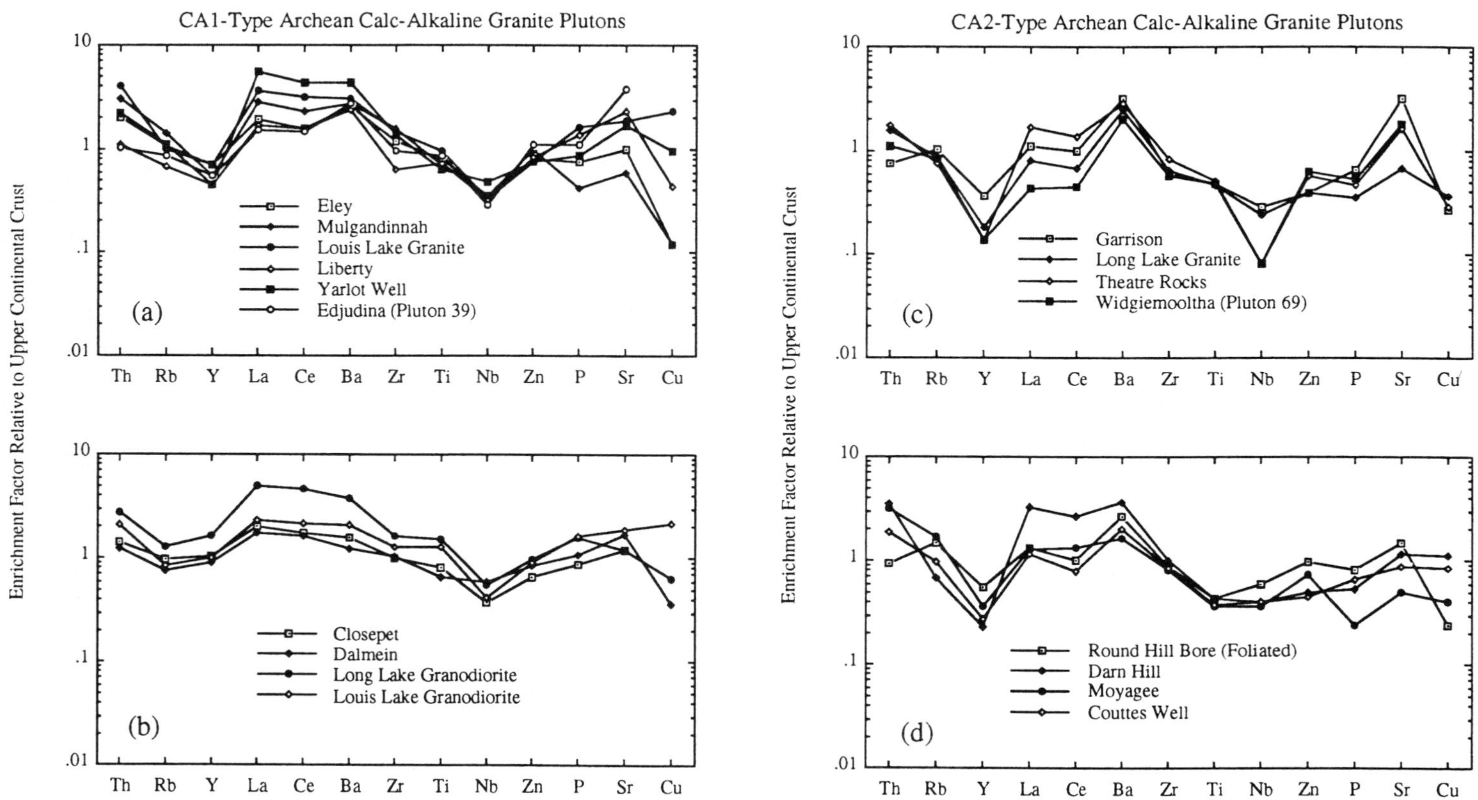

Fig. 6. Upper continental crust-normalized trace element diagram for (a, b) CA1-Type and (c, d) CA2-Type Archean calc-alkaline granite plutons. Sources of data for the plutons are listed in Table 1. Composition of the upper continental crust is given in the caption to Fig. 2.

These observations suggest that the secular evolution of the composition of calc-alkaline granite is linked intimately to the evolution of the composition of the TTG suite through time.

Not all of the differences between the Archean and Phanerozoic plutons, however, can be explained by appealing to the secular evolution of tonalitic source compositions. For instance, Ba concentrations (Table 1) and Ba/Zr ratios (Fig. 6) of both CA1- and CA2-type Archean plutons tend to be high relative to those of Phanerozoic plutons, but Archean TTG have, according to Condie (1993), a lower mean Ba concentration (660 ppm) and Ba/Zr ratio (4.1) than do Phanerozoic TTG (Ba = 800 ppm, Ba/Zr = 5.7). There is no obvious reason why the Phanerozoic plutons should have suffered more extensive fractional crystallization of a Ba-rich phase such as alkali feldspar than did the Archean plutons. Thus, rather than being the result of differences in source composition or crystallization histories, the different Ba concentrations and Ba/Zr ratios of the two plutons probably reflect different amounts of residual biotite in their source regions. Biotite is probably the dominant sink for Ba in restite formed in equilibrium with calc-alkaline granite melt because, according to the experiments of Skjerlie and Johnston (1993), alkali feldspar is not particularly stable. If we assume that the partition coefficient for Ba in biotite is 10 (Arth, 1976) and that both Phanerozoic and Archean plutons are 50% melts of TTG with Ba concentrations of 800 and 660 ppm, respectively, then we can calculate, assuming batch melting (equation 1 of Arth, 1976), that biotite comprises ~15% of the Phanerozoic restite but no more than ~1% of the Archean restite. This is an interesting result because biotite is progressively consumed with increasing temperature in the melting experiments of Skjerlie and Johnston (1993). It follows that the magmas which formed Archean calc-alkaline plutons, both those derived in the middle and lower crust, separated from their source regions at higher temperatures than did the magmas which formed Phanerozoic calc-alkaline plutons. The plutons thus provide evidence that, during episodes of crustal melting, geothermal gradients were steeper in Archean continents than in Phanerozoic continents.

Evidence that continental geothermal gradients have decreased through time is of some importance. Although there is little doubt that the Archean sub-lithospheric mantle was hotter than its modern counterpart, it has been argued that Archean continents were underlain by thick (>150 km), refractory, low-density, lithospheric mantle roots and both the continental crust and mantle root were relatively cool during the Archean and have remained so since (see Bickle, 1986, for a review). In contrast, Grotzinger and Royden (1990) presented sedimentological evidence that a thick mantle root had not formed beneath the Slave Province by 1900 Ma and thus thermal gradients, at least there, have decreased by at least a factor of two since the Archean. The fact that Archean calc-alkaline granites derived from both the middle and lower crust in several of the shield regions studied here seem to have formed at higher temperatures than their Phanerozoic counterparts suggests that steep geothermal gradients may have been more common in Archean continents than has been generally recognized.

CHEMICAL COMPOSITIONS OF STRONGLY PERALUMINOUS GRANITE PLUTONS

Phanerozoic Plutons

Two kinds of strongly peraluminous granite plutons are found in large granite-dominated Phanerozoic provinces. For each type, a mean composition is presented in Table 3 and upper continental crust-normalized minor and trace element enrichment patterns of individual plutons are plotted in Fig. 7. "SP1-type" plutons have higher mean concentrations of CaO, TiO_2, Sr, Y, Zr and Ba and lower mean concentrations of Al_2O_3, FeO_t, MgO, Na_2O, K_2O, P_2O_5, Sn and Rb than do "SP2-type" plutons. SP1s thus have particularly high CaO/Na_2O ratios and low Rb/Y ratios compared to SP2s. Of the provinces used to compile the mean pluton compositions of Table 3, the Lachlan Fold Belt seems to contain many more SP1-type plutons than SP2s, whereas in the British Caledonides and European Hercynides, the reverse seems to be true.

On the basis of their elevated isotopic ratios and distinctive chemical compositions, it is generally agreed that SP1-type plutons were derived by partial melting of sedimentary rocks, as was first suggested by Chappell and White (1974). The plutons have much lower concentrations of CaO, Na_2O and Sr than do their calc-alkaline (I-type) counterparts which, as discussed by Chappell and White (1992), can be attributed to the fact that greywackes and pelites (shales) have lower concentrations of CaO, Na_2O and Sr than do tonalites (Condie, 1993), the probable sources of calc-alkaline granites. There has been some discussion as to whether greywacke or pelite is a more appropriate source rock for SP1-type plutons. Experiments by Patino Douce and Johnston (1991) have shown that while melts with compositions similar to those of SP1 granites can be generated from pelites, such source rocks are not particularly fertile, having too much Al_2O_3, TiO_2, FeO and MgO. Greywackes possess less Al_2O_3, TiO_2 and MgO than do pelites (Condie, 1993) and hence may be more fertile but, unlike pelites, greywackes have lower concentrations of K_2O and Rb than do tonalites (Condie, 1993). The K_2O and Rb concentrations of SP1-type plutons, in contrast, are higher than those of calc-alkaline granite plutons (Tables 1 and 3), which are thought to be derived from tonalites. It seems likely therefore that SP1-type plutons formed by partial melting of rocks with a bulk composition transitional between greywacke and pelite, i.e., a lithologic package containing, in decreasing abundance, biotite, quartz, plagioclase and aluminosilicates, as described by Patino Douce and Johnston (1991).

Somewhat less certainty attends the origin of SP2-type plutons. Because they seem to be largely isolated in a geographic sense from SP1-type plutons, as noted above, SP2s were probably not derived from SP1-type magmas by fractional crystallization. SP2s have less CaO, Na_2O and Sr than do Phanerozoic calc-alkaline granite plutons and, hence, like SP1s, could have been derived from greywackes and pelites by partial melting. Because there is much more Na_2O and much

TABLE 3

Mean chemical compositions of strongly peraluminous granite plutons

	Phanerozoic		Archean	
	SP1-type	SP2-type	SP3-type	SP4-type
wt%				
SiO_2	71.6±1.1 (5)	72.25±0.20 (7)	73.71±0.25 (7)	74.39±0.31 (14)
TiO_2	0.46±0.08 (5)	0.24±0.02 (7)	0.21±0.02 (9)	0.12±0.02 (13)
Al_2O_3	13.75±0.31 (5)	14.65±0.14 (7)	14.49±0.22 (9)	14.15±0.17 (14)
FeO_t	3.24±0.47 (5)	1.60±0.22 (7)	1.53±0.22 (9)	1.10±0.10 (14)
MnO	0.05±0.00 (5)	0.04±0.01 (7)	0.03±0.00 (9)	0.03±0.01 (12)
MgO	1.20±0.29 (5)	0.46±0.11 (7)	0.41±0.07 (9)	0.26±0.06 (14)
CaO	1.44±0.25 (5)	0.61±0.10 (7)	1.12±0.08 (9)	0.57±0.06 (14)
Na_2O	2.20±0.32 (5)	3.37±0.17 (7)	2.94±0.22 (9)	3.91±0.12 (14)
K_2O	4.14±0.15 (5)	4.83±0.34 (7)	4.69±0.21 (9)	4.69±0.12 (14)
P_2O_5	0.13±0.02 (5)	0.28±0.03 (7)	0.09±0.01 (8)	0.10±0.03 (13)
LOI	1.53±0.28 (5)	1.23±0.16 (7)	0.79±0.09 (2)	0.65±0.08 (11)
ppm				
Li		160±70 (5)	7.0±0.0 (1)	120±40 (9)
Be		11±3 (5)	1.0±0.0 (1)	4.5±0.4 (9)
B		21±2 (5)	12±0 (2)	19±9 (2)
F		240±50 (4)	240±40 (2)	100±50 (2)
Sc	10±1 (5)	4±0 (1)	2.8±0.3 (5)	1.8±0.2 (5)
V	46±9 (5)	30±7 (6)	9.0±0.0 (1)	3.1±0.4 (3)
Cr	29±11 (5)	12±3 (7)	9.9±2.4 (6)	6.3±1.4 (5)
Co	11±2 (5)	11±3 (6)	2.63±0.92 (4)	1.87±0.52 (5)
Ni	13±4 (5)	10±2 (7)	5.0±0.0 (1)	6.3±1.5 (3)
Cu	9±2 (5)	15±3 (7)	60±0 (1)	12±9 (5)
Zn	64±10 (5)	63±12 (7)	29±7 (2)	27±4 (10)
Ga	17±1 (5)	21±2 (3)	14±0 (1)	23±3 (6)
Rb	190±15 (5)	371±51 (7)	140±11 (9)	292±25 (14)
Sr	124±21 (5)	72.2±9.0 (7)	189±19 (9)	72.9±7.6 (14)
Y	36±3 (5)	13±2 (7)	7.3±1.7 (3)	21±4 (9)
Zr	181±20 (5)	92±10 (7)	116±16 (6)	90±13 (13)
Nb	11±2 (5)	13±2 (7)	3.3±1.2 (3)	16±3 (9)
Sn	5.8±1.7 (5)	22±3 (6)	1±0 (1)	6.5±0.6 (4)
Cs		32±10 (4)	2.4±0.5 (5)	12±6 (6)
Ba	650±50 (5)	232±20 (7)	740±59 (9)	372±55 (14)
La	29.4±2.5 (5)	21.2±2.4 (7)	38.1±5.2 (7)	20.3±4.8 (9)
Ce	67.2±4.5 (5)	45.8±5.7 (7)	76.4±8.6 (8)	41.9±8.2 (9)
Nd		21.1±2.7 (6)	28.0±3.4 (8)	19.0±4.8 (9)

(continued)

	Phanerozoic		Archean	
	SP1-type	SP2-type	SP3-type	SP4-type
Sm		4.91±0.53 (5)	4.62±0.61 (8)	3.80±0.70 (9)
Eu		0.55±0.11 (4)	0.813±0.089 (8)	0.307±0.040 (9)
Gd		4.05±0.29 (4)	3.78±0.67 (5)	2.76±0.29 (3)
Tb			0.43±0.06 (6)	0.57±0.05 (6)
Dy		2.6±0.2 (4)	2.2±0.5 (5)	4.4±2.4 (3)
Yb		0.86±0.18 (4)	0.82±0.15 (8)	2.05±0.64 (9)
Lu		0.16±0.04 (4)	0.14±0.02 (7)	0.32±0.11 (6)
Hf			3.9±0.5 (6)	3.3±0.8 (6)
Ta			0.48±0.09 (4)	3.8±1.8 (5)
Pb	31±2 (5)	30±3 (7)	32±7 (2)	36±1 (3)
Th	20±2 (5)	12±2 (7)	22±5 (8)	21±5 (7)
U	4.0±0.5 (5)	9.1±1.6 (7)	11±7 (8)	5.4±1.0 (7)
ratios				
Al_2O_3/TiO_2	33±4 (5)	66±8 (7)	77±8 (9)	157±23 (13)
CaO/Na_2O	0.67±0.08 (5)	0.18±0.02 (7)	0.40±0.04 (9)	0.15±0.02 (14)
Mg#	38±3 (5)	32±3 (7)	32±3 (9)	29±3 (13)
[La/Yb]n		27±4 (4)	36±6 (7)	8±2 (9)
Eu/Eu*		0.37±0.05 (4)	0.66±0.06 (8)	0.30±0.04 (8)
Rb/Y	5.6±0.8 (5)	34±7 (7)	23±5 (3)	17±3 (9)
Gd/Yb		5.2±0.8 (4)	6.1±1.6 (5)	2.5±0.3 (3)

Error on the mean is calculated as $\sigma_n\sqrt{(n-1)}$ where n, the number of plutons, is given in parentheses. FeO_t is total iron expressed as FeO. Mg# = 100[molecular MgO/(MgO + FeO_t)]. [La/Yb]n is the chondrite-normalized La/Yb ratio. Eu/Eu* is the observed Eu concentration divided by the Eu concentration calculated assuming that the chondrite-normalized REE pattern had no Eu anomaly. Data sources: *Phanerozoic SP1-Type plutons* — Cooma granodiorite (i = number of individual samples = 1), Happy Jacks granodiorite (i = 1), Dalgety monzogranite (i = 1), Strathbogie granite (i = 1) and Bethanga granite (i = 1) (Chappell and White, 1992). *Phanerozoic SP2-Type plutons* — Leinster Non-Porphyritic Granite (i = 9) (Sweetman, 1987); Threlkeld granite (i = 13) (O'Brien et al., 1985); Granya monzogranite (i = 1) (Chappell and White, 1992); St. Sylvestre granite (i = 2), St. Austell granite (i = 1), Carnmenellis granite (i = 1) and Barruecopardo granite (i = 1) (Shaw and Guilbert, 1990). *Archean SP3-type plutons* — white muscovite leucogranite, western Sturgeon Lake Batholith (i = 8) (Percival, 1989); units 5 and 7 of Giants Range Batholith (i = 2) (Arth and Hanson, 1975); Lankin Dome Granite (i = 29) (Stuckless and Miesch, 1981); Malley Rapids Granite (i = 4) (Frith and Fryer, 1985); Bears Ears Granite, northern pluton (i = 9) (Stuckless, 1989); Dudhsagar Granite (i = 3) and Chandranath Granite (i = 7) (Dhoundial et al., 1987); unit GLB-1, Ghost Lake Batholith (i = 5) (Breaks and Moore, 1992); Kabetogama Lake leucogranites, Vermilion Granite Complex (i = 6) (Day and Weiblen, 1986). *Archean SP4-type plutons* — Glacier Lake pluton (i = 3), Barbara Lake stock (i = 2) and MNW stock (i = 2) of Georgia Lake (Kissin and Zayachkivsky, 1985); Preissac (i = 5), Lacorne (i = 2) and LaMotte (i = 5) plutons (Feng and Kerrich, 1991; Boily et al., 1990); Pluton H, Mt. Edgar Batholith (i = 4) (Davy and Lewis, 1986); Bear Mountain Granite (i = 6) (Gosselin et al., 1990); leucogranite of Lac du Bonnet Batholith (i = 17) (Cerny et al., 1987); Sparrow pluton (i = 27) (Kretz et al., 1989); Prosperous Lake pluton (i = 3) (Drury, 1979; Green et al., 1968); Kuuk Granite (i = 1) and Pluton #29 (i = 1) (Frith and Fryer, 1985); Garden Creek Adamellite (i = 10) (Bickle et al., 1989).

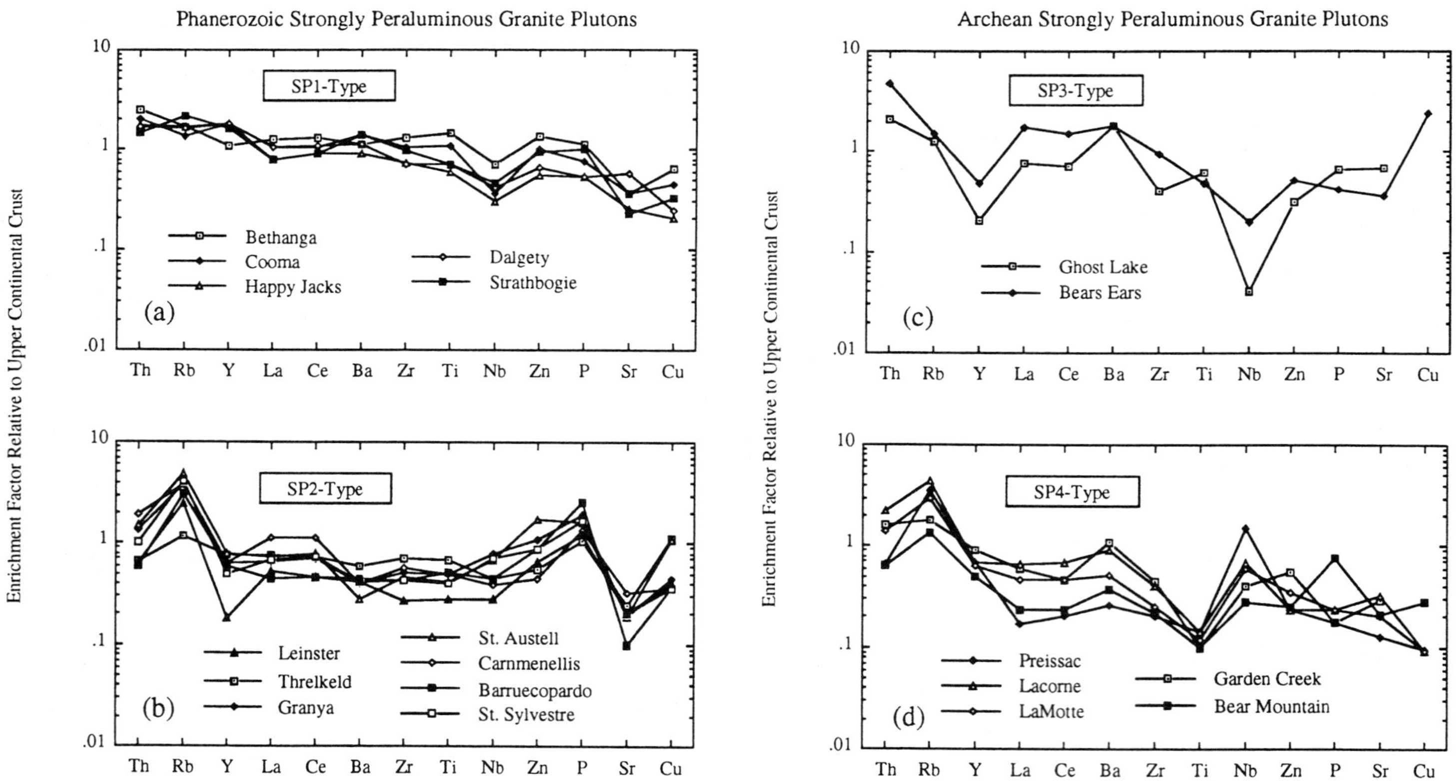

Fig. 7. Upper continental crust-normalized trace element diagram for (a) Phanerozoic SP1-Type, (b) Phanerozoic SP2-Type, (c) Archean SP3-Type and (d) Archean SP4-Type strongly peraluminous granite plutons. Sources of data for the plutons are listed in Table 3. Composition of the upper continental crust is given in the caption to Fig. 2.

less CaO and Sr in SP1s than in SP2s (Table 3), however, felsic volcanic rocks probably formed a significant part of the source of the SP2s. Felsic volcanic rocks have higher concentrations of Na_2O than do greywackes and pelites, and lower concentrations of CaO and Sr than do greywackes (Condie, 1993). Moreover, the close spatial association of felsic volcanic rocks and sedimentary rocks in many supracrustal successions is well-established, making a mixed volcanic–sedimentary source quite plausible for SP2-type plutons.

Another difference between the origins of SP1- and SP2-type plutons is that garnet was probably a residual phase during the melting history of the SP2s. This is because SP2s have much higher Gd/Yb and Rb/Y ratios (means of ~5 and 34, respectively, Table 3) than do Phanerozoic greywackes, pelites, or felsic volcanic rocks (means of ~2 and 3, ~1.8 and 5, and ~1.6 and 3, respectively, Condie, 1993), and garnet strongly partitions Yb and Y over Gd and Rb (Arth, 1976, Sisson and Bacon, 1992). The experimental study of Patino Douce and Johnston (1991) showed that biotite reacts with aluminosilicate and quartz to form residual garnet in pelitic source rocks melted at pressures of 7–13 kbar and suggested that, at lower pressures, residual cordierite would form instead. Because cordierite does not preferentially incorporate Y over Rb as it crystallizes (Harris et al., 1992), low-pressure melts of pelitic rocks would presumably possess low Rb/Y ratios, as are seen in SP1-type plutons (Table 3). Thus, a model can be constructed in which SP1-type plutons formed from partial melts of a mixed greywacke–pelite source at mid-crustal depths, whereas SP2-type plutons formed from partial melts of a mixed greywacke–pelite–felsic volcanic source at somewhat greater depths.

The foregoing discussion must be ended on a word of caution: some plutonic and meta-plutonic rocks have compositions similar to those of greywackes, pelites and felsic volcanics and such rocks cannot be excluded as sources of some strongly peraluminous granite plutons. For instance, experiments by Holtz and Johannes (1991) have shown that partial melting of a strongly peraluminous tonalite gneiss produces melts with compositions similar to those of strongly peraluminous granite plutons.

Archean plutons

As in the Phanerozoic, the spectrum of compositions of strongly peraluminous granite plutons formed during the Archean can be subdivided into two types. For each kind, a mean composition is presented in Table 3 and upper continental crust-normalized minor and trace element enrichment patterns of individual plutons are plotted in Fig. 7. "SP3-type" plutons have higher mean concentrations of CaO, TiO_2, FeO_t, MgO, Sr, Ba, La, Ce, Nd and Eu, lower mean concentrations of Na_2O, Rb, Y, Nb, Cs, Tb, Dy, Yb, Lu and Ta, and higher mean CaO/Na_2O and Gd/Yb ratios than do "SP4-type" plutons. As can be seen in Fig. 8, chondrite-normalized REE patterns of SP3-type plutons tend to decrease more steeply from La to Lu than do patterns of SP4-type plutons. Mean chondrite-normalized La/Yb ratios are 36 in the SP3s and 8 in the SP4s (Table 3). The patterns of most plutons of both types have negative Eu anomalies, but those anomalies tend to be larger in the SP4s (mean Eu/Eu* = 0.66 in the SP3s vs 0.30 in the SP4s, Table 3), possibly indicating that the SP4s suffered more extensive feldspar fractionation than did the SP3s.

Strongly peraluminous plutons have yet to be documented in the Yilgarn Block, the Kaapvaal Craton or the North Atlantic Craton, but those in the Dharwar Craton are SP3s, those in the Pilbara Block are SP4s, and those in the Slave, Superior and Wyoming Provinces are both types. Strongly peraluminous plutons of the Slave and Superior Provinces seem to be particularly abundant, and it is probably no coincidence that sedimentary rocks are also particularly voluminous. Worldwide, SP3-type plutons seem to be more voluminous than SP4-type plutons. Of the 23 plutons used in compiling Table 3, the mean area of the SP3s is 370 ± 180 km^2, whereas for SP4s, it is only 54 ± 14 km^2. Seven of nine SP3-type plutons, but only two of 14 SP4-type plutons, have areas greater than 100 km^2.

Age relationships between SP3- and SP4-type plutons found within a single craton are poorly known but, within the Superior Province, where age documen-

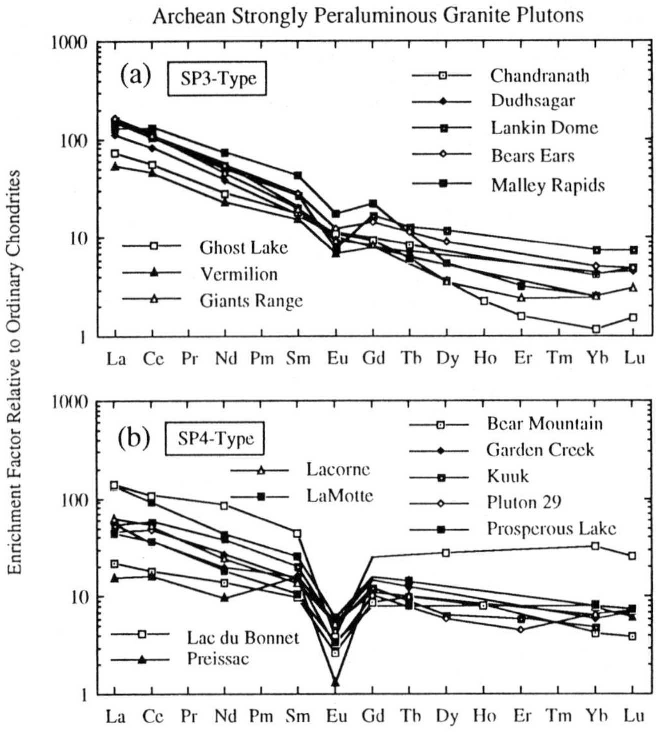

Fig. 8. Ordinary chondrite-normalized REE plots for (a) SP3-Type and (b) SP4-Type Archean strongly peraluminous granite plutons. Sources of data for the plutons are listed in Table 3. REE concentrations for ordinary chondrites as in Fig. 3.

tation is most advanced, some SP3s are known to be somewhat older than some SP4s. For instance, the SP3 "GLB-1 unit" of the Ghost Lake Batholith formed ~2685 Ma (Davis, unpublished data reported in Breaks and Moore, 1992) and the SP3 "white muscovite leucogranite" of the western Sturgeon Lake Batholith formed 2670 ± 2 Ma (Percival and Sullivan, 1988), whereas the SP4 Lacorne pluton formed 2643 ± 4 Ma (Feng and Kerrich, 1991). Some SP3-type plutons are located quite close to SP4-type plutons — the SP3 Malley Rapids Granite and SP4 Kuuk Granite of the Slave Province are separated by less than 25 km, for instance — but other SP3s are found without any SP4s nearby, and *vice versa*.

Because, as noted in the Introduction, little is known about the undisturbed isotopic compositions of Archean granite plutons and their potential source rocks, information about the sources of SP3- and SP4-type plutons must be inferred from their chemical compositions. Thus, it is of interest that both SP3s and SP4s have low CaO and Sr concentrations compared to Archean calc-alkaline granite plutons. Only SP3s, however, have lower Na_2O contents than do both kinds of calc-alkaline plutons. Hence, following the same reasoning used above for Phanerozoic strongly peraluminous plutons, SP3s could have been formed from partial melts of a greywacke–pelite source, whereas SP4s could have been derived from a greywacke–pelite–felsic volcanic source. As SP3-type plutons have much higher Gd/Yb and Rb/Y ratios (means of ~6 and 23, respectively, Table 3) than do Archean greywackes and pelites (means of ~3 and 3, and ~2 and 4, respectively, Condie, 1993), garnet was probably a residual phase in their source regions. In contrast, although the Rb/Y ratios of SP4-type plutons (mean of ~17, Table 3) are much larger than those of Archean greywackes, pelites and felsic volcanics (mean of ~1 for the felsic volcanics, Condie, 1993), the Gd/Yb ratios of SP4s (mean of ~2.5, Table 3) are not much different than those of their probable source rocks (mean of ~1.5 for the felsic volcanics, Condie, 1993). Thus, unlike in the sources of SP3-type plutons, garnet was probably not a restitic phase in the sources of SP4-type plutons.

The reason that the Rb/Y ratios of SP4-type plutons are so much higher than their probable source rocks is largely because SP4s have much higher Rb concentrations (mean of ~290 ppm, Table 3) than do greywackes, pelites or felsic volcanic rocks (means of ~70, 110 and 50 ppm, respectively, Condie, 1993). This suggests that, during crustal melting, Rb was strongly partitioned into the parental melts of SP4-type plutons and, hence, biotite, the major Rb-bearing phase present in the restites of many strongly peraluminous granites, was not present in the restites of SP4-type plutons. The lack of both residual biotite and garnet in the sources of SP4-type plutons can be attributed to their precursor melts having formed at unusually high temperatures. The experiments of Patino Douce and Johnston (1991) showed that if a pelite is melted at temperatures high enough to exhaust biotite, then garnet and aluminosilicate, the phases produced from that biotite, will melt incongruently, producing more granitic liquid and spinel, until all of the garnet is consumed. In those experiments, for example, residual biotite and garnet were both no longer present at 975°C and 7 kbar.

Archean plutons compared to their Phanerozoic counterparts

In both the Archean and Phanerozoic, we find strongly peraluminous granite plutons with low CaO and Sr contents, reflecting their derivation from source rocks that consisted solely or largely of greywacke and pelite. The sources of some of these plutons, the SP1s of the Phanerozoic and the SP3s of the Archean, seem to have included a felsic volcanic rock component, but this does not seem to have been the case for the other plutons, the SP2s of the Phanerozoic and the SP4s of the Archean. Phanerozoic strongly peraluminous SP1 plutons seem to have formed at somewhat shallower depths than all other Archean and Phanerozoic strongly peraluminous plutons because there is no evidence that garnet ever existed in their source regions. Similarly, Archean strongly peraluminous SP4 plutons seem to have formed at somewhat higher temperatures than all other Archean and Phanerozoic strongly peraluminous plutons because there is evidence that garnet formed in their source regions as a result of the complete breakdown of biotite and then, through further heating, was melted completely itself. Thus, some Archean strongly peraluminous plutons, like the Archean calc-alkaline plutons mentioned earlier, provide evidence that geothermal gradients were steeper in Archean continents than in more modern continents.

Given the aforementioned variations in source composition, melting depth, and melting temperature found among the four kinds of Archean and Phanerozoic strongly peraluminous plutons, it should not be surprising that their compositions, as represented by upper continental crust-normalized multi-element enrichment patterns (Fig. 7) are so different from one another. The variations in source lithology and melting regime make it quite difficult to identify differences in composition between Archean and Phanerozoic strongly peraluminous plutons that may have resulted from secular changes in the compositions of their source rocks. For instance, concentrations of Sc, V, Cr, Co and Ni are much higher in both shales and felsic volcanic rocks of Archean age than in their Phanerozoic counterparts (Condie, 1993). Thus, it might be expected that concentrations of these elements in Archean strongly peraluminous granite plutons would be much higher than in Phanerozoic strongly peraluminous plutons. In fact, the opposite tends to be true (Table 3). The reason that the Archean SP3-type plutons have lower concentrations of Sc, V, Cr, Co and Ni than do the Phanerozoic SP1-type plutons may be that partition coefficients for these elements are larger in the residual garnet of SP3s than in the residual cordierite of SP1s. The reason that the Archean SP4s have lower concentrations of Sc, V, Cr, Co and Ni than do the Phanerozoic SP2s may be that residual spinel, which is produced from the breakdown of garnet, and into which large amounts of Sc, V, Cr, Co and Ni may substitute, was more abundant in the source regions of the SP4s than in those of the SP2s. Spinel would have been more abundant in the SP4s source regions because garnet seems to have reacted out completely there, in contrast to in the SP2 source regions, where some residual garnet seems to have remained.

CHEMICAL COMPOSITIONS OF ALKALINE GRANITE PLUTONS

Phanerozoic plutons

The range of compositions of alkaline granite plutons found in large granite-dominated Phanerozoic provinces can be split into two types, referred to here as "ALK1" and "ALK2". A mean composition of each is presented in Table 4. Upper continental crust-normalized minor and trace element enrichment patterns of individual plutons are plotted in Fig. 9. ALK1 plutons seem to be more abundant than ALK2 plutons in the Pan-African of Arabia, whereas in the British Caledonides and the Lachlan Fold Belt, the opposite is observed.

ALK1- and ALK2-type plutons have mean major element compositions that are very similar to each other, except for somewhat higher FeO_t and TiO_2 concentrations and lower Mg numbers for the ALK1s. There are several significant trace element differences between the two kinds of plutons, however. ALK1-type plutons have higher mean concentrations of Sc, Cu, Zn, Y, Zr, Nb, light REEs, Eu and Hf and lower mean concentrations of Rb and Th than do ALK2-type plutons. ALK1s thus have particularly low Rb/Y and Th/Zr ratios compared to ALK2s.

By comparing Tables 1 and 4, one can see that alkaline granite plutons have much more evolved compositions than do calc-alkaline granite plutons. What Tables 1 and 4 do not show, but what was noted by Sylvester (1989), is that alkaline granite plutons that are less evolved than average tend to possess higher concentrations of FeO_t, Na_2O and K_2O and lower concentrations of Al_2O_3 and CaO than do their calc-alkaline counterparts. There are trace element differences between alkaline and calc-alkaline granite plutons: both ALK1s and ALK2s have higher mean concentrations of Rb, Y, Sn, heavy REEs, Pb, Th and U and lower mean concentrations of V, Cr, Co and Sr. In addition, ALK1s have higher mean F, Zn, Zr, Nb and light REE contents than do the calc-alkaline plutons. As can be seen in Table 4, the fluorine contents of ALK2-type plutons are not well-known. In summarizing unpublished data from the Lachlan Fold Belt, however, King (1993) noted that while the fluorine contents of ALK1-type plutons tend to be higher than those of calc-alkaline plutons, it is ALK2-type plutons that possess some of the highest fluorine contents of all.

The origin of Phanerozoic alkaline granite plutons is unresolved. Collins et al. (1982) and Whalen et al. (1987) referred to ALK1-type plutons as "A-type granites" and argued that they were derived by partial melting of a calc-alkaline granite melt-depleted restite. In contrast, Sylvester (1989) showed that the A-type granites of the Lachlan Fold Belt have too much Ba, Rb, Ce, Y and Sc to be derived from restites of the region. Cullers et al. (1981) and Anderson (1983) suggested that ALK1-type plutons formed by partial melting of tonalites, as did calc-alkaline granites, but by much smaller degrees of melting (10–30% vs. >30%). In the experiments of Skjerlie and Johnston (1993), however, partial melting of a tonalite produced melts with major element compositions and high fluorine contents like

TABLE 4

Mean chemical compositions of alkaline granite plutons

	Phanerozoic		Archean	
	ALK1-type	ALK2-type	ALK3-type	ALK4-type
wt%				
SiO_2	74.62±0.57 (11)	75.75±0.27 (12)	73.33±0.69 (17)	74.20±0.12 (11)
TiO_2	0.25±0.04 (11)	0.13±0.02 (12)	0.29±0.05 (17)	0.14±0.01 (11)
Al_2O_3	12.29±0.20 (11)	12.80±0.13 (12)	13.08±0.24 (17)	13.52±0.14 (11)
FeO_t	2.29±0.30 (11)	1.09±0.09 (12)	2.35±0.26 (17)	1.32±0.08 (11)
MnO	0.06±0.01 (11)	0.04±0.01 (12)	0.04±0.01 (17)	0.04±0.00 (11)
MgO	0.19±0.05 (11)	0.17±0.03 (12)	0.39±0.06 (17)	0.23±0.02 (11)
CaO	0.68±0.12 (11)	0.69±0.07 (12)	1.21±0.13 (17)	0.94±0.05 (11)
Na_2O	3.72±0.12 (11)	3.91±0.14 (12)	3.26±0.14 (17)	3.74±0.10 (11)
K_2O	4.43±0.12 (11)	4.51±0.15 (12)	4.78±0.12 (17)	4.68±0.09 (11)
P_2O_5	0.05±0.01 (11)	0.03±0.00 (12)	0.09±0.02 (16)	0.06±0.01 (10)
LOI	1.05±0.13 (10)	0.80±0.06 (12)	0.55±0.06 (15)	0.75±0.07 (9)
ppm				
Li	41±17 (2)	8.5±2.5 (2)	26±2 (4)	66±24 (6)
Be	3±0 (2)	2.8±0.2 (2)	4±0 (1)	4.7±1.4 (3)
B			2±0 (1)	4±1 (2)
F	1200±300 (2)	950±0 (1)	650±125 (6)	1270±360 (5)
Sc	11±2 (11)	4.7±0.6 (11)	3.2±0.5 (12)	2.6±0.2 (7)
V	6±1 (10)	4±1 (11)	19±5 (3)	9±1 (7)
Cr	2.1±0.1 (7)	1.8±0.7 (11)	48±28 (9)	37±21 (9)
Co	2.6±0.4 (10)	2.4±0.3 (10	11±3 (11)	19±7 (5)
Ni	8±5 (6)	1.2±1.0 (10)	14±4 (5)	10±3 (7)
Cu	6±1 (9)	2.1±0.8 (11)	12±3 (7)	11±3 (8)
Zn	143±8 (10)	38±7 (12)	47±9 (8)	44±2 (10)
Ga	22±2 (8)	19±2 (11)	16±1 (4)	20±1 (4)
Rb	180±10 (11)	300±50 (12)	211±21 (17)	339±35 (11)
Sr	90±20 (11)	50±10 (12)	172±38 (17)	102±15 (11)
Y	93±6 (10)	60±15 (12)	58±13 (8)	31±4 (10)
Zr	430±80 (11)	140±15 (12)	276±30 (17)	134±11 (11)
Nb	33±6 (10)	22±3 (12)	19±2 (8)	19±3 (10)
Sn	19±3 (3)	12±4 (8)	6.5±3.5 (2)	7.6±1.4 (3)
Cs	6.0±0.9 (8)	5.7±0.3 (2)	6.0±1.0 (9)	10±3 (6)
Ba	500±90 (11)	310±80 (12)	830±160 (17)	505±40 (11)
La	65±4 (11)	27±4 (16)	98±14 (16)	44±8 (11)
Ce	135±10 (11)	62±8 (12)	179±26 (17)	86±15 (11)
Nd	66±5 (9)	29±4 (4)	81±11 (13)	28±5 (8)
Sm	16±2 (9)	10±1 (2)	12±1 (13)	5.3±0.9 (8)
Eu	2.0±0.3 (9)	0.7±0.6 (2)	1.6±0.2 (13)	0.45±0.06 (8)
Gd	12.7±1.0 (8)	11±2 (2)	5.4±2 (2)	3.0±0.9 (3)
Tb	2.4±1.0 (2)	1.9±0.6 (2)	1.8±0.2 (12)	0.85±0.20 (6)

(continued)

	Phanerozoic		Archean	
	ALK1-type	ALK2-type	ALK3-type	ALK4-type
Dy		11±4 (2)		3.1±1.1 (3)
Yb	8.7±0.4 (9)	7±3 (2)	5.5±0.8 (12)	1.9±0.5 (7)
Lu	1.29±0.04 (8)		0.83±0.11 (13)	0.37±0.09 (7)
Hf	9.6±1.0 (8)	3.2±0.6 (2)	8.0±0.7 (12)	4.3±0.5 (7)
Ta	3.3±0.2 (9)		2.4±0.2 (12)	2.3±1.0 (4)
Pb	31±1 (8)	28±4 (11)	29±2 (7)	41±5 (10)
Th	21±1 (11)	29±6 (12)	31±3 (16)	41±3 (11)
U	5.7±0.3 (11)	9±2 (12)	3.9±0.5 (17)	6.8±1.5 (10)
ratios				
Al_2O_3/TiO_2	63±9 (11)	156±40 (12)	72±11 (17)	116±20 (11)
CaO/Na_2O	0.18±0.03 (11)	0.18±0.02 (12)	0.39±0.05 (17)	0.25±0.01 (11)
Mg#	11±3 (12)	21±3 (12)	21±2 (17)	24±2 (11)
[La/Yb]n	5.2±0.3 (9)	7±5 (2)	24±5 (12)	22±5 (7)
Eu/Eu*	0.42±0.06 (9)	0.2±0.2 (2)	0.46±0.05 (13)	0.36±0.08 (8)
Rb/Y	2.0±0.2 (10)	5.6±0.5 (12)	3.8±0.6 (8)	13±2 (10)
Gd/Yb	1.46±0.13 (8)	1.79±0.34 (2)	2.45±0.49 (2)	2.05±0.59 (3)
Tb/Yb	0.26±0.05 (2)	0.28±0.02 (2)	0.32±0.02 (11)	0.49±0.07 (5)
Th/Zr	0.07±0.02 (11)	0.23±0.04 (12)	0.12±0.01 (16)	0.31±0.02 (11)

Error on the mean is calculated as $\sigma_n\sqrt{(n-1)}$ where n, the number of plutons, is given in parentheses. FeO_t is total iron expressed as FeO. Mg# = 100[molecular MgO/(MgO + FeO_t)]. [La/Yb]n is the chondrite-normalized La/Yb ratio. Eu/Eu* is the observed Eu concentration divided by the Eu concentration calculated assuming that the chondrite-normalized REE pattern had no Eu anomaly. Data sources: *ALK1-Phanerozoic plutons* — Mumbulla (i = number of individual samples = 6), Dr George (i = 2), Watergums (i = 3), Naghi (i = 1), Nagha (i = 2), Howe Range (i = 2) and Gabo Island (i = 1) (Collins et al., 1982); Narraburra Granite (i = 3) (Wormald and Price, 1988); Granite pluton of Meatiq Dome (i = 21) (Sultan et al., 1986; M. Sultan, pers. comm.); Bayda (i = 6) and Al Yoob (i = 4) Granites (unpublished analytical database from Jackson et al., 1984). *ALK2-Phanerozoic plutons* —Hamamah (i = 2) (unpublished analytical database from Jackson et al., 1984); Ennerdale (i = 18) (O'Brien et al., 1985); Bodalla (i = 1), Maffra (i = 1), Thologolong (i = 1) and Coles Bay (i = 1) monzogranites (Chappell and White, 1992); Mt. Mittamatite (i = 1) and Pine Mountain (i = 2), leucogranites (Price et al., 1983); Yeoval Adamellite (i = 1), Grenfell Granite (i = 1), Scammells Adamellite (i = 1) and Knocker Granite (i = 1) (Wyborn et al., 1987). *ALK3-type Archean plutons* — Chitradurga Granite (i = 12) (Taylor et al., 1984); Mondana Adamellite (i = 10) (Davy, 1988); Pluton F, Mt. Edgar Batholith (i = 7) (Davy and Lewis, 1986); Benchmark 627 Granite (i = 8) (Tyler et al., 1992; I. Tyler, pers. comm.); Little Elk Granite (i = 10) (Gosselin et al., 1990); Koolanooka Pluton (i = 5), Damperwah Pluton (i = 6) and Pluton 00 (i = 1) (Watkins and Hickman, 1990); Mooihoek (i = 4), Mhlosheni (i = 5), Spekboom (i = 5), Godlwayo (i = 5), Mpageni (i = 4), Mbabane (i = 5), Ngwempisi (i = 6), Sicunusa (i =4) and Kwetta (i = 5) Plutons (Meyer et al., 1992; Condie and Hunter, 1976). *ALK4-type Archean plutons* — Cooglegong Adamellite (i = 19) (Blockley, 1980; Bickle et al., 1989); Moolyella Adamellite (i = 22) (Blockley, 1980; Jahn et al., 1981; Davy and Lewis, 1986); Pluton I, Mt. Edgar Batholith (i = 7) (Davy and Lewis, 1986); Pluton 7 (i = 1), Pluton 29 (i = 2) and Pluton 17 (i = 1) (Watkins and Hickman, 1990); Mungari Leucogranite (i = 9) (Cassidy et al., 1991; Hill et al., 1992a; A. Thom, pers. comm.); Pluton 76 syenogranite (i = 1) (Cassidy et al., 1991); Goodia Dome (i = 1) and Dundas (i = 2) Plutons (Hill et al., 1992a; A. Thom, pers. comm.); Granite of Owl Creek Mountains (i = 25) (Stuckless et al., 1986).

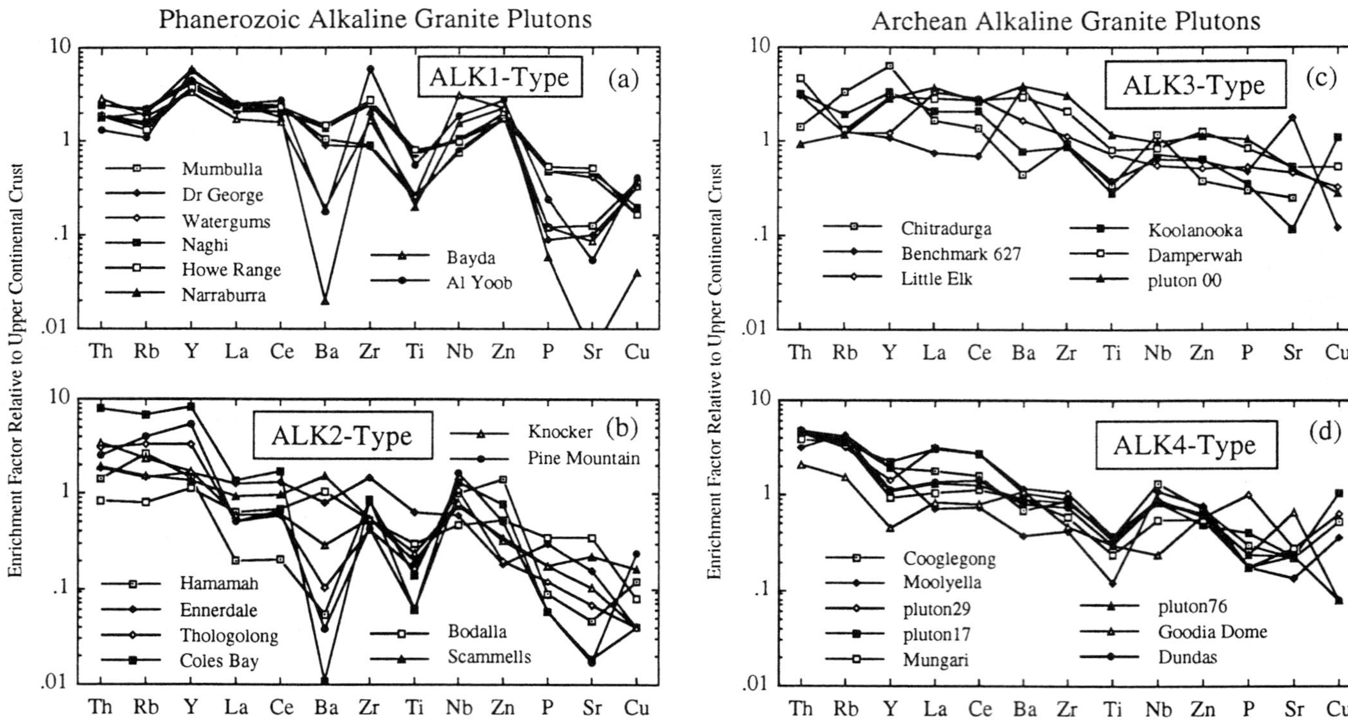

Fig. 9. Upper continental crust-normalized trace element diagram for (a) Phanerozoic ALK1-Type, (b) Phanerozoic ALK2-Type, (c) Archean ALK3-Type and (d) Archean ALK4-Type alkaline granite plutons. Sources of data for the plutons are listed in Table 4. Composition of the upper continental crust is given in the caption to Fig. 2.

those of ALK1-type plutons only after *larger* degrees of melting than that necessary to produce low-fluorine, calc-alkaline granite melts. Also, Sylvester (1989) showed that the Lachlan A-type granites have too much Y and Sc to be derived from the same tonalites that produced the Lachlan I-type granites. He suggested that the source of the A-type granites had more Y and Sc than did the tonalite source of the I-type granites, possibly because it was a more mafic, amphibole-rich tonalite. Creaser et al. (1991), on the other hand, argued that ALK1-type granites, because their compositions tend to be more evolved than calc-alkaline granites, formed from a felsic tonalite or granodiorite, i.e., sources more evolved than those of calc-alkaline granites. Patino Douce and Johnston (1991) have pointed out, however, that granodiorites are too anhydrous to be fertile sources of granites.

The origin of ALK2-type granite plutons has received much less attention than that of the ALK1s, and is equally unclear. Wyborn et al. (1987) and Chappell and White (1992) noted that ALK2-type granite plutons in the Lachlan Fold Belt are associated closely with I-type plutons. They further showed that the distribution of major elements and many trace elements in the ALK2-type plutons are consistent with an origin involving fractional crystallization of orthopyroxene, plagioclase and zircon from an I-type granite magma. Hence, they considered ALK2-type plutons to be fractionated I-type granites and referred to them as such. Fractional crystallization, however, may not explain the distribution of all trace elements in these rocks. For instance, Chappell and White (1992) noted that the concentration of Ba in their fractionated I-type granites (our ALK2-type granites) is lower than in their unfractionated I-type granites. This relationship is difficult to explain by fractional crystallization without alkali-feldspar or biotite being among the removed phases, yet if biotite was separated, one would expect the concentration of Rb, which is strongly partitioned into that phase, to be lower in their fractionated I-type granites than in their unfractionated I-type granites. In fact, the concentration of Rb is much higher. Given such difficulties, and the fact that ALK2-type granite plutons are much more similar in composition to ALK1-type plutons than to calc-alkaline plutons, it is possible that ALK2s were derived by partial melting of sources similar to those of ALK1s rather than by fractional crystallization of calc-alkaline magmas.

The formidable problems encountered in trying to explain the derivation of alkaline granites from common igneous source rocks or magmas solely by melt-crystal equilibria suggest that fluid-transport of elements played a role in their origin. In particular, the high concentrations of Rb, Y, Sn, REEs, Pb, Zn, Nb, Th and U in alkaline granites compared to their calc-alkaline counterparts could in part be the result of enrichment of these elements in their sources by chlorine- and/or fluorine-rich aqueous fluids before or during melting. Experiments by Webster and Holloway (1990) and Keppler and Wyllie (1991) have shown that aqueous fluid/melt partition coefficients for Rb, Y, Sn, REEs, Pb, Zn, Nb and U are high if the fluid contains chlorine, while strong partitioning of Th into the fluid occurs in the presence of fluorine.

Archean plutons

Each of the two types of Phanerozoic alkaline granite plutons seems to have a counterpart in the Archean. A mean composition for each Archean type is presented in Table 4. Upper continental crust-normalized minor and trace element enrichment patterns of individual Archean alkaline plutons are plotted in Fig. 9. "ALK3-type" plutons have higher mean concentrations of TiO_2, FeO_t, V, Sr, Y, Zr, Ba, REEs and Hf and lower mean concentrations of F, Rb, Pb, Th and U than

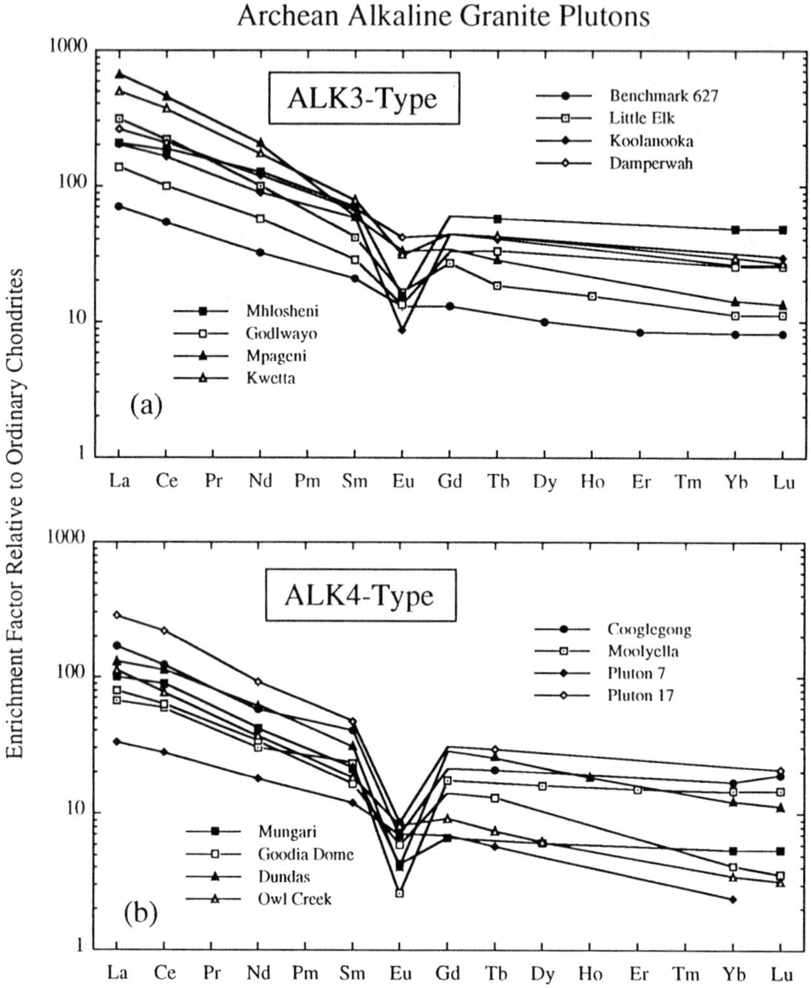

Fig. 10. Ordinary chondrite-normalized REE plots for (a) ALK3-Type and (b) ALK4-Type Archean alkaline granite plutons. Sources of data for the plutons are listed in Table 4. REE concentrations for ordinary chondrites as in Fig. 3.

do "ALK4-type" plutons. ALK3s thus have lower mean Al_2O_3/TiO_2, Rb/Y and Th/Zr ratios than do ALK4s, mirroring the relationship observed between ALK1s and ALK2s. As can be seen in Fig. 10, chondrite-normalized REE patterns of ALK3-type plutons are similar in shape to those of ALK4-type plutons. Mean chondrite-normalized La/Yb ratios are 24 in the ALK3s and 22 in the ALK4s; mean Eu/Eu* ratios are ~0.4 in both types of plutons (Table 4).

Alkaline granite plutons have yet to be well-documented in the Superior and Slave Provinces or in the North Atlantic Craton, but those in the Dharwar and Kaapvaal Cratons are ALK3s, while those in the Yilgarn Block, Pilbara Block and Wyoming Province are both types. There is not much difference in the average sizes of the two kinds of plutons. Of the 28 used in compiling Table 4, the mean area of the ALK3s is 180 ± 50 km^2 and that of the ALK4s is 130 ± 40 km^2. Age relationships between ALK3- and ALK4-type plutons are poorly known but, in the Yilgarn Block, the ALK3 Damperwah pluton formed 2641 ± 5 Ma (Wiedenbeck and Watkins, 1993) and was followed by the ALK3 Koolanooka (Wiedenbeck and Watkins, 1993) and ALK4 Mungari (Hill et al., 1992a) plutons at ~2600 Ma. In the Wyoming Province, the ALK4 Granite of Owl Creek Mountains has an age of 2730 ± 35 My (Stuckless et al., 1986), whereas the ALK3 Little Elk Granite has been dated at 2549 ± 11 My (Gosselin et al., 1990). In cratons where alkaline and calc-alkaline granite plutons are both present, the alkaline plutons typically formed tens of million years after the calc-alkaline plutons. There are exceptions, however. For instance, in the Dharwar Craton, intrusion of the ALK3 Chitradurga Granite (Taylor et al., 1984) preceded emplacement of the calc-alkaline Closepet Granite (Friend and Nutman, 1991) by almost 100 million years.

The fact that, in general, ALK3- and ALK4-type plutons did not form alongside co-magmatic calc-alkaline granite plutons suggests that they are not related to the latter by fractional crystallization. Instead, because concentrations of Rb, Nb, Y, Sn, Yb, Lu and Th are enriched in ALK3s and ALK4s relative to the Archean calc-alkaline CA1s and CA2s, the Archean alkaline granites, like Phanerozoic alkaline granites, may have been formed by partial melting of metasomatized source rocks.

Archean plutons compared to their Phanerozoic counterparts

Because the precise nature of the source rocks of both Archean and Phanerozoic alkaline granites is not known, it is not possible, of course, to identify secular changes in the chemical composition of those sources or in the thermal conditions under which the granites formed. It is noteworthy, however, that there are distinct differences in composition between Archean and Phanerozoic alkaline granites. For instance, while both Archean ALK4s and Phanerozoic ALK2s have larger Rb/Y and Th/Zr ratios than do either Archean ALK3s or Phanerozoic ALK1s, those ratios are largest in the ALK4s (Table 4). Furthermore, ALK3s seem to have lower concentrations of F than not only ALK4s, but also ALK1s and ALK2s, although the database is admittedly small. These differences may represent somewhat different origins for Archean and Phanerozoic alkaline granites.

HEAT SOURCES

An understanding of the mechanism by which continents become hot enough to melt is surely the key to understanding the origin of granites. Four possible mechanisms are advection, conduction, decompression and metasomatism.

Advection

In the advection model, magmas that are usually thought to be basaltic are intruded into continental crust as sills and their heat of crystallization melts the country rock, producing granite magma. Melting is essentially instantaneous so that the basalt sills and granites form at very nearly the same time (Huppert and Sparks, 1988). It is unlikely that this model is applicable to the origin of most Archean granite plutons because significant volumes of co-magmatic basalt are not found with them.

Conduction

In the conduction model, the upper mantle beneath continental crust becomes anomalously hot and heat is transferred into the crust by conduction, causing melting. The upper mantle may become hot as a result as a thermal plume rising from the deeper mantle (Hill et al., 1992a,b) or delamination of the mantle lithosphere (and perhaps the lower crust) and upwelling of the subjacent asthenosphere (Kay and Kay, 1991). In either case, basalts formed from the hot mantle are either ponded at the base of the crust or intrude the crust as dikes rather than sills, thereby limiting amounts of advective heat transfer. Because heat transfer by conduction is much slower than by advection, crustal melting and the consequent production of granites occur sometime after mantle heating and basalt formation. Hill et al. (1992b) calculated that a 1300°C mantle layer located ~30 km beneath a crustal layer will raise the temperature of that crustal layer from 500 to 850°C in ~30 My by conduction alone.

The conductive heat-transfer model is an attractive explanation for Archean granite plutons in that it is consistent with them being emplaced typically tens of millions of years after the formation of basalts in nearby Archean greenstone belts. The model, however, conflicts with other observations. For instance, there are *no* basalts associated with many granite plutons, as described earlier in this chapter, and seismic-velocity models suggest that few mafic rocks now reside in the lower parts of Archean cratons (Durrheim and Mooney, 1991). This suggests that extensive mantle melting was not occurring when many Archean plutons formed.

Also, as pointed out by Hill et al. (1992a,b), conductive heat transfer from the mantle predicts that granites derived by melting the lower crust should be older than those formed by melting the middle crust because of the extra time required to transfer heat from the lower to middle crust. In fact, what was shown earlier in this chapter was that granite plutons derived from the lower crust did not form any earlier than those from the middle crust.

Decompression

In the decompression model, deep-seated crustal rocks that are just below their solidus temperatures are uplifted to shallower levels where they begin to melt, producing granites. In order to bring crustal rocks close to their soldii in the first place, and to explain their subsequent uplift, crustal thickening, usually thought to be the result of tectonic stacking and/or interleaving of thrust sheets, is a necessary prerequisite for decompression melting (England and Thompson, 1986). Temperatures in a crustal column increase with thickening because cool, shallow rocks are buried to hotter, deeper regions and because radiogenic heating increases. Beneath the large overburden of a thickened crust, radiogenic heat is trapped more easily, and will be greater anyhow, because the stacking of thrust sheets significantly increases the bulk concentration of heat-producing elements (K, U, Th) in the crust. Crustal melting that results from thickening and uplift seems to be a slow process; Zen (1988) calculated that melting will follow thickening by about 20 to 60 My. He also showed that melting begins at the base of a thickened crust but quickly moves to shallower levels.

The decompression model explains the time lag observed in many Archean cratons between compressional tectonism and granite plutonism. It is also consistent with near-simultaneous melting of the middle and lower crust. It is uncertain, however, whether the high melting temperatures (>900°C) inferred for some Archean granites can be attained by this model. There is also some question as to whether, just before emplacement of granite plutons, Archean cratons were as thick as that commonly assumed in models of decompression melting (~60 km). Seismic-velocity models suggest a present-day thickness of ~35 km for many Archean cratons (Durrheim and Mooney, 1991) and post-Archean erosion must have thinned them significantly, but because greenschist-facies assemblages are widespread, it is doubtful that as much as 25 km, on average, were lost. Given the likelihood that the formation of Archean granites required high fusion temperatures and that Archean crustal thickening was only moderate, processes other than decompression melting must have been important.

Metasomatism

In the metasomatic model, crustal rocks are not heated *per se*. Instead, a mobile, water-rich fluid phase lowers their solidus temperatures sufficiently to cause melting. Newton (1990) invoked a variant of this model to explain the remarkable coincidence of the lower boundary of the Closepet Granite occurring at precisely the same paleodepth at which orthopyroxene, formed by the dehydration reaction of hornblende and biotite, first appears in lower crust. He suggested that a mantle-derived fluid phase scavenged potassium from the breakdown of lower crustal biotite and deposited it in the middle crust. Potassium, like water, can significantly lower the solidus temperatures of potential source rocks of granites.

The origin of granites by metasomatic-induced melting is not a popular idea because, as noted by Skjerlie et al. (1993), a water-saturated granite melt cannot ascend far from its source region without solidifying. There are also questions about how fluids could move from the mantle to the lower crust and on to the middle crust, given the presumably low porosity and permeability of rock in those regions. Nonetheless, it is possible that mantle- or crustal-derived fluids stripped lower-crustal rocks of heat-producing elements, deposited those elements in mid-crustal rocks before moving on to the upper crust and, in this way, led to radiogenic heating and melting of the middle crust and the formation of some Archean granites. As we have seen, many Archean plutons are associated with large transcurrent shear zones that, as noted by Newton (1990), could have provided ample pathways for fluid migration. Also, granulite-facies rocks from the lower crust are depleted in heat-producing elements, particularly U, relative to lower-grade, upper-crustal rocks (Rudnick and Presper, 1990). The experiments of Keppler and Wyllie (1991) showed that aqueous fluids containing chloride and/or CO_2 could have efficiently leached U from these granulites.

Towards a model of Archean granite formation

It seems likely that radiogenic heating due to crustal thickening and metasomatic enrichment of heat-producing elements played a much more important role in bringing about the crustal melting that formed Archean granite plutons than did heating from basaltic magmas. At the same time, because Archean cratons were on average probably never as much as 60 km thick, and because there is no obvious mechanism for enriching the Archean *lower* crust in heat-producing elements, it may seem doubtful that radiogenic heating alone could have been responsible for partially melting the deep-seated source regions of some Archean granites. It must be remembered, however, that many of the rocks in those source regions, namely TTGs, probably did not form much earlier than the Archean granites and, hence, would have been still quite warm at the time the granites were produced. Also, because the mantle heat flux has steadily decreased with the decay of radiogenic heat-producing elements through time, the Archean continental crust should have been somewhat warmer than is the case today, arguments for thick refractory mantle root zones in the Archean (Bickle, 1986) notwithstanding. Thus, perhaps only small increases in temperature were sufficient to induce crustal melting and thereby form Archean granite plutons.

TECTONIC ENVIRONMENT

As has been stressed throughout this chapter, the characteristics of most Archean granite plutons are similar to those of granites formed ~600 Ma in the Pan-African orogeny, ~400 Ma ago in the British Caledonides and the Lachlan Fold Belt, and ~300 Ma ago in the Hercynides of Europe. The tectonic environment in which the Phanerozoic granites formed is not without uncertainty but most models involve crustal melting following collision and amalgamation of island arcs and continental fragments (see reviews by Pitcher, 1987 and Sylvester, 1989). As in the Archean regions, the heat sources for the Phanerozoic granites are not obvious: basalts are rare and the crust into which the granites intruded was probably never as much as 60 km thick on average. The latter characteristic contrasts with the well-known Himalayan collision, during which the crust doubled in thickness as a result of one large, thick continental mass overriding another. In the other Phanerozoic collisions, which were probably much more common than the Himalayan case, comparatively small and thin pieces of crust seem to have been thrust together.

Based on analogy with their Phanerozoic counterparts, collision models can be invoked for the tectonic setting of Archean granite plutons and many recent studies have done just that (Kusky, 1989; Card, 1990; Gosselin et al., 1990). Formation of Archean granites in a collisional setting is consistent with the thrusting and folding that preceded and, in some cases, accompanied their emplacement (King and Helmstaedt, 1989; Card, 1990), the paucity of co-magmatic mantle-derived rocks, intrusion of the granites over broad areas, and their association with older greenstone belt volcanics (Sylvester et al., 1987; Condie, 1989) and TTG plutons (Martin, 1986) that can be plausibly explained as being subduction-related. It is also probably true that collisions are the only tectonic setting in which, as is observed in the Archean, large volumes of calc-alkaline, strongly peraluminous and alkaline granites are all produced at more-or-less the same time (Harris et al., 1986; Sylvester, 1989). This characteristic of collision-related granites should dissuade anyone from attempting to discern the tectonic setting of any granite, Archean or otherwise, on the basis of trace element diagrams that imply, as in those of Pearce et al. (1984), that all calc-alkaline granites are subduction-related, all alkaline granites are anorogenic, and all collision-related granites are strongly peraluminous. In fact, in the Archean granite provinces, many of the early, somewhat deformed plutons, which are most commonly calc-alkaline and strongly peraluminous in composition, may be described as "syn-collisional" whereas many of the late, undeformed plutons, most commonly alkaline in composition, may be "post-collisional".

Not all of the Archean granite plutons necessarily formed in a collisional setting, of course. A good case can be made, for instance, for an anorogenic setting for the 3075 My-old alkaline granites of the Kaapvaal Craton, which formed 150 My after a major thrust faulting event and are associated with major coeval basaltic volcanism. These granites may have been produced as a result of crustal melting induced by the intrusion of basalt sills, perhaps above the hot head of a rising mantle plume (Hill et al., 1992a,b).

SUMMARY

Large volumes of calc-alkaline, strongly peraluminous and alkaline granite plutons were emplaced across broad areas of Archean cratons several million years or more after episodes of greenstone belt volcanism and TTG plutonism. In general, compressional deformation occurred before and continued through the beginning of granite plutonism but then quickly waned. Hence, the earliest plutons, most commonly calc-alkaline and strongly peraluminous granites, tend to be somewhat deformed, whereas the later plutons, commonly alkaline granites, tend to be undeformed.

The Archean granite plutons were formed by partial melting of igneous and sedimentary rocks in the middle and lower crust, in most cases with little involvement from mantle-derived magmas. Similar granites occur in the Phanerozoic in what are thought to be collision-related orogenic provinces, and the same tectonic setting may be appropriate for many of the Archean granites. Thus, many of the early, calc-alkaline and strongly peraluminous Archean plutons may be "syn-collisional" and many of the late, alkaline Archean plutons may be "post-collisional".

There are differences in composition between Archean and Phanerozoic granite plutons. Many of these differences are probably the result of changes in the composition of the continental crust through time. For instance, high La/Y and Sr/P ratios and low Mg numbers of Archean calc-alkaline granite plutons, as compared to their Phanerozoic counterparts, may reflect compositional differences between Archean and Phanerozoic TTGs. Other differences between Archean and Phanerozoic granites probably reflect steeper geothermal gradients in Archean continents than in the continents of today, at least during episodes of crustal melting. Thus, high Ba concentrations and Ba/Zr ratios in calc-alkaline Archean plutons suggest that most of the biotite in their source regions was consumed under unusually high temperatures. The combination of high Rb concentrations and low Gd/Yb ratios in some strongly peraluminous Archean granite plutons suggests that both biotite and garnet were consumed in their source regions, again as a result of high-temperature melting.

Although the basic characteristics of Archean granite plutons are now fairly clear, and their importance can be doubted no longer, they present many unresolved issues that deserve further work. Five problems stand out in particular. First, and possibly foremost, is the mechanism by which Archean crustal rocks became partially molten, thereby forming the granite plutons. Were the source rocks already so warm that addition of a volatile phase or hyperfusible element such as potassium was alone sufficient to induce melting? Or was an enhanced heat flux from the mantle required? Was mantle-derived heat carried by fluids and how could have such fluids infiltrated the crust so efficiently? What role did collisional tectonism have in bringing about crustal melting? Second is the question of the nature of the source rocks of alkaline granite plutons. Did alkaline granites form simply by melting "normal" crustal rocks under unusual conditions such as at very high temperatures or were the sources themselves unusual in some way, such as being extremely enriched in incompatible trace elements? Again, was a fluid phase important in modifying the composition of the source rocks?

Third, it would be useful to pin down as precisely as possible the temperatures and pressures under which Archean calc-alkaline and strongly peraluminous granite plutons formed in order to determine just how much hotter the continental crust may have been in the Archean than it is today. If the granites require Archean continents to have been *much* hotter, can this be reconciled with evidence that some of those continents were underlain by thick lithospheric mantle roots? Fourth, it should be determined whether any of the compositional characteristics that distinguish Archean granites from Phanerozoic granites are common in Proterozoic granites. If Proterozoic granites have compositions more similar to Archean granites than to Phanerozoic granites, how does this impact on the often-made case for changes in processes of crustal evolution across the Archean–Proterozoic boundary?

Finally, there is the problem of why most Archean granites seem to have formed after ~3100 Ma rather than before. If the mantle has cooled progressively through time should not crustal melting have been more widespread and granite plutonism more abundant early in the Archean rather than later? Is it possible that large volumes of early Archean granites have been lost from the geologic record through erosion or have otherwise gone unrecognized? Perhaps a related issue is the curious observation that almost all of the few granites that did form before 3100 Ma occur now as strongly metamorphosed gneisses rather than as plutons. If these gneisses were emplaced originally as plutons, the question arises as to why it was not until 3100 Ma that granite plutons could escape strong deformation and metamorphism? What new properties did continents begin to acquire 3100 Ma?

Answers to most of these questions will no doubt come slowly and then only through inputs from many sub-disciplines of geology. The reward however will be nothing less than a solid understanding of Archean crustal evolution.

ACKNOWLEDGEMENTS

I thank Andrea Thom and Ian Tyler for supplying unpublished chemical data collected at the ANU and the Chemistry Centre of W.A., respectively, and used in this study, Thom and Kevin Cassidy for their thoughts about the Yilgarn granites, and Alison Leitch for assistance with data compilation. Fred Breaks, Anton Brown, Richard Davy, F.M. Meyer, John Percival, Laurence Robb, John Stuckless and Keith Watkins supplied reprints and preprints of pertinent publications that I otherwise would not have been able to acquire. Charlotte Allen, Kent Condie, Laurence Robb and John Rogers provided very helpful comments on the manuscript.

REFERENCES

Aleinikoff, J.N., Williams, I.S., Compston, W., Stuckless, J.S. and Worl, R.G., 1989. Evidence for an early Archean component in the middle to late Archean gneisses of the Wind River Range, west-central Wyoming; conventional and ion microprobe U-Pb data. Contrib. Mineral. Petrol., 101: 198–206.

Allen, P., Condie, K. and Bowling, G.P., 1986. Geochemical characteristics and possible origins of the southern Closepet batholith, South India. J. Geol., 94: 283–299.

Anders, E. and Grevesse, N., 1989. Abundances of the elements: Meteoritic and solar. Geochim. Cosmochim. Acta, 53: 197–214.

Anderson, J.L., 1983. Proterozoic anorogenic granite plutonism of North America. In: L.G. Medaris, C.W. Byers, D.M. Mickelson and W.C. Shanks (Eds.), Proterozoic Geology: Selected Papers from an International Proterozic Symposium. Geol. Soc. Am. Mem., 161: 133–154.

Armstrong, R.A., Compston, W., Retief, E.A., Williams, I.S. and Welke, N.J., 1991. Zircon ion-microprobe studies bearing on the age and evolution of the Witwatersrand triad. Precambrian Res., 53: 243–266.

Arth, J.G., 1976. Behavior of trace elements during magmatic processes — a summary of theoretical models and their applications. J. Res. U.S. Geol. Survey, 4: 41–47.

Arth, J.G. and Hanson, G.N., 1975. Geochemistry and origin of the early Precambrian crust of northeastern Minnesota. Geochim. Cosmochim. Acta, 39: 325–362.

Baadsgaard, H., 1976. Further U-Pb dates on zircons from the early Precambrian rocks of the Godthåbsfjord area, West Greenland. Earth Planet. Sci. Lett., 33: 261–267.

Balakrishnan, S. and Rajamani, V., 1987. Geochemistry and petrogenesis of granitoids around the Kolar Schist Belt, South India: constraints for the evolution of the crust in the Kolar area. J. Geol., 95: 219–240.

Barker, F., 1979. Trondhjemite: Definition, environment and hypotheses of origin. In: F. Barker (Ed.), Trondhjemites, Dacites and Related Rocks. Elsevier, Amsterdam, pp. 1–12.

Barley, M.E. and Groves, D.I., 1990. Deciphering the tectonic evolution of Archaean greenstone belts: the importance of contrasting histories to the distribution of mineralization in the Yilgarn Craton, Western Australia. Precambrian Res., 46: 3–20.

Barton, J.M. and van Reenen, D.D., 1992. When was the Limpopo Orogeny? Precambrian Res., 55: 7–16.

Beakhouse, G.P., McNutt, R.H. and Krogh, T.E., 1988. Comparative Rb-Sr and U-Pb zircon geochronology of late- to post-tectonic plutons in the Winnipeg River Belt, northwestern Ontario, Canada. Chem. Geol., 72: 337–351.

Beakhouse, G.P. and McNutt, R.H., 1991. Contrasting types of Late Archean plutonic rocks in northwestern Ontario: implications for crustal evolution in the Superior Province. Precambrian Res., 49: 141–165.

Bevier, M.L. and Gebert, J.S., 1991. U-Pb geochronology of the Hope Bay–Elu Inlet area, Bathurst Block, northeastern Slave Structural Province, Northwest Territories. Can. J. Earth Sci., 28: 1925–1930.

Bickle, M.J., 1986. Implications of melting for stabilisation of the lithosphere and heat loss in the Archean. Earth Planet. Sci. Lett., 80: 314–324.

Bickle, M.J., Bettenay, L.F., Barley, M.E., Chapman, H.J., Groves, D.I., Campbell, I.H. and de Laeter, J.R., 1983. A 3500 Ma plutonic and volcanic calc-alkaline province in the Archaean East Pilbara Block, Western Australia. Contrib. Mineral. Petrol., 84: 25–35.

Bickle, M.J., Bettenay, L.F., Chapman, H.J., Groves, D.I., McNaughton, H.J., Campbell, I.H. and de Laeter, J.R., 1989. The age and origin of younger granitic plutons of the Shaw Batholith in the Archaean Pilbara Block, Western Australia. Contrib. Mineral. Petrol. 101: 361–376.

Bickle, M.J., Bettenay, L.F., Chapman, H.J., Groves, D.I., McNaughton, H.J., Campbell, I.H. and de Laeter, J.R., 1993. Origin of the 3500–3300 Ma calc-alkaline rocks in the Pilbara Archaean: isotopic and geochemical constraints from the Shaw Batholith. Precambrian Res., 60: 117–149.

Blockley, J.G., 1980. The tin deposits of Western Australia. Geol. Surv. West. Aust. Mineral Res. Bull. 12: 1–184.

Boily, M., Williams-Jones, A.E. and Mulja, T., 1990. Rare element granitic pegmatites in the Abitibi Greenstone Belt: a case study of the Preissac–Lacorne Batholith. In: M. Rive, P. Verpaelst, Y. Gaanou, J.M. Lulin, G. Riveriu and A. Simard (Eds.), The northwestern Quebec polymetallic belt. Canadian Institute of Mining and Metallurgy, Special Volume 43, pp. 299–311.

Breaks, F.W., Cherry, M.E. and Janes, D.A., 1985. Metallogeny of Archaean granitoid rocks of the English River Subprovince, Superior Province, Ontario, Canada: a review. In: High Heat Production Granites, Hydrothermal Circulation and Ore Genesis. Institute of Mining and Mineralogy, Chameleon Press, pp. 9–31.

Breaks, F.W. and Moore, J.M., 1992. The Ghost Lake Batholith, Superior Province of northwestern Ontario: a fertile, S-type, peraluminous granite–rare-element pegmatite system. Can. Mineral., 30: 835–875.

Brown, M., Friend, C.R.L., McGregor, V.R. and Perkins, W.T., 1981. The late Archaean Qôrqut Granite Complex of southern West Greenland. J. Geophys. Res., 86: 10617–10632.

Burke, K. and Kidd, W.S.F., 1978. Were Archean continental geothermal gradients much steeper than those of today? Nature, 272: 240–241.

Card, K.D., 1990. A review of the Superior Province of the Canadian Shield, a product of Archean accretion. Precambrian Res., 48: 99–156.

Cassidy, K.F., Barley, M.E., Groves, D.I., Perring, C.S. and Hallberg, J.A., 1991. An overview of the nature, distribution and inferred tectonic setting of granitoids in the late-Archaean Norseman–Wiluna Belt. Precambrian Res., 51: 51–83.

Cerny, P. and Meintzer, R.E., 1988. Fertile granites in the Archean and Proterozoic fields of rare-element pegmatites: crustal environment, geochemistry and petrogenetic relationships. Can. Inst. Min. Metal., Spec. Vol. 39: 170–207.

Cerny, P., Fryer, B.J., Longstaff, F.J. and Tammemagi, H.Y., 1987. The Archean Lac du Bonnet batholith, Manitoba: Igneous history, metamorphic effects, and fluid overprinting. Geochim. Cosmochim. Acta, 51: 421–438.

Chadwick, B., Vasudev, V.N., Krishna Rao, B. and Hedge, G.V., 1992. The Dharwar Supergroup: basin development and implications for late Archaean tectonic setting in western Karnataka, southern India. Univ. West. Australia, Publ. 22: 3–15.

Chappell, B.W. and White, A.J.R., 1974. Two contrasting granite types. Pacific Geology, 8: 173–174.

Chappell, B.W. and White, A.J.R., 1992. I- and S-type granites in the Lachlan Fold Belt. Trans. Royal Soc. Edinburgh Earth Sci., 83: 1–26.

Collins, W.J., Beams, S.D., White, A.J.R. and Chappell, B.W., 1982. Nature and origin of A-type granites with particular reference to southeastern Australia. Contrib. Mineral. Petrol, 80: 189–200.

Compton, P., 1978. Rare earth evidence for the origin of the Nûk gneisses Buksefjorden Region, southern West Greenland. Contrib. Mineral. Petrol., 66: 283–293.

Condie, K.C., 1981. Geochemical and isotopic constraints on the origin and source of Archaean granites. Spec. Publs. geol. Soc. Aust. 7: 469–479.

Condie, K.C., 1989. Geochemical changes in basalts and andesites across the Archean–Proterozoic boundary: identification and significance. Lithos, 23: 1–18.

Condie, K.C., 1993. Chemical composition and evolution of the upper continental crust: Contrasting results from surface samples and shales. Chem. Geol., 104: 1–37.

Condie, K.C. and Hunter, D.R., 1976. Trace element geochemistry of Archean granitic rocks from the Barberton Region, South Africa. Earth Planet. Sci. Lett., 29: 389–400.

Creaser, R.A., Price, R.C. and Wormald, R.J., 1991. A-type granites revisited: Assessment of a residual-source model. Geology , 19: 163–166.

Cullers, R.L., Koch, R.J. and Bickford, M.E., 1981. Chemical evolution of magmas in the Proterozoic terrane of the St. Francois Mountains, southeastern Missouri 2. Trace element data. J. Geophys. Res., 86: 10388–10401.

Davy, R., 1988. Geochemical patterns in granitoids of the Corunna Downs Batholith, Western Australia. Geol. Survey West. Australia, Prof. Pap. 23: 51–84.

Davy, R. and Lewis, J.D., 1986. The Mount Edgar Batholith Pilbara Area, Western Australia geochemistry and petrography. Geol. Survey West. Australia, Report 17: 43pp.

Day, W.C., 1990. Petrology of the Rainy Lake area, Minnesota, USA — implications for petrotectonic setting of the Archean southern Wabigoon subprovince of the Canadian Shield. Contrib. Mineral. Petrol., 105: 303–321.

Day, W.C. and Weiblen, P.W., 1986. Origin of late Archean granite: geochemical evidence from the Vermilion Granitic Complex of northern Minnesota. Contrib. Mineral. Petrol., 93: 283–296.

de Wit , M.J., Roering, C., Hart, R.J., Armstrong, R.A., de Ronde, C.E.J., Green, R.W.E., Tredoux, M., Peberdy, E. and Hart, R.A., 1992. Formation of an Archaean continent. Nature, 357: 553–562.

Deer, W.A., Howie, R.A. and Zussman, J. 1966. An introduction to the rock-forming minerals. Longman Press, London, 528 pp.

Dhoundial, D.P., Paul, D.K., Amitabha, S., Trivedi, J.R., Gopalan, K., and Potts, P.J., 1987. Geochronology and geochemistry of Precambrian granitic rocks of Goa, SW India. Precambrian Res., 36: 287–302.

Drury, S.A., 1979. Rare-earth and other trace element data bearing on the origin of Archaean granitic rocks from Yellowknife, Northwest Territories. Can. J. Earth Sci., 16: 809–815.

Durrheim, R.J. and Mooney, W.D., 1991. Archean and Proterozoic crustal evolution: evidence from crustal seismology. Geology, 19: 606–609.

England, P.C. and Thompson, A.B., 1986. Some thermal and tectonic models for crustal melting in continental collision zones. In: M.P. Coward and A.C. Ries (Eds.), Collision Tectonics. Geol. Soc. Spec. Pub. 19: 83–94.

Feng, R. and Kerrich, R., 1991. Single zircon age constraints on the tectonic juxtaposition of the Archean Abitibi greenstone belt and Pontiac subprovince, Quebec, Canada. Geochim. Cosmochim. Acta, 55: 3437–3441.

Feng, R. and Kerrich, R., 1992. Geochemical evolution of granitoids from the Archean Abitibi Southern Volcanic Zone and the Pontiac subprovince, Superior Province, Canada: Implications for tectonic history and source regions. Can. J. Earth Sci., 29: 2266–2286

Friend, C.R.L. and Nutman, A.P., 1991. SHRIMP U-Pb geochronology of the Closepet granite and Peninsular gneiss, Karnataka, South India. J. Geol. Soc. India, 38: 357–368.

Frith, R.A. and Fryer, B.J., 1985. Geochemistry and origin of the Regan Intrusive Suite and other granitoids in the northeastern Slave Province, northwest Canadian Shield. Can. J. Earth Sci., 22: 1048–1065.

Glikson, A.Y., 1976. Trace element geochemistry and origin of early Precambrian acid igneous series, Barberton Mountain Land, Transvaal. Geochim. Cosmochim. Acta, 40: 1261–1280.

Gosselin, D.C., Papike, J.J., Shearer, C.K., Peterman, Z.E. and Laul, J.C., 1990. Geochemistry and origin of Archean granites from the Black Hills, South Dakota. Can. J. Earth Sci., 27: 57–71.

Gower, C.F., Crocket, J.H. and Kabir, A., 1983. Petrogenesis of Archean granitoid plutons from the Kenora area, English River subprovince, northwest Ontario, Canada. Precambrian Res., 22: 245–270.

Green, D.C., Baadsgaard, H. and Cumming, G.L., 1968. Geochronology of the Yellowknife area, Northwest Territories, Canada. Can. J. Earth Sci., 5: 725–735.

Green, T.H. and Pearson, N.J., 1987. An experimental study of Nb and Ta partitioning between Ti-rich minerals and silicate liquids at high pressure and temperature. Geochim. Cosmochim. Acta, 51: 55–62.

Griffin, T.J., White, A.J.R. and Chappell, B.W., 1978. The Moruya Batholith and geochemical contrasts between the Moyura and Jindabyne Suites. J. Geol. Soc. Austr., 25: 235–247.

Grotzinger, J. and Royden, L, 1990. Elastic strength of the Slave craton at 1.9 Gyr and implications for the thermal evolution of the continents. Nature, 347: 64–66.

Harris, N.B.W., Pearce, J.A. and Tindle, A.G., 1986. Geochemical characteristics of collision-zone magmatism. In: M.P. Coward and A.C. Ries (Eds.), Collision Tectonics. Geol. Soc. Spec. Pub. 19: 67–81.

Harris, N.B.W., Gravestock, P. and Inger, S., 1992. Ion-microprobe determinations of trace-element concentrations in garnets from anatectic assemblages. Chem. Geol., 100: 41–49.

Henderson, J.B., van Breemen, O. and Loveridge, W.D., 1987. Some U-Pb zircon ages from Archean basement, supracrustal and intrusive rocks, Yellowknife–Hearne Lake area, District of Mackenzie. In: Radiogenic Age and Isotopic Studies: Report 1, Geol. Survey Canada, Paper 87-2: 111–121.

Hickman, A.H., 1983. The geology of the Pilbara Block and its environs. West. Australia Geol. Survey, Bull. 127: 268 pp.

Hill, R.I., Chappell, B.W. and Campbell, I.H., 1992a. Late Archaean granites of the southeastern Yilgarn Block, Western Australia: age, geochemistry, and origin. Trans. Royal Soc. Edinburgh Earth Sci., 83: 211–226.

Hill, R.I., Campbell, I.H. and Chappell, B.W., 1992b. Crustal growth, crustal reworking, and granite genesis in the southeasten Yilgarn Block, Western Australia. Univ. West. Australia, Publ. 22: 203–212.

Holtz, F. and Johannes, W., 1991. Genesis of peraluminous granites I. Experimental investigation of melt compositions at 3 and 5 kb and various H_2O activities. J. Petrol., 32: 935–958.

Hunter, D.R., 1991. Crustal processes during Archean evolution of the southeastern Kaapvaal province. J. Afr. Earth Sci ., 13: 13–26.

Huppert, H.E. and Sparks, R.S.J., 1988. The generation of granitic magmas by intrusion of basalt into continental crust. J. Petrol., 29: 599–624.

Isachsen, C.E., Bowring, S.A. and Padgham, W.A., 1991. U-Pb zircon chronology of the Yellowknife volcanic belt, NWT, Canada: new constraints on the timing and duration of greenstone belt magmatism. J. Geol., 99: 55–67.

Jackson, N.J., Walsh, J.N. and Pegram, E., 1984. Geology, geochemistry and petrogenesis of late Precambrian granitoids in the Central Hijaz Region of the Arabian Shield. Contrib. Mineral. Petrol., 87: 205–219.

Jahn, B.-M., Glikson, A.Y., Peucat, J.J. and Hickman, A.H., 1981. REE geochemistry and isotopic data of Archean silicic volcanics and granitoids from the Pilbara Block, Western Australia: implications for the early crustal evolution. Geochim. Cosmochim. Acta, 45: 1633–1652.

Jayananda, M. and Mahabaleswar, B., 1991a. Relationship between shear zones and igneous activity: The Closepet granite of southern India. Proc. Indian Acad. Sci. (Earth Planet. Sci.) 100: 31–36.

Jayananda, M. and Mahabaleswar, B., 1991b. The generation and emplacement of Closepet granite during late Archaean granulite metamorphism in Southeastern Karnataka. J. Geol. Soc. India, 38: 418–426.

Kamo, S. and Davis, D., 1991. A review of geochronology from the Barberton Mountain Land. In: L.D. Ashwal (Editor), Two Cratons and an Orogen — Excursion Guidebook and Review Articles for a Field Workshop through Selected Archaean Terranes of Swaziland, South Africa and Zimbabwe. IGCP Project 280, Dept. of Geology, Univ. Witwatersrand, Johannesburg, pp. 59–68.

Kay, R.W. and Mahlburg-Kay, S., 1991. Creation and destruction of lower continental crust. Geol. Rundschau, 80/2: 259–278.

Keppler, H. and Wyllie, P.J., 1991. Partitioning of Cu, Sn, Mo, W, U, and Th between melt and aqueous fluid in the systems haplogranite–H_2O–HCl and haplogranite–H_2O–HF. Contrib. Mineral Petrol., 109: 139–150.

King, J.E. and Helmstaedt, H., 1989. Deformation history of an Archean fold belt, eastern Point Lake area, Slave Structural Province, N.W.T. Can. J. Earth Sci., 26: 106–118.

King, J.E., Davis, W.J., Relf, C. and Van Nostrand, T., 1990. Geology of the Contwoyto–Nose lakes map area, central Slave Province, District of MacKenzie, N.W.T. In: Current Research, Part C, Geol. Survey Canada, Paper 90-1C: 177–187.

King, P.L., 1993. A-type granites from the Lachlan Fold Belt. A case study and reassessment. B.Sc. (Honours) Thesis, The Australian National University, Canberra, 80 pp. + appendices.

Kissin, S.A. and Zayachkivsky, B., 1985. Genesis of Pegmatites in the Quetico gneiss belt of northwestern Ontario - rare-element pegmatites and associated granitoids of the Georgia Lake Pegmatite Field. Ontario Geol. Surv. Misc. Paper 127: 186–199.

Koesterer, M.E., Frost, C.D., Frost, B.R., Hulsebosch, T.P., Bridgwater, D. and Worl, R.G., 1987. Development of the Archean crust in the Medina Mountain area, Wind River Range, Wyoming (U.S.A.). Precambrian Res., 37: 287–304.

Krapez, B. and Barley, M.E., 1987. Archaean strike-slip faulting and related ensialic basins: evidence from the Pilbara Block, Australia. Geol. Mag., 124: 555–567.

Kretz, R., Loop, J. and Hartree, R., 1989. Petrology and Li-Be-B geochemistry of muscovite–biotite granite and associated pegmatite near Yellowknife, Canada. Contrib. Mineral. Petrol., 102: 174–190.

Krogh, T.E. and Turek, A., 1982. Precise U-Pb zircon ages from the Gamitagama greenstone belt, southern Superior Province. Can. J. Earth Sci., 19: 859–867.

Krogstad, E.J., Balakrishnan, S., Mukhopadhyay, D.K., Rajamani, V. and Hanson, G.N., 1989. Plate tectonics 2.5 billion years ago: evidence at Kolar, South India. Science, 243: 1337–1340.

Kusky, T.M., 1989. Accretion of the Archean Slave Province. Geology, 17: 63–67.

Martin, H., 1986. Effect of steeper Archean geothermal gradient on geochemistry of subduction-zone magmas. Geology, 14: 753–756.

Martin, H., 1987. Petrogenesis of Archaean trondhjemites, tonalites and granodiorites from eastern Finland: major and trace element geochemistry. J. Petrol., 28: 921–953.

McGregor, V.R., 1979. Archean gray gneisses and the origin of the continental crust: evidence the Godthåb region, West Greenland. In: F. Barker (Ed.), Trondhjemites, Dacites and Related Rocks. Elsevier, Amsterdam, pp. 169–204.

McGregor, V.R., Bridgwater, D. and Nutman, A.P., 1983. The Qârusuk dykes: post-Nûk, pre-Qôrqut granitoid magmatism in the Godthåb region, southern West Greenland. Rapp. Grønlands geol. Unders., 112: 101–112.

McGregor, V.R., Nutman, A.P. and Friend, C.R.L., 1986. The Archean geology of the Godthabsfjord region, southern west Greenland. In: L.D. Ashwal (Ed.), Workshop on Early Crustal Genesis: The

World's Oldest Rocks. LPI Tech. Rpt. 86-04, Lunar and Planetary Institute, Houston, pp. 113–169.

Meen, J.K., Rogers, J.J.W. and Fullagar, P.D., 1992. Lead isotopic compositions of the western Dharwar Craton, southern India: Evidence for distinct Middle Archean terranes in a Late Archean craton. Geochim. Cosmochim. Acta, 56: 2455–2470.

Meyer, F.M., Robb, L.J., Reimold, W.U. and de Bruiyn, H.D., 1992. S- and I-type granites during late-stage magmatism in the Barberton Mountain Land, Southern Africa. Univ. Witwatersrand Econ. Geology Res. Unit Information Circular no. 257, 18 pp.

Miller, C.F., 1985. Are strongly peraluminous magmas derived from pelitic sedimentary sources? J. Geol., 93: 673–689.

Moore, M., Davis, D.W., Robb, L.J., Jackson, M.C. and Grobler, D.F., 1993. Archean rapakivi granite–anorthosite–rhyolite complex in the Witwatersrand basin hinterland, southern Africa. Geology, 21: 1031–1034.

Mortensen, J.K., Thorpe, R.I., Padgham, W.A., King, J.E. and Davis, W.J. 1988. U-Pb zircon ages for felsic volcanism in the Slave Province, N.W.T. In: Radiogenic age and isotopic studies: Report 2. Geol. Survey Canada, Paper 88–2: 85–95.

Mueller, P.A., Shuster, R.D., Graves, M.A., Wooden, J.L. and Bowes, D.R., 1988. Age and composition of a late Archean magmatic complex, Beartooth Mountains, Montana–Wyoming. Montana Bur. Mines and Geology, Spec. Publ., 96: 7–22.

Myers, J.S., 1976. Acid and intermediate intrusions, deformation and gneiss formation, north-east of Fiskenæsset. Rapp. Grønlands geol. Unders., 73: 7–15.

Naha, K., Srinivasan, R. and Jayaram, S., 1991. Sedimentattional, structural and migmatitic history of the Archaean Dharwar tectonic province, southern India. Proc. Indian Acad. Sci. (Earth Planet. Sci.), 100: 413–433.

Newton, R.C., 1990. Fluids and melting in the Archean deep crust of southern India. In: J. Ashworth and M. Brown (Eds.), High Temperature Metamorphism and Crustal Anatexis. Mineral. Soc. Ser. 2, pp. 149–179.

Nutman, A.P., Friend, C.R.L., Baadsgaard, H. and McGregor, V.R., 1989. Evolution and assembly of Archean gneiss terranes in the Godthabsfjord region, southern West Greenland: structural, metamorphic and isotopic evidence. Tectonics, 8: 573–589.

Nutman, A.P., Chadwick, B., Ramakrishnan, M. and Viswanatha, M.N., 1992. SHRIMP U-Pb ages of detrital zircon in Sargur Supracrustal rocks in western Karnataka, southern India. J. Geol. Soc. India, 39: 367–374.

Nutman, A.P., Friend, C.R.L., Kinny, P.D. and McGregor, V.R., 1993. Anatomy of an early Archean gneiss complex: 3900 to 3600 Ma crustal evolution in southern West Greenland. Geology, 21: 415–418.

O'Brien, C., Plant, J.A., Simpson, P.R. and Tarney, J., 1985. The geochemistry, metasomatism and petrogenesis of the granites of the English Lake District. J. Geol. Soc. London, 142: 1139–1157.

Padgham, W.A., 1985. Observations and speculations on supracrustal successions in the Slave Structural Province. In: L.D. Ayres, P.C. Thurston, K.D. Card and W. Weber (Eds.), Evolution of Archean Supracrustal Sequences. Geol. Assoc. Canada Spec. Paper 28, pp. 133–151.

Patino Douce, A.E. and Johnston, A.D., 1991. Phase equilibria and melt productivity in the pelitic system: implications for the origin of peraluminous granitoids and aluminous granulites. Contrib. Mineral. Petrol., 107: 202–218.

Pearce, J.A., Harris, N.B.W. and Tindle, A.G., 1984. Trace element discrimination diagrams for the tectonic interpretation of granitic rocks. J. Petrol., 25: 956–983.

Percival, J.A., 1989. A regional perspective of the Quetico metasedimentary belt, Superior Province, Canada. Can. J. Earth Sci., 26: 677–693.

Percival, J.A. and Sullivan, R.W., 1988. Age constraints on the evolution of the Quetico Belt, Superior Province, Ontario. Report 2, Geol. Surv. Can., Paper 88-2: 97–107.

Pidgeon, R.T., Aftalion, M. and Kalsbeek, F., 1976. The age of the Ilivertalik granite in the Fiskenæsset area. Rapp. Grønlands geol. Unders., 73: 31–33.

Pidgeon, R.T., Wilde, S.A., Compston, W. and Shield, M.W., 1990. Archaean evolution of the Wongan Hills Greenstone Belt, Yilgarn Craton, Western Australia. Australian J. Earth Sci., 37: 279–292.

Pitcher, W.S., 1987. Granites and yet more granites forty years on. Geol. Rundschau, 76/1: 51–79.

Price, R.C., Brown, M.W. and Woolard, C.A., 1983. The geology, geochemistry and origin of the late-Silurian high-Si igneous rocks of the Upper Murray Valley, NE Victoria. J. Geol. Soc. Australia, 30: 443–459.

Ridley, J.R., 1992. The thermal causes and effects of voluminous, Late Archean monzogranite plutonism. Univ. West. Australia, Publ. 22: 275–285.

Robb, L.J., 1983. Geological and chemical characteristics of late granite plutons in the Barberton Region and Swaziland with an emphasis on the Dalmein Pluton — a review. Spec. Publ. geol. Soc. S. Afr., 9: 153–167.

Robb, L.J., Meyer, F.M., Ferraz, M.F. and Drennan, G.R., 1990. The distribution of radioelements in Archaean granites of the Kaapvaal Craton, with implications for the source of uranium in the Witwatersrand Basin. S. Afr. J. Geol., 93: 5–40.

Robb, L.J., Davis, D.W. and Kamo, S.L., 1991. Chronological framework for the Witwatersrand Basin and environs: towards a time-contrained depositional model. S. Afr. J. Geol., 94: 86–95.

Robb, L.J., Meyer, F.M., Kroener, A., Trumbull, R.B., Reimold, W.U., de Bruiyn, H., walraven, F. and Toulkeridis, T., 1993. Late-stage granite plutons in the Barberton region and Swaziland: an update. Extended abstract, 16th Int. Symp. African Geology, Mbabane, Swaziland.

Rogers, J.J.W., 1988. The Arsikere granite of Southern India: magmatism and metamorphism in a previously depleted crust. Chem. Geol., 67: 155–163.

Rottura, A., Bargossi, G.M., Caironi, V., D'Amico, C. and Maccarrone, E., 1989. Petrology and geochemistry of late-hercynian granites from the Western Central System of the Iberian Massif. Eur. J. Mineral., 1: 667–683.

Rudnick, R.L. and Presper, T., 1990. Geochemistry of intermediate- to high-pressure granulites. In: D. Vielzeuf and Ph. Vidal (Eds.), Granulites and Crustal Evolution. Kluwer Academic Publishers, Netherlands, pp. 523–550.

Shaw, A.L. and Guilbert, M., 1990. Geochemistry and metallogeny of Arizona peraluminous granitoids with reference to Appalachian and European occurrences. In: H.J. Stein and J.L. Hannah (Eds.), Ore-bearing granite systems; Petrogenesis and mineralization processes. Geol. Soc. Amer. Spec. Paper 246: 317–352.

Shirey, S.B. and Hanson, G.N., 1985. Mantle heterogeneity and crustal recycling in Archean granite-greenstone belts: evidence from Nd isotopes and trace elements in the Rainy Lake area, Superior Province, Ontario, Canada. Geochim. Cosmochim. Acta, 50: 2631–2651.

Sisson, T.W., 1991. Pyroxene-high silica rhyolite trace element partition coefficients measured by ion microprobe. Geochim. Cosmochim. Acta, 55: 1575–1585.

Sisson, T.W. and Bacon, C.R., 1992. Garnet/high silica rhyolite trace element partition coefficients measured by ion microprobe. Geochim. Cosmochim. Acta, 56, 2133–2136.

Skjerlie, K.P. and Johnston, A.D., 1993. Fluid-absent melting behavior of an F-rich tonalitic gneiss at mid-crustal pressures: implications for the generation of anorogenic granites. J. Petrol., 34: 785–815.

Skjerlie, K.P., Patino Douce, A.E. and Johnston, A.D., 1993. Fluid absent melting of a layered crustal protolith: implications for the generation of anatectic granites. Contrib. Mineral. Petrol., 114:

365–378.

Smith, T.E., Huang, C.H., Riddle, C. and Choudry, A.G., 1985. The geochemistry of Archean igneous rocks and metaigneous rocks, Gamitagama area, Wawa (Shebandowan) Sub-province, Ontario. Neues Jahrbuch Miner. Abh., 151: 53–86.

Sultan, M., Batiza, R. and Sturchio, N.C., 1986. The origin of small-scale geochemical and mineralogic variations in a granite intrusion. Contrib. Mineral. Petrol., 93: 513–523.

Streckeisen, A.L., 1976. To each plutonic rock its proper name. Earth Sci. Rev., 12: 1–33.

Stuckless, J.S., 1989. Petrogenesis of two contrasting, late Archean granitoids, Wind River Range, Wyoming. USGS Prof. Paper 1491: 38 pp.

Stuckless, J.S. and Meisch, A.T., 1981. Petrogenetic modeling of a potential uranium source rock, Granite Mountains, Wyoming. USGS Prof. Paper 1225: 34 pp.

Stuckless, J.S., Miesch, A.T. and Wenner, D.B., 1986. Geochemistry and petrogenesis of an Archean Granite from the Owl Creek Mountains, Wyoming. USGS Prof. Paper 1388-A: 1–21.

Sutcliffe, R.H., Barrie, C.T., Burrows, D.R. and Beakhouse, G.P., 1993. Plutonism in the southern Abitibi subprovince: A tectonic and petrogenetic framework. Econ. Geol., 88: 1359–1375.

Sweetman, T.M., 1987. The geochemistry of the Blackstairs Unit of the Leinster Granite, Ireland. J. Geol. Soc. London, 144: 971–984.

Sylvester, P.J., 1989. Post-collisional alkaline granites. J. Geol., 97: 261–280.

Sylvester, P.J., Attoh, K. and Schulz, K.J., 1987. Tectonic setting of Late Archean bimodal volcanism in the Michipicoten (Wawa) greenstone belt, Ontario. Can. J. Earth Sci., 24: 1120–1134.

Tarney, J., Weaver, B. and Drury, S.A., 1979. Geochemistry of Archean trondhjemitic and tonalitic gneisses from Scotland and East Greenland. In: F. Barker (Ed.), Trondhjemites, Dacites and Related Rocks. Elsevier, Amsterdam, pp. 275–299.

Taylor, P.N., Chadwick, B., Moorbath, S., Ramakrishnan, M. and Viswanatha, M.N., 1984. Petrography, chemistry and isotopic ages of Peninsular gneiss, Dharwar acid volcanic rocks and the Chitradurga granite with special reference to the late Archaean evolution of the Karnataka craton, Southern India. Precambrian Res., 23: 349–375.

Taylor, S.R. and McLennan, S.M., 1985. The Continental Crust: Its Composition and Evolution. Blackwell, Oxford, 312 pp.

Thompson, R.N., Morrison, M.A., Hendry, G.L. and Parry, S.J., 1984. An assessment of the relative roles of crust and mantle in magma genesis: an elementary approach. Philos. Trans. R. Soc. London, 310: 549–590.

Tindle, A.G. and Pearce, J.A., 1981. Petrogenetic modelling of in situ fractional crystallization in the zoned Loch Doon Pluton, Scotland. Contrib. Mineral. Petrol., 78: 196–207.

Tyler, I.M., 1991. The geology of the Sylvania Inlier and the Southeast Hamersley Basin Geol. Surv. West. Aust. Bull. 138: 120 pp.

Tyler, I.M., Fletcher, I.R., de Laeter, J.R., Williams, I.R. and Libby, W.G., 1992. Isotope and rare earth element evidence for a late Archaean terrane boundary in the southeastern Pilbara Craton, Western Australia. Precambrian Res., 54: 211–229.

van Breeman, O. and Henderson, J.B., 1988. U-Pb zircon and monazite ages from the eastern Slave Province and Thelon Tectonic Zone, Artillery Lake area, N.W.T. In: Radiogenic age and isotopic studies. Report 2, Geol. Survey Canada, Paper 88–2: 73–83.

van Breeman, O., Henderson, J.B., Sullivan, R.W. and Thompson, P.H., 1987. U-Pb, zircon and monazite ages from the eastern Slave Province, Healy Lake area, N.W.T. In: Radiogenic age and isotopic studies. Report 1, Geol. Survey Canada, Paper 87-2: 101–110.

van Breeman, O., King, J.E. and Davis, W.J., 1989. U-Pb zircon and monazite ages from plutonic rocks in the Contwoyto–Nose lakes map area, central Slave Province, District of MacKenzie, Northwest Territories. In: Radiogenic age and isotopic studies. Report 3, Geol. Survey Canada,

Paper 89-2: 29–37.

Watkins, K.P. and Hickman, A.H., 1990. Geological evolution and mineralization of the Murchison Province, Western Australia. Geol. Surv. West. Aust. Bull. 137.

Watkins, K.P., Fletcher, I.R. and de Laeter, J.R., 1991. Crustal evolution of Archaean granitoids in the Murchison Province, Western Australia. Precambrian Res., 50: 311–336.

Webster, J.D. and Holloway, J.R., 1990. Partitioning of F and Cl between magmatic hydrothermal fluids and highly evolved granitic magmas. In: H.J. Stein and J.L. Hannah (Eds.), Ore-bearing granite systems; Petrogenesis and mineralizing processes. Geological Society of America Special Paper 246: 21–34.

Wells, P.R.A., 1979. Chemical and thermal evolution of Archaean sialic crust, southern West Greenland. J. Petrol., 20: 187–226.

Whalen, J.B., Currie, K.L. and Chappell, B.W., 1987, A-type granites: geochemical characteristics, discrimination and petrogenesis. Contrib. Mineral. Petrol., 95: 407–419.

White, A.J.R., Clemens, J.D., Holloway, J.R., Silver, L.T., Chappell, B.W. and Wall, V.J., 1986. S-type granites and their probable absence in southwestern North America. Geology, 14: 115–118.

Wickham, S.M., 1987. Crustal anatexis and granite petrogenesis during low-pressure regional metamorphism: the Trois Seigneurs Massif, Pyrenees, France. J. Petrol., 28: 127–169.

Wiedenbeck, M. and Watkins, K.P., 1993. A time scale for granitoid emplacement in the Archean Murchison Province, Western Australia, by single zircon geochronology. Precambrian Res., 61: 1–26.

Wilde, S.A. and Pidgeon, R.T., 1986. Geology and geochronology of the Saddleback Greenstone Belt in the Archaean Yilgarn Bock, southwestern Australia. Austr. J. Earth Sci., 33: 491–501.

Williams, H.R., 1990. Subprovince accretion tectonics in the south-central Superior Province. Can. J. Earth Sci., 27: 570–581.

Williams, I.R., 1989. Explanatory notes on the Balfour Downs 1:250,000 geological sheet, Western Australia (2nd edition). Geol. Surv. West. Aust. Record 1989/2.

Winther, K. T. and Newton, R. C., 1991. Experimental melting of hydrous low-K tholeiite: evidence on the origin of Archean cratons. Bull. Geol. Soc. Denmark, 39: 213–228.

Witt, W.K. and Swager, C.P., 1989. Structural setting and geochemistry of the Archaean I-type granitoids in the Bardoc–Coolgardie area of the Eastern Goldfields Province, Western Australia. Precambrian Res., 44: 323–351.

Wooden, J.L., Mueller, P.A., Hunt, D.K. and Bowes, D.R., 1982. Geochemistry and Rb-Sr geochronology of Archean Rocks from the interior of the southeastern Beartooth Mountains, Montana and Wyoming. Montana Bur. Mines and Geology, Spec.Publ. 84: 45–55.

Wormald, R.J. and Price, R.C., 1988. Peralkaline granites near Temora, southern New South Wales: tectonic and petrological implications. Austr. J. Earth Sci., 35: 209–221.

Wyborn, D., Turner, B.S. and Chappell, B.W., 1987. The Boggy Plain Supersuite: a distinctive belt of I-type igneous rocks of potential economic significance in the Lachlan Fold Belt. Austr. J. Earth Sci., 34: 21–34.

Wyborn, L.A.I., Wyborn, D., Warren, R.G. and Drummond, B.J., 1992. Proterozoic granite types in Australia: implications for lower crust composition, structure and evolution. Trans. Royal Soc. Edinburgh Earth Sci., 83: 201–209.

Zen, E., 1988. Thermal modelling of stepwise anatexis in a thrust-thickened sialic crust. Trans. R. Soc. Edinburgh Earth Sci., 79: 223–235.

Zorpi, M.J., Coulon, C. and Orsini, J.B., 1991. Hybridization between felsic and mafic magmas in calc-alkaline granitoids — a case study in northern Sardinia, Italy. Chem. Geol., 92: 45–86.

Chapter 8

ARCHEAN ANORTHOSITES

L.D. ASHWAL and J.S. MYERS

INTRODUCTION

Anorthosite forms a minor but prominent component of Archean crust (Fig. 1), and can be found associated with mafic intrusives and extrusives of many (but not all) greenstone belts. These anorthosites are characterized by megacrysts of equidimensional calcic plagioclase (usually $An_{80\pm5}$), and form extensive sheet-like bodies, mostly emplaced at shallow depths into basaltic volcanic rocks. Original thicknesses of the anorthositic sheets are difficult to determine because of deformation. Archean anorthosites were derived from fractionated basaltic magmas (Fe-rich tholeiites), and related basaltic rocks with calcic plagioclase megacrysts also form dikes and flows.

The major anorthosite intrusions comprise layered components of anorthosite, leucogabbro, gabbro, melanogabbro and ultramafic rocks. They contain rocks, textures and structures similar to those of younger layered basic intrusions, but are generally distinguished by the abundance of anorthosite and leucogabbro over gabbro and ultramafic rocks, and the equidimensional megacrysts of calcic plagioclase. They differ in overall structure and composition from Proterozoic massif-type anorthosites, which commonly form large steep-sided plutons and are characterized by tabular crystals of intermediate plagioclase. The calcic megacrystic anorthosites, therefore, are distinct from other anorthosite types. Their almost complete restriction to the Archean is comparable to the temporality of komatiitic magmatism, to which they may be related.

Most Archean anorthosites have been deformed and metamorphosed. All stages of deformation can be seen from undeformed rocks to intensely deformed rocks in which igneous layering is streaked out into thin discontinuous layers, and igneous textures are completely replaced by metamorphic minerals with either tectonic or granoblastic, post-tectonic, fabrics. Metamorphic grades range from low greenschist to granulite facies.

In this paper we summarize the features of Archean anorthosites, provide descriptions of selected occurrences, and speculate on the possible origin of these anorthosites. Interested readers are referred to previous review papers by Windley (1973), Windley et al. (1981), Leelanandam (1987), and Phinney et al. (1988a).

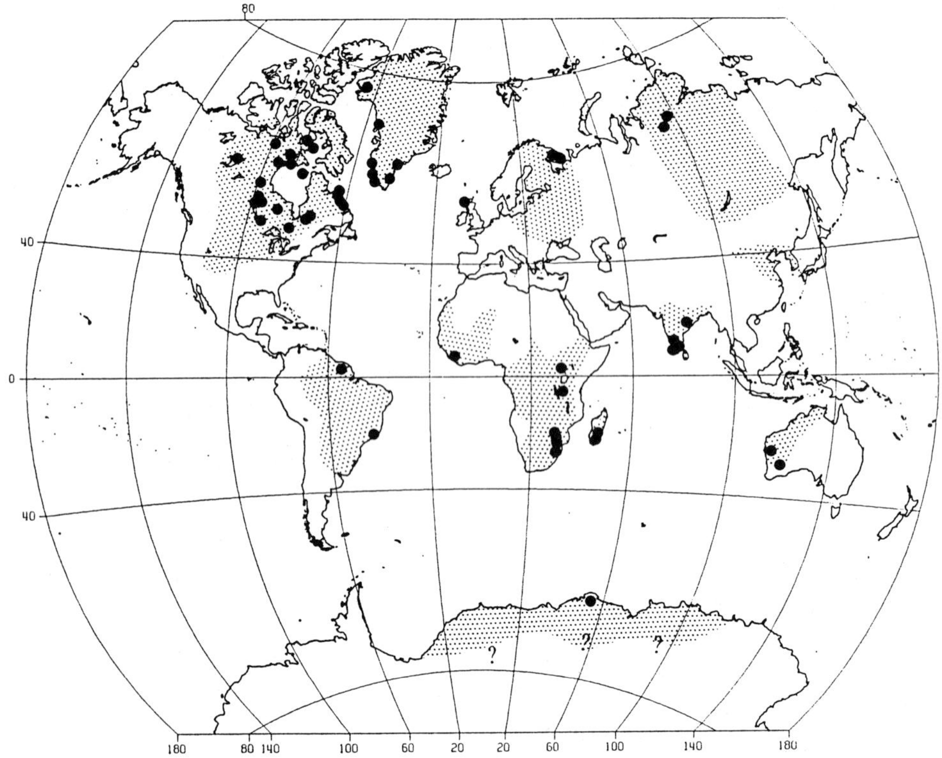

Fig. 1. World map (van der Grinten projection) showing locations of known Archean anorthosites (black dots) and areas known or suspected to be underlain by Archean rocks and reworked equivalents (shaded). Base map modified from Condie (1981) and de Wit et al. (1988).

FIELD RELATIONS

Most primary contact relations between anorthosite complexes and their original host rocks are obscured by the subsequent intrusion of sheets of granitoid rocks along the contacts, and by deformation. Primary relations are seen in the Bad Vermilion Lake complex (Ontario) where the anorthosite intruded mafic volcanic rocks of the Wabigoon greenstone belt (Wood et al., 1980), and the upper contact of the Fiskenæsset complex (West Greenland) where amphibolite with deformed pillow lava structure and metasedimentary rocks are preserved. At the latter locality the host rocks also occur as rafts within the uppermost part of the anorthosite complex (Myers, 1985).

Although most Archean anorthosite occurrences are associated directly or indirectly with mafic volcanics (or metamorphosed equivalents), the Sittampundi (India) and Messina (southern Africa) complexes are both associated with marbles, quartzites and schists, possibly implying a geologic setting different from the majority of complexes of "greenstone belt" affinity (e.g. Phinney et al., 1988a,b). However, inasmuch as both complexes are highly deformed and metamorphosed to granulite facies, the possibility must be considered that these occurrences have been tectonically intercalated with associated supracrustals rather than originally emplaced into them. The apparent absence of low-grade occurrences with similar association may be significant.

MAGMATIC TEXTURES AND STRUCTURES

Large plagioclase megacrysts of equant, euhedral to subhedral, calcic plagioclase surrounded by a mafic matrix are the most distinctive feature of Archean

Fig. 2. Undeformed, partly recrystallized equant igneous plagioclase of the Upper Leucogabbro Unit at Majorqap qâva, Fiskenæsset complex, West Greenland. Note the large plagioclase cluster with radial structure to the right of the hammer, possibly indicating rapid growth (Myers, 1985).

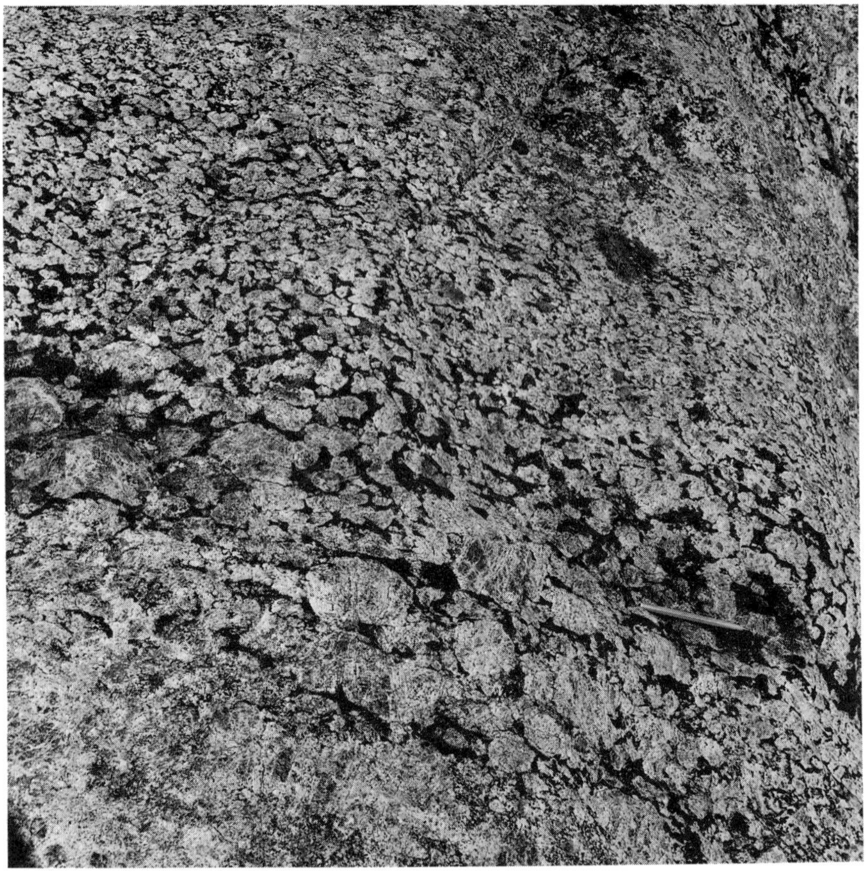

Fig. 3. Undeformed, partly recrystallized, size-graded cumulus plagioclase in leucogabbro of the Lower Leucogabbro Unit at Majorqap qâva, Fiskenæsset complex, West Greenland. Original way up is towards the top right. The pen is 13 cm (Myers, 1981).

anorthosites (Figs. 2 and 3; see also Figs. 10, 12, 13, 18) (Phinney, 1982; Phinney et al., 1981a,b, 1986a,b). This texture has variously been referred to as "football or baseball anorthosite", "leopard rock", "cat rock", and "ovoid, blotchy, mottled or glomeroporphyritic gabbro".

The megacrysts range in size from 0.5 to 30 cm in diameter, but most are 1–5 cm across. The anorthosites may have uni- or polymodal size distributions of megacrysts. The matrix may be either fine or coarse grained, and is generally basaltic or gabbroic in composition. In some occurrences many megacrysts are enclosed by large, continuous amphiboles, and in a few cases pyroxene remnants within this matrix suggest that the amphibole was derived from igneous pyroxene. The matrix is extensively preserved as pyroxene in the Windimurra

Fig. 4. Mineral-graded layers within a large trough layer in the Middle Gabbro Unit at Majorqap qâva, Fiskenæsset complex, West Greenland. The hornblende-rich layer immediately above the hammer forms a slump structure that is truncated by the overlying mineral-graded layer. The rocks now consists of metamorphic plagioclase and green hornblende.

Intrusion in Australia, and forms giant crystals poikilitically enclosing the plagioclase megacrysts.

In the Bad Vermilion Lake and Fiskenæsset complexes the megacrysts locally define size-graded layering (Fig. 3) (Ashwal et al., 1983; Morrison et al., 1987, 1988; Myers, 1985). A variety of other magmatic structures occur in the Fiskenæsset complex, including mineral-graded layering (Fig. 4), trough layers (Fig. 5), cross-bedding (Fig. 6), slump structures (Fig. 4; see also Fig. 22), snowflake structures (Fig. 7) and pipe-like intrusions (Fig. 8) (Myers, 1976, 1978, 1985).

Fig. 5. Peridotite trough layer (center) in leucogabbro, sub-unit 5.1 of the Upper Leucogabbro Unit (Fiskenæsset complex, West Greenland), overlain by slumped blocks of hornblende chromitite. The leucogabbro now consists entirely of plagioclase and green hornblende. Note the 4 cm thick dark hornblende rim around the peridotite. Original way up is towards the right (Myers, 1985).

Fig. 6. Cross-bedded mineral-graded layers in gabbro of the Middle Gabbro Unit at Majorqap qâva, Fiskenæsset complex, West Greenland. Original way up is towards the right (Myers, 1985).

Fig. 7. "Snowflake" structures, clusters of both large equant plagioclase crystals in radial chains and large radial wedge-shaped crystals, in gabbro sub-unit 4.3 of the Middle Gabbro Unit at Majorqap qâva, Fiskenæsset complex, West Greenland. Original way up is towards top of photograph (Myers, 1985).

Fig. 8. Pipe of hornblende pegmatite (P) cutting sharply across mineral-graded layering in gabbro of the Middle Gabbro Unit at Majorqap qâva, Fiskenæsset complex, West Greenland. Note displaced blocks of leucogabbro (L) within the pipe (Myers, 1985).

Fig. 9. Intrusive sheets of tonalite disrupting the stratigraphy of the Fiskenæsset complex, West Greenland, seen on a 300 m high cliff, 40 km east of Fiskenæsset. The upper part of the cliff marked ULG consists of leucogabbro of the Upper Leucogabbro Unit, with compositional igneous layering. The middle part of the cliff comprises black layers of gabbro and ultramafic rocks of the Middle Gabbro Unit, disrupted by intrusive sheets of tonalite. The lower part of the cliff consists of leucogabbro and gabbro of the Lower Leucogabbro Unit (LLG) and strongly deformed tonalite gneiss (GN).

TECTONIC FABRICS AND METAMORPHIC TEXTURES

Many Archean anorthosites have been disrupted by thrusting and/or the intrusion of granitoids (Fig. 9), and have been repeatedly folded and metamorphosed. Deformation was generally heterogeneous at both large and small scales, and various stages of progressive deformation and recrystallization can be seen in individual complexes (Figs. 10 and 11). Many outcrop patterns reflect complex fold structures. Large scale examples can be seen in maps of the anorthosite complexes of Messina (Barton et al., 1979), Sittampundi (Ramadurai et al., 1975), Shawmere (Percival, 1981) and Fiskenæsset (see Figs. 19 and 21) (Myers, 1985).

PETROLOGY AND GEOCHEMISTRY

No Archean anorthosites are known to have survived in a pristine state. Most were at least partly recrystallized during metamorphic episodes, and many were completely recrystallized. The most abundant primary minerals that survive are

Fig. 10. Shear zone with pronounced schistosity developing from partly deformed relict igneous plagioclase with chromite and hornblende in a 1-m thick leucogabbro-chromitite layer in anorthosite of the Anorthosite Unit at Majorqap qâva, Fiskenæsset complex, West Greenland (Myers, 1981).

plagioclase, followed by spinel, chromite, magnetite, and in a few cases pyroxene and olivine.

Most igneous plagioclase has a composition of $An_{80\pm5}$ (Fig. 12). A substantial amount of metamorphic plagioclase also has a similar composition, but much shows significant departures from igneous compositions and ranges below An_{75} and above An_{90}. Compositions more sodic than about An_{75} can be attributed to removal of An component by epidote group minerals or calcite during low grade metamorphism. Very calcic plagioclase (An_{90-100}) is far more common in high grade occurrences (e.g. Fiskenæsset, Shawmere, Sittampundi); in these cases Ab component may have been removed by partial melts. Most igneous crystals do not show any marked zonation, even in large megacrysts, and there is little cryptic variation within the layered complexes.

Igneous plagioclase crystals generally contain abundant inclusions that impart a pale green color. The most abundant inclusions are acicular amphibole, with lesser amounts of pyroxenes and rutile. During metamorphic recrystallization, large igneous plagioclase crystals were replaced by granoblastic aggregates of clear white grains that purged themselves of inclusions (Fig. 13). Many low and high grade anorthosites have been affected by late hydrothermal alteration which converted both igneous and metamorphic plagioclase to epidote group minerals

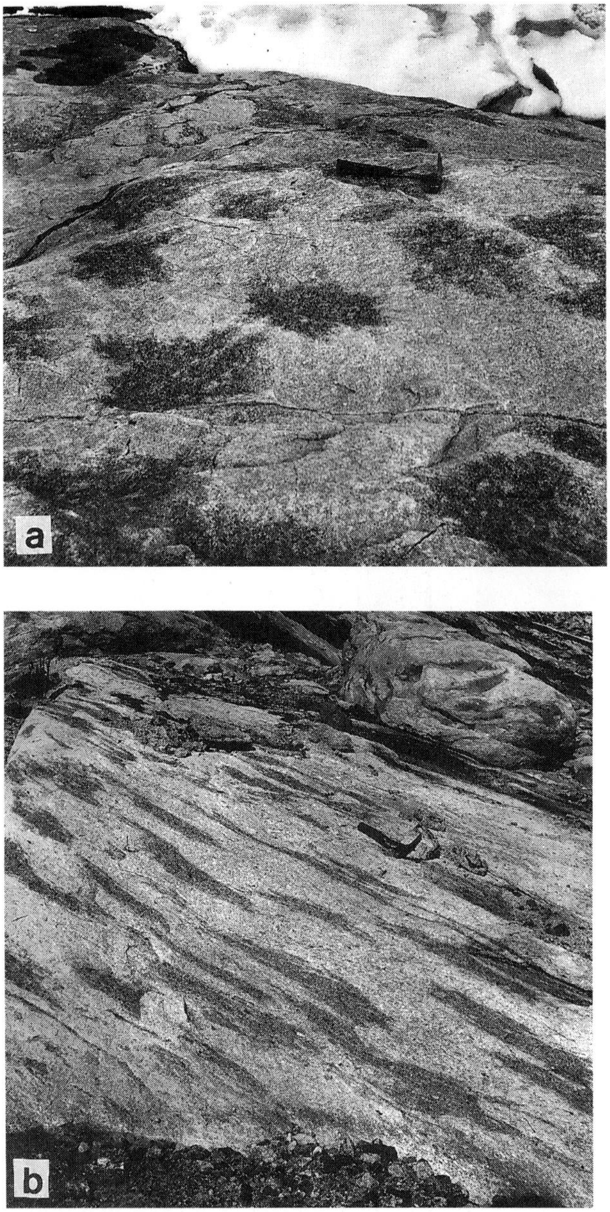

Fig. 11. (A) Undeformed, recrystallized sub-spherical patches of leucogabbro in anorthosite, typical of the least deformed anorthosite of the Anorthosite Unit at Majorqap qâva, Fiskenæsset complex, West Greenland. (B) Recrystallized, deformed anorthosite with schlieren of leucogabbro derived from sub-spherical patches, in the Anorthosite Unit at Majorqap qâva, Fiskenæsset complex, West Greenland. *Opposite, top*: (C) Recrystallized, strongly deformed anorthosite with leucogabbro patches streaked out to form a tectonic banding in the Anorthosite Unit at Majorqap qâva, Fiskenæsset complex, West Greenland (Myers, 1985).

Fig. 11. (C) Caption opposite.

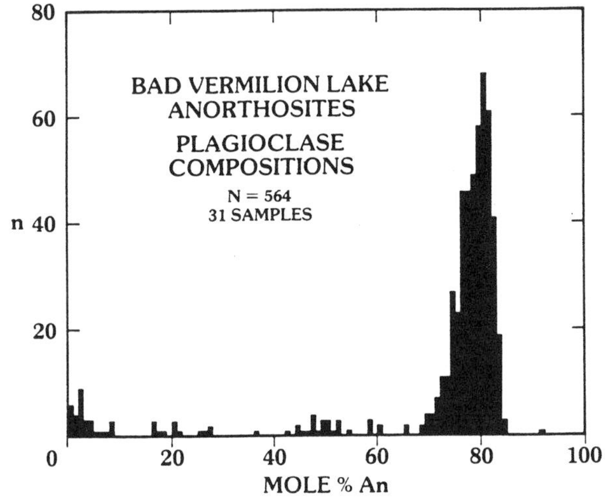

Fig. 12. Histogram of plagioclase compositions determined by microprobe for anorthositic rocks of the Bad Vermilion Lake anorthosite complex, Ontario. The large peak at about An_{80} probably represents primary compositions; more sodic compositions result from low-grade metamorphic effects (Ashwal et al., 1983).

Fig. 13. Photomicrographs (crossed polarizers) of anorthosites from the Majorqap qâva area, Fiskenæsset complex, West Greenland. Top: Large, relict plagioclase megacryst (An_{85}) surrounded by recrystallized, granoblastic plagioclase (also An_{85}). Sample GGU 159415. Long dimension of photograph is 3 cm. Bottom: Granoblastic texture with well developed 120° grain boundary intersections. Sample GGU 125632. Long dimension of photograph is 2.6 cm.

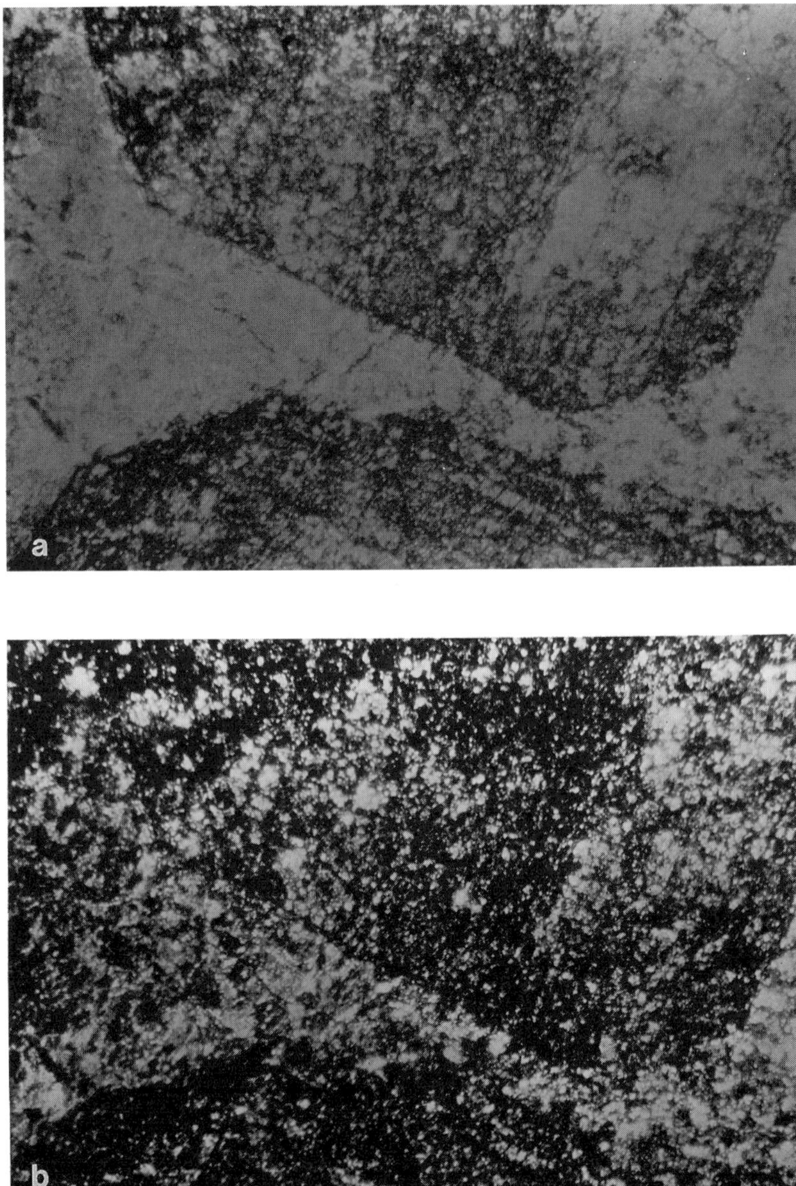

Fig. 14. Photomicrographs of megacrystic anorthosite from the Bad Vermilion Lake anorthosite complex, Ontario. Euhedral outlines of primary plagioclase crystals have been preserved despite having been completely altered to epidote group minerals. Interstitial mafic matrix is now composed of chlorite, amphibole, albite, calcite, and quartz. Plane polarized light (a), crossed polarizers (b). Long dimension of photographs is 2.5 cm.

(Fig. 14). This is a prominent feature in the complexes at Dore Lake, Quebec (Allard, 1970), Bad Vermilion Lake, Ontario (Ashwal et al., 1983) and Manfred, Western Australia (Myers, 1988).

Metamorphic amphibole is the main mafic mineral of most Archean anorthosites. In a few cases it encloses remnants of igneous pyroxene from which it seems to have been derived. This, and other evidence (see Ashwal, 1993, pp. 33–34) argues against the suggestion (e.g. Windley et al., 1973) that amphibole represents a primary igneous phase. The amphiboles in low grade rocks are generally actinolite or actinolitic hornblende in the classification scheme of Leake (1978). At high grades, pargasitic and/or tschermakitic hornblende is predominant (Windley et al., 1981).

Bulk whole-rock geochemistry of anorthosites (Table 1) with well preserved igneous stratigraphy indicates upward iron enrichment trends (Windley et al., 1973). Similar trends were found in associated metavolcanic rocks and were considered to reflect igneous melt fractionation in which the bulk composition was controlled by the precipitation of cumulus plagioclase and pyroxene (Rivalenti, 1976; Weaver et al., 1981; Ashwal et al., 1983).

TABLE 1

Modal and whole rock chemical compositions of typical rock types from the Fiskenæsset Complex

GGU sample	Anorthosite 125763	Leucogabbro 159415	Leucogabbro 159452	Leucogabbro 125632	Gabbro 159448	Peridotite 151012
Plagioclase	98	85	85	85	50	
Hornblende	1	15	15	15	45	30
Biotite	1					
Olivine						60
Pyroxene					5	4
Magnetite						4
Spinel						2
SiO_2	48.97	46.39	47.30	44.95	48.44	39.55
TiO_2	0.12	0.02	0.11	0.03	0.20	0.14
Al_2O_3	31.41	33.44	29.11	31.61	15.82	2.68
Fe_2O_3	0.25	0.15	0.65	0.32	1.55	3.92
FeO	0.75	0.46	2.09	1.43	5.65	16.89
MnO	0.03	0.00	0.05	0.01	0.16	0.39
MgO	0.90	0.71	2.70	2.27	10.20	32.60
CaO	14.63	16.69	14.57	16.64	13.80	2.67
Na_2O	1.79	1.17	1.78	0.80	0.95	0.27
K_2O	0.46	0.14	0.19	0.14	0.24	0.16
P_2O_5	0.10	0.10	0.10	0.01	0.11	0.09
H_2O	0.44	0.56	0.64	0.90	1.32	0.56
	99.85	99.83	99.29	99.11	98.44	99.92

GGU sample	Anorthosite 125763	Leucogabbro 159415	Leucogabbro 159452	Leucogabbro 125632	Gabbro 159448	Peridotite 151012
Li	16	15	44		52	20
Ba	33	38	19	34	26	22
Rb	26	14	4.3	12	2.1	14
Sr	83	128	82	160	30	0.8
Pb	5	1	1	4	1	1
Th	3	1	1	1	1	1
U	0.53	0.06	0.73	0.2	0.16	0.02
Zr	6		3	0.27	1	6
Nb	0.05	1	1	1	1	1
Mo	2		2	2	2	2
Sn	0.05	0.05	10.5	0.5	0.5	1
Y	2	1.9	4	1	5	4
La	11	5	3	1	2	10
Ce	18	2	2	3	6	6
Nd	6	2	2	2	2	2
Sc	3	5	10	2	51	26
V	28	20	48	20	171	56
Cr	232	35	55	37	1490	312
Co	13	16	24	23	51	174
Ni	25	18	100	76	273	1370
Cu	17	27	11	28	27	107
Zn	8		12	1	70	93
Ga	21	26	19	18	13	1
Ge	0.9	1.1	1.2	0.9	1.8	1.7
Pd (ppb)						7.4
Pt (ppb)						7.8
Rh (ppb)						0.2
S	190	70	10	70	60	2400
Cl	25	26	90	150	180	180

Modal and major element analyses in percentages; major elements by XRF, I. Sorensen, analyst. Trace elements in ppm by XRF, J.C. Bailey, analyst; except for U by DNA, R. Gwozdz, analyst. Pd, Pt, Rh in ppb from Page et al. (1980). Analyses were selected from over 150 whole rocks analysed.

Anorthosites typically show light REE enrichment, strong positive Eu anomalies, and flat or slightly depleted heavy REE abundances (Fig. 15). They have a range in REE abundances but these are generally less than 10× chondrites.

Phinney and Morrison (1990) used a multiple aliquot technique to overcome effects of alteration and contamination. By careful analysis of megacrysts and matrices from basaltic dikes, sills and flows from the Superior Province, they were able to determine plagioclase/liquid partition coefficients (K_d) for REE and several other trace elements, including Co, Sc and Ta. When applied to megacrysts from the Bad Vermilion Lake Complex, these partition coefficients predict tholeiitic equilibrium melts with about 13 wt% FeO_t, 0.15 wt% K_2O, 1.5 wt% TiO_2, and REE

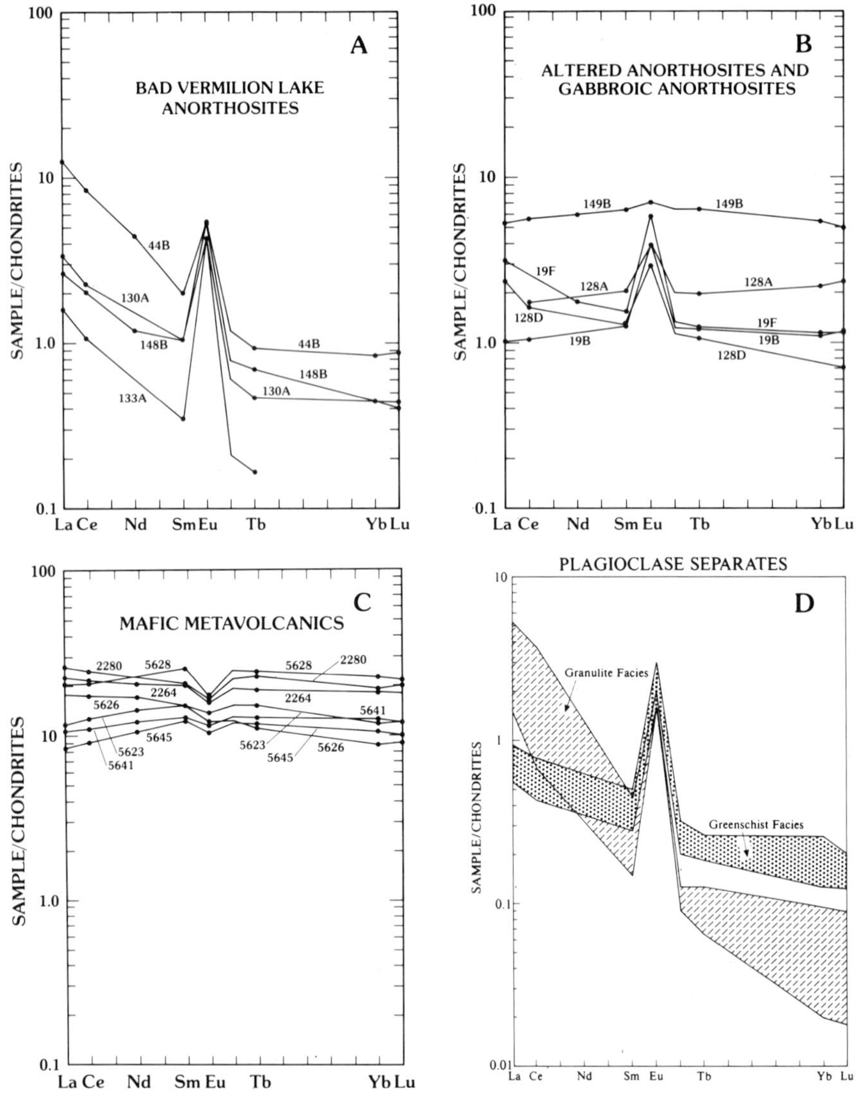

Fig. 15. Chondrite-normalized REE patterns for Archean anorthositic rocks (A, B) and associated mafic metavolcanics (C) from Bad Vermilion Lake, Ontario (Ashwal et al., 1983). D: Comparison of REE patterns in plagioclase separates from granulite vs. greenschist grade Archean anorthosite occurrences (after Phinney and Morrison, 1990).

patterns with 10–20× chondrites, showing slight LREE depletion. These results indicate that such moderately Fe-rich tholeiites may be parental to other similar Archean anorthosite complexes. This evidence supports the previous suggestion of Ashwal et al. (1983) that the similarity in the REE patterns of the mafic matrices

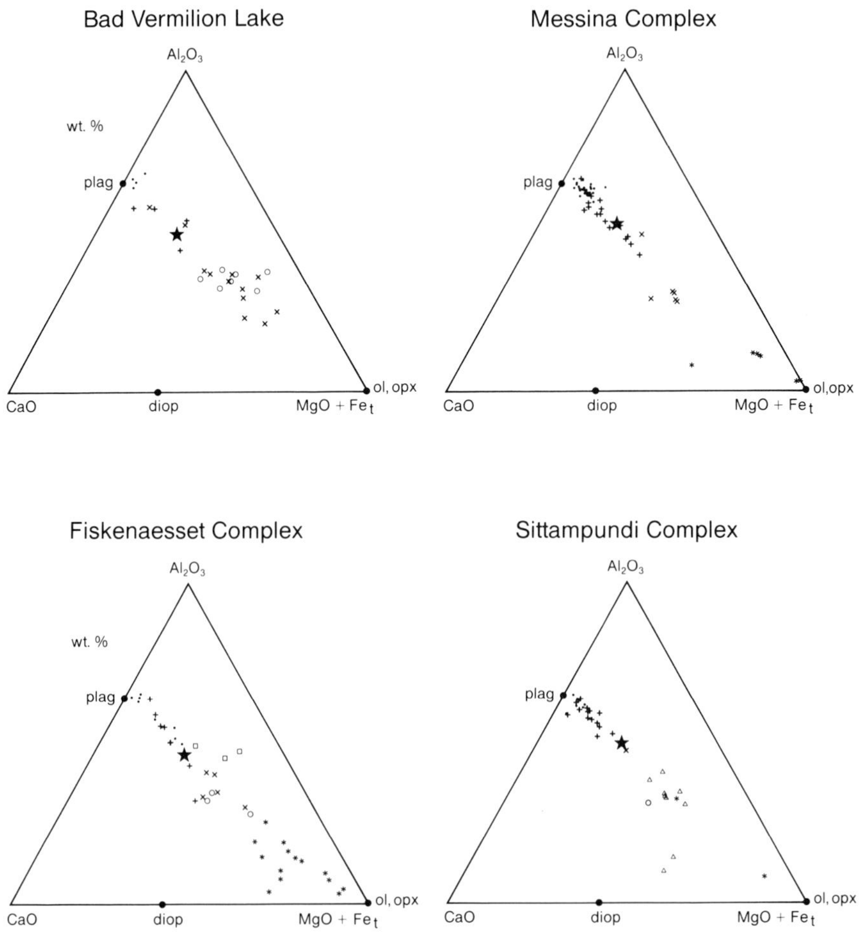

Fig. 16. ACF diagrams showing variations of whole-rock compositions of major lithologies for four Archean anorthosite complexes. Data sources: Bad Vermilion Lake (Ashwal et al., 1983); Messina (Hor et al., 1975; Barton et al., 1979); Fiskenæsset (Windley, 1969a, Windley et al., 1973; Henderson et al., 1976; Weaver et al., 1981; Myers, 1985); Sittampundi (Subramaniam, 1956; Janardhan and Leake, 1975). Symbols: • = anorthosite; + = leucogabbro; × = gabbro; ∗ = ultramafic; o = amphibolite; open square = chromitite; open triangle = garnet granulite; solid star = bulk composition for each complex (Ashwal, 1993, Fig. 2.13).

of the plagioclase megacrysts and associated metabasalts may indicate a genetic connection. They concluded that the Bad Vermilion Lake complex may represent a cumulate body derived from a subvolcanic intrusion that fed mafic lavas during its crystallization. Some associated metabasalts at this locality and at Fiskenæsset contain plagioclase megacrysts similar to those in the adjacent anorthosite complexes.

Average bulk compositions of well exposed anorthosite complexes are remarkably similar, with high Al_2O_3 (23–25 wt%) and CaO (13–14 wt%) abundances

(Fig. 16, Table 1). Bulk trace element abundances show flat REE patterns (Fig. 15) with positive Eu anomalies and low abundance levels (1–10× chondrites) (Phinney et al., 1981a). The average bulk composition is that of leucogabbro. These complexes are therefore much more plagioclase-rich (69–74% normative plagioclase) than typical basaltic melt compositions, and should not be considered analogous to anorthosite-bearing layered mafic intrusions, which have an overall basaltic composition (Ashwal, 1993).

AGES AND ISOTOPIC COMPOSITIONS

Direct ages of Archean anorthosites have been obtained for a few occurrences, using a variety of techniques (Table 2), but additional isotopic work on these and other complexes would be most welcome.

Mafic matrix contains sufficiently higher Sm/Nd and Rb/Sr than coexisting plagioclase such that isochrons can be attempted. Sm-Nd whole rock or mixed

TABLE 2

Compilation of available isotopic data for Archean anorthosites

	Age (Ma)	I_{Sr}	ε_{Nd}	μ_1	References
Fiskenæsset, West Greenland	2860±50 (Sm-Nd)	0.70108–0.70162	+2.9±0.4	4–8	Ashwal et al (1989); Gancarz (1976); Black et al. (1973); Taylor et al. (1980)
Bad Vermilion Lake, Ontario	2747±58 (Sm-Nd)	0.70068–0.70117	+2.0±1.4		Ashwal et al. (1985)
Shawmere, Ontario	>2765*	0.70061–0.70130			Simmons et al. (1980); Percival and Krogh (1983)
Dore Lake, Quebec	>2717±2*	0.70003–0.70096			Jones et al. (1974); Krogh (1982)
Messina, South Africa	3153±47 (Rb–Sr)	0.70308–0.70364			Barton et al. (1979)
Rooiwater, South Africa	>2646*	0.70005–0.70031		7.34±0.27	Vearncombe et al. (1988)
Manfred, Western Australia	3730±6 (U-Pb)	0.69729–0.70096	−0.2±0.7	10.23±0.35	Kinney et al. (1988); Myers (1988); Fletcher et al. (1988)
Windimurra, Western Australia	>2669**	0.70091–0.70122			Ahmat and De Laeter (1982)
Holenarasipur, India	3095±58 (Rb-Sr)	0.7015–0.7017			Kutty et al (1984)

*From U–Pb zircon age of crosscutting granitoid.

**From Rb–Sr age of crosscutting granitoid.

whole rock–mineral isochrons have been obtained for Bad Vermilion Lake (Ashwal et al., 1985), Fiskenæsset (Ashwal et al., 1989), and Manfred (Fletcher et al., 1988). These appear to record primary ages, although some visibly altered or metamorphosed samples show disturbance of the Sm-Nd isotopic system.

Some anorthosite complexes show unusually large ranges in Rb/Sr that cannot be accounted for by igneous fractionation. The concordance of Rb-Sr and Sm-Nd ages at Bad Vermilion Lake suggests possible Rb introduction by hydrothermal fluids or low grade metamorphism soon after primary crystallization (Ashwal et al., 1985). Similar conclusions were reached for Messina (Barton et al., 1979) and Holenarasipur (Kutty et al., 1984).

Plagioclases in anorthositic rocks of the Manfred complex contain abundant euhedral zircons which have been analysed by ion probe, yielding an age of 3730 $\pm$6 Ma (Kinny et al., 1988). This age is equivalent to that obtained from the Sm-Nd (3680 $\pm$ 70 Ma) and Pb-Pb (3689 $\pm$ 146 Ma) whole-rock methods (Fletcher et al., 1988), although the Rb-Sr whole-rock and Sm-Nd internal isotopic systems have been disturbed by younger metamorphic events.

Initial isotopic ratios of Archean anorthosites are summarized in Fig. 17. Fiskenæsset and Bad Vermilion Lake both show positive epsilon Nd values, indicating derivation from depleted mantle. The Manfred complex, however, with $\varepsilon_{Nd} = 0.2 \pm 0.7$ was either derived from chondritic mantle or was contaminated with early Archean crustal components, a conclusion supported by μ_1 values > 10 (Table 2).

Initial Sr ratios of Archean anorthosites cluster near the bulk Earth Sr evolution line (Fig. 17), but some complexes show significant departures. Rooiwater and Dore Lake show low I_{Sr}, indicating possible derivation from depleted mantle. Fiskenæsset, Holenarasipur, and Messina all show elevated I_{Sr}, suggesting either derivation from enriched sources, contamination with older crust, or disturbance of the Rb-Sr isotopic system during amphibolite or granulite grade metamorphism.

DESCRIPTIONS OF SELECTED OCCURRENCES OF ARCHEAN ANORTHOSITES

Megacrystic anorthosites are best known from the Archean granite–greenstone terranes of the Superior Province of Canada and the Archean high-grade gneiss complexes of Greenland, Canada, India and South Africa (Fig. 1). They range in age from some of the oldest known terrestrial rocks in Labrador, Greenland and Western Australia at 3.8–3.7 Ga, to the late Archean at 2.7–2.6 Ga. Some typical examples are summarized below. They are described in more detail in Ashwal (1993) and in the references cited below.

For ease of description, anorthosite is generally used in a broad sense to include both leucogabbro with 10–30% mafic minerals as well as anorthosite with less

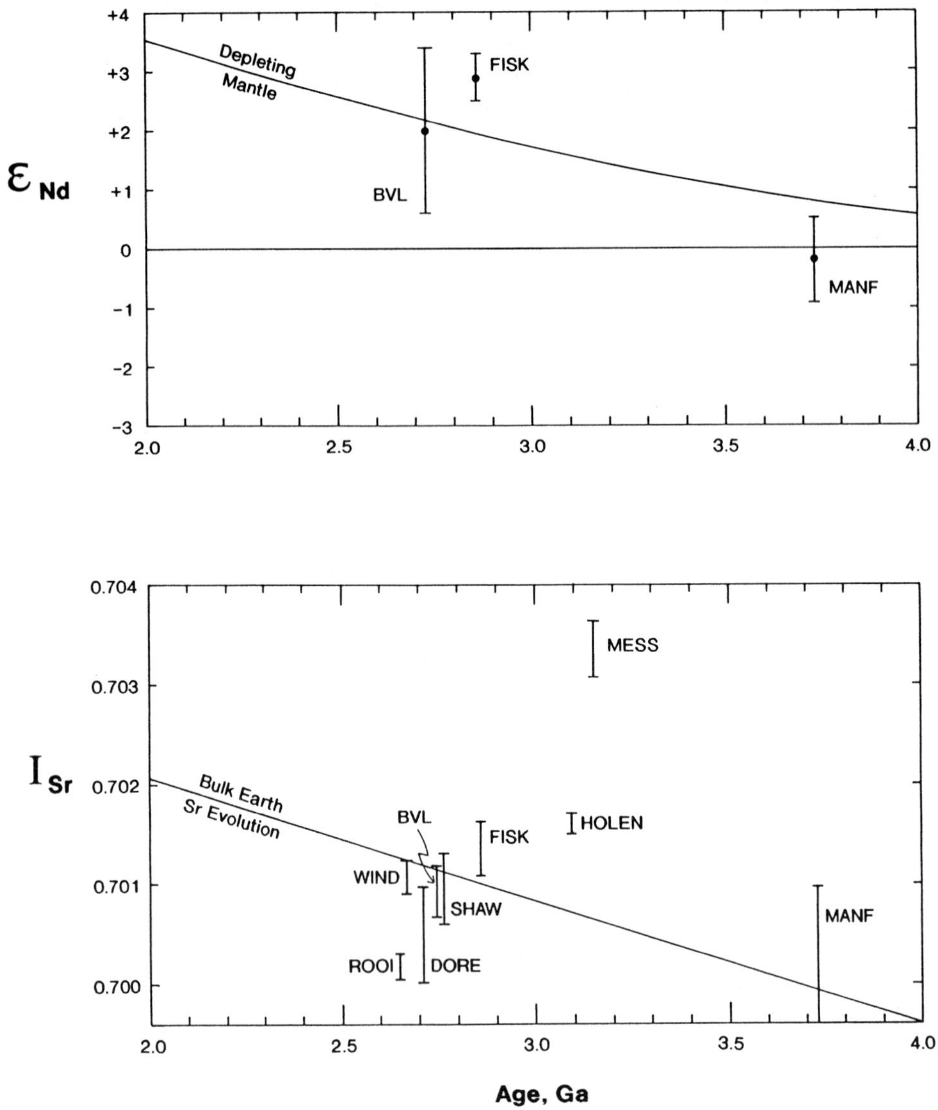

Fig. 17. Top. Available initial Nd ratios of Archean anorthosites expressed as ε_{Nd} calculated from whole-rock and mixed mineral–whole-rock isochrons. Depleting mantle curve from DePaolo (1981). Bottom. Initial Sr ratios of Archean anorthosites calculated at best estimates of their crystallization ages. In some cases these represent minimum ages as determined from U-Pb zircon ages of cross-cutting granitoids. Bulk Earth Sr evolution from DePaolo and Wasserburg (1976). Age uncertainties are not shown. Data sources as in Table 2. Abbreviations: BVL = Bad Vermilion Lake, Ontario; DORE = Dore Lake, Quebec; FISK = Fiskenæsset, West Greenland; HOLEN = Holenarasipur, India; MANF = Manfred, Western Australia; MESS = Messina, South Africa; SHAW = Shawmere, Ontario; WIND = Windimurra, Western Australia.

than 10% mafics. Anorthosite is only used in a strict sense when it is specifically distinguished from leucogabbro. All the rocks described are metamorphosed but for simplicity, the prefix meta- is generally omitted.

North Atlantic Craton

Anorthosites with plagioclase megacrysts (Fig. 2) are abundant and widespread in the Superior and Nain Provinces of Canada and Greenland as thin (up to 1 km thick), folded and disrupted sheets associated with gabbro and ultramafic rocks. They are also prominent in some Proterozoic orogens, such as the Torngat Orogen, Labrador, where they were further deformed and recrystallized. They occur in both low grade granite–greenstone terrains, and in high grade gneiss terrains. In both cases they appear to have initially been associated with mafic volcanic rocks, to which they were probably genetically related (Phinney et al., 1986a,b; Ashwal et al., 1983, 1985). Similar plagioclase megacrysts are also found in dikes such as the Matachewan swarm of the Canadian Shield (Ernst, 1982), and in sills and flows (Phinney et al., 1986a,b; 1988a) (Fig. 18).

Fiskenæsset Complex, West Greenland

The Fiskenæsset complex is one of the best preserved examples of anorthosites that are widespread throughout the Archean gneiss complex of Greenland and Labrador (Bridgwater et al., 1976). These anorthosites form thin layers generally less than 50 m thick, and trains of inclusions in tonalitic gneiss. The types of relict igneous texture and bulk lithology of the anorthosites are surprisingly similar throughout the region. Most rocks are strongly deformed and recrystallized to amphibolite or granulite facies, but as they were generally more competent than their host rocks, relict igneous textures are widely preserved, even in small inclusions less than 20 cm across.

The Fiskenæsset complex (Figs. 19 and 20) comprises seven major lithostratigraphic units which are, in ascending order: lower gabbro (50 m), ultramafic (40 m), lower leucogabbro (50 m), middle gabbro (40 m), upper leucogabbro (60 m), anorthosite (250 m) and upper gabbro (50 m) units (Myers, 1985). Cyclic mineral-graded and modal layering occur intermittently in all rock types. The complex was derived from tholeiitic magmas intruded into basic volcanic rocks (now amphibolites) (Weaver et al., 1981). The intrusion was fragmented by the emplacement of concordant granitoid sheets (now gneisses) (Fig. 9), which form 80% of the region (Fig. 19), and by contemporaneous thrusting.

These rocks were folded together into large scale recumbent folds (Fig. 21) and then refolded into dome-and-basin interference structures by two sets of folds with steep axial surfaces at high angles to each other (Fig. 19). Deformation was heterogeneous, and all stages can be seen from undeformed to intensely deformed rocks (Fig. 10). The ultramafic unit contains cyclic mineral-graded and size-graded layers of dunite, peridotite and hornblendite; igneous olivine, pyroxene and

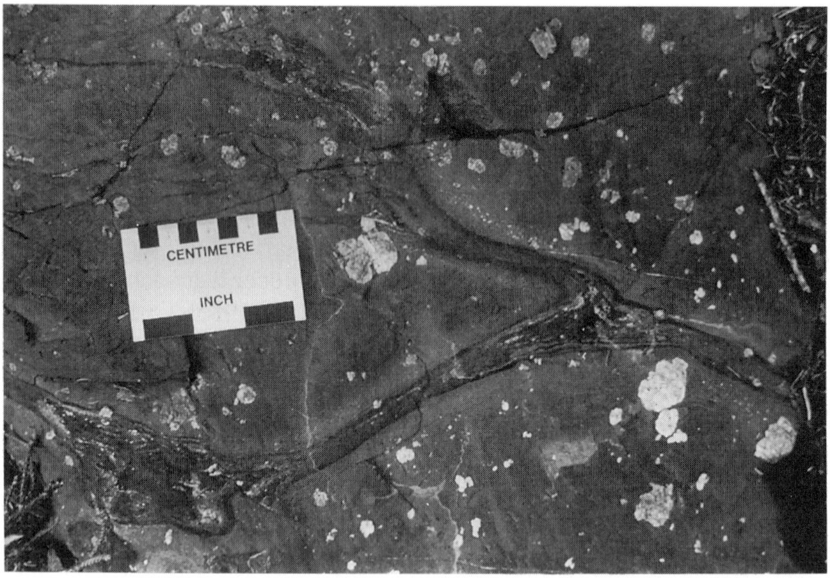

Fig. 18. Top: Calcic plagioclase megacrysts in glacial boulder near Fishtail, Montana, probably derived from mafic dike of Matachewan swarm. U.S. quarter dollar for scale. Bottom: Altered plagioclase megacrysts in basaltic pillow lavas, Bird River area, Manitoba (Photo courtesy of W.C. Phinney).

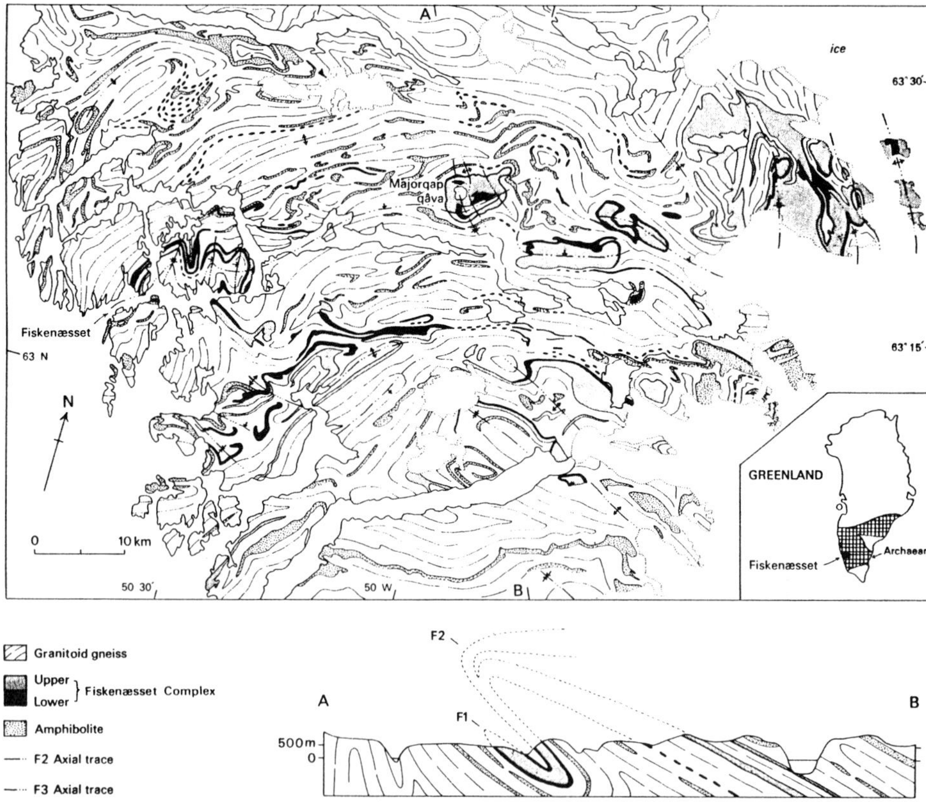

Fig. 19. Geological map and cross-section showing the extent and structure of the Fiskenæsset complex, West Greenland (Myers, 1981).

spinel are locally preserved in these rocks. Both the lower and upper leucogabbro units contain megacrysts of calcic plagioclase, and well preserved igneous textures and structures, including mineral- and size-graded layering, scoured channels and trough layering.

The middle gabbro unit contains columns of trough shaped mineral-graded layers as well as extensive, centimetre scale, mineral-graded layers (Fig. 8). A horizon of giant clusters of wedge shaped radiating supercooled plagioclase crystals occurs at the base of the overlying upper leucogabbro unit, and these clusters are locally deposited within the uppermost part of the middle gabbro unit (Fig. 7, see also Fig. 2). Individual clusters are generally 20 cm in diameter and comprise wedge shaped plagioclase crystals up to 10 cm long, that form a layer up to 10 m thick.

The lower part of the upper leucogabbro unit contains lenticular mineral-graded ultramafic trough layers (Fig. 5). They are interpreted as deposits from turbidity

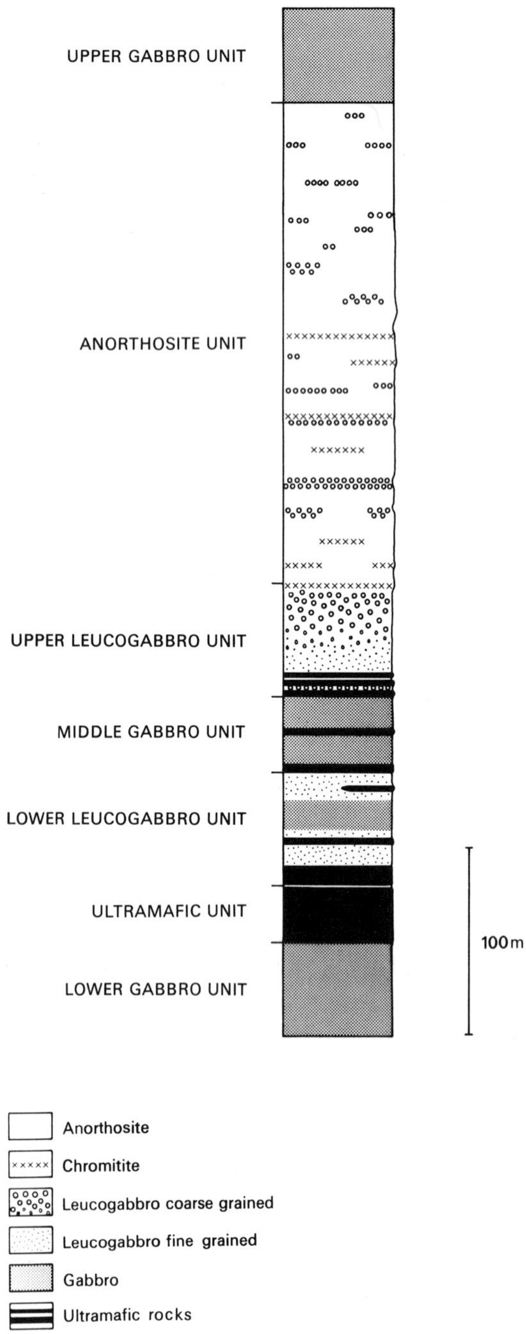

Fig. 20. General stratigraphic column of the Fiskenæsset complex, West Greenland (Myers, 1981).

Fig. 21. Recumbent isoclinal fold of amphibolite (black) and anorthosite (white) of the Anorthosite Unit of the Fiskenæsset complex, West Greenland. View to the northeast of a 400 m high cliff on the edge of the inland ice cap 55 km east of Fiskenæsset.

currents in scoured channels (Myers, 1976). Some are overlain by plagioclase-chromite cumulates. There is an upward increase in the size of plagioclase megacrysts within the upper leucogabbro unit from 2 to 10 cm.

The relatively thick anorthosite unit lacks marked internal stratigraphy and the most pronounced layering is tectonic. The unit is characterized by irregular sub-spherical patches of leucogabbro, typically 30 cm in diameter, in a matrix of anorthosite (Fig. 11). They appear to represent large oikocrysts of pyroxene, poikilitically enclosing equidimensional plagioclase.

Chromitite occurs as discontinuous layers and lenses throughout the anorthosite and upper leucogabbro units. Locally it forms massive mafic layers or lenses of chromite–hornblende–biotite rock up to a metre thick, but chromite generally occurs evenly distributed with hornblende and biotite between igneous or relict igneous plagioclase 1–2 cm in diameter in layers or lenses 1–2 m thick where cumulus plagioclase makes up about 80% of the rock. Some of the chromitite layers were disrupted by slumping during the accumulation of the cumulates (Fig. 22).

The upper gabbro unit consists of gabbro with layers of peridotite, dunite, melanogabbro, leucogabbro and anorthosite. Magnetite and ilmenite are prominent and locally form massive layers up to 1 m thick, or occur with equant relict igneous plagioclase in magnetite-rich leuogabbro.

Ashwal et al. (1989) obtained a five point mixed whole rock and mineral Sm-Nd isochron age of 2860 ± 50 Ma. This may either represent the igneous age of the complex, or disturbance during the high-grade metamorphism at about 2800 Ma.

Fig. 22. Fragments of a chromitite layer disrupted by slumping in anorthosite, Qeqertarssuatsiaq, Fiskenæsset complex, West Greenland (Ashwal, 1993).

Bad Vermilion Lake Complex, Ontario

This complex forms an elongate body of anorthosite and gabbro, emplaced into mafic and felsic volcanic rocks of the Wabigoon greenstone belt (Fig. 23). It was intruded by sheets of tonalite and metamorphosed at greenschist to amphibolite grade. Most plagioclase has been converted to epidote group minerals, and the mafic matrix to chlorite, amphiboles, albite, calcite and quartz (Fig. 14). Most rocks are undeformed or little deformed and relict igneous textures are well preserved. The most prominent relict igneous texture consists of equant plagioclase megacrysts 1–30 cm in diameter.

Mafic flows and dikes, chemically similar to the mafic volcanic rocks, contain plagioclase megacrysts. The mafic matrix of the anorthosite has a composition similar to the associated basaltic volcanics, suggesting that the anorthosite complex may have accumulated from a subvolcanic magma chamber which fed lavas to the surface (Ashwal et al., 1983). These authors obtained a Sm-Nd isochron of 2747 ± 58 Ma from combined anorthosite, gabbro and basalt, which they interpreted as the time of igneous crystallization.

Shawmere Complex, Ontario

This is one of the largest known Archean anorthosites, and forms two bodies with areas of 750 and 75 km^2 (Thurston et al., 1977; Simmons et al., 1980; Percival, 1981; Riccio, 1981). It occurs in the Kapuskasing Structural Zone, a narrow belt of high grade gneiss, a middle crustal portion of the granite–green-

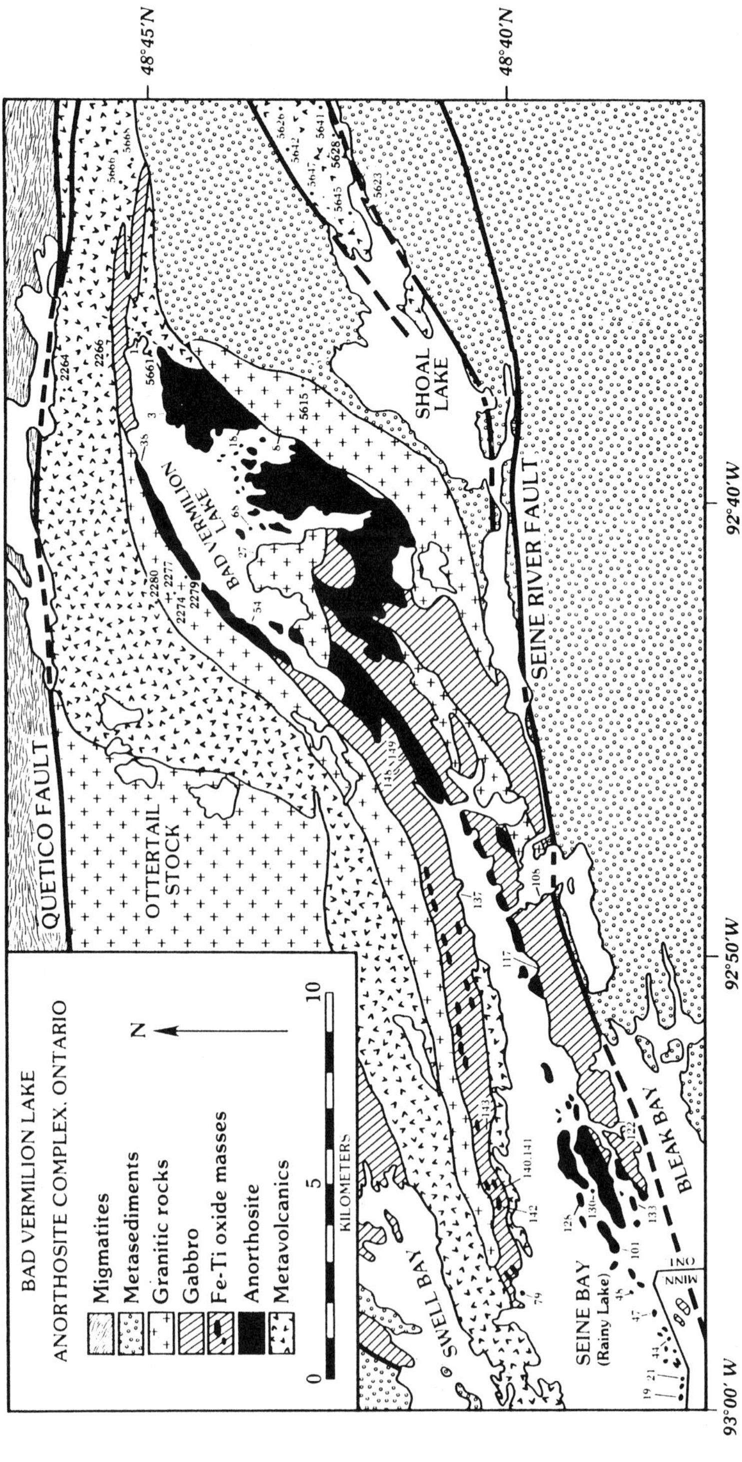

Fig. 23. Geologic map of the Bad Vermilion Lake anorthosite complex, Ontario (Ashwal et al., 1983).

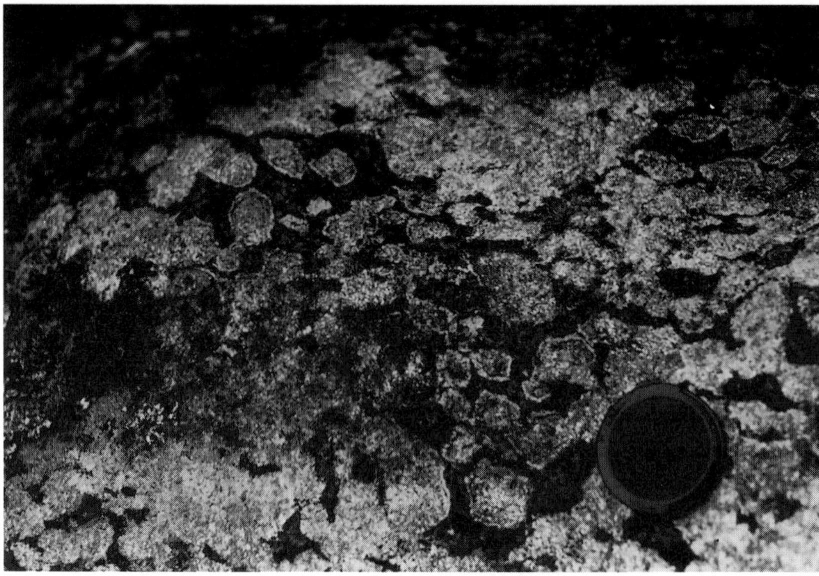

Fig. 24. Megacrystic leucogabbro from the Shawmere complex, Ontario. Granulite facies metamorphism has produced reaction rims of garnet between calcic plagioclase and mafic matrix.

stone terrain, uplifted along a thrust (Percival and Card, 1983; Cook, 1985). It is surrounded by high grade gneissic amphibolite, tonalite and minor paragneiss (Simmons et al., 1980; Percival, 1981; Riccio, 1981). The contacts are tectonised and conformable with the host rocks, as a result of deformation during granulite grade metamorphism (Percival, 1981).

Most of the complex is strongly deformed, and igneous textures and minerals are only well preserved in the interior of the larger body (Fig. 24). Part of this body comprises units of: leucogabbro with equant plagioclase crystals up to 50 cm in diameter; anorthosite with 1–30 cm thick layers of anorthosite, and minor gabbro and ultramafic rocks; and a border zone 50–100 m thick of garnet amphibolite (Riccio, 1981). The layered rocks are cut by dikes of anorthosite and leucogabbro (Simmons et al., 1980).

Dore Lake Complex, Quebec

The Dore Lake complex was intruded into volcanic rocks of the Abitibi greenstone belt. Allard (1970) distinguished a basal anorthosite zone (2500–3700 m), a middle ultramafic zone (500–1000 m), and an upper tonalite zone (1600–2000 m). The tonalitic rocks constitute the intrusive Chibougamou Lake pluton, and are genetically unrelated to the anorthosites and ultramafics. The rocks are generally little deformed, and are metamorphosed at greenschist grade. Cu-Au mineralization (associated with late shear zones) is present in the anorthosite. The anorthosites are characterized by plagioclase megacrysts, but they show a great range in grain size and texture. The complex is cut by the tonalitic Chibougamou pluton which has given a U-Pb zircon age of 2712 ± 2 Ma (Krogh et al., 1982).

Dharwar Craton, India

Anorthosites are widespread in the Archean Dharwar Craton. Most occur in association with the Sargur supracrustal rocks, considered to be older than 3000 Ma (Ramakrisnan et al., 1984; Taylor et al., 1984), or as fragments in the heterogeneous Peninsular Gneiss. Most are strongly deformed and metamorphosed at amphibolite or granulite grade. They range from small layers and lenses to large complexes such as Chimalpahad and Sittampundi which contain layers of anorthosite up to 150 m thick.

Sittampundi Complex, Salem District

The Sittampundi Complex (Subramaniam, 1956a) forms layers with a strike length of 20–25 km that have been repeatedly folded together with adjacent quartzofeldspathic gneiss containing layers of marble, quartzite and banded iron formations (Ramadurai et al., 1975). The complex is dominated by anorthosite with layers and lenses of chromitite, garnet–pyroxene rocks, gabbro, hornblendite, pyroxenite and anthophyllite rocks (Janardhan and Leake, 1975). The rocks are intensely deformed, igneous textures have been almost totally obliterated, and the rocks have granoblastic metamorphic textures that formed under high pressures (Chappell and White, 1970). It has not been possible to determine a reliable igneous stratigraphy, but Ramadurai et al. (1975) estimated a maximum thickness of 975 m.

Kalahari Craton, Southern Africa

Messina Complex, South Africa, Zimbabwe, and Botswana

The Messina Complex forms discontinuous layers of calcic anorthosite, gabbro, and minor ultramafics that crop out over several hundred kilometres in the late Archean Limpopo Mobile Belt (Hor et al., 1975; Barton et al., 1979). They are associated with the heterogeneous Sand River Gneiss and supracrustal rocks including quartzite, marble, pelitic gneiss, banded iron formations and amphibolite. The rocks have been repeatedly folded, intensely deformed, and recrystallized in granulite facies. The original relationships with the associated supracrustal rocks and basement gneisses are uncertain.

Pilbara and Yilgarn Cratons, Australia

Archean anorthosites occur in the granite-greenstone terrains of both the Pilbara and Yilgarn Cratons, and in high grade gneiss in the northwest part of the Yilgarn Craton. Thin mafic–ultramafic sills are abundant in the late Archean

Fig. 25. Partly deformed and recrystallized coarse-grained leucogabbro with some remnants of cumulus plagioclase crystals (grey) enclosed by metamorphic plagioclase (white), in a matrix of metamorphic hornblende (black), Manfred Complex, Western Australia. The pen is 13 cm long. (Myers, 1988).

greenstone sequences of the Yilgarn Craton and a few contain layers of megacrystic anorthosite, locally known by the old miners name 'cat rock', after a spotted cat-like native marsupial.

Manfred Complex

The Manfred Complex occurs in the high grade Narryer Terrane of the Yilgarn Craton (Myers, 1988). It comprises thin layers and lenses of anorthosite, gabbro and ultramafic rocks. It was intruded by granite at 3.4–3.3 Ga (Dugel Gneiss) and again at 2.75–2.60 Ga, and repeatedly folded and metamorphosed at granulite facies. The anorthosite is characterized by relict plagioclase megacrysts typically 2–5 cm in diameter (Fig. 25), although some occur up to 30 cm across. Some igneous minerals and layering structures are preserved, although in many anorthosites plagioclase was extensively replaced by epidote group minerals.

Zircon and apatite are abundant as inclusions within igneous remnants of plagioclase megacrysts. These zircons have given U-Pb ages of 3730 ± 6 Ma (Kinny et al., 1988), in agreement with Sm-Nd (3680 ± 70 Ma) and Pb-Pb (3689 ± 146 Ma) ages (Fletcher et al., 1988). If these ages represent the emplacement age of the Manfred complex then this complex is the oldest of known terrestrial anorthosites. Some anorthosites occur in Labrador and West Greenland that are thought to have formed at ca. 3.8 Ga.

Windimurra Complex

The Windimurra complex along the eastern margin of the Murchison Terrane of the Yilgarn Craton is a large (22 × 103 km) body of leucogabbro and leucotroctolite with minor anorthosite and ultramafic rocks (Ahmat, 1990; Ahmat and Mathesen, in prep). It is in tectonic contact with granitoid rocks and volcanic rocks. Most of the intrusion is internally undeformed or little deformed, and the body is only partly recrystallized in greenschist facies. Igneous textures and minerals are well preserved, and plagioclase megacrysts, typically up to 2 cm diameter, are generally enclosed by giant poikilitic pyroxenes. Igneous layering occurs sporadically.

Anabar Shield, Siberia

The Anabar Shield of Siberia contains several large lensoid anorthosite bodies enclosed by Archean granitoid gneiss, amphibolite and minor quartzite (Lutts, 1974; Sukhanov, 1984a). They are divided into the Kotuykan and Magan Groups on the basis of their geographic location, and collectively contain about 900 and 350 km^2 of anorthositic masses, respectively. The rocks are strongly deformed and recrystallized in granulite facies.

Baltic Shield

Archean anorthosites occur in the Kola Peninsula of Russia. The Kolmozero complex (also called the Patchemvarak complex) is a lensoid body of gabbro, anorthosite and ultramafic rocks within granitoid gneiss (Yudin, 1974; Sharkov, 1984). The anorthosite contains plagioclase megacrysts up to 4 cm in diameter and some rhythmic layering. The complex has been metamorphosed to amphibolite grade, and is intruded by pegmatites with a U-Pb zircon age of 2750 ± 50 Ma. The nearby Achinsk complex forms layers extending over 100 km that have been repeatedly deformed.

ORIGIN OF ARCHEAN ANORTHOSITES

Analogies and comparisons

In considering the origin of Archean anorthosites, many authors have attempted to draw analogies with more familiar or better understood geological features. These anorthosites have been interpreted as: metamorphosed equivalents of layered mafic intrusions such as Bushveld (Harpum, 1957; Windley, 1969a), tectonic slices of lower crust (Windley, 1969b), fragments of the Earth's putative primordial anorthositic crust (Windley, 1970), Archean analogues to modern ophiolite complexes (Sutton and Windley, 1974), and analogues to anorthositic components of cordilleran granitoid batholiths (Windley and Smith, 1976). Most of these ideas can now be eliminated, as discussed below (see also Ashwal, 1993, pp. 74–75). It was not until detailed studies were made of occurrences at low metamorphic grade (e.g. Phinney et al., 1988a) that the prevailing association with Archean greenstone belts was recognized.

Although some magmatic features present in Archean anorthosite complexes resemble those found in mafic layered intrusions, the megacrystic textures, lack of cryptic variation in mineral compositions, and highly feldspathic bulk compositions of these complexes effectively distinguish them from most, if not all layered mafic complexes such as Bushveld, Stillwater, or Dufek. The common association of Archean anorthosites with supracrustal rocks (even those now at high metamorphic grade) argues against deep crustal emplacement, and therefore they cannot represent uplifted pieces of lower crust. Isotopic data demonstrate that Archean anorthosites are not fragments of Earth's earliest crust, and cannot be considered analogous to the far more calcic (An_{94-99}) lunar anorthosites, which probably formed from a magma ocean shortly after the Moon accreted. Although there is some merit in an analogy between Archean anorthosites and those present in Phanerozoic ophiolite complexes (e.g. Bay of Islands, Newfoundland), in the latter case, the anorthositic rocks occur as thin (<1 m) layers between gabbroic cumulates, and are therefore much smaller in scale than the large masses of megacrystic plagioclase found in the Archean complexes. The very minor anorthositic rocks present in calc-alkaline plutons of modern continental arcs (e.g. Peninsular Ranges batholith, California) differ from Archean anorthosites in terms of texture, mineralogy and lithologic association, and are therefore an unsuitable analog.

We may also consider the differences between Archean anorthosites and those of the large Proterozoic massifs. Massif-type anorthosites are larger (up to 17,000 km^2), younger (~1.65–1.0 Ga), and more sodic (An_{40-60}) than their Archean counterparts. They were emplaced into continental rather than oceanic crust, although a two-stage model involving deep crustal ponding of mafic magmas, followed by shallow emplacement of plagioclase-rich mushes may be applicable to both Archean and massif-type anorthosites. An oceanic setting for Archean anorthosite emplacement is considered below.

The presence of continental crust can, at least in part, account for some of the differences between massif-type and Archean anorthosites by a combination of effects (Ashwal, 1993, p. 347). One effect involves the effect of pressure upon the composition of liquidus plagioclase, such that Ab content of crystals is greater at higher pressure, all other things being equal. The second effect involves assimilation of Na-rich continental crustal components by massif-type anorthosite parental magmas. Archean anorthosites might be more calcic, therefore, partly because of the absence of continental crust, both as a potential assimilant and as a barrier to shallow crystallization. Differences in primary magma compositions, however, may also play an important role, as discussed below.

Chemical, mineralogical, textural and geological features of Archean anorthosites, therefore, distinguish them from most, if not all other types of anorthosite. Some of the cogent aspects relating to the origin of Archean anorthosites are discussed below.

Parental magmas and petrogenesis

Most Archean anorthosites are spatially associated with rocks of basaltic origin and minor metasedimentary rocks. The few surviving primary contact relations show that Archean anorthosite complexes were intruded into basaltic volcanic rocks and associated supracrustal rocks. The primary textures of most Archean anorthosites are cumulate. Although it is difficult to determine the compositions of melts from which cumulates were derived, several studies suggest that the parental magma was basaltic (Windley et al., 1973; Barton et al., 1979; Weaver et al., 1981; Wiener, 1981; Ashwal et al., 1983). Geochemical evidence also suggests that the intrusions were genetically linked with basaltic dikes and volcanic rocks.

These features suggest that many Archean anorthosites were emplaced at shallow depths into consanguineous volcanic rocks, either in an oceanic environment or into flood basalt sequences on sialic crust. High grade metamorphic features are secondary and indicate the many of the complexes subsequently experienced conditions at middle to lower crustal depths.

Although some Archean anorthosites show upward iron enrichment trends, plagioclase, the only widely preserved cumulus mineral, is uniformly calcic. This contrasts with younger layered basic intrusions in which plagioclase and other cumulus minerals generally show pronounced cryptic variation with stratigraphic height. Younger layered intrusions generally contain mafic and ultramafic portions that, when combined with felsic portions, equate to a bulk basaltic composition. In contrast, Archean anorthosites are much more feldspathic, as their mafic and ultramafic portions do not generally complement the feldspathic portions.

As noted above, the melts in equilibrium with the calcic plagioclase megacrysts of Archean anorthosites appear to have been Fe-rich tholeiitic basalts. Such compositions are probably derivative in nature, having fractionated from more primitive melts or magmas such as picrite or komatiite. A genetic connection between megacrystic anorthosites and komatiites is, therefore, tempting, given their equivalent temporal restriction to the Archean, and the association of both with Archean greenstone belts. Petrologic evidence supporting a link between komatiites and Archean anorthosites comes from observations that early plagioclases crystallizing after mafic silicates from ultramafic magmas tend to be very calcic (An_{80-90}) (Phinney, 1982).

Anorthosite emplacement

A two-stage emplacement model that could link anorthosites and komatiites was proposed by Phinney (1982) (Fig. 26). He suggested that mantle-derived mafic melts were ponded at or near the crust/mantle boundary where they reached isostatic equilibrium. Some may have become contaminated by partial melting and assimilation of lower crustal rocks. During cooling, the ponded melt began to crystallize mafic silicates that sank and accumulated as layered ultramafic rocks. Some melt may have been erupted as komatiites at this stage. With continued fractionation, the residual melt became enriched in Ca and Al and crystallized calcic plagioclase. The long cooling times allowed large, homogeneous plagioclase crystals to develop. These were separated by flotation, and rose as crystal mushes with various amounts of fractionated melts. They spread out as sills when they reached levels at which they were in isostatic equilibrium, and with further cooling they crystallized as layered anorthosite complexes. Some of the plagioclase megacrysts were carried up to higher crustal levels with Fe-rich basaltic liquids and crystallized in dikes or flows.

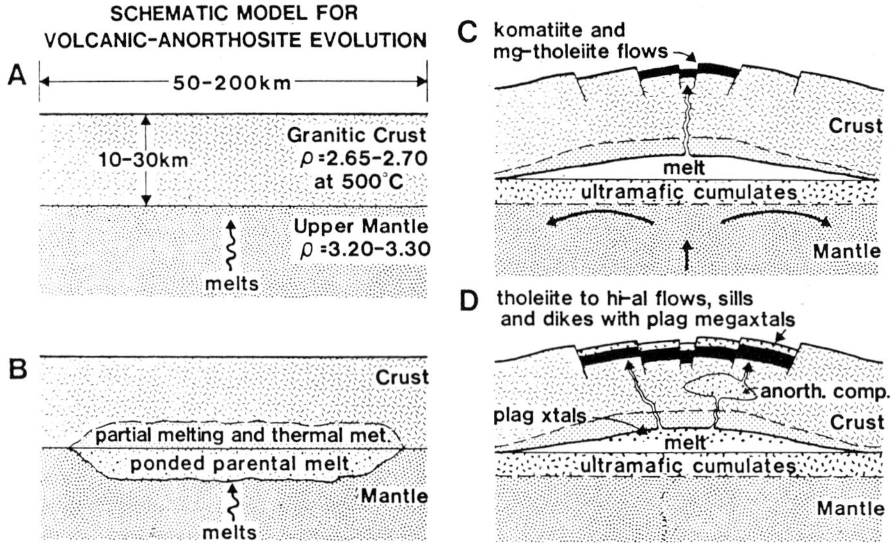

Fig. 26. Model for genesis of Archean megacrystic anorthosites proposed by Phinney (1982). Mantle-derived melts (A) ascend and pond at or near the crust-mantle boundary (B), where they induce partial melting and metamorphism of lower crustal rocks. Mafic silicates crystallize and sink, forming ultramafic cumulates, while some residual melt is tapped to the surface forming Mg-rich flows of komatiite or Mg-tholeiite. C. Calcic plagioclase precipitates and floats due to high compressibility of silicate melts at lower crustal pressures (D). Large, compositionally uniform, calcic plagioclase megacrysts form in this environment, and are emplaced to near-surface depths as flows, sills, dikes, and anorthosite complexes with various crystal/melt ratios (Phinney, 1982).

A modification of this model by Phinney et al. (1988a), based on experimental work by Morrison et al. (1985), suggests that plagioclase megacrysts could also have formed in high-level magma chambers. Moderate pressure (5–10 kbar) fractionation and precipitation of olivine and orthopyroxene from melts ponded near the crust/mantle boundary could have decreased the density of the residual melt sufficiently for it to rise further. The melt then spread out at a high crustal level, and calcic plagioclase would have nucleated and crystallized in these conditions of reduced isostatic pressure.

Repeated injections of relatively hot magmas into this magma chamber may have induced supercooling and the abundant nucleation of plagioclase. Evidence of this can be seen in the Fiskenæsset complex (Myers, 1976, 1985), where injections of ultramafic magma into the upper part of the crystallizing anorthosite complex led to downward surges of dense clouds of olivine, pyroxene, spinel and chromite. In some cases these surges entrained plagioclase megacrysts from the magma through which they passed, and these megacrysts were fractured, mechanically rounded, and then partly overgrown by shells of pyroxene during their transport. When they reached the floor of the magma chamber, the dense surges scoured channels 100–500 m wide in the underlying cumulates, and travelled for tens of kilometres. They deposited their load of crystals in these scoured troughs, forming both mineral-graded and massive layers of peridotite 1–3 m thick (Fig. 5). These deposits immediately overlie a prominent horizon up to 5 m thick of giant, 20 cm diameter, snowflake like clusters of radiating wedge-shaped plagioclase (Fig. 7).

Phinney and Morrison suggest that the equidimensional, compositionally uniform, plagioclase megacrysts typical of Archean anorthosites could have originated as snowflake-type crystals (Figs. 2 and 7) generated during supercooling episodes induced by intermittent injection of relatively hot mafic magmas, that were subsequently infilled during periods of slow cooling. Some of these hot mafic magmas may have been potential komatiites that did not reach the Earth's surface.

Tectonic setting

The lack of obvious modern analogues and the poor state of preservation of many Archean anorthosite complexes prevents a convincing determination of tectonic setting. Many analogues have been proposed, as discussed above, but most of these are not sufficiently convincing.

The two-stage emplacement models described above for Archean anorthosites are comparable to those proposed for the origin of massif-type anorthosites (Morse, 1968; Barker et al., 1975; Emslie, 1978; Longhi and Ashwal, 1985; Ashwal, 1993), and imply the existence of continental crust to act as an initial barrier to rising melts. Such models, therefore, are applicable only to those Archean anorthosite occurrences which might be associated with cratonic settings.

These include megacryst-bearing dike swarms and possibly those anorthosite complexes associated with cratonic shelf sequences, such as Sittampundi and Messina.

However, the association of many calcic megacrystic anorthosites with voluminous mafic pillow lavas, and the geochemical links between anorthosites and mafic volcanics permit, if not suggest, an oceanic crustal environment. It is not clear if a two stage petrogenetic model requiring initial moderate pressure magmatic ponding can be accommodated without a suitable density barrier such as continental crust to initially prevent shallow emplacement. It could be argued, therefore, that the anorthositic complexes formed during an initial rifting stage of continental break-up. Alternatively, mechanisms of deep magmatic ponding in oceanic crust must be considered. The suggestion of increased oceanic crustal thickness during the Archean (e.g. Sleep and Windley, 1982) may be pertinent to this problem.

Among those workers who favor an oceanic setting for Archean greenstone belts, there is little agreement as to whether mid-ocean ridge, back-arc basin, or oceanic plateau environments are most applicable in specific cases, or whether modern-day analogues are appropriate at all (e.g. de Wit and Ashwal, 1986, 1994). If we are prepared to accept an oceanic environment for these anorthosite complexes, then their present locations in continental crust may mark sites of ancient ocean closures, and in this sense they may be used as an indicator of Archean ophiolite assemblages. If so, then we must explain the significant differences between Archean anorthosites and those found as minor components in Phanerozoic ophiolite complexes (Ashwal, 1993). A primary komatiitic magma for Archean anorthosites may be part of the answer.

SUMMARY

Archean anorthositic rocks are characterized by distinctive megacrystic textures, consisting of equidimensional, calcic (>An_{80}) plagioclase crystals 0.5–30 cm in diameter, in a mafic groundmass. Associated rocks commonly include mafic metavolcanics and other supracrustals of Archean greenstone belts. Ages range between 3.7 Ga (Manfred Complex, Australia) and 2.75 Ga (examples from Superior Province, Canada). Metamorphic grade of known occurrences varies from sub-greenschist to granulite. Low grade examples are usually extensively altered, and high grade ones commonly deformed, but in both cases magmatic textures, structures and mineralogy can be partly preserved.

A genetic link between these anorthosites and mafic volcanics of greenstone belts comes from occurrences of basaltic flows, sills, and dikes that contain similar calcic plagioclase megacrysts, and chemical similarities between the basalts and the mafic groundmass surrounding the plagioclase megacrysts of the anorthosites. The anorthosite complexes, therefore, can be interpreted as subvolcanic magma chambers that fed mafic lavas to the surface during their crystallization. Parental magmas were Fe-rich tholeiites that may represent fractionation products from primitive picrites or komatiites.

Two-stage crystallization/emplacement models involving ponding and fractionation of mantle-derived melts or magmas in deep crust, followed by shallow emplacement of plagioclase-rich mushes or plagioclase-supersaturated melts can account for many features of Archean anorthosite complexes. If continental crust is required as a density filter to initiate ponding, then this might best apply to megacryst-bearing dike swarms and possibly those few complexes associated with cratonic shelf sequences. A continental rift environment might be appropriate to account for these. However, most Archean anorthosite complexes and associated greenstone belts appear to have formed in oceanic crustal environments. If this is the case, then it is not clear how two-stage petrogenetic models can be accommodated unless Archean oceanic crust was thicker than existing Phanerozoic equivalents..

ACKNOWLEDGMENTS

LDA wishes to thank NASA and the Lunar and Planetary Institute (Houston) for support of research activities leading to this paper, and colleagues W.C. Phinney, D.A. Morrison, and D.E. Maczuga for initial opportunities and continued collaboration. JSM thanks the Geological Survey of Greenland for supporting his work on the Fiskenæsset complex as part of the Survey's Systematic Mapping Program, and publishes the results of this work with permission. Critical reviews by S.A. Morse, R.A. Wiebe and K.C. Condie improved the manuscript substantially and are much appreciated.

REFERENCES

Ahmat, A.L. and de Laeter, J.R., 1982. Rb-Sr isotopic evidence for Archaean–Proterozoic crustal evolution of part of the central Yilgarn Block, Western Australia: constraints on the age and source of the anorthositic Windimurra Gabbroid. J. Geol. Soc. Aust., 29: 177–190.

Ahmat, R.L. and Mathesen, C.I., in preparation. Windimurra layered gabbroid complex, Yilgarn Craton, Western Australia.

Allard, G.O., 1970. The Dore Lake complex, Chibougamou, Quebec — a metamorphosed Bushveld-type layered intrusion. In: D.J.L. Visser and G. von Gruenewaldt (Eds.), Symposium on the Bushveld Igneous Complex and Other Layered Intrusions. Geol. Soc. S. Afr. Spec. Publ. No. 1, Johannesburg, pp. 477–491.

Ashwal, L.D., 1993. Anorthosites. Springer-Verlag, New York, Berlin, Heidelberg, 422 pp.

Ashwal, L.D., Morrison, D.A., Phinney, W.C., and Wood, J., 1983. Origin of Archean anorthosites: evidence from the Bad Vermilion Lake complex, Ontario. Contrib. Mineral. Petrol., 82: 259–273.

Ashwal, L.D., Wooden, J.L., Phinney, W.C., and Morrison, D.A., 1985. Sm-Nd and Rb-Sr isotope systematics of an Archean anorthosite and related rocks from the Superior Province of the Canadian Shield. Earth Planet. Sci. Lett., 74: 338–346.

Ashwal, L.D., Jacobsen, S.B., Myers, J.S., Kalsbeek. F., and Goldstein, S.J., 1989. Sm-Nd age of the Fiskenæsset Anorthosite Complex, West Greenland. Earth Planet. Sci. Lett., 91: 261–270.

Barker, F., Wones, D.R., Sharp, W.N., and Desborough, G.A., 1975. The Pikes Peak Batholith, Colorado Front Range, and a model for the origin of gabbro–anorthosite–syenite–potassic granite suite. Precambrian Res., 2: 97-150.

Barton, J.M., Jr., Fripp, R.E.P., Horrocks, P.C., and McLean, N., 1979. The geology, age, and tectonic setting of the Messina Layered Intrusion, Limpopo Mobile Belt, southern Africa. Am. J. Sci., 279: 1108–1134.

Black, L.P., Moorbath, S., Pankhurst, R.J., and Windley, B.F., 1973. $^{207}Pb/^{206}Pb$ whole rock age of the Archean granulite facies metamorphic event in west Greenland. Nature, 244: 50–53.

Bridgwater, D.D., Keto, L., McGregor, V.R., and Myers, J.S., 1976. Archaean gneiss complex of Greenland. In: A. Escher and W.S. Watt (Eds.), Geology of Greenland. Grønl. geol Unders., Copenhagen, pp. 18–75.

Chappell, B.W., White, A.J.R., 1970. Further data on an "eclogite" from the Sittampundi complex, India. Mineral. Mag., 37: 555–560.

Condie, K.C., 1981. Archean Greenstone Belts. Elsevier, New York, 434 pp.

DePaolo, D.J., 1981. Neodymium isotopes in the Colorado Front Range and crust-mantle evolution in the Proterozoic. Nature, 291: 193–196.

DePaolo, D.J. and Wasserburg, G.J., 1976. Inferences about magma sources and mantle structure from variations in $^{143}Nd/^{144}Nd$. Geophys. Res. Lett., 3: 743–746.

de Wit, M.J. and Ashwal, L.D. (Eds.), 1986. Workshop on Tectonic Evolution of Greenstone Belts. Lunar Planet. Inst. Tech. Rept. 86-10, Lunar Planet. Inst., Houston, 227 pp.

de Wit, M.J. and Ashwal, L.D. (Eds.), 1994. Tectonic Evolution of Greenstone Belts. Oxford Univ. Press, in preparation.

de Wit, M.J., Jeffery, M., Bergh, H., Nicolaysen, L., 1988. Geological map of sectors of Gondwana reconstructed to their disposition ~150 Ma. Am. Assoc. Petrol. Geol., Tulsa, Scale 1:10,000,000.

Emslie, R.F., 1978. Anorthosite massifs, rapakivi granites, and late Proterozoic rifting of North America. Precambrian Res., 7: 61–98.

Ernst, R.E., 1982. Structural and chemical studies of mafic dike swarms in northern Ontario. Ontario Geol. Surv. Misc. Pap. 106, pp. 53–56.

Fletcher, I.R., Rosman, K.J.R., and Libby, W.G., 1988. Sm-Nd, Pb-Pb and Rb-Sr geochronology of the Manfred Complex, Mount Narryer, Western Australia. Precambrian Res., 38: 343–354.

Gancarz, A.J., 1976. Isotopic systematics in Archean rocks, west Greenland. Ph.D. Thesis, Calif. Inst. Tech., 349 pp.

Harpum, J.R., 1957. Discussion of Boulanger, J., 1957, Les Anorthosites de Madagascar. Comm. Tech. Coop. Africa South of the Sahara, Conf. de Tananarive, Prem. Vol., pp. 71–92.

Henderson, P., Fishlock, S.J., Laul, J.C., Cooper, R.L., Conard, R.L., Boynton, W.V., and Schmitt, R.A., 1976. Rare earth element abundances in rocks and minerals from the Fiskenæsset Complex, West Greenland. Earth Planet. Sci. Lett., 30: 37–49.

Hor, A.K., Hutt, D.K., Smith, J.V., Wakefield, J., Windley, B.F., 1975. Petrochemistry and mineralogy of early Precambrian anorthositic rocks of the Limpopo Belt, southern Africa. Lithos, 8: 297–310.

Janardhan, A.S. and Leake, B.E., 1975. The origin of meta-anorthositic gabbros and garnetiferous granulites of the Sittampundi complex, Madras, India. J. Geol. Soc. India, 16: 391–408.

Jones, L.M., Walker, R.L., and Allard, G.O., 1974. The rubidium–strontium whole-rock age of major units of the Chibougamou greenstone belt, Quebec. Can. J. Earth Sci., 11: 1550–1561.

Kinny, P.D., Williams, I.D., Froude, D.O., Ireland, T.R., and Compston, W., 1988. Early Archean zircon ages from orthogneisses and anorthosites at Mount Narryer, Western Australia. Precambrian Res., 38: 325–341.

Krogh, T.E., 1982. Improved accuracy of U-Pb zircon ages by the creation of more concordant systems using an air abrasion technique. Geochim. Cosmochim. Acta, 46: 637–649.

Krogh, T.E., Davis, D.W., Nunes, P.D., and Corfu, F., 1982. Archean evolution from precise U-Pb isotopic dating. Geol. Assoc. Can. - Mineral. Assoc. Can. Abstr., 7: 61.

Kutty, T.R.N., Anantha Iyer, G.V., Ramakrishnan, M., and Verma, S.P., 1984. Geochemistry of meta-anorthosites from Holenarasipur, Karnataka, South India. Lithos, 17: 317–328.

Leake, B.E., 1978. Nomenclature of amphiboles. Am. Mineral., 63: 1023–1052.

Leelanandam, C., 1987. Archaean anorthosite complexes: An overview. In: Saha, A.K. (Ed.), Geological Evolution of Peninsular India — Petrological and Structural Aspects. Recent Researches in Geology, 13: 108–116.

Longhi, J. and Ashwal, L.D., 1985. Two-stage models for lunar and terrestrial anorthosites: Petrogenesis without a magma ocean. Proc. Lunar Planet. Sci. Conf. 15th, Part 2, J. Geophys. Res (Suppl), C571–C584.

Lutts, B.G., 1974. Anorthosites of the Anabar Shield. In: O.A. Bogatikov O.A. (Ed.), Anorthosites of the USSR. Nauka, Moscow, pp. 70–84 (in Russian).

Morrison, D.A., Haskin, L.A., Qiu, Y.Z., Phinney, W.C., and Maczuga, D.E., 1985. Alteration in Archean anorthosite complexes. Lunar and Planet. Sci. XVI. Lunar Planet. Inst., Houston, pp. 589–590.

Morrison, D.A., Phinney, W.C., and Maczuga, D.E., 1987. Archean anorthosites: constraints on the accumulation process. Lunar and Planet. Sci. XVIII. Lunar Planet. Inst., Houston, pp. 670–671.

Morrison, D.A., Phinney, W.C., and Maczuga, D.E., 1988. The petrogenetic significance of plagioclase megacrysts in Archean rocks. In: L.D. Ashwal (Ed.), Workshop on the Deep Continental Crust of South India. Lunar Planet. Inst. Tech. Rep. 88-06, Lunar Planet. Inst., Houston, pp. 112–114.

Morse, S.A., 1968. Layered intrusions and anorthosite genesis. In Y.W. Isachsen (Ed.), Origin of Anorthosite and Related Rocks. N.Y. State Mus. Sci. Serv. Mem., 18: 175–187.

Myers, J.S., 1976. Channel deposits of peridotite, gabbro and chromitite from turbidity currents in the stratiform Fiskenæsset anorthosite complex, southwest Greenland. Lithos, 9: 281–291.

Myers, J.S., 1978. Formation of banded gneisses by deformation of igneous rocks. Precambrian Res., 6: 43–64.

Myers, J.S., 1981. The Fiskenæsset anorthosite complex: a stratigraphic key to the tectonic evolution of the West Greenland gneiss complex 3000–2800 m.y. ago. Spec. Publ. Geol. Soc. Austr., 7: 351–360.

Myers, J.S., 1985. Stratigraphy and structure of the Fiskenæsset Complex, West Greenland. Grønl. Geol. Unders. Bull., 150, 72 pp.

Myers, J.S., 1988. Oldest known terrestrial anorthosite at Mount Narryer, Western Australia. Precambrian Res., 38: 309–323.

Page, N.J., Myers, J.S., Haffty, J., Simon, F.O., and Aruscavage, P.J., 1980. Platinum, palladium, and rhodium in the Fiskenæsset complex, southwestern Greenland. Econ. Geol., 75: 907–915.

Percival, J.A., 1981. Geological evolution of part of the central Superior Province based on relationships among the Abitibi and Wawa subprovinces and the Kapuskasing Structural Zone. Ph.D. Thesis, Queen's Univ., 300 pp.

Percival, J.A. and Krogh, T.E., 1983. U-Pb zircon geochronology of the Kapuskasing structural zone and vicinity in the Chapleau–Foleyet area. Can. J. Earth Sci., 20: 830–843.

Phinney, W.C., 1982. Petrogenesis of Archean anorthosites. In: D. Walker and I.S. McCallum (Eds.), Workshop on Magmatic Processes of Early Planetary Crusts: Magma Oceans and Stratiform Layered Intrusions. Lunar Planet. Inst. Tech. Rep. 82-01, Lunar Planet. Inst., Houston, pp. 121–124.

Phinney, W.C., Morrison, D.A., 1990. Partition coefficients for calcic plagioclase: Implications for Archean anorthosites. Geochim. Cosmochim. Acta, 54: 1639–1654.

Phinney, W.C., Morrison, D.A., and Ashwal, L.D., 1981a. Archean anorthosites and plagioclase megacrysts: evidence for early crustal formation processes. EOS, 62: 420.

Phinney, W.C., Morrison, D.A., and Ashwal, L.D., 1981b. Implications of Archean anorthosites for crust–mantle evolution. Lunar Planet. Sci. XII. Lunar Planet. Inst., Houston, pp. 830–832.

Phinney, W.C., Morrison, D.E., and Maczuga, D.E., 1986a. Archean megacrystic plagioclase units and the tectonic setting of greenstones. In: M.J. de Wit and L.D. Ashwal (Eds.), Workshop on Tectonic Evolution of Greenstone Belts. Lunar Planet. Inst. Tech. Rep. 86-10. Lunar Planet. Inst., Houston, pp. 174–176.

Phinney, W.C., Morrison, D.E., and Maczuga, D.E., 1986b. Petrogenesis of calcic plagioclase megacrysts in Archean rocks. In: L.D. Ashwal (Ed.), Workshop on Early Crustal Genesis: the World's Oldest Rocks. Lunar Planet. Inst. Tech. Rep. 86-04. Lunar Planet. Inst., Houston, pp. 90–92.

Phinney, W.C., Morrison, D.A., and Maczuga, D.E., 1988a. Anorthosites and related megacrystic units in the evolution of Archean crust. J. Petrol., 29: 1283–1323.

Phinney, W.C., Morrison, D.A., and Maczuga, D.E., 1988b. Tectonic implications of anorthosite occurrences. In: L.D. Ashwal et al. (Eds.), Workshop on the Deep Continental Crust of South India. Lunar Planet. Inst. Tech. Rep. 88-06. Lunar Planet. Inst., Houston, pp. 135–137.

Ramadurai, S., Sankaran, M., Selvan, T.A., and Windley, B.F., 1975. The stratigraphy and structure of the Sittampundi complex, Tamil Nadu, India. J. Geol. Soc. India, 16: 409–414.

Ramakrishnan, M., Moorbath, S., Taylor, P.N., Anatha Iyer, G.V., and Viswanatha, M.N., 1984. Rb-Sr and Pb-Pb whole-rock isochron ages of basement gneisses in Karnataka craton. J. Geol. Soc. India, 25: 20–34.

Riccio, L., 1981. Geology of the northeastern portion of the Shawmere anorthosite complex, District of Sudbury. Ontario Geol. Surv. Open File Rep. 5338, 101 pp.

Rivalenti, G., 1976. Geochemistry of metavolcanic amphibolites from south-west Greenland. In: B.F. Windley (Ed.), The Early History of the Earth. Wiley, London, pp. 213–224.

Sharkov, E.V., 1984. Anorthosite massifs of the Kola peninsula (in Russian). In: M.S. Markov and O.A. Bogatikov (Eds.), Anorthosites of the Earth and Moon. Nauka, Moscow, pp. 5–61.

Simmons, E.C., Hanson, G.N., and Lumbers, S.B., 1980. Geochemistry of the Shawmere anorthosite complex, Kapuskasing structural zone, Ontario. Precambrian Res., 11: 43–71.

Sleep, N.H. and Windley, B.F., 1982. Archaean plate tectonics: constraints and inferences. J. Geol., 90: 363–379.

Subramaniam, A.P., 1956. Mineralogy and petrology of the Sittampundi complex, Salem District, Madras State, India. Geol. Soc. Am. Bull., 67: 327–379.

Sukhanov, M.K., 1984. Anorthosite association of the Anabar Shield (in Russian). In: M.S. Markov and O.A. Bogatikov (Eds.), Anorthosites of the Earth and Moon. Nauka, Moscow, pp. 61–86.

Sutton, J., Windley, B.F., 1974. The Precambrian. Sci. Progr., 61: 401–420.

Taylor, P.N., Moorbath, S., Goodwin, R., and Petrykowski, A.C., 1980. Crustal contamination as an indicator of the extent of early Archean continental crust: Pb isotopic evidence from the late Archean gneisses of west Greenland. Geochim. Cosmochim. Acta, 44: 1437–1453.

Taylor, P.N., Moorbath, S., Chadwick, B., Ramakrishna, M., and Viswanatha, M.N., 1984. Petrography, chemistry, and isotopic ages of Peninsular gneiss, Dharwar acid volcanic rocks, and the Chitradurga granite with special reference to the late Archean evolution of the Karkataka craton. Precambrian Res., 23: 349–375.

Thurston, P.C., Siragusa, G.M., and Sage, R.P., 1977. Geology of the Chapleau area, Districts of Algoma, Sudbury, and Cochrane. Ontario Div. Mines Geosci. Rep. 157, 293 pp.

Vearncombe, J.R., Barton, J.M., and Walsh, K.L., 1988. The Rooiwater Complex and associated rocks, Murchison granitoid–greenstone terrane, Kaapvaal Craton. Trans. Geol. Soc. S. Afr., 90: 361–377.

Weaver, B.L., Tarney, J., and Windley, B., 1981. Geochemistry and petrogenesis of the Fiskenæsset complex, southern west Greenland: nature of the parent magma. Geochim. Cosmochim. Acta, 45: 711–725.

Wiener, R.W., 1981. Tectonic setting, rock chemistry, and metamorphism of an Archean gabbro–anorthosite complex, Tessiuyakh Bay, Labrador. Can. J. Earth Sci., 18: 1409–1421.

Windley, B.F., 1969a. Anorthosites of southern west Greenland. Am. Assoc. Petrol. Geol., Mem. 12, pp. 899–915.

Windley, B.F., 1969b. Evolution of the early Precambrian basement complex of southern west Greenland. Geol. Assoc. Canada, Spec. Pap. 5, pp. 155–161.

Windley, B.F., 1973. Archean anorthosites: a review with the Fiskenæsset Complex, West Greenland as a model for interpretation. Spec. Publ. Geol. Soc. So. Africa 3, pp. 319–322.

Windley, B.F., 1970. Anorthosites in the early crust of the Earth and on the Moon. Nature, 226: 333–335.

Windley, B.F. and Smith, J.V., 1970. Archaean high grade complexes and modern continental margins. Nature, 260: 671–675.

Windley, B.F., Bishop, F.C. and Smith, J.V., 1981. Metamorphosed layered igneous complexes in Archean granulite–gneiss belts. Ann. Rev. Earth Planet. Sci., 9: 175–198.

Windley, B.F., Herd, R.K., and Bowden, A.A. 1973. The Fiskenæsset Complex, West Greenland, Part I. A preliminary study of the stratigraphy, petrology, and whole rock chemistry from Qeqertarssuatsiaq. Grønl. Geol. Unders. Bull., 106, 80 pp.

Wood, J., Dekker, J., Jansen, J.G., Keay, J.P. and Panagapko, D., 1980. Mine Centre Area, District of Rainy River. Ontario Geol. Surv. Prelim. Maps P2201 and P2202, Geol. Series.

Yudin, B.A., 1974. Gabbro-labradorites from the Kola peninsula. In: O.A. Bogatikov (Ed.), Anorthosites of the USSR. Nauka, Moscow, pp. 21–29 (in Russian).

Chapter 9

ARCHEAN HIGH-GRADE METAMORPHISM

J.A. PERCIVAL

INTRODUCTION

It is a common perception that high-grade metamorphism typifies rocks of Archean age. While many large granulite complexes are indeed Archean, there also exist many well-preserved low-grade Archean sequences, implying a range of exhumation magnitude and tectonic controls. The abundance of high-grade rocks in regions of Archean age, along with the widespread occurrence of Archean komatiites, taken to indicate hotter mantle conditions, have given rise to speculation that the early Earth was characterized by considerably higher geothermal gradients than today's. There is growing recognition that many granulites of Archean and younger age formed in continental marginal settings, along highly perturbed geotherms, making estimates of geothermal gradient from metamorphic assemblages relevant only to transient conditions (e.g. Bohlen, 1987).

This chapter analyzes processes involved in high-grade metamorphism through a survey of Archean granulites. "High-grade metamorphism" is generally considered to correspond to regional granulite facies, although migmatites in the upper amphibolite facies are common associates of granulites. Granitoid rocks with primary igneous orthopyroxene (charnockite in the original definition of Holland, 1900), are common components of high-grade metamorphic regions and where crustally derived, bridge the gap between metamorphic and igneous processes (Percival, 1991a; Kilpatrick and Ellis, 1992).

Modern analytical tools have been increasingly applied to high-grade metamorphic belts over the past two decades. Notably, quantitative estimates of metamorphic conditions are available for many regions owing to calibration of thermobarometers applicable to equilibria in the granulite facies (e.g. Newton and Perkins, 1982; Bohlen et al., 1983; Berman, 1991). Prograde paths are generally not preserved in rocks of granulite facies owing to peak conditions beyond homogenization temperatures for garnet (e.g. Ganguly and Chakraborty, 1991). Furthermore, extraction of retrograde *P-T-t* paths from granulites is subject to many pitfalls (e.g. Frost and Chacko, 1989; Harley, 1989; Selverstone and Chamberlain, 1990; Spear and Florence, 1992), but through careful treatment, a variety of paths has been extracted from Archean granulite-facies rocks (e.g. Mezger et al., 1990a; Harley, 1992). Path information can be used in light of lithological, structural and geochronological constraints to derive integrated models of tectonic setting.

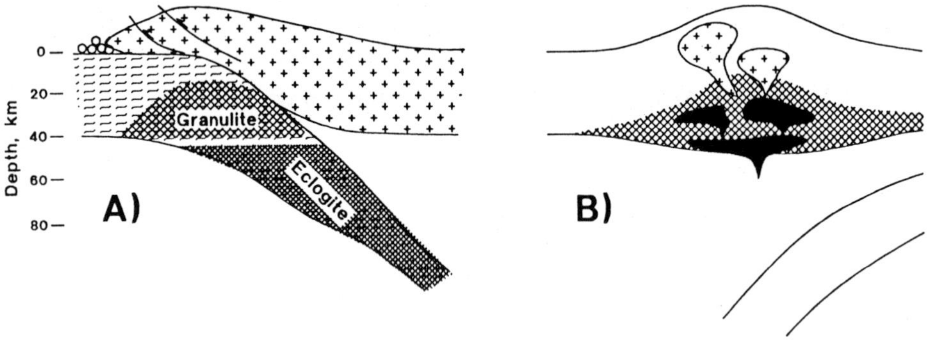

Fig. 1. End-member tectonic environments for the formation of granulites. (A) Continent–continent collision zone in which the lower plate is buried to great depth and heated through thermal relaxation. Zones of potential granulite and eclogite-facies metamorphism are indicated. Clockwise *P-T* paths are predicted for granulites formed in this setting. (B) Magmatic arc setting, showing under- and intra-plated mantle-derived basaltic magmas (black) which upon crystallization produce granulite-facies metamorphism and partial melts of country rock (shaded zone). Magmas of granitoid composition rise and crystallize as charnockitic and granitic plutons (cross pattern). Many high-grade regions have elements of both end members.

From analytical studies, two end-member granulite types can be distinguished in rocks of Archean age (Fig. 1): (1) those formed by granulite-facies metamorphism of existing crust during collisional orogeny, followed by return to normal crustal levels through isostatic rebound; and (2) those formed in the deep crust as an integral part of juvenile crustal production. High-grade metamorphism cannot be isolated from its lower grade equivalents and several regions, including the Pikwitonei domain and Kapuskasing uplift of the Superior Province (see below), have lithological, structural and temporal linkages across a range of grade. Such gradational relationships permit an assessment of differences and analysis of depth-related and lateral variability.

In contrast to views prevailing only a decade ago (e.g. Condie, 1981; Kröner and Greiling, 1984), there is general consensus that some form of plate tectonics existed during the Archean (for a minority view, see Hamilton in Reed et al., 1993). Geodynamic models range from those that differ only slightly from the current situation of thick lithospheric plates on convecting asthenosphere (Burke et al., 1976; Bickle, 1978; 1986; Davies, 1979; England, 1979; Sleep and Windley, 1982; Nisbet and Fowler, 1983; Arndt, 1983; Jarvis and Campbell, 1983; England and Bickle, 1984; Richardson et al., 1984; Richter, 1985; Wilks, 1988), to models that accommodate higher rates of plate motion (Abbott and Hoffman, 1984; Bickle, 1986; Nelson and Forsythe, 1989), to those involving a more chaotic regime dominated by plume/hot spot activity (Fyfe, 1978; Campbell et al., 1989; Hill, 1993). Accordingly, recent interpretations of Archean high-grade metamorphism are set within a plate-tectonic context.

ARCHEAN METAMORPHISM: GENERAL CONCEPTS

Studies of Archean metamorphism initially focused on low-grade greenstone belts as a result of their association with gold and other mineral deposits. Unbroken transitions to regions of high metamorphic grade are rare (Windley and Bridgwater, 1973); most commonly, faults bound both greenstone belts and high-grade regions. Early syntheses of Archean metamorphism, based mainly on studies of granite–greenstone belts, noted the prevalence of high-T, low-P conditions and led to interpretations of Archean geothermal gradients on the order of 100°C/km (e.g. Fyfe, 1973; Collerson and Fryer, 1978; Condie, 1984). In many regions the apparent geothermal gradient is not representative, having been perturbed at the time of metamorphism by granitoid plutonism (e.g. Windley and Smith, 1976; Watson, 1978). Data from deeper crustal levels, in the zone of melt generation rather than pluton emplacement (Grambling, 1981; Newton and Perkins, 1982), indicate elevated geotherms related to major orogenic events, conditions not substantially different from those in more recent orogens (Burke and Kidd, 1978; Collerson and Fryer, 1978; England and Bickle, 1984). In addition to observations from metamorphism, geodynamic constraints (e.g. Bickle, 1978; 1986; England, 1979; Jarvis and Campbell, 1983; Nisbet and Fowler, 1983; England and Bickle, 1984; Richter, 1985) and the occurrence of Archean diamonds (Richardson et al., 1984; Boyd et al., 1985) suggest average continental geotherms not greatly different from modern continental geotherms.

The underlying control of metamorphic conditions and *P-T-t* path in orogens of any age is the tectonic setting and history (Percival, 1990a; Garde, 1990). Because deep crustal rocks are presently exposed at the surface, collisional tectonic models commonly have been applied to explain burial and metamorphism of older protoliths, followed by isostatic uplift (e.g. England and Bickle, 1984; Wilks, 1988). Granulites formed in this setting are characterized by isothermal decompression and clockwise *P-T* paths (Fig. 2a) (Harley, 1989). However, another major environment, recognized through associations between plutonic rocks and high-grade metamorphism (e.g. Wells, 1980a; Mezger et al., 1989; Nutman et al., 1991), is the deep crust of magmatic arcs (Windley and Smith, 1976). Metamorphic rocks formed in this environment underwent isobaric cooling and are characterized by counterclockwise *P-T* paths (Fig. 2b) (e.g. Bohlen, 1987).

The following descriptions of high-grade regions (Fig. 3) are intended to illustrate the diversity in processes and products of Archean high-grade metamorphism. The compilation does not include all known examples and draws heavily on many recent studies of Archean granulites from North America. On the global scale, most Archean granulites date from the Late Archean, although some older metamorphic events have been recognized geochronologically in southwest Greenland, Antarctica, and Siberia. Several additional examples contain Early Archean protoliths that were not metamorphosed until the Late Archean. Pressure, temperature and age estimates are reported from the original sources, without

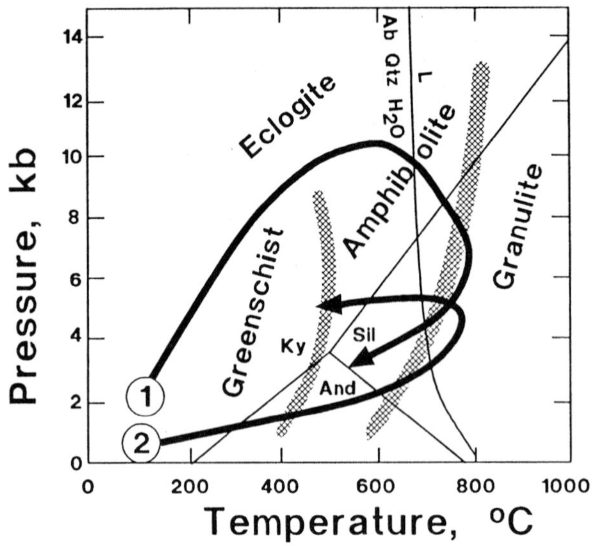

Fig. 2. Schematic *P-T* diagram showing fields of major regional metamorphic facies and typical *P-T* paths for granulites. Path 1 describes rocks affected by collisional thickening (see Fig. 1 (A), leading to a clockwise *P-T* trajectory including isothermal decompression. Path 2 shows the counterclockwise trajectory of rocks metamorphosed in a magmatic arc environment (Fig. 1 (B)), including a late stage of isobaric cooling. Later events are required to exhume granulites formed in this environment.

recalculation to common standards. Most of the studies cited used modern techniques of thermobarometry and U-Pb geochronology, which are summarized below.

HIGH-GRADE METAMORPHISM: TOOLS

Techniques for determining critical parameters of high-grade metamorphism, including pressure, temperature, age, *P-T* path, and fluid composition, have been developed recently and widely applied to Archean regions. However, complete data sets are not yet available for most regions, due at least partly to the nature of assemblages available for study.

Pressure estimates provide control on the depth of metamorphism, and recorded pressure variations within rocks yield constraints on vertical crustal trajectories during metamorphism. Geobarometers developed for common granulite assemblages include garnet–pyroxene–plagioclase–quartz (Newton and Perkins, 1982; Bohlen et al., 1983; Moecher et al., 1988), with a relatively low dP/dT. Estimates from these barometers (using either ortho- or clinopyroxene) are generally consistent with independent constraints from petrogenetic grids (e.g. Hensen and Harley, 1990), owing partly to internally consistent thermochemical databases (e.g. Berman,

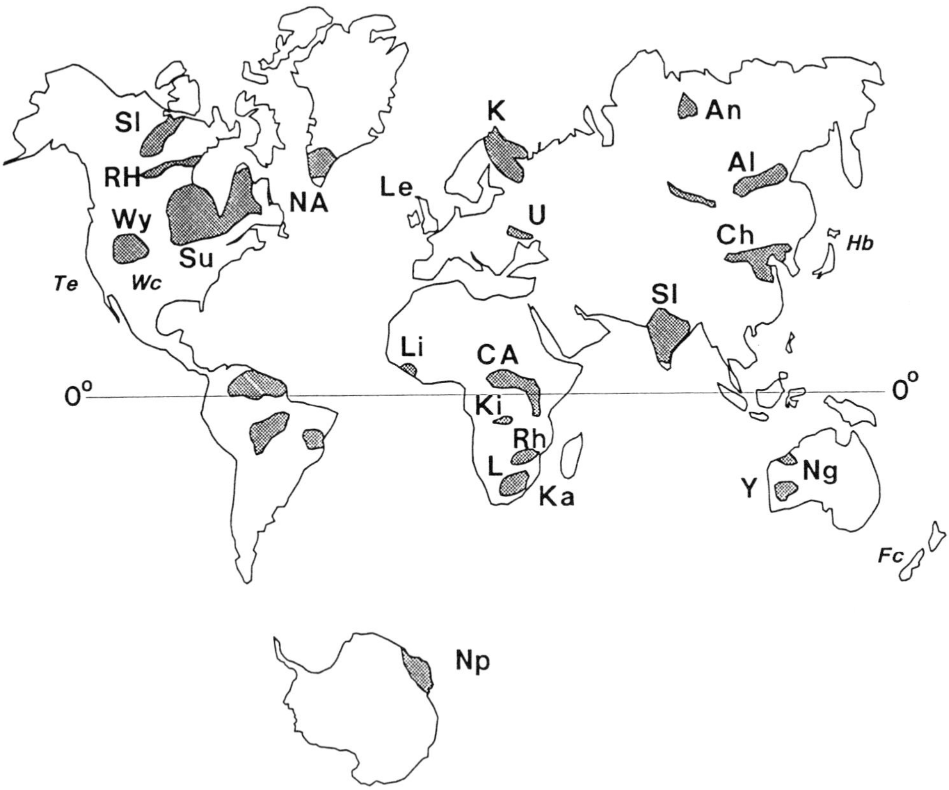

Fig. 3. Global distribution of Archean provinces (shaded) (after Condie, 1981). Granulite localities referred to in the text are labelled: An: Anabar shield; Al: Aldan shield; CA: Central African craton; Ch: Chinese craton; K: Kola shield; Ka: Kaapvaal craton; Ki: Kasai craton; L: Limpopo belt; Le: Lewisian Complex; Li: Liberian craton; NA: North Atlantic craton; Ng: Narryer Gneiss Complex; Np: Napier complex; RH: Rae-Hearne province; Rh: Rhodesion craton; SI: South Indian craton; Sl: Slave province; Su: Superior province; U: Ukranian shield; Wy: Wyoming province; Y: Yilgarn block. Comparative Phanerozoic examples include the Fiordland Complex (*Fc*), Hidaka belt (*Hb*), Tehachapi Complex (*Te*) and Waterman Metamorphic Complex (*Wc*).

1991). Pressure trajectories may be recorded as chemical zonation within minerals, leading to different pressure estimates from mineral cores and rims (e.g. Anovitz, 1991). More evident are mineral reaction rims, coronas and symplectites, recording pressure-induced recrystallization. For example, garnet–clinopyroxene coronas between orthopyroxene and plagioclase suggest isobaric cooling, whereas orthopyroxene–plagioclase overgrowths may indicate isothermal decompression or renewed heating (see Harley, 1989 for review). Caution is necessary in interpretation since similar textures may result from superimposed, unrelated metamorphic events.

Temperature is known to be generally above 700°C in granulites, based on the presence of *in situ* leucosome in rocks of appropriate bulk composition and knowledge of solidus temperatures for granitic rocks. The high temperatures are a disadvantage in extraction of peak conditions, because intra- and intergranular diffusion are efficient at elevated temperature (Ganguly and Chakraborty, 1991), with the result that mineral thermometers are commonly reset during cooling (e.g. Frost and Chacko, 1989; Harley, 1989; Chamberlain and Selverstone, 1990). Evidence of high temperature may be preserved through the cooling process by immobile elements such as Al in pyroxene (Anovitz, 1991; Pattison and Bégin, 1994).

Fluid compositions can be estimated through mineral equilibria involving volatile species, particularly H_2O and CO_2, or through fluid inclusion studies. These two independent techniques commonly provide divergent estimates of metamorphic fluid compositions (e.g. Lamb et al., 1987). It is likely that calculations based on hydrous equilibria of metamorphic assemblages including hornblende or biotite yield fluid activities relevant to peak metamorphic conditions, whereas fluid inclusions probably correspond to compositions trapped at some point on the cooling/uplift path (Lamb, 1990).

Age of metamorphism can generally be determined through U-Pb analysis of zircon, monazite or garnet. Closure temperatures for zircon may be >1000°C (Black et al., 1986), for garnet it is on the order of >800°C (Mezger et al., 1989), and for monazite it is approximately 700°C (Parrish, 1990), making these minerals useful for recording granulite-facies events. In Archean rocks, precision of 1 to 5 m.y. is common. Garnet can be used to date reactions that produced it (Mezger et al., 1989), although low U and Pb contents make analysis challenging, and bulk grains, rather than specific zones, of garnet are required to obtain enough material. Garnet-whole rock Sm-Nd isochrons also hold promise of precise ages, including individual garnet zones (Burton and O'Nions, 1991). Zircon is a common metamorphic mineral, although specific reactions to produce it and their P-T-X_{fluid} conditions have not been defined. New metamorphic overgrowths on older cores or recrystallized zones may be dated by ion probe, or conventionally by removing and dating growth rims. Monazite probably dates cooling through the closure temperature in most granulite areas. Additional, lower-temperature geochronometers, such as titanite (~600°C) and rutile (~400°C) (e.g. Mezger et al., 1990b) can be used to determine cooling rates and thereby draw inferences on uplift rate and tectonic process controls.

P-T-t paths can be determined through a variety of techniques. The form of the retrograde path provides information on the cooling and exhumation history in many high-grade areas (e.g. Bohlen, 1987; Harley, 1989; Fig. 2). Prograde paths are generally not preserved in granulite-facies minerals, owing to high-temperature intragranular diffusion, but mineral inclusions may provide a qualitative sense of prograde paths (e.g. Harley and Hensen, 1990). Integrated studies, using geochronology to define depositional and metamorphic ages, provide more complete information. Where possible, data from high structural levels can and should be incorporated into integrated models of crustal evolution (e.g. Mezger et al., 1990a; Percival and West, 1994).

EARLY ARCHEAN (>3.5 Ga) METAMORPHISM

Some parts of the North Atlantic craton (Fig. 3) experienced granulite-facies metamorphism prior to 3.5 Ga. In southern West Greenland, metamorphism of the 3.87–3.70 Ga, mainly tonalitic rock assemblage was predominantly to amphibolite facies at 3.65 Ga (Nutman et al., 1989; 1993); small occurrences of higher metamorphic grade are widespread. Granulite-facies mafic gneiss associated with this event has been dated by U-Pb on metamorphic zircon at ~3.60 Ga (Baadsgaard et al., 1984) and yields *P-T* conditions of 630–700°C at 8–10 kb (Griffin et al., 1980). Younger Archean (~2800 Ma) granulite-facies metamorphism in the same area (cf. Wells, 1979; 1980b; see below) has probably altered mineral compositions used in the thermobarometry to produce the anomalously low-temperature results.

MID-ARCHEAN (3.5–3.0 Ga) METAMORPHIC COMPLEXES

The Narryer Gneiss Complex of Western Australia (Fig. 3; Myers, 1988; 1993) contains Early Archean protoliths including 3.73 Ga anorthosite and 3.68 Ga monzogranite gneiss (Kinny et al., 1988), as well as Mid-Archean (3.49–3.38 Ga) syenogranite (Nutman et al., 1991). High-grade (750–850°C, 7–10 kb) metamorphism (Muhling, 1990) occurred at 3.30 Ga (Kinny et al., 1988) in association with emplacement of granites and pegmatites (3.30–3.28 Ga; Nutman et al., 1991) and was followed by isobaric cooling (Muhling, 1990). Both the *P-T* path and ages support a close association between widespread felsic magmatism and high-grade metamorphism.

Early components of the Stanovik complex of the Aldan Shield, eastern Siberia (Fig. 3), also underwent granulite-facies metamorphism in the Mid-Archean (~3.1 Ga; Moskovchenko et al., 1993). Similarly, in the Novopavlovsk complex of the Ukranian Shield (Fig. 3), enderbite dated at 3.44 Ga (Bibikova et al., 1990) may correspond to metamorphism of rocks as old as 3.65 Ga at ~3.4 Ga.

SUPERIOR PROVINCE: THREE TYPES OF GRANULITE IN A SINGLE CRATON

The Late Archean represents a time of vigorous crustal growth and reworking of older Archean lithosphere. High-grade metamorphism is commonly associated with these processes and exhibits a wide range of styles and products.

The Superior Province of the Canadian Shield (Figs. 3 and 4) is the largest preserved Archean craton in the world and contains a diverse range of belt types that are generally interpreted in a plate tectonic context (e.g. Card, 1990; Williams et al., 1992; Kimura et al., 1993). Known for both its low-grade belts and granulites, it has the unique distinction of exposing low- to high-grade transitions through greenstone belts in oblique crustal cross section (Fountain and Salisbury, 1981). The Kapuskasing uplift occurs within the core of the craton and the Pikwitonei domain at its margin (Percival et al., 1992a). In addition to the examples of exhumed deep crust, the Superior Province has two of Earth's largest massif granulite complexes, the Minto block and Ashuanipi complex (Fig. 4). A third granulite type, consisting of older rocks reworked during a Late Archean collision, occurs in the Minnesota River Valley (Fig. 4).

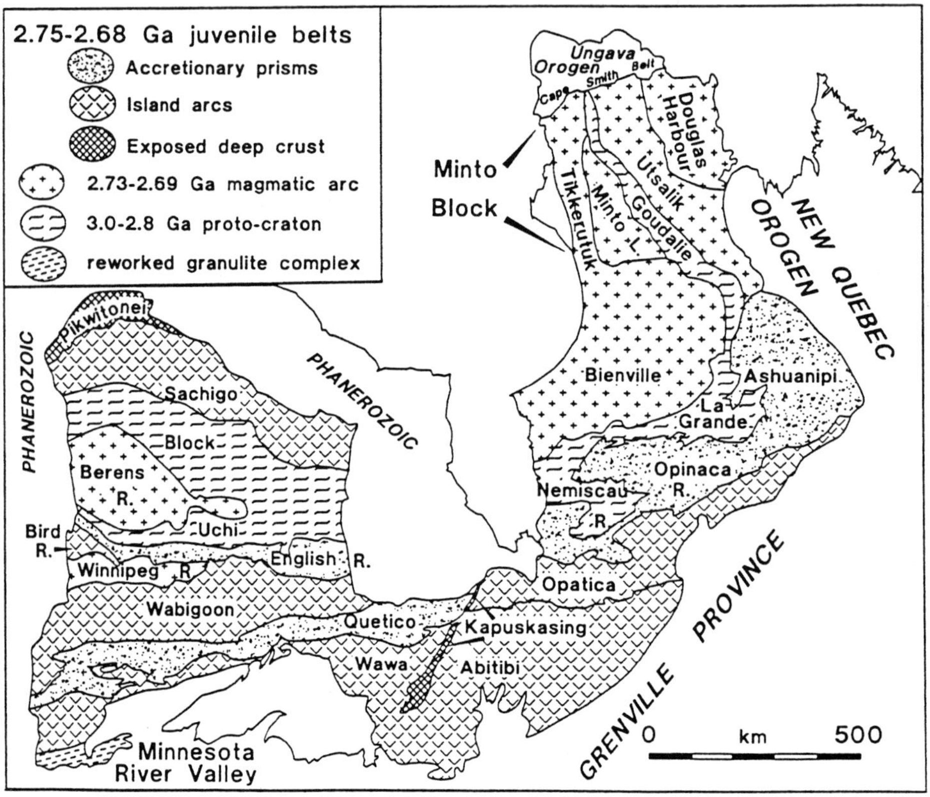

Fig. 4. Tectonic map of Superior Province (modified after Card, 1990; Percival et al., 1992), showing location of granulites discussed in the text.

Exhumed deep crust

Several Late Archean granulite occurrences of the Superior Province represent exhumed parts of the deep crust (Percival, 1990a), exposed in oblique crustal cross sections (Percival et al., 1992a). These regions have relatively simple geological histories, having evolved during single magmatic–metamorphic episodes at ~2.7 Ga. The examples of the Pikwitonei domain and Kapuskasing uplift nevertheless illustrate considerable structural–metamorphic complexity and a protracted period of high-temperature metamorphism that may be typical of granulites with prolonged deep crustal residence times.

The Pikwitonei domain occurs at the northwestern margin of the Superior Province (Fig. 4), adjacent to the 1.9–1.8 Ga Trans–Hudson orogen which probably was responsible for its exhumation. Pikwitonei granulites may represent the deep crust beneath the Cross Lake granite–greenstone subprovince based on common rock units of similar age, including tonalite and pillowed mafic volcanic rocks (Weber and Mezger, 1990). Lithological proportions vary regionally and the Pikwitonei domain is dominated by felsic components, with some anorthosite, paragneiss and mafic gneiss. Age data indicate that metamorphism at amphibolite and granulite facies lasted from 2744 to 2590 Ma, punctuated by zircon and garnet growth at discrete times (2744–2738, 2700–2687, 2660–2637, and 2605–2591 Ma; Mezger et al., 1989) (Fig. 5). The 2700 Ma event corresponds to emplacement of hot (>1100°C) orthopyroxene-bearing tonalitic magma (enderbite), a probable important heat source for the regional metamorphism (Mezger et al., 1990a). Leucosome formation at 2695 and 2637 Ma (Krogh et al., 1986) may correspond to development of dilatant sites during structural events. The metamorphic peak, at 2648–2641 Ma, varied with structural level in the section, from 575°C at 3 kb, through 750°C at 7 kb, to 830°C at 7.5–8 kb (Mezger et al., 1990a). Isobaric cooling was followed by late granite emplacement in the interval 2629–2591 Ma and continued slow cooling into the Early Proterozoic (Fig. 5; Mezger et al., 1990b).

Mezger (1992) summarized the *P-T-t* path for the Pikwitonei domain and proposed tectonic causes for the complex history (Fig. 5). Evidence from Sm/Nd studies suggests a ~2.9 Ga crustal prehistory for the domain, interpreted as pre-existing continental crust. The Late Archean metamorphism (2744–2590 Ma) relates to construction of a continental magmatic arc, with periodic injection of felsic and anorthositic intrusions under high-grade metamorphic conditions. Crustal thickening associated with mantle-derived magmatism led to a counterclockwise *P-T* path, followed by tectonic stability that yielded an isobaric cooling regime. Late, unrelated uplift was probably caused by collision between the Churchill and Superior Provinces at ~1.8–1.9 Ga (Weber, 1990).

In the central Superior Province, the intracratonic Kapuskasing uplift exposes the deep crust of the Abitibi–Wawa granite–greenstone belt (Fig. 4; Percival and Card, 1983). Supracrustal rocks of the high-level (2–3 kb) greenschist-facies belts

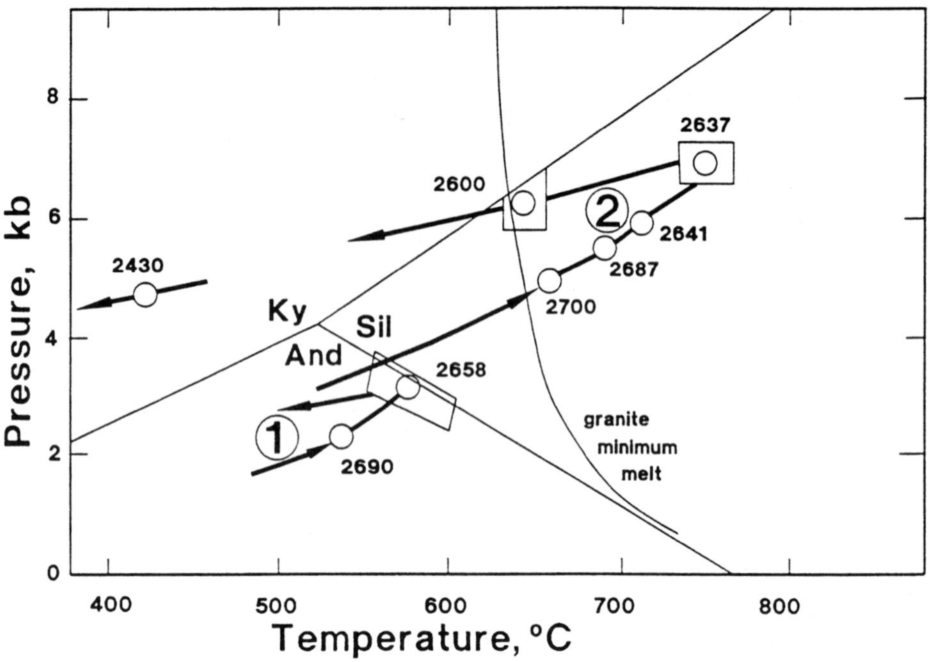

Fig. 5. *P-T-t* path for metamorphism in the Pikwitonei domain, Superior Province, Manitoba (modified after Mezger et al., 1990a). Boxes represent defined *P-T* conditions based on metamorphic assemblages and circles indicate U-Pb ages on metamorphic minerals. *P-T* paths for both shallow (1) and deep (2) crustal levels are counterclockwise, suggesting metamorphism in a magmatic arc environment followed by later, unrelated exhumation.

have U-Pb zircon ages mainly in the range 2.75–2.70 Ga (Corfu, 1993), and are cut by several suites of plutonic rock (Sutcliffe et al., 1993). Polyphase tonalitic gneisses, representing intermediate crustal levels (4–6 kb igneous hornblende barometer pressures) were emplaced mainly between 2.72 and 2.66 Ga (Moser, 1994). Structurally beneath a mid-crustal discontinuity (Percival, 1986; Fountain et al., 1989) are high-grade rocks of the Kapuskasing structural zone, metamorphosed to upper amphibolite and granulite facies (Percival, 1983; Mäder et al., 1994). Protoliths are mainly igneous rock types of tonalitic, dioritic and anorthositic composition. Ages of relict supracrustal units are known only from Nd model ages on mafic gneiss and paragneiss (2.75–2.70 Ga; McNutt and Dickin, 1989). Structural style and orientation of the dominant foliation vary systematically with structural level. Upright folds and cleavage at high levels change downward to a pervasive, late ductile re-orientation of gneissosity into subhorizontal orientations (Bursnall et al., 1994). Both supracrustal and intrusive rocks are migmatitic, with leucosome of tonalitic composition (Fig. 6). Thermobarometry based on garnet–clinopyroxene–hornblende–plagioclase–quartz in mafic gneiss and garnet–ortho-

Fig. 6. Mafic gneiss from the Kapuskasing structural zone with characteristic modal layering of garnet–clinopyroxene–hornblende–plagioclase–quartz assemblages (Percival, 1990b). Metamorphic conditions of 10–11 kb, 730–780°C are indicated for this outcrop. Note abundant tonalitic leucosome in concordant and dilatant orientations. Zircons from different structural sites in this outcrop produced a range of U-Pb ages. A prograde garnet–clinopyroxene–plagioclase–hornblende zone gave 2640 Ma, a retrograde hornblende-rich equivalent 2630 Ma, and a boudin-fill pegmatite 2585 Ma (Krogh, 1993). Hammer handle is 30 cm long.

pyroxene–biotite–plagioclase–quartz in paragneiss indicates 750–790°C at 8–11 kb, probably close to peak conditions (Mäder et al., 1994) (Fig. 7). Metamorphic fluids probably contained both water, based on evidence for partial melting (Percival, 1983) and CO_2, present in "primary" fluid inclusions (Rudnick et al., 1984). Graphite grain-boundary films (Mareschal et al., 1992) present in all Kapuskasing rock types may have precipitated from late CO_2-rich fluids (cf. Frost et al., 1989).

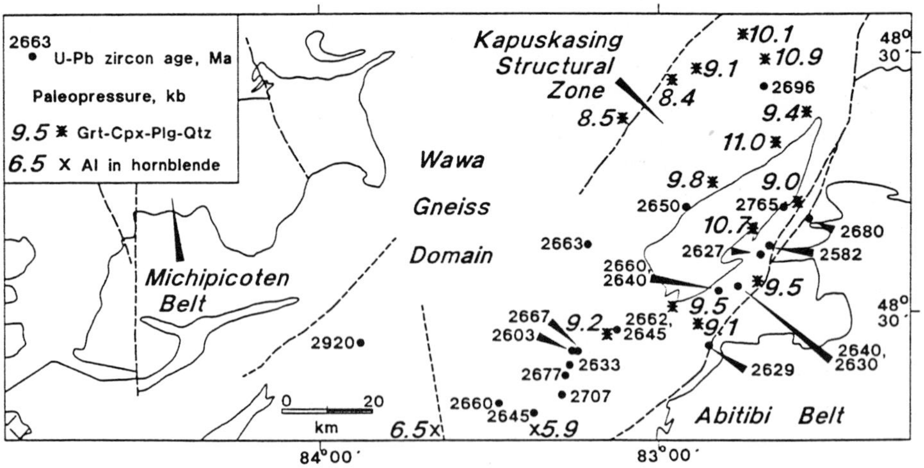

Fig. 7. Generalized geological map, with *P-T* conditions and U-Pb ages of metamorphic zircon from garnet–clinopyroxene–hornblende–plagioclase–quartz granulites of the Kapuskasing structural zone, Superior Province, Ontario (after Percival, 1983; Mäder et al., 1994; Krogh, 1993). U-Pb zircon ages from Percival and Krogh (1983) and Krogh (1993). Pressures from Al content of hornblende in calc-alkaline tonalites calculated with calibration of Schmidt (1992).

Little path information is preserved in the chemically homogeneous high-grade minerals. Zircon of metamorphic origin occurs within mafic gneiss (Percival and Krogh, 1983) and provides estimates of the timing and duration of the high-grade event (Krogh, 1993) (Figs. 7 and 8). Ages in a wide range between 2695 and 2625 Ma occur in low-uranium zircons from high-grade gneisses with 10±1 kb signatures and appear to record prograde metamorphism, whereas younger, high-uranium zircons and overgrowths (2640–2580 Ma) are associated with local amphibolite-facies retrogression and late extensional deformation (Krogh, 1993; Moser, 1994). The wide (110 Ma) range of zircon ages, in addition to a large spread of titanite dates (2680–2493 Ma) from the Kapuskasing zone indicates a protracted period of high-grade metamorphism and plutonism (Corfu, 1987) followed by slow cooling, from which an extended period of residence in the deep crust can be inferred (Fig. 8).

The lower crust of the Abitibi greenstone belt 150 km to the east was sampled by Cretaceous kimberlite plugs. Deep crustal xenoliths consist mainly of garnet–clinopyroxene–plagioclase mafic granulite with zircon ages in the range 2580–2490 Ma (D.E. Moser, personal communication, 1993; Moser and Heaman, 1994). Rock types are similar to Kapuskasing mafic gneisses, although the xenoliths are less hydrous and may come from deeper levels. Together with the Kapuskasing oblique crustal profile, representing upper and middle crust, the xenoliths provide age control on metamorphism throughout the crust of an Archean greenstone belt (Fig. 9). A striking feature is the apparent downward decrease in the age of

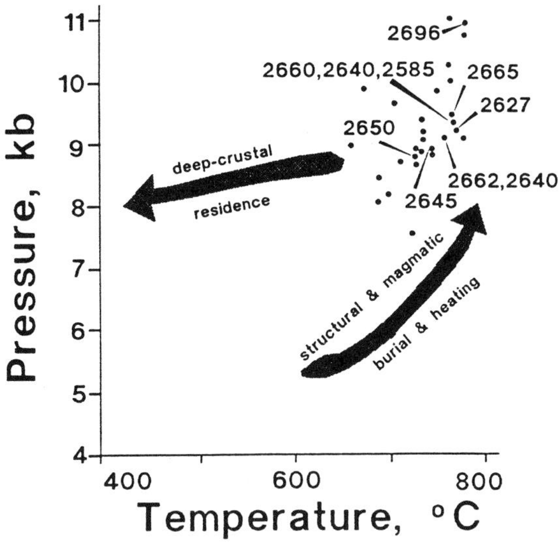

Fig. 8. *P-T* diagram showing range of apparent metamorphic conditions based on garnet–clinopyroxene–plagioclase–quartz equilibria (Mäder et al., 1994), along with U-Pb dates of metamorphic zircon (Percival and Krogh, 1983; Krogh, 1993) from representative Kapuskasing granulites. The *P-T-t* data describe a deep crustal (8–11 kb) environment for granulites of the Kapuskasing zone, from 2696 to at least 2585 Ma. Still younger titanite ages indicate continued elevated temperatures until ~2500 Ma, and Early Proterozoic Rb-Sr biotite ages (Percival and Peterman, 1994) suggest continued deep-crustal residence until ~1900 Ma. The long-term high-temperature metamorphism, association with voluminous tonalitic intrusive rocks, and retarded cooling are consistent with a postulated counterclockwise path and evolution in a deep magmatic arc setting.

metamorphism, as recorded in zircon in mafic rocks, from near 2680 Ma at high crustal levels, to 2660–2620 in the middle crust, to 2580 and 2490 Ma at the deepest levels represented. The pattern is difficult to explain by the normal upward prograde movement of heat from the mantle, which should affect the deepest levels earliest. There is no indication that metamorphic temperatures approached those at which zircon is reset, and therefore it appears that zircon grew at progressively deeper levels late during the metamorphic evolution. Models that account for this phenomenon involve either late tectonic injection of slabs of mafic rock into the lower crust, which then underwent metamorphism (tectonic underplating; Krogh, 1993), or episodic late zircon growth at progressively deeper structural levels in a cooling, ductile crust in response to structurally-induced recrystallization (Percival and West, 1994).

The slow rate of cooling was used by Oxburgh (1990) to derive an estimate of lithospheric thickness of >80 km in this part of the Superior Province at ~2.65 Ga. The value is remarkably consistent with that (80–90 km) derived by Windley and Davies (1978) on the basis of the spacing (corrected for deformation) of 2.72–2.70

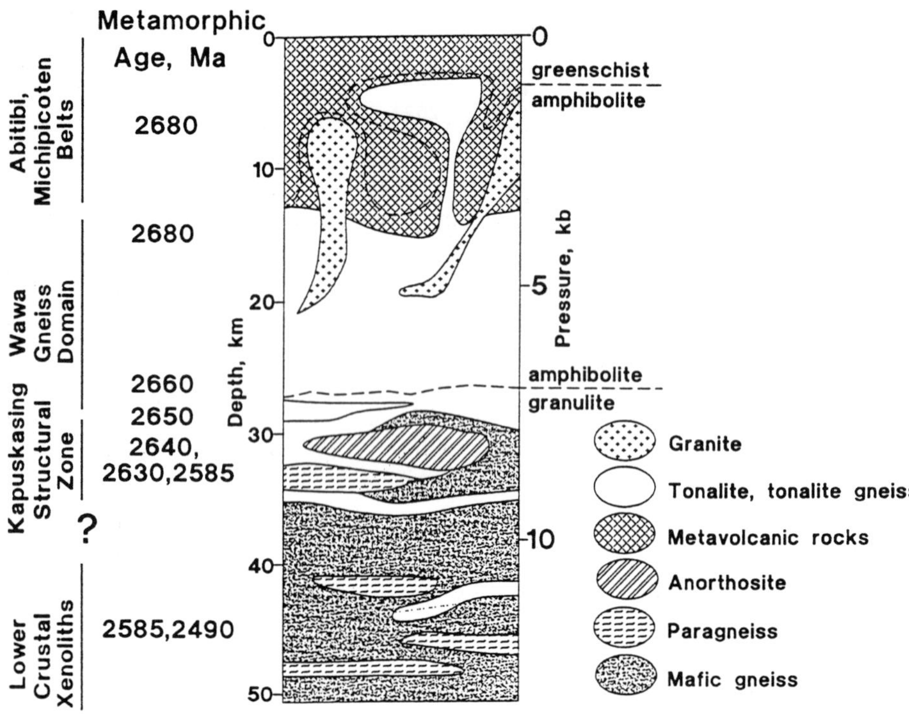

Fig. 9. Composite crustal cross section of central Superior Province showing age progression of metamorphism with crustal level. The section incorporates data from greenstone belt (2–3 kb), exposed granulites (8–11 kb) and lower crustal xenolith (10–15 kb?) levels. The section shows progressively younger metamorphic ages with increasing depth.

Ga volcanoes in the same region. However, recent views on the assembly of the Abitibi belt from distinct lithotectonic terranes (e.g., Kimura et al., 1993; Ludden et al., 1993) may invalidate the volcano spacing analysis. The presence of thick Archean mantle lithosphere beneath the Superior Province has been inferred on the basis of mantle anisotropy (Silver and Chan, 1988) and seismic tomography (Grand, 1987; Hoffman, 1990).

Uplift of the Kapuskasing structure along a brittle thrust fault has been attributed to Early Proterozoic (~1.9 Ga) tectonism elsewhere in the Canadian shield (Percival and McGrath, 1986). Evidence for the timing of uplift includes independent studies of the cooling history (Percival and Peterman, 1994), structural evolution (Bursnall, 1990; Bursnall et al., 1994), seismic reflection data (Percival et al., 1989), and analysis of mafic dyke swarms of 2.45–2.04 Ga (West and Ernst, 1991; Bates and Halls, 1991: Percival et al., 1994a). The brittle uplift was coupled with ductile shortening in the lower crust to produce a crustal root 5–10 km thick (Boland and Ellis, 1989; Parphenuk et al., 1994; Percival and West, 1994).

A long period of deep-crustal residence followed by late unrelated uplift is consistent with a magmatic arc setting during the Archean. The large volume of tonalite emplaced at mid- to upper-crustal depths was derived from lower crustal or slab sources (Rudnick and Taylor, 1986; Truscott and Shaw, 1990). Heat advected by plutons into the mid-crust would have augmented normal orogenic thermal relaxation, leading to a long and complex metamorphic history, as well as to a blurred isotopic character as predicted by Wells (1980a) and supported by subsequent workers (Bohlen, 1987; Mezger et al., 1990a; Harley, 1989).

Similar oblique exposures that traverse from greenstone belts to coeval deep-crustal granulites occur in other regions of Archean age, and display comparable crustal structure. For example, Glikson and Lambert (1976) applied a similar interpretation to the western Yilgarn craton of Australia (Fig. 3) and the Kasila Group of the Liberian craton (Fig. 3) of Sierra Leone, has many similar features (Rollinson, 1982; Williams, 1988). In the predominantly greenschist– to amphibolite-facies Slave Province (Fig. 3) of northwestern Canada, granulites metamorphosed at 4–6 kb (Farquhar et al., 1993) occur locally in a Proterozoic uplift (Henderson and Schaan, 1993).

Granulites from the deep crust beneath greenstone belts, particularly the Superior Province examples, resemble the root zones of Cenozoic magmatic arcs of the southwestern U.S.A. (Ross, 1985; 1989) in terms of metamorphic–magmatic evolution and uplift history. Batholithic infrastructural levels are locally exposed in oblique cross section adjacent to late faults (Saleeby, 1990).

Giant granulite complexes

In the northeastern Superior Province, the ~200,000 km^2 Minto block and ~90,000 km^2 Ashuanipi complex are two of the largest Archean granulite occurrences on Earth (Figs. 3 and 4). Although of similar Late Archean age, their character and geological history are distinct. Both have elements of the juvenile and reworked end-member granulite types. The Minto block had a crustal pre-history that culminated in metamorphism and charnockitic magmatism at 2725 Ma (Percival et al., 1992b), whereas the Ashuanipi Complex represents dominantly sedimentary protoliths deposited shortly before high-grade metamorphism and crustal magmatism at ~2690 Ma (Percival, 1991a).

The Minto block consists dominantly of variably pyroxene-bearing plutonic rocks of charnockitic affinity (Figs. 10 and 11), dated at 2725, 2712 and 2690 Ma, with only rare supracrustal and older gneissic enclaves (Percival et al., 1992b; Stern et al., 1994). Tonalitic gneisses, present as enclaves and in map-scale domains, give U-Pb zircon ages in the range 2.9–3.1 Ga. Nd isotopic values of the plutonic rocks indicate crustal precursors in the range 2.8–3.1 Ga (Stern et al., 1994). Sparse enclaves of granulite-facies paragneiss contain a record of the metamorphic history. Peak metamorphic conditions, based on garnet–orthopyroxene assemblages, varied regionally from 950–1000°C, 7–10 kb, to 750–800°C,

Fig. 10. Massive, homogeneous, medium-grained granodiorite from the eastern Minto block, with orthopyroxene, clinopyroxene, hornblende and biotite of igneous origin. Similar rocks underlie large parts of the Minto block and have consistent 2725 Ma zircon ages (Percival et al., 1992b; Stern et al., 1994). Lens cap is 5 cm in diameter.

5–6 kb (Bégin and Pattison, 1994) and relate to emplacement of pyroxene-bearing granitoid rocks, which occurred at elevated temperatures and 4–6 kb pressure, based on igneous hornblende barometry (Percival et al., 1992b). Retrograde *P-T* paths indicate near isobaric cooling, consistent with the dominantly magmatic thermal history (Bégin and Pattison, 1994). The high-grade metamorphism can be directly related to the presence of hot, dry plutonic rocks (Figs. 10 and 11), although the ultimate source of heat that produced crustal melting is likely to have been mantle-wedge-derived mafic magmas that mixed with older crust (Stern et al., 1994; Hildreth and Moorbath, 1988). A tectonic model for the Minto block (Fig. 12) involves construction of 2.72-Ga continental magmatic arcs on older (2.9–3.1) felsic crust, followed by back-arc rifting (Skulski et al., 1994), collision and renewed metamorphism and magmatism (Percival et al., 1994b).

Protoliths in the Ashuanipi Complex (Fig. 4) are mainly greywacke, based on local relict rhythmic layering in garnet–biotite–orthopyroxene paragneiss, with some iron formation (Lapointe and Chown, 1993). The composition, character and age of the paragneiss resemble those of metagreywackes which occur in lower grade belts west of the complex (Fig. 4) (Percival, 1989; 1990). Early (pre-metamorphic) tonalites (Percival, 1991b) occur as sills and plutons. Large parts of the complex consist of plutons and batholiths of enclave-laden orthopyroxene ± garnet-bearing granodiorite (charnockite) (Fig. 13), whose composition resembles

Fig. 11. Massive, homogeneous, medium-grained granodiorite with igneous pyroxenes and hornblende, from the western Minto block. The mafic enclave has a hornblende-rich interior and orthopyroxene-rich reaction selvage inferred to have been produced by dehydration in an anhydrous magma (after Percival et al., 1992b). Hammer head is 17 cm long.

that of metasedimentary and tonalitic country rocks (Percival, 1991a). *P-T* conditions for both igneous and metamorphic rock types were in the range 700–850°C at 3.5–6.5 kb (op. cit.). Early CO_2-rich fluid inclusions indicate a clockwise retrograde *P-T* path for the complex (Moritz and Chevé, 1992). A relatively short history for the complex is implied by detrital zircon as young as 2.70 Ga and plutonic rocks of 2.68–2.66 Ga (Percival et al., 1992b). Although little mineralogical *P-T* path information is preserved, U-Pb monazite dates of 2.67–2.63 Ga suggest rapid initial cooling to <700°C (Percival et al., 1992b), presumably due to exhumation immediately following metamorphism. Rapid (<20 Ma) heating of protoliths was necessary to produce high degrees of crustal melting. Both the time-scale and implied high crustal temperatures appear inappropriate for thermal relaxation processes following collisional thickening (cf. England and Thompson, 1984; 1986); mantle-derived mafic magmas are required as the ultimate heat source. Like the linear metasedimentary belts to the west, which also have high-temperature, low-pressure granulite occurrences (Percival, 1989), the Ashuanipi Complex has been interpreted as part of an accretionary prism (Percival and Williams, 1989; cf. Platt, 1986), deposited and deformed rapidly in a trench setting. The subsequent high-temperature metamorphism occurred following an arc collision which placed the thick sedimentary wedge in a back-arc environment (Fig. 14).

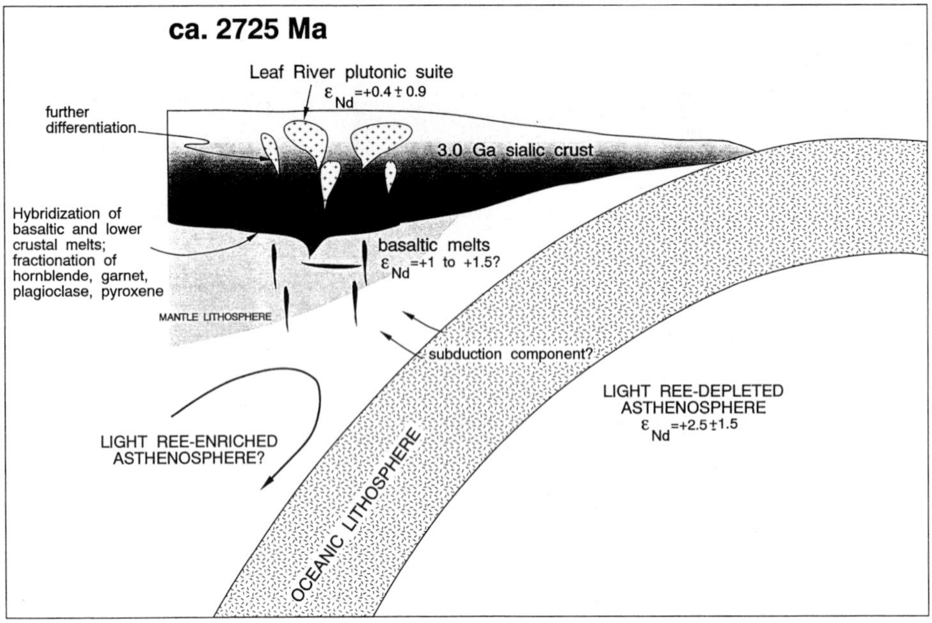

Fig. 12. Tectonic model for evolution of the Minto block in magmatic arcs built on ~3 Ga continental crust (after Stern et al., 1994). Basaltic melts originating in a slab-fluxed mantle wedge pond at the base of the crust, producing and mixing with lower-crustal melts. The high-temperature, anhydrous melts rise to crystallize as charnockitic and more hydrous granitic plutons (Leaf River plutonic suite). Granulite-facies metamorphism occurs in association with emplacement of the charnockitic plutonic rocks.

Areas reworked in the granulite facies

The Minnesota River Valley gneiss complex (Fig. 4) contrasts with other granulites of the Superior Province in having developed from significantly older protoliths. The oldest components of the complex are orthogneisses of 3.6, 3.3 and 3.0 Ga age that were reworked by high-grade metamorphism and granite plutonism at 2.6 Ga (Goldich et al., 1980). Metamorphic conditions, based on thermobarometry of garnet–orthopyroxene assemblages in mafic gneiss and paragneiss, were of the order of 6 kb, 700°C (Perkins and Chipera, 1985; Moecher et al., 1986). The occurrence of old protoliths is rare in the Superior Province and led to the suggestion that the Minnesota River Valley complex is an allochthonous terrane that collided with the southern Superior Province at ~2.6 Ga (Sims et al., 1980).

Parts of the Winnipeg River subprovince (Fig. 4) also consist of older orthogneiss protoliths (to 3.2 Ga; Corfu, 1988) metamorphosed at ~2.7 Ga. Peak metamorphic conditions reached 630–800°C at pressures of 4–7 kb, during a period of crustal thickening (Beakhouse, 1991), and were followed by cooling to below the titanite closure temperature (~600°C) by 2650 Ma (Corfu, 1988).

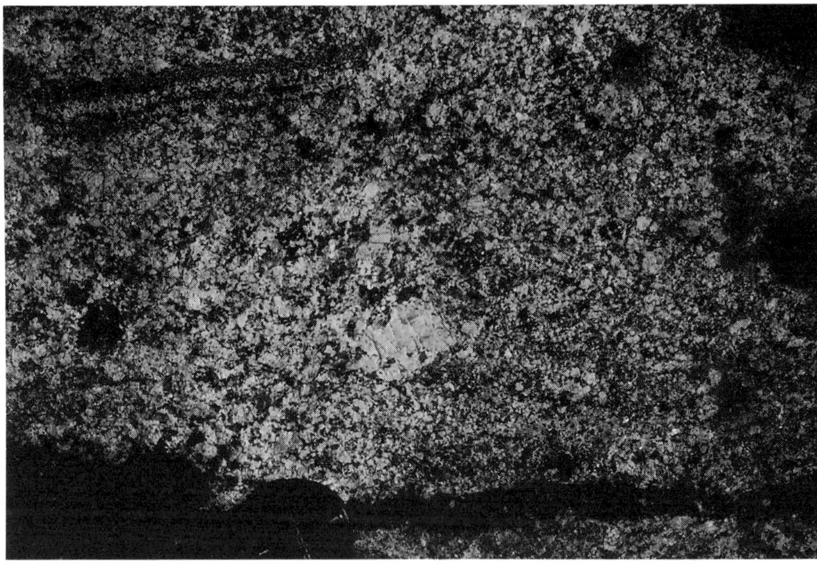

Fig. 13. Massive, homogeneous, garnet–orthopyroxene–biotite granodiorite with characteristic blocky white feldspars (lower center) and biotite-rich schlieren (upper left). These charnockitic igneous rocks underlie large parts of the Ashuanipi Complex and resemble country-rock paragneiss in composition. The plutons, which crystallized at 5–7 kb, are inferred to have been derived from paragneiss at deep crustal levels, through wholesale melting at high temperature and low water activity (Percival, 1991a). Lens cap is 5 cm in diameter.

Although the derived cooling rate is slower than that at high crustal levels, it is relatively fast compared to that recorded in examples of exhumed deep crust. The older rocks may represent microcontinental fragments, incorporated during Late Archean accretionary events (Williams et al., 1992).

HIGH-PRESSURE METAMORPHISM

A distinguishing feature of most Archean metamorphism is its generally high-temperature, low-pressure nature. Various authors have cited the lack of Archean blueschists or eclogites as supporting high Archean geothermal gradients (Baer, 1981; Condie, 1981). However, their absence could also be due to lack of preservation or insufficient detailed information, as exemplified by recent discoveries of eclogite in rocks of both Proterozoic and Archean age (Davidson, 1990; M.L. Williams, pers. comm., 1993).

Recent mapping in the Snowbird tectonic zone separating the Rae and Hearne provinces of Canada (Fig. 3) has identified an association of granulite and eclogite-facies rocks at least 2.6 Ga old. The Snowbird zone consists of lozenges of

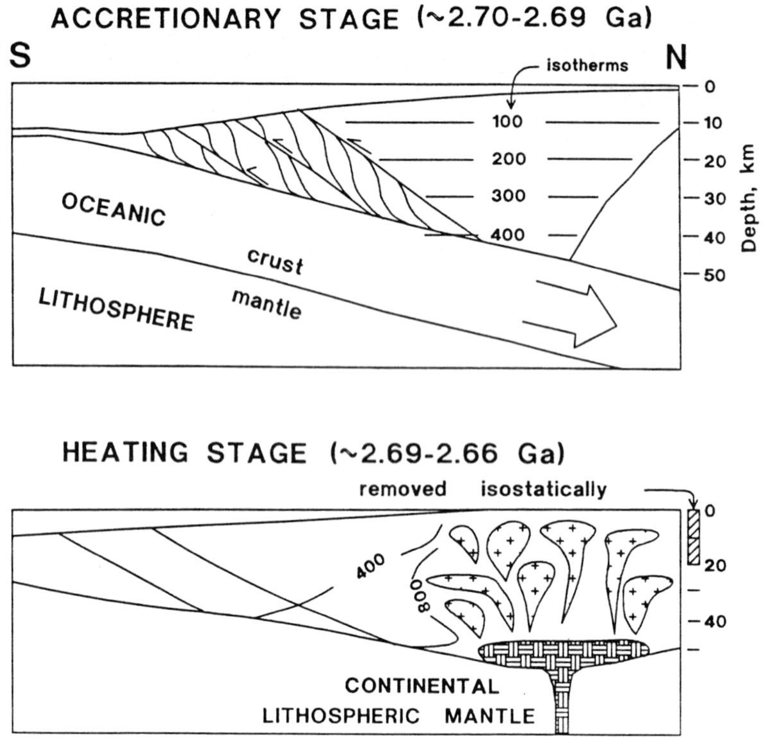

Fig. 14. Model for origin of the Ashuanipi Complex in an accretionary prism setting: (a) deposition and deformation of greywacke in a forearc prism to form sedimentary crust up to 55 km thick; (b) high-grade metamorphism, melting, magmatism and isostatic recovery of the accretionary prism. The rapid heating to conditions in the granulite facies can be attributed to arc collision, subduction zone retreat, and placement of the former forearc in a back-arc setting (after Percival et al., 1992b).

high-grade mafic and anorthositic rocks separated from amphibolite-facies wall rocks by ultramylonites in the granulite facies (Hanmer and Kopf, 1993). Lenses and layers of mafic eclogite occur in association with high-pressure quartzite and diatexite (migmatitic rock containing >50% mobilisate component; Brown, 1973) at the base of a thrust slab near the tip of one shear-bounded lozenge. The mafic rocks contain Mg-rich garnet (50–60 mol% pyrope) and omphacitic pyroxene (~20 mol% jadeite with plagioclase rims). Textural relationships indicate that the eclogitic assemblages developed from granulite-facies precursors and were subsequently variably retrograded, through orthopyroxene–plagioclase symplectites, to granulite facies. Estimates based on mineral equilibria suggest initial granulite-facies conditions of ~1000°C, 10 kb, with an excursion to >20 kb (eclogite facies), and final equilibration at 850°C, 12 kb (Snoeyenbos and Williams, 1994). The high-pressure rocks appear to represent a tectonic flake, rapidly transported from mid- to lower and back to mid-crustal levels within a transcurrent fault zone under high-temperature, ductile conditions. The extent of rocks of similar character is unknown owing to lack of recent mapping in granulite-facies regions northwest of the Snowbird tectonic zone.

HIGH-TEMPERATURE METAMORPHISM

Whereas high-pressure rocks are rare in Archean regions, high-temperature metamorphism is common (e.g. Sandiford and Powell, 1991). Examples of sapphirine–quartz assemblages, indicating conditions in the 900°C range, are known from several cratons, and examples of granulite facies at pressures of <5 kb are also widespread.

Napier Complex, Enderby Land, Antarctica

Evidence of high metamorphic temperatures is preserved in a variety of gneissic rock types of the Napier Complex, Antarctica (Fig. 3) (Dallwitz, 1968; Ellis, 1980; Grew, 1980). The complex is dominated by orthogneisses of tonalitic to granitic composition, containing paragneiss, mafic and ultramafic enclaves. Some tonalites have ages in excess of 3800 Ma. Younger magmatic rocks, generated from crustal precursors in the range 3930–3800 Ma, were emplaced at 3070–3000 Ma during a high-temperature metamorphic event, accompanied by major deformation that may have extended from 3070 to 2900 Ma (Black et al., 1986). High metamorphic temperatures are indicated by a number of observations. Sapphirine–quartz assemblages indicate conditions beyond the stability field of garnet–cordierite (>1050°C; Harley and Hensen, 1990), as do occurrences of osumilite (Ellis, 1980; Grew, 1982), which also indicate very low water activity. An isograd based on the presence of exsolved and inverted pigeonite indicates temperatures in excess of 980°C (Harley, 1987), consistent with garnet–orthopyroxene thermometry indicating 900–1000°C (Harley, 1985). The high temperatures occurred at pressures that vary both regionally and with time, from ~11 to 6 kb (Harley and Hensen, 1990) (Fig. 15). *P-T-t* paths, based on a variety of reaction textures and chemical zonation, suggest a multi-stage cooling and uplift history (Harley et al., 1990). The initial stage locally involved 2–3 kb of decompression at temperatures above 900°C, but the main record involves approximately 300°C of later, near-isobaric cooling (to 5–9 kb) prior to renewed metamorphism in the Late Archean (~2.5 Ga) at 5–8 kb, 650–700°C. Further isobaric cooling is inferred for the interval 2.5–1.1 Ga, at which time Proterozoic metamorphism associated with the Rayner complex brought the deep-crustal granulites toward the surface in an isothermal decompression event (Harley and Hensen, 1990).

A variety of tectonic models has been suggested to account for the high metamorphic temperatures in the Napier Complex and subsequent cooling and uplift history. Because only the post-peak-metamorphic history is recorded in the

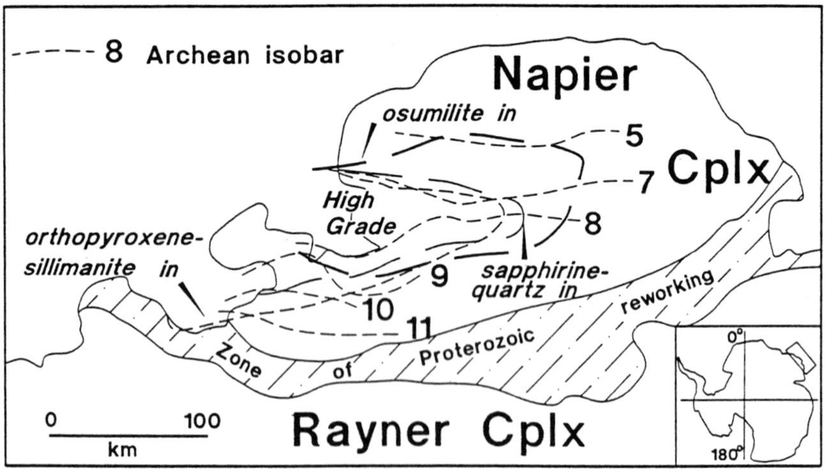

Fig. 15. Metamorphic map of the Napier complex, Antarctica, showing Archean isobars and location of diagnostic high-temperature mineral assemblages (modified after Harley and Hensen, 1990).

P-T paths, the critical clockwise vs anticlockwise nature of the heating phase of the history is not known. Models generally involve a collisional event followed by extensional relaxation (e.g. Ellis, 1987) or collapse (Sandiford, 1989a), with or without accompanying magmatism (Harley et al., 1990). Further geochronology is required to assess the extent and exact timing of magmatism with respect to the metamorphic peak. Clearly, a special combination of circumstances operated to produce the very high temperature granulites of the Napier Complex (Harley and Black, 1987; Harley and Hensen, 1990).

The Lewisian Complex of Scotland

The Lewisian Complex (Fig. 3) contains an additional example of metamorphic temperatures in the 850–920°C range, (Barnicoat, 1983; Sills and Rollinson, 1987; Cartwright, 1992). Corresponding pressures during the 2.7 Ga metamorphism are also high, generally considered to be in the range 8–11 kb. Estimates are based on assemblages in mafic and paragneisses within an area dominated by tonalite (for a review see Cartwright, 1990). Geochemical depletion of U, Th, K, Rb, Cs, and associated elements characterizes gneisses of the Scourie Complex (Weaver and Tarney, 1983), and may be related to anatexis and removal of granitoid components (e.g. Pride and Muecke, 1980; Cartwright, 1990) or flushing by carbonic fluids (Weaver and Tarney, 1981; 1983). A clockwise *P-T-t* path has been constructed from a variety of assemblages and textures (Cartwright, 1990) and is consistent with a tectonic model involving crustal thickening and thermal relaxation (Barnicoat, 1987), as well as input from magmatic sources (Holland and Lambert, 1975).

Labwor Hills, Uganda

In the Labwor Hills of the Central African Craton (Fig. 3), similarly high temperatures are recorded in rocks of probable Archean age (Sandiford et al., 1987). Peak metamorphic conditions of >1000°C, 7–9 kb are indicated by sapphirine–cordierite–quartz assemblages. Retrograde sapphirine–hypersthene–K–feldspar–quartz symplectites suggest isobaric cooling. Exhumation of the lower-crustal granulites is inferred to have resulted from late, unrelated Pan-African tectonism (Sandiford et al., 1987). In the adjacent Kasai craton (Fig. 3), 2.8-Ga granulites experienced conditions of 720°C at 6.7 kb, followed by renewed metamorphism at ~2.4 Ga (Bingen et al., 1988).

CARBONIC FLUIDS IN HIGH-GRADE METAMORPHISM

The role of desiccating fluids in promoting granulite-facies metamorphism has received wide attention over the past decade (e.g. Collerson and Fryer, 1978; Newton et al., 1980; Newton, 1987; 1992a). Evidence for CO_2 streaming was first recognized in the Archean Dharwar craton of southern India, where veinlets and patches of charnockite transect amphibolite-facies tonalites (Fig. 16). Discussion of the relative importance of vapour-absent partial melting and fluid infiltration in dehydration of granulites and geochemical depletion continues to this day (e.g. Vielzeuf and Vidal, 1990; Newton, 1991; Stevens and Clemens, 1993).

Dharwar Craton

The Dharwar craton (Fig. 3) consists of two supracrustal sequences, the older (>3.0 Ga) Sargur Group and younger (3.0–2.5 Ga) Dharwar Supergroup, that occur within the dominantly tonalitic Peninsular Gneiss (3.4–2.6 Ga) (Pichamuthu and Srinivasan, 1984). Supracrustal rocks decrease in abundance from north to south over a 200 km distance, as their metamorphic grade increases from greenschist facies (3 kb) to granulite facies (8 kb) with charnockite massifs (Janardhan et al., 1982; Raith et al., 1983; Hansen et al., 1984; Srikantappa, 1993; Eckert and Newton, 1993). Early isothermal uplift of the granulites to levels corresponding to ~4.5 kb is indicated by symplectitic textures (Mohan and Windley, 1993). In quarry exposures near the orthopyroxene isograd, hornblende–biotite tonalitic gneisses are transformed in patches and along shears (Fig. 16), with subtle geochemical modification (Condie et al., 1982; Stahle et al., 1987), to coarse orthopyroxene-bearing assemblages (e.g. Friend, 1981; 1985). Channelized fluids rich in CO_2 are inferred to have decreased water activity in these domains to the point where hornblende and biotite became unstable at sub-granulite-facies conditions. Massif charnockites to the south have been interpreted as a pervasively CO_2-flushed region (Harris et al., 1982; Newton, 1991).

(a)

Fig. 16. (a) Outcrop of tonalitic gneiss (pale grey) with orthopyroxene-bearing (dark grey) patches and seams (scale card is 8 cm long). (b) Orthopyroxene developed along a minor dextral shear zone. Kabbal Durga quarry, South India. The presence of orthopyroxene suggests that CO_2-rich fluids infiltrated along structurally-controlled sites, dehydrating tonalite at upper amphibolite-facies metamorphic conditions.

High-grade metamorphism is coeval (2510 Ma) with emplacement of the Closepet granite (Friend and Nutman, 1991), a body that extends across the low- to high-grade transition (Friend, 1983). Several workers have postulated genetic links between the metamorphism and granite generation. Friend (1981; 1983) suggested that aqueous fluids, flushed out of granulites by CO_2-rich fluids, promoted partial melting of amphibolite-facies rocks to produce the granite, whereas Newton (1991) proposed that the Closepet granite is a metasomatic body which developed in a mega-shear zone that channeled K-rich fluids. Recent geochronological evidence suggests that a suite of juvenile (mantle-derived) tonalite–granodiorite plutons was emplaced (2.55–2.53 Ga zircon) immediately prior to or

(b)

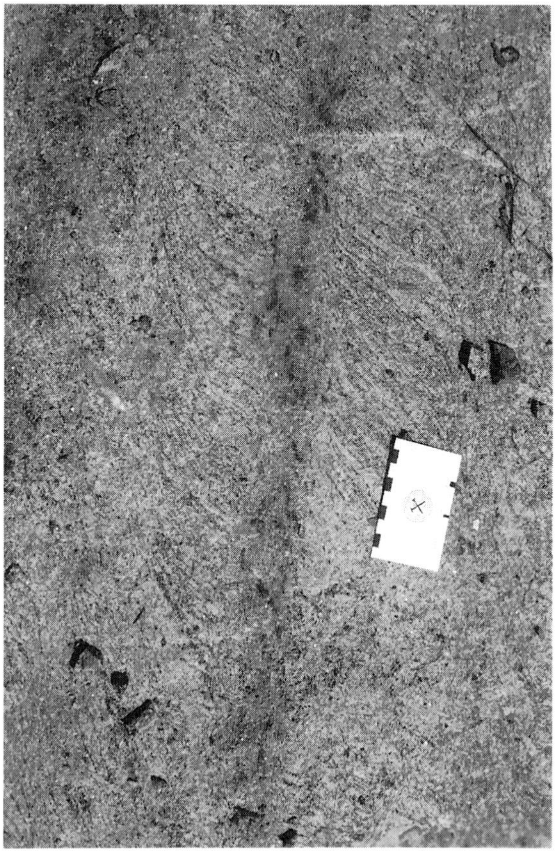

Fig. 16. (b) Caption oppposite.

during the high-grade metamorphism (2.51 Ga monazite) (Peucat et al., 1993). The plutons represent the probable immediate heat source for the metamorphism, although their relationship to the source of carbonic fluids is not known.

Textures and structures characteristic of "incipient charnockitization" are distinctive and widespread in southern India, both in the Dharwar craton and in rocks of Pan-African (~550 Ma) age to the south (e.g. Chacko et al., 1987). The process of carbonic metamorphism therefore appears to be more than a local, Archean phenomenon. However, on a global scale, Archean high-grade metamorphism generally involves migmatite production, and structures characteristic of influence by carbonic fluids are rare. An additional example on a minor scale occurs in 2.8 Ga granulites of West Greenland (Friend et al., 1987).

Granulites of the Dharwar craton contain evidence for several different processes leading to high-grade metamorphism. The presence of old crust (3.3 Ga Peninsular Gneiss) and evidence for rapid decompression suggest collisional tectonics, whereas coeval magmatism and metamorphism argue for evolution in a magmatic arc (Peucat et al., 1993). The origin of carbonic fluids and their role in the metamorphism cannot be linked directly to underlying tectonic causes.

Wind River Mountains, Wyoming

In Archean granulites of the Wind River Range of the Wyoming craton (Fig. 3) (Koesterer et al., 1987), local dehydration effects have also been documented (Frost and Frost, 1987). In these examples, granulitic assemblages are developed adjacent to thin charnockitic dykes and along their extensions (Frost et al., 1989). This phenomenon implies the presence of a mixed CO_2–H_2O volatile component in granitic melt that evolves a CO_2-rich phase early in the crystallization history.

REPEATED GRANULITE-FACIES METAMORPHISM

Several high-grade metamorphic complexes contain evidence for more than one Archean granulite-facies event, and some have an additional high-grade overprint of Early Proterozoic age. Metamorphic events generally correspond to regional orogenic activity and associated emplacement of granitoid rocks (cf. Stüwe et al., 1993). These regions appear to have remained in tectonically active settings for extended periods of time.

Napier Complex, Antarctica

The Napier Complex of Enderby Land (Fig. 3) experienced a very high temperature metamorphic event at ~3 Ga (see above), then cooled isobarically through the remainder of the Archean (Harley and Hensen, 1990). In the latest Archean or Early Proterozoic (2.5–2.46 Ga), the still deeply buried mid-Archean granulites were deformed and subjected to renewed granulite to upper amphibolite facies metamorphism at 650–750°C, 5–8 kb (Black et al., 1983). Coeval emplacement of a suite of A-type granites and high-grade pegmatites may have re-elevated temperatures in the deep crust into the granulite field.

Aldan Shield, Siberia

In the Aldan Shield (Fig. 3), several high-grade Archean metamorphic events have been recognized in the Stanovik complex (Rosen et al., 1994). The earliest event, at ~3.1 Ga, is recorded in a dominantly mafic sequence with components as old as 3.5 Ga (Dook et al., 1989). Conditions reached 1000°C, 10 kb locally, prior to a second granulite-facies event at ~2.7 Ga in which conditions associated with charnockite and enderbite emplacement attained 700–950°C at 9–12 kb (Moskovchenko et al., 1993). Further deformation, metamorphism and magmatism in

Early Proterozoic events at ~2.1 and 1.9 Ga are probably responsible for uplift and exhumation of the metamorphic complex (see Chapter 10, this volume).

Similar events are recognized in the Anabar Shield (Fig. 3) where mafic, sedimentary and tonalitic protoliths of ~3.1 Ga were metamorphosed to the granulite facies (820–950°C, 8.5–11 kb; Rosen, 1990) at ~2.76 Ga (Rosen et al., 1994). Renewed metamorphism and tectonism in the Early Proterozoic (~1.9 Ga) is probably responsible for exhumation of the Archean granulites.

Narryer Gneiss Complex, Western Australia

The Narryer Gneiss Complex (Fig. 3) exhibits a complex history involving at least two high-grade Archean metamorphic events (Myers, 1988). The dominantly granitoid magmatic complex of 3.73–3.4 Ga age was affected by an early (~3.35 Ga) granulite-facies event at 750–850°C, 7–10 kb (Muhling, 1990). The isobaric *P-T* path for this event (op. cit.) suggests that the complex remained at depth until later (~2.62 Ga) reworking, accompanied by leucogranite plutonism, at conditions of 650–750°C, 7.5–8.5 kb. This event is also recognized in the 3.0–2.6 Ga granite–greenstone belts of the Yilgarn craton to the east and is thought to result from juxtaposition of the two terranes (Myers, 1988). Isobaric cooling following this event was followed by uplift in an unrelated late event, the ~1.6–1.9 Ga Capricorn Orogeny (Myers, 1993).

Hebei Province, China

Rocks of Early to Mid-Archean age occur in Late Archean granulites in the eastern Hebei Province of the Chinese craton (Fig. 3) (Jahn and Zhang, 1984; Liu et al., 1990). Supracrustal rocks, including quartzite, carbonate, iron formation and volcanic rocks, were probably deposited prior to 3 Ga (Xuan et al., 1986). The main metamorphic events occurred at ~2.7 Ga, accompanied by major plutonism of granitic and charnockitic rocks (Kaiyi et al., 1990; Liu et al., 1990), and at 2.5 Ga, when metamorphic conditions reached 700°C, 7–8 kb (Sills et al., 1987), and in adjacent areas, ~800°C at 8–10 kb on a clockwise *P-T* path (Lu and Jin, 1993) and 850°C, 9.5 kb on a counterclockwise path (Liu et al., 1993). Early Proterozoic (2.0–1.9 Ga) metamorphism had variable effects on the Archean complex, from low-grade retrogression to renewed granulite facies (Lu and Jin, 1993; Liu et al., 1993).

Wind River Range, Wyoming province

The Wind River Range contains evidence for three Archean metamorphic events. The first, undated event affected rocks possibly as old as 3.4 Ga, and attained granulite facies at approximately 750°C, 5 kb, although subsequent high-grade conditions may have altered mineral compositions (Koesterer et al.,

1987). The second, amphibolite– to granulite-facies event predates a set of plutons 2.7 Ga old (Stuckless et al., 1985), and the third is a granulite-facies contact-metamorphic event at about 750°C, 5 kb, associated with emplacement of charnockites of crustal derivation (Koesterer et al., 1987) at 2.63–2.5 Ga (Stuckless et al., 1985).

West Greenland

Several distinct Archean high-grade metamorphic events, at ~3.6, 3.0 and 2.8 Ga, have been recognized in western Greenland. Recent work (Friend et al., 1987; 1988) suggests that the individual events are restricted to distinct tectonic blocks that were juxtaposed after 2.8 Ga. Therefore, the events are not superimposed. Synmetamorphic calc-alkaline tonalitic intrusions (Wells, 1979; 1980b) appear to be the main heat source in both the 3.0 Ga (800°C, 7.9 kb) and 2.8 Ga (780°C, 8.9 kb) metamorphic events (Riciputi et al., 1990).

Concluding statement

The polymetamorphic complexes cited above experienced at least two distinct events, separated in time by 300–700 Ma. In examples such as the Napier and Narryer Complexes, the rocks appear to have remained deeply buried until the later event, whereas other regions may have been eroded to some extent in the interval. The Napier, Narryer, Eastern Hebei and Stanovik complexes have ancient protoliths affected by two high-grade Archean events, as well as strong Proterozoic overprints, attesting to continued reworking. The recurrence of granulite-facies metamorphism in these areas suggests repeated involvement in continental margin processes including deformation and juvenile additions.

ARCHEAN LOWER-CRUSTAL GRANULITE XENOLITHS

An important question in Archean high-grade metamorphism concerns how representative exposed granulites are as samples of the lower crust (e.g. Bohlen and Mezger, 1989; Rudnick and Presper, 1990; Rudnick, 1992). Xenolith suites generally indicate mafic compositions for the present lower crust, whereas exposed granulites have intermediate to felsic bulk compositions (Rudnick and Presper, 1990). It is possible that xenolith suites are biased toward mafic compositions as a result of mechanical or thermal instability of felsic rocks in the host mafic magma. The mafic rocks may also be the intrusive equivalents of the younger magmas carrying the xenoliths and therefore not representative of unaltered lower Archean crust. Xenolith-bearing volcanoes rarely occur within Archean cratons, making comparison difficult between exposed Archean granulites and their contemporary lower crusts. Recent work on North American examples sheds light on this problem.

Bearpaw Mountains, Montana

In the Archean granulite-facies region exposed in the Bearpaw Mountains of the Wyoming province (Fig. 3), Eocene volcanic rocks carry lower crustal xenoliths (Collerson et al., 1988). The mainly mafic suite comprises high- and intermediate-pressure granulites, noritic anorthositic rocks, and quartzofeldspathic and pelitic granulites that indicate conditions of 750–1050°C, 9–16 kb. Some xenoliths have Archean ages whereas others are compositionally distinct and are probably cumulates derived from younger, underplated basaltic magmas equivalent in age to the volcanics.

Abitibi belt, Superior province

Xenoliths from the Abitibi belt were discussed previously. A Cretaceous kimberlite pipe cuts low-grade supracrustal rocks in the Kirkland Lake area of Ontario. Its xenolith suite includes anhydrous mafic granulites of probable lower crustal provenance, as well as ultramafic rocks from mantle depths. The mafic rocks, consisting of garnet–clinopyroxene–plagioclase assemblages, are similar to amphibole-bearing mafic granulites exposed in the Kapuskasing structure 150 km to the west, which have an average laboratory-measured seismic P-wave velocity of 7.23 km/s (Fountain et al., 1989). Regional refraction studies (e.g. Boland and Ellis, 1989) show P-wave velocities of 6.9–7.5 km/s in the 30–40 km depth range of the central Superior Province. Large parts of the lower crust probably consist of anhydrous mafic granulite.

The crustal xenoliths contain complex populations of metamorphic zircon that indicate deep-crustal events at 2580 and 2490 Ma. Based on geochronology in rocks from higher structural levels, both periods significantly postdate all crust-forming events. The young ages may reflect slow decay in the lower crust of the thermal anomaly related to crustal growth at 2730–2650 Ma. Representatives of the Cretaceous magmatism have not been identified in the suite of xenoliths, consistent with the minor extent of magmatism of this age. It is unlikely that there is an extensive magmatic underplate of Cretaceous age beneath the Superior Province.

ARCHEAN GRANULITES AND CONTINENTAL COLLISION

Probably the most important processes leading to crustal reworking occur in continental collision zones (e.g. England and Bickle, 1984; Wilks, 1988). England and Bickle (1984), noting large Archean granulite complexes with uniform pressures of ~8 kb (Newton and Perkins, 1982), postulated the existence of Himalayan-scale collisions, including the presence of 10-km-high mountains, inferred to have been supported by a strong lithosphere.

The Limpopo belt of southern Africa (Fig. 3) is a 200-km-wide zone of high-grade gneisses that records a 2.65–2.70 Ga collision (Barton and van Reenen, 1992) between the Zimbabwe craton of granite–greenstone belts (3.5–2.6 Ga; Wilson, 1990) to the north and Kaapvaal craton (3.5–2.9 Ga; de Wit et al., 1992) to the south (Fig. 3). Metamorphic grade increases from greenschist to granulite facies through northern and southern marginal zones. The high-grade Central Zone contains units inferred to be of platformal sedimentary origin, lithologically and chronologically exotic with respect to the greenstone belts. Granulite-facies metamorphism attained 790–830°C, 11–13 kb in the Central Zone (Droop, 1989; Tsunogae et al., 1992) and 750–800°C, 7.5–8 kb, in both marginal zones, with clockwise *P-T* paths (Stevens and van Reenen, 1992; Tsunogae et al., 1992). Widespread charnockitic magmatism accompanied metamorphism in the Northern Marginal Zone (Ridley, 1992) and occurred to a lesser extent in the Southern Marginal Zone (Bohlender et al., 1992). A tectonic model for the region involves an early accretionary history during which the granite–greenstone belts were assembled from oceanic and continental precursors, followed by collisions (Treloar et al., 1992) which resulted in formation of granulites within the Limpopo orogen (Ashwal et al., 1992). Continued compression led to tectonic expulsion of the orogenic roots onto the cratons (van Reenen et al., 1987; Roering et al., 1992).

The Limpopo example may serve as a general model for the production of granulites through reworking of pre-existing continental crust. Many granulites occur within fault-bounded blocks, separated from their original forelands by later faults or cover, requiring interpretation of tectonic setting based solely on the information provided by the high-grade rocks themselves.

ARCHEAN GRANULITE METAMORPHISM AND MAGMATISM

Direct links can be made between regional metamorphism and magmatism in both low- and high-grade metamorphic belts (e.g. Lux et al., 1986; Wells, 1980a), and underplated mafic magmas have been implicated as heat sources in many high-grade metamorphic regions (e.g. Bohlen, 1987; Bohlen and Mezger, 1989). Close temporal and causative relationships are evident between charnockitic magmas and high-grade metamorphism (Newton, 1992b).

Basaltic magmas

In most regions the link between metamorphism and magmas of basaltic composition is circumstantial, based on the requirement to explain crustal temperatures higher than those normally attainable through the processes of crustal thickening and thermal relaxation (England and Thompson, 1986). The common occurrence of large volumes of granitoids of crustal origin in high-grade regions requires a deep crustal heat source that may have been underplated mafic magmas derived from a mantle wedge in a subduction environment (e.g. Wyllie et al., 1976; Wyllie, 1977; Kay et al., 1992). As well as providing heat and fluids (e.g. Newton, 1989; Thompson, 1990; Clemens, 1990; Bohlen, 1991; Fyfe, 1993) to promote high-grade deep-crustal metamorphism, the mafic underplate might cool through the eclogite facies (e.g. Herzberg et al., 1983; Furlong and Fountain, 1986; Griffin and O'Reilly, 1987), inhibiting topographic rise and thereby contributing to isobaric cooling paths in granulites formed in magmatic arc settings. Alternatively, the exhumation of regional high-grade complexes could be partly linked to epeirogenic rise associated with non-eclogitic underplates (McKenzie, 1984). Delamination of eclogitic "sinkers" (e.g. Houseman et al., 1981; Kay and Kay, 1991; Bohlen, 1991) would cause late uplift of high-grade regions, although the application to belts of Archean age requires further assessment (e.g. Ellis, 1992). The consequences of mafic underplating for crustal magmatism are discussed further in the following sections.

Granitic magmas

Two main types of processes link granitoids and granulites: (1) Large bodies of granite in some areas are temporally related to metamorphism and therefore may represent significant heat sources. An example is the Dugel granite of the Narryer Gneiss Complex of Western Australia, emplaced at 3375 Ma and metamorphosed at 3300 Ma (Nutman et al., 1991). These rocks apparently crystallized hydrous assemblages and have been subsequently metamorphosed in a regional event. (2) Granitoid melts are generated through vapour-absent dehydration melting in the granulite facies (Powell, 1983; Ellis and Thompson, 1986; Vielzeuf and Holloway, 1988; Clemens, 1990; Vielzeuf et al., 1990). The melts, if present in large enough volume to migrate (Wickham, 1987; Clemens and Mawer, 1992; Brown, 1994), provide a vehicle for geochemical depletion observed in granulite complexes such as the Scourie (Pride and Muecke, 1982; Cartwright and Barnicoat, 1986). The derived granites rise, advecting heat and LILE-enriched material (e.g. Condie et al., 1985) to higher crustal levels.

Intermediate magmas

The process of magmatic heat advection into mid-crustal levels to promote granulite-facies metamorphism has been modeled by Wells (1980a). To provide metamorphic temperatures in the 800°C range, magmatic compositions more mafic than granite with liquidus temperatures ~1000°C, are required. Assuming normal water contents, Wells assumed compositions in the mafic tonalite to diorite range, somewhat more mafic than those observed in most granulite occurrences. Similar temperatures could be achieved through emplacement of anhydrous granitic magmas of the charnockite suite.

Charnockitic magmas

Considerable variability exists in the literature regarding use of the term "charnockite" (see Newton (1992b) for a brief review). Most Indian workers use the term to describe orthopyroxene-bearing igneous or meta-igneous rocks formed at high metamorphic grade; elsewhere it is generally used to refer to felsic igneous rocks containing magmatic orthopyroxene. Igneous charnockites are generally, but not exclusively, associated with rocks of the granulite facies. In this discussion, "charnockite" refers to magmatic rocks of granitic (charnockite) to tonalitic (enderbite) composition that contain igneous orthopyroxene, with the implication of anhydrous igneous crystallization.

Clear examples of undeformed magmatic charnockites have been documented in geological provinces of diverse age (see Kilpatrick and Ellis (1992) for a review); more controversial are foliated rocks whose origin may be either igneous or metamorphic (e.g. Condie and Allen, 1984; Bohlender et al., 1992). Although magmatic charnockites generally occur in granulite complexes, several examples of pigeonite-bearing volcanic rocks have been linked with charnockites on the basis of mineralogical and geochemical similarity (Kilpatrick and Ellis, 1992).

Magmatic charnockitic rocks are common components of Archean high-grade complexes, where they are intimately associated with the metamorphism. For example, in the Wind River Range of Wyoming, synmetamorphic charnockite dykes have granulite-facies reaction selvages (Frost and Frost, 1987). In the Ashuanipi Complex of Quebec, large plutons of peraluminous charnockite intrude granulite-facies migmatitic paragneiss. The chemistry of the Ashuanipi charnockitic plutons suggests derivation from paragneiss sources. The bodies appear to have carried heat and fluids into mid-crustal (4–6 kb) levels (Percival, 1991a). Several observations indicate igneous crystallization under low PH_2O. First, orthopyroxene, occurs as blocky, randomly-oriented, centimetre-scale crystals in massive, coarse-grained granodiorite (Fig. 13). Second, mafic enclaves containing interior amphibole have dehydration (orthopyroxene–plagioclase) rinds where in contact with host granodiorite similar to that illustrated in Fig. 11. As there was not likely a temperature gradient within ~30 cm xenoliths, the dehydration zones suggest that amphibole breakdown was induced in the contact zone by low water activity in the host magma. In the Minto block, several compositional suites of charnockitic rocks of different age have been distinguished (Percival et al., 1992b). Intrusions of progressively younger age, from 2725 to 2690 Ma, show a systematic increase in crustal involvement in their bulk geochemistry and Nd composition (Stern et al., 1994). The small volume of metamorphic rock present as screens in plutons reached granulite facies during the repeated high-temperature magmatism (Bégin and Pattison, 1994). Dykes of charnockite cutting country rock show dehydration selvages (Percival et al., 1992b) similar to those in the Wind River Range. Variably deformed synmagmatic mafic dykes within charnockitic bodies may represent late fractionates of contemporaneous basaltic magma that caused melting in the lower crust (Stern et al., 1994).

The preceding examples show that relatively anhydrous charnockitic magmas, as well as being agents for heat advection within the crust, also may have contained CO_2 (see Frost et al., 1989). Experimental support for the solubility of CO_2 in biotite-bearing granitic rocks has been provided by Peterson and Newton (1989; 1990) and challenged by Clemens (1993a,b). A general model for the role of igneous charnockite in regional metamorphism could involve underplating by CO_2-bearing basaltic melts that on crystallizing, release heat and fluids to melt fertile lower-crustal rock types (Stevens and Clemens, 1993). The hot (>1000°C; Kilpatrick and Ellis, 1992) charnockitic magmas rise within the crust, crystallizing at mid-crustal levels, and carrying sufficient heat and CO_2 to cause granulite-facies metamorphism (Holland and Lambert, 1975; Frost et al., 1987).

Charnockitic magmas may crystallize early anhydrous phases, fractionating water-rich derivative magmas that rise to higher crustal levels (Fig. 17). Supporting field evidence is derived from metasedimentary belts of the Superior Province, exposed at a range of structural levels. Deep (5–7 kb) levels have peraluminous charnockitic plutons of 2668 Ma age whereas shallower levels (2–3 kb) have fractionated peraluminous granitic plutons of 2667 Ma age (Percival, 1990a). Charnockitic magmas represent a link between granulite metamorphism and granite plutonism.

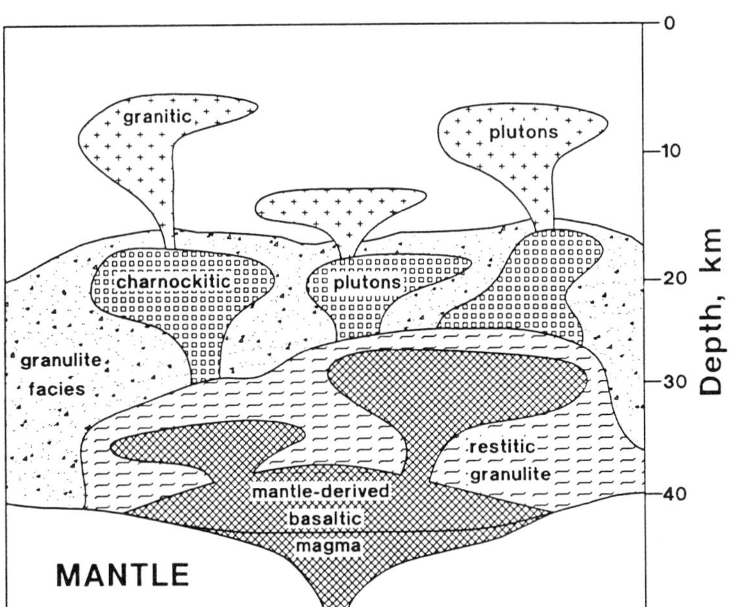

Fig. 17. Model for heat advection through mafic and granitoid magmas. Mantle-derived magmas crystallize in the deep crust, releasing heat and fluids into fertile crustal country rocks. Dehydration melting, leaving restitic granulites, produces granitoid magmas which rise to mid-crustal levels and fractionate anhydrous (charnockitic) plutons. Further fractionation of water-richer compositions yields hydrous granitoid magmas which rise to higher levels of the crust.

TECTONIC SETTINGS OF ARCHEAN HIGH-GRADE METAMORPHISM

The preceding examples of regions and processes demonstrate that diverse tectonic settings, from collisional to extensional, may account for granulite-facies metamorphism. All types of available information need to be incorporated into an assessment of environment: rock types and associations, chronology, structural history, as well as prograde and retrograde *P-T* paths. Collisional settings have been inferred for granulite complexes containing reworked rocks with clockwise *P-T* paths and geochronological evidence for synmetamorphic exhumation (Figs. 1 and 2) (e.g. Van Reenen et al., 1987). These regions contain evidence for large crustal thicknesses (60–70 km) such as currently exist beneath the Himalayas. Mantle-wedge-derived magmatic rocks and their derivatives may be important components of these complexes. Also developed at convergent margins are granulites within the deep parts of magmatic arcs. These complexes are made up mainly of synmetamorphic magmatic rocks, display isobaric cooling paths (Figs. 1 and 2), and owe their uplift to younger, unrelated events, or to slow exhumation of magmatically thickened crust such as that currently beneath the Andes. A variety of igneous compositions may be present, from mantle-wedge-derived basalt, to tonalite including a slab component, to granite and charnockite of crustal origin.

Extensional settings may also produce deep-crustal granulites, characterized by subhorizontal structures and isobaric cooling paths (Ellis, 1987; Sandiford, 1989a). Heating to granulite temperatures may occur through ponding of asthenospheric melts, or exposure of the lower crust to asthenosphere through lithospheric thinning or delamination. A period of extension, caused by crustal collapse, may be a normal late feature of contractional orogens (Sandiford and Powell, 1986; Dewey, 1988; Sandiford, 1989a).

ARCHEAN GEOTHERMAL GRADIENTS

The geothermal regime during the Archean has been the subject of considerable discussion. Because radiogenic elements were approximately twice as abundant in the Archean as presently, mantle temperatures may have been 50–100°C higher (Davies, 1979; 1992; 1993; Nisbet et al., 1993). This effect does not necessarily translate into high continental geothermal gradients for several reasons. Owing to its lower viscosity at higher temperature, the mantle was convecting more rapidly, producing more magma at oceanic ridges. Therefore, as it is today, the mantle heat flow was partitioned into the oceans (Bickle, 1978; 1986; Sleep and Windley, 1982; Nisbet and Fowler, 1983; Bickle and England, 1984; Richter, 1985; Warren, 1984). A consequence of higher mantle temperatures may have been numerous plumes rather than linear ridges (Fyfe, 1978; Campbell et al., 1989; Campbell and Griffiths, 1990; Hill, 1993), with consequences for the origin of Archean granite–greenstone belts (Hill et al., 1992).

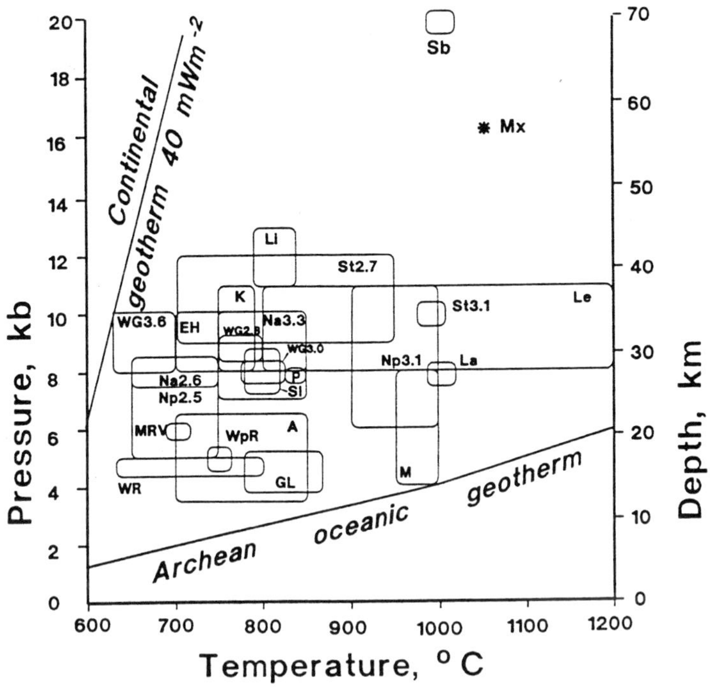

Fig. 18. *P-T* diagram showing estimated metamorphic conditions for Archean granulites: A: Ashuanipi (Percival, 1991); EH: eastern Hebei Province (Sills et al., 1987); GL: Ghost Lake, Slave Province (Farquhar et al., 1993); K: Kapuskasing uplift (Mäder et al., 1994); La: Labwor Hills (Sandiford et al., 1987); Le: Lewisian complex (Cartwright, 1990); LI: Limpopo belt (Droop, 1989; Tsunogae et al., 1992); M: Minto block (Bégin and Pattison, 1994); MRV: Minnesota River Valley (Perkins and Chipera, 1985; Moecher et al., 1986); Mx: Montana xenoliths (Collerson et al., 1988); Na: Narryer gneiss complex (Muhling, 1990) (3.3, 2.6 Ga metamorphic ages); Np: Napier complex (Harley and Hensen, 1990) (3.1, 2.5 Ga: metamorphic ages); P: Pikwitonei (Mezger et al., 1990b); Sb: Snowbird (M. Williams, personal communication, 1993); SI: Southern India (Eckert and Newton, 1993); St: Stanovik complex (Moscovchenko et al., 1992) (3.1, 2.7 Ga metamorphic ages); WG: West Greenland (Riciputi et al., 1990) (3.6, 3.0, 2.8 Ga metamorphic ages); WpR: Winnipeg River (Beakhouse, 1991); WR: Wind River range (Koesterer et al., 1987). Archean oceanic geotherm from Davies (1992), based on mantle heat generation three times present value and lithosphere heat generation two times present value. Continental geotherm (after Boyd et al., 1985) required to maintain diamond stability. Most Archean high-grade metamorphism probably occurred in continental marginal settings, either continent–continent collisional zones, or marginal magmatic arcs. Magmatism in both environments, as well as tectonically transient geotherms, prohibits direct estimation of ambient geothermal gradient.

That the Archean mantle geotherm beneath continents was not greatly different from today's (Fig. 18) (e.g. Burke and Kidd, 1978) is illustrated by the presence of diamonds of Archean age from the Kaapvaal craton, indicating the presence of lithosphere on the order of 150–200 km thick by 3.3 Ga ago (Boyd et al., 1985).

It is likely that high-grade metamorphism occurred in thermally disturbed continental margins as illustrated by *P-T* conditions of Archean granulites, which span the range of conditions bounded by Archean oceanic and continental geotherms (Fig. 18).

Metamorphic assemblages developed during high-grade events are probably unsuitable for estimation of ambient geothermal gradients in the Archean or at any subsequent time. Metamorphism is generally intimately associated with tectonism, and in any setting the geotherm must be disturbed by tectonic processes. For example, in a magmatic arc, temperatures of 900°C can be achieved at virtually any crustal level through emplacement of plutons of appropriate composition. In collisional environments, the pressure at the time of peak temperature is controlled as much by the exhumation rate as by the ambient geotherm (England and Richardson, 1977). Therefore, inferences on geothermal gradient from high-grade metamorphic rocks apply only to transient and not ambient conditions.

Sandiford (1989a) noted the dichotomy between the existence of thick lithosphere and common occurrence of high-temperature, low-pressure metamorphic belts (Grambling, 1981) in the Archean and suggested a solution in mantle lithospheric delamination at collisional margins (Houseman et al., 1981). Rather than global heat flow evolving over time, more vigorous Archean mantle convection would make lithospheric roots more susceptible to delamination, creating high-temperature conditions and deep-crustal extension (Sandiford, 1989b).

COMPARISON WITH YOUNGER HIGH-GRADE METAMORPHIC BELTS

High-grade belts of Proterozoic age show many similarities to the Archean examples discussed above, particularly the reworked type resulting from collisional processes (e.g. Windley, 1981; Harley, 1992). The interpretation of these belts in terms of tectonic processes is subject to many of the same uncertainties that characterize the Archean examples. Therefore, the following discussion focuses on relatively young geological areas in which the tectonic setting can be readily constrained. Several examples of Cretaceous granulite-facies metamorphism have been well documented. Most occurrences formed in deep magmatic arc settings that have been exhumed through subsequent tectonism associated with convergent margin processes.

The Tehachapi Complex of the southernmost Sierra Nevada (*Te*, Fig. 3) appears to represent a relatively mafic and high-grade metamorphic basal part of an oblique cross section through a dominantly granitic Cretaceous (~100 Ma) batholith (Sams and Saleeby, 1988). Granulites of the Tehachapi complex formed at 6–8 kb from mafic, dioritic, tonalitic and rare metasedimentary protoliths (Ross, 1985; 1989), in contrast to higher structural levels where variably crustally contaminated calc-alkaline granitoid rocks were emplaced at 1–5 kb (Ague and Brimhall, 1988). Oblique exposure of the batholith was probably a consequence of transpressional deformation and uplift along discrete faults between 80 and 40 Ma (Saleeby et al., 1987; Sams and Saleeby, 1988). In granulites of similar age and setting to the south, Barth and May (1992) determined an isobaric cooling path between 108 and 88 Ma, followed by rapid tectonic uplift between 88 and 78 Ma.

Complex uplift mechanisms are responsible for exhumation of the Cretaceous Fiordland Complex of New Zealand (*Fc*, Fig. 3) (Gibson et al., 1988). Granulite-facies rocks metamorphosed at 650–700°C, 9–12.5 kb (Newton and Perkins, 1982) are interpreted to be the lower crust of a 120–130 Ma magmatic arc (Oliver and Coggon, 1979). Rock types include mafic rocks containing metasedimentary and metaplutonic enclaves, cut by noritic, anorthositic and ultramafic dykes and sills. Uplift occurred during an early extensional unroofing event (~90 Ma; Gibson et al., 1988), followed by transpressional deformation associated with movement on the Alpine fault. These examples show many similarities to Archean high-grade complexes formed in magmatic arc environments. They include an association with igneous rocks, igneous and minor supracrustal protoliths, isobaric cooling paths, and exhumation related to later tectonism.

Tertiary granulites are exposed in northern Japan in the imbricate thrust stack of the Hidaka metamorphic belt (*Hb*, Fig. 3) (Osanai et al., 1991). Protoliths include mafic and felsic meta-igneous rocks, pelite and S-type tonalite, which reached peak conditions of 870°C, 7 kb about 56 Ma (Osanai et al., 1992), probably in the basal part of an island arc (Komatsu et al., 1989). An important metamorphic heat source appears to be peraluminous, garnet-, orthopyroxene-bearing tonalites of crustal origin, intruded during high-grade metamorphism at temperatures >800°C (Shimura et al., 1992). A prograde *P-T* path, including early retrograde isothermal decompression, has been deduced for the complex (Osanai et al., 1991). The high-grade metamorphism and charnockitic plutonism developed rapidly in a basal arc environment from protoliths 80–65 Ma, and was rapidly exhumed. Final exposure probably occurred ~20 Ma in response to late faulting (Osanai et al., 1991; Shimura, 1992).

The Archean Ashuanipi Complex has many lithological and chronological similarities to the Hidaka belt. Both have dominantly metasedimentary protoliths, rapidly metamorphosed to produce peraluminous charnockitic magmas emplaced at ~7 kb levels into granulite-facies migmatitic paragneiss, quickly followed by cooling and exhumation. Major differences between the regions include their size, crustal distribution and exhumation mechanism. The Hidaka rocks occur in a 140 by 10–20-km belt, whereas the Ashuanipi Complex is a 300-km-diameter massif. Hidaka granulites are the lower part of an island arc crust, exposed in a series of thrust imbricates; Ashuanipi granulites are areally extensive and presumably extend to the base of the ~35–40 km thick crust. Uplift of the Hidaka rocks is partly isostatic, aided by later faults (Shimura, 1992), whereas the Ashuanipi complex appears to have been exposed entirely isostatically (Percival et al., 1992b) (Fig. 14).

A few granulite complexes are exposed in young extensional settings, including the Fiordland complex mentioned above. In the Mojave desert of the southwestern

United States, the Waterman Metamorphic Complex (*Wc*, Fig. 3) was exposed on Early Miocene extensional detachment faults (Dokka, 1989). The complex metamorphic and deformational history of the granulite-facies supracrustal and intrusive protoliths involved an early high-pressure granulite-facies event (800–850°C at 7.5–9 kb) and retrogression at 750–800°C, 10–12 kb, followed by early extension, all presumably in a 112 Ma magmatic arc environment (Henry and Dokka, 1992). Isobaric cooling followed until extensional unroofing in the Miocene (~22 Ma), which produced mylonites and a mild greenschist-facies overprint. These relatively superficial metamorphic effects do not correspond to the granulite-facies lower-crustal conditions expected during high heat-flow regimes associated with extension and mafic underplating. The Waterman Complex probably represents crustal levels exposed from 10–17 km depths (op. cit.), and the present lower crust probably experienced renewed granulite conditions during the extensional event.

To date, examples of granulites exhumed in Archean metamorphic core complexes have not been documented, although core complexes may have formed locally in the Archean. In the Sleepy Dragon Complex of the Slave Province of northwestern Canada, juxtaposed upper amphibolite and greenschist facies domains are related by a late, low-angle extensional fault (James and Mortensen, 1992). The geometry, resembling Cordilleran metamorphic core complexes, may have resulted from extensional collapse of the orogen following thickening, perhaps caused by lithospheric delamination (Davis and Hegner, 1992).

SUMMARY

Over the past two decades, advances in geothermobarometry, geochronolgy and isotope geochemistry have allowed quantitative estimates of P-T-X_{fluid} and age for Archean high-grade metamorphic domains. Recent attention to the P-T-t path of granulites to constrain tectonic evolution has succeeded in defining retrograde portions and hence uplift trajectories. Although this information is instructive, an outstanding problem remains elucidation of the prograde path of granulites. This will involve documentation of the age of key events in the evolution of a metamorphic complex through mineral systems robust enough to retain information through high-grade metamorphism. Critical constraints include the age of deposition of rocks at surface or pluton emplacement, the timing of deformation(s), as well as the age and duration of high-grade metamorphic conditions. Combined with information on cooling/uplift rates available for many regions, the prograde constraints will permit more accurate analogies between the evolution of ancient and younger orogens.

Two end-member tectonic settings led to widespread high-grade metamorphism during the Archean: (1) the deep crust of magmatic arcs; and (2) tectonic reworking in collisional orogeny. Many combinations of processes exist, as in the examples of magmatic arcs built on older crust, and juvenile intrusions emplaced within reworked gneiss complexes. Metamorphism in both environments is controlled tectonically, as evidenced by coeval deformation and the presence of arc-related magmas. Even the examples of CO_2-infiltration-driven dehydration are demonstrably associated with deformation and probably linked to mantle degassing through tectonic processes.

High-grade metamorphism appears to be widespread in regions of Archean age, leading to the suggestion that higher Archean heat production and consequent elevated geothermal gradients are responsible. However, many Archean granulites are characterized by isobaric cooling paths and development in magmatic arcs, suggesting that late, unrelated events are responsible for their exhumation. It is probable that high-grade metamorphism is a common trait of the deep crust of all ages, but that because of their age, Archean rocks have a high probability of being exposed in younger, including Proterozoic, tectonic "accidents". Massif granulites, with no evidence of younger uplift structures, appear to be more common in Archean cratons than in younger belts. Perhaps equivalent complexes lie at depth within thick crusts beneath the Andes and Himalayas and will be exhumed once orogeny ceases.

Magmatism commonly accompanies high-grade metamorphism, but only in rare examples can the two be directly linked. Underplated basaltic magmas are commonly implicated as a heat source to elevate crustal temperatures into the 800–1000°C range, in order to produce high-temperature granulites and promote crustal melting. The commonly observed associates are granitoid and charnockitic plutons, mainly of crustal derivation. While these magmas are obvious advectors of heat and fluids, the ultimate driving force is generally cryptic.

A variety of tectonic settings has been proposed to explain granulite-facies metamorphism in different Archean regions. Convergent margins account for both the reworked (collisional) type of granulite belt, characterized by older protoliths, clockwise *P-T* paths and syntectonic exhumation; and deep magmatic arcs, dominated by plutonic rocks and isobaric cooling paths. Granulites may also develop in extensional settings with thinned lithosphere and associated elevated geotherms, although Archean examples are rare. Late extensional collapse is a common structural feature of compressional orogens and has been recognized in some high-grade regions of Archean age such as the Kapuskasing uplift.

Various lines of evidence indicate that continental lithosphere as thick or thicker than present thicknesses existed beneath some Archean cratons (Hoffman, 1990). Higher mantle heat production during the Archean was dissipated through the oceans by hotter and/or larger volumes of basaltic magma. The consequent faster, hotter subduction environments can account for the relatively rapid growth of continental crust and granulite production during the Archean in widespread magmatic arcs.

ACKNOWLEDGMENTS

I wish to thank Kent Condie for his invitation to contribute to this volume, as well as for guidance. Discussions over the past decade with K.D. Card, K.D. Collerson, D.M. Fountain, R.C. Newton, D.R.M. Pattison and R.A. Stern have contributed to the synthesis presented here. Critical reviews by M. Brown, K.C. Condie, T. Frisch and S.L. Harley materially improved the manuscript. Geological Survey of Canada contribution number 35894.

REFERENCES

Abbott, D.H., and Hoffman, S.E., 1984. Archean plate tectonics revisited 1: Heat flow, spreading rate and the age of subducting oceanic lithosphere and their effects on the origin and evolution of continents. Tectonics, 3: 429–448.

Ague, J.J., and Brimhall, G.H., 1988. Magmatic arc asymmetry and distribution of anomalous plutonic belts in the batholiths of California: Effects of assimilation, crustal thickness, and depth of crystallization. Geol. Soc. Am. Bull., 100: 912–927.

Anovitz, L.M. 1991. Al-zoning in pyroxene and plagioclase: Window on late prograde to early retrograde P-T paths in granulite terranes. Am. Mineral., 76: 1328–1343.

Arndt, N.T., 1983. Role of a thin, komatiite-rich oceanic crust in the Archean plate-tectonic process. Geology, 11: 372–375.

Ashwal, L.D., Morgan, P., and Hoisch, T.D., 1992. Tectonics and heat sources for granulite metamorphism of supracrustal-bearing terranes. Precambrian Res., 55: 525–538.

Baadsgaard, H., Nutman, A.P., Bridgwater, D., Rosing, M., McGregor, V.R., and Allaart, J.H., 1984. The zircon geochronology of the Akilia association and Isua supracrustal belt, West Greenland. Earth Planet. Sci. Lett., 68: 221–228.

Baer, A.J., 1981. Geotherms, evolution of the lithosphere and plate tectonics. Tectonophysics, 72: 203–227.

Barnicoat, A.C., 1983. Metamorphism of the Scourian complex, NW Scotland. J. Metamorph. Geol., 1: 163–182.

Barth, A.P., and May, D.J., 1992. Mineralogy and pressure-temperature–time path of Cretaceous granulite gneisses, southeastern San Gabriel Mountains, southern California. J. Metamorph. Geol., 10: 529–544.

Barton, J.M. Jr., and van Reenen, D.D. 1992. When was the Limpopo orogeny? Precambrian Res., 55: 7–16.

Bates, M.P., and Halls, H.C., 1991. Broad-scale Proterozoic deformation of the central Superior Province revealed by paleomagnetism of the 2.45 Ga Matachewan dyke swarm. Can. J. Earth Sci., 28: 1780–1796.

Beakhouse, G.P., 1991. Winnipeg River subprovince. In: P.C. Thurston, H.R. Williams, R.H. Sutcliffe and G.M. Stott (Eds.), Geology of Ontario. Ont. Geol. Surv., Spec. Vol. 4, Pt. 1, pp. 279–301.

Bégin, N.J., and Pattison, D.R.M., 1994. Metamorphic evolution of granulites in the Minto block, northern Quebec: extraction of peak P-T conditions taking account of late Fe–Mg exchange. J. Metamorph. Geol., 31: 1134–1145.

Berman, R.G., 1991. Thermobarometry using multiequilibrium calculations: a new technique with petrological applications. Can. Mineral., 29: 833–855.

Bibikova, E.V., and Williams, I.S., 1990. Ion microprobe U-Th-Pb isotopic studies of zircons from three early Precambrian areas in the U.S.S.R. Precambrian Res., 48: 203–221.
Bickle, M.J., 1978. Heat loss from the earth: A constraint on Archaean tectonics from the relation between geothermal gradients and the rate of plate production. Earth Planet. Sci. Lett., 40, 301–315.
Bickle, M.J., 1986. Implications of melting for stabilization of the lithosphere and heat loss in the Archaean. Earth Planet. Sci. Lett., 80: 314–324.
Bingen, B., Demaiffe, D., and Delhal, J., 1988. Aluminous granulites of the Archean craton of Kasai (Zaire): petrology and P-T conditions. J. Petrol., 29: 899–919.
Black, L.P., James, P.R., and Harley, S.L., 1983. Geochronology and geological evolution of metamorphic rocks in the Field Islands area, East Antarctica. J. Metamorph. Geol., 1: 277–303.
Black, L.P., Williams, I.S., and Compston, W., 1986. Four zircon ages from one rock: the history of a 3900 Ma-old granulite from Mount Sones, Enderby Land, Antarctica. Contrib. Mineral. Petrol., 94: 427–437.
Bohlen, S.R., 1987. Pressure–temperature–time paths and a tectonic model for the evolution of granulites. J. Geol., 95: 617–632.
Bohlen, S.R., 1991. On the formation of granulites. J. Metamorph. Geol., 9: 223–229.
Bohlen, S.R., and Mezger, K., 1989. Origin of granulite terranes and the formation of the lowermost continental crust. Science, 244: 326–329.
Bohlen, S.R., Wall, V.J., and Boettcher, A.L., 1983. Experimental investigation and application of garnet granulite equilibria. Contrib. Mineral. Petrol., 28: 310–318.
Bohlender, F., van Reeneen, D.D., and Barton, J.M., 1992. Evidence for metamorphic and igneous charnockites in the Southern Marginal Zone of the Limpopo belt. Precambrian Res., 55: 429–449.
Boland, A.V. and Ellis, R.M., 1989. Velocity structure of the Kapuskasing uplift, northern Ontario, from seismic refraction studies. J. Geophys. Res., 94: 7189–7204.
Boyd, F.R., Gurney, J.J., and Richardson, S.H., 1985. Evidence for a 150–200-km thick Archaean lithosphere from diamond inclusion thermobarometry. Nature, 315: 387–389.
Brown, M., 1973. The definition of metatexis, diatexis and migmatite. Proc. Geol. Assoc. 84, 371–382.
Brown, M., 1994. The generation, segregation, ascent and emplacement of granite magma: the migmatite-to-crustally-derived granite connection in thickened orogens. Earth Sci. Rev., in press.
Burke, K., and Kidd, W.S.F., 1978. Were Archaean continental geothermal gradients much steeper than those of today? Nature, 262: 240–241.
Burke, K., Dewey, J.F., and Kidd, W.S.F., 1976. Dominance of horizontal movements, arc and microcontinental collisions during the later permobile regime. In: B.F. Windley (Ed.), The Early History of the Earth. John Wiley, New York, pp. 113–129.
Bursnall, J.T., 1990. Deformation sequence in the southeastern Kapuskasing Structural Zone, Ivanhoe Lake, Ontario, Canada. In: M.H. Salisbury and D.M. Fountain (Eds.), Exposed Cross Sections of the Continental Crust. Kluwer, Dordrecht, pp. 469–484.
Bursnall, J.T., Leclair, A.D., Moser, D.E., and Percival, J.A., 1994. Structural correlation within the Kapuskasing uplift. Can. J. Earth Sci., in press.
Burton, K.W., and O'Nions, R.K., 1991. High-resolution garnet chronometry and the rates of metamorphic processes. Earth Planet. Sci. Lett. 107, 649–671.
Campbell, I.H., and Griffiths, R.W., 1990. Implications of mantle plume structure for the origin of flood basalts. Earth Planet. Sci. Lett., 99: 79–93.
Campbell, I.H., Griffiths, R.W., and Hill, R.I., 1989. Melting in an Archean mantle plume: heads it's

basalts, tails it's komatiites. Nature, 339: 697–699.

Card, K.D. 1990. A review of the Superior Province of the Canadian Shield, a product of Archean accretion. Precambrian Res., 48: 99–156.

Cartwright, I., 1990. Prograde metamorphism, anatexis, and retrogression of the Scourian complex, north-west Scotland. In: J.R. Ashworth and M. Brown (Eds.), High-temperature Metamorphism and Crustal Anatexis. Mineral. Soc. Ser., 2. Unwin Hyman, London, pp. 371–399.

Cartwright, I., 1992. Archaean granulite facies metamorphism of the Lewisian of Tiree, Inner Hebrides, northwest Scotland. J. Metamorph. Geol., 10: 727–744.

Cartwright, I., and Barnicoat, A.C., 1986. The generation of quartz-normative melts and corundum-normative restites by crustal anatexis: petrogenetic modelling based on an example from the Lewisian of north-west Scotland. J. Metamorph. Geol., 4: 79–99.

Chacko, T., Ravindra Kumar, G.R., and Newton, R.C., 1987. Metamorphic P-T conditions of the Kerala (South India) khondalite belt, a granulite facies supracrustal terrain. J. Geol., 95: 343–358.

Clemens, J.D., 1990. The granulite–granite connexion. In: D. Vielzeuf and P. Vidal (Eds.), Granulites and Crustal Evolution. Kluwer, Dordrecht, pp. 25–36.

Clemens, J.D., 1993a. Carbonic metamorphism and new experiments in the system $KAlO_2$–MgO–SiO_2–H_2O–CO_2: no melt fluxing by CO_2 or production of magnesian lamprophyres from sialic crust. EOS Trans. Am. Geophys. Un., 74: 657.

Clemens, J.D., 1993b. Experimental evidence against CO_2-promoted deep crustal melting. Nature, 363: 336–338.

Clemens, J.D., and Mawer, C.K., 1992. Granite magma transport by fracture propagation. Tectonophysics, 204: 339–360.

Collerson, K.D., and Fryer, B.J., 1978. The role of fluids in the formation and subsequent development of early continental crust. Contrib. Mineral. Petrol., 67: 1151–167.

Collerson, K.D., Hearn, B.C., Macdonald, R.A., Upton, B.G.J., and Park, J.G., 1988. Granulite xenoliths from the Bearpaw Mountains, Montana: constraints on the character and evolution of lower continental crust. Terra Cognita, 8: 270.

Condie, K.C., 1981. Archean Greenstone Belts. Elsevier, Amsterdam, 434 p.

Condie, K.C., 1984. Archean geotherms and supracrustal assemblages. Tectonophysics, 105: 29–41.

Condie, K.C., and Allen, P., 1984. Origin of Archean charnockites from southern India. In: A. Kröner (Ed.), Archaean Geochemistry. Springer, Berlin, pp. 182–293.

Condie, K.C., Allen, P., and Narayana, B.L., 1982. Geochemistry of the Archean low- to high-grade transition zone, southern India. Contrib. Mineral. Petrol., 81: 157–167.

Condie, K.C., Bowling, G.P., and Allen, P., 1985. Missing Eu anomaly and Archaean high-grade granites. Geology, 13: 633–636.

Corfu, F., 1987. Inverse age stratification in the Archean crust of the Superior Province: evidence for infra- and subcrustal accretion from high-resolution U-Pb zircon and monazite ages. Precambrian Res., 36: 259–275.

Corfu, F., 1988. Differential response of U-Pb systems in coexisting accessory minerals, Winnipeg River subprovince, Canadian Shield: implications for Archean crustal growth and stabilization. Contrib. Mineral. Petrol., 98: 312–325.

Corfu, F., 1993. The evolution of the southern Abitibi greenstone belt in light of precise U-Pb geochronology. Econ. Geol., 88: 1323–1340.

Dallwitz, W.B., 1968. Coexisting sapphirine and quartz in granulite from Enderby Land, Antarctica. Nature, 219: 476–477.

Davies, G.F., 1979. Thickness and thermal history of continental crust and root zones. Earth Planet. Sci. Lett., 44: 231–238.

Davies, G.F., 1992. On the emergence of plate tectonics. Geology, 20: 963–966.
Davies, G.F., 1993. Conjectures on the thermal and tectonic evolution of the Earth. Lithos, 30: 281–289.
Davidson, A., 1990. Evidence for eclogite metamorphism in the southwest Grenville Province, Ontario. Current Research, Part C. Geol. Surv. Can. Pap., 90-1C: 113–118
Davis, W.J., and Hegner, E., 1992. Neodymium isotopic evidence for the tectonic assembly of Late Archean crust in the Slave Province, northwest Canada. Contrib. Mineral. Petrol., 111: 493–504.
Dewey, J.F., 1988. Extensional collapse of orogens. Tectonics, 7: 1123–1139.
de Wit, M.J., Roering, C., Hart, R.J., Armstrong, R.A., de Ronde, C.E.J., Green, R.W.E., Tredoux, M., Peberdy, E. and Hart, R.A., 1992. Formation of an Archaean continent. Nature, 357: 553-562.
Dokka, R.K., 1989. The Mojave extensional belt of southern California. Tectonics, 8: 363–390.
Dook, V.L., Neymark, L.A., and Rudnick, V.A. (Eds.), 1989. The oldest rocks of the Aldan–Stanovik Shield, eastern Siberia, USSR. Excursion guide for geological field trip to the Aldan–Stanovik Shield, Leningrad–Mainz, 121 p.
Droop, G.T.R., 1989. Reaction history of garnet–sapphirine granulites and conditions of high-pressure granulite-facies metamorphism in the central Limpopo Belt, Zimbabwe. J. Metamorph. Geol., 7: 383–403.
Eckert, J.O., and Newton, R.C., 1993. Palaeopressures of South Indian two-pyroxene garnet granulites from thermochemically calibrated CMAS barometers. J. Metamorph. Geol., 11: 845–854.
Ellis, D.J., 1980. Osumilite–sapphirine–quartz granulites from Enderby Land, Antactica: P-T conditions of metamorphism, implications for garnet–cordierite equilibria and the evolution of the deep crust. Contrib. Mineral. Petrol., 74: 201–210.
Ellis, D.J., 1987. Origin and evolution of granulites in normal and thickened crusts. Geology, 15: 167–170.
Ellis, D.J., 1992. Precambrian tectonics and the physicochemical evolution of the continental crust. II. Lithosphere delamination and ensialic orogeny. Precambrian Res., 55: 507–524.
Ellis, D.J., and Thompson, A.B., 1986. Subsolidus and partial melting reactions in the quartz-excess $CaO + MgO + Al_2O_3 + SiO_2 + H_2O$ system under water-excess and water-deficient conditions to 10 kb: some implications for the origin of peraluminous melts from mafic rocks. J. Petrol., 27: 91–121.
England, P.C., 1979. Continental geotherms during the Archaean. Nature, 277, 556–558.
England, P.C., and Bickle, M.J., 1984. Continental thermal and tectonic regimes during the Archaean. J. Geol., 92: 353–367.
England, P.C., and Richardson, S.W., 1977. The influence of erosion upon the mineral facies of rocks from different metamorphic environments. J. Geol. Soc. Lond., 134: 201–213.
England, P.C., and Thompson, A.B., 1984. Pressure–temperature–time paths of regional metamorphism I. Heat transfer during the evolution of regions of thickened continental crust. J. Petrol., 25: 894–928.
England, P.C., and Thompson, A.B., 1986. Some thermal and tectonic models for crustal melting in continental collision belts. In: M.P. Coward and A.C. Reis (Eds.), Collision Tectonics. Geol. Soc. Lond., pp. 83–94.
Farquhar, J., Staveley, J.A., and Chacko, T., 1993. Granulite facies metamorphism near Ghost Lake, Slave Province, N.W.T. Geol. Assoc. Can., Abstr. Prog., 18: A28.
Fountain, D.M. and Salisbury, M.H., 1981. Exposed cross sections through the continental crust: Implications for crustal structure, petrology and evolution. Earth Planet. Sci. Lett., 5: 263–277.
Fountain, D.M. and Salisbury, M.H., and Percival, J.A., 1989. Seismic structure of the continental

crust based on rock velocity measurements from the Kapuskasing uplift. J. Geophys. Res., 95: 1167–1186.

Friend, C.R.L., 1981. Charnockite and granite formation and influx of CO_2 at Kabbaldurga. Nature, 294: 550–552.

Friend, C.R.L., 1983. The link between charnockite formation and granite production: evidence from Kabbaldurga, Karnataka, southern India. In: M.P. Atherton and C.D. Gribble (Eds.), Migmatites, Melting and Metamorphism. Shiva, Nantwich, UK, pp. 264–276.

Friend, C.R.L., 1985. Evidence for fluid pathways through Archean crust and the generation of the Closepet granite, Karnataka, South India. Precambrian Res., 27: 239–250.

Friend, C.R.L., and Nutman, A.P., 1991. SHRIMP U-Pb geochronology of the Closepet granite and Peninsular gneiss, Karnataka, South India. J. Geol. Soc. India, 38: 357–368.

Friend, C.R.L., Nutman, A.P., and McGregor, V.R., 1987. Late Archaean tectonics in the Faeringhavn–Tre Brodre area, south of Buksefjorden, southern West Greenland. J. Geol. Soc. Lond., 144: 369–376.

Friend, C.R.L., Nutman, A.P., and McGregor, V.R., 1988. Late Archaean terrane accretion in the Godthab region, southern West Greenland. Nature, 335: 535–538.

Frost, B.R., and Chacko, T., 1989. The granulite uncertainty principle: limitations on thermobarometry in granulites. J. Geol., 97: 435–450.

Frost, B.R., and Frost, C. 1987. CO_2, melts and granulite metamorphism. Nature, 327: 503–506.

Frost, B.R., Frost, C., and Touret, J.R.L., 1989. Magmas as a source of heat and fluids in granulite metamorphism. In: D. Bridgwater (Ed.), Fluid Movements–Element Transport and the Composition of the Deep Crust. Kluwer, Dordrecht, pp. 1–18.

Frost, B.R., Fyfe, W.S., Tazaki, K., and Chan, T., 1989. Grain-boundary graphite in rocks and implications for high electrical conductivity in the lower crust. Nature, 340: 143–146.

Furlong, K.P., and Fountain, D.M., 1986. Continental crustal underplating: thermal considerations and seismic–petrologic consequences. J. Geophys. Res., 91: 8285–8294.

Fyfe, W.S., 1973. The granulite facies, partial melting and the Archaean crust. Phil. Trans. Roy. Soc. Lond., A273: 457–461.

Fyfe, W.S., 1978. The evolution of the earth's crust: modern plate tectonics to ancient hot spot tectonics? Chem. Geol., 23: 89–114.

Fyfe, W.S., 1993. Hot spots, magma underplating, and modification of continental crust. Can. J. Earth Sci., 30: 908–912.

Garde, A., 1990. Thermal granulite facies metamorphism with diffuse retrogression in Archaean orthogneisses, Fiskefjord, southern West Greenland. J. Metamorph. Geol., 6: 663–682.

Ganguly, J., and Chakraborty, S., 1991. Compositional zoning and cation diffusion in garnets. In: J. Ganguly (Ed.), Diffusion, Atomic Ordering and Mass Transport. Springer-Verlag, New York, pp. 120–175.

Gibson, G.M., McDougall, I, and Ireland, T.R., 1988. Age constraints on metamorphism and development of a metamorphic core complex in Fiordland, southern New Zealand. Geology, 16: 405–408.

Glikson, A.Y., and Lambert, I.B., 1976. Vertical zonation and petrogenesis of the early Precambrian crust in western Australia. Tectonophysics, 30: 55–89.

Goldich, S.S., Hegde, C.E., Stern, T.W., Wooden, J.L., Bodkin, J.B., and North, R.M., 1980. Archean rocks of the Granite Falls area, southwestern Minnesota. Geol. Soc. Am. Spec. Pap., 182: 19–43.

Grambling, J.A., 1981. Pressures and temperatures in Precambrian metamorphic rocks. Earth Planet. Sci. Lett., 53: 63–68.

Grand, S.P., 1987. Tomographic inversion for shear velocity beneath the North American plate. J.

Geophys. Res. 92, 14,065–14,090.
Grew, E.S., 1980. Sapphirine + quartz association from Archean rocks in Enderby Land, Antarctica. Am. Mineral., 65: 821–836.
Grew, E.S., 1982. Osumilite in the sapphirine–quartz terrane of Enderby Land, Antarctica; implications for osumilite perogenesis in the granulite facies. Am. Mineral., 67: 762–787.
Griffin, W.L., and O'Reilly, S.Y., 1987. Is the continental Moho the crust-mantle boundary? Geology, 15: 241–244.
Griffin, W.L., McGregor, V.R., Nutman, A., Taylor, P.N., and Bridgwater, D., 1980. Early Archean granulite-facies metamorphism south of Ameralik, West Greenland. Earth Planet. Sci. Lett., 50: 59–74.
Hanmer, S., and Kopf, C., 1993. The Snowbird tectonic zone in District of Mackenzie, Northwest Territories. Current Research Part C. Geol. Surv. Can. Pap., 93-1C: 41–52.
Hansen, E.C., Newton, R.C., and Janardhan, A.S., 1984. Fluid inclusions in rocks from the amphibolite-facies gneiss to charnockite progression in southern Karnataka, India: direct evidence concerning the fluids of granulite metamorphism. J. Metamorph. Geol., 2: 249–264.
Harley, S.L, 1985. Garnet–orthopyroxene bearing granulites from Enderby Land, Antarctica: metamorphic pressure–temperature–time evolution of the Archaean Napier Complex. J. Petrol., 26: 819–856.
Harley, S.L., 1987. A pyroxene-bearing meta-ironstone and other pyroxene granulites from Tonagh Island, Enderby Land, Antactica: further evidence for very high temperature (>980°C) Archaean regional metamorphism in the Napier Complex. J. Metamorph. Geol., 5: 341–356.
Harley, S.L., 1989. The origins of granulites: a metamorphic perspective. Geol. Mag., 126: 215–247.
Harley, S.L., 1992. Proterozoic granulite terranes. In: K.C. Condie (Ed.), Proterozoic Crustal Evolution. Elsevier, Amsterdam, pp. 301–359.
Harley, S.L., and Black, L.P., 1987. The Archaean geological evolution of Enderby Land, Antarctica. In: R.G. Park and J. Tarney (Eds.), The Evolution of the Lewisian and Comparable Precambrian High-grade Terrains. Geol. Soc. Lond. Spec. Publ., 27: 285–296.
Harley, S.L, and Hensen, B.J., 1990. Archean and Proterozoic high-grade terranes of East Antarctica (40–80°E): a case study of diversity in granulite facies metamorphism. In: J.R. Ashworth and M. Brown (Eds.), High-temperature Metamorphism and Crustal Anatexis. Mineral. Soc. Ser., 2. Unwin Hyman, London, pp. 320–370.
Harley, S.L., Hensen, B.J., and Sheraton, J.W., 1990. Two-stage decompression in orthopyroxene–sillimanite granulites from Forefinger Point, Enderby Land, Antarctica: implications for the evolution of the Archaean Napier complex. J. Metamorph. Geol., 8: 591–613.
Harris, N.B.W., Holt, R.W., and Drury, S.A., 1982. Geobarometry, geothermometry, and late Archean geotherms from the granulite facies terrain of South India. J. Geol., 90: 509–528.
Henderson, J.B., and Schaan, S.E., 1993. Geology of the Wijinnedi Lake area: a transect into mid-crustal levels in the western Slave Province, District of Mackenzie, Northwest Territories. In Current Research, Part C. Geol. Surv. Can. Pap., 93–1C: 83–91.
Henry, D.J., and Dokka, R.K., 1992. Metamorphic evolution of exhumed middle to lower crustal rocks in the Mojave extensional belt, southern California, USA. J. Metamorph. Geol., 10: 347–364.
Hensen, B.J., and Harley, S.L., 1990. Graphical analysis of P-T-X relations in granulite facies metapelites. In: J.R. Ashworth and M. Brown (Eds.), High-temperature Metamorphism and Crustal Anatexis. Mineral. Soc. Ser., 2. Unwin Hyman, London, pp. 19–56.
Herzberg, C.T., Fyfe, W.S., and Carr, M.J., 1983. Density constraints on the formation of the continental Moho and crust. Contrib. Mineral. Petrol., 84: 1–5.
Hildreth, W., and Moorbath, S. 1988. Crustal contributions to arc magmatism in the Andes of central

Chile. Contrib. Mineral. Petrol., 98: 455–489.

Hill, R.T., 1993. Mantle plumes and continental tectonics. Lithos, 30: 193–206.

Hill, R.T., Chappell, B.W., and Campbell, I.H., 1992. Late Archaean granites of the southeastern Yilgarn Block, Western Australia: age, geochemistry, and origin. Trans. R. Soc. Edinburgh: Earth Sci., 83: 211–226.

Hoffman, P.F., 1990. Geological constraints on the origin of the mantle root beneath the Canadian shield. Phil. Trans. R. Soc. Lond. A 331, 523–532.

Holland, J.G., and Lambert, R. St. J., 1975. The chemistry and origin of the Lewisian gneisses of the Scottish mainland: the Scourie and Inver assemblages and sub-crustal accretion. Precambrian Res., 2: 161–188.

Holland, T.H., The charnockite series, a group of Archaean hypersthenic rocks in peninsular India. Geol. Surv. India Mem., 2, 192–249.

Houseman, G.A., McKenzie, D.P., and Molnar, P., 1981. Convective instability of a thickened boundary layer and its relevance for the thermal evolution of continental convergent belts. J. Geophys. Res., 86: 6115–6132.

Jahn, B.M., and Zhang, Z.Q., 1984. Archaean granulite gneisses from eastern Hebei Province, China: rare earth geochemistry and tectonic implications. Contrib. Mineral. Petrol., 85: 224–243.

James, D.T., and Mortensen, J.K., 1992. An Archean metamorphic core complex in the southern Slave Province: basement-cover structural relations between the Sleepy Dragon Complex and the Yellowknife Supergroup. Can. J. Earth Sci., 29: 2133–2145.

Janardhan, A.S., Newton, R.C., and Hansen, E.C., 1982. The transformation of amphibolite-facies gneiss to charnockite in southern Karnataka and northern Tamil Nadu, India. Contrib. Mineral. Petrol.. 79: 130–149.

Jarvis, G.T., and Campbell, I.H., 1983. Archean komatiites and geotherms: solution to an apparent contradiction. Geophys. Res. Lett., 10: 1133–1136.

Kaiyi, W., Windley, B.F., Sills, J.D., and Yuehua, Y., 1990. The Archaean gneiss complex in E. Hebei Province, North China: geochemistry and evolution. Precambrian Res., 48: 245–265.

Kay, R.W., and Kay, S. Mahlburg, 1991. Creation and destruction of lower continental crust. Geol. Rundsch., 80: 259–278.

Kay, R.W., Kay, S. Mahlburg, and Arculus, R.J., 1992. Magma genesis and crustal processing. In: D.M. Fountain, R. Arculus, and R.W. Kay (Eds.), Continental Lower Crust. Elsevier, Amsterdam, pp. 423–445.

Kilpatrick, J.A., and Ellis, D.J., 1992. C-type magmas: igneous charnockites and their extrusive equivalents. Trans. R. Soc. Edinburgh: Earth Sci., 83: 155–164.

Kimura, G., Ludden, J.N., Desrochers, J-P., and Hori, R., 1993. A model of ocean–crust accretion for the Superior province, Canada. Lithos, 30: 337–355.

Kinny, P.D., Williams, I.S., Froude, D.O., Ireland, T.R., and Compston, W., 1988. Early Archaean zircon ages from orthogneisses and anorthosites at Mount Narryer, Western Australia. Precambrian Res., 38: 325–341.

Koesterer, M.E., Frost, C.D., Frost, B.R., Hulsebosch, T.P., Bridgwater, D., and Worl, R.G., 1987. Development of the Archean crust in the Medina Mountain area, Wind River Mountains, Wyoming (U.S.A.). Precambrian Res., 37: 287–304.

Komatsu, M., Osanai, Y., Toyoshima, T., and Miyashita, S., 1989. Evolution of the Hidaka metamorphic belt, northern Japan. In: J.S. Daly, R.A. Cliff and B.W.D. Yardley (Eds.), Evolution of Metamorphic Belts. Geol. Soc. Spec. Publ., 43: 487–493.

Krogh, T.E. 1993. High precision U-Pb ages for granulite metamorphism and deformation in the Archean Kapuskasing structural zone, Ontario: implications for structure and development of the lower crust. Earth Planet. Sci. Lett., 119: 1–18.

Krogh, T.E., Heaman, L., Machado, N., Davis, D., and Weber, W., 1986. U-Pb geochronology program, northern Superior Province. Rep. Field Act., 1986. Manitoba Dep. Mines Energy, 1986, 178–180.

Kröner, A., and Greiling, R. (editors), 1984. Precambrian Tectonics Illustrated. E. Schweizerbart'sche Verlagsbuchhandlung, Stuttgart, Germany,

Lamb, W.M., 1990. Fluid inclusions in granulites: peak vs. retrograde formation. In: D. Vielzeuf and P. Vidal (Eds.), Granulites and Crustal Evolution. Kluwer, Dordrecht, pp. 419–433.

Lamb, W.M., Valley, J.W., and Brown, P.E., 1987. Post-metamorphic CO_2-rich inclusions in granulites. Contrib. Mineral. Petrol., 96: 485–495.

Lapointe, B., and Chown, E.H., 1993. Gold-bearing iron formation in a granulite terrane of the Canadian Shield: a possible deep-level expression of an Archean gold-mineralizing system. Mineral. Deposita, 28, 191–197.

Liu, D.Y., Shen, Q.H., Zhang, Z.Q., Jahn, B.M., and Auvray, B., 1990. Archean crustal evolution in China: U-Pb geochronology of the Qianxi Complex. Precambrian Res., 48: 223–244.

Liu, X., Jin, W., Li, S., and Xu, X., 1993. Two types of Preambrian high-grade metamorphism, Inner Mongolia, China. J. Metamorph. Geol., 11: 499–510.

Lu, L., and Jin, S., 1993. P-T-t paths and tectonic history of an early Precambrian granulite facies terrane, Jining district, south-east Inner Mongolia, China. J. Metamorph. Geol., 11: 483–498.

Ludden, J.N., Hubert, C., Barnes, A., Milkereit, B., and Sawyer, E., 1993. A three dimensional perspective on the evolution of Archaean crust: LITHOPROBE seismic reflection images in the southwestern Superior province. Lithos, 30: 357–372.

Lux, D.R., De Yoreo, J.J., Guidotti, C.V., and Decker, E.R., 1986. Role of plutonism in low-pressure metamorphic belt formation. Nature, 323: 794–797.

Mäder, U., Percival, J.A., and Berman, R.G., 1994. Thermobarometry of garnet–clinopyroxene–hornblende granulites from the Kapuskasing structural zone. Can. J. Earth Sci., 31: 1134–1145.

Mareschal, M., Fyfe, W.S., Percival, J.A., and Chan, T., 1992. Grain-boundary graphite in Kapuskasing gneisses and implications for lower-crustal conductivity. Nature, 357: 674–676.

McKenzie, D., 1984. A possible mechanism for epeirogenic uplift. Nature, 307: 616–618.

McNutt, R.H., and Dickin, A.P., 1989. Rb/Sr and Sm/Nd studies of four units in the vicinity of the Kapuskasing structural zone. Geol. Assoc. Can. Prog. Abstr., 14: A124.

Mezger, K., 1992. Temporal evolution of regional granulite terranes: Implications for the formation of lowermost continental crust. In: D.M. Fountain, R. Arculus, and R.W. Kay (Eds.), Continental Lower Crust. Elsevier, Amsterdam, pp. 447–478.

Mezger, K., Hanson, G.N., and Bohlen, S.R., 1989. U-Pb systematics of garnet: dating the growth of garnet in the late Archean Pikwitonei granulite domain at Cauchon and Natawahunan Lakes, Manitoba, Canada. Contrib. Mineral. Petrol., 101: 136–148.

Mezger, K., Bohlen, S.R., and Hanson, G.N., 1990a. Metamorphic history of the Archean Pikwitonei granulite domain and the Cross Lake subprovince, Superior Province, Manitoba. J. Petrol., 31: 483–517.

Mezger, K., Hanson, G.N., and Bohlen, S.R., 1990b. U-Pb ages of metamorphic rutiles: application to the cooling history of high grade terranes. Earth Planet. Sci. Lett., 96: 106–118.

Moecher, D.P., Essene, E.J., and Anovitz, L.M., 1988. Calibration and application of clinopyroxene–garnet–plagioclase–quartz geobarometers. Contrib. Mineral. Petrol., 100: 92–106.

Moecher, D.P., Perkins, D., Leier-Englehardt, P.J., and Medaris, L.G., 1986. Metamorphic conditions of late Archean high-grade gneisses, Minnesota River Valley, U.S.A. Can. J. Earth Sci., 23: 633–645.

Mohan, A., and Windley, B.F., 1993. Crustal trajectory of sapphirine-bearing granulites from Ganguvarpatti, South India: evidnce for an isothermal decompression path. J. Metamorph. Geol.,

11: 867–878.

Moritz, R.P., and Chevé, S.R., 1992. Fluid-inclusion studies of high-grade metamorphic rocks of the Ashuanipi complex, eastern Superior Province: constraints on the retrograde P-T path and implications for gold metallogeny. Can. J. Earth Sci., 29: 2309–2327.

Moser, D., 1994. The geology and structure of the mid-crustal Wawa gneiss domain — a key to understanding tectonic variation with depth and with time in an Archean orogen. Can. J. Earth Sci., 31: 1064–1080.

Moser, D.E., and Heaman, L.M., 1994. A U-Pb zircon study of lower-crustal xenoliths from the Archean Superior Province: evidence for two anomalously "young" metamorphic events at 2.58 and 2.49 Ga. Geol. Assoc. Can. Abstr. Prog., 19: A79.

Moskovchenko, N.I., Ovchinnikova, G.V., and Kastrykina, V.M., 1993. High-pressure granulites of East Siberia in terms of Archaean and Proterozoic evolution. Precambrian Res., 62: 473–491.

Muhling, J.R., 1988. The nature of Poterozoic reworking of early Archaean gneisses, Mukalo Creek area, southern Gascoyne Province, Western Australia. Precambrian Res., 38: 297–307.

Muhling, J.R., 1990. The Narryer gneiss complex of the Yilgarn Block, Western Australia: a segment of Archaean lower crust uplifted during Proterozoic orogeny. J. Metamorph. Geol., 8: 47–64.

Myers, J.S., 1988. Early Archaean Narryer gneiss complex, Yilgarn craton, Western Australia. Precambrian Res., 38: 297–307.

Myers, J.S., 1993. Precambrian history of the west Australian craton and adjacent orogens. Annu. Rev. Earth Planet. Sci., 21: 453–485.

Nelson, E.P., and Forsythe, R.D., 1989. Ridge collision at convergent margins: implications for Archean and post-Archean crustal growth. Tectonophysics, 161: 307–315.

Newton, R.C., 1987. Late Archean/Early Proterozoic CO_2 streaming through the lower crust and geochemical segregation. Geophys. Res. Lett., 14: 287–290.

Newton, 1989. Metamorphic fluids in the deep crust. Ann. Rev. Earth Planet. Sci., 17: 385–412.

Newton, R.C., 1990. Fluids and melting in the Archaean deep crust of southern India. In: J.R. Ashworth and M. Brown (Eds.), High-temperature Metamorphism and Crustal Anatexis. Mineral. Soc. Ser. 2. Unwin Hyman, London, pp. 149–179.

Newton, R.C., 1992a. Charnockitic alteration: evidence for CO_2 infiltration in granulite facies metamorphism. J. Metamorph. Geol., 10: 383–400.

Newton, R.C., 1992b. An overview of charnockite. Precambrian Res., 55: 399–405.

Newton, R.C., and Perkins, D., 1982. Thermodynamic calibration of geobarometers based on the assemblages garnet–plagioclase–orthopyroxene (clinopyroxene)–quartz. Am. Mineral., 67: 203–222.

Newton, R.C., Smith, J.V., and Windley, B.F., 1980. Carbonic metamorphism, granulites and crustal growth. Nature, 288: 45–50.

Nisbet, E.G., and Fowler, C.M.R., 1983. Model for Archean plate tectonics. Geology, 11: 375–379.

Nisbet, E.G., Cheadle, M.J., Arndt, N.T., and Bickle, M.J., 1993. Constraining the potential temperature of the Archaean mantle: a review of the evidence from komatiites. Lihos, 30: 291–307.

Nutman, A.P., Friend, C.R.L., Baadsgaard, H., and McGregor, V.R., 1989. Evolution and assembly of of Archean gneiss terranes in the Godthabsfjord region, southern west Greenland: structural, metamorphic and isotopic evidence. Tectonics, 8: 573–589.

Nutman, A.P., Kinny, P.D., Compston, W., and Williams, I.S., 1991. SHRIMP U-Pb zircon geochronology of the Narryer gneiss complex, Western Australia. Precambrian Res., 52: 275–300.

Nutman, A.P., Friend, C.R.L., Kinny, P.D., and McGregor, V.R., 1993. Anatomy of an Early

Archean gneiss complex: 3900 to 3600 Ma crustal evolution in southern West Greenland. Geology, 21: 415–418.

Oliver, G.J.H., and Coggon, J.H., 1979. Crustal structure of Fiordland, New Zealand. Tectonophysics, 54: 253–292.

Osanai, Y., Komatsu, M., and Owada, M., 1991. Metamorphism and granite genesis in in the Hidaka metamorphic belt, Hokkaido, Japan. J. Metamorph. Geol., 9: 111–124.

Osanai, Y., Owada, M., and Kawasaki, T., 1992. Tertiary deep crustal ultrametamorphism in the Hidaka metamorphic belt, northern Japan. J. Metamorph. Geol., 10: 401–414.

Oxburgh, E.R., 1990. Some thermal aspects of granulite history. In: D. Vielzeuf and P. Vidal (Eds.), Granulites and Crustal Evolution. Kluwer, Dordrecht, pp. 569–580.

Parphenuk, O.I., Dechoux, V, and Mareschal, J-C., 1994. Finite element models of evolution for the Kapuskasing structural zone. Can. J. Earth Sci., 31: 1227–1234.

Parrish, R.R., 1990. U-Pb dating of monazite and its application to geological problems. Can. J. Earth Sci., 27: 1431–1450.

Pattison, D.R.M., and Bégin, N.J., 1994. Complex zoning in orthopyroxene and garnet in granulites: implications for geothermometry. J. Metamorph. Geol., 12: 387–410.

Percival, J.A., 1983. High-grade metamorphism in the Chapleau–Foleyet area, Ontario. Am. Mineral., 68: 667–686.

Percival, J.A., 1986. A possible exposed Conrad discontinuity in the Kapuskasing uplift, Ontario. In: M. Barazangi and L.D. Brown (Eds.), Reflection Seismology: The Continental Crust. Am. Geophys. Un. Geodynam. Ser., 14: 135–141.

Percival, J.A., 1989. A regional perspective of the Quetico metasedimentary belt, Superior Province, Canada. Can. J. Earth Sci., 26, 677–693.

Percival, J.A., 1990a. Archean tectonic settings of granulite terranes of the Superior Province, Canada: A view from the bottom. In: D. Vielzeuf and P. Vidal (Eds.), Granulites and Crustal Evolution. Kluwer, Dordrecht, pp. 171–193.

Percival, J.A., 1990b. A field guide to the Kapuskasing uplift, a cross section through the Archean Superior Province. In: M.H. Salisbury and D.M. Fountain (Eds.), Exposed Cross Sections of the Continental Crust. Kluwer, Dordrecht, pp. 227–283.

Percival, J.A., 1991a. Granulite-facies metamorphism and crustal magmatism in the Ashuanipi complex, Labrador–Quebec, Canada. J. Petrol., 32, 1261–1297.

Percival, J.A., 1991b. Orthopyroxene–poikilitic tonalites from the Desliens igneous suite, Ashuanipi granulite complex, Labrador–Quebec, Canada. Can. J. Earth Sci., 28, 743–753.

Percival, J.A., and Card, K.D., 1983. Archean crust as revealed in the Kapuskasing uplift, Superior Province, Canada. Geology, 11: 323–326.

Percival, J.A., and Krogh, T.E., 1983. U-Pb zircon geochronology of the Kapuskasing structural zone and vicinity in the Chapleau–Foleyet area, Ontario. Can. J. Earth Sci., 20: 830-843.

Percival, J.A., and McGrath, P.H., 1986. Deep crustal structure and tectonic history of the northern Kapuskasing uplift of Ontario: An integrated petrological–geophysical study. Tectonics, 5: 553–572.

Percival, J.A., and Peterman, Z.E., 1994. Rb-Sr biotite and whole-rock data from the Kapuskasing uplift and their bearing on the cooling and exhumation history. Can. J. Earth Sci., 31: 1172–1181.

Percival, J.A., and West, G.F., 1994. The Kapuskasing uplift: a geological and geophysical synthesis. Can. J. Earth Sci., 31: 1256–1286.

Percival, J.A., and Williams, H.R., 1989. Late Archean Quetico accretionary complex, Superior Province, Canada. Geology, 17: 23–25.

Percival, J.A., Green, A.G., Milkereit, B., Cook, F.A., Geis, W. and West, G.F., 1989. Seismic reflection profiles across deep continental crust exposed in the Kapuskasing uplift structure.

Nature, 342: 416–420.

Percival, J.A., Fountain, D.M., and Salisbury, M.H., 1992a. Exposed crustal cross sections as windows on the lower crust. In: D.M. Fountain, R.J. Arculus and R.W. Kay (Eds.), Continental Lower Crust. Elsevier, Amsterdam, pp. 317–362.

Percival, J.A., Mortensen, J.K., Stern, R.A., Card, K.D., and Begin, N.J., 1992b. Giant granulite terranes of northeastern Superior Province: The Ashuanipi complex and Minto block. Can. J. Earth Sci., 29: 2287–2308.

Percival, J.A., Palmer, H.C. and Barnett, R.L., 1994a. Quantitative estimates of emplacement level of post-metamorphic mafic dykes and subsequent erosion magnitude in the southern Kapuskasing uplift. Can. J. Earth Sci., 31: 1218–1226..

Percival, J.A., Stern, R.A., Skulski, T., Card, K.D., Mortensen, J.K., and Bégin, N.J., 1994b. Minto block, Superior province: missing link in deciphering assembly of the craton at 2.7 Ga. Geology, 22: 839–842.

Perkins, D., and Chipera, S.J., 1985. Garnet–orthopyroxene–plagioclase–quartz barometry: refinement and application to the English River subprovince and Minnesota River valley. Contrib. Mineral. Petrol., 89: 69–80.

Peterson, J.W., and Newton, R.C. 1989. Reversed experiments on biotite–quartz–feldspar melting in the system KMASH: Implications for crustal anatexis. J. Geol., 97: 465–485.

Peterson, J.W., and Newton, R.C. 1990. Experimental biotite–quartz melting in the KMASH–CO_2 system and the role of CO_2 in the petrogenesis of granites and related rocks. Am. Mineral., 75: 1029–1042.

Peucat, J.J., Mahabaleswar, B., and Jayananda, M., 1993. Age of younger tonalitic magmatism and granulitic metamorphism in the South Indian transition zone (Krishnagiri area); comparison with older Peninsular gneisses from the Gorur–Hassan area. J. Metamorph. Geol., 11: 879–888.

Pichamuthu, C.S., and Srinivasan, R., 1984. The Dharwar craton. Indian Nat. Sci. Acad. Persp. Rep. Ser., 7: 3–34.

Platt, J.P., 1986. Dynamics of orogenic wedges and the uplift of high-pressure metamorphic rocks. Geol. Soc. Am. Bull., 97: 1037–1053.

Powell, R., 1983. Processes in granulite-facies metamorphism. In: M.P. Atherton and C.D. Gribble (Eds.), Migmatites, Melting and Metamorphism. Shiva, Nantwich, pp. 127–139.

Pride, C., and Muecke, G.K., 1980. Rare earth element geochemistry of the Scourie complex, NW Scotland — evidence for the granite–granulite link. Contrib. Mineral. Petrol., 73: 403–412.

Pride, C., and Muecke, G.K., 1982. Geochemistry and origin of granitic rocks, Scourian complex, NW Scotland. Contrib. Mineral. Petrol., 80: 379–385.

Raith, M., Raase, P., Ackermand, D., and Lal, R.K., 1983. Regional geothermobarometry in the granulite facies terrane of South India. Trans. R. Soc. Edinburgh, Earth Sci., 73: 221–244.

Reed, J.C., Ball, T.T., Farmer, G.L., and Hamilton, W.B., 1993. A broader view. In: J.C. Reed Jr., M.E. Bickford, R.S. Houston, P.K. Link, D.W. Rankin, P.K. Sims, and W.R. Van Schmus (Eds.), Precambrian: Conterminus U.S. Geol. Soc. Amer. The Geology of North America, C-2, 597–636.

Richardson, S.H., Gurney, J.J., Erlank, A.J., and Harris, J.W., 1984. Origin of diamonds in old enriched mantle. Nature, 310: 198–202.

Richter, F.M., 1985. Models for the Archaean thermal regime. Earth Planet. Sci. Lett., 73: 350–360.

Riciputi, L.R., Valley, J.W., and McGregor, V.R., 1990. Conditions of Archean granulite metamorphism in the Godthab–Fiskenaesset region, southern West Greenland. J. Metamorph. Geol., 8: 171–190.

Ridley, J., 1992. On the origins and significance of the charnockite suite of the Archaean Limpopo Belt, Northern Marginal Zone, Zimbabwe. Precambrian Res., 55:, 407-427.

Roering, C., Van Reenen, D.D., Smit, C.A., Barton, J.M., Jr., De Beer, J.H., de Wit, M.J., Stettler, E.H., Van Schalkwyk, J.F., Stevens, G. and Pretorius, S., 1992. Tectonic model for the evolution of the Limpopo Belt. Precambrian Res., 55: 539-552.

Rollinson, H.R., 1982. P-T conditions in coeval greenstone belts and granulites from the Archaean of Sierra Leone. Earth Planet. Sci. Lett., 59: 177–191.

Rosen, O.M., 1990. (Compiler) Metamorphic rocks of the Anabar Shield. In: K.C. Condie and N.A. Bogdanov (Eds.), International Field Conference on Archean Geology and Geochemistry of the Anabar Shield. Acad. Sci. USSR Inst. Lithosphere, 102 p.

Rosen, O.M., Condie, K.C., Natapov, L.M., and Nozhkin, A.D., 1994. Precambrian tectonic development of the Siberian Plate: A preliminary assessment. In: K.C. Condie (Ed.), Archean Crustal Evolution, Chap. 10, pp. 411–459.

Ross, D.C., 1985. Mafic gneiss complex (batholithic root?) in the southernmost Sierra Nevada, California. Geology, 13: 288–291.

Ross, D.C., 1989. The metamorphic and plutonic rocks of the southernmost Sierra Nevada, California, and their tectonic framework. U.S. Geol. Surv. Prof. Pap., 1381, 159 p.

Rudnick, R.L., 1992. Xenoliths — Samples of the lower continental crust. In: D.M. Fountain, R. Arculus, and R.W. Kay (Eds.), Continental Lower Crust. Elsevier, Amsterdam, pp. 269–316.

Rudnick, R.L., and Presper, T., 1990. Geochemistry of intermediate- to high-pressure granulites. In: D. Vielzeuf and P. Vidal (Eds.), Granulites and Crustal Evolution. Kluwer, Dordrecht, pp. 523–550.

Rudnick, R.L., and Taylor, S.R., 1986. Geochemical constraints on the origin of Archaean tonalitic–trondhjemitic rocks and implications for lower crustal composition. In: J.B. Dawson, D.A. Carswell, J. Hall, and K.H. Wedepohl (Eds.), The Nature of the Lower Continental Crust. Geol. Soc. Spec. Publ., 24: 179–191.

Rudnick, R.L., Ashwal, L.D. and Henry, D.J., 1984. Fluid inclusions in high-grade gneisses of the Kapuskasing structural zone, Ontario: metamorphic fluids and uplift/erosion path. Contrib. Mineral. Petrol., 87: 399–406.

Saleeby, J.B., 1990. Progress in tectonic and petrotectonic studies in an exposed cross section of young (100 Ma) continental crust, southern Sierra Nevada, California. In: M.H. Salisbury and D.M. Fountain (Eds.), Exposed Cross Sections of the Continental Crust. Kluwer, Dordrecht, pp. 137–158.

Saleeby, J.B., Sams, D.B., and Kistler, R.W., 1987. U/Pb zircon, strontium and oxygen isotopic and geochronological study of the southernmost Sierra Nevada batholith, California. J. Geophys. Res., 92: 10,443–10,466.

Sams, D.B., and Saleeby, J.B., 1988. Geology and petrotectonic significance of crystalline rocks of the southernmost Sierra Nevada, California. In: W.G. Ernst (Ed.), Metamorphism and Crustal Evolution of the Western United States. Prentice-Hall, Englewood Cliffs, pp. 865–893.

Sandiford, M., 1989a. Horizontal structures in granulite terrains: A record of mountain building or mountain collapse? Geology 17: 449–452.

Sandiford, M., 1989b. Secular trends in the thermal evolution of metamorphic belts. Earth Planet. Sci. Lett., 95: 85–96.

Sandiford, M., and Powell, R., 1986. Deep crustal metamorphism during continental extension: modern and ancient examples. Earth Planet. Sci. Lett., 79: 151–158.

Sandiford, M., and Powell, R., 1991. Some remarks on high-temperature–low-pressure metamophism in convergent orogens. J. Metamorph. Geol., 9: 333–340.

Sandiford, M., Neall, F.B., and Powell, R., 1987. Metamorphic evolution of aluminous granulites from Labwor Hills, Uganda. Contrib. Mineral. Petrol., 95: 217–225.

Schmidt, M.W., 1992. Amphibole composition as a function of pressure: an experimental calibration of the Al-in-hornblende barometer. Contrib. Mineral. Petrol., 110: 304–310.

Selverstone, J., and Chamberlain, C.P., 1990. Apparent isobaric cooling paths from granulites: two counterexamples from British Columbia and New Hampshire. Geology, 18: 307–310.

Shimura, T., 1992. Intrusion of granitic magma and uplift tectonics of the Hidaka Metamorphic Belt, Hokkaido. J. Geol. Soc. Japan, 98: 1–20.

Shimura, T., Komatsu, M., and Iiyama, J.T., 1992. Genesis of the lower crustal garnet–orthopyroxene tonalites (S-type) of the Hidaka Metamorphic Belt, northern Japan. Trans. R. Soc. Edinburgh: Earth Sci., 83: 259–268.

Sills, J.D., and Rollinson, H.R., 1987. Metamorphic evolution of the mainland Lewisian complex. In: R.G. Park and J. Tarney (Eds.), The Evolution of the Lewisian and Comparable Precambrian High-grade Terrains. Geol. Soc. Lond. Spec. Publ., 27: 81–92.

Sills, J.D., Wang, K.Y., Yan, Y.H., and Windley, B.F., 1987. The Archean high-grade gneiss terrane in E. Hebei Province, NE China: geological framework and conditions of metamorphism. In: R.G. Park and J. Tarney (Eds.), The Evolution of the Lewisian and Comparable Precambrian High-grade Terrains. Geol. Soc. Lond. Spec. Publ., 27: 297–305.

Silver, P.G., and Chan, W.W., 1988. Implications for continental structure and evolution from seismic anisotropy. Nature, 335: 34–39.

Sims, P.K., Card, K.D., Morey, G.B., and Peterman, Z.E., 1980. The Great Lakes tectonic zone — a major Precambrian crustal structure in central North America. Geol. Soc. Am. Bull., 91: 690–698.

Skulski, T., Percival, J.A., and Stern, R.A., 1994. Oceanic allochthons in an Archean continental margin sequence, Vizien greenstone belt, northern Quebec. Current Research, Geol. Surv. Can. Pap., 1994-C: 311–320.

Sleep, N.H., and Windley, B.F., 1982. Archean plate tectonics: constraints and inferences. J. Geol., 90: 363–379.

Snoeyenbos, D.R., and Williams, M.L., 1994. An Archean eclogite facies terrane from the Snowbird tectonic zone, northern Saskatchewan. EOS, Trans. Am. Geophys. Un., 75: 355.

Spear, F.S., and Florence, F.P., 1992. Thermobarometry in granulites: pitfalls and new approaches. Precambrian Res., 55: 209–241.

Srikantappa, C., 1993. High prssure charnockites of the Nilgiri Hills, southern India. In: B.P. Radhakrishna (Ed.), Continental Crust of South India. Geol. Soc. India Mem., 25: 95–110.

Stähle, H.J., Raith, M., Hoernes, S., and Delfs, A., 1987. Element mobility during incipient granulite formation at Kabbaldurga, southern India. J. Petrol., 28: 803–843.

Stern, R.A., Percival, J.A., and Mortensen, J.K., 1994. Geochemical evolution of the Minto block: A 2.7 Ga continental magmatic arc built on the Superior proto-craton. Precambrian Res., 65: 115–153.

Stevens, G. and Van Reenen, D.D. 1992. Constraints on the form of the P-T loop in the Southern Marginal Zone of the Limpopo Belt, South Africa. Precambrian Res., 55: 279-296.

Stevens, G., and Clemens, J.D., 1993. Fluid-absent melting and the roles of fluids in the lithosphere; a slanted summary? Chem. Geol., 108: 1–17.

Stuckless, J.G., Hedge, C.E., Worl, R.G., Simmons, K.R., Nkomo, I.T., and Wenner, D.B., 1985. Isotopic studies of Late Archean plutonic rocks in the Wind River Range, Wyoming. Geol. Soc. Am. Bull., 96: 850–860.

Stüwe, K., Sandiford, M., and Powell, R., 1993. Episodic metamorphism and deformation in low-pressure, high-temperature terranes. Geology, 21: 829–832.

Sutcliffe, R.H., Barrie, C.T., Burrows, D.R., and Beakhouse, G.P., 1993. Plutonism in the southern Abitibi subprovince: a tectonic and petrogenetic framework. Econ. Geol., 88: 1359–1375.

Thompson, A.B., 1990. Heat, fluids and melting in the granulite facies. In: D. Vielzeuf and P. Vidal

(Eds.), Granulites and Crustal Evolution. Kluwer, Dordrecht, pp. 37–57.
Treloar, P.J., Coward, M.P., and Harris, N.B.W., 1992. Himalayan–Tibetan analogies for the evolution of the Zimbabwe Craton and Limpopo Belt. Precambrian Res. 55: 571–587.
Truscott, M.G. and Shaw, D.M., 1990. Average composition of lower and intermediate continental crust, Kapuskasing structural zone, Ontario. In: M.H. Salisbury and D.M. Fountain (Eds.), Exposed Cross Sections of the Continental Crust. Kluwer, Dordrecht, pp. 421–436.
Tsunogae, T., Miyano, T. and Ridley, J., 1992. Metamorphic P-T profiles from the Zimbabwe Craton to the Limpopo Belt, Zimbabwe. Precambrian Res., 55: 259-277.
Van Reenen, D.D., Barton, J.M., Jr., Roering, C., Smit, C.A. and Van Schalkwyk, J.F., 1987. Deep-crustal response to continental collision: The Limpopo Belt of southern Africa. Geology, 15: 11-14.
Vielzeuf, D., and Holloway, J.R., 1988. Experimental determination of the fluid-absent melting relations in the pelitic system. Consequences for crustal differentiation. Contrib. Mineral. Petrol., 98: 257–276.
Vielzeuf, D., and Vidal, P. (Eds.), 1990. Granulites and crustal evolution. Kluwer, Dordrecht, 585 pp.
Vielzeuf, D., Clemens, J.D., Pin, C., and Moinet, E., 1990. Granites, granulites, and crustal differentiation. In: D. Vielzeuf and P. Vidal (Eds.), Granulites and Crustal Evolution. Kluwer, Dordrecht, pp. 59–85.
Warren, P.H., 1984. Primordial degassing, lithosphere thickness, and the origin of komatiites. Geology 12: 335–338.
Watson, J.V., 1978. Precambrian thermal regimes. Phil. Trans. R. Soc. Lond. Ser. A, 288: 431–440.
Weaver, B.L., and Tarney, J., 1981. Lewisian gneiss geochemistry and Archaean crustal development models. Earth Planet. Sci. Lett., 55: 171–180.
Weaver, B.L., and Tarney, J., 1983. Elemental depletion in Archaean granulite-facies rocks. In: M.P. Atherton and C.D. Gribble (Eds.), Migmatites, Melting and Metamorphism. Shiva, Nantwich, UK, pp. 250–263.
Weber, W., 1990. The Churchill–Superior boundary zone, southeast margin of the Trans-Hudson orogen: A review. In: J.F. Lewry and M.R. Stauffer (Eds.), The Early Proterozoic Trans-Hudson Orogen. Geol. Assoc. Can. Spec. Pap., 37: pp. 41–55.
Weber, W., and Mezger, K., 1990. An oblique cross section of Archean continental crust at the northwestern margin of Superior Province, Manitoba, Canada. In: M.H. Salisbury and D.M. Fountain (Eds.), Exposed Cross Sections of the Continental Crust. Kluwer, Dordrecht, pp. 327–341.
Wells, P.R.A., 1979. Chemical and thermal evolution of Archaean sialic crust, southern West Greenland. J. Petrol., 20: 187–226.
Wells, P.R.A., 1980a. Thermal models for the magmatic accretion and subsequent metamorphism of continental crust. Earth Planet. Sci. Lett., 46: 253–265.
Wells, P.R.A., 1980b. Late Archaean metamorphism in the Buksefjorden region, southwest Greenland. Contrib. Mineral. Petrol., 56: 229–242.
West, G.F. and Ernst, R.E., 1991. Evidence from aeromagnetics on the configuration of Matachewan dykes and the tectonic evolution of the Kapuskasing structural zone, Canada. Can. J. Earth Sci., 28: 1797–1811.
Wickham, S.M., 1987. The segregation and emplacement of granitic magmas. J. Geol. Soc. Lond., 144: 281–297.
Wilks, M.E., 1988. The Himalayas — a modern analogue for Archaean crustal evolution. Earth Planet. Sci. Lett., 87: 127–136.
Williams, H.R., 1988. The Archaean Kasila Group of western Sierra Leone: geology and relations

with adjacent granite–greenstone terrane. Precambrian Res., 38: 201–213.
Williams, H.R., Stott, G.M., Thurston, P.C., Sutcliffe, R.H., Bennett, G., Easton, R.M., and Armstrong, D.K., 1992. Tectonic evolution of Ontario: summary and synthesis. In: P.C. Thurston, H.R. Williams, R.H. Sutcliffe and G.M. Stott (Eds.), Geology of Ontario. Ont. Geol. Surv. Spec. Vol. 4, Pt. 2, pp. 1255–1332.
Wilson, J.F., 1990. A craton and its cracks: some of the behaviour of the Zimbabwe block from the Late Archaean to the Mesozoic in response to horizontal movements, and the significance of some of its mafic dyke patterns. J. Afr. Earth Sci., 10: 483–501.
Windley, B.F., 1981. Phanerozoic granulites. J. Geol. Soc. Lond., 138: 745–751.
Windley, B.F., and Bridgwater, D., 1971. The evolution of Archaean low- and high-grade terrains. Spec. Publ. Geol. Soc. Australia, 3: 33–46.
Windley, B.F., and Davies, F.B., 1978. Volcano spacings and lithospheric/crustal thickness in the Archaean. Earth Planet. Sci. Lett., 38: 291–297.
Windley, B.F., and Smith, J.V., 1976. Archaean high-grade complexes and modern continental margins. Nature, 260: 671–675.
Wyllie, P.J., 1977. Crustal anatexis: an experimental review. Tectonophysics, 43: 41–71.
Wyllie, P.J., Huang, W.L., Wuu-liang, Stern, C.R., and Maaloe, S., 1976. Granitic magmas: possible and impossible sources, water contents, and crystallization sequences. Can. J. Earth Sci., 13: 1007–1019.
Xuan, H., Ziwei, B., and DePaolo, D.J., 1986. Sm-Nd study of early Archean rocks, Qianan, Hebei Province, China. Geochim. Cosmochim. Acta, 50: 625–631.

Chapter 10

ARCHEAN AND EARLY PROTEROZOIC EVOLUTION OF THE SIBERIAN CRATON: A PRELIMINARY ASSESSMENT

O.M. ROSEN, KENT C. CONDIE, LEV M. NATAPOV and A.D. NOZHKIN

INTRODUCTION

The Siberian craton extends from Lake Baikal on the south for over 2500 km to the Arctic Ocean on the north, and from the Yenisey River on the west to the Sea of Okhotsk on the east (Fig. 1) The borders of the craton are major Phanerozoic suture zones that formed during the aggregation of Pangea in the late Paleozoic and early Mesozoic (Zonenshain et al., 1989; Parfenov, 1991). The oldest suture is between the southwestern margin of the Siberian craton and the Central Asia Paleozoic orogenic belt (Fig. 1). This orogenic belt formed by accretion of Late Proterozoic and early Paleozoic arc systems to the southwestern margin of the Siberian craton beginning in the Late Riphean and continuing into the mid-Paleozoic. During this collision in the Devonian, the Barguzin terrane composed of Paleozoic and Late Proterozoic rocks was thrust onto the southern margin of the Siberian craton (Fig. 1). During the mid to late Mesozoic, terranes collided with Siberia along the southeastern and eastern borders (directions based on present position) forming the Mongolia–Okhotsk and Verkhoyansk orogenic belts. Although the western border of the Siberian craton is covered with Mesozoic cratonic sediments, geophysical data suggest the presence of a tectonic boundary in the subsurface west of Noril'sk (Khain, 1985; Yanshin and Borukaev, 1988) (Fig. 1). The Barentsia Massif, a hypothetical landmass of Late Proterozoic age, may have collided with Siberia along this boundary in the early or middle Paleozoic (Zonenshain et al., 1989). The Taymyr orogenic belt along the northern margin of the Siberian craton is a late Paleozoic–early Mesozoic orogenic belt.

Beneath the Siberian craton, the Moho is 35–60 km deep and is characterized by P_n seismic wave velocities of 8.2–8.6 km/sec beneath the northern part of the craton and 7.8–8.2 km/sec beneath the southern part (Belousov et al., 1991). The lowest velocities (7.6–7.8 km/sec) occur beneath the Baikal rift and the Barguzin terrane. Results from deep seismic sounding indicate a thickness of the lithosphere of 150–200 km for the eastern part of Siberia, which is also characterized by low heat flow (25–40 mW/m^2). The lithosphere thins to 130–140 km in the western northwestern parts of the craton (Belousov et al., 1991; Fotiadi, 1989), perhaps in response to a mantle plume from which the Siberian traps were erupted in the late

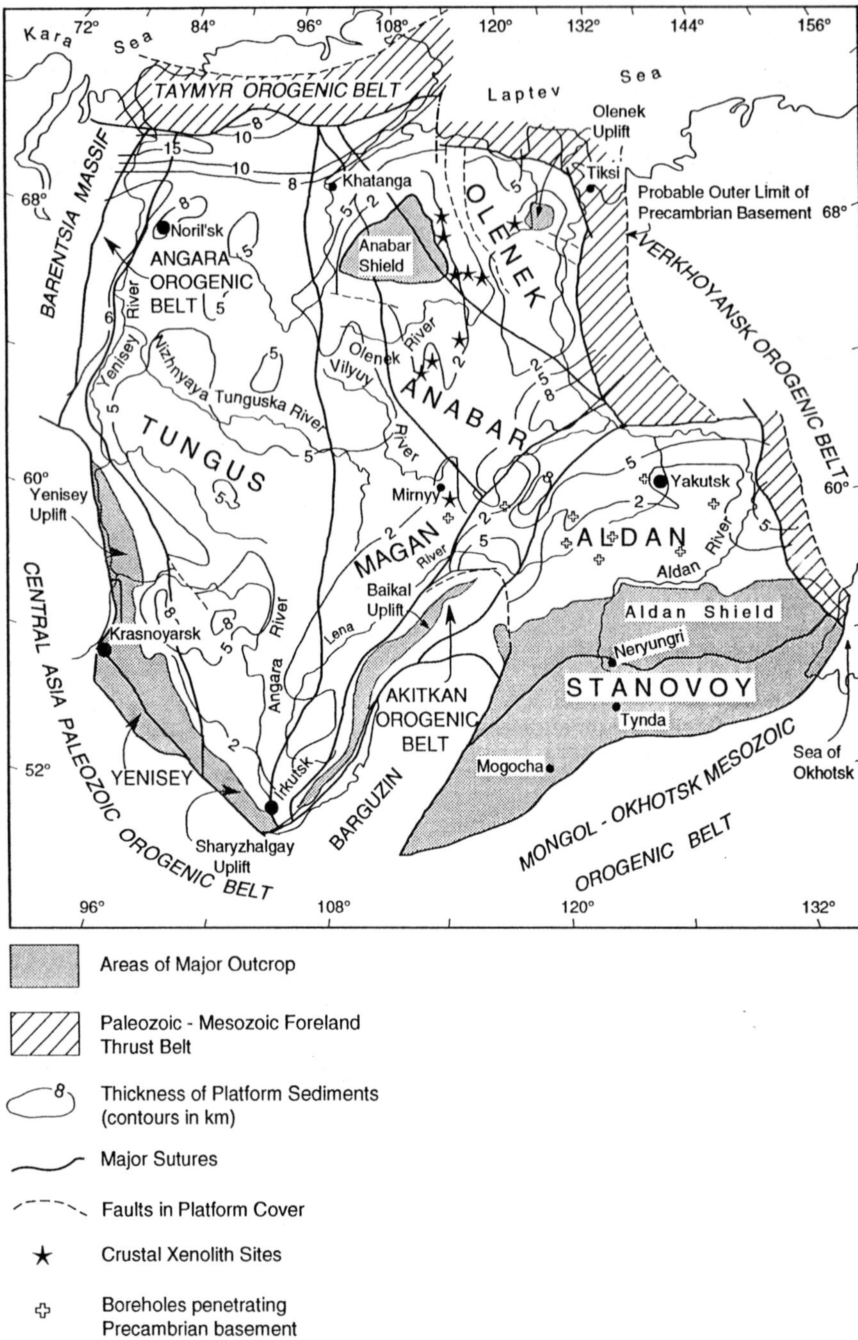

Fig. 1. Tectonic map of the Siberian craton showing Precambrian crustal provinces in the basement. Stars indicate kimberlite fields with crustal xenoliths and pluses are boreholes reaching Precambrian basement.

Paleozoic. It is notable that these lithospheric thicknesses are considerably less than those reported for Archean cratons in North America from shear wave velocity studies (Grand, 1987; Anderson et al., 1992) and from South Africa from mantle xenolith studies (Richardson et al., 1984). Pre-Riphean (>1650 Ma) basement is exposed in three large areas in the Siberian craton: the Aldan shield on the southeast, the Anabar shield in the north, and the Yenisey uplift along the southwestern margin (Fig. 1). In addition, smaller Precambrian exposures are found in the Baikal uplift west of Lake Baikal, in the Sharyzhalgay uplift west of Irkutsk, and in the Olenek uplift southwest of Tiksi on the Laptev Sea. Collectively, only about 25% of the basement of the Siberian craton is exposed at the surface. The remainder is covered by Riphean, Vendian, and Phanerozoic sediments typically 2–5 km thick, and locally in rifts reaching thicknesses in excess of 10 km (Fig. 1). Flood basalts of the Siberian traps erupted near the Permian–Triassic boundary reach thicknesses of 5 km in the western part of Siberia.

Structural features as well as constraints on lithology and age of covered portions of the Siberian craton can be deduced from recently released gravity and magnetic anomaly maps of Russia (Litvinova et al., 1978; Samkov and Potapyev, 1986) (Fig. 2), as well as from boreholes that penetrate the basement, and crustal xenoliths in kimberlite pipes (Shatsky et al., 1990; Neymark et al., 1993a). From a combination of results from these three sources, together with outcrop and recent geochronologic data, we present a preliminary subdivision of the basement of the Siberian craton into terranes, crustal provinces, and orogenic belts. Terranes are fault-bounded segments of continental crust with distinctive rock assemblages and tectonic histories, whereas crustal provinces are collages of terranes accreted during a limited time interval (Jones et al., 1983; Condie, 1992). Orogenic belts are linear terranes or crustal provinces that have undergone one or more periods of intense deformation. Boundaries of these subdivisions are defined from outcrop and geophysical anomalies in the basement. Seven crustal provinces are recognized in the Siberian craton, and these are bounded by large faults or/and large geophysical anomalies. Some terrane boundaries are poorly defined, and at this stage rather speculative. In all but the Akitkan belt, Archean basement is known to be present in Siberian orogenic belts. Differences in our subdivisions from those proposed before (Khiltova, 1993; Rundkvist and Sokolov, 1993 and others) are probably due to a combination of factors including the following: (1) we did not use seismic refraction data, which are not sensitive enough to define province boundaries; (2) we did not model the magnetic anomaly maps, but used the data as published; and (3) we had access to recent geologic mapping and high-precision U/Pb zircon ages published in the last few years.

In the Aldan shield, two crustal provinces are exposed: the Aldan on the north and the Stanovoy on the south. These provinces are separated by the Kalar shear zone, a major E–W trending shear zone some 500 km south of Yakutsk (Fig. 3). From geologic mapping and geochronology in these provinces, four terranes and two orogenic belts are recognized in the Aldan province and three terranes and one

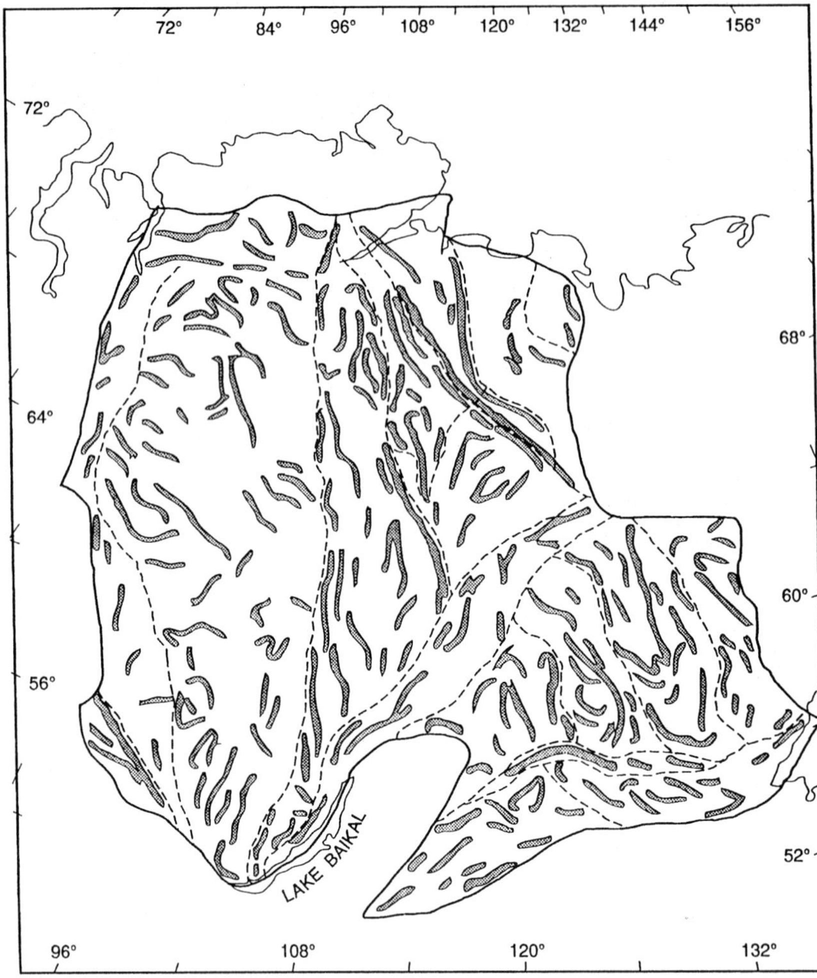

Fig. 2. Simplified map showing distribution of positive magnetic anomalies in the Siberian craton (after Litvinova et al., 1978). Anomalies shown have magnitudes of >1 mE. Dashed lines are borders of Precambrian crustal provinces and terranes as defined in this study.

orogenic belt in the Stanovoy province. The northern boundary of the Aldan province, herein called the Lena fault zone, is defined by a disruption of geophysical anomalies and rapid thickening of the overlying Riphean sediments. Three provinces are exposed in the Anabar shield. The Anabar province is a triangular-shaped province bounded by the Vilyuy, Billyakh, and Kotuykan fault zones, and it comprises two terranes (Fig. 7). The Olenek province on the extreme NE corner of the Siberian craton includes three terranes. The Magan province in central Siberia is bounded on the west by the Sayan–Taymyr fault zone, defined by a

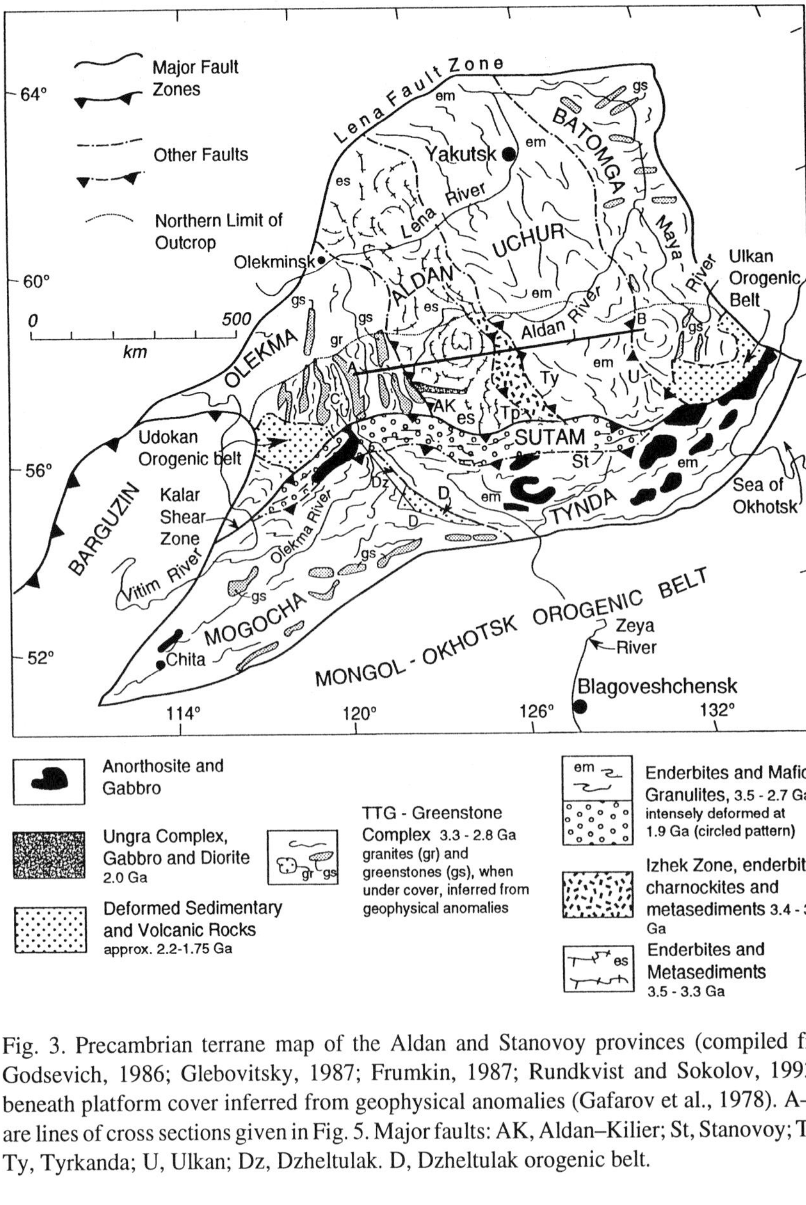

Fig. 3. Precambrian terrane map of the Aldan and Stanovoy provinces (compiled from data in Godsevich, 1986; Glebovitsky, 1987; Frumkin, 1987; Rundkvist and Sokolov, 1993). Features beneath platform cover inferred from geophysical anomalies (Gafarov et al., 1978). A–B and C–D are lines of cross sections given in Fig. 5. Major faults: AK, Aldan–Kilier; St, Stanovoy; Tp, Timpton; Ty, Tyrkanda; U, Ulkan; Dz, Dzheltulak. D, Dzheltulak orogenic belt.

major north-trending break in magnetic and gravity anomaly patterns (Figs. 2 and 7). West of this break, anomalies are randomly oriented or NW-trending, whereas to the east they are chiefly north-trending. The large Tungus province is exposed only in a small area in the Sharyzhalgay uplift, and the Yenisey province is exposed in the Yenisey uplift southeast of Krasnoyarsk (Fig. 1).

In this study, we present a summary of the geologic and geochronologic results from Archean and Early Proterozoic rocks of the Siberian craton, most of which are published in Russian and not readily available to western scientists. We also present a preliminary interpretation of these data as they relate to the tectonic history of the Siberian craton during the Precambrian, as well as point out problem areas for future research.

ALDAN PROVINCE

Olekma terrane

The Olekma terrane, which is one the most studied areas in the Aldan province, comprises a typical Archean TTG (tonalite–trondhjemite–granodiorite)-greenstone assemblage in which major greenstone belts trend in a northerly direction (Rundkvist and Sokolov, 1993) (Fig. 3). Two domains are recognized within the Olekma terrane, the Olondo and Tungurcha domains, bounded by the Temulyakit fault (Fig. 4). Although most of the terrane reflects low to middle metamorphic grades, some regions have been metamorphosed to granulite grade. Greenstone belts are up to 20 km wide and some exceed 100 km in length. As with many other Archean greenstone successions, Olekma greenstones are composed chiefly of basalts and komatiites at lower stratigraphic levels, with andesites, felsic volcanics, and volcaniclastic sediments (including some carbonate) becoming more abundant at higher levels. Late syntectonic to post-tectonic granitoids are also found in parts of the Olondo domain (Popov et al., 1990; Puchtel, 1992).

The geochronology of the Olekma terrane is discussed in Nutman et al. (1992a) and Glebovitsky and Drugova (1993). Although the oldest component in the Tungurcha domain defined by SHRIMP zircon ages from tonalitic gneisses is about 3250 Ma, the most widespread ancient component in this domain is 3212 ± 8 Ma (Nutman et al., 1992a), an age that agrees within the range of the large uncertainty of a previously reported Sm/Nd isochron age of 3235 ± 144 Ma (Puchtel et al., 1989). There are several zircon and Sm/Nd isochron ages at 3000–2950 Ma in both the Tungurcha and Olondo domains (Neymark et al., 1990; Baadsgaard et al., 1990; Nutman et al., 1992a) suggesting this was an important time of TTG production. ε_{Nd} values at 3 Ga range from +1 to +2 (Jahn et al., 1990; Puchtel, 1992) suggesting a short (<100 Ma) crustal residence time for these rocks. Greenstone volcanism is also recorded by zircon SHRIMP and Sm/Nd isochron ages in the Olondo domain at 3018–3000 Ma and TTG plutonism at about 2860 Ma (Bibikova, 1989a; Baadsgaard et al., 1990). Zircon ages from two syntectonic granitoid plutons are 2984 ± 22 Ma and 2999 ± 51 Ma (Neymark et al., 1993b). Less reliable Sm/Nd isochron ages in the Tungurcha domain suggest TTG production at about 2850 Ma (Nemchin et al., 1989; Puchtel et al., 1989). Two SHRIMP zircon ages (2738 ± 8 Ma and 2751 ± 8 Ma) from granites, which are interpreted by Russian geologists as syntectonic with motion along the Temulyakit fault, suggest juxtaposition of the Olondo and Tungurcha

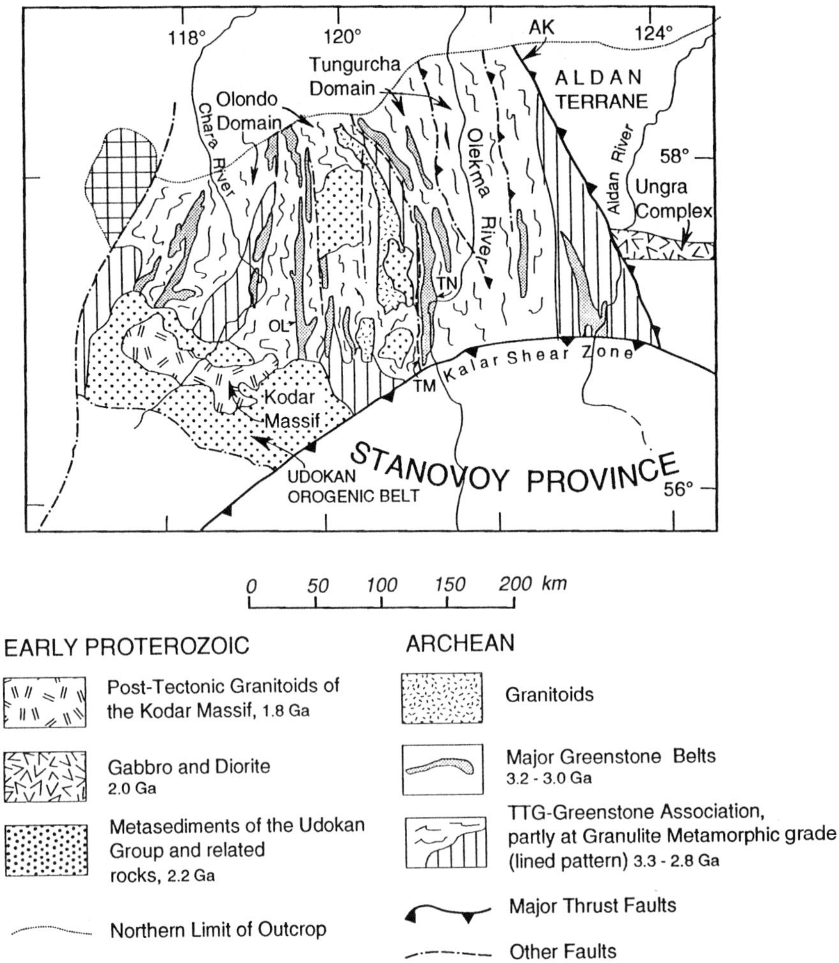

Fig. 4. Generalized geologic map of Precambrian rocks of the Olekma terrane (after Fedorovsky, 1985; Kitsul and Dook, 1985; Dook, 1989a). Important greenstone belts: OL, Olondo; TN, Tungurcha. Faults: TM, Temulyakit; AK, Aldan–Kilier.

domains at 2750–2740 Ma. SHRIMP ages of euhedral zircon overgrowths indicate partial melting and granulite-grade metamorphism at 1895 ± 4 Ma along the eastern border of the Olekma terrane (Nutman et al., 1992a).

Aldan terrane

The Aldan terrane is a small high-grade terrane bounded on both the east and west by east-dipping thrust faults (Figs. 3 and 5a) (Glebovitsky, 1987; Dook, 1989b; Kitsul, 1986). The southern boundary is a major suture with the Stanovoy

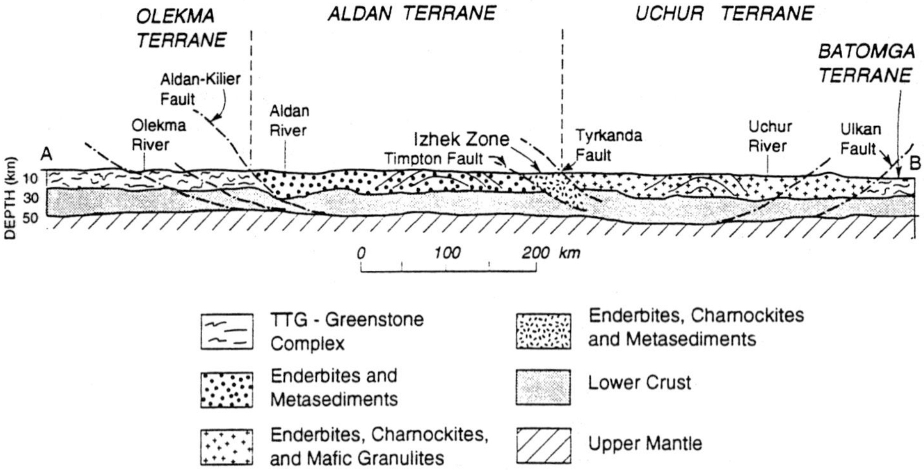

Fig. 5. Schematic cross sections (after Petrov et al., 1985; Dook, 1989a; Popov et al., 1989). (a) eastern Magocha terrane to the Batomga terrane, and (b) (*opposite*)central Olekma terrane to eastern Magocha terrane across the Sutam terrane. Inclination of faults based on geophysical data and isostatic gravity profile after data given Pismenny and Alakshin (1984) and Alakshin and Karsakov (1985). Lines of sections shown on Fig. 3.

province. The terrane is composed chiefly of a TTG complex including charnockites, granites, and metasediments. These rocks are multiply deformed and metamorphosed to the granulite facies. Although the metasediments are generally thought to be Archean in age like associated enderbites, the possibility they are infolded Early Proterozoic sediments cannot be eliminated (Nutman et al., 1992a). Enderbites are the most widespread felsic components, and field relationships indicate that some are older and some younger than metasediments. Metasediments are chiefly pelites, quartzites, and carbonates, and appear to reflect a cratonic suite. Estimates of the relative rock abundances are as follows (Rosen and Zlobin, 1990): orthopyroxene–plagioclase enderbites and charnockites 52%; mafic and ultramafic rocks 6%; metapelites 24%; quartzites 9%; calc-silicates and marbles 9%. The north-central part of the Aldan terrane is a broad domal structure (Moralev, 1986) in which some of the oldest rocks in the terrane exposed (Drugova, 1989). Quartzites are widespread on the western slope of this dome (Kitsul, 1986) and appear to be younger than felsic gneisses composing the core of the dome. Deformed remnants of mafic dikes are widespread throughout the Aldan terrane. Supracrustal rocks in the Izhek zone, an upthrusted slice between the Timpton and Tyrkanda faults, contain as much as 70% metasediments, and include also the large intrusive Izhek charnockite massif. The Ungra Complex, which is composed of gabbro, diorite, tonalite, and trondhjemite, is intruded into the basement gneisses and extends for over 100 km in strike length (Figs. 3 and 4). It crosscuts older N–NW striking folds and redeforms adjacent

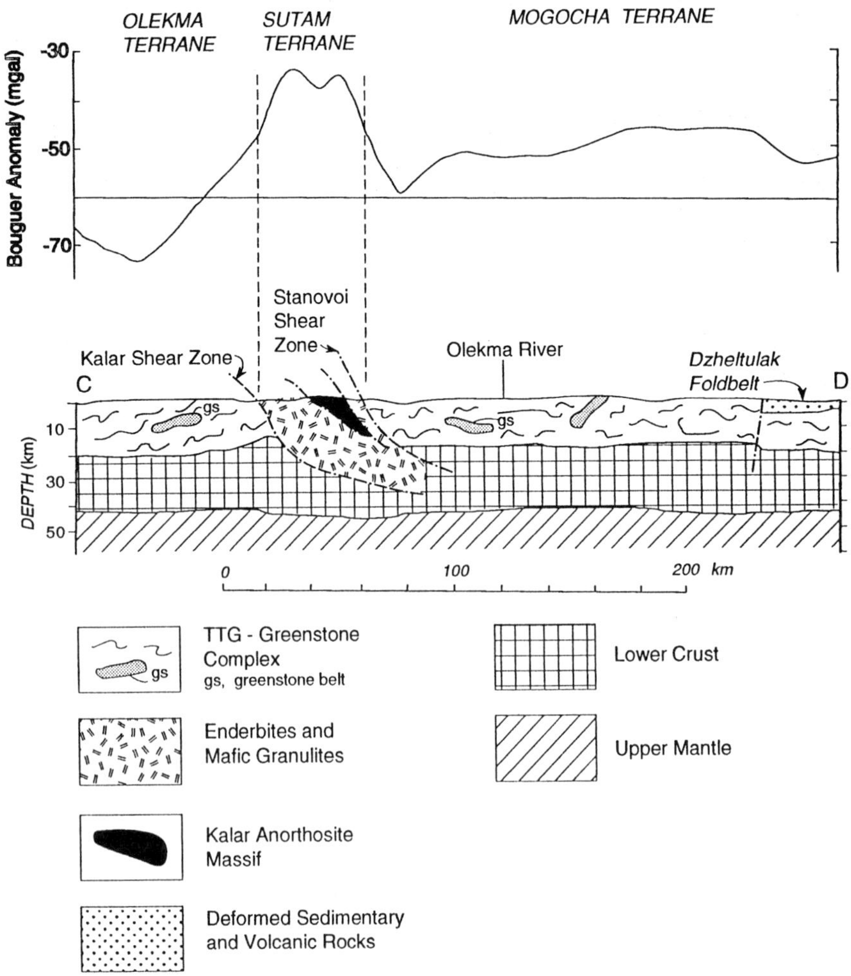

Fig. 5 (continued). (b) Caption on opposite page.

rocks into E–W trending folds with axes parallel to the long axis of the intrusion (Dook, 1989b; Glebovitsky, 1987). Structural and metamorphic studies along the Aldan–Kilier thrust fault (Fig. 5a) suggest the Aldan terrane was thrust over the Olekma terrane (Dook, 1989a), and thus is the hinterland in the deformation belt. Mapping has revealed two stages of deformation and plutonism in the Aldan terrane, both under granulite-facies conditions (Kotov et al., 1989). The first stage is characterized by syntectonic intrusion of tonalite and diorite into an isoclinally folded migmatite complex. The second stage involves open folding, partial melting leading to more migmatites, followed by intrusion of biotite–two feldspar aplite dikes.

In the central and eastern part of the Aldan terrane, thermobarometric studies using orthopyroxene–garnet, cordierite–garnet, and garnet–sillimanite–K-feldspar assemblages suggest minimum metamorphic temperatures of 800–825°C and pressures of 8–8.2 kbar (Kitsul, 1986). In the western part of the terrane, similar studies show temperatures of 690–780°C and pressures of 6–7.4 kbar, and in the Izhek zone, temperatures of 840–970°C and pressures of 8.6–10.7 kbar. During the terminal stages of metamorphism, the Izhek charnockite was one of several large granitoids intruded into the metamorphic rocks in the Izhek zone, in part contemporaneous with motion on the Timpton and Tyrkanda shear zones.

As evident from U/Pb zircon ages, the oldest known rocks in the Aldan terrane occur along the northwestern side of the large central domal structure (Figs. 3 and 5a). The upper intercept of discordia defined by a population of clear, zoned zircons from both biotite granite and trondhjemite from along the Aldan River (near the mouth of the Lesser Nimnyr River) yield an age of 3386 ± 100 Ma (Morozova et al., 1989). Downstream along the Aldan River, nebulitic felsic granulites contain two populations of zircons. The most abundant is a group of prismatic, cloudy to nearly opaque zircons that yield an upper discordia SHRIMP intercept age of 3335 ± 3 Ma, whereas a less common group of clear equant grains yields an age of 1929 ± 9 Ma (Nutman et al., 1992a). Some components in this zircon population must have a minimum age of 3500 Ma. TTG from the Iengra Group along the Aldan River have yielded a U/Pb zircon age of 3570 ± 50 Ma (Glebovitsky and Drugova, 1993), which is currently the oldest reported zircon age from the Siberian craton. A leucosome from a migmatite in this area contains zircons with rounded dark cores that yield a SHRIMP age of 3328 Ma, and clear euhedral overgrowths yield a mean age of 1947 ± 5 Ma (Nutman et al., 1992a). These zircon overgrowths appear to have grown during metamorphism associated with major motion on the Aldan–Kilier thrust, and thus may record collision of the Olekma and Aldan terranes. Numerous K/Ar ages from phlogopites in skarns in the Aldan central dome are around 1840 Ma and probably are cooling ages (Shepel, 1980). A discordia upper intercept age of 2037 ± 20 Ma has been obtained from an equant population of transparent zircons from a leucogabbro in the Ungra complex (Fig. 4) in the southwestern part of the Aldan terrane (Bibikova et al., 1984a). This age is a lower limit for the age of the Aldan–Kilier thrust (which cuts the Ungra complex) and an upper age limit for metasediments, which are intruded by the complex.

An upper discordia intercept age of 1916 ± 10 Ma has been obtained from a prismatic transparent zircon population from a charnockite massif in the Izhek zone (Bibikova et al., 1986). Because motion on the Timpton fault and charnockite intrusion are largely syntectonic (Kitsul and Dook, 1985), this should also be the approximate age of collision of the Aldan and Uchur terranes. A Rb/Sr whole rock isochron age with a large uncertainty from the Izhek charnockite of 1770 ± 130 Ma (i = 0.7124 ± 0.0033) (Dook et al.,1989) suggests that the Izhek zone remained relatively hot (> 550°C) for 100–200 Ma.

In summary, results from the Aldan terrane suggest protolith ages for some and probably most rocks of about 3350 Ma with some components ≥ 3500 Ma. One or more granulite-facies metamorphic events coincident with collisions of the Olekma, Aldan, and Uchur terranes at 1950–1920 Ma, was followed by slow uplift and cooling of the Aldan terrane for another 100–200 Ma.

Uchur terrane

The western border of the Uchur terrane is the Tyrkanda fault, along which this terrane has been thrust westward over the Aldan terrane (Figs. 3 and 5a). Although not as well known, the eastern boundary is a west- dipping fault zone, the Ulkan fault, along which the transport direction appears to be to the east (Fig. 5a). The Uchur terrane is composed chiefly of an charnockite–mafic granulite association, which likely represents a TTG–greenstone association metamorphosed to the granulite facies. Granulite-grade metasediments (metapelites, marbles, quartzites) occur sporadically around the periphery of the province where they compose up to 30% of the supracrustal successions (Frumkin, 1987). Geochemical studies of metavolcanics from the Uchur terrane reveal a complete suite from mafic to felsic compositions, with mafic components dominating in most successions (Popov et al., 1989). Plutonic components in the central domal structure of the terrane, however, are bimodal in composition, with the felsic end member (charnockite) dominating.

Thermobarometric studies of high-grade mineral assemblages in the Kholbolokh and Kyurikan suites in the Uchur terrane record granulite-facies temperatures of 820–830°C and pressures of 8.3–8.7 kbar (Kitsul, 1986). In addition, retrograde metamorphism at amphibolite-facies conditions (T = 650°C, P = 5.7–7.0 kbar) is widespread in the terrane.

Although there are no reliable isotopic ages from the Uchur terrane, similarity of rock associations and structure to other high-grade Archean terranes in Siberia suggest an Archean age for most or all of this terrane.

Batomga terrane

Very little is known about the Batomga terrane in the easternmost part of the Aldan province. Exposures in the eastern Aldan shield indicate that it is a TTG-greenstone terrane of probable Archean age. Along its western margin, the Batomga terrane is thrust beneath the Uchur terrane along the Ulkan fault (Fig. 5a). Magnetic anomalies are interpreted to reflect greenstone belts in the covered basement, suggesting a northward extension of this terrane to the fault that offsets the Verkhoyansk orogenic belt (Figs. 1 and 3). The Batomga terrane comprises chiefly a TTG association with supracrustal enclaves composed of mafic igneous rocks, pelites, quartzites, and carbonates. Locally, all rocks are metamorphosed to granulite grade. Some low grade supracrustal successions include komatiite, basalt, and intermediate volcanics with smaller amounts of pelite, conglomerate, quartzite, and carbonate (Glebovitsky, 1987). These supracrustals are typically intruded by mafic to ultramafic bodies and by late tectonic granites and granodiorites.

The similarity of both TTG and supracrustal rocks to those found in other Archean TTG–greenstone associations suggests an Archean age for the much of the basement rock in the Batomga terrane. The terrane must be older than the unconformably overlying Early Proterozoic Ulkan supracrustal rocks.

Udokan orogenic belt

Several remnants of deformed and metamorphosed Early Proterozoic sedimentary successions unconformably overlie Archean basement in the Olekma terrane, the largest and best preserved of which is the Udokan orogenic belt (Fig. 4). The Udokan Group contains quartz arenites (in part copper-bearing), mudstones, and rarely marine carbonates, and up to 10 km of section is preserved (Sochava, 1986). Limited sedimentological studies of these rocks, which commonly contain cross-bedding and ripple marks, show they are derived from the Archean basement (Feoktistov, 1986). Quartz arenites are especially common in the lower part of the Udokan Group and associated conglomerates are composed almost entirely of quartz and quartzite clasts (Fedorovsky, 1985). In the upper part of the succession, detrital quartz decreases chiefly at the expense of feldspars, and the sandstones become more arkosic. The upper conglomerates also contain a larger proportion of granite and rhyolite clasts. Mudstones in the lower part of the succession are commonly associated with cherty dolomites, have high sodium contents, and contain widespread mudcracks, features that suggest a partially closed marine basin (Fedorovsky, 1985). It would appear that the Udokan succession reflects a cratonic basin that evolved with time into a rift basin.

A conventional multigrain zircon suite from a tuffaceous sandstone in the Udokan Group yields an upper concordia U/Pb intercept age of 2180 ± 50 Ma, which is interpreted to reflect the depositional age (Berezhnaya et al., 1988). A Rb/Sr whole rock isochron age of 1950 ± 110 Ma (i = 0.7075 ± 0.002) from an Udokan mudstone may represent the metamorphic age of this succession (Gorokhov et al., 1989). A post-tectonic granite intruding the Udokan Group yields a whole-rock Rb/Sr isochron age of 1800 Ma (i = 0.7065) (Tauson et al., 1983), and a similar granite from the Kodar massif has a Pb/Pb zircon age of 1780 ± 50 Ma (Rublev et al., 1981). The existing data base suggests that the Udokan Group was deposited at about 2.2 Ga, deformed and metamorphosed at about 1.95 Ga during collision of the Olekma and Aldan terranes, and intruded with post-tectonic granites at 1.8 Ga.

Ulkan orogenic belt

The Ulkan orogenic belt of Early Proterozoic age occurs along the southern edge of the Batomga terrane (Fig. 3). The best exposures are in the Uchur and Maya River basins. In the southern part of the orogenic belt, deformation is characterized by simple east-striking folds, whereas in the west one-sided grabens are common. Deformation is complex in the eastern part of the belt, which is bounded by a westward-dipping thrust zone and characterized by north-striking isoclinal folds (Drugova et al., 1993). Supracrustal rocks in the orogenic belt, which rest unconformably on older Archean basement, include in ascending stratigraphic order, red quartz sandstones (200 m thick), subaerial trachybasalt, trachyte and alkali rhyolite accompanied by volcaniclastic sediments (5 km thick) (Petrov et al., 1985). These rocks in the western part of the belt are intruded with rapakivi granites and quartz syenites (Gamaleya, 1968).

Upper intercept zircon ages from an intrusive granite (1703 ± 18 Ma) and hornblende syenite (1727 ± 6 Ma) provide a minimum age for the supracrustal succession (Neymark et al., 1992b). A zircon from a rhyolite in the Ulkan succession yields an age of 1727 ± 18 Ma. These results suggest that deposition, volcanism, and associated plutonism were approximately coeval at 1730–1700 Ma in the Ulkan orogenic belt. Although the timing of deformation is unknown, later thermal overprinting of the Ulkan granites is recorded by K/Ar ages of 755 ± 31 Ma and 453 ± 16 Ma (Gamaleya, 1968).

STANOVOY PROVINCE

Mogocha terrane

The Mogocha terrane comprises chiefly an Archean TTG–greenstone complex, in which major greenstone belts occur in the southern part (Figs. 3 and 6). Remnants of mafic and ultramafic intrusions are also common in the south-central part of the terrane. Rocks are metamorphosed to amphibolite grade, reaching granulite grade in the central part of the terrane (Karsakov, 1983; Velikoslavinsky et al., 1993). The Mogocha terrane is bounded on the west and south by Phanerozoic sutures, on the east by the Dzheltulak fault, and on the north by the Stanovoy fault, where it is thrust over the Sutam terrane (Fig. 6). Both the Mogocha and Tynda terranes are intruded by large Phanerozoic batholiths that comprise about 30% of the exposed rocks.

The proportion of greenstones relative to TTG is greater in the Mogocha terrane than in most Archean granite–greenstone provinces. Geochemical characteristics of both the TTG and greenstone components are similar to those of other Archean TTG–greenstone associations (Sedova and Glebovitsky, 1985; Velikoslavinsky et al. 1993). Several greenstone belts, usually associated with shear zones, have been mapped along the southern margin of the terrane (Rasskozov and Yalynichev, 1972). These are composed of basalts, basaltic komatiite, basaltic andesite, graywacke, quartzite, and BIF (in decreasing order of abundance), and they are intruded by basaltic andesite dikes and layered dunite–troctolite–gabbro intrusions (Kastrykina and Elyanov, 1985; Moskovchenko et al., 1987).

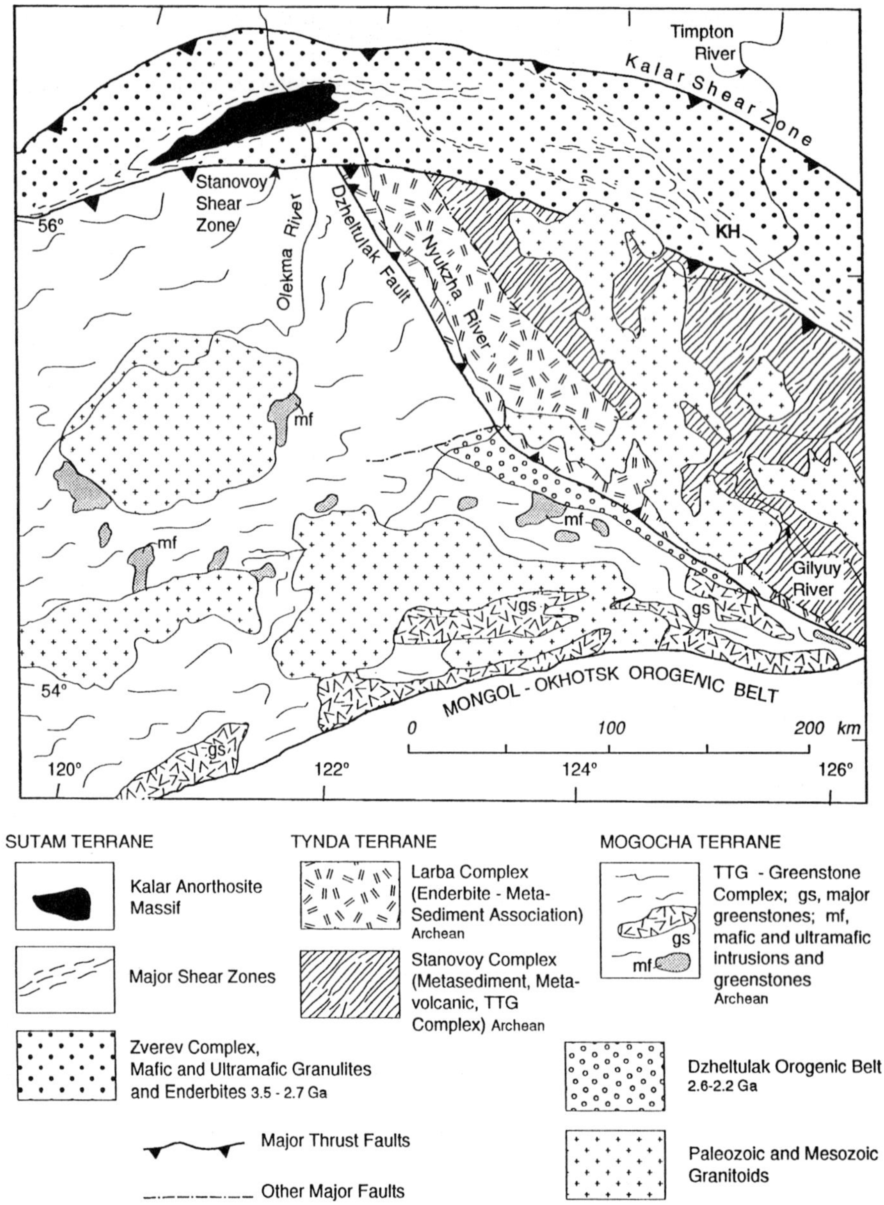

Fig. 6. Generalized geologic map of Precambrian rocks in the central part of the Stanovoy province (after Panchenko, 1985; Kastrykina and Elyanov, 1985; Moskovchenko et al., 1987). KH, Kholodnikan volcanic belt.

Although Archean zircon ages have not yet been reported from the Mogocha terrane, K/Ar and Rb/Sr isotopic ages indicate this terrane is also Archean in age (Manuilova, 1968; Iskanderova et al., 1980). Zircons from amphibolites from a sequence of amphibolites and cordierite–sillimanite paragneisses in the Archean basement from near the town of Mogocha were analyzed with SHRIMP and yield an age of 1873 ± 6 Ma (Nutman et al., 1992b). Based on the euhedral shapes and uniformity of these zircons, Nutman et al. (1992b) interpret this age as a metamorphic age. Multi-domained zircons from paragneisses in the same sequence are discordant with Pb/Pb ages of about 1950 Ma (Bibikova et al., 1989a). According to Nutman et al. (1992b), this age may result from analyzing a zircon population with older cores and younger metamorphic overgrowths. Migmatitic granite gneisses from the southeastern part of the Mogocha terrane (in the Mogocha and Amazar River basins) have yielded a discordia upper intercept zircon age of 1930 Ma (Iskanderova et al., 1980), which is probably the age of metamorphism/anatexis. Thus, it appears that one or two high grade metamorphic events accompanied by migmatite formation occurred in the Mogocha terrane at about 1900 Ma.

Dzheltulak orogenic belt

A small remnant of deformed Early Proterozoic sediments, the Dzheltulak orogenic belt, occurs adjacent to the Dzheltulak fault along the eastern boundary of the Mogocha terrane (Fig. 6). These sediments (the Dzheltulak and Odolo Groups), which are similar to those in the Udokan orogenic belt, have well preserved primary textures and rest unconformably on Archean crust. They have been metamorphosed to low or middle grades, and show zonal arrangement of metamorphic facies (Kastrykina, 1983; Panchenko, 1985). Thermobarometric studies show maximum metamorphic temperatures around 600–650°C and pressures chiefly between 5 and 7 kbar. These rocks appear to have been metamorphosed when the Tynda terrane was thrust over the Mogocha terrane from the northeast.

Pelites, quartz arenites, and dolomites (often enriched in carbonaceous matter) predominate in the lower part of the Dzheltulak succession, whereas arkosic sandstones, conglomerates, and sedimentary breccias are typical of the upper part (Kastrykina, 1983). Basalts also occur interlayered with the upper sediments and these successions are cut by diabase dikes. Clasts in the conglomerates appear to come from the adjacent Archean TTG–greenstone complex, as well as from the reworking of older parts of the sedimentary succession (Kastrykina, 1983; Panchenko, 1985). The Dzheltulak sediments are deformed and intruded with syntectonic aegerine syenites and alkali granites (Elyanov et al., 1985). In terms of tectonic setting, the Dzheltulak succession appears to represent a transition from a craton or passive continental margin into a continental rift, in much the same way as does the Udokan Group.

An approximate upper limit on the age of the Dzheltulak Group is provided by a K/Ar date of 2208 ± 44 Ma from intrusive pegmatites (Kastrykina, 1983). Thus it would appear that the Dzheltulak sediments were deposited sometime between 2.6 Ga (the youngest age from the Archean basement) and 2.2 Ga.

Tynda terrane

The Tynda terrane is bounded on the north by the Stanovoy fault system and appears to have been thrust northward over the Sutam terrane (Fig. 6). The southern and eastern boundaries are Phanerozoic sutures and the western boundary is the Dzheltulak thrust fault along which the Tynda terrane was thrust westward over the Mogocha terrane, after deposition of the Dzheltulak Group. From west to east, the Tynda terrane comprises three granulite-grade domains (known as the Larba Complex) alternating with two amphibolite-grade domains (known as the Stanovoy Complex) (Godsevich, 1986). Boundaries between the Larba and Stanovoy Complexes are always faults or shear zones (Fig. 6). Also exposed in the Tynda terrane are numerous large anorthosite bodies (Fig. 3). Thermobarometric results suggest metamorphic temperatures of 650–700°C and pressures of 5–7 kbar in the eastern parts of the Tynda terrane, and temperatures exceeding 900°C at pressures in excess of 9 kbar adjacent to the Dzheltulak fault. Relative proportions of rock types in the Larba Complex (estimated from the mapping of Moskovchenko and Kastrykina, 1989) are as follows: enderbite and hornblende–biotite gneiss, 37%; ultramafic schists with interlayered felsic pyroxene gneisses and kinzigite, 36%; calc-silicates, marbles, quartzite (metachert), and BIF, 20%; anorthosite, 4%; and gabbro, 3%. In contrast, the Stanovoy Complex is composed chiefly of diorite–tonalite–trondhjemite (55%), basalt and gabbro (15–20%), and paragneisses (metagraywacke, metapelite) (25%), as estimated from the mapping of Godsevich (1986). In addition, harzburgite and basalt dikes intrude this complex. Mafic to felsic components of the Stanovoy Complex show continuous calc-alkaline or tholeiitic trends (Sedova and Glebovitsky, 1985). The striking difference of these associations suggests different tectonic settings: perhaps a TTG–greenstone setting for the Stanovoy Complex and a cratonic setting for the Larba Complex. If so, these two very different tectonic groups are now tectonically interleaved in the Tynda terrane.

Composite zircons from tonalitic gneisses with relict granulite-facies mineral assemblages from along the Gilyuy River in the southwestern part of the Tynda terrane (Fig. 6) have been dated by the U/Pb SHRIMP method. Zircon cores, which are thought to record the time of tonalite intrusion, yield an age of 2785 ± 9 Ma, while the zircon overgrowths give an age of 1960 ± 25 Ma, probably the age of granulite-facies metamorphism (Nutman et al., 1992b). Zircons separated from garnets in granulites from the Larba Complex give a nearly concordant U/Pb age of 2585 ± 20 Ma, which is interpreted as the age of granulite grade metamorphism (Bibikova et al., 1984; Shuldiner and Panchenko, 1985), which may be the age of some component in the source of these metasediments. Zircons, which grew in ultramafic and mafic dikes during retrogressive amphibolite facies metamorphism yield SHRIMP ages of 1929 ± 13 Ma and 1924 ± 24 Ma, respectively (Nutman et al., 1992b). Zircons from an anorthosite in the Gheran intrusion of the Zhugzhur massif yield an upper intercept age of 1734 ± 12 Ma and probably reflect the age of emplacement of this body (Neymark et al., 1992a).

Sutam terrane

The Sutam terrane is tectonically sandwiched between the Aldan province on the north and the Stanovoy province on the south (Figs. 3, 5b and 6). Two major thrust faults, the Kalar and Stanovoy faults, border the Sutam terrane on the north and south, respectively. These faults, which are manifest by gravity and magnetic anomalies, are major crustal structures perhaps penetrating the entire crust and extending into the mantle lithosphere (Fig. 5b) (Alakshin and Karsakov, 1985). The Kalar fault truncates structures within the Aldan province as well as boundary faults between terranes of the Aldan province, and thus its age must be <1.92 Ga, the age of the Timpton fault between the Uchur and Aldan terranes. If the Sutam terrane is an upthrusted piece of lower crust as suggested in the cross section (Fig. 5b), the Stanovoy and Kalar faults are probably of similar age. The Sutam terrane, which is composed chiefly of an enderbite–mafic granulite association known as the Zverev Group, includes large intrusions of anorthosite such as the Kalar massif (Fig. 6), as well as major shear zones with extensive mylonites. High-grade supracrustal rocks of the Zverev Group have been correlated over large parts of the Sutam terrane, based on similar stratigraphic sequences. Rock proportions in the Zverev Group estimated from the mapping and cross sections of Godsevich (1986) are as follows: mafic granulites 37%, enderbites 25%, charnockites 15%, metapelites 10%, quartzites and marbles 10%, and anorthosites 3%. Anorthosite bodies are composed of two-pyroxene gabbroic anorthosites and anorthosites with well preserved cumulus textures, and they are commonly associated with monzodiorite and pyroxene syenite (Sukhanov and Panskikh, 1981). Xenoliths of surrounding rock types in the anorthosites, as well as contact metamorphic zones, clearly indicate an intrusive origin. Also widespread in the Sutam terrane are high grade mafic to ultramafic massifs up to 50 km in size and generally conformable with the structural fabric (Bereskin, 1986). These massifs are composed of gabbro, diorite, lherzolite and pyroxenite, and these rocks generally define a single tholeiitic geochemical trend within a given massif.

As indicated by thermobarometric studies (Kitsul, 1986), the Sutam terrane appears to represent a slice of the middle to lower crust. Temperatures recorded by metamorphic mineral assemblages are typically in the range of 820–920°C at pressures of 8–11 kbar, and textural studies suggest at least two periods of granulite metamorphism. Retrograde metamorphic minerals are common in these high-pressure granulites. Also supporting a deep crustal origin for rocks of the

Sutam terrane are the high densities (2.95–3.5 gm/cm^3) and P-wave velocities (6.36–7.2 km/sec) measured in rock samples from the terrane (Glukhovsky et al., 1977). In many respects, the Sutam terrane is similar to the Kapuskasing uplift in eastern Canada, where a slice of lower Archean crust has been brought to the surface between two thrust faults during an Early Proterozoic collision (Percival and Card, 1983).

Some low-grade metavolcanic rocks are preserved in shear zones in the central Sutam terrane, as for instance the Kholodnikan belt (KH in Fig. 6). These volcanics contain basalt, andesite, komatiite, quartzite, and carbonate and are cut by synvolcanic diabase dikes with well preserved diabasic textures (Bereskin, 1986). Deformational fabrics in the Kholodnikan belt are similar to those in the Aldan terrane to the north, and this belt may represent a fragment of the Aldan terrane tectonically isolated in the Sutam terrane when the Sutam terrane collided with the Aldan terrane.

Isotopic ages from the Sutam terrane are few in number. Zircon xenocrysts from a metagabbro in the western part of the terrane (Kurulta Group) yield an upper intercept U/Pb discordia age of 3460 ± 16 Ma and a lower intercept age of 2200 ± 20 Ma (Bibikova et al., 1989b). The 2200 Ma date is generally considered the age of granulite-facies metamorphism. Zircons from enderbite/charnockite surrounding the Kalar anorthosite yield an upper intercept discordia age of 2660 ± 60 Ma with a lower intercept age of 1800 Ma (Levchenkov et al., 1987). These are thought to represent the ages of the protolith and granulite-facies metamorphism, respectively. A Sm/Nd mineral isochron age from the Zverev Group of 1950 Ma, and another from a low grade mafic volcanic rock of 2150 Ma (Dook, 1989b) are probably metamorphic ages. Results indicate that both Early and Late Archean rocks occur in the Sutam terrane and that several periods of high grade metamorphism are recorded in the terrane.

OLENEK PROVINCE

Birekte terrane

The Birekte terrane in the northeastern part of the Siberian craton is defined entirely from geophysical data, as there are no surface exposures (Figs. 1 and 7). Up to 10 km of Riphean and Phanerozoic sediments cover the basement rocks of this terrane. The western terrane boundary is equated with a major linear magnetic anomaly (Fig. 2) that may represent a fault zone. The northeastern boundary is not well constrained, but from magnetic and gravity data it also may be a tectonic contact with the Aekit orogenic belt. Three large negative gravity anomalies in the Birekte terrane have been interpreted as major granitic batholiths (Gafarov et al., 1978; Khoreva, 1987). Because the pattern of geophysical anomalies in the Birekte basement is similar to that of Archean granite–greenstone terrains in the

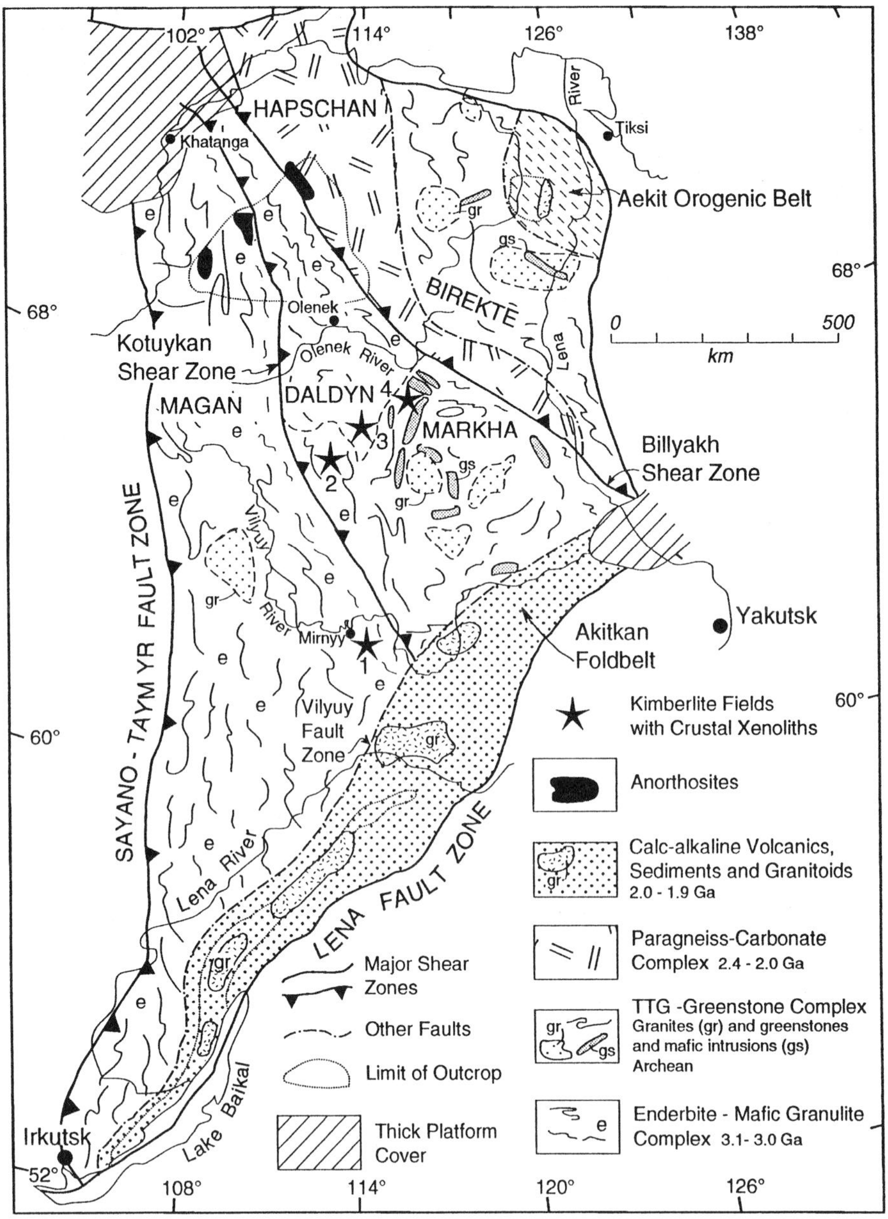

Fig. 7. Precambrian terrane map of the Magan, Anabar and Olenek Provinces. After data in Fedorovsky (1985), Petrov et al. (1985), and Rosen (1989). Features beneath platform cover inferred from geophysical anomalies (Gafarov et al., 1978; Khoreva, 1987; Samkov and Potapyev, 1986). Kimberlite fields: 1, Mirnyy; 2, Alakit; 3, Daldyn; 4, Muna.

Aldan province where basement is exposed, the Birekte terrane may be a similar rock association. Supporting this interpretation are common clasts of tonalite and amphibolite in overlying Late Proterozoic sediments that appear to have been derived from Birekte basement sources (Mokshantsev, 1979; Petrov et al., 1985). As with the covered basement in the Aldan province, the structural trends and lithologic units in the Birekte terrane (Fig. 7) are defined from geophysical data employing both magnetic and gravity data.

Although an Archean age is likely for the Birekte terrane, isotopic ages are not available from basement rocks to confirm or deny this inference.

Aekit orogenic belt

The Aekit orogenic belt, which is exposed in the Olenek uplift southwest of Tiksi (Fig. 7), is bounded on the east by the Mesozoic Verkhoyansk orogenic belt and on the west by a poorly defined tectonic contact with the Birekte terrane. Where exposed, the Aekit Group is composed chiefly of felsic volcanics (rhyolites and dacites) interbedded with smaller amounts of pelite and sandstone (Mokshantsev, 1979). Primary textures and structures are well preserved in these rocks, which are metamorphosed only to prehnite–pumpellyite or greenschist grade. The sediments are chiefly volcanogenic and commonly contain clastic pyroxenes, amphiboles, garnet, epidote, and volcanic rock fragments. The Aekit Group is deformed into linear isoclinal folds and intruded with syn- to post-tectonic granitoids and late trachyte and diabase dikes. Geochemical data from these rocks suggest affinities to modern continental-margin arc systems.

Although no recent isotopic ages are available from the Aekit orogenic belt, older K/Ar dates from metamorphic micas are about 1980 Ma (Krylov et al., 1963), and similar dates from granitoid micas lie in the range of 2080 to 1850 Ma (Mokshantsev, 1979). These results suggest extensive metamorphism and plutonism in the Aekit orogenic belt about 2000 Ga. The Aekit Group is probably Early Proterozoic in age by analogy with other dated successions in the Siberian craton.

Hapschan terrane

The Hapschan terrane is exposed in the eastern Anabar shield where it is thrust SW over the adjacent Daldyn terrane along the Billyakh shear zone (Fig. 8). In the exposed area, rocks occur in tight isoclinal folds with northwestern strikes and they are commonly overturned to the southwest, perhaps reflecting SW-thrusting of the Hapschan terrane (Luts and Oskman, 1990). A similar structural pattern can be traced into adjacent covered areas with geophysical data (Vishnevsky and Turchenko, 1986).

The terrane is composed of the Hapschan Group, which comprises chiefly garnet paragneisses, calc-silicates and marbles, of which carbonate metasediments appear to have originally composed up to 40% of the succession (Rosen, 1989,

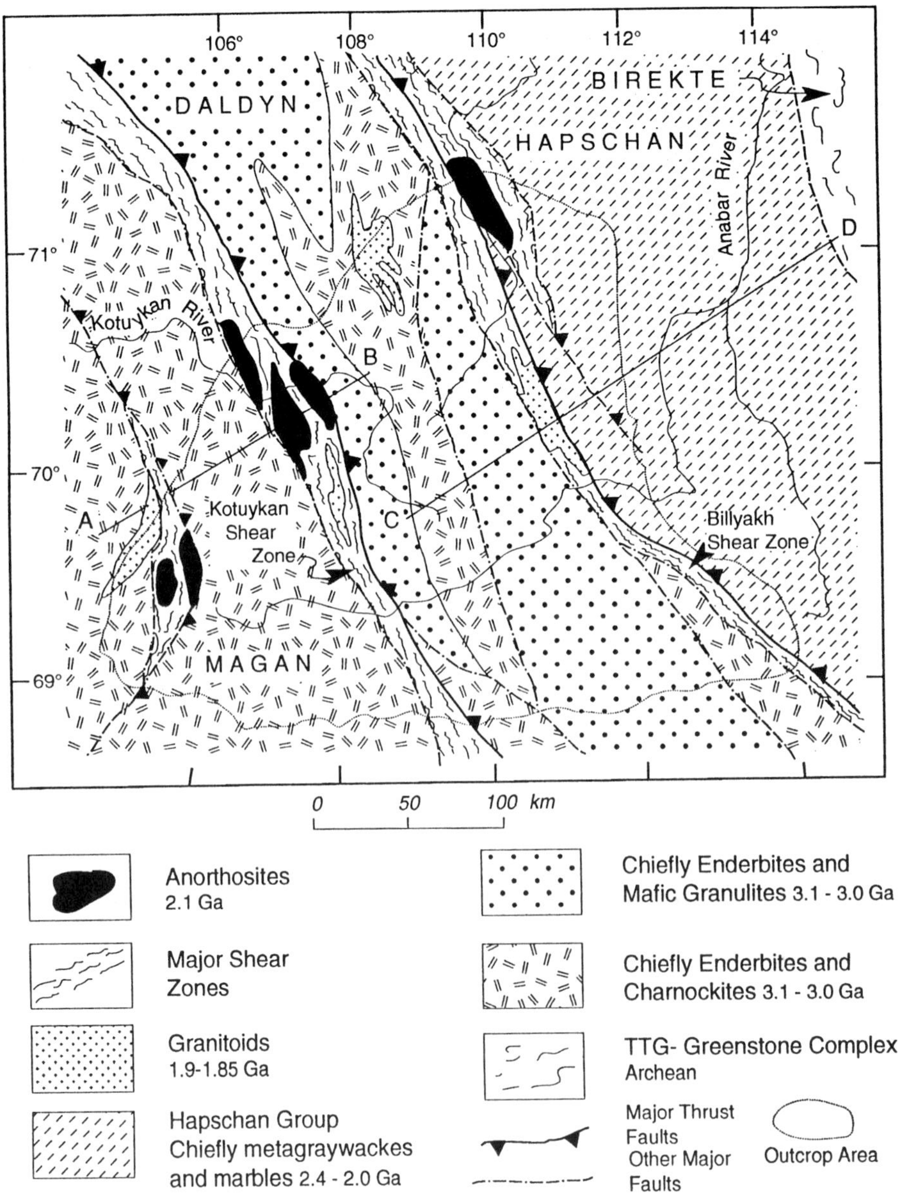

Fig. 8. Generalized geologic map and cross section of Precambrian rocks in the Anabar shield and surrounding areas. After Samkov and Potapyev (1986), Rosen (1989, 1992), and Rosen et al. (1990). LC – lower crust, M – mantle.

1992). Enderbites and mafic granulites are minor components in this group. The Hapschan terrane is metamorphosed from amphibolite to granulite grade, with maximum temperatures of 750–820°C and pressures of 5.5–7 kbar, as deduced

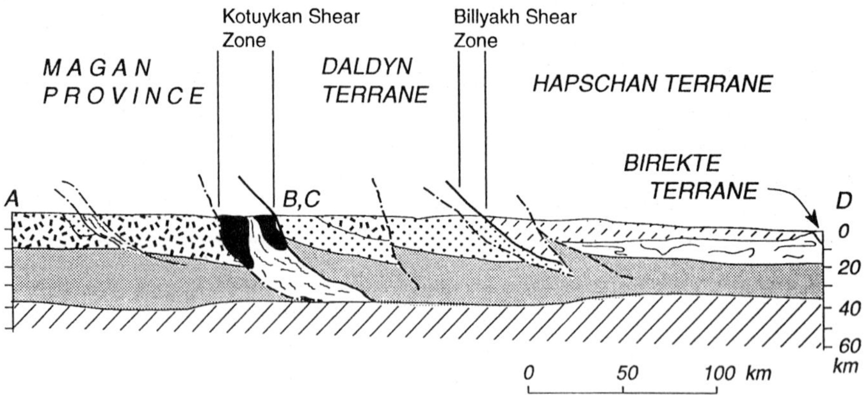

Fig. 8 (continued). Caption on previous page.

from thermobarometric mineral studies (Vishnevsky and Turchenko, 1986). Geochemical studies of the paragneisses suggest graywacke protoliths with felsic sources (Condie et al., 1991). The Hapschan Group appears to have been deposited in a cratonic basin or passive continental margin. The nature and age of the basement on which the group was deposited is unknown, although it could be the Birekte terrane as suggested in the cross section in Fig. 8.

Few isotopic ages are available from the Hapschan terrane. Paragneisses and calcsilicates with a tight grouping of both Nd isotopic ratio and Sm/Nd ratio yield a T_{DM} model age of about 2400 Ma (Rosen et al., 1991). Although the interpretation of this age is ambiguous, it is likely that the depositional age of the metasediments is not greater than 2400 Ma, and hence, they are not Archean. Transparent equant zircons from a garnet paragneiss yield a conventional multigrain U/Pb, nearly concordant, age of 1970 ± 20 Ma (Rosen et al., 1991). This has been interpreted by Rosen et al. as the age of granulite facies metamorphism and widespread deformation that may be associated with movement along the Billyakh shear zone. Thus, the depositional age of the Hapschan Group appears to lie between 2.4 and 2.0 Ga.

ANABAR PROVINCE

Daldyn terrane

The Daldyn terrane is a slice of high grade rocks thrust over the Magan province on the west along the Kotuykan shear zone (Figs. 7 and 8). Although now covered with Phanerozoic cratonic sediments, this terrane appears to be transected by the Taymyr orogenic belt on the north. Where exposed in the Anabar shield, the terrane is composed of enderbite–mafic granulite and enderbite–charnockite complexes. Rocks in these complexes are folded into dominantly linear isoclinal folds with northeastern strikes and general dips to the northeast (Rosen, 1989). In outcrop, fold closures are common on scales of hundreds of meters. With magnetic and gravity anomalies, these rocks can be traced into the basement both north and south of the Anabar shield (Fig. 2). Thermobarometric studies of Daldyn metamorphic rocks indicate granulite temperatures in the range of 820–950°C and very high pressures of 8.5–11 kbar (Luts, 1974).

Supracrustal rocks (including dikes), referred to as the Daldyn Group, are dominantly two-pyroxene mafic to felsic granulites with minor amounts of metasediment. These are intruded by widespread enderbites and subordinate charnockites, partially melted at granulite grade (Rosen, 1992). Geochemical studies indicate most of the supracrustals had igneous protoliths with rocks of andesitic to tonalitic composition dominating (46%) followed by felsic (23%) and mafic compositions (21%) (Rosen, 1989, 1992). Marbles, calc-silicates, quartzites, and BIF also are recognized in small amounts in the Daldyn Group. The igneous components fall into three distinct geochemical series, low-alkali, calc-alkaline, and high-alkali, similar to those characteristic of arc volcanics. From the K, Th, and U distributions in Daldyn rocks, the weighted heat production of the terrane is about 0.4×10^{-6} W/m^3, the average density 2.83 g/cm^3, and the calculated average P-wave velocity 6.3 km/sec (Rosen, 1992). These parameters are consistent with the Daldyn terrane representing and elevated slice of the mid- to lower continental crust. It is possible, if not probable, that the Daldyn protoliths were components in an Archean TTG–greenstone complex.

Crustal xenoliths of Daldyn basement are available in kimberlite pipes in the southern part of the terrane (Fig. 7). They are chiefly of two types: garnet, two-pyroxene mafic granulite and garnet–scapolite–zoisite–plagioclase granulite, the latter perhaps representing metacarbonate (Gerasimchuk and Serenko, 1988). Thermobarometric studies of these mineral assemblages indicate metamorphic temperatures of 800–950°C and pressures of 8 kbar to perhaps as high as 13 kbar (Luts, 1974; Neymark et al., 1993a). The mafic granulite xenoliths have major and trace element distributions similar to normal MORB, indicating the presence of a LIL-element depleted mantle source for their protoliths (Shatsky et al., 1990).

The Billyakh shear zone is a major terrane boundary in northeastern Siberia. Where exposed in the eastern Anabar shield, it ranges from 10–20 km wide (Fig. 8). With magnetic and gravity anomalies it can be traced for over 1200 km in a northeasterly direction (Fig. 7). Within the shear zone, the Daldyn Group is fragmented into slices and lenses bounded by mylonite zones, and in places the mylonites are tectonically mixed with fragments of the adjacent Hapschan terrane. Migmatization is widespread in the shear zone, and locally anorthosites and granites occur as large tectonic boudins. Mylonites generally dip to the NE and most motion indicators show upthrusting to the southwest.

Prismatic, pink zircons from Daldyn enderbites in the Anabar shield yield nearly concordant U/Pb SHRIMP ages of 3164 ± 32 Ma, and the same suite of zircons yields a multigrain conventional upper-intercept discordia age of 3000 ± 20 Ma (Rosen et al., 1991). A Sm/Nd whole rock isochron age of 3063 ± 80 Ma (ε_{Nd} = 3.1 ± 1.8) is also reported from mafic granulites from the same region (Spiridonov et al., 1991). These results suggest a protolith age for igneous rocks of about 3100 Ma. A suite of transparent, equant zircons from one Daldyn enderbite is interpreted as metamorphic in origin, and yields an upper-intercept discordia U/Pb age of 2760 ± 10 Ma, probably the age of major deformation and granulite-facies metamorphism (Rosen et al., 1991). In addition, a Sm/Nd mineral isochron age of 1884 ± 5 Ma (ε_{Nd} = –6.3) has been obtained from a mafic granulite xenolith from the Udachnaya kimberlite pipe (Daldyn kimberlite field, Fig. 7), and probably represents the age of the last metamorphic event in the southern part of the Daldyn terrane (Neymark et al., 1993a).

In summary, isotopic ages from the Daldyn terrane suggest a crustal formation age of about 3.1 Ga followed by major deformation and granulite-grade metamorphic events at 2.76 and 1.9 Ga, the latter of which may be a cooling age associated with the collision of the Daldyn terrane with the Magan province along the Kotuykan shear zone.

Markha terrane

Little is known of the Markha terrane south of the Daldyn terrane (Fig. 7). This terrane is not exposed at the surface and is defined entirely from geophysical and crustal xenolith data. The eastern and western borders are the Billyakh and Kotuykan shear zones, respectively, as reflected by major magnetic and gravity anomalies contiguous with the exposed portions of these shear zones (Fig. 2). The southern boundary is also a sharp break in geophysical anomalies that may be a fault zone. The northwestern boundary of the terrane is poorly defined and is tentatively placed where magnetic and gravity anomalies change trend from dominantly NW in the Daldyn terrane, to dominantly NE (Fig. 2). Wide, rather equant negative gravity anomalies in the Markha terrane have been interpreted as large granitoid complexes, and smaller sublinear positive anomalies, either as greenstone belts or large mafic intrusions (Gafarov et al., 1978).

Common crustal xenoliths reported from both the Muna and Alakit kimberlite fields (Fig. 7) are two-pyroxene–garnet–hornblende mafic granulites and high-MgO, two-pyroxene–plagioclase–olivine xenoliths, the latter perhaps of komatiitic parentage (Gerasimchuk and Serenko, 1988). In addition, the Muna kimberlites contain plagioclase–garnet–orthopyroxene–quartz granulite xenoliths of intermediate composition and the Alakit kimberlites contain high-alumina, garnet–sillimanite felsic to intermediate granulite xenoliths of probable sedimentary origin. These xenolith assemblages suggest that basement protoliths in this region are dominantly mafic (and perhaps komatiitic) to intermediate volcanics and intrusives with some metasediments (graywackes?).

A garnet amphibolite xenolith from the Novinka kimberlite pipe in the Muna field has a Sm/Nd whole rock isochron age of 1756 ± 6 Ma ($\varepsilon_{Nd} = -0.2$), a possible age for the granulite metamorphism, and a T_{DM} model age 2920 Ma (Neymark et al., 1993a). Other garnet gneiss and amphibolite xenoliths from the Muna field have Sm/Nd model ages ranging from 3.3 to 3.1 Ga, which are minimum ages of their protoliths. Although zircon ages are not yet available from the Markha terrane, the Sm/Nd ages are similar to those from the Daldyn terrane, and it is possible that the Markha and Daldyn terranes are really part to the same terrane.

MAGAN PROVINCE

The Magan province, exposed in the western part of the Anabar shield, appears to extend to Lake Baikal on the south (Fig. 7). The eastern boundary is the Kotuykan shear zone, and the western boundary, defined entirely from geophysical data, is interpreted to be tectonic and will be referred to as the Sayan–Taymyr fault zone. Magnetic and gravity anomalies along and east of this inferred fault are typically north-trending, whereas west of the fault they are more irregular and often trend northwesterly (Fig. 2). The rather abrupt change in orientation of these anomalies is equated with the fault trace. Gravity models suggest the Sayan–Taymyr fault dips to the east, and if so the Magan province appears to be thrust over the adjacent Tungus province. On the north, the Magan province plunges beneath thick Phanerozoic cover where it is truncated by the Taymyr orogenic belt, and on the south it appears to be truncated by the Vilyuy fault zone as defined by a change in trend of geophysical anomalies from northerly to northeasterly in the Akitkan orogenic belt.

Where exposed in the Anabar shield (Fig. 8), the Magan province includes the Upper Anabar Group, which is a composed of highly deformed enderbites and charnockites, with variable amounts of mafic to intermediate granulites of probable supracrustal origin, and rare metasediments (Rosen, 1992). Crustal xenoliths from the Mirnyy kimberlite field (Fig. 7) are chiefly two-pyroxene mafic to intermediate granulites. Isoclinal folds and associated faults in the Upper Anabar Group trend chiefly to the north or northwest. As with the Daldyn Group, geochemical studies of Upper Anabar supracrustals indicate three geochemical series (low-alkali, calc-alkaline, high-alkali) in igneous protoliths, similar to those characteristic of arc systems (Rosen, 1992). Calculated average heat production of the Upper Anabar Group (0.56×10^{-6} W/m^3) is somewhat higher and the average rock density (2.76 g/cm^3) somewhat lower than corresponding values of the Daldyn Group. Such values suggest that at least the northern part of the Magan province represents a shallower crustal level than the adjacent Daldyn terrane.

The Kotuykan shear zone along the eastern border of the Magan province is one of the largest and best known province boundaries in the Siberian craton. It ranges from 10–30 km in width and can be traced with magnetic and gravity anomalies for at least 1300 km in a northwesterly direction (Fig. 7). Although most strain indicators within the shear zone are consistent with thrusting to the southwest, rotated feldspars locally indicate left lateral motion. Within the shear zone, rocks are typically retrograded from granulite to amphibolite grade and large tectonic blocks of such diverse rock types as granite, anorthosite and jotunite are dispersed among an extensive array of mylonites (Rosen et al., 1990). In addition, syntectonic migmatites are widespread in the shear zone. Anorthosite massifs up to 60 km in size within the shear zone are elongated parallel to the tectonic fabric. In addition to anorthosite, these massifs contain pyroxenites and gabbros and are often associated with monzodiorites (Sukhanov and Rachkov, 1988). Although the borders of the anorthosite massifs are strongly deformed, relict primary layering and cumulus textures are locally preserved in their interiors.

A clinopyroxene amphibolite xenolith from the Mir kimberlite pipe in the Mirnyy field yields a Sm/Nd T_{DM} model age of about 3000 Ma (Neymark et al., 1993a), which provides a minimum age for at least some of the basement rocks in this region. In the Kotuykan shear zone, zircons yield an upper intercept age of about 2 Ga for one of the anorthosite bodies, and an age of 2.7 Ga for associated monzodiorites (Sukhanov and Rachkov, 1988). The discrepancy in these ages may be due to a xenocrystic origin for the monzodiorite zircons, which may have been entrained in the magma as it intruded Archean granulites. A Sm/Nd mineral isochron from the same monzodiorite yields an age of 2180 ± 20 Ma ($\varepsilon_{Nd} = -6.1$) (Sukhanov et al., 1990) in agreement with the zircon age, and is interpreted as the age of crystallization of the anorthosite–monzodiorite complex. Monazites from a syntectonic migmatite in the Kotuykan shear zone yield an upper intercept U/Pb discordia age of 1925 ± 100 Ma and genetically related uraninites have reported Pb/Pb ages of 2000–1900 Ma (Stepanov, 1974). The youngest rocks in the Kotuykan shear zone are biotite granites that appear to have been emplaced during late stages (or renewed stages) of shearing. These have U/Pb zircon upper intercept discordia ages of 1870–1840 Ma (Stepanov, 1974).

Available isotopic ages suggest the Magan province has a crustal formation age similar to the Daldyn terrane of 3.1–3.0 Ga, with one or more periods of anorthosite intrusion and granulite-grade metamorphism at 2000–1900 Ma. The Daldyn terrane may have collided with the Magan province along the Kotuykan shear zone at about 1.93 Ga.

TUNGUS PROVINCE

The Tungus province, which is over 2000 km long and 200–800 km wide, is exposed only in the Sharyzhalgay uplift west of Irkutsk (Figs. 9 and 10). It is covered by great thicknesses of Riphean to Mesozoic sediments and in the north by the Siberian traps. In contrast to the Magan province, it is characterized by a rather irregular distribution of magnetic and gravity anomalies. The western boundary with the Angara orogenic belt is placed where the generally NW-trending

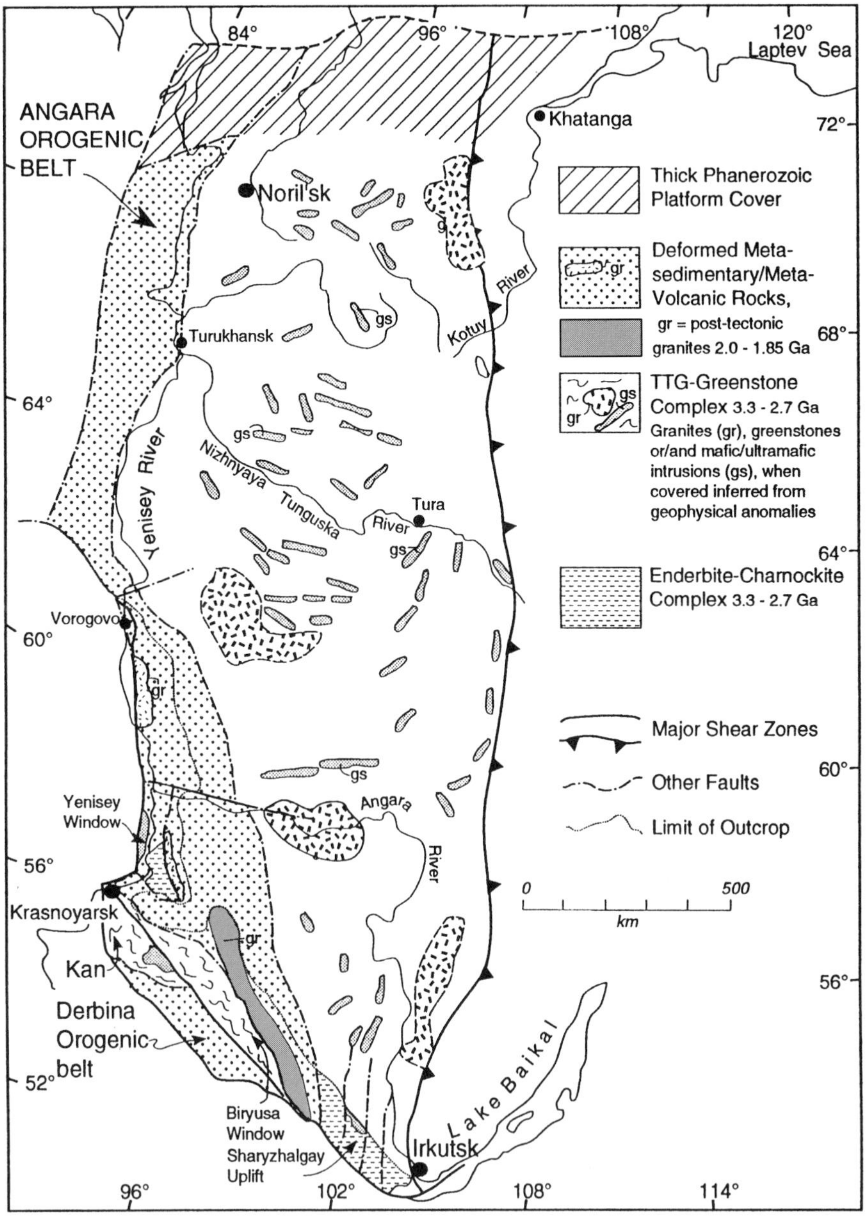

Fig. 9. Precambrian terrane map of the Western Siberian craton. After Postelnikov and Museibov (1992), Rundkvist and Sokolov (1993), and Sezko (1988). Features beneath platform cover inferred from geophysical anomalies (Gafarov et al., 1978; Khain, 1985; Khoreva, 1987).

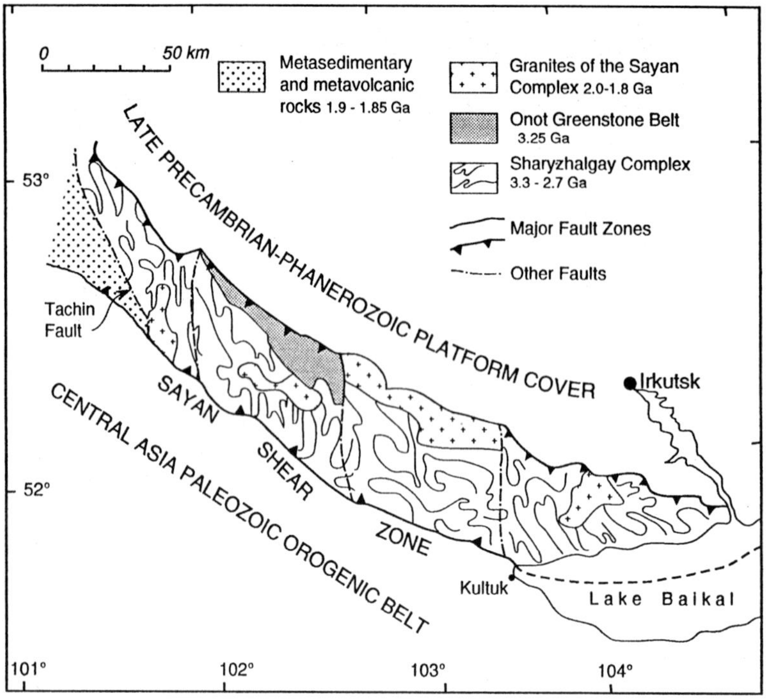

Fig. 10. Generalized geologic map of Precambrian rocks in the Sharyzhalgay uplift west of Irkutsk, Siberia (after Melnikov, 1991).

positive magnetic anomalies of the Tungus province terminate against the northerly-trending anomalies of the Angara belt (Fig. 2). Overall, the Tungus province is characterized by positive magnetic anomalies and the Angara by negative anomalies. The Tachin fault separates the two terranes in the western part of the Sharyzhalgay uplift (Fig. 10). The northern boundary of the province, which is probably a suture with the Taymyr orogenic belt, is covered with thick Mesozoic sediments. Along the southern border, the Tungus province is thrust southwards over the Central Asiatic orogenic belt along the Sayan shear zone.

Where exposed in the Sharyzhalgay uplift, the Tungus province comprises Archean high grade metamorphic rocks and locally, a TTG–greenstone association. Several north-trending fault zones cut the exposures (Fig. 10). The terrane is also intruded by large Proterozoic granitoid bodies of the Sayan Complex. Interpretive studies of gravity and magnetic data suggest that the positive anomalies in the unexposed part of the province are underlain by greenstone belts or mafic/ultramafic intrusions (Gafarov et al., 1978). The Onot greenstone belt about 100 km west of Irkutsk includes a variably metamorphosed and deformed supracrustal suite of tholeiitic amphibolite, orthogneisses formed from felsic volcanics, quartzite, BIF and dolomites with magnesite-rich horizons (Rundkvist and Sokolov,

1993). Remnants of Archean greenstones in other places in the uplift contain also conglomerates with clasts of mafic granulites, gneisses, and carbonates. Field studies suggest that TTG locally intrusive into to Onot greenstone belt is remobilized basement rocks.

Most of the Sharyzhalgay uplift is underlain by Archean high grade rocks assigned to the Sharyzhalgay Complex. This complex is composed chiefly of charnockites and biotite granites with associated migmatites. Found within the plutonic bodies are remnants of mafic volcanics, metapelites, quartzites, and carbonates, ranging in metamorphic grade from granulite in the eastern part of the uplift to amphibolite in the west (Petrova and Levitsky, 1990; Melnikov, 1991). Detailed structural and metamorphic studies of exposures along the southwestern shore of Lake Baikal indicate the existence of two major tectonic events (Hopgood and Bowes, 1990). The first, which is thought to be collision-related, is characterized by isoclinal and asymmetrical folds and tectonically emplaced ultramafic bodies, followed by tight to open folds and minor granite intrusion. The second is related to crustal extension and is characterized by folding with syntectonic development of patchy charnockites, probably resulting from the passage of CO_2 through the crust. During the late stages of this event, upright folds and shear zones formed, followed by intrusion with various granitoids and pegmatites.

The oldest isotopic ages from the Tungus province come from tonalitic gneisses that are intrusive into the Onot greenstone belt. These gneisses yield a U/Pb upper-intercept discordia zircon age of 3250 ± 50 Ma (Bibikova et al., 1982). Similar zircon ages from high grade felsic gneisses and charnockites from the Sharyzhalgay Complex range from 2778–2710 Ma and are interpreted as the time of widespread intrusion (Bibikova et al., 1990b; Aftalion et al., 1991). Infolded tectonic slices of kinzigite and high-Al gneiss have what appear to be metamorphic zircon suites yielding upper intercept ages of 2568–2560 Ma, which probably record a major period of deformation and granulite-facies metamorphism (Bibikova et al., 1990a; Aftalion et al., 1991). Three Rb/Sr whole rock isochron ages from high grade granitoids range from 2540–2530 Ma and are probably cooling ages related to this event (Krylov et al., 1980). Zircon ages from Early Proterozoic granites and charnockites are 1965 ± 4 and 1950 ± 5, Ma (Bibikova et al., 1990a; Aftalion et al., 1991), and Sm/Nd T_{DM} (2 Ga) model ages from similar rocks are in the range of 2.4–2.3 Ga (Aftalion et al., 1991). From these results, an age of about 1960 Ma has been assigned to the first Early Proterozoic deformation, metamorphism and plutonism recognized by Hopgood and Bowes (1990). This event may have been caused by collision of the Tungus and Magan provinces (Zonenshain et al., 1989; Aftalion et al., 1991). The youngest ages from the Tungus province are lower-intercept zircon discordia ages of 1873 ± 4 and 1817 ± 30 Ma from Archean granitoids and a concordant U/Pb monazite age of 1862 ± 4 Ma from a garnet gneiss (Aftalion et al., 1991). These ages are thought to date the second period of deformation, metamorphism, and plutonism in the Sharyzhalgay uplift at about 1.87–1.86 Ga.

YENISEY PROVINCE

The Yenisey province is located on the extreme southwestern edge of the Siberian craton south of Krasnoyarsk (Figs. 1 and 11). It is bounded on the northeast by the Sayan shear zone, an Early Proterozoic structure that was reactivated in the Late Proterozoic and Paleozoic (Postelnikov, 1990). The other margins of this province are major thrusts associated with Late Proterozoic and Paleozoic deformation and collision with the Central Asia orogenic belt. The Yenisey province has been divided into two tectonic subdivisions: the Kan terrane on the north comprising Archean rocks, and the Early Proterozoic Derbina orogenic belt on the south (Khiltova, 1993) (Figs. 9 and 11). Geophysical anomalies in the Derbina orogenic belt show a strong easterly trend compared to the NW trend of anomalies in the Kan terrane (Fig. 2). Much of the Archean basement of the Kan terrane is either covered with Devonian volcanics and sediments or intruded with Cambrian to Devonian granitoids (Fig. 11). The boundary of the Kan terrane and the Derbina orogenic belt is a tectonic contact defined by several W- to NW-striking faults.

The Kan terrane is a typical Archean TTG–greenstone terrane of which greenstones comprise about 50% by area. Most greenstone belts include, in the lower parts of stratigraphic successions, dominantly tremolite schists (with relics of olivine, clinopyroxene, and spinel), amphibolites (basalts and diabases), and associated ultramafic/mafic intrusives, with biotite gneisses (meta felsic volcanics), metapelites, quartzites (metachert), and carbonate becoming important at higher stratigraphic levels (Nozhkin and Smaguin, 1988; Tsypukov et al., 1993). Most successions are multiply deformed and metamorphosed to amphibolite grade. Intruded into some greenstones are plutonic complexes up to 40 km long composed of pyroxenite, anorthosite, and gabbro.

The Derbina orogenic belt comprises a deformed Early Proterozoic succession of chiefly metasediments, which were deposited unconformably on Archean basement. At lower stratigraphic levels, paragneisses and quartzites are common in most successions, with marbles dominating at middle stratigraphic levels. The Derbina supracrustals probably represent Early Proterozoic platform cover on a small Archean craton. These successions were later deformed into NE- and E-trending folds, variably metamorphosed to amphibolite or granulite grade, and intruded with syntectonic granitoids.

High quality isotopic ages from the Yenisey province are few. Earlier reported K/Ar and Rb/Sr ages clearly indicate a Late Archean (2.8–2.7 Ga) age for Archean basement in both tectonic subdivisions. Tonalitic gneisses from the Kan terrane basement yield a Rb/Sr whole-rock isochron age of 3260 ± 50 Ma (i = 0.7126) (Romanov and Gerasimov, 1987). Various granitoids from the Kan terrane have reported Pb/Pb zircon and allanite ages of 1900 ± 65, 1830 ± 250, and 1750 ± 340, and 650 ± 100 Ma (Volobuev et al., 1980). A U/Pb zircon age from metavolcanics overlying the basement is 2200 Ma, and tonalites intruding these volcanics yield a zircon age of 1860 Ma (Nozhkin et al., 1989). Relatively undeformed Riphean

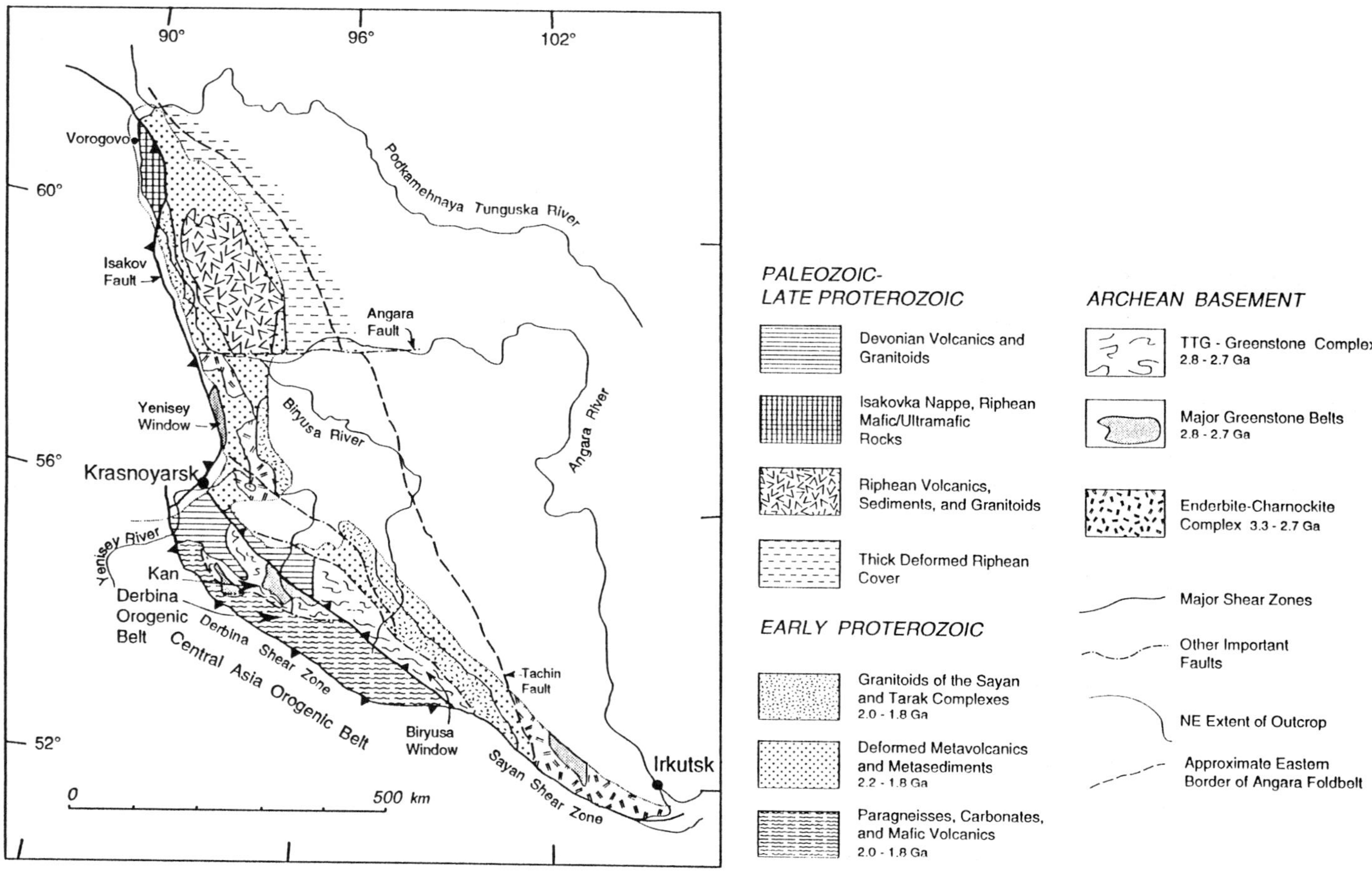

Fig. 11. Generalized geologic map of Precambrian rocks of the Yenisey Uplift and adjacent areas along the southwestern border of the Siberian craton. After Nozhkin (1986), Nozhkin and Smaguin (1988), and Postelnikov and Museibov (1992).

basal clastic sediments unconformably overlie the Derbina orogenic belt and provide an upper limit for the age of the belt of about 1150 Ma based on older K/Ar ages (Volobuev et al., 1980). Whole rock Pb/Pb isochron ages from Derbina marbles are 1570 ± 220 Ma (Derbina Suite) and 1050 ± 50 Ma (Alygzher Suite), and may record regional metamorphic/deformation events (Volobuev et al., 1980). The timing of collision between the Yenisey and Tungus provinces along the Sayan shear zone is poorly constrained. It must be older than Riphean (1.65 Ga) and either synchronous or younger than 1.9–1.8 Ga granitoids.

AKITKAN OROGENIC BELT

The Akitkan orogenic belt, which is exposed only in the Baikal uplift, is a major Proterozoic orogenic belt in the Siberian craton (Yanshin and Borukaev, 1988). Magnetic and gravity anomalies of exposed rocks in the Baikal uplift can be traced along strike into the basement and suggest that the belt is more than 1500 km long and from 50 to 250 km wide (Fig. 7). The southeastern boundary with the Aldan province is defined by an abrupt change in gravity and magnetic anomalies from typically NW-trending in the Aldan province to NE-trending in the Akitkan orogenic belt (Fig. 2). This boundary is herein referred to as the Lena fault zone and is regarded as a major Precambrian province boundary. Farther south, the orogenic belt is separated from the Barguzin terrane by a major high-angle thrust fault, the Abchad fault, which is exposed north of Lake Baikal (Fig. 12). The northern boundary of the Akitkan orogenic belt is defined by a break in basement geophysical anomalies interpreted as a major fault zone (Vilyuy fault zone).

West of the Davan fault (Fig. 12), the Akitkan orogenic belt includes the Sarma Group, which comprises metasediments (pelite, arkosic sandstones, conglomerates) and metavolcanics (rhyolite, dacite) intruded with diabase sills (Fedorovsky, 1985). These rocks, which contain well preserved primary textures and structures, are metamorphosed from greenschist to amphibolite grade, deformed into narrow folds of northeastern strike, and intruded with syntectonic granitoids of the Kocherikovo Complex. These relationships are well exposed on Olkhon Island in Lake Baikal. Also in the domain west of the Davan fault is the Akitkan Group, which includes low grade, chiefly felsic volcanics (rhyolite, trachyte, andesite), clastic sediments (sandstone, conglomerate), and locally, basalt flows (Bukharov, 1987). Comagmatic plutonic and volcanic rocks of the Irel Complex (granites and syenites, some with rapakivi textures) are closely associated with the Akitkan felsic volcanics (Sryvtsev, 1986; Bukharov, 1987). Between the Davan and Abchad faults are the Olkhon and Chuya Groups, which are high grade, but strongly retrogressed mafic granulites, marbles and paragneisses (metapelites), with minor amounts of BIF. These groups are intruded with syntectonic monzodiorites, syenites, granites and granodiorites, chiefly near the Abchad or Davan faults. These granulite-grade, multiply deformed and metamorphosed rocks may represent an exposed slice of Archean basement of the Magan province (Rundkvist and Sokolov, 1993).

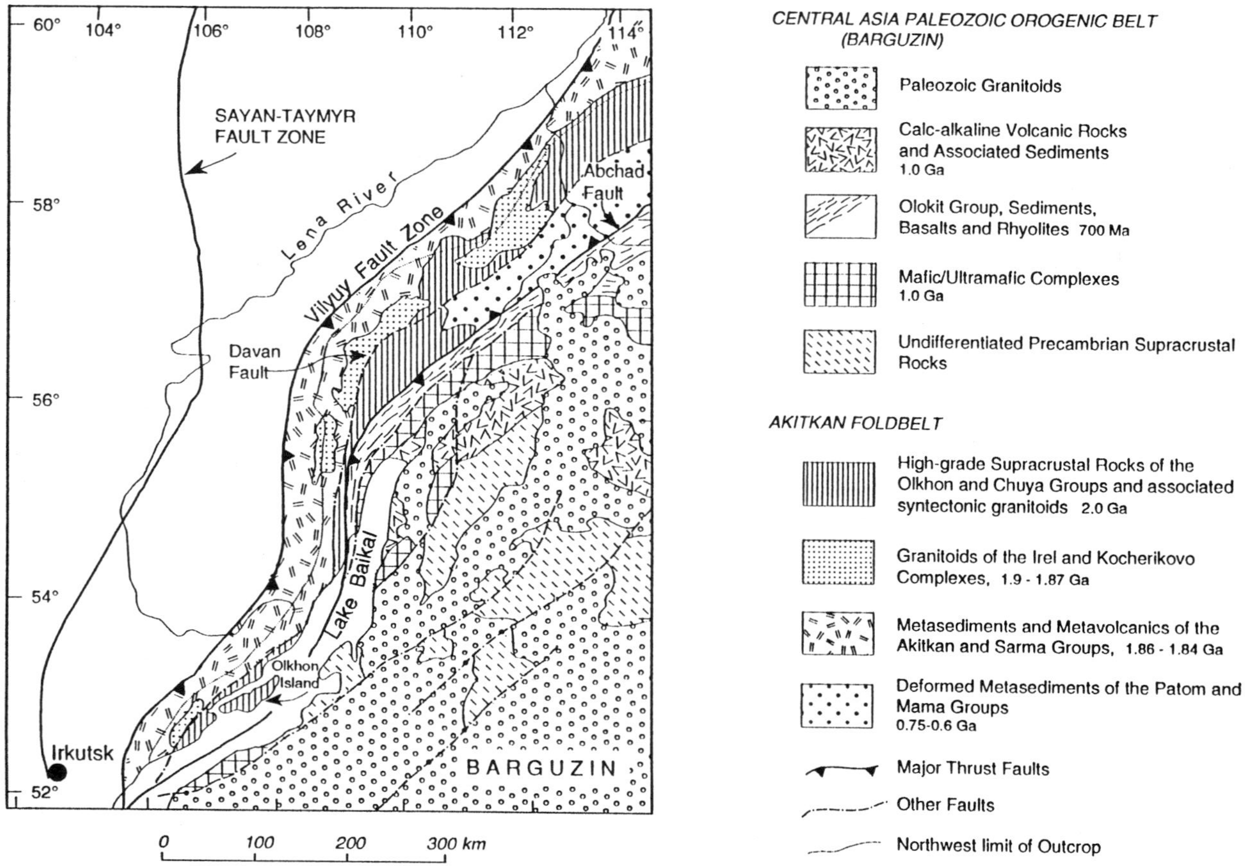

Fig. 12. Generalized geologic map of Precambrian rocks of the Baikal uplift and adjacent a parts of the Barguzin terrane, southern Siberia. After Fedorovsky (1985) and Rundkvist and Sokolov (1993).

U/Pb upper-intercept discordia zircon ages from granitoids of the Kocherikovo Complex are 1910 ± 3 and 1890 ± 25 Ma, and 1866 ± 7 Ma from granitoids of Irel Complex; 1861 ± 8 Ma for granites from the Abchad Massif; 1863 ± 5 Ma for felsic volcanics of Ilovir suite; and 1835 ±8 Ma from felsic volcanics of the Sarma Group (Bibikova et al., 1987; Neymark et al., 1993a). Syntectonic gneiss domes were also formed concurrently with the earliest period of granitoid production. A concordant U/Pb zircon age of 1880 Ma from a garnet–cordierite granulite from the west coast of Lake Baikal (Bibikova et al., 1990a) may reflect a period of high grade metamorphism at this time. Results suggest several periods of granitoid intrusion, volcanism, metamorphism and deformation in the Akitkan orogenic belt between 1.9 and 1.84 Ga. A post-tectonic A-type monzogranodiorite intrusive into the Chuya Group yields a Rb/Sr isochron age of 2034 Ma (Sryvtsev, 1986), which is an upper limit for the age of deposition of this group. In addition, zircon xenocrysts from one of the Kocherikovo granitoids yield a discordia age of 2180 Ma (Bibikova et al., 1990a), which is a lower limit for the age of the oldest basement in this part of the Akitkan belt. Lower-intercept discordia ages (from the upper intercept ages cited above) range chiefly from 500 to 450 Ma and may record times of widespread early Paleozoic deformation associated with the collision of Barguzin with the Siberian craton. Granitoids from the northernmost part of the Akitkan belt have reported U/Pb upper intercept zircon ages ranging from 735 to 322 Ma (Neymark et al., 1993a).

Although modern isotopic ages from the Akitkan orogenic belt are still few in number, it is important that no Archean ages have been reported from the belt. If this is substantiated by future work, the Akitkan orogenic belt may be the only major segment of post-Archean juvenile Early Proterozoic crust in the Siberian craton.

ANGARA OROGENIC BELT

The Angara orogenic belt extends for over 2000 km along the western margin of the Siberian craton. It is bounded on the west by major Late Proterozoic and Paleozoic shear zones and on the southwest by a large Early Proterozoic shear zone (Fig. 9). The eastern boundary of the Angara belt is probably a series of faults or lithologic breaks marked by changes in geophysical anomalies in the basement. The only exposed part of this boundary is the Tachin fault in the Sharyzhalgay uplift (Figs. 10 and 11).

The exposed part of the Angara belt, which extends from the western Sharyzhalgay uplift on the south to the northern end of the Yenisey uplift near Vorogovo (Fig. 11), comprises chiefly a deformed Early Proterozoic metavolcanic and metasediment association intruded with granitoids. The supracrustal rocks lie unconformably on Archean basement, of which two windows of this basement are exposed in the Yenisey uplift. Early Proterozoic supracrustal rocks in the southern part of the belt (upper Biryussa River basin) include a thick succession of dominantly submarine tholeiitic and calc-alkaline volcanics (ranging in composition from basalt to rhyolite) with associated volcaniclastic sediments and carbonates (Sezko, 1988; Bukharov, 1987). These are unconformably overlain by subaerial sandstones and polymictic conglomerates with associated basalts and rhyodacites. Both sequences are deformed, with major fold axes trending northwest, and metamorphosed to the amphibolite facies (Konikov and Travin, 1986). In addition, syn- to post-tectonic granitoids of the Sayan Complex intrude these rocks along a strip many hundreds of kilometers in length (Fig. 11). Near the mouth of the Angara River, supracrustal rocks of the Yenisey Group include dacites, basalts, pelites, quartzites, and carbonates, all metamorphosed to amphibolite grade. These are intruded by granitoids of the Tarak Complex, which are similar to granitoids of the Sayan Complex. In the region between the Angara River and Vorogovo, the succession includes pelites, quartzites, and felsic volcanics overlain by conglomerates, quartzites, and carbonates interlayered with basalts of the Teya Group. Intrusive granitoids in this region include plagiogranites and associated migmatites, syenites, and rapakivi granites (Nozhkin, 1986; Postelnikov and Museibov, 1992).

The Biryusa Archean window southeast of Krasnoyarsk comprises a supracrustal sequence composed of a TTG–greenstone association overlain by metapelites and amphibolites (Nozhkin and Smaquin, 1988; Sezko, 1988). Ultramafic and mafic bodies within the TTG appear to represent remnants of greenstones. Although amphibolite-grade mineral assemblages typify the Biryusa Archean, relicts of orthopyroxene occur in some rocks and together with locally preserved eclogites, suggest earlier high-grade metamorphism. The Biryusa rocks are multiply deformed and the last deformation is characterized by open folds with shallow axial plunges and NNE trends (Rundkvist and Sokolov, 1993).

The Yenisey Archean window northeast of Krasnoyarsk includes a basement complex (the Kuzeev and Atamanov Groups) composed of felsic garnet-two pyroxene gneisses, mafic granulites, metapelites, and charnockites and associated migmatites (Nozhkin, 1986). Intruded into this complex are norites and pyroxenites accompanied by a small amount of anorthosite and dunite. This Archean basement is deformed into subvertical isoclinal folds with a northerly strike, and multiply metamorphosed to granulite grade. Thermobarometric studies indicate the first metamorphism at 900°C and about 8 kbar and a second at 550–800°C and 4–7 kbar (Perchuk et al., 1989). The Yukseevo greenstone belt occurs in the western part of the window and includes a thick succession of basalts, dacites and rhyolites associated with graywacke turbidites, and quartzites. Rocks in this greenstone belt are metamorphosed to greenschist grade and intruded with widespread syntectonic TTG and later mafic, andesitic and felsic dikes (Nozhkin, 1985; 1986).

Isotopic ages from the Angara orogenic belt are relatively few in number and many are of questionable quality. An upper intercept zircon age from orthopyroxene gneiss in the Yenisey Archean window is 2730 ± 180 Ma, recording perhaps the approximate age of the felsic protolith (Bibikova et al., 1993). The lower intercept age of the same zircon suite is 1900 ± 20 Ma and may reflect the timing of granulite-grade metamorphism. This is similar to an upper intercept zircon age of 1840 ± 40 Ma from a charnockitic intrusion in the Archean basement (Bibikova et al., 1993). Zircons from a felsic volcanic in the Yukseevo greenstone belt yield a TE (thermo-ionic emission) Pb/Pb age of 2750 Ma, which is interpreted as the age of the volcanic protolith (Nozhkin et al., 1989). An upper intercept zircon age of 1880 ± 40 Ma is reported from a biotite–hornblende gneiss from the Yenisey Group, and is probably a protolith age. A lower intercept age from these rocks of 1780 ± 20 Ma may be a metamorphic overprinting related to post-tectonic intrusion of the widespread Tarak granites (Bibikova et al., 1993). Zircons from the post-tectonic Garyovo rapakivi granite yield an upper intercept zircon age of 1760 Ma and supracrustal rocks in this area yield K/Ar cooling ages in the range of 1670 to 1650 Ma.

A tonalite from the Archean basement of the Biryusa window yields a TE Pb/Pb zircon age of 2670 Ma, interpreted as the protolith age, and intrusive charnockites have a concordant U/Pb zircon age of 1926 ± 26 Ma. A granite of the Sayan Complex intruding both the Archean basement and Early Proterozoic supracrustal rocks has a U/Pb zircon age of 1900 ± 15 Ma and a Rb/Sr mineral isochron age of 2014 ± 75 Ma from associated pegmatites (Pakholchenko, 1980). An age of 1900 Ma may represent the approximate time of intrusion of these granites. Another group of granites intruding basement in the Biryusa window yield thermal ionic emission Pb/Pb ages in the range 1760–1740 Ma and pegmatites cutting associated Proterozoic supracrustal rocks yield K/Ar mica ages of 1830 and 1730 Ma, recording perhaps late thermal events at these times (Ovchinnikov et al., 1980).

A preliminary assessment of the above ages suggests the following sequence of events in the Angara orogenic belt. The Archean basement in the Yenisey and Biryusa windows formed between about 2.75 and 2.7 Ga and is probably part of the Tungus province on the east. Early Proterozoic volcanic and sedimentary rocks were deposited on this basement and deformed, intruded with granitoids, and metamorphosed several times between 2 and 1.8 Ga, with the most intense and perhaps most widespread activity between 1.9 and 1.85 Ga. Volcanism, plutonism, and deformation in the Angara belt at this time may have been associated with subduction along the western margin of the Tungus terrane.

DISCUSSION

A summary of zircon isotopic ages from Precambrian rocks of the Siberian craton is given in Fig. 13, together with comparable data from the northern parts

---

Opposite: Fig. 13. Summary of the pre-1.6 Ga U/Pb zircon geochronology of the Siberian plate compared to similar results from the northern part of the Canadian and Baltic shields. Results from Canada and Scandinavia compiled from many sources. Key: 1, Plutonism, deformation, metamorphism, and in some instances volcanism; 2, plutonism; 3, sedimentation; and 4, deformation and metamorphism.

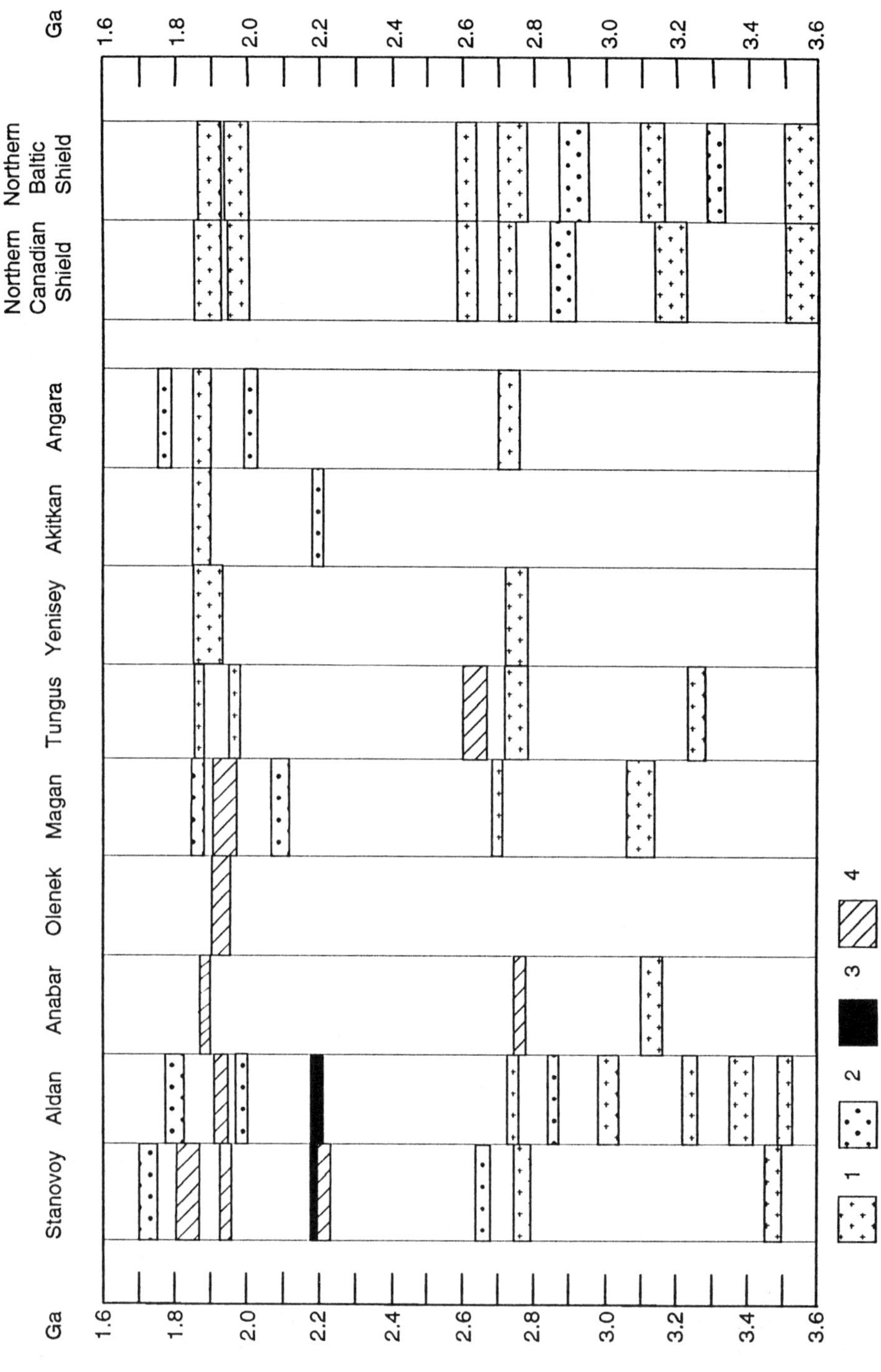
Ga
1.6
1.8
2.0
2.2
2.4
2.6
2.8
3.0
3.2
3.4
3.6
Northern Baltic Shield
Northern Canadian Shield
Angara
Akitkan
Yenisey
Tungus
Magan
Olenek
Anabar
Aldan
Stanovoy
1
2
3
4

of the Canadian and Baltic shields. Of the nine major tectonic subdivisions shown, only the Aldan province has an adequate number of zircon ages to characterize most or all Precambrian events. Four major periods of Archean plutonism and orogenesis are recorded in the Aldan province at 3.5, 3.35, 3.0, and 2.75 Ga. The relative importance of each of these in terms additions of juvenile crust, however, is not well known. One or two of these episodes is also recognized in the Stanovoy, Anabar, Magan, Yenisey, and Tungus provinces and in the Angara orogenic belt. The absence of Archean ages from the Olenek province probably reflects an inadequate data base. Based on the currently available zircon and Sm/Nd isotopic data summarized in this paper, it would appear that ≥90% of the Precambrian basement of the Siberian craton represents crust extracted from the mantle during the Archean.

The tectonic juxtaposition of Archean TTG–greenstone and high grade terrains, such as found in the Aldan, Stanovoy, and Tungus (Sharyzhalgay uplift) provinces, is a common feature in the Siberian craton. The similarity in chemical compositions and lithologic proportions (greenstone/TTG ratios in low grade areas and mafic granulite/enderbite ratios in high grade areas) suggests that many of these Archean terrains represent shallow and deep exposure levels of the same crust. Even within the same terrane, such as within the Olekma and the Tynda terranes, low grade and high grade slices of TTG–greenstone are tectonically juxtaposed. Only two examples of Archean terranes with cratonic components (i.e., quartzite, metapelite, carbonate) are recognized: the Aldan terrane and the Larba Complex within the Tynda terrane. Even in these cases, the possibility exists that the cratonic supracrustals are Early Proterozoic in age and are tectonically infolded into the Archean basement. If exposed areas today are representative of buried portions of the Siberian craton, Archean high grade terrains could comprise more than 70% of the Siberian basement. Much of the high grade metamorphism of these terranes, however, may have occurred in the Early Proterozoic. If so, this requires considerable uplift and erosion prior to deposition of Riphean sediments on the craton. Perhaps the high grade metasediments in the Hapschan terrane are remnants of sediments produced during this uplift.

If the seismic sounding results are correct, and the lithosphere beneath the Siberian craton is of normal thickness (100–200 km), the Siberian Archean crust differs from that in North America and South Africa where seismic shear wave studies and crustal xenolith studies indicate the presence of anomalously thick (up to 400 km) mantle lithosphere beneath Archean crust (Richardson et al., 1984; Grand, 1987; Anderson et al., 1992). Because of extensive Early Proterozoic reworking of the Siberian craton, however, it is not clear if this thin lithosphere developed during the Archean or during the Proterozoic. Attesting to the fact that thick Archean lithosphere can be preserved during later reworking is the thick lithosphere beneath parts of the Archean Canadian shield that were extensively reworked in the Proterozoic (such as Baffin Island, the Ungava peninsula, and North Greenland) (Grand, 1987). The fact that Proterozoic craton is preferentially buried with younger sediments compared to Archean craton in North America has been related to a thick, depleted and thus more buoyant lithospheric root beneath the Archean crust (Hoffman, 1989). Consistent with thinner Archean lithosphere in Siberia is the widespread, often thick, Riphean sedimentary cover over much of the craton. Although a large mantle plume associated with eruption of the Siberian traps in the late Paleozoic could have destroyed a thick Archean lithosphere, it cannot explain the thick Riphean cover, which appears to record a normal thickness for the Archean lithosphere. If the lithosphere is not thickened beneath Siberia, more than one mechanism for the formation of continental crust and associated lithosphere may have operated in the Archean. Clearly, the thickness of the Siberian lithosphere needs to be accurately estimated from shear wave velocity studies to answer this very important question related to early continental growth.

The most widespread Early Proterozoic orogenic events in the Siberian craton occurred between 2.0 and 1.8 Ga, with perhaps the major event or events centered at 1.95–1.90 Ga (Fig. 13). It is not yet clear if the lack of Archean ages from the Akitkan orogenic belt is real, but if it is, this orogenic belt may be the only Early Proterozoic juvenile crust in the Siberian craton. Widespread deformation and granulite-facies metamorphism is recorded in all provinces and orogenic belts at 2.0–1.9 Ga, and plutonism was also important at this time in the Magan, Tungus, and Yenisey provinces and in the Akitkan and Angara orogenic belts. From the available constraints on ages of major fault zones (Table 1), it would appear that 1.95–1.90 Ga was the time of assembly of the Siberian craton. This is also the assembly time for most of northern North America and the northern part of the Baltic shield (Hoffman, 1989), and it is likely that Siberia was part of this assembling landmass.

TABLE 1

Constraints on ages of major Precambrian fault zones in the Siberian Plate

Fault/shear zone	Age limits (Ga)	Probable age of major deformation (Ga)
1. Aldan–Kilier	> 1.8, < 2.0	1.95
2. Tyrkanda/Timpton	–	1.92
3. Ulkan	>1.9, < 2.4	–
4. Dzheltulak	>1.8, < 2.2	1.95 ?
5. Kalar/Stanovoy	>1.8, <1.92	1.90
6. Lena	>1.8, <1.95	1.90 ?
7. Vilyuy	>1.8, <1.93	1.90–1.87
8. Billyakh	–	1.97
9. Kotuykan	>1.8, <2.0	1.93 (1.87–1.84)
10. Sayan–Taymyr	>1.8, <2.0	1.96 (1.87–1.84)
11. Sayan	–	1.9–1.8

( ) = Renewed orogenic activity.

The number of plates and the distances they moved during the Early Proterozoic growth of Siberia are not well constrained. Zonenshain et al. (1989) suggest the existence of convergent boundaries along the Ulkan and Akitkan orogenic belts during the Early Proterozoic. We would add to this list the Aekit and Angara orogenic belts. Supracrustal and plutonic rocks in all four of these orogenic belts are similar to those characteristic of modern continental arc systems. Because the eastern margin of the Siberian craton is now disrupted and partially covered by thrusts of the Mesozoic Verkhoyansk orogenic belt, the original lengths of the Aekit and Ulkan orogenic belts is unknown, and it is possible they were part of the Akitkan orogenic belt. Also, because of lack of exposure, foreland and hinterland relationships along these boundaries during collision are not known. The overall similarity of Archean rock associations and geochronology of adjacent provinces, such as the Stanovoy and Aldan provinces, is consistent with, but does not demand these provinces being part of the same crustal segment when they formed. If so, it would necessitate a period of continental fragmentation prior to the 1.95–1.90 Ga collisions, but after or during deposition of the Udokan, Ulkan, Dzheltaulak and similar Early Proterozoic sediments at about 2.2 Ga. Perhaps the changes from craton-related to rift-related sediments with time recorded in some of these successions (such as the Udokan) reflect rifting and partial or complete fragmentation of Siberian crust in the earliest Proterozoic. This question needs to be further evaluated with paleomagnetic studies.

Because the histories of motion on the 1.95–1.90 Ga boundary faults of provinces within the Siberian craton are not well known, the nature of the collisional events at this time remains obscure. However, the widespread distribution of granulite-facies metamorphism and accompanying deformational styles clearly suggest compressive forces and crustal thickening of most of the provinces now comprising the Siberian craton. The apparent absence, however, of pre- to syntectonic Early Proterozoic intrusive and volcanic activity along some of the province boundaries (such as the Billyakh and Kalar faults) may indicate largely transcurrent motion along these faults.

Similar Precambrian zircon isotopic histories of the Siberian craton, the northern Canadian shield, and the northern Baltic shield are consistent with a Siberia–Laurentia connection in the Arctic region, probably between the Akitkan orogenic belt in Siberia and the Thelon orogenic belt in Canada (Condie and Rosen, 1994). Zircon ages in both belts fall in the 2.0–1.9 Ga range and appear to record additions of juvenile crust. Also supporting this fit is the match between the Archean Slave province in Canada and the Aldan province in Siberia. Common zircon ages in both provinces are >3.5 to 3.2 Ga, 3.1–2.9 Ga, and 2.8–2.6 Ga. Fragmentation of Siberia and Laurentia appears to have occurred in the Late Riphean, perhaps coincident with intrusion of mafic dike swarms in both areas between 780 and 615 Ma (Maschak, 1989). These dike swarms may have been derived from a mantle plume centered beneath the Queen Elizabeth Islands in the Canadian Arctic (Condie and Rosen, 1994).

SUMMARY

Geologic mapping, gravity and magnetic anomaly distributions, and U/Pb zircon isotopic ages suggest the existence of seven Archean crustal provinces and two major and five minor Early Proterozoic orogenic belts in the Siberian craton. Four major periods of plutonism and orogenesis are recorded in the Aldan province at ≥3.5, 3.35, 3.0, and 2.75 Ga and one or two of these is also recognized in the Stanovoy, Anabar, Magan, Yenisey, and Tungus provinces and in the Angara orogenic belt. Zircon and Sm/Nd isotopic data suggest that ≥90% of the Precambrian basement of the Siberian craton was extracted from the mantle during the Archean.

Seismic sounding results indicate that the lithosphere beneath Siberia is of normal thickness (100–150 km), unlike the thickened Archean lithosphere in North America and South Africa (300–400 km thick). If these differences are real and do not reflect Early Proterozoic reworking of the lithosphere, more than one mechanism for the formation of continental lithosphere may have operated in the Archean.

Widespread Early Proterozoic orogenic events including remobilization of Archean basement occurred between 2.0 and 1.9 Ga in the Siberian craton, and constraints on ages of major terrane boundaries suggest that the craton was assembled at this time. The Early Proterozoic Akitkan orogenic belt in south-central Siberia may be the only significant post-Archean juvenile crust in the Siberian craton.

ACKNOWLEDGMENTS

This research effort was supported largely by U. S. National Science Foundation grant EAR-9201722 to KCC. The grant paid also for a 5-month visit of OMR to New Mexico Tech where most of the synthesis and map compilation was undertaken. We would also like to acknowledge the "unsung heroes and heroines" that made this compilation and synthesis possible: the many Russian scientists who over the last two decades have contributed an enormous field and geochronologic data base, published chiefly in Russian, and thus not readily available to western scientists. Much of the synthesis and some of the interpretations presented have benefited from first-hand discussions between OMR and many of the Russian scientists, who over the years have contributed to our understanding of the evolution of the Siberian craton. J. Barton Jr., D. Bridgwater, T. Frisch, H. Jackson, and N.V. Popov made available to us unpublished material, and discussion and comments by E.V. Bibikova, N.L. Dobretsov, V.S. Fedorovsky, V.A. Glebovitsky, E.S. Postelnikov, M.A. Semikhatov, and B.R. Shpount were useful in improving our data base and in constraining our interpretations. An earlier version of the manuscript was improved by in-depth reviews of Allen Nutman, Brian Windley, and Kevin Burke.

REFERENCES

Aftalion, M., Bibikova, E.V., Bowes, D.R., Hopgood, A.M. and Perchuk, L.L., 1991. Timing of Early Proterozoic collisional and extensional events in the granulite–gneiss–charnockite–granite complex, Lake Baikal, USSR: A U/Pb, Rb/Sr, and Sm/Nd isotopic study. J. Geol., 99: 851–862.

Alakshin, A.M. and Karsakov, L.P., 1985. Deep-seated structure of the Stanovoy fault zone. Tikhookeanskaya Geol., 3: 76–86 (in Russian).

Anderson, D.L., Zhang, Y.S. and Tanimoto, T., 1992. Plume heads, continental lithosphere, flood basalts and tomography. Geol. Society London, Spec. Publ., 68: 99–124.

Baadsgaard, H., Nutman, A.P. and Samsonov, A.V., 1990. Geochronology of the Olondo greenstone belt. Abstract Volume, Fifth Intern. Conference on Geochronology, Cosmochronology, and Isotope Geology, Geol. Soc. Australia, p. 6.

Belousov, V.V., Pavlenkova, N.I. and Kvyatkovskaya, G.N. (Editors), 1991. Deep structure of the USSR territory. Nauka Press, Moscow, 222 pp. (in Russian).

Berezhnaya, N.G., Bibikova, E.V., Sochava, A.V., Kirnozova, T.I., Makarov, V.A.and Bogomolov, E.S., 1988. Isotopic age of the Chiney Subseries of the Udokan Group in the Kodaro–Udokan Depression. Doklady, Russian Acad. Sci., 302, 5: 1209–1212 (in Russian).

Bereskin, V.I., 1986. Magmatic rocks. In: N.L. Dobretsov (Editor), Early Precambrian of South Yakutia. Nauka Press, Moscow, pp. 138–151, 204–206, 246–250 (in Russian).

Bibikova, E.V., Khiltova, U. Yu., Gracheva, T.V. and Makarov, V.A., 1982. Age of greenstone belts in the Sayan region. Doklady, Russian Acad. Sci., 267 (5): 1171–1174 (in Russian).

Bibikova, E.V., Gracheva, T.V., Dook, V.L., Kitsul, V.I. and Makarov, V.A., 1984. Isotopic age of the Ungra Magmatic Complex in the Aldan Shield. Doklady, Russian Acad. Sci., 276 (4): 206–209 (in Russian).

Bibikova, E.V., Drugova, G.M., Dook, V.L., Levsky, L.K., Levchenkov, D.A. and Morozova, I.M., 1986. Geochronology of the Vitim–Aldan Shield. In: Methods of Isotopic Geochronology and the Geochronological Scale. Nauka Press, Moscow, pp. 135–189 (in Russian).

Bibikova, E.V., Korikovski, S.P., Kirnosova, T.I., Sumin, L.V., Arakelyants, M.M., Fedorovsky, V.S. and Petrova, Z.U., 1987. Isotopic age determination of rocks in the Baikal–Vitim greenstone belt. In: Isotopic Dating of Metamorphic and Metasomatic Processes. Nauka Press, Moscow, pp. 154–164 (in Russian).

Bibikova, E.V., Gavrikova, S.N., Kirnozova, T.V. and Fedchuk, V., 1989a. Early Proterozoic granites in the southern Aldan Shield: Isotopes and isotopic ages. In: Isotopes in Nature. German Democratic Republic Academy of Sciences, Central Institute of Isotope and Radiation Research, Leipzig, pp. 13–14.

Bibikova, E.V., Morozova, I.M., Gracheva, T.V. and Makarov, V.A., 1989b. U/Pb age of granulites from the Kurulta Complex. In: V.A. Rudnik (Ed.), The Oldest Rocks of the Aldan–Stanovik Shield, Eastern Siberia, USSR. Excursion Guide, IGCP 280, Leningrad–Mainz, pp. 89–91.

Bibikova, E.V., Karpenko, S.F., Sumin, L.V., Bogdanovsky, D.G., Kirnozova, T.I., Lyalikov, A.V., Makarov, V.A., Arakalyants, M.M., Korikovsky, S.P., Fedorovsky, V.S., Petrova, Z.I. and Levitsky, V.I., 1990a. U/Pb, Sm/Nd and K/Ar age of metamorphic and magmatic rocks in the Olkhon area, western Baikalia. In: Precambrian Geology and Geochronology of the Siberian Platform and Surroundings. Nauka Press, Leningrad, pp. 171–183 (in Russian).

Bibikova, E.V., Kirnozova, T.I. and Makarov, V.A., 1990b. Age boundaries in the evolution of the Sharyzhalgay Complex in western Baikalia (U/Pb system in zircons). In: Precambrian Geology and Geochronology of the Siberian Platform and Surrounding Areas. Nauka Press, Leningrad, pp. 162–169.

Bibikova, E.V., Gracheva, T.V., Makarov, V.A. and Nozhkin, A.D., 1993. Age boundaries of the early Precambrian geological evolution of the Yenisey Ridge. Stratigraphy and Geological Correlation, 1: 29–34.

Bukharov, A.A., 1987. Protoactivated zones of the ancient platforms. Nauka Press, Novosibirsk, 202 pp. (in Russian).

Condie, K.C., 1992. Proterozoic terranes and continental accretion in southwestern North America. In: K.C. Condie (Ed.), Proterozoic Crustal Evolution. Elsevier, Amsterdam, pp. 447–480.

Condie, K.C. and Rosen, O.M., 1994. The Laurentia–Siberia Connection Revisited. Geology (in press).

Condie, K.C., Wilks, M., Rosen, O.M. and Zlobin, V.L., 1991. Geochemistry of metasediments from the Precambrian Hapschan Series, eastern Anabar Shield, Siberia. Precambrian Res., 50: 37–47.

Dook, V.L., 1989a. Geological framework of the Aldan–Stanovik Shield. In: V.A. Rudnik (Ed.), The Oldest Rocks of the Aldan–Stanovik Shield, Eastern Siberia, USSR. Excursion Guide, IGCP 280, Leningrad–Mainz, pp. 2–3.

Dook, V.L., 1989b. Precambrian of the Aldan Shield: structural evolution and geological history. Geol. Inst., Russian Acad. Sci. Publ., Moscow, 43 pp. (in Russian).

Dook, V.L., Gorokhov, I.M., Kitsul, V.I., Kutyavin, E.P. and Varshavshkaya, E.S., 1989. Rb/Sr age and genesis of charnockites of the Ust–Idzhek Massif, central Aldan Shield. In: Isotopic Geochronology of the Precambrian. Nauka Press, Leningrad, pp. 126–135 (in Russian).

Drugova, G.M., 1989. Tonalites from probable basement of the central Aldan tectonic domain. In: V.A. Rudnik (Ed.), The Oldest Rocks of the Aldan–Stanovik Shield, Eastern Siberia, USSR. Excursion Guide, IGCP 280, Leningrad–Mainz, pp. 91–96.

Drugova, G.M., Dook, V.L. and Sochava, A.V., 1993. Batomga granite–greenstone terrain. In: D.V. Rundqvist and F.P. Mitrofanov (Eds.), Precambrian Geology of the USSR. Elsevier, Amsterdam, pp. 339–344.

Elyanov, A.A., Kastrykina, V.M. and Kastrykin, Yu.P., 1985. New data on the geology and metallogeny of the Dzheltaulak trough. In: Precambrian Rifts in the Baikal–Amur Region and their Metallogeny. Nauka Press, Novosibirsk, pp. 151–167 (in Russian).

Fedorovsky, V.S., 1985. Lower Proterozoic of the Baikal Mountains. Trudy Geol. Inst., Russian Acad. Sci. Nauka Press, Moscow, Vol. 400, 200 pp. (in Russian).

Feoktistov, V.P., 1986. Copper-bearing Precambrian sediments of the Uguy zone. Geol. Rudnykh Mestor., 1: 65–72 (in Russian).

Fotiadi, E.E., 1987. Deep heat field of the Siberia. Nauka Press, Novosibirsk, 194 pp. (in Russian).

Frumkin, I.M., 1987. Tectonic regimes in the Archean of the Aldan Shield. In: F.P. Mitrofanov (Ed.), Geology and Ore Prospecting of the Ancient Platform Basement. Nauka Press, Moscow, pp. 305–318 (in Russian).

Gafarov, R.A., Leites, A.M., Fedorovsky, V.S., Prozorov, I.P, Savinskaya, M.S. and Savinsky, K.A., 1978. Tectonic subdivisions of the Siberian platform basement and stages of continental crust formation. Geotektonika, 1: 43–58 (in Russian).

Gamaleya, Yu.N., 1968. Formation analysis of the southeastern Siberian Platform. Geotektonika, 6: 34–45 (in Russian).

Gerasimchuk, A.V. and Serenko, V.P., 1988. Compositional and petrophysical data to basement subdivision of the Daldyn–Alakit region. Sovetskaya Geol., 11: 74–80 (in Russian).

Glebovitsky, V.A., editor., 1987. Metamorphic Facies of the Baikal–Amur Zone. Nauka Press, Leningrad, 77 pp. (in Russian).

Glebovitsky, V.A. and Drugova, G.M., 1993. Tectonothermal evolution of the western Aldan shield, Siberia. Precambrian Res., 62: 493–506.

Glukhovsky, M.Z., Moralev, V.M. and Kuzmin, M.I., 1977. Tectonics and petrogenesis of the

Kata–Archean complex of the Aldan Shield in connection with the proto-ophiolite problem. Geotektonika, 6: 103–117 (in Russian).

Godsevich, B.L., 1986. Archean stratigraphy of the southern Aldan–Stanovoy Shield. In: Problems of Early Precambrian Stratigraphy in Middle Siberia. Nauka Press, Moscow, pp. 127–136 (in Russian).

Gorokhov, I.M., Timofeev, V.F., Bizunok, M.B., Berezkin, V.I., Dook, V.L., Krylov, I.N., Kutyavin, E.P., Melnikov, N.N. and Smelov, A.P., 1989. Rb/Sr systems in metasediments of the Khani graben, Olekma greenstone belt. In: Isotopic Geochronology of the Precambrian. Nauka Publ. House, Leningrad, pp. 110–125 (in Russian).

Grand, S.P., 1987. Tomographic inversion for shear velocity beneath the North American plate. J. Geophys, Res., 92: 14065–14090.

Hoffman, P.F., 1989. Precambrian geology and tectonic history of North America. In: The Geology of North America — An Overview. Geol Soc. Am., DNAG. Vol. A, 447–511.

Hopgood, A.M. and Bowes, D.R., 1990. Contrasting structural features in the granulite–gneiss–charnockite–granite complex, Lake Baikal, USSR: Evidence for diverse geotectonic regimes in Early Proterozoic times. Tectonophysics, 174: 274–299.

Iskanderova, A.D., Mirkina, S.L., Neymark, L.A., Chukhonia, A.P. and Khoreva, B.Ya., 1980. New data on lead isotopic studies of Archean metamorphic rocks and granitic gneisses in the Stanovoy area of the Aldan Shield. In: Geochronology of Eastern Siberia and the Far East. Nauka Press, Moscow, pp. 132–153 (in Russian).

Jahn, B.M., Gruau, G., Bernard-Griffiths, J., Carnichet, J., Kroner, A. and Wendt, I., 1990. The Aldan Shield, Siberia: Geochemical characterization, ages, petrogenesis and comparison with the Sino–Korean craton. In: Third Intern. Archean Symposium Abstract Volume, Univ. Western Australia, Perth, pp. 179–181.

Jones, D.L., Howell, D.G., Coney, P.J. and Monger, J.W.H., 1983. Recognition, character and analysis of tectonostratigraphic terranes in western North America. In: M. Hashimoto and S. Nyeda (Eds.), Advances in Earth and Planetary Sciences. Terra Scientific Publ. Co., Tokyo, pp. 21–35.

Karsakov, L.P., 1983. Metamorphic complexes of the Amur River area. In: A.N. Neelov (Ed.), Precambrian Metamorphism in the Baikal–Amur Railway Area. Nauka Publ. House, Leningrad, pp. 66–96 (in Russian).

Kastrykina, V.M., 1983. Metamorphism of the central Dzhugdzhur–Stanovik fold area. In: A.N. Neelov (Ed.), Precambrian Metamorphism in the Baikal–Amur Railway Area. Nauka Press, Leningrad, pp. 140–163 (in Russian).

Kastrykina, V.M. and Elyanov, A.A., 1985. Stanovoy Complex stratigraphy in the area between Nyukzha and Tynda Rivers and the problem of structural zonality of the Dzhugdzhur–Stanovoy folded area. In: Early Precambrian of the Aldan Massif and Vicinity. Nauka Press, Leningrad, pp. 69–721 (in Russian).

Khain, V.E., 1985. Geology of the USSR. Berlin–Stuttgart, 385 pp.

Khiltova, U.Ya, 1993. The Kan Pre-Sayan terrain. In: D.V. Rundqvist and F.P. Mitrofanov (Eds.), Precambrian Geology of the USSR. Elsevier, Amsterdam, pp. 365–387.

Khoreva, B.Ya. (Ed.), 1987. Map of Metamorphic and Granitic Rocks of the USSR, 1:10,000,000. All Union Geol. Inst. Ministry Geology Map Factory, Leningrad (in Russian).

Kitsul, V.I., 1986. Metamorphism. In: Early Precambrian of South Yakutia. Nauka Press, Moscow, pp. 152–193 (in Russian).

Kitsul, V.I. and Dook, V.L., 1985. Endogenic formation regimes and evolution stages of the Early Precambrian lithosphere in the Vitim–Aldan Shield. In: Endogenic Formation Regimes of the Earth's Crust and Ore Deposits in the Early Precambrian. Nauka Press, Leningrad, pp. 217–235 (in Russian).

Konikov, A.Z., and Travin, L.V., 1986. Lower Proterozoic stratigraphy of the Urik–Iya graben, eastern Sayan area. In: Problems of Early Proterozoic Stratigraphy in Middle Siberia. Nauka Press, Moscow, pp. 21–29 (in Russian).

Kotov, A.B., Zagornaya, N.Yu. and Berezhnaya, N.G., 1989. Succession of granite forming events. In: V.A. Rudnik (Ed.), The Oldest Rocks of the Aldan–Stanovik Shield, Eastern Siberia, USSR. Excursion Guide, IGCP 280, Leningrad–Mainz, pp. 96–98.

Krylov, A.Ya., Vishnevsky, A.N. and Silin, B.I., 1963. Absolute age of rocks in the Anabar Shield. Geokhimia, 12: 1140–1144 (n Russian).

Krylov, I.N., Gorokhov, I.M., Kutyavin, E.P., Melnikov, N.N. and Varshavskaya, E.S., 1980. Rb/Sr dating of polymetamorphic rocks in the Sharyzhalgay Series (SW Baikalia). In: Geochronology of Eastern Siberia and the Far East. Nauka Press, Moscow, pp. 80–94 (in Russian).

Levchenkov, D.A., Morozova, I.M., Drugova, G.M., Dook, V.L. and Levsky, L.K., 1987. U/Pb dating of the oldest rocks in the Aldan Shield. In: Isotopic Dating of Metamorphic and Metasomatic Processes. Nauka Press, Moscow, pp. 116–138 (in Russian).

Litvinova, T.P., Shmiyarova, N.P. and Yermoshko, L.V., 1978. Map of the Magnetic Anomaly Field of the USSR and Surrounding Areas, 1:10,000,000. Aerogeologia Map Factory, Leningrad (in Russian).

Luts, B.G., 1974. Petrology of the Crustal Deep seated zones and the Upper Mantle. Nauka Press, Moscow, 304 pp. (in Russian).

Luts, B.G. and Oksman, V.S., 1990. Deep seated fault zones of the Anabar shield. Nauka Press, Moscow, 260 pp. (in Russian).

Manuilova, M.M. (Ed.), 1968. Precambrian geochronology of the Siberian Platform and its surrounding folded areas. Nauka Press, Leningrad, 331 pp. (in Russian).

Maschak, M.S., 1989. Structural position of Riphean dolerite dikes in the southern Anabar shield, in Basic Magmatism in the Siberian platform and its metallogeny. Nauka Press, Yakutsk, pp. 29–31 (in Russian).

Melnikov, A.I., 1991. Sharyzhalgay structural zone in the south Siberian Platform. In: Granulitic complexes of the lower continental crust (regional overviews of the USSR). Institute of the Lithosphere Publ., Russian Acad. Sci., Moscow, pp. 71–92 (in Russian).

Moralev, V.M., 1986. Early Stages of the Evolution of the Continental Lithosphere. Nauka Press, Moscow, 185 pp. (in Russian).

Mokshantsev, K.B., 1979. Proterozoic of the northeastern Siberian Platform. Nauka Press, Novosibirsk, 215 pp. (in Russian).

Morozova, I.M., Bogomolov, E.S., Jakovleva, S.Z., Belyatsky, B.V. and Berezhnaya, N.G., 1989. U/Pb dating of granite–gneiss and tonalite–trondhjemite gneisses. In: V.A. Rudnik (Ed.), The Oldest Rocks of the Aldan–Stanovik Shield, Eastern Siberia, USSR. Excursion Guide, IGCP 280, Leningrad–Mainz, pp. 105–107.

Moskovchenko, N.I. and Kastrykina, V.M., 1989. The Larba Block. In: V.A. Rudnik (Ed.), The Oldest Rocks of the Aldan–Stanovik Shield, Eastern Siberia, USSR. Excursion Guide, IGCP 280, Leningrad–Mainz, pp. 82–85.

Moskovchenko, N.I., Shustova, L.E., Krasnikov, N.N., Semenov, A.P, Shemyakin, V.M. and Kastrykina, V.M., 1987. Stanovoy Megablock: tectonic subdivisions, structural complexes, and metamorphism. In: F.P. Mitrofanov (Ed.), Evolution of Early Precambrian lithosphere in the Aldan–Olekma–Stanovoy region. Nauka Press, Leningrad, pp. 146–200 (in Russian).

Nemchin, A.A., Kovach, V.P., Kotov, A.B. and Vinogradov, D.P., 1989. Geochemistry and Sm/Nd geochronology. In: V.A. Rudnik (Ed.), The Oldest Rocks of the Aldan–Stanovik Shield, Eastern Siberia, USSR. Excursion Guide, IGCP 280, Leningrad–Mainz, pp. 48–55.

Neymark, L.A., Kovach, V.P., Nemchin, A.A., Morozova, I.M., Kotov, A.B., Vinogradov, D.P.,

Gorokhovshy, B.M., Ovchinnikova, G.V., Bogomolova, L.M. and Smelov, A.P., 1990. Late Archean intrusive complexes in the Olekma granite–greenstone terrane, Eastern Siberia: geochemical and isotopic study. In: Third Intern. Archean Symposium, Abstr. Vol., Univ. Western Australia, Perth, pp. 173–174.

Neymark, L.A., Larin, A.M., Ovchinnikova, G.V. and Yakovleva, S.Z., 1992a. U/Pb age of the Zhugzhur anorthosite. Doklady, Russian Acad. Sci., 323: 514–518 (in Russian).

Neymark, L.A., Larin, A.M. and Yakovleva, S.Z., 1992b. U/Pb age of magmatic rocks in the Ulkan graben, south eastern Aldan Shield. Doklady, Russian Acad. Sci., 323: 1152–1156 (in Russian).

Neymark, L.A., Nemchin, A.A., Rosen, O.M., Serenko, V.P., Spetsius, Z.V. and Shulesko, I.K., 1993a. Sm/Nd isotopic systems in lower crustal xenoliths of Yukutian kimberlites. Doklady, Russian Acad. Sci., 327: 374–378 (in Russian).

Neymark, L.A., Kovach, V.P., Nemchin, A.A., Morozova, I.M., Kotov, A.B., Vinogradov, D.P., Gorokhovsky, B.M., Ovchinnikova, G.V., Bogomolova, L.M. and Smelov, A.P., 1993b. Late Archean intrusive complexes in the Olekma granite–greenstone terrain eastern Siberia: geochemical and isotopic study. Precambrian Res, 62: 453–472.

Nozhkin, A.D., 1985. Geochemical peculiarities of Early Precambrian trough complexes in Middle Siberia. In: Geology and Radiogeochemisty of Middle Siberia. Nauka Press, Novasibirsk, pp. 118–140 (in Russian).

Nozhkin, A.D., 1986. Mean features of the composition and structure of the Precambrian sequences of the Yenisey Mountain Range. In: V.V. Reverdatto and V.V. Khlestov (Eds.), Precambrian Crystalline Complexes of the Yenisey Mountain Range. Russian Acad. Sci. Publ., Novosibirsk, pp. 9–18 (in Russian).

Nozhkin, A.D. and Smaguin, A.N., 1988. New subdivision scheme for Precambrian metamorphic complexes of the Kan black. Geologia i Geophysica, 12: 3–121 (in Russian).

Nozhkin, A.D., Malyshev, V.I., Sumin, L.V., Ostapenko, E.I. and Gerya, T.V., 1989. Geochronological study of metamorphic complexes in the SW Siberian platform. Geologia i Geofisika, 1: 26–33 (in Russian).

Nutman, A.P., Chernyshev, I.V., Baadsgaard, H. and Smelov, A.P., 1992a. The Aldan Shield of Siberia, USSR: the age of its Archean components and evidence for widespread reworking in the mid-Proterozoic. Precambrian Res., 54: 195–210.

Nutman, A.P., Gavrikova, S.N. and Chernyshev, I.V., 1992b. Late Archean crust formation and mid-Proterozoic reworking in the Stanovik block of the Aldan Shield, USSR. In: J.E. Glover and S.E. Ho (Eds.), The Archean: Terrains, Processes and Metallogeny. Univ. Western Australia Publ., 22: 89–92.

Ovchinnikov, L.N., Voronovsky, S.N. and Ovchinnikova, L.B., 1980. Absolute age of pegmatites and near-pegmatite metasomatic rocks in the Eastern Sayan Mountain Range. In: Geochronology of Eastern Siberia and the Far East. Nauka Press, Moscow, pp. 111–120 (in Russian).

Pakholchenko, Yu.A., 1980. Rb/Sr age of rare metal pegmatites. In: Geochronology of Eastern Siberia and the Far East. Nauka Press, Moscow, pp. 30–47 (in Russian).

Panchenko, I.V., 1985. Geology and metamorphic evolution of the lower Precambrian in the Stanovoy Mountain Range. Far East Science Center, Russian Acad. Sci., Vladivostok, 152 pp. (in Russian).

Parfenov, L.M., 1991. Tectonics of the Verkhoyansk–Kolyma Mosozoides in the context of plate tectonics. Tectonophys., 199: 319–342.

Perchuk, L., Gerya, T. and Nozhkin, A., 1989. Petrology and retrograde P-T path in granulites of the Kanskaya Formation, Yenisey Range, Eastern Siberia. J. Metamorphic Geol., 7: 599–617.

Percival, J.A. and Card, K.D., 1983. Archean crust as revealed in the Kapuskasing uplift, Superior province, Canada. Geology, 11: 323–326.

Petrov, A.F., Gusev, G.S., Tretyakov, F.F. and Oksman, V.S., 1985. The Archean (Aldanian) and Early Proterozoic (Karelian) megacomplexes. In: V.V. Kovalsky (Ed.), Structure and Evolution of the Earth's Crust in Yakutia. Nauka Press, Moscow, pp. 9–39 (in Russian).

Petrova, Z.I. and Levitsky, V.I., 1990. General regularities in the granulite–gneiss complex evolution in the basement and folded belts of the Siberian Platform. In: Precambrian Geology and Geochronology of the Siberian Platform and Surrounding Areas. Nauka Press, Leningrad, pp. 49–57 (in Russian).

Pismenny, B.M. and Alakshin, A.M., 1984. Structure of the Earth's crust in the western Aldan Shield and Stanovoy folded area. Tekhookeanskaya Geol., 5: 86–98 (in Russian).

Popov, N.V., Zedgeniziov, A.N. and Berezkin, V.I., 1989. Petrochemistry of Archean metavolcanics in the Sunnagin block of the Aldan Massif. Nauka Press, Novosibirsk, 25 pp. (in Russian).

Popov, N.V., Smelov, A.P., Dobretsov, N.N., Bogomolva, L.M., Kartavchenko, V.G., 1990. Olondo greenstone belt. Russian Acad. Sci. Publ., Yakutsk, 170 pp. (in Russian).

Postelnikov, E.S., 1990. Late Proterozoic structures and sequences along the eastern slope of the Yenisey Mountains. Bull. MOIP, 65 (1): 14–30.

Postelnikov, E.S. and Museibov, N.I., 1992. Baikalian orogeny basement structure in the SW Siberian platform. Geotectonics, 6: 37–51.

Puchtel, I.S., 1992. Mafic-ultramafic rocks and crust-mantle evolution in the Early Precambrian: the Olekma gneiss–greenstone area as an example, Publ. Inst. Petrography, Mineralogy and Ore Deposits of Russia. Russian Acad. Sci., Moscow, 20 pp. (in Russian).

Puchtel, I.S., Samsonov, A.V., Simon, A.K. and Zhuravlev, D.Z., 1989. Geology, geochemistry, and Sm/Nd geochronology of the oldest rocks from the Olekma granite–greenstone terrain. In: V.A. Rudnik (Ed.), The Oldest Rocks of the Aldan–Stanovik Shield, Eastern Siberia, USSR. Excursion Guide, IGCP 280, Leningrad–Mainz, pp. 57–63.

Rasskazov, Yu.P. and Yalynichev, E.B., 1972. Ust–Gilyuy trough — a special type of structure in the Precambrian of the Stanovoy Mountains. In: Geology of Far East, ITG, Far East Science Center. Russian Acad. Sci., Khabarovsk, pp. 120–131 (in Russian).

Richardson, S.H., Gurney, J.J., Erlank, A.J. and Harris, J.W., 1984. Origin of diamonds in old enriched mantle. Nature, 310: 198–202.

Romanov, I.A. and Gerasimov, N.S., 1987. Rb/Sr dating of the Precambrian Malotogul deposit (Eastern Sayan). In: F.A. Letnikov (Ed.), Geology, Ore Deposits and Geochronology of the Precambrian in the Siberian Platform and Surrounding Areas. Inst. Earth Crust Publ., Irkutsk, pp. 226–227 (in Russian).

Rosen, O.M., 1989. Two geochemically different types of Precambrian crust in the Anabar Shield, North Siberia. Precambrian Res., 45: 129–142.

Rosen, O.M., 1992. Earth crust formation in the Anabar Shield. Russian Acad. Sci., Inst. Lithosphere Publ., Moscow, 48 pp. (in Russian).

Rosen, O.M. and Zlobin, V.L., 1990. Carbonate sediments of Early Archean granulite and greenstone belts. Int. Geol. Revs., 32 (6): 539–550.

Rosen, O.M., Bibikova, E.V. and Zhuravlev, A.B., 1991. Early crust of the Anabar Shield: age and formation models. In: Early Earth's Crust: the Composition and Age. Nauka Press, Moscow, pp. 199–224 (in Russian).

Rosen, O.M., Rachkov, V.S. and Sonyushkin, V.E., 1990. Metasomatism and partial melting of tectonites and origin of granites in shear belts of the Anabar Shield (North Siberia). Geologica Carpatica, Bratislava, 4 (6): 693–708.

Rublev, A.G., Chukhonin, A.D., Neymark, A.P. and Zaitzev, V.S., 1981. Age of the Kodar Massif, in Geology and Mineralogy of the Precambrian in the Baikal–Amur Railway area. Trudy All Union Geol. Institute, Ministry Geology, Nov. ser., 278: 54–60 (in Russian).

Rundkvist, D.V. and Sokolov, Yu.M., 1993. Main features of Precambrian metallogeny. In: D.V. Rundqvist and F.P. Mitrofanov (Ed.), Precambrian Geology of the USSR. Elsevier, Amsterdam, pp. 1–6.

Samkov, V.V. and Potapyev, S.V., 1986. Interpretation of the gravity field and deep seismic sounding. In: V.M. Moralev (Ed.), The Crustal Structure of the Anabar Shield. Nauka Press, Moscow, pp. 130–155 (in Russian).

Sedova, I.S. and Glebovitsky, V.A., 1985. Granite formation at amphibolite facies conditions in the Stanovoy Complex. In: Early Precambrian of the Aldan Massif and Surroundings. Nauka Press, Leningrad, pp. 92–112 (in Russian).

Sezko, A.L., 1988. Main stages of continental crust formation in Eastern Sayan Area. In: Precambrian and Paleozoic Evolution of the Earth's Crust. Nauka Press, Novosibirsk, pp. 7–37 (in Russian).

Shatsky, V.S., Rudnick, R.L., Jagoutz, E., 1990. Mafic granulite xenoliths from Udachnaya pipe, Yakutia: Samples of Archean lower crust? In: Deep Seated Magmatism and Evolution of Lithosphere of the Siberian Platform. Intern. Seminar Abstracts, Inst. Geol. & Geophys, Siberian Branch, Russian Acad. Sci., Novosibirsk, pp. 35–38.

Shepel, A.B., 1980. K/Ar age of phlogopites in scarn–magnetite deposits of the Aldan Shield and its geological interpretation. In: Geochronology of East Siberia. Nauka Press, Moscow, pp. 38–47 (in Russian).

Shuldiner, V.I. and Panchenko, I.V., 1985. Isotopic age of granulites in the western Stanovoy area and its geological meaning. Geologia i Geofisika, 11: 39–45 (in Russian).

Sochava, A.V., 1986. Petrochemistry of the Late Archean and Proterozoic in the western Vitimo–Aldan Shield. Nauka Press, Leningrad, 142 pp (in Russian).

Spiridonov, V.G., Sukhanov, M.K., Karpenko, S.F. and Lyalikov, A.V., 1991. Sm-Nd isotope system of the oldest granulites in the Anabar shield. Doklady, Russian Acad. Sci., 319 (5): 1209–1212.

Sryvtsev, N.A., 1986. Structure and geochronology of the Akitkan Series in western Baikalia. In: Problems of the Early Precambrian Stratigraphy of Middle Siberia. Nauka Press, Moscow, pp. 28–35 (in Russian).

Stepanov, L.L., 1974. Radiogenic age of polymetamorphic rocks in the Anabar Shield. In: Early Precambrian Rocks and Related Ore Deposits of the Central Arctic. Nauka Press, Leningrad, pp. 77–84.

Sukhanov, M.K. and Panskikh, E.A., 1981. Geological structure, petrology, and ore deposits of the Kalar Anorthosite Massif. In: Petrography and Petrochemistry of Ore-Bearing Magmatic Formation. Nauka Press, Moscow, pp. 189–300 (in Russian).

Sukhanov, M.K. and Rachkov, V.S., 1988. Anorthosites and related rocks. In: M.S. Markov (Ed.), Archean of the Anabar Shield and Problems of Early Evolution of the Earth. pp. 62–83 (in Russian).

Sukhanov, M.K., Spiridonov, V.G. and Karpenko, S.F., 1990. First results of Sm/Nd dating of the anorthosites in the Anabar Shield. Doklady, Russian Acad. Sci., 310: 448–453.

Tauson, L.V., Sobachenko, B.N. and Plyusnin, G.S., 1983. Rb/Sr age of rapakivi-like granites and metasomatic rocks in Katugin–Ayan Zone, northeastern Trans–Baikalia. Doklady, Russian Acad. Sci., 273 (5): 1233–1236 (in Russian).

Tsypukov, M.Yu., Nozhkin, A.D., Bobrov, V.A., Schipitsyn, Yu.G., 1993. Komatiite–basaltic association in the Kan greenstone belt. Geologia i Geofisika, 8: 98–108 (in Russian).

Velikoslavinsky, S.D., Tolmacheva, E.V., Dook, V.L., Melkevich, R.I. and Rudnik, V.A., 1993. Geochemical mapping of basic complexes in the early Precambrian Aldan–Stanovoy shield of Siberia. Precambrian Res., 62: 507–525.

Vishnevsky, A.N. and Turchenko, C.I., 1986. Granulite complex structure in the Anabar Shield. In: V.M. Moralev (Ed.), The Crustal Structure of the Anabar Shield. Nauka Press, Moscow, pp. 17–38 (in Russian).

Volobuev, M.I., Zykov, S.I. and Stupnikova, N.I., 1980. Pb isotope geochronology of Precambrian metamorphic complexes along the southwestern margin of the Siberian Platform. In: Geochronology of Eastern Siberia and the Far East. Nauka Press, Moscow, pp. 14–30 (in Russian).

Yanshin, A.L. and Borukaev, Ch.B. (Eds.), 1988. Tectonics and evolution of the earth's crust in Siberia. Nauka Press, Novosibirsk, 174 pp. (in Russian).

Zonenshain, L.P., Kuzmin, M.I. and Natapov, L.M., 1989. Geology of the USSR: A Plate Tectonic Synthesis. Am. Geophys. Union, Geodynamics Series, vol. 21, 242 pp.

Chapter 11

ARCHEAN MINERALIZATION

D.I. GROVES AND M.E. BARLEY

INTRODUCTION

Archean terrains are some of the most richly mineralized on Earth, both in terms of the mineral deposits which formed during the Archean and also in terms of the deposit styles which developed on or adjacent to stable Archean cratons following their stabilization (e.g. diamonds in kimberlites, copper and REE in carbonatites and other alkaline rocks, bauxites, beach sands). As such, the metallogeny of the Archean terrains has been the subject of several reviews (Hutchison et al., 1971; Anhauesser, 1981; Hutchinson, 1981; Franklin and Thorpe, 1982; Groves and Batt, 1984; Thurston and Chivers, 1990; Barley et al., 1992) that integrate the data from numerous publications that describe individual deposits or deposit styles: refer to Anhaeusser and Maske (1986), Hughes (1990), and Thurston et al. (1992) for descriptions and exhaustive references on Archean mineral deposits for South Africa, Australia and Canada, respectively. This review attempts not only to describe the most important Archean deposit styles, but to place them in a modern tectonic context.

Emphasis is placed on the metallogeny of the granitoid-greenstone terrains (see below), particularly with respect to their Late-Archean evolution, as this was a metallogenetic period matched only by that of the post-Paleozoic metallogeny of the Pacific (and proto-Pacific) Rim. Each of the major deposit styles is described with respect to their common characteristics, and rock associations where appropriate, in terms of their timing relative to volcanism and subsequent orogeny and associated plutonism. It is immediately evident that there is heterogeneous distribution of the major ore styles on both a craton- and global scale (Fig. 1). Thus, emphasis is placed on the major metallogenic provinces that contain clusters of a specific deposit style. In addition, as the geochronology of the terrains and mineral deposits is best constrained in published data from Canada, Australia, and southern Africa, deposit styles from these regions dominate the discussion.

Following discussion of the nature of each major Archean ore style, there is a short review of analogous post-Archean deposit styles and their modern or inferred tectonic settings, and discussion of genetic models. Finally, there is an attempt to integrate the data on the deposits into a holistic tectonic and metallogenic model for Archean granitoid-greenstone terrains.

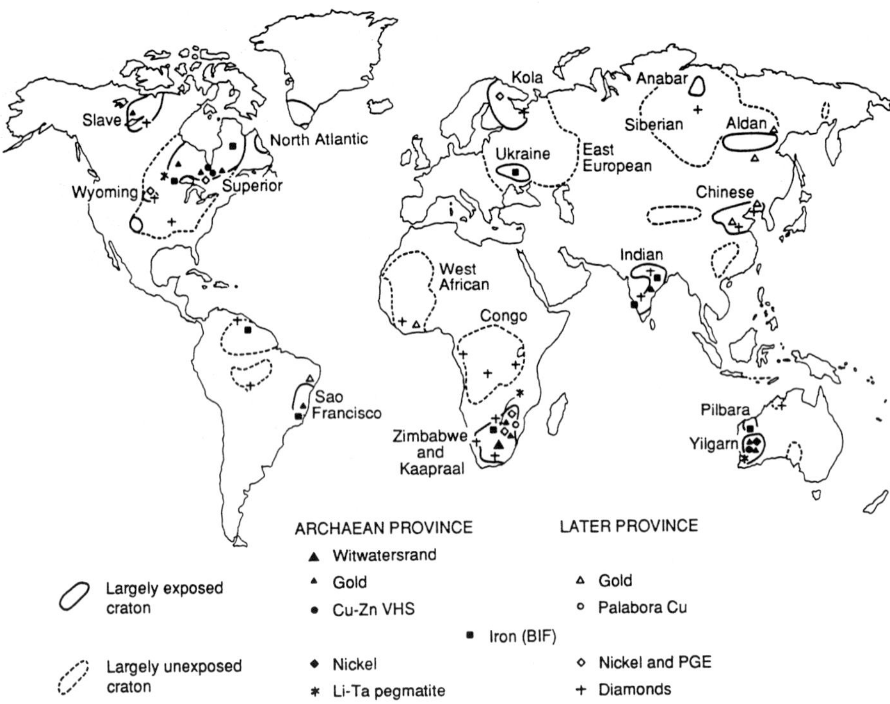

Fig. 1. World map showing the distribution of major Archean cratons and contained world-class ore deposits divided into those that formed in the Archean cratons and those that formed subsequent to cratonization. Positions are generalized to show major deposit clusters or camps. The major Precambrian iron ores are shown, although it is realised that the host BIF's may be Archean or early Proterozoic and the enriched iron ores may be Proterozoic in age. Other major deposits, such as laterites and beach sands, that are developed on or around Archean cratons are not shown due to space restrictions.

Although there has always been controversy over precise tectonic models for Archean granitoid-greenstone terrains (Windley, 1976; Condie, 1981), with both unique (*in situ* history) and non-unique plate-tectonic (accretionary history) models being proposed, there has been increasing support for plate tectonic models of greenstone belt evolution (Talbot, 1973; Tarney et al., 1976; Dimroth et al., 1983; de Wit et al., 1983; Barley et al., 1989; Kerrich and Wyman, 1990; Thurston and Chivers, 1990; de Wit et al., 1992; Williams et al., 1992), albeit modified to some extent (Barley and Groves, 1992) by the evidence for a hotter Archean Earth and by the need to explain komatiites as very deep mantle melts (Campbell et al., 1989). The approach taken in this paper is to use mineral deposits as a critical parameter additional to the more commonly used constraints, such as volcanic and sedimentary associations and structural style and sequence, that help define the tectonic setting of the highly mineralized greenstone belts (Garson and Mitchell, 1981; Hutchinson, 1981; Barley and Groves 1990; 1992 Barley et al., 1992).

CLASSIFICATION OF ARCHEAN TERRAINS

If 2500 Ma is accepted as the end of the Archean, there are a number of types of Archean terrain that must be considered in terms of their metallogeny. These are discussed below in approximate order of the crustal depth, largely defined by their metamorphic grade, that they are interpreted to represent.

(1) Completely deformed and metamorphic high-grade gneiss terrains include some of the oldest rocks (>3700 Ma) and minerals (>4200 Ma) on Earth (Froude et al., 1983). They appear to represent the high-grade metamorphic equivalents of both greenstone and metasedimentary belts, described below, and are commonly affected by the younger magmatic and deformational events recorded in these belts. These terrains are poorly mineralized, in general, although there are some rare-metal pegmatites, including the giant Greenbushes pegmatite (Partington, 1990), and other styles of mineralization described by Katz (1988).

(2) So-called plutonic provinces (Fyon et al., 1992) are dominated by tonalite plutons and complexly deformed tonalitic gneisses with small remnants of highly deformed greenstone-belt lithologies, probably representing the immediate root zones of granitoid-greenstone terrains. These are, in general, poorly mineralized.

(3) Granitoid-greenstone provinces are the most widespread and best documented Archean terrains, comprising deformed and largely near-vertical supracrustal sequences of ultramafic, mafic and felsic volcanic and volcaniclastic rocks and metasedimentary rocks intruded by a variety of variably deformed tonalitic to granitic plutons and batholiths, and variably metamorphosed from sub-greenschist to upper amphibolite facies. Well documented examples in Australia, Canada, India and South Africa range in age from *ca* 3500 Ma to *ca* 2700 Ma, although there may be younger belts both in Brazil and Russia. It is these provinces, particularly the Late-Archean (<2800 Ma) greenstone belts within them, that make the Archean terrains so prospective for mineral exploration. They contain the plethora of mineral deposits, including very large to giant Cu-Zn volcanogenic massive sulfide (VMS) deposits, the largest known komatiite-associated Ni-Cu deposits, very large to giant "mesothermal" lode-gold deposits, and giant rare-metal pegmatites.

(4) Metasedimentary subprovinces are mainly recognised from the Superior Province (Fyon et al., 1992), where they occur as elongate belts of deformed (largely near-vertical orientation) and metamorphosed turbidites cut by granitoid intrusions that are separated from adjacent elongate granitoid-greenstone belts by strike-parallel deformation zones. They resemble accretionary prisms (Fyon et al., 1992). Overall, these terrains are relatively poorly mineralized, although the granitoids contain anatectic two-mica granites, and rare-element and U-bearing pegmatite deposits are characteristic mineralization styles in them (Fyon et al., 1992).

(5) Intracratonic or continental-margin successions, widely regarded as Proterozoic in age (Tankard et al., 1982) until precise Archean ages were obtained in

the last decade, sit unconformably on one or more of the terrain types described above. These successions are shallowly dipping and at low metamorphic grade over much of their extent, but may be severely deformed around basin margins and/or adjacent to regional deformation zones representing sites of subsequent collisional orogeny. The oldest of these successions is the Witwatersrand Supergroup in South Africa, which is constrained to have formed between 3000 and 2800 Ma (Armstrong et al., 1991) This largely sedimentary succession contains several giant goldfields with numerous "reefs" producing about 50% of world gold and a significant proportion of its uranium production. Although there has been long debate about the origin of this Au-U mineralization, and epigenetic concepts have recently been revived (Phillips et al., 1987), a modified placer origin appears most likely (Minter et al., 1993). A balanced description and discussion of the Witwatersrand ores would involve a paper of the total length of this review, and hence they are not described in detail further here, although they are considered in the metallogenic overview.

Overlying the Witwatersrand Basin in South Africa, and directly overlying granitoid-greenstone belts in the Pilbara Craton of Australia, are *ca* 2780 Ma and younger continental rift sequences (Blake, 1993) of clastic sedimentary rocks and tholeiitic basalts (Ventersdorp Supergroup, South Africa; Fortescue Group, Australia) overlain by continental margin or platform sequences of clastic sedimentary rocks, dolomites, and Superior type banded iron formations (BIFs), typified by the Hamersley Basin in Western Australia (Trendall and Blockley, 1970; Blake and Barley, 1989). The BIF units were subsequently enriched in the Proterozoic (Morris et al., 1980) to form some of the world's giant iron-ore deposits. As for the Witwatersrand ores, these are not described further here, although their tectonic setting is discussed below in terms of its importance to Late-Archean evolution.

SYNVOLCANIC DEPOSITS IN GREENSTONE BELTS

Introduction

The two major styles of synvolcanic mineral deposit in Archean greenstone belts are volcanogenic massive Cu-Zn±Pb deposits and komatiite-associated Ni-Cu deposits. These are heterogeneously distributed, with a major VMS province in the Superior Province, particularly the Abitibi Belt, Canada, some important deposits in the Slave Province, and only minor, commonly isolated, deposits elsewhere in the world, and a major komatiite-associated nickel province in the Yilgarn Craton, particularly the Norseman–Wiluna Belt, Western Australia, with similarly scattered deposits on a global scale.

Other deposit styles include enriched iron ores in Algoma type BIF, which are widespread but small relative to those in Superior type BIF of overlying Late-

Archean to early Proterozoic basins, and minor porphyry Cu ± Mo ± Au deposits that are scattered throughout greenstone belts, worldwide. Small, intrusion-hosted Ni-Cu deposits may be related to the more important komatiite-hosted deposits mentioned above.

Volcanogenic massive sulfide deposits

Deposit characteristics and associations

As noted above, discussion of this deposit style focuses on deposits in the Superior Province, and largely on deposits in the Abitibi Belt, with comparisons made with other deposits where appropriate. Australian examples are reviewed by Barley (1992).

In accord with many other deposit styles, the VMS deposits occur in clusters or "camps" (Fig. 2), commonly with one or more large, very large or giant deposits in association with numerous smaller deposits (Hutchinson, 1973). The deposits tend to occur at specific stratigraphic horizons, commonly boundaries between contrasting lithologies (e.g. andesite and rhyolite) in the volcanic successions (Fig. 3). Most economic deposits are in the range of 1 to 10 million tonnes (mt) of ore with >1% Cu and >3% Zn, with a few in the range 10 to 50 mt, and rare examples

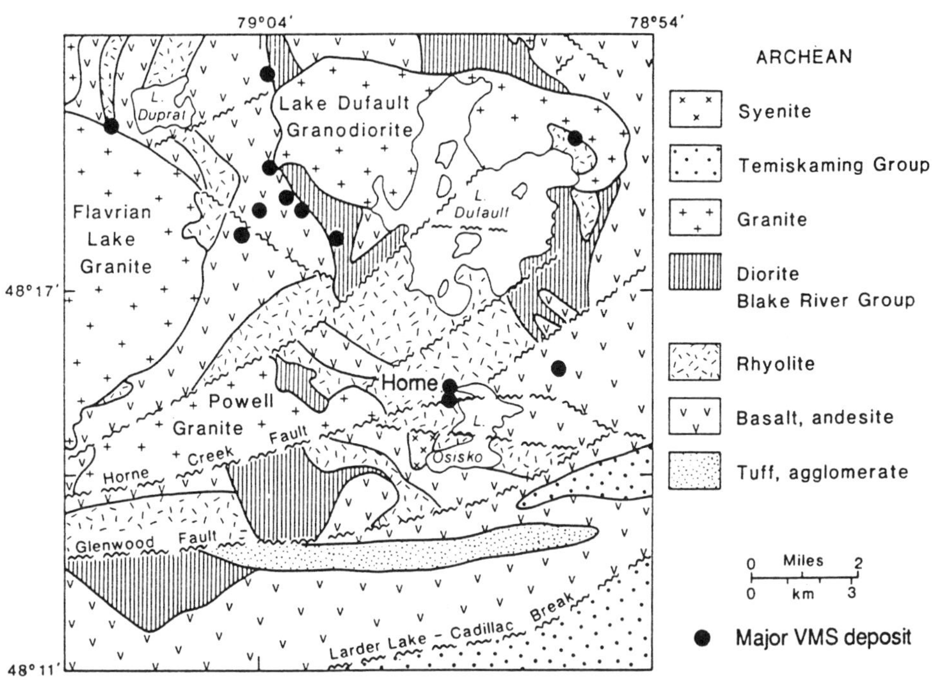

Fig. 2. Volcanogenic massive sulfide (VMS) deposits of the Noranda district, Quebec, showing clustering of major deposits. From Franklin (1990), as adapted from earlier workers.

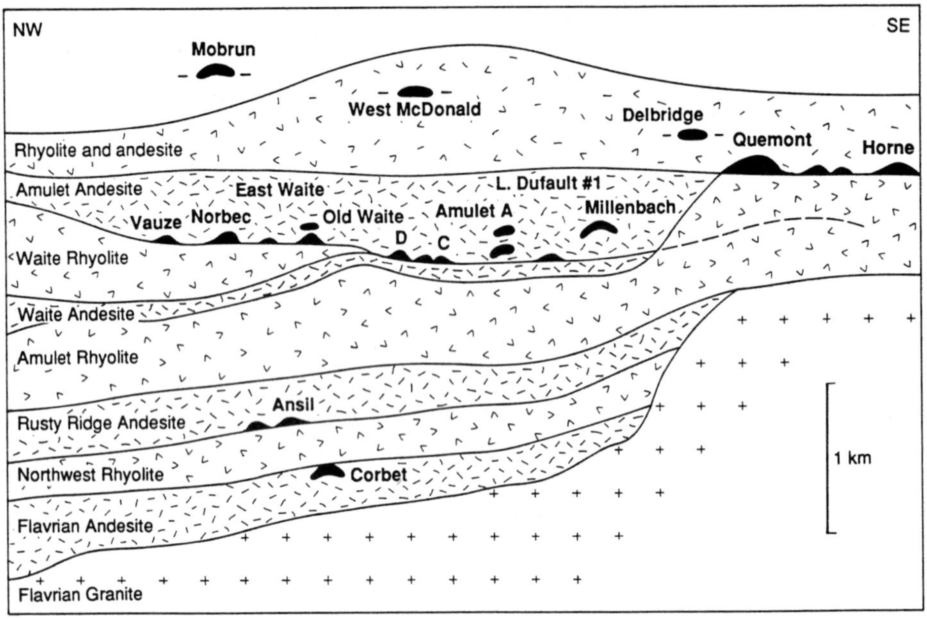

Fig. 3. Longitudinal section of the Noranda district, Quebec, showing the stratigraphic position of all the major VMS deposits over a horizontal distance of about 10 km. Adapted from Spence and Rosen-Spence (1975).

above this size, with the largest being the Kidd Creek deposit at about 130 mt of about 2.7% Cu and 6.5% Zn. The deposits have been classified, on the basis of Cu/Zn ratio and style and intensity of alteration (Morton and Franklin, 1987), into two main styles, the Noranda- and Mattabi-type deposits.

An idealized deposit, commonly taken as one of those in the Noranda camp (e.g. Fig. 4), comprises a lenticular or pod-like body containing more than 60% massive sulfides and zoned from a sphalerite-, pyrite-, and/or galena-rich top to a pyrite- and chalcopyrite-rich base (Franklin et al., 1981). There may be cherty sedimentary rocks or "exhalites" along strike from the massive sulfide bodies. Commonly, the massive or semi-massive sulfides are underlain, normally gradationally, by a discordant, hydrothermal alteration zone (normally chlorite ± quartz ± sericite ± carbonate) containing chalcopyrite–pyrite stringer mineralization, and interpreted as the zone of fluid discharge (Sangster and Scott, 1976). In addition (Noranda-type), or instead (Mattabi-type), there may be regionally extensive semi-conformable alteration zones below the ore zone (e.g. Sturgeon Lake camp; Morton and Franklin, 1987; Galley 1993). At low metamorphic grade, alteration assemblages close to Mattabi-type mineralization include sericite–quartz–calcite or dolomite, whereas those at some depth below Noranda-type deposits reflect silicification and spilitization.

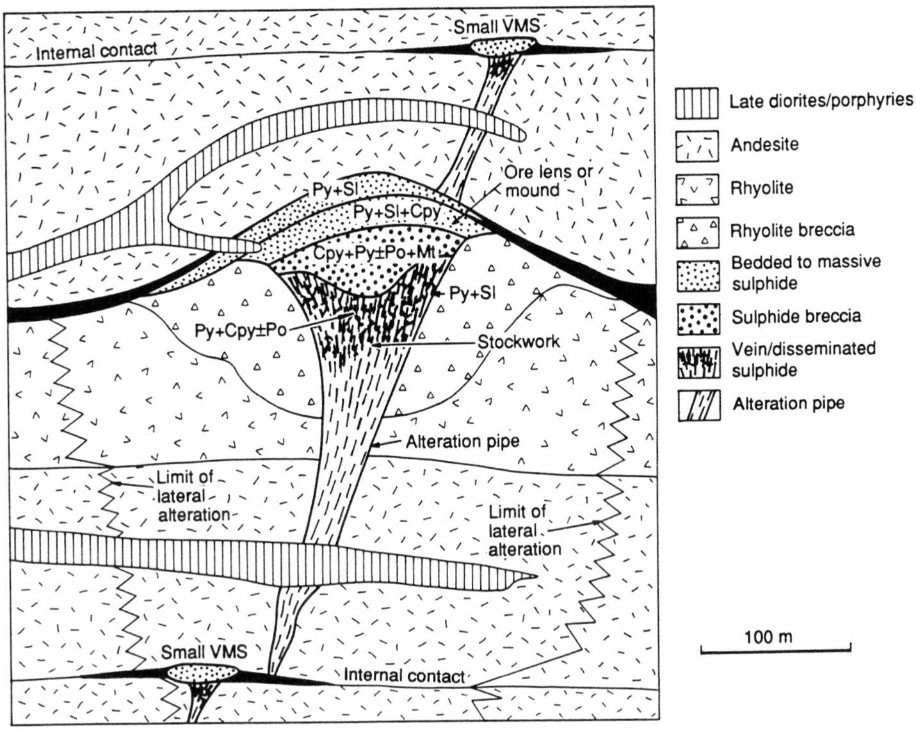

Fig. 4. Schematic cross section through a Noranda-type VMS deposit illustrating the position of sulfide ores, alteration, and geometry of pre-deformation ore body and rock relations. Adapted from Franklin (1990).

Many of the VMS clusters are associated with felsic volcanic rocks which have distinctive trace-element geochemistry (Condie, 1981), the so-called F III type volcanics of Lesher et al. (1986), which are typified by moderate to pronounced Eu anomalies. Such volcanic rocks are believed to have erupted from high-level, subvolcanic, zoned basic-felsic magma chambers, (Lesher et al., 1986), which may be represented by the subvolcanic intrusions which commonly occur one to several kilometres stratigraphically below the VMS deposits (Franklin et al., 1981). These subvolcanic intrusions are interpreted to be heat sources ("heaters") and possibly metal sources for the VMS deposits (Campbell et al., 1981). Recent oxygen isotope studies by Cathles (1993) at Noranda clearly demonstrate that the intrusions are the "heaters" driving convective hydrothermal activity and mineralization.

Most of the VMS camps are associated with bimodal volcanic assemblages of submarine intermediate to felsic flows and pyroclastic rocks, interpreted to represent stratovolcanoes (Thurston et al., 1978), central volcanic complexes (Dimroth et al., 1985), or large composite cones (Gibson and Watkinson, 1990). Many authors stress a caldera setting, cauldron subsidence and resurgence, and localization of volcanic vents and alteration pipes along synvolcanic faults (Gibson and Watkinson, 1990). Fewer deposits are associated with komatiite–tholeiite assemblages consisting of various combinations of komatiites, tholeiitic basalts, icelandite (Macdonald et al., 1990), F III-type rhyolites and calc-alkaline andesite and rhyolite. However, importantly, the giant Kidd Creek deposit occurs in this association.

All major VMS deposits in the Superior Province occur in volcanic assemblages that range in age from 2740 to 2700 Ma (Fyon et al., 1992), although younger hydrothermal activity has probably reset the system at Kidd Creek (Schandl et al., 1990). There are few reliable age data on VMS deposits elsewhere in the world, although deposits in the Yilgarn Block of Western Australia (Barley, 1992) may be both *ca* 2700 Ma (e.g. Teutonic Bore) and 3000 Ma (Scuddles). The oldest (*ca* 3450–3200 Ma) small VMS deposits in the Pilbara Craton are quite distinct from their younger Archean counterparts, but more like Phanerozoic equivalents, in that they commonly contain barite and galena and are thus Cu-Zn-Pb-Ba rather than Cu-Zn or Zn-Cu deposits. Stratiform barite deposits in these older greenstone belts, both in the Pilbara and at Barberton, may be related to VMS deposits, but are more likely to be replaced evaporites (Buick and Dunlop, 1990).

Younger analogues

Most post-Archean VMS deposits are Paleozoic (e.g. Iberian Pyrite Belt, Norwegian Caledonides, Tasman Orogenic Belt of Australia) or younger (Cyprus deposits, Kuroko deposits of Japan), although Proterozoic examples occur at Flin Flon–Snow Lake, Manitoba (Syme and Bailes, 1993), and Jerome, Arizona (Gustin, 1990). Excellent general reviews of the Australian Paleozoic deposits and Kuroko deposits are provided by Large (1992) and Ohmoto and Skinner (1983), respectively.

All post-Archean deposits have been ascribed to subduction-related plate-tectonic settings (Sawkins, 1990). The deposit types are commonly subdivided into Cu-rich Cyprus-type deposits formed at oceanic or back-arc spreading ridges, Beshi-type Cu-Zn deposits formed in mafic volcanic sequences in the early stages of volcanic arc formation, and Kuroko-type Cu-Zu-Pb ± Ag ± Au deposits formed in felsic volcanic sequences in the later stages of volcanic ore development. Today, hydrothermal vents producing modern equivalents of the VMS deposits occur at many sites along spreading centres (Rona, 1988; Hannington and Scott, 1988), and these provide the basis for current genetic models, although many of the physical and chemical parameters of ore formation were already well known before their discovery (Stanton, 1972; Solomon and Walshe, 1979). Most modern seafloor hydrothermal systems occur at mid-ocean ridges or transform zones, and their products are unlikely to be preserved, but recent discoveries in the back-arc environments of the Woodlark Basin, Papua New Guinea (Binns and Scott, 1993) are more likely to be better analogues of ancient VMS deposits.

Genetic models

Due to the occurrence of modern analogues, there is now reasonable consensus on genetic models for VMS deposits, as reviewed by Lydon (1988). Most models invoke the convective circulation of seawater, as proposed by Spooner and Fyfe (1973) and modified by Cathles (1983), with strong evidence from oxygen isotope studies for high fluid/rock ratios around subvolcanic heaters (Cathles, 1993). Most authors invoke leaching of metals from the volcanic pile, although metal contributions from magmas in subvolcanic plutons can not be ruled out (Sawkins, 1990). In cases where there are extensive originally near-vertical alteration pipes below the VMS deposits, advection of fluids by synvolcanic fault zones appears likely, although lateral flow along more permeable aquifers is indicated in some cases by extensive zones of stratiform alteration. The final form and compositions of the VMS deposits are probably controlled by such factors as the depth of seawater, composition of volcanic rocks, and temperature, fS_2, and fO_2 of ore fluids, as discussed by Large (1992) The rapid oxidation of sulfide chimneys precipitated by the so-called black smokers of the seafloor today (Alt et al., 1987) requires that either massive sulfides formed below a carapace of early-precipitated hydrothermal products, the mound theory (see Large, 1992, for discussion), and/or were rapidly covered by subsequent volcanic flows or sediments.

Komatiite-associated nickel deposits

Deposit characteristics and associations

As noted above, the major nickel camps associated with komatiites in Archean terrains occur in the Yilgarn Block of Western Australia, and the following review is based primarily on these deposits which represents a significant proportion of world total identified nickel sulfide resources in deposits with ≥0.8% Ni, although similar deposits occur in the Superior Province (summarised by Fyon et al., 1992), in Zimbabwe (Williams, 1979), and in Brazil (Brenner et al., 1990). Previous summaries of these deposits have been provided by Groves and Hudson (1981), Martin et al. (1981), Naldrett (1981), and Ross and Travis (1981), and Marston et al. (1981), and Lesher (1989) provides the most recent exhaustive review.

Lesher (1989) subdivided the deposits into two major groups, komatiitic peridotite-hosted deposits and komatiitic dunite-hosted deposits. The former are mainly stratiform deposits with generally low tonnages ($1–5 \times 10^6$ tonnes) of high-grade (2–4% Ni head grades) massive and matrix sulfides at the base of komatiitic peridotite flows, although rare stratabound deposits of disseminated or blebby sulfides may be associated with them. Most of the mined deposits in Western Australia, for example at Kambalda, Widgiemooltha and Windarra, are of this type, as are the massive ores at Agnew (Barnes et al., 1988a) and probably the Forrestania deposits (Perring et al., 1993), and the deposits in Brazil, Canada and Zimbabwe. The komatiitic dunite-hosted deposits, on the other hand, are large (up to 3×10^8 tonnes) stratabound deposits of low-grade (<1% Ni) finely disseminated sulfides within komatiitic dunites interpreted to be thick dunitic flows representing lava channels or lakes (Hill et al., 1987; Barnes et al., 1988a). Examples include the disseminated ores at Agnew and the giant Mt. Keith deposit, both in the northern part of the Norseman-Wiluna Belt (Fig. 5).

Komatiite-hosted nickel deposits are heterogeneously distributed on a craton scale (as discussed above), and on an intracratonic scale as shown by the selective concentrations of deposits within the Norseman–Wiluna Belt of the Yilgarn Block. They are also heterogeneously distributed in terms of deposit styles, with komatiitic peridotite-hosted deposits distributed throughout the Norseman–Wiluna Belt, although concentrated in the south, whereas the komatiitic dunite-associated nickel deposits are virtually restricted to the northern part of the belt (Fig. 5). As with the VMS deposits, where precise dating is available, the majority of deposits occur in greenstone belts that are younger than 2750 Ma, although the age of the Forrestania deposits is still not clear. The older >3000 Ma greenstone belts contain komatiites but show a remarkable lack of komatiite-hosted nickel deposits, with the Ruth Well deposit of the western Pilbara Block (Nisbet and Chinner, 1981) being the only significant deposit.

Most mineralized komatiite sequences have a systematic upwards symmetry, varying from thick (10s–100s m) komatiitic peridotites or dunites at the base through thinner, aphyric and spinifex-textured komatiite flows to massive or pillowed komatiitic basalt flows at the top. There may also be lateral variations from komatiitic dunites into komatiitic peridotites or porphyritic komatiites. The komatiite sequence is commonly more complex above ore zones, implying a volcanic control on ore localisation (Gresham and Loftus-Hills, 1981), and lacks interflow sedimentary units which are common in flanking environments (Fig. 6). No unequivocal volcanic feeder dykes have been recorded in the footwall rocks, which vary between different deposits, with mafic to felsic volcanic rocks or BIF present in specific cases.

Host komatiitic peridotites and dunites are highly magnesian, olivine-enriched rocks which are normally confined to footwall embayments and may show complex intraflow relationships within these structures (Fig. 7). The footwall embayments vary from broad shallow depressions through discontinuous re-entrant troughs (the classic Kambalda deposits: Fig. 6) to deep transgressive troughs (Agnew). In combination, the nature of the hosting komatiite units and the embayments are taken to indicate that the host komatiites represent dynamic lava channels (Lesher et al., 1984; Lesher and Groves, 1986; Donaldson et al., 1986; Barnes et al., 1988b; Lesher and Campbell, 1993) confined to lows in the topography (Lesher et al., 1984; Evans et al., 1989) accentuated by thermal erosion (Huppert et al., 1984), and subsequently modified by deformation (Barrett et al., 1977; Cowden and Archibald, 1987).

Typical profiles of stratiform ores consist of a thin, commonly discontinuous, massive sulfide (>80% sulfides) layer which directly overlies footwall rocks and is overlain by a thick, more continuous layer of matrix sulfides (40–80% sulfides)

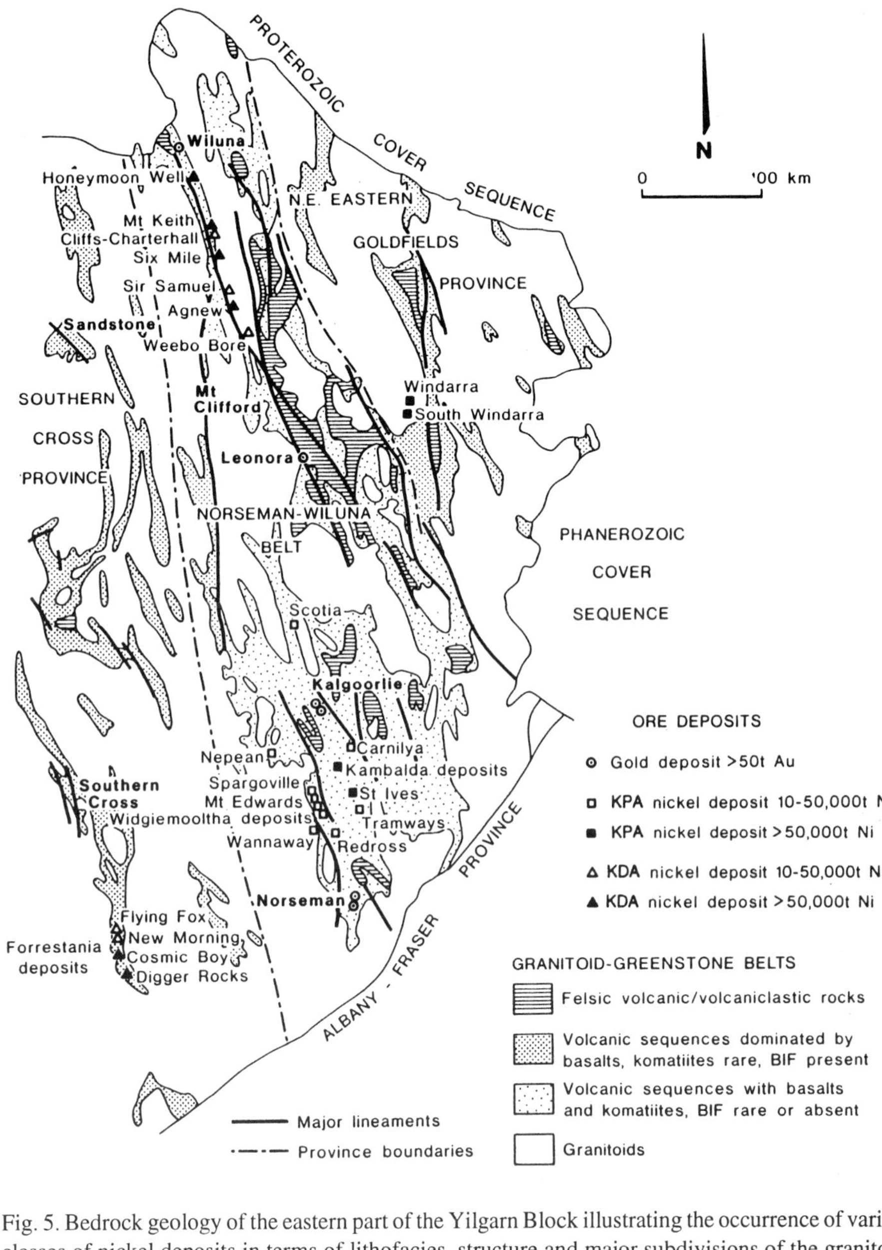

Fig. 5. Bedrock geology of the eastern part of the Yilgarn Block illustrating the occurrence of various classes of nickel deposits in terms of lithofacies, structure and major subdivisions of the granitoid-greenstone terrains. Adapted from Groves et al. (1984).

and disseminated sulfides (10–40%). The sulfide ores comprise pyrrhotite, pentlandite, pyrite, chalcopyrite, ferrochromite and magnetite. Massive ores may be layered and show mineral zonation or may be truly massive. Most of the

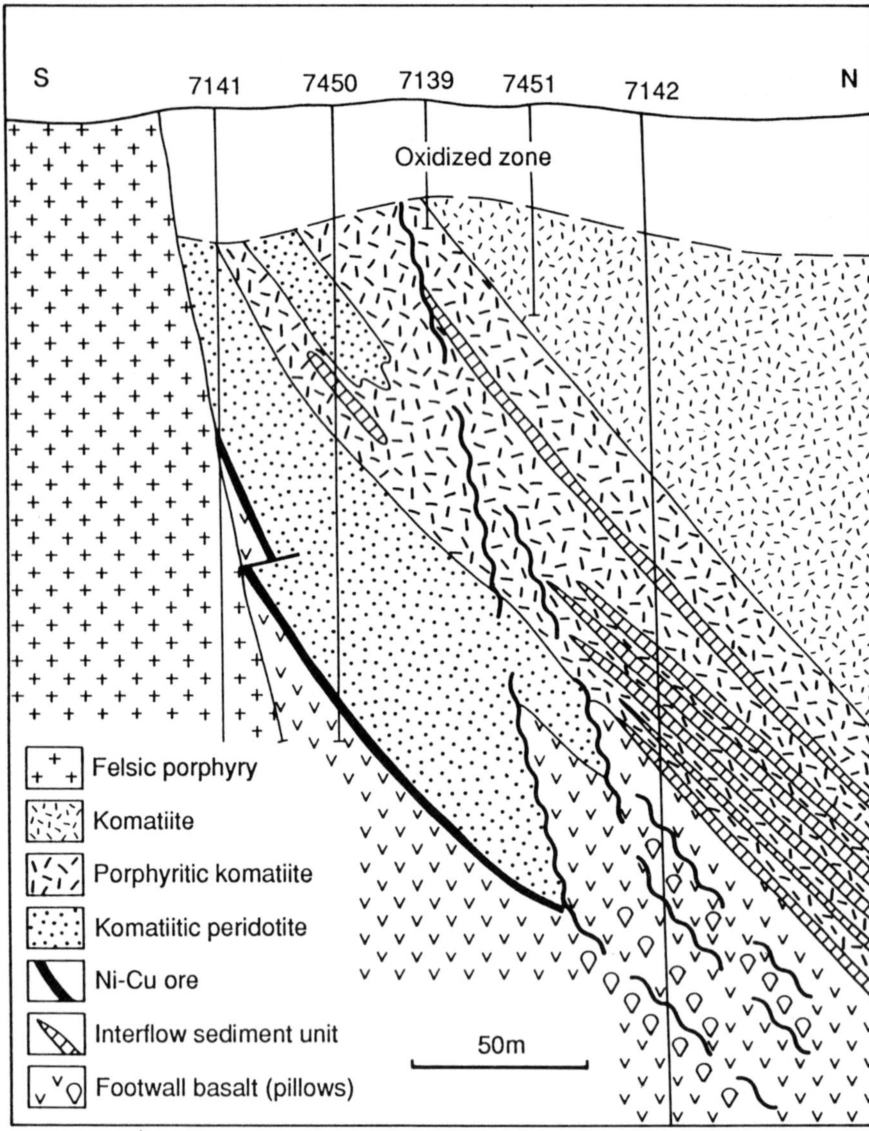

Fig. 6. Cross section of Durkin Shoot, Kambalda, Western Australia. After Lesher (1989).

structures and textures are due to metamorphism and deformation (Barrett et al., 1977). In contrast, the stratabound mineralization in komatiitic dunites comprises extensive disseminated sulfides, commonly interstitial to olivine or olivine pseudomorphs, which may form zones over 1 km long, at least 1 km deep, and 100 to 300 m wide of >1% Ni grade (Fig. 8). In these deposits, the ore mineralogy

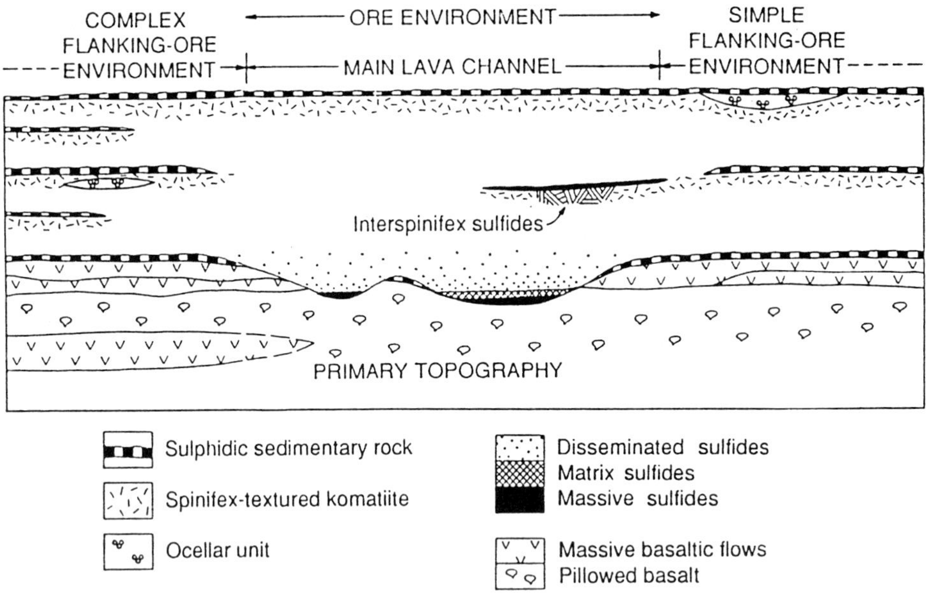

Fig. 7. Schematic section showing the geometry and facies distribution of the Kambalda ore environment at the time of ore formation. Adapted from Lesher et al. (1984), Groves et al. (1986), and Frost and Groves (1989).

everywhere has been severely modified by physiochemical changes imposed by reactions between metamorphic fluids and the dominant silicate rocks to produce serpentinites or talc–carbonate rocks (Eckstrand, 1975; Groves and Keays, 1979).

Despite the metamorphic overprint, most geochemical parameters of the ores are consistent with a magmatic origin (Keays et al., 1981; Cowden et al., 1986), although adjacent ore shoots may show different average compositions and there may be variations within individual shoots (Marston and Kay, 1980; Cowden and Woolrich, 1987). Sulfur isotopes and S/Se ratios (Seccombe et al., 1978; Groves et al., 1979) suggest links between the sulfur sources for the interflow sulfide-rich sedimentary units and the nickel ores. Direct evidence for assimilation of the sulfidic sedimentary rocks is provided by the occurrence of ocellar komatiites which represent unmixing of immiscible felsic melts derived from melting of the sedimentary rocks (Frost and Groves, 1989). A reconstruction of the channel environment is shown in Fig. 7.

Younger analogues

Unlike the VMS deposits, there are no modern, or even Phanerozoic, volcanic analogues of the komatiite-hosted nickel deposits, and the komatiites of Gorgona Island (e.g. Echeverria, 1982) are the only well documented Phanerozoic examples of this magma series.

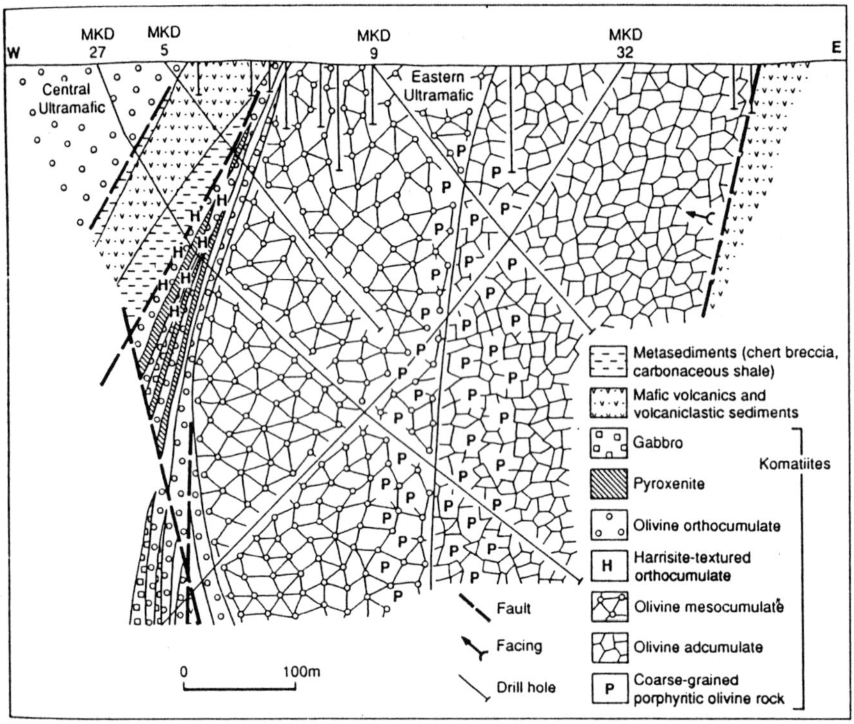

MKD 27
MKD 5
MKD 9
MKD 32
W
E
Central Ultramafic
Eastern Ultramafic
Metasediments (chert breccia, carbonaceous shale)
Mafic volcanics and volcaniclastic sediments
Gabbro
Pyroxenite
Olivine orthocumulate
Harrisite-textured orthocumulate
Olivine mesocumulate
Olivine adcumulate
Coarse-grained porphyritic olivine rock
Komatiites
Fault
Facing
Drill hole
0 100m

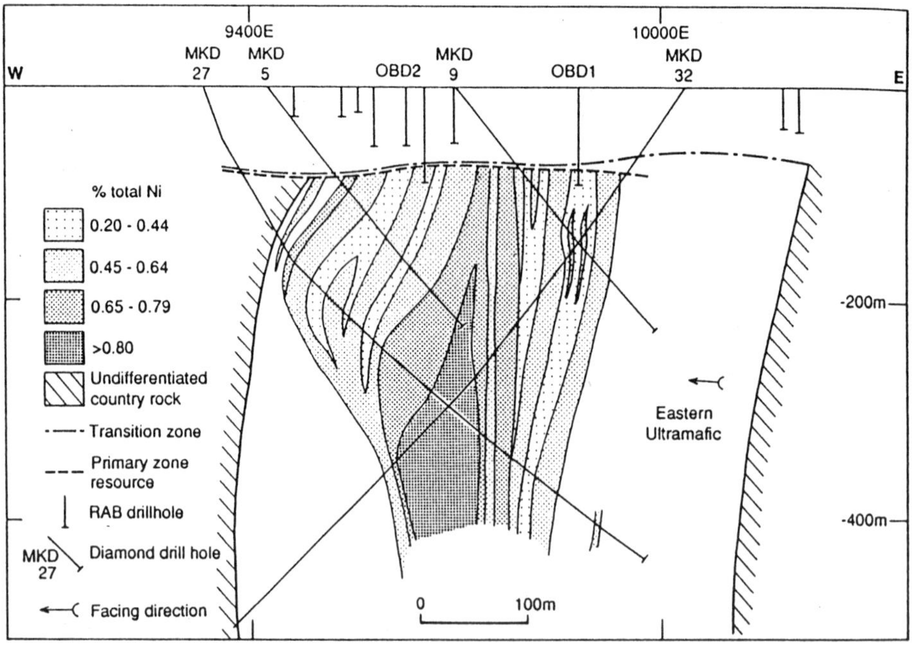

9400E
10000E
MKD 27
MKD 5
OBD2
MKD 9
OBD1
MKD 32
W
E
% total Ni
0.20 - 0.44
0.45 - 0.64
0.65 - 0.79
>0.80
Undifferentiated country rock
Transition zone
Primary zone resource
RAB drillhole
MKD 27
Diamond drill hole
Facing direction
Eastern Ultramafic
-200m
-400m
0 100m

Proterozoic analogues occur in the Thompson, Manitoba, and Cape Smith (Quebec) belts of Canada. The deposits in the Thompson Belt show some similarities to the Archean komatiite-associated nickel ores, but are highly deformed in most cases and, although of major economic importance, contribute little to better understanding of this deposit class. However, the well-exposed Cape Smith deposits, which are at low metamorphic grade and low strain, show many features that strengthen the interpretation that olivine-enriched komatiite hosts to nickel mineralization formed in dynamic lava channels that focused large volumes of komatiite magma. These *ca* 1.96–1.92 Ga sequences (Parrish, 1989) represent rift sequences that have subsequently been thrust faulted but have not suffered the crustal melting typical of the Archean granitoid-greenstone terrains, and hence field evidence for lava channels is less equivocal (Gillies and Lesher, 1992).

Genetic models

Although hydrothermal models have been suggested (Lusk, 1976; Hopwood, 1981), most authors now accept a magmatic origin for the komatiite-hosted nickel ores. The most widely accepted model involves voluminous rapid eruption of turbulently convecting komatiites forming large to giant channelized flows that assimilate variable amounts of sulfide-rich footwall rocks during eruption. The lava channels are thought to have been focused in pre-existing topographic lows, possibly sea-floor grabens, in the substrate, but these were deepened and modified by thermal erosion due to continuous or periodic flow-through of very hot, turbulently convecting komatiite lava.

The originally sulfur-undersaturated flows are interpreted to have assimilated sulfide-rich sediments, thereby achieving early sulfur saturation. Immiscible sulfide-oxide liquid droplets scavenged chalcophile elements from the komatiite liquid and settled rapidly to the base of the flow where dynamic flow segregation and static buoyancy processes produced observed massive-matrix ore profiles: Lesher and Campbell (1993) present an integrated model to explain ore tenor variations in the resultant sulfides. In other cases, the flows assimilated smaller amounts of their substrate or more sulfur-deficient footwall rocks, reaching sulfur saturation later in their crystallization history and resulting in the stratabound disseminated ores. In all cases, exsolution of sulfide-oxide liquid and its subsequent concentration occurred downstream from magma conduits.

Opposite: Fig. 8. Cross section through the Mt Keith komatiitic dunite showing geology (A) and nickel distribution (B). Adapted from Dowling and Hill (1993).

Other mineralization styles

Iron deposits

Banded iron-formations (BIFs) are common components of Archean greenstone belts and metasedimentary terrains worldwide. They are generally of Algoma type and comprise a number of facies including oxide (magnetite), carbonate (siderite/ankerite), sulfide (pyrite/pyrrhotite), and silicate (greenalite/minnesotaite) facies (James, 1954), that formed in a variety of sedimentary environments (Kimberley, 1978). Although all facies may be present in greenstone belts of any age, there is a tendency for oxide-facies iron-formations to be more common in 3000 to 2900 Ma greenstone belts in Western Australia (Groves and Batt, 1984) and Canada (Fyon et al., 1992).

Banded iron-formations are virtually restricted to the Precambrian, but are generally thicker and more laterally extensive in Late Archean to early Proterozoic basins (Klein and Beukes, 1992) than in greenstone belts. The origin of the oxide-facies BIF units is contentious (Holland, 1984), although there is general agreement that they formed by chemical precipitation in sediment-starved environments under less oxidising atmosphere/hydrosphere conditions than in the Phanerozoic. The delicate banding, and its correlation over large distances (Trendall, 1983) in Late Archean to early Proterozoic, largely Superior type equivalents, suggests that it was related to cyclic sedimentation.

Most of the iron ore production from BIF comes from the giant Superior type deposits in Late Archean to early Proterozoic basins unconformably overlying basement rocks that may include the granitoid-greenstone terrains (e.g. Western Australia, Brazil), but iron ores have been mined from fold hinges where the BIF units are either thickened (unenriched ore; e.g. Bruce Lake, Ontario; Shklanka, 1970) and/or brecciated and enriched by dissolution of silica (e.g. Steep Rock and Wawa, Ontario; Shklanka, 1972: Koolyanobbing, Western Australia). Economically important ores include the Adams, Sherman and Moose Mountain deposits of the Superior Province.

Porphyry copper–molybdenum–gold deposits

Porphyry-style mineralization is best developed in Mesozoic to Recent convergent margins in association with shallow-level to subvolcanic tonalitic to granodioritic plutons (Beane and Titley, 1981), but there are numerous small or low-grade sub-economic or non-economic deposits developed in Archean greenstone belts: these are of geologic interest but are normally only economic where they lie within gold camps (e.g. McIntyre, Abitibi Belt) or have experienced supergene enrichment (e.g. Boddington, Western Australia).

Barley (1982) described a number of low-grade (<0.2% Cu, <0.1% Mo) porphyry-style deposits from the 3450–3300 Ma greenstone belts of the Pilbara Block. These are hosted by calc-alkaline felsic intrusions and associated with altered subaerial volcanics, and include the large tonnage but subeconomic grade

Coppins Gap deposit. However, the majority of the porphyry-style deposits appear to be associated with greenstone belts that are younger than 2750 Ma. In the Superior Province, there are more than ten examples that are mostly associated with synvolcanic to epizonal plutons of tonalite, trondjhemite or granodiorite emplaced into volcanic-plutonic complexes (Ayres and Cerny, 1982). However, some examples (e.g. McIntyre deposit) are associated with post-tectonic plutons. All deposits have stockwork mineralization of chalcopyrite- or molybdenite-bearing quartz veins in zones of potassic, phyllic and porpylitic alteration similar to those of Phanerozoic analogues. Minor porphyry-style mineralization may also occur in some VMS camps (Noranda; Goldie et al., 1979).

In recent years, there has been the recognition that the hypogene mineralization from which the giant laterite-hosted Boddington gold deposit was derived may be a porphyry Cu-Au-Mo system (Symons et al., 1988; Roth et al., 1991). This deposit, which is dated at *ca* 2690 Ma, appears to correspond to the diorite model of Hollister (1975) in terms of its metal associations and alteration style.

Sulfide-oxide deposits in mafic–ultramafic intrusions

As noted above, in addition to the komatiite-associated nickel deposits in flow sequences, there may also be small nickel deposits in both broadly coeval komatiitic and tholeiitic layered intrusions in Western Australia (Marston et al., 1981). In the Superior Province, chromite–sulfide layers may also occur in similar layered intrusions (e.g. Shebandowan; Whittaker, 1986), while elsewhere zoned gabbro-anorthosite sills may contain lower sulfide layers and upper magnetite–apatite–rutile layers (e.g. Dore Lake; Poulsen and Hodgson, 1984).

In addition, the *ca* 3500 Ma greenstone sequences at Sherugwe (Selukwe) in Zimbabwe contain a 1 km thick ultramafic unit which hosts lenses of chromite up to 300 m long and 10 m thick (Cotterill, 1969). These deposits, which are coarse-grained and podiform in nature and hence resemble podiform chromite deposits in ophiolites, have been Africa's most important chrome producers in the past, although they are now supplanted by the deposits of the Bushveld Complex and the Great Dyke.

SYN- TO POST-OROGENIC DEPOSITS IN GREENSTONE BELTS

Introduction

The major deposit style associated with the late orogenic stage of Archean greenstone evolution is lode-gold mineralization, which is more uniformly distributed in Australia, Brazil, Canada, India and Africa than either VMS or komatiite-hosted nickel mineralization, but is still relatively poorly developed in some Archean greenstone terrains such as those of Finland and the former USSR. The other main deposit style that developed late in greenstone belt evolution was pegmatite-associated rare-metal mineralization which is scattered over many Archean cratons.

Minor deposits that occur, at least in part, within Archean greenstone belts, but that are probably related to deformational and metamorphic processes during post-Archean events, include the Co-Ni-Ag ores of Cobalt, Ontario (Smyk and Watkinson, 1990), largely sited in Neoproterozoic sequences, and the Cu-Au deposits of Chibougamau, Quebec (Guha, 1984). These are not discussed further in this review.

Lode-gold deposits

Introduction

Due to the boom in gold mining in the past decade, there have been numerous papers devoted to individual deposits as well as to published reviews on Archean lode-gold deposits generally (Colvine et al., 1988; Colvine, 1989; Foster, 1989; Groves et al., 1989, 1992; Groves and Foster, 1991; Foster and Piper, 1993) that provide far more data than can be condensed into this short review. Thus, only a brief overview of the major features of this diverse group of deposits is presented below, with particular emphasis on the Yilgarn Block where the authors have the most personal experience.

The Archean VMS and komatiite-associated deposits (discussed above) are clearly identified deposit types and genetic groups on the basis of similar host rocks, mineralization styles, and ore and alteration mineralogy. The Archean lode-gold deposits, in contrast, vary widely in their structural setting, host rocks, mineralization styles, and ore and alteration mineralogy, leading to a plethora of classification schemes and genetic models and to the real possibility that they may include several distinct types of deposit and genetic groups. However, there are enough similarities in the characteristics of the deposits to treat the majority of them as a coherent deposit group. These consistent features include: (i) their constant Au-Ag $\pm$ As $\pm$ Sb $\pm$ Te $\pm$ W $\pm$ B association with generally low Cu-Zn-Pb, (ii) the similar *chemistry* (addition of CO_2-K $\pm$ Na) of their proximal alteration zones, (iii) the common low-salinity H_2O-CO_2 $\pm$ CH_4 fluid from which they were deposited, (iv) their ubiquitous structural control, and (v) their late-orogenic timing. For the purposes of this review, the deposits are considered as a coherent group, such that the characteristics of individual deposits can be integrated into an all-embracing model that reveals the four-dimensional architecture of the deposit group in a comparable way to the initial advance in understanding the porphyry–copper deposits of the continental USA created by the integrated model of Lowell and Guilbert (1970).

Deposit characteristics and associations

The lode-gold deposits occur in greenstone belts of all ages, although, like the VMS and komatiite-associated nickel deposits, the most highly mineralized belts containing the giant deposits, for example the Abitibi and Norseman–Wiluna belts, are mainly younger than 2750 Ma old. The greenstone belts of the Barberton Mountain Land are the oldest to host world-class gold deposits (de Ronde et al., 1992).

The major gold provinces are broadly associated with elongate granitoid-greenstone terrains which have a high density of craton- or terrain-scale deformation zones that are high-strain flattening or shear zones in otherwise low-strain greenstone terrains: examples include the Abitibi and Norseman–Wiluna Belts. Terrains dominated by more domal granitoids and intervening stellate greenstones with more widely spaced deformation zones (e.g. Pilbara Block) tend to be more poorly mineralized, although the Zimbabwe greenstone terrains are relatively well mineralized (Foster, 1989).

On the goldfield (or camp) scale (Fig. 9), the lode-gold deposits show a strong structural control by lower order faults or shear zones that in places show a geometrical relationship to first-order deformation zones (Eisenlohr et al., 1989). There are no unique structural controls, with strike-slip, oblique-slip, normal, thrust and high-angle faults or shear zones all hosting gold mineralization in different terrains at different crustal levels (Hodgson, 1989). Reactivation of pre-existing faults is probably common (Sibson et al., 1988), and heterogeneities, such as irregular granitoid contacts or earlier fold or fault geometries, locally controlled fluid focusing and gold mineralization (Ojala et al., 1993).

Similarly, there are no unique host rocks to mineralization, with all components of the granitoid-greenstone terrains being mineralized in particular provinces or goldfields. Despite this, particular host rocks or lithological contacts between well defined host rocks are commonly selectively mineralized within specific greenstones or their component goldfields. For example, layered mafic intrusions, and particularly the granophyric units within them, are selectively mineralized in the Kambalda–Kalgoorlie segment of the Norseman–Wiluna Belt (Phillips and Groves, 1983), whereas BIFs are selectively mineralized in the Geraldton area of Ontario (Macdonald, 1988). Both these hosts have high Fe contents and high Fe/Fe $\pm$ Mg ratios, implying chemical controls for these rock types. Rheological contrasts between lithological units also appear to be important controls on mineralization, and thus different units may control mineralization in contrasting lithostratigraphic packages and at different crustal levels.

Considerable interest has been generated in the potential association of specific igneous rocks with the lode-gold deposits, and associations with felsic porphyry intrusions or small tonalite–trondjhemite plutons are commonly stressed (Burrows and Spooner, 1986; Hodgson and Troop, 1988). In many mineralized environments representing relatively high crustal levels, there are swarms of felsic porphyry dykes and sills (Perring et al., 1988), with or without shoshonitic lamprophyre intrusions (Rock and Groves, 1988; Wyman and Kerrich, 1989). However, there is no unique association between specific igneous rock suites and gold deposits, and none of the suites shows intrinsic gold enrichment (Wyman and

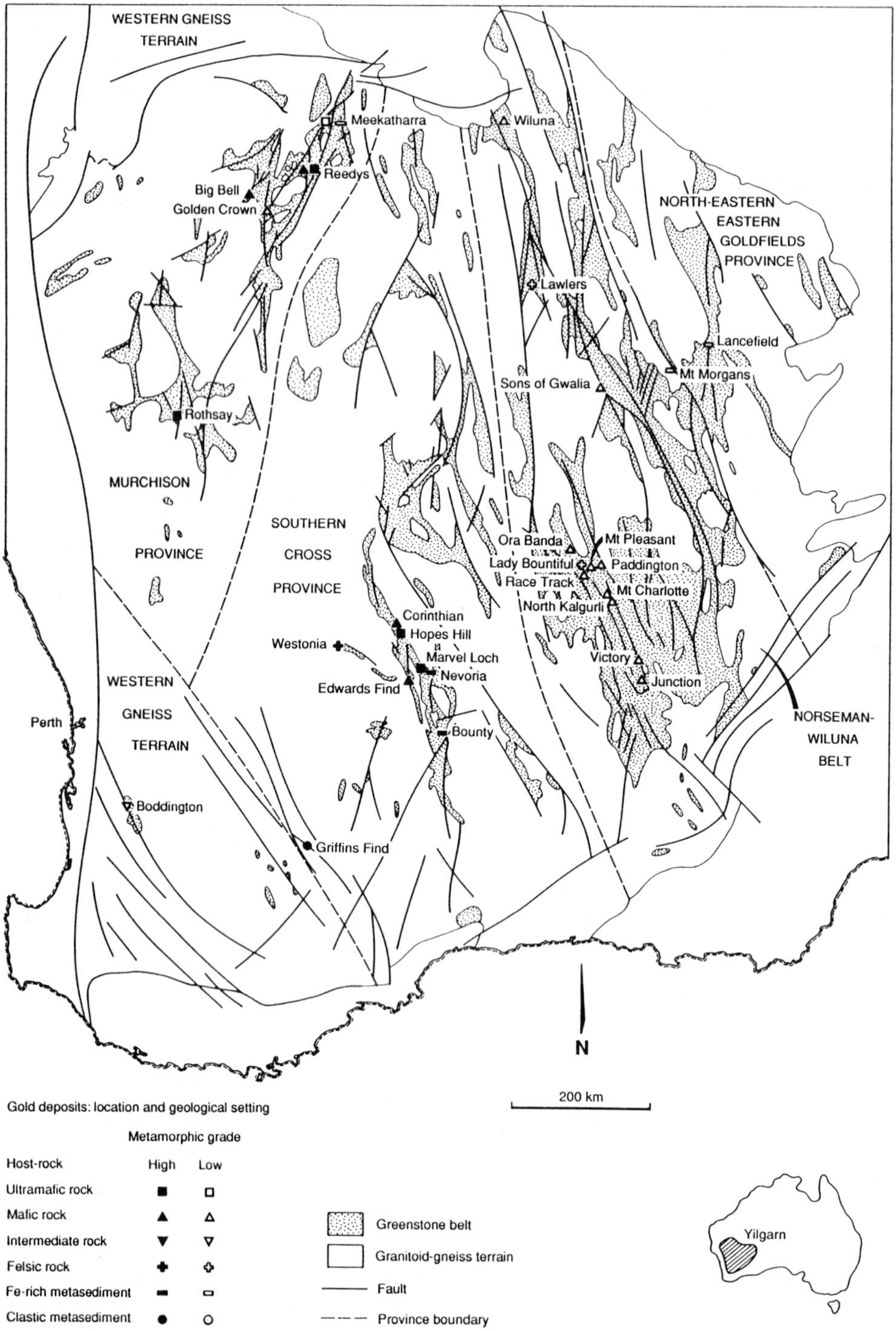
WESTERN GNEISS TERRAIN
Meekatharra
Wiluna
Reedys
Big Bell
Golden Crown
NORTH-EASTERN
EASTERN
GOLDFIELDS
PROVINCE
Lawlers
Lancefield
Mt Morgans
Sons of Gwalia
Rothsay
MURCHISON
PROVINCE
SOUTHERN
CROSS
PROVINCE
Ora Banda
Mt Pleasant
Lady Bountiful
Paddington
Race Track
Mt Charlotte
North Kalgurli
Corinthian
Hopes Hill
Westonia
Marvel Loch
Nevoria
Edwards Find
Victory
Junction
WESTERN
GNEISS
TERRAIN
Perth
Bounty
NORSEMAN-
WILUNA
BELT
Boddington
Griffins Find
N
200 km
Gold deposits: location and geological setting
Metamorphic grade
Host-rock
High
Low
Ultramafic rock
Mafic rock
Intermediate rock
Felsic rock
Fe-rich metasediment
Clastic metasediment
Greenstone belt
Granitoid-gneiss terrain
Fault
Province boundary
Yilgarn

Kerrich, 1989; Taylor et al., 1994). Pegmatites instead of porphyries are the common associates of gold mineralization at deeper crustal levels.

Most gold production has come from middle to upper greenschist facies metamorphic terrains which contain many of the very large to giant deposits (e.g. Timmins, Ontario; Kalgoorlie, Western Australia), and descriptions of these so-called mesothermal deposits dominate the literature. This metamorphic regime corresponds broadly to the brittle–ductile transition (Sibson et al., 1975), where gold deposits may be selectively developed. However, recent research has shown that Archean lode-gold deposits occur in terrains that range from sub-greenschist facies (e.g. Wiluna, Western Australia; Commonor, Zimbabwe) to lower granulite facies (e.g. Griffins Find, Western Australia; Renco, Zimbabwe), and that it is partly this variation in the P-T conditions of fluid advection and gold deposition that results in the extreme variability of deposit styles within the group.

Following the initial models of Colvine (1989) and Barnicoat et al. 1991), Groves et al. (1992) proposed that Archean lode-gold deposits represent a crustal continuum with deposits deposited over a range of P-T conditions from about 200°C at <1 kb to about 700°C at 5 kb. The deposit styles show a corresponding gradational change from brittle vein sets and/or breccias at shallow levels to more deformed vein systems hosted by more ductile shear zones at deep levels (Fig. 10). Proximal wallrock alteration assemblages show a complementary change from white mica-ankerite, through biotite (or phlogopite)-ankerite ± albite and biotite-amphibole to biotite (or K feldspar)-clinopyroxene ± garnet assemblages with increasing depth (Mueller and Groves, 1991; Witt, 1991). At high crustal levels, the ore fluids are well-documented as H_2O-CO_2 ± CH_4 fluids with 10–25 mole % CO_2, low salinity, and moderate density (Ho, 1987; Robert and Kelly, 1987; Clark et al., 1989), and reconnaissance studies indicate that similar, although possibly more complex, fluids deposited the vein minerals formed under higher P-T conditions.

Most lode-gold deposits in greenschist facies terrains formed after peak metamorphism of the greenstone belts (Clark et al., 1989), but deposits in higher grade metamorphic environments normally formed at, or close to, the metamorphic peak (Barnicoat et al., 1991). In the Yilgarn Block, available precise geochronology suggests that gold mineralization was broadly contemporaneous at *ca* 2640 to 2630 Ma irrespective of the age of hosting greenstone sequences (Wang et al., 1993). However, in the Superior Province, it was apparently diachronous (see summary in Fyon et al., 1992), with gold mineralization related to orogenic events between about 2860 Ma and 2600 Ma in different provinces. A major problem in defining the precise age of mineralization in any one deposit or goldfield is the

---

Opposite: Fig. 9. Schematic map of the Yilgarn Block, Western Australia, showing the distribution of the major provinces, major deformation zones, and well studied gold deposits. The gold deposits are shown in terms of their host rocks and the metamorphic grade of host sequences (low = greenschist–amphibolite transition and below : high = lower amphibolite to lower granulite facies) to show the extreme variations in the environments of deposition of the gold deposits.

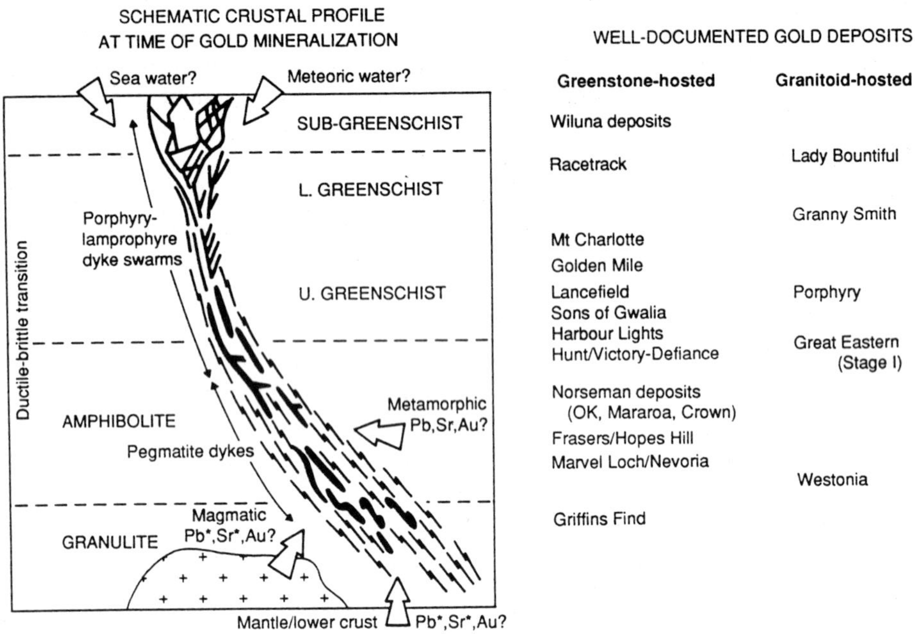

Fig. 10. Schematic reconstruction of a hypothetical continuous auriferous hydrothermal system extending over a crustal range of >20 km, showing potential fluid and solute sources. Schematic hydrothermal systems for Archean lode-gold deposits is taken from Groves et al. (1992). Note that the continuous section is derived from a study of deposits in discrete areas and that the deposits may not all occur in one vertical profile.

variable ages given by various techniques such as U-Pb in rutile, Pb-Pb mineral isochrons, and $^{40}Ar/^{39}Ar$ ratios of micas. The interpretation of zircons in hydrothermal veins as hydrothermal or relict phases is also a problem (Jemielita et al., 1990). Despite this uncertainty, most authors conclude that gold mineralization occurred during the waning stages of accretionary orogenic events (Barley et al., 1989, Kerrich and Wyman, 1990; Fyon et al., 1992), and most of the major gold provinces were developed in the Late Archean, broadly synchronous with deformation and metamorphism in both the upper and lower crust (Colvine, 1989) prior to final cratonization.

Younger analogues

Similar deposit styles occur in early Proterozoic greenstone belts such as those in Ghana (Eisenlohr, 1989), which host the very large Ashanti deposit, and in parts of Brazil where there are still uncertainties on the ages of greenstone belts and mineralization. Archean lode-gold deposits also show similarities to those sited in Paleozoic turbidite or slate belts, where structurally controlled mesothermal deposits also formed from H_2O-$CO_2 \pm CH_4$ fluids during deformation (Peters, 1993). Analogies to these deposits can help constrain the processes important to fluid advection and gold deposition.

However, potentially more can be gained from the viewpoint of understanding the broad tectonic setting and ore-forming processes by comparison with the gold province of the Pacific Rim, as this contains the largest repositories of gold since the Late Archean mineralization: this is discussed further below. Although the Pacific Rim mainly hosts large porphyry-style to epithermal gold (± copper ± silver) deposits (Sillitoe, 1989), there are also mesothermal deposits in the North American Cordillera which are strikingly similar to the Archean lode-gold deposits (Nesbitt et al., 1986). In these terrains, gold mineralization was superimposed on deformed and metamorphosed sedimentary and igneous rocks synchronously with subduction or accretion of oceanic crust, and probably coincided with a change in plate motion which caused a shift from convergent to partly transcurrent tectonics (Goldfarb et al., 1991). Such a change may be recorded in Archean lode-gold deposits by the reactivation of earlier structures (Sibson et al., 1988) and/or by the change from one structural regime to another during mineralization (e.g. at Kalgoorlie; Mueller et al., 1988).

Genetic models

Following earlier syngenetic or modified syngenetic models for the Archean lode-gold deposits, particularly those hosted by BIF, there is now general consensus that the deposits are epigenetic and that the fluid and solute sources are external to the depositional environment, although pre-enriched source rocks are still invoked in some models: see the review of this problem by Hutchinson (1993).

The Late-Archean hydrothermal systems were clearly very large, with advection of overpressured, hot, low salinity H_2O-$CO_2 \pm CH_4$ fluids up vertically extensive shear zones and/or fault-induced permeability, more-or-less synchronously over areally extensive terrains. Gold, transported as reduced sulphur complexes (Phillips and Groves, 1983), was deposited via fluid/rock reactions (e.g. sulfidation) or phase separation in structural traps, possibly zones of low mean stress (Ridley, 1993), in rocks or contact zones of suitable rheology and/or chemistry. Many large to giant lode-gold deposits were deposited close to the brittle–ductile transition, but gold was deposited from >20 km to <5 km depths at P-T conditions between 700°C and 5 Kb and 200°C at <1 Kb. At upper crustal levels, advecting ore fluids derived from a deep source apparently mixed with surface water (Gebre-Marian et al., 1993), with fluid mixing as a potential gold-deposition mechanism.

The vertically extensive fluid conduits tapped several solute reservoirs, such that stable isotope compositions of ore-related minerals, and the inferred fluids that deposited them, show a wide spread and do not reflect a single source region (Kerrich, 1987). Radiogenic isotopes (King and Kerrich, 1989; McNaughton et al., 1990; Mueller et al., 1991) rule out the greenstone belts, I-type granitoids or porphyry/lamprophyre complexes as the *sole* sources of solutes, but imply that anatectic granitoids and/or their parent rocks were the sources of some solutes. Available data thus favour a sub-greenstone source for at least some of the ore fluid and solute components, but the precise nature of this source (or sources) is debated. The gold depositional environments are commonly characterised by the broad conjunction of reactivated deformation zones, granitic plutons, felsic porphyry and lamprophyre dyke swarms, as well as repeated hydrothermal activity (Kerrich and Wyman, 1990), thus obscuring the precise source and association(s) of the ore fluids. As suggested by Fyon et al. 1989) for the Superior Province, and in agreement with radiogenic isotope data and timing relationships in the Yilgarn Craton (Barnicoat et al., 1991), however, gold mineralization appears linked to metamorphic–magmatic processes in the lower to middle crust.

Rare element pegmatites

Rare element pegmatites are widely distributed through space (e.g. Superior Province and Slave Province, Canada; Yilgarn Block, Australia; Sao Francisco Craton, Brazil; Zimbabwe Craton) and time, and some occur in even the oldest greenstone belts (e.g. Pilbara Block; Blockley, 1980). but the first important and widespread pegmatite formation occurred between 2800 and 2600 Ma (Cerny, 1982). This is broadly the same Late-Archean period in which the world class VMS, komatiite-associated nickel and lode-gold deposits developed.

Most of the economically important Li ± Cs ± Ta ± Sn pegmatites occur in low-grade (greenschist facies) greenstone or metasedimentary belts; Fyon et al. (1992) provide a comprehensive review of such pegmatites in Ontario, Canada. The pegmatites are complex, with well-developed zoning, and contain an abundance of different minerals (Cerny, 1982). The Tanco deposit, near the Manitoba-Ontario border, is one of the larger and more typical examples, and is a subhorizontal 1.5 × 1.0 × 0.1 km body with significant resources of Ta (mainly tantalite), Li (spodumene) and Cs (pollucite). It is one of the most fractionated pegmatites known, and apparently crystallized from the walls inwards (Cerny, 1982). There appears to be a genetic relationship with pegmatitic granites in the area. Similar, shallowly dipping pegmatites with 2 km strike length and up to 60 m thickness occur at Bikita, Zimbabwe. These irregularly zoned pegmatites are worked for Li (petalite, lepidolite, spodumene) and Cs (pollucite) with minor Be (beryl), and have greisenized margins containing tantalite and cassiterite (Cooper, 1964). Rare-element pegmatites are also described for the dominantly metasedimentary terrains of the Slave Province of Canada, where the pegmatites, as in Ontario, appear to be related to two-mica granites (Poulsen et al., 1992).

The Greenbushes Pegmatite (Fig. 11), in a probably high-grade greenstone remnant in the Western Gneiss Terrain of the Yilgarn Block, is arguably the largest deposit of this type with resources of 13.5 m tonnes of 590 ppm Ta_2O_5, 400 ppm Nb_2O_5 and 0.15% Sn (up to 40 mt at lower cut-off grades) and 5.9 m tonnes at 4.0 % Li_2O in spodumene within 28.3 m tonnes at 2.8% Li_2O (Hatcher and

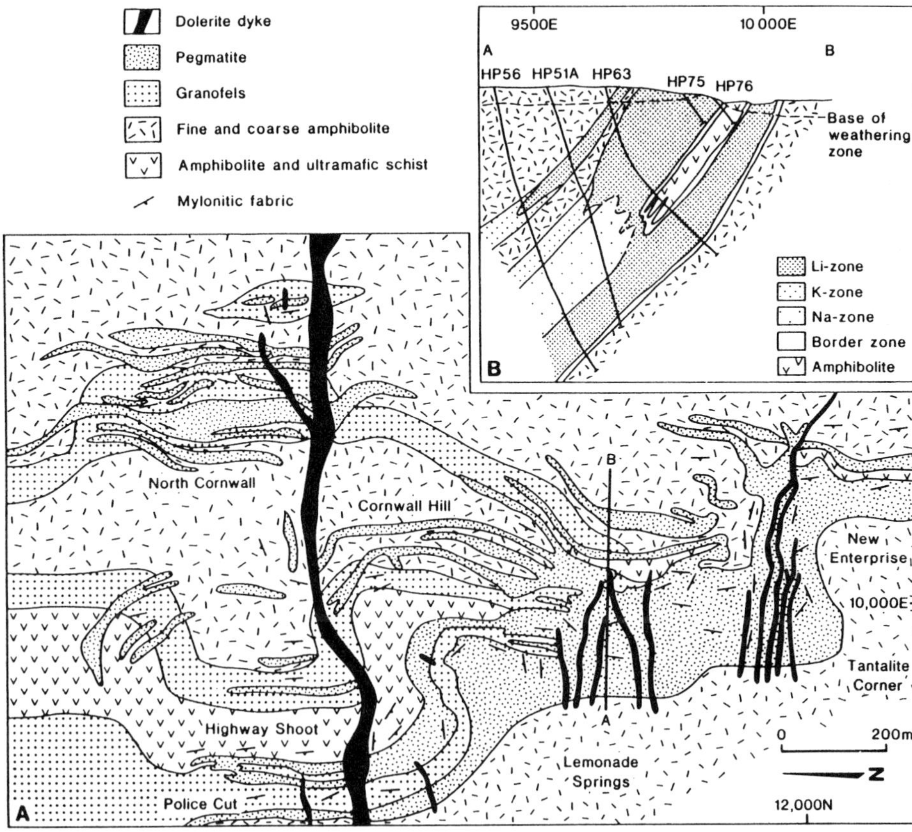

Fig. 11. Greenbushes pegmatite, Balingup Complex, approximately 250 km south of Perth, Western Australia. A. geological plan of the northern part of the pegmatite showing its complex shape and relationship to country rocks. B. Section at 12380N showing zoning in the pegmatite. Modified from Hatcher and Elliot (1986).

Elliot, 1986). Like the other pegmatites, the body is shallowly dipping, very large (at least $3 \times 0.4 \times 0.2$ km), and is quite strongly zoned (Partington, 1990). However, it is anomalous in a number of respects: (i) at *ca* 2530 Ma, it is younger than most of the other Late-Archean pegmatites, (ii) its ore-grade Li-Ta-Sn association and the occurrence of spodumere as the major Li mineral are unusual, (iii) there is no association with contemporaneous specialized granites, (iv) it is synkinematic rather than postkinematic in timing, and (v) it occurs in a high-grade polymetamorphic gneiss terrain rather than a lower-grade greenstone belt.

Fyon et al. (1992) presented a synthesis of depositional and tectonic models for the rare-element pegmatites in Ontario, suggesting a collisional environment for the generation of parent granites, but it is unclear how the giant Greenbushes Pegmatite fits into such a model.

METALLOGENIC SYNTHESIS

Tectonics of Late Archean terrains

The nature of the global tectonic regime and its relationship to metallogeny during the Archean is still the subject of debate. Most current models for Archean tectonics focus on the evolution of granitoid-greenstone terrains, and favour either convergent plate tectonic settings or a non-plate tectonic regime for their development. However, any model for Archean tectonics must be capable of explaining the features of the full range of Archean terrains and their mineralization. In order to achieve this it is necessary to place magmatic, orogenic, and mineralization "events" in a variety of environments in widely-spaced terrains in a better chronostratigraphic framework than has been possible previously. Sequence stratigraphy (Vail et al., 1977; van Wagoner et al., 1990) is the method used by the oil industry for global correlation of events in sedimentary basins. Krapez (1993) and Blake (1993) have recently demonstrated that it is possible to successfully apply sequence stratigraphic techniques to Archean terrains in Western Australia.

If applied carefully, sequence stratigraphy combined with precise geochronology can provide a superior method of unravelling the tectonic history of Archean basins and greenstone belts, and thus establish the chronostratigraphic correlations necessary for tectonic and metallogenic syntheses. In this section, the tectonic and metallogenic evolution of some key, well-mineralized, Late-Archean terrains is analysed using a sequence stratigraphic approach, and a unified model for Late-Archean metallogeny is developed. The metallogeny of earlier Archean terrains is then examined in terms of this model.

Graphs of the distribution of metallic mineralization through time (Fig. 12) show that several types of metal deposits have a peak in abundance or preservation between 3.0 and 2.5 Ga during the Late Archean (Meyer, 1988; Barley and Groves, 1992). As discussed above, there is a particular concentration of major ore deposits in the latter part of this period (Fig. 13). Some of these metal deposit types such as VMS and mesothermal gold, which form at convergent plate margins and in back-arc basins, also have peaks in the Paleozoic and Mesozoic–Cenozoic. As shown in previous sections, the Archean deposits are similar in most respects to their modern counterparts. The Late Archean also marks the first appearance, or preservation, of large intercontinental and continental margin basins such as the Hamersley and Witwatersrand Basins, in the rock record. The tectonic setting of these basins is easier to constrain than that of granitoid-greenstone terrains and, as they also contain important gold and iron mineralization, they are of critical importance to Archean tectonic and metallogenic syntheses. In order to discuss the tectonic setting of Late-Archean terrains, it is necessary to concentrate on those with adequate high-precision geochronology. For this reason, the Hamersley Basin and Yilgarn Craton of Western Australia and the Superior Province of Canada are described below (see Fig. 14).

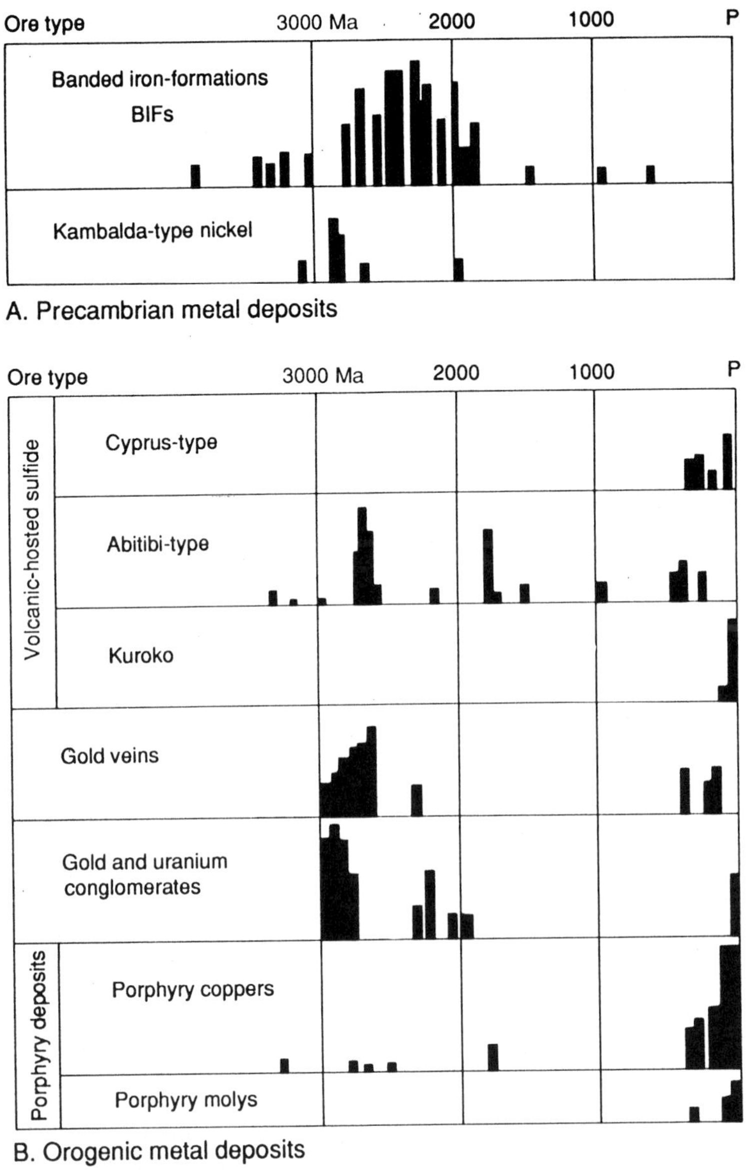

Fig. 12. Distribution through time of styles of mineral deposits discussed in the text (from Barley and Groves, 1992). Width of each vertical bar represents interval of ~50 m.y. as compared with total estimated tonnage for that style of deposit through geologic time. Data have been taken from Meyer (1988); ages of Archean gold mineralization are corrected to reflect new geochronological data.

The Hamersley Basin in northwestern Australia developed on the Pilbara granitoid-greenstone province in the Late Archean. It thus formed at the same time as some granitoid-greenstone terrains that have abundant mineralization. Blake

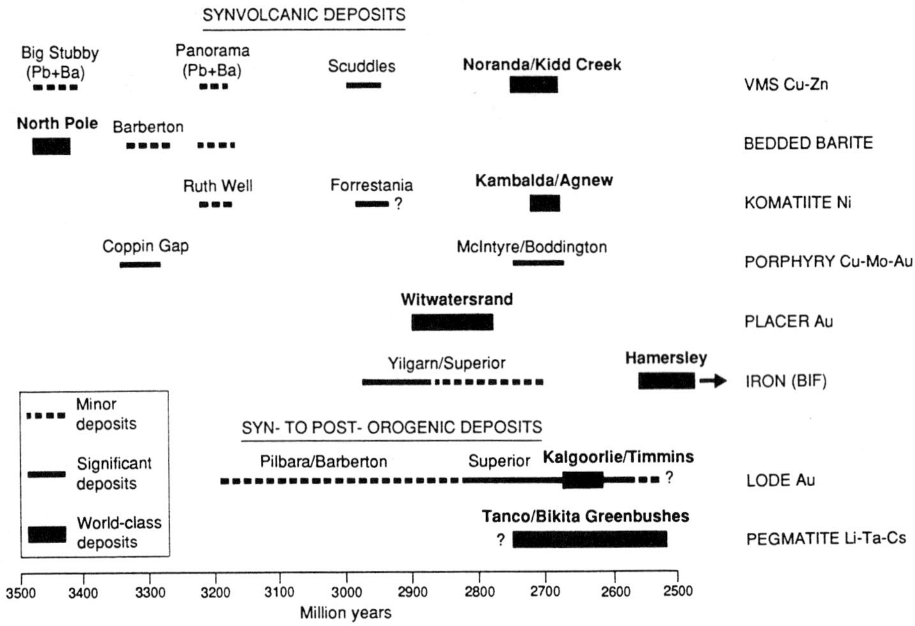

Fig. 13. Timing of major classes of Archean mineral deposits showing major metallogeny at 2700 ± 100 Ma. Note that iron ores (BIF) are included although it is realised that many are post-Archean and that enriched iron ores developed in the Proterozoic.

and Barley (1992) have recently interpreted a sequence stratigraphy for the fill of the Hamersley Basin, the Mt Bruce Megasequence Set (see terminology in van Wagoner et al. 1990). It contains two supersequences of flood basalts and terrigenous sedimentary rocks (at *ca* 2.78–2.76 and *ca* 2.74 to 2.7 Ga), overlain by a passive margin supersequence (at *ca* 2.69 to 2.6 Ga). These supersequences comprise a megasequence that represents the rifting of a continent which contained the Pilbara Craton and the opening of an ocean to the west and south. Open ocean is represented by a lacuna or condensed section (from *ca* 2.6 to 2.49 Ga), followed by a continental back-arc basin supersequence (*ca* 2.49 to 2.44 Ga) containing the most important iron formations and a foreland basin supersequence. These supersequences comprise a megasequence which records the closure of an ocean to the north (prior to 2.49 Ga) and development of a convergent margin to the south of the present craton. Thus, the sequence stratigraphy of the Hamersley Basin suggests that a continent rifted in the Late Archean and that the Pilbara Craton crossed an ocean in much the same way that India has migrated out of Gondwanaland and into Asia since the Mesozoic. Importantly, there are continental basalts of the same age as those in the Hamersley Basin in Brazil and South Africa.

The tectonic evolution of the granitoid-greenstone terrain of the Superior Province in Canada is better constrained by high precision geochronology than

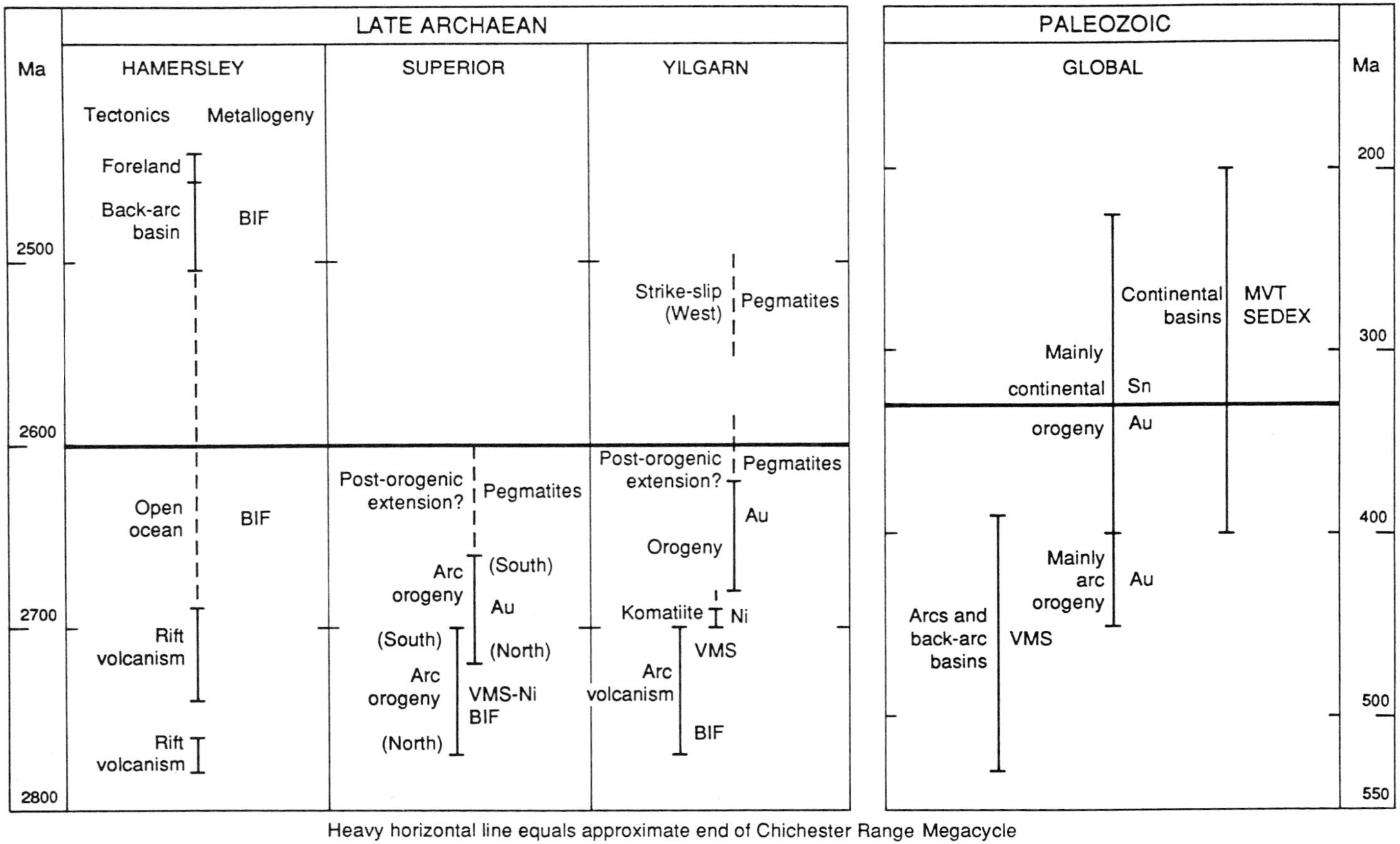

Fig. 14. Late Archean tectonic and metallogenic summary of the Hamersley Basin, Superior Province, and Yilgarn Craton compared to the Paleozoic.

any other Late-Archean granitoid-greenstone terrain. It's tectonic evolution is also interpreted by comparison with modern environments by most workers (Card 1990; Poulsen et al., 1992). The Superior Province records intensive magmatism between 2.77 and 2.66 Ga, which starts with early episodes (between 2.77 and 2.70 Ga) of arc-type magmatism with associated Cu-Zn VMS and porphyry Cu-Mo mineralization, and komatiite volcanism associated with nickel mineralization. Volcanism was diachronous, and peaked at *ca* 2.74 Ga in the Uchi and Wabigoon Subprovinces (north and centre) and *ca* 2.71 Ga in the Abitibi Subprovince (south). The tectonic evolution of the province then progressed to an orogenic stage characterized by voluminous calc-alkaline magmatism, deposition of turbidite sequences, polyphase folding, thrusting, and local alluvial sedimentation and alkaline magmatism. These orogenic phases are interpreted to be the combined response to a regime of north–south compression that led to the accretion of individual arcs and earlier greenstone belts to a craton. This occurred diachronously, prior to 2.71 Ga in the north, between 2.71 and 2.69 Ga in the centre, and between 2.70 and 2.68 Ga in the south. Mesothermal gold mineralization is also apparently diachronous, mainly related to orogenic events between 2.71 and 2.66 Ga. This interpreted tectonic and metallogenic evolution of the Superior Province is comparable with that of southwestern Pacific island arcs and back-arc basins.

The Yilgarn Craton in southwestern Australia contains a well mineralized granitoid-greenstone terrain. Its Late Archean evolution is less well constrained and apparently more complex than that of the Superior Province. It involved 2.76 to 2.70 Ga mafic through intermediate to silicic arc-type magmatism in the western, central, and eastern greenstone belts, with Cu-Zn VMS mineralization at Teutonic Bore, and porphyry-style Cu-Au mineralization at Boddington. This was followed by an important period of deep-water komatiite and tholeiite volcanism, which hosts the important nickel sulfide deposits in the Norseman-Wiluna Belt at 2.70 to 2.69 Ga. The komatiite and tholeiite volcanism is interpreted by Barley et al. (1989) as either occurring in a back-arc basin or a rift within a pre-existing arc. Komatiitic to basaltic volcanism in the Norseman–Wiluna Belt was followed by intermediate to silicic volcanism and sedimentation (mainly turbidites), thrusting, and granitoid emplacement at *ca* 2.68 Ga followed by further sedimentation, strike-slip to compressive deformation, metamorphism and episodic extension with Yilgarn-wide granitoid emplacement between 2.66 and 2.63 Ga. This granitoid magmatism was more extensive than that in the Superior Province. From the sparse geochronological data available, mesothermal gold was deposited in structural sites at *ca* 2.63 Ga. Volumetrically minor granitoid and mineralized pegmatite emplacement, regional extension and uplift, and strike-slip deformation occurred until after 2.6 Ga.

Because the overall rock assemblages, depositional and deformational histories, and mineralization of the greenstone belts resemble those of Phanerozoic greenstone complexes, most workers have interpreted the evolution of the Yilgarn granitoid-greenstone terrain as resulting from some form of convergent tectonic regime (Barley et al., 1989; Swager et al., 1990). Barley et al. (1993) interpreted this complex history as a continental microplate in a complex convergent environment such as in the Indonesian Archipelago, Kalimantan and the Philipines where opposing subduction zones have affected a collage of microplates since the early Mesozoic.

In contrast, Campbell et al. (1989) and Hill et al. (1992) have interpreted high-temperature komatiite volcanism and subsequent granitoid magmatism as resulting from a mantle plume in a non-plate tectonic regime. However, the two models need not be mutually exclusive. It is likely that periods of plume activity induce magmatism and rifting in, or near, existing convergent plate margins as well as within continental and oceanic settings. Some oceanic plateaus (Mahoney and Spencer, 1991) are the best modern analogues for submarine volcanic provinces generated by Archean mantle plumes. The distribution of modern oceanic plateaus (Ben-Avraham et al., 1981) indicates that some of these are within marginal seas of the Pacific basin and that most will eventually be accreted to continental crust at convergent plate margins. The Cretaceous komatiites of Gorgona Island have also recently been interpreted as forming at an ocean plateau (Storey et al., 1990), and subsequently to have been juxtaposed with convergent-margin rocks. Closure of a basin containing a mantle thermal anomaly may also provide a convenient explanation for apparently synchronous Yilgarn-wide crustal melting and anatectic granitoid emplacement at 2.66 to 2.63 Ga, preceding cratonization.

It is apparent from this brief summary that the tectonic evolution of the three Late-Archean terrains with the best coverages of high-precision geochronological data can be explained in broad terms by comparison with modern environments and processes, and that, in general, the metallogeny of these terrains is similar to that expected for younger analogues.

Tectonics related to Late Archean metallogeny

The tectonics and metallogeny of these three Late Archean provinces is summarized and compared to that of analogous mineral deposits during the Paleozoic in Fig. 14. The first feature that is evident is that, although volcanism and orogeny are diachronous in detail, there is a remarkable synchroneity between pulses of magmatism and mineralization in the Yilgarn Craton and Superior Province and the supersequences (supercycles) of the Hamersley Basin. All three provinces start with volcanism at 2.78 to 2.76 Ga, with further pulses of magmatism at 2.74 to 2.72 Ga and 2.70 to 2.68 Ga, with the cycle of volcanism, orogeny and cratonization in the Yilgarn Craton and Superior Province over by 2.60 Ga, the end of the first Hamersley Basin megacycle. This supports the interpretation (Blake and Barley, 1992) that the supersequences and megasequences mapped in the Hamersley Basin may indeed represent the rock record of global Late-Archean supercycles and megacycles which were similar in duration and style to their

Phanerozoic equivalents. Progressive changes in both the nature of the dominant mineralization styles during a single megacycle and from the first to second megacycle in a megacycle set are also evident.

These observations can be combined to produce an integrated tectonic and metallogenic model for the Late Archean. Mineralization related to rifting or plume activity (komatiite-associated Ni), arc and back-arc magmatism (VMS, porphyry Cu-Mo-Au) were most abundant in the Yilgarn Craton and Superior Province synchronous with the early rifting of the Hamersley Basin. This is followed by mesothermal gold mineralization in the Yilgarn Craton and Superior Province and mineralized pegmatites in the Superior Province in the second half of the first megacycle as marginal seas closed and arcs and oceanic plateaus were accreted to nascent or pre-existing continental plates and microplates. The thick Superior type iron formations of the Hamersley Basin and mineralized pegmatites in the Yilgarn Craton are interpreted to have formed during a second megacycle, which was dominated by intracratonic tectonics.

This parallels tectonic and metallogenic patterns during the Phanerozoic (Titley, 1991; 1993; Barley and Groves, 1992). These involve early Paleozoic peaks in VMS deposit abundance in Cambrian, Ordovician and Silurian submarine volcanic rocks in Caledonian orogenic belts. Some of these volcano-sedimentary assemblages resemble those in Archean greenstone belts, and are interpreted to result from pulses in oceanic, arc, and back-arc magmatism during an early Paleozoic megacycle as the Neoproterozoic supercontinent disaggregated. This overlapped, and was followed by, peaks in mesothermal gold mineralization and tin granites as well as sediment-hosted mineralization in continental basins as continental fragments amalgamated to form Pangea during a second megacycle in the Devonian Carboniferous and Permian. The pattern of early Mesozoic "plume related" Ni, Cr and PGE mineralization (e.g. Noril'sk), with Mesozoic to Cenozoic VMS (e.g. Kuroko), and Mesozoic to Cenozoic gold mineralization (e.g. Pacific Rim) during and following the breakup of Pangea is also broadly similar.

Older Archean metallogeny

Both Australia and Canada contain granitoid-greenstone terrains formed between 3.2 Ga and 2.8 Ga with similar rock assemblages to the younger greenstone belts, but which in general are less well mineralized (Barley and Groves 1990; Fyon et al.,1992) as shown in Fig. 13. These terrains most likely represent the remains of older Archean tectonic cycles (Krapez, 1993) that were either not as well preserved, or were not as well mineralized as their Late Archean counterparts. However, the Witwatersrand Basin in southern Africa, which formed between 3.0 and 2.8 Ga, contains some of the worlds largest gold deposits. This basin is interpreted as a continental foreland basin which received sediment from intensely mineralized magmatic arc rocks in much the same way that giant Mesozoic–Cenozoic placer deposits formed surrounding the Pacific ocean.

Older 3.5 to 3.2 Ga granitoid-greenstone terrains occur in the Pilbara Craton and in southern Africa. These terrains also contain similar assemblages and mineralization styles to their younger counterparts (Fig. 13). They locally contain unusual (for the Archean) sulphate-rich VMS deposits and, with the exception of the Barberton Mountainland which contains major *ca* 3.0 to 2.9 Ga mesothermal gold mineralization (De Ronde et al., 1992), are not intensely mineralized. However, the similarity of rock assemblages and mineral deposit types with Late Archean terrains indicates that relationships between mineralization and tectonics were essentially the same between 3.5 and 2.5 Ga.

SUMMARY

Archean terrains are some of the most richly mineralized on Earth, both in terms of mineral deposits that are an integral part of Archean tectonic evolution and also those that formed in Archean cratons at a subsequent time. These Archean terrains comprise: (1) poorly mineralized high-grade gneiss belts including the oldest known terrains, (2) poorly mineralized plutonic provinces, (3) heterogeneous, but commonly highly mineralized granitoid-greenstone terrains, (4) poorly mineralized metasedimentary provinces, and (5) intracratonic or continental-margin successions which include the highly mineralized (Au-U) Witwatersrand Basin and several iron-ore provinces.

Synvolcanic mineral deposits in Archean greenstone belts are unevenly distributed. Copper-Zn VMS districts are most abundant in the Superior Province of Canada, but only scattered examples are developed elsewhere. Similarly, komatiite-associated nickel districts are best developed in the Yilgarn Block of Western Australia, but only scattered examples are known from other provinces. Porphyry Cu-Mo-Au deposits are scattered throughout greenstone belts, but are nowhere as well developed as in Phanerozoic terrains. By analogy with younger examples of similar mineralization styles, most of the synvolcanic deposits suggest an arc environment in a convergent margin setting, although the komatiites and associated nickel deposits may form on the equivalent of oceanic plateaus.

Syn- to post-orogenic mineral deposits are dominated by the mesothermal group of gold deposits, although rare-metal pegmatites are also locally important. The gold deposits are interpreted to have been deposited in a convergent margin setting based on the nature of controlling structures and associated igneous activity. A mid- to lower-crustal event of craton scale is thought to have triggered hydrothermal activity late in the orogenic cycle.

There was a major peak in metallogeny in the Late Archean, with synvolcanic Cu-Zn VMS and komatiite-associated Ni deposits and syn- to post-orogenic gold deposits and rare-metal pegmatites all being best developed in the period 2.75 to 2.6 Ga. Although outside this range, the Witwatersrand deposits (3.0–2.8 Ga) and the major iron ores (BIF at ~2.5 Ga) are still Late Archean in age. Mineralization in older Archean terrains is less well developed, and in some cases anomalous (e.g. sulphate rich VMS, stratiform barite) relative to the Late Archean greenstone belts.

Analysis of Late Archean greenstone terrains with well constrained chronology using sequence stratigraphy confirms the opinion of most researchers that the tectonic evolution of these Late Archean terrains can be broadly explained in terms of modern tectonic environments and processes, in agreement with the analogies between early Paleozoic, post-Mesozoic and Late Archean metallogeny. There appears to be a Late-Archean tectonic and metallogenic cycle recorded by granitoid-greenstone terrains such as the Superior Province and Yilgarn Craton. This starts with pulses of magmatism and associated mineralization at *ca* 2.78 Ga followed by orogeny with associated mineralization, and ending with cratonization before 2.6 Ga. This broadly parallels the tectonic and metallogenic evolution of the early Paleozoic or Mesozoic to Cenozoic with the breakup of supercontinents, formation of volcanic arcs, and opening and closure of marginal seas.

ACKNOWLEDGMENTS

We are grateful to our colleagues in the Key Centre for Strategic Mineral Deposits, particularly Tim Blake, Brian Krapez, Neal McNaughton and John Ridley for concepts developed and discussed over the past decade. We are also indebted to the late Nick Rock and Rob Kerrich for stimulating discussions. The research on which this paper is based was funded by ARC, AMIRA, MERIWA, UWA and numerous mining companies. We are most grateful for this support. We also acknowledge the most useful reviews by Kent Condie, Dick Hutchinson and Mike Lesher.

REFERENCES

Alt, J.C., Lonsdale, P., Haymon, R. and Muehlenbachs, K., 1987. Hydrothermal sulfide and oxide deposits on seamounts near 21°N, East Pacific Rise. Geol. Soc. Am. Bull., 98: 157–168.

Anhaeusser, C.R., 1981. The relation of mineral deposits to early crustal evolution. Econ. Geol., 75th Anniv. Vol., 42–62.

Anhaeusser, C.R. and Maske, S. (Eds.), 1986. Mineral Deposits of South Africa, Vols. I & II. Geol. Soc. S. Africa, Johannesburg, 2376 pp.

Armstrong, R.A., Compston, W., Retief, E.I., Williams, I.S. and Welke, H.J., 1991. Zircon ion microprobe studies bearing on the age and evolution of the Witwatersrand Triad. Precambrian Res., 53: 243–266.

Ayres, L.D. and Cerny, P., 1982. Metallogeny of granitoid rocks in the Canadian Shield. Can. Mineral., 20: 439–536.

Barley, M.E., 1982. Porphyry-style mineralization associated with early Archean calc-alkaline igneous activity, Eastern Pilbara, Western Australia. Econ. Geol., 77: 1230–1235.

Barley, M.E., 1992. A review of Archean volcanic-hosted massive sulfide and sulfate mineralization in Western Australia. Econ. Geol., 87: 855–872.

Barley, M.E. and Groves, D.I., 1990. Deciphering the tectonic evolution of Archaean greenstone belts: the importance of contrasting histories to the distribution of mineralization in the Yilgarn Craton, Western Australia. Precambrian Res., 46: 3–20.

Barley, M.E. and Groves, D.I., 1992. Supercontinent cycles and the distribution of metal deposits through time. Geology, 20: 291–294.

Barley, M.E., Eisenlohr, B., Groves, D.I., Perring, C.S. and Vearncombe, J.R., 1989. Late Archaean convergent margin tectonics and gold mineralization: a new look at the Norseman–Wiluna Belt, Western Australia. Geology, 17: 826–829.

Barley, M.E., Krapez, B. and Groves, D.I., 1993. Resolving conflicting messages from granitoids and greenstones: the key to understanding Archaean tectonics. Aust. Geol. Surv. Organization Recd. 1993/54: 57–61.

Barnes, S.J., Gole, M.J. and Hill, R.E.T., 1988a. The Agnew nickel deposit, Western Australia. Econ. Geol., 83: 524–550.

Barnes, S.J., Hill, R.E.T. and Gole, M.J., 1988b. The Perseverance ultramafic complex, Western Australia: the product of a komatiite lava river. J. Petrol., 29: 305–331.

Barnicoat, A.C., Fare, R.J., Groves, D.I. and McNaughton, N.J., 1991. Synmetamorphic lode-gold deposits in high-grade Archean settings. Geology, 19: 921–924.

Barrett, F.M., Binns, R.A., Groves, D.I., Marston, R.J. and McQueen, K.G., 1977. Structural history and metamorphic modification of Archean volcanic-type nickel deposits, Yilgarn block, Western Australia. Econ. Geol., 72: 1195–1223.

Ben-Avraham, Z., Nur, A., Jones, D. and Cox, P., 1981. Continental accretion and orogeny. From oceanic plateaus to allochthonous terranes. Science, 213: 47–54.

Binns, R.A. and Scott, S.D., 1993. Actively forming polymetallic sulfide deposits associated with felsic volcanic rocks in the eastern Manus back-arc Basin, Papua New Guinea. Econ. Geol., 88: 2226–2236.

Blake, T.S., 1993. Late Archaean crustal extension, sedimentary basin formation, flood basalt volcanism and continental rifting: the Nullagine and Mount Jope Supersequences, Western Australia. Precambrian Res., 60: 185–241.

Blake, T.S. and Barley, M.E., 1992. Tectonic evolution of the Late Archaean to Early Proterozoic Mount Bruce Megasequence Set, Western Australia. Tectonics, 11: 1415–1425.

Blockley, J.G., 1980. The tin deposits of Western Australia. Geol. Surv. West. Aust., Min. Res. Bull., 12, 184 pp.

Brenner, T.L., Teixeira, N.A., Oliveira, J.A.L., Franke, N.D. and Thompson, J.F.H., 1990. The O'Toole nickel deposit, Morro do Ferro greenstone belt, Brazil. Econ. Geol., 85: 904–920.

Buick, R. and Dunlop, J.S.R., 1990. Evaporitic sediments of Early Archaean age from the Warrawoona Group, North Pole, Western Australia. Sedimentology, 37: 247–277.

Burrows, D.R. and Spooner, E.T.C., 1986. The McIntyre Cu-Au deposit, Timmins, Ontario, Canada. In: A.J. Macdonald (Ed.), Proceedings of Gold '86, An International Symposium on the Geology of Gold. Konsult International Inc., Toronto, pp. 23–39.

Campbell, I.H., Franklin, J.M., Gorton, M.P., Hart, T.R. and Scott, S.D., 1981. The role of subvolcanic sills in the generation of massive sulfide deposits. Econ. Geol., 76: 2248–2253.

Campbell, I.H., Griffiths, R.W. and Hill, R.I., 1989. Melting in an Archaean mantle plume: heads it's basalts, tails it's komatiites. Nature, 339: 679–699.

Card, K.D., 1990. A review of the Superior Province of the Canadian Shield, a product of Archaean accretion. Precambrian Res., 48: 99–156.

Cathles, L.M., 1983. An analysis of the hydrothermal systems responsible for massive sulphide deposition in the Hokuroko Basin of Japan. In: H. Ohmoto and B.J. Skinner (Eds.), Kuroko and Related Volcanogenic Massive Sulphide Deposits. Econ. Geol. Mono., 5: 439–487.

Cathles, L.M., 1993. Oxygen isotope alteration in the Noranda mining district, Abitibi greenstone belt, Quebec. Econ. Geol.,88: 1483–1511.

Cerny, P., 1982. Granitic pegmatites in science and industry. Mineral. Ass. Can., Short Course Handbook, 8, 555pp.

Clark, M.E., Carmichael, D.M., Hodgson, C.J. and Fu, M., 1988. Wall-rock alteration, Victory Gold Mine, Kambalda, Western Australia: Processes and P-T-XCO_2 conditions of metasomatism. In: R.R. Keays, W.R.H. Ramsay and D.I. Groves (Eds.), The Geology of Gold Deposits: The Perspective in 1988. Econ. Geol. Mono., 6: 445–459.

Colvine, A.C., 1989. An empirical model for the formation of Archean gold deposits: Products of final cratonization of the Superior Province, Canada. Econ. Geol. Mono., 6: 37–53.

Colvine, A.C., Fyon, J.A., Heather, K.B., Marmont, S., Smith, P.M. and Troop, D.G., 1988. Archean lode gold deposits in Ontario. Ontario Geol. Surv. Misc. Pap., 139, 136pp.

Condie, K.C., 1981. Archean Greenstone Belts. Elsevier Scientific Publishing, Amsterdam, 434pp.

Cooper, D.G., 1964. The geology of the Bikita pegmatite. In: S.H. Haughton (Ed.), The Geology of some Ore Deposits in Southern Africa, Vol. II. Geol. Soc. South Africa, Johannesburg: 441–462.

Cotterill, P., 1969. The chromite deposits of Selukwe, Rhodesia. Econ. Geol. Mono., 4: 154–186.

Cowden, A. and Archibald, N.T., 1987. Massive-sulfide fabrics at Kambalda and their relevance to the inferred stability of monosulfide solid-solution. Can. Mineral., 25: 37–50.

Cowden, A. and Woolrich, P., 1987. Geochemistry of the Kambalda iron–nickel sulfides: implications for models of sulfide–silicate partitioning. Can. Mineral., 25: 21–36.

Cowden, A., Donaldson, M.J., Naldret, A.J. and Campbell, I.H., 1986. Platinum-group elements and gold in the komatiite-hosted Fe-Ni-Cu sulfide deposits at Kambalda, Western Australia. Econ. Geol., 81: 1226–1235.

De Ronde, C.E.J., Spooner, E.T.C., De Wit, M.J. and Bray, C.J., 1992. Shear zone-related, Au quartz vein deposits in the Barberton greenstone belt, South Africa: Field and petrographic characteristics, fluid properties, and light stable isotope geochemistry. Econ. Geol., 87,: 366–402.

De Wit, M.J., Armstrong, R., Hart, R.J. and Wilson, A.H., 1987. Felsic igneous rocks within the 3.3 to 3.5 Ga Barberton greenstone belt: High crustal level equivalents of the surrounding tonalite–trondhjemite terrain, emplaced during thrusting. Tectonics, 6: 529–549.

De Wit, M.J., Roering, C., Hart, R.J., Armstrong, R.A., De Ronde, C.E.J., Green, R.W.E., Tredoux, M., Peperdy, E. and Hart, R.A., 1992. Formation of an Archaean continent. Nature, 357: 553–562.

Dimroth, E., Imreh, L., Cousineau, P., Leduc, M. and Sanschagrin, Y., 1985. Paleogeographic analysis of massive submarine flows and its use in the exploration for massive sulphide deposits. In: L.D. Ayres, P.C. Thurston, K.D. Card and W. Weber (Eds.), Evolution of Archean Supracrustal Sequences. Geol. Assoc. Can., Spec. Paper, 28. Meeting of Geol. Soc. Can., Winnipeg, 1982, pp. 203–222

Dimroth, E., Imreh, L., Goulet, N. and Rocheleau, M., 1983. Evolution of the south-central segment of the Abitibi Belt, Quebec. Part II: tectonic evolution and geomechanical model. Can. J. Earth Sci., 20: 1355–1373.

Donaldson, M.J., Lesher, C.M., Groves, D.I. and Gresham, J.J., 1986. Comparison of Archaean dunites and komatiites associated with nickel mineralization in Western Australia: Implications for dunite genesis. Mineral. Deposita, 21: 296–305.

Dowling, S.E., and Hill, R.E.T., 1993. The Mount Keith ultramafic complex and the Mount Keith nickel deposit. Aust. Geol. Surv. Organisation Recd., 1993/54, pp. 165–169.

Echeverria, L.M., 1982. Komatiites from Gorgona Island, Columbia. In: N.T. Arndt and E.G. Nisbet

(Eds.), Komatiites. George Allen and Unwin, London, pp. 199–209.

Eckstrand, O.R., 1975. The Dumont serpentinite: A model for control of nickeliferous opaque assemblages by alteration products in ultramafic rocks. Econ. Geol., 70: 183–201.

Eisenlohr, B.N., 1989. The structural geology of Birimian and Tarkwaian rocks in Ghana. BGR Rept. No. 106448: 66 pp.

Eisenlohr, B., Groves, D.I. and Partington, G.A., 1989. Crustal-scale shear zones and their significance to Archean gold mineralization. Mineral. Deposita, 24: 1–8.

Evans, D.M., Cowden, A. and Barratt, R.M., 1989. Deformation and thermal erosion at the Foster nickel deposit, Kambalda–St. Ives, Western Australia. In: M.D. Prendergast and M.J. Jones (Eds.), Magmatic Sulphides — the Zimbabwe Volume, Proceedings of the Fifth Magmatic Sulphide Field Conference. Inst. Min. Metall., London, pp. 215–219.

Foster, R.P., 1989. Archean gold mineralization in Zimbabwe: Implications for metallogenesis and exploration. Econ. Geol. Mono., 6: 54–70.

Foster, R.P. and Piper, D.P., 1993. Archaean lode gold deposits in Africa: Crustal setting, metallogenesis and cratonization. Ore Geol. Rev., 8: 303–347.

Franklin, J.M., 1990. Volcanic-associated massive sulphide deposits. In: S.E. Ho, F. Robert and D.I. Groves (Eds.), Gold and Base Metal Mineralization in the Abitibi Subprovince, Canada, with Special Emphasis on the Quebec Segment. Geol. Dept. (Key Centre) & Univ. Extension, Univ. West. Aust., Publ., 24: 211–241.

Franklin, J.M. and Thorpe, R.I., 1982. Comparative metallogeny of the Superior, Slave, and Churchill provinces. In: R.W. Hutchinson, L.D. Spence and J.M. Franklin (Eds.), Precambrian Sulphide Deposits. Geol. Ass. Can. Spec. Pap., 25: 3–90.

Franklin, J.M., Lydon, J.W. and Sangster, D.F., 1981. Volcanic-associated massive sulfide deposits. Econ. Geol., 75: 485–627.

Frost, K.M. and Groves, D.I., 1989. Ocellar units at Kambalda: evidence for sediment assimilation by komatiite lavas. In: M.D. Pendergast and M.J. Jones (Eds.), Magmatic Sulphides — The Zimbabwe Volume. Proceedings of the Fifth Magmatic Sulphide Field Conference, Harare, Zimbabwe, pp. 207–214.

Froude, D.O., Ireland, T.R., Kinny, P.D. Williams, I.S., Compston, W., Williams, I.R. and Myers, J.S., 1983. Ion microprobe identification of 4100–4200 M yr-old terrestrial zircons. Nature, 304: 616–618.

Fyon, J.A., Breaks, F.W., Heather, K.B., Jackson, S.L., Muir, T.L., Stott, G.M. and Thurston, P.C., 1992. Metallogeny of metallic mineral deposits in the Superior Province of Ontario. Ontario Geol. Surv. Spec. Vol., 4 (2): 1091–1174.

Fyon, J.A., Troop, D.G., Marmont, S. and Macdonald, A.J., 1989. Introduction of gold into Archean crust, Superior Province, Ontario — coupling between mantle-initiated magmatism and lower crustal thermal maturation. Econ. Geol. Mono., 6: 479–490.

Galley, A.G., 1993. Characteristics of semi-conformable alteration zones associated with volcanogenic massive sulphide districts. J. Geochem. Expl., 48: 175–200.

Garson, M.S. and Mitchell, A.H.G., 1981. Precambrian ore deposits and plate tectonics. In: A. Kröner (Ed.), Precambrian Plate Tectonics. Inst. Geowiss., Johannes Gutenberg Univ. Mainz, Federal Republic of Germany: 689–731.

Gebre-Mariam, M., Groves, D.I., McNaughton, N.J., Mikucki, E.J. and Vearncombe, J.R., 1993. Archaean Au-Ag mineralisation at Racetrack, near Kalgoorlie, Western Australia: a high crustal-level expression of the Archaean composite lode-gold system. Mineral. Deposita, 28: 375–387.

Gibson, H.L. and Watkinson, D.H., 1990. Volcanogenic massive sulphide deposits of the Noranda cauldron and shield volcano, Quebec. In: M. Rive (Ed.), The Northwestern Quebec Polymetallic Belt: A Summary of 60 Years of Mining Exploration. Can. Inst. Min. Metall. Spec. Vol., 43: 119–132.

Gillies, S.L. and Lesher, C.M., 1992. Lava channelization in the Katinniq Peridotite Complex, Cape Smith belt, New Quebec. Geol. Soc. Am., Abstracts with Programs, 24 (7): A267.

Goldfarb, R.J., Snee, L.W., Miller, L.D. and Newberry, R.J., 1991. Rapid dewatering of the crust deduced from ages of mesothermal gold deposits. Nature, 354: 296–298.

Goldie, R.J., Kotila, B. and Seward, D., 1979. The Don Rouyn Mine: an Archaean porphyry copper deposit near Noranda, Quebec. Econ. Geol., 74: 1680–1684.

Gresham, J.J. and Loftus-Hills, G.N., 1981. The geology of the Kambalda nickel field, Western Australia. Econ. Geol., 76: 1373–1416.

Groves, D.I. and Batt, W.D., 1984. Spatial and temporal variations of Archaean metallogenic associations in terms of evolution of granitoid-greenstone terrains with particular emphasis on Western Australia. In: A. Kröner, G.M. Hanson and A.M. Goodwin (Eds.), Archaean Geochemistry. Springer-Verlag, Berlin, pp. 73–98.

Groves, D.I. and Foster, R.P., 1990. Archaean lode gold deposits. In: R.P. Foster (Ed.), Gold Metallogeny and Exploration. Blackie and Son Ltd., pp. 63–103.

Groves, D.I. and Hudson, D.R., 1981. The nature and origin of Archaean strata-bound volcanic-associated nickel–iron–copper sulfide deposits. In: K.H. Wolf (Ed.), Handbook of Strata-Bound and Stratiform Ore Deposits, 9. Elsevier, Amsterdam, pp. 305–410.

Groves, D.I. and Keays, R.R., 1979. Mobilization of ore-forming elements during alteration of dunites, Mt. Keith-Betheno, Western Australia. Can. Mineral., 17: 373–389.

Groves, D.I., Barley, M.E. and Ho, S.E., 1989. The nature, genesis and tectonic setting of mesothermal gold mineralization in the Yilgarn Block, Western Australia. Econ. Geol. Mono., 6: 71–85.

Groves, D.I., Barrett, F.M. and McQueen, K.G., 1979. The relative roles of magmatic segregation, volcanic exhalation and regional metamorphism in the generation of volcanic-associated nickel ores of Western Australia. Can. Mineral., 17: 319–336.

Groves, D.I., Barley, M.E., Barnicoat, A.C., Cassidy, K.F., Fare, R.J., Hagemann, S.G., Ho, S.E., Hronsky, J.M.A., Mikucki, E.J., Mueller, A.G., McNaughton, N.J., Perring, C.S., Ridley, J.R. and Vearncombe, J.R., 1992. Sub-greenschist to granulite-hosted Archaean lode-gold deposits of the Yilgarn Craton: A depositional continuum from deep-sourced hydrothermal fluids in crustal-scale plumbing systems. In: J.E. Glover and S.E. Ho (Eds.), The Archaean: Terrains, Processes and Metallogeny. Geol. Dept. (Key Centre) & Univ. Extension, Univ. Western Australia Publ., 22, pp. 325–337.

Guha, J., 1984. Hydrothermal systems and correlations of mineral deposits in the Chibougamau mining district — an overview. In: J. Guha and E.H. Chown (Eds.), Chibougamau — Stratigraphy and Mineralization. Can. Inst. Min. Metall. Spec. Vol., 34: 517–534.

Gustin, M.S., 1990. Stratigraphy and alteration of the host rocks, United Verde massive sulphide deposit, Jerome, Arizona. Econ. Geol., 85: 29–49.

Hannington, M.D. and Scott, S.D., 1988. Gold and silver potential of polymetallic sulphide deposits on the seafloor. Mar. Min., 7: 271–282.

Hatcher, M.I. and Elliot, A., 1986. Greenbushes — a new world resource of lithium. 7th Indust. Min. Int. Cong., pp. 217–232.

Hill, R.E.T., Gole, M.J. and Barnes, S.J., 1987. Physical volcanology of komatiites: A field guide to the komatiites between Kalgoorlie and Wiluna, Eastern Goldfields Province, Yilgarn Block, Western Australia. Excursion Guide Book No 1, Geol. Soc. Aust., W.A. Div., Perth, 74 pp.

Hill, R.I., Chappell, B.W. and Campbell, I.H., 1992. Late Archaean granites of the southeastern Yilgarn Block: age, geochemistry and origin. Trans. Royal Soc. Edinburgh, Earth Sci., 83: 211–226.

Ho, S.E., 1987. Fluid inclusions: Their potential as an exploration tool for Archaean gold deposits. In: S.E. Ho and D.I. Groves (Eds.), Recent Advances in Understanding Precambrian Gold

Deposits, Vol. 1. Geol. Dept. & Univ. Extension, Univ. West. Aust., Publ., 11: 239–264.
Hodgson, C.J. and Troop, D.G., 1988. A new computer-aided methology for area selection in gold exploration: a case study from the Abitibi greenstone belt. Econ. Geol., 83: 952–977.
Hodgson, C.J., 1989. The structure of shear-related vein-type gold deposits: A review. Ore Geol. Rev., 4: 231–273.
Holland, H.D., 1984. The Chemical Evolution of the Atmosphere and Oceans. Princeton Univ. Press, Princeton, 582 pp.
Hollister, V.F., 1975. An appraisal of the nature of some porphyry-copper deposits. Min. Sci. Engng., 7: 225–233.
Hopwood, T., 1981. The significance of pyritic black shales in the genesis of Archean nickel sulphide deposits. In: K.H. Wolf (Ed.), Handbook of Strata-Bound and Stratiform Ore Deposits, 9. Elsevier, Amsterdam, pp. 411–468.
Hughes, F.E. (Ed.), 1990. Geology of the Mineral Deposits of Australia and Papua New Guinea, Vols. I and II. Australas. Inst. Min. Metall., Melbourne, 1828 pp.
Huppert, H.H., Sparks, S.J., Turner, J.S. and Arndt, N.T., 1984. Emplacement and cooling of komatiite lavas. Nature, 309: 19–22.
Hutchinson, R.W., 1973. Volcanogenic massive sulfides and their metallogenic significance. Econ. Geol., 68: 1223–1246.
Hutchinson, R.W., 1981. Mineral deposits as guides to supracrustal evolution. In: R.J. O'Connel and W.S. Fyfe (Eds.), Evolution of the Earth. Am. Geophys. Union–Geol. Soc. Am., Geodynamic Series, 5: 120–141.
Hutchinson, R.W., 1993: A multi-stage, multi-process genetic hypothesis for greenstone-hosted gold lodes. Ore Geol. Rev., 8: 349–382.
Hutchinson, R.W., Ridler, R.H. and Suffel, G.G., 1971. Metallogenic relationships in the Abitibi belt, Canada: A model for Archean metallogeny. Can. Min. Metall. Bull., 64: 48–57.
James, H.L., 1954. Sedimentary facies of iron formation. Econ. Geol., 49: 235–293.
Jemielita, R.A., Davies, D.W. and Krogh, T.E., 1990. U-Pb evidence for Abitibi gold mineralization postdating greenstone magmatism and metamorphism. Nature, 346: 831–834.
Katz, M.B., 1988. Metallogeny of Early Precambrian granulite facies terrains. Precambrian Res., 39: 77–84.
Keays, R.R., Ross, J.R. and Woolrich, P., 1981. Precious metals in volcanic peridotite-associated nickel sulfide deposits in Western Australia, II: Distribution within ores and host rocks at Kambalda. Econ. Geol., 76: 1645–1674.
Kerrich, R., 1987. The stable isotope geochemistry of Au-Ag vein deposits in metamorphic rocks. In: T. Kyser (Ed.), Stable Isotope Geochemistry of Low Temperature Processes. Assoc. Can. Short Course Handbook, 13, Univ. Saskatoon, Saskatoon, pp. 287–336.
Kerrich, R. and Wyman, D., 1990. The geodynamic setting of mesothermal gold deposits: an association with accretionary tectonic régimes. Geology, 18: 882–885.
Kimberley, M.M., 1978. Paleoenvironmental classification of iron formations. Econ. Geol., 73: 215–229.
King, R.W. and Kerich, R.W., 1989. Strontium isotope compositions of tourmaline from lode gold deposits of the Archean Abitibi greenstone belt (Ontario–Quebec, Canada): implications for source reservoirs. Chem. Geol., 79: 225–240.
Klein, C. and Beukes, N.J., 1992. Proterozoic iron-formations. In: K.C. Condie (Ed.), Proterozoic Crustal Evolution. Elsevier, Amsterdam, pp. 383–418.
Krapez, B., 1993. Sequence stratigraphy of the Archaean supracrustal belts of the Pilbara Block, Western Australia. Precambrian Res., 60: 1–45.
Large, R.R., 1992. Australian volcanic-hosted massive sulfide deposits: Features, styles, and genetic

models. Econ. Geol., 87: 471–510.

Lesher, C.M., 1989. Komatiite-associated nickel sulfide deposits. Econ. Geol. Rev., 4: 45–101.

Lesher, C.M. and Campbell, I.H., 1993. Geochemical and fluid dynamic modelling of compositional variations in Archean komatiite-hosted nickel sulfide ores in Western Australia. Econ. Geol., 88: 804–816.

Lesher, C.M. and Groves, D.I., 1986. Controls on the formation of komatiite-associated nickel-copper sulfide deposits. In: G.H. Friedrich, A.J. Genkin, A.J. Naldrett, J.D. Ridge, R.H. Sillitoe and F.M. Vokes (Eds.), Geology and Metallogeny of Copper Deposits. Proceedings of the Twenty-Seventh International Geological Congress, Moscow. Springer-Verlag, Berlin, pp. 43–62.

Lesher, C.M., Goodwin, A.M., Campbell, I.H. and Gorton, M.P., 1986. Trace-element geochemistry of ore-associated and barren, felsic metavolcanic rocks in the Superior Province, Canada. Can. J. Earth Sci., 23: 222–237.

Lowell, J.D. and Guilbert, J.M., 1970. Lateral and vertical alteration–mineralization zoning in porphyry ore deposits. Econ. Geol., 65: 373–408.

Lusk, J., 1976. A possible volcanic-exhalative origin for lenticular nickel deposits of volcanic association, with special reference to those in Western Australia. Can. J. Earth Sci., 13: 451–458.

Lyndon, J.W., 1988. Volcanogenic massive sulphide deposits, Part 2: Genetic models. Geosci. Can., 15: 43–65.

Macdonald, A.J., 1988. The Geraldton gold camp: the role of banded iron formation. Ontario Geol. Surv., Open File Rept., 5694, 173 pp.

Macdonald, R., McGravie, D.W., Pinkerton, H. Smith, R.L. and Palacz, Z.A., 1990. Petrogenetic evolution of the Torfajokull volcanic complex, Iceland; I Relationship between the magma types. J. Petrol., 31: 429–459.

Mahoney, J.J. and Spencer, J.J., 1991. Isotopic evidence for the origin of the Manihiki and Ontong Java oceanic plateaus. Earth Planet. Sci. Lett., 104: 196–210.

Marston, R.J. and Kay, B.D., 1980. The distribution, petrology and genesis of nickel ores at the Juan complex, Kambalda, Western Australia. Econ. Geol., 75: 546–565.

Marston, R.J., Groves, D.I., Hudson, D.R. and Ross, J.R., 1981. Nickel sulfide deposits in Western Australia: A review. Econ. Geol., 76: 1330–1363.

McNaughton, N.J., Cassidy, K.F., Dahl, N., Groves, D.I., Perring, C.S. and Sang, J.H., 1990. Source of ore fluids and ore components. In: S.E. Ho, D.I. Groves and J.M. Bennett (Eds.), Gold Deposits of the Archaean Yilgarn Block, Western Australia: Nature, Genesis and Exploration Guides. Geol. Dept. (Key Centre) & Univ. Extension, Univ. West. Aust. Publ., 20: 226–236.

Meyer, C., 1988. Ore deposits as guides to geologic history of the Earth. Ann. Rev. Earth Planet. Sci., 16: 147–171.

Minter, W.E.L., Goedhart, M., Knight, J. and Frimmel, H.E., 1993. Morphology of Witwatersrand gold grains from the basal reef: Evidence for their detrital origin. Econ. Geol., 88: 237–248.

Morris, R.C., Thornber, M.R. and Ewers, W.E., 1980. Deep–seated iron ores from banded iron-formation. Nature, 288: 250–252.

Morton, R.L. and Franklin, J.M., 1987. Two-fold classification of Archean volcanic-associated massive sulfide deposits. Econ. Geol., 82: 1057–1063.

Mueller, A.G. and Groves, D.I., 1991. The classification of Western Australian greenstone-hosted gold deposits according to wallrock-alteration mineral assemblages. Ore Geol. Rev., 6: 291–331.

Mueller, A.G., De Laeter, J.R. and Groves, D.I., 1991. Strontium isotope systematics of hydrothermal minerals from epigenetic Archean gold deposits in the Yilgarn Block, Western Australia. Econ. Geol., 86: 780–809.

Mueller, A.G., Harris, L.B. and Lungan, A., 1988. Structural control of greenstone-hosted gold mineralization by transcurrent shearing: a new interpretation of the Kalgoorlie Mining District,

Western Australia. Ore Geol. Rev., 3: 359–387.
Naldrett, A.J., 1981. Nickel sulfide deposits: Classification, composition and genesis. Econ. Geol, 75: 628–685.
Nesbitt, B.E., Murochick, J.B. and Muehlenbachs, K., 1986. Dual origins of lode gold deposits in the Canadian Cordillera. Geology, 14: 506–509.
Nisbet, E.G. and Chinner, G.A., 1981. Controls on the eruption of mafic and ultramafic lavas, Ruth Well Ni-Cu prospect, West Pilbara. Econ. Geol., 76: 1719–1735.
Ohmoto, H. and Skinner, B.J. (Eds.), 1983. Kuroko and related volcanogenic massive sulphide deposits. Econ. Geol. Mono., 5: 439–487.
Ojala, V.J., Ridley, J.R., Groves, D.I. and Hall, G.C., 1993. The Granny Smith Gold Deposit: role of heterogeneous stress distribution at an irregular granitoid contact in a greenschist facies terrane. Mineral. Deposita, 28: 409–419.
Parrish, R.R., 1989. U-Pb geochronology of the Cape Smith Belt and Sugluk block, northern Quebec. Geoscience Can., 16: 126–130.
Partington, G.A., 1990. Environment and structural controls on the intrusions of the giant rare metal Greenbushes Pegmatite, Western Australia. Econ. Geol., 85: 437–456.
Perring, C.S., Barley, M.E., Cassidy, K.F., Groves, D.I., McNaughton, N.J., Rock, N.M.S., 1989. The association of linear orogenic belts, mantle-crustal magmatism and Archaean gold mineralization in the Eastern Yilgarn Block of Western Australia. In: R.R. Keays, W.R.H. Ramsay and D.I. Groves (Eds.), The Geology of Gold Deposits: The Perspective in 1988. Econ. Geol. Mono., 6: 571–584.
Perring, C.S., Barnes S.J. and Hill R.ET., 1983. The physical volcanology of Archaean komatiitic sequences from Forrestania, Southern Cross Province. IAVCEI, Canberra 1993, Abst., 86.
Peters, S.G., 1993. Formation of oreshoots in mesothermal gold–quartz vein deposits: Examples from Queensland, Australia. Ore Geol. Rev., 8: 277–301.
Phillips, G.N. and Groves, D.I., 1983. The nature of Archaean gold-bearing fluids as deduced from gold deposits of Western Australia. J. Geol. Soc. Aust., 30: 25–39.
Phillips, G.N., Myers, R.E. and Palmer, J.A., 1987. Problem with the placer model for Witwatersrand gold. Geology, 15: 1027–1030.
Poulsen, K.H., Card, K.D. and Franklin, J.M., 1992. Archean tectonic and metallogenic evolution of the Superior Province of the Canadian Shield. Precambrian Res., 58: 25–54.
Poulsen, K.H. and Hodgson, C.J., 1984. Mineralization associated with Archean gabbro-anorthosite intrusions in the Rainy Lake area, northwestern Ontario. In: J. Guha and E.H. Chown (Eds.), Chibougamau; Stratigraphy and Mineralization. Spec. Vol., Can. Inst. Min. Metall., 34: 329–344.
Poulsen, K.H., Card, K.D. and Franklin, J.M., 1992. Archaean tectonic and metallogenic evolution of the Superior Province of the Canadian Shield. Precambrian Res., 58: 25–54.
Ridley, J.R., 1993. The relations between mean rock stress and fluid flow in the crust: With reference to vein- and lode-style gold deposits. Ore Geol. Rev., 8: 23–38.
Robert, F. and Kelly W.C., 1987. Ore-forming fluids in Archaean gold-bearing quartz veins at the Sigma Mine, Abitibi greenstone belt, Quebec, Canada. Econ. Geol., 82: 1464–1482.
Rock, N.M.S. and Groves, D.I., 1988. Can lamprophyres resolve the genetic controversy over mesothermal gold deposits? Geology, 16: 538–541.
Ross, J.R. and Travis, G.A., 1981. The nickel sulfide deposits of Western Australia in global perspective. Econ. Geol., 76: 1291–1329.
Roth, E., Groves, D.I., Anderson, G., Daley, L. and Staley, R., 1991. Primary mineralization at the Boddington Gold Mine, Western Australia: an Archaean porphyry Cu-Au-Mo deposit. In: E.A. Ladeira (Ed.), Brazil Gold '91: The Economics, Geology, Geochemistry, and Genesis of Gold Deposits. A.A. Balkema, Rotterdam, pp. 481–488.

Sangster, D.F. and Scott, S.D., 1976. Precambrian, strata-bound, massive Cu-Zn-Pb sulphide ores of North America. In: K.H. Wolf (Ed.), Handbook of Strata-Bound and Stratiform Ore Deposits, 6. Elsevier, Amsterdam, pp. 130–221.

Sawkins, F.J., 1990. Integrated tectonic–genetic model for volcanic-hosted massive sulfide deposits. Geology, 18: 1061–1064.

Schandl, E.S., Davis, D.W. and Krogh, T.E., 1990. Are the alteration haloes of massive sulphide deposits syngenetic? Evidence from U-Pb dating of hydrothermal rutile at the Kidd volcanic centre, Abitibi subprovince, Canada. Geology, 18: 505–508.

Seccombe, P.K., Groves, D.I., Binns, R.A. and Smith, J.W., 1978. A sulfur isotopic study to test a genetic model for Fe-Ni sulphide mineralization at Mt. Windarra, Western Australia. In: B.W. Robinson (Ed.), Stable Isotopes in the Earth Sciences. New Zealand Dept. Sci. Indust. Res. Bull., 220: 187–200.

Shklanka, R., 1970. Geology of the Bruce Lake area. Ontario Dept. Mines, Geol. Rept. 82, 27 pp.

Shklanka, R., 1972. Geology of the Steep Rock Lake area, District of Rainy River. Ontario Dept. Mines and Northern Affairs, Geol. Rept. 93, 114 pp.

Sibson, R.H., Moore, R.M. and Rankin, A.H., 1975. Seismic pumping — a hydrothermal transport mechanism. J. Geol. Soc. London, 131: 653–659.

Sibson, R.H., Robert, F. and Poulsen, K.H., 1988. High-angle reverse faults, fluid-pressure cycling, and mesothermal gold–quartz deposits. Geology, 16: 551–555.

Sillitoe, R.H., 1989. Gold deposits in western Pacific island arcs: The magmatic connection. Econ. Geol. Mono., 6: 274–291.

Smyk, M.C. and Watkinson, D.H., 1990. Sulfide remobilization in Archean volcano-sedimentary rocks and its significance in Proterozoic silver vein genesis, Cobalt, Ontario. Can. J. Earth Sci., 27: 1170–1181.

Solomon, M. and Walshe, J.L., 1979. The formation of massive sulfide deposits on the seafloor. Econ. Geol., 74: 797–813.

Spence, C.D. and de Rosen-Spence, A.F., 1975. The place of sulfide mineralization in the volcanic sequence of Noranda, Quebec. Econ. Geol., 70: 90–101.

Spooner, E.T.C. and Fyfe, W.S., 1973. Sub-sea floor metamorphism, heat and mass transfer. Contrib. Mineral. Petrol., 42: 287–304.

Stanton, R.L., 1972. Ore Petrology. McGraw Hill Book Co., New York, 713 pp.

Storey, M., Mahoney, J.J., Kroenke, L.W., and Saunders, A.D., 1990. Are oceanic plateaus sites of komatiite formation? Geology, 19: 376–379.

Swager, C.P., Griffin, T.J., Witt, W.K., Wyche, S., Ahmat, A.L. and Hunter, W.M., 1990. Geology of the Archaean Kalgoorlie Terrane: an explanatory note. Geol. Surv. Western Australia, Record 1990/12.

Syme, E.C. and Bailes, A.H., 1993. Stratigraphic and tectonic setting of early Proterozoic volcanogenic massive sulfide deposits, Flin Flon, Manitoba. Econ. Geol., 88: 566–589.

Symons, P.M., Anderson, G., Beard, T.J., Hamilton, L.H., Reynolds, G.D., Robinson, J.M. and Staley, R.W., 1988. The Boddington gold deposit. Bicentennial Gold '88, Ext. Abst. Oral Program. Geol. Soc. Aust. Inc. Abst. Ser., 22: 56–61.

Talbot, C.J., 1973. A plate tectonic model for the Archaean crust. Phil. Trans. Roy. Soc. London Ser. A, 273: 413–427.

Tankard, A.J., Jackson, M.P.A., Eriksson, K.A., Hobday, D.K., Hunter, D.R. and Minter, W.E.L., 1982. Crustal Evolution of Southern Africa: 3.8 Billion Years of Earth History. Springer-Verlag, New York, 523 pp.

Tarney, J., Dalziel, I.W.D. and De Wit, M.J., 1976. Marginal basin "Rocas Verdes" complex from S. Chile: A model for Archaean greenstone belt formation. In: B.F. Windley (Ed.), The Early

History of the Earth, John Wiley and Sons, London, pp. 131–146.

Taylor, W.R., Rock, N.M.S., Groves, D.I., Perring, C.S. and Golding, S.D., 1994. Geochemistry of Archean shoshonitic lamprophyres from the Yilgarn Block, Western Australia: Au abundance and association with gold mineralization. Appl. Geochem., 9: 197–222.

Thurston, P.C. and Chivers, K.M., 1990. Secular variation in greenstone sequence development emphasizing Superior Province, Canada. Precambrian Res., 46: 21–58.

Thurston, P.C., Wan, J., Squair, H.S., Warburton, A.F. and Wiersbick, V.W., 1978. Volcanology and Mineral Deposits of the Uchi-Confederation Lakes Area, Northwestern Ontario. In: Field Guides, Joint Annual Meeting, Geol. Soc. Am.–Geol. Ass. Can.: 302–324.

Thurston, P.C., Williams, H.R., Sutcliffe, R.H. and Stott, G.M. (Eds.), 1992. Geology of Ontario. Ontario Geol. Surv. Spec. Vol., 4, 1525 pp.

Titley, S.R., 1991. Phanerozoic ocean cycles and sedimentary rock hosted gold ores. Geology, 19: 645–648.

Titley, S.R., 1993. The relationship of stratabound ores with tectonic cycles of the Phanerozoic and Precambrian. Precambrian Res., 61: 295–322.

Trendall, A.F., 1983. The Hamersley Basin. In: A.F. Trendall and R.C. Morris (Eds.), Iron Formation Facts and Problems. Elsevier Scientific Publishing, Amsterdam, pp. 69–130.

Trendall, A.F. and Blockley, J.G., 1970. The iron formations of the Precambrian Hamersley Group, Western Australia, with special reference to the associated crocidolite. Geol. Surv. West. Australia Bull., 119, 366 pp.

Vail, P.R., Mitchum, R.M. Jr., Todd, R.G., Widmier, J.M., Thompson, S. III, Sangree, J.B., Bubb, J.N. and Hatllid, W.G., 1977. Seismic stratigraphy and global changes of sea level. Am. Assoc. Petrol. Geol. Mem., 26: 49–212.

van Wagoner, J.C., Mitchum, R.M., Campion, K.M. and Rahmanian, V.D., 1990. Siliciclastic sequence stratigraphy in well logs, cores and outcrops: concepts for high resolution correlation of time and facies. Am. Assoc. Petrol. Geol. Methods in Exploration Series, 7.

Wang, L.G., McNaughton, N.J. and Groves, D.I., 1993. An overview of the relationships between granitoid intrusions and gold mineralization in the Archaean Murchison Province, Western Australia. Mineral. Deposita, 28: 482–494.

Whittaker, P.J., 1986. Gold deposits of the southwestern Abitibi Subprovince of Ontario. In: A.M. Chater (Ed.), Gold '86, an International Symposium on the Geology of Gold Deposits; Poster Paper Abstracts. Newmont Explor. Can., Toronto, Canada, pp. 172–173.

Williams, D.A.C., 1979. The association of some nickel sulfide deposits with komatiitic volcanism in Rhodesia. Can. Mineral., 17: 337–350.

Williams, H.R., Stott, G.M., Thurston, P.C., Sutcliffe, R.H., Bennet, G., Easton, R.M. and Armstrong, D.K., 1992. Tectonic evolution of Ontario: Summary and synthesis. Ontario Geol. Surv. Spec. Vol., 4: 1255–1332.

Windley, B.F. (Ed.), 1976. The Early History of the Earth. John Wiley and Sons, New York, 619 pp.

Witt, W.K., 1991. Regional metamorphic controls on alteration associated with gold mineralization in the Eastern Goldfields Province, Western Australia: Implications for the timing and origin of Archean lode-gold deposits. Geology, 19: 982–985.

Wyman, D. and Kerrich, R., 1989. Archean shoshonitic lamprophyres associated with Superior Province gold deposits: distribution, tectonic setting, noble metal abundances, and significance for gold mineralization. Econ. Geol. Mono., 6: 661–667.

Chapter 12

THE ARCHEAN ATMOSPHERE: ITS COMPOSITION AND FATE

DAVID J. DES MARAIS

The composition of Earth's early atmosphere must be inferred both through its interactions with the oceans and sediments, and by other independently-deduced constraints on the early environment. Fortunately the interactions between the atmosphere, ocean and sediments can be rapid and pervasive, thus clues about atmospheric composition were indeed recorded. Furthermore, the atmosphere responded to those processes that altered Earth's crust during the Archean and Early Proterozoic. These changes in turn altered climate and probably influenced the evolution of the biosphere. This review will summarize evidence both for the composition of the Archean atmosphere between 3.8 and 3.0 Ga, and also about how the atmosphere changed between 3.0 and 2.0 Ga.

ORIGIN OF THE ATMOSPHERE

Earth's earliest atmosphere probably derived from multiple sources. Nonradiogenic rare gases that might have been inherited from the solar nebula, and that could have been retained gravitationally in the atmosphere since the time of planetary accretion, are extremely depleted relative to their cosmic abundances (Anders and Owen, 1977). Thus much of the early atmosphere was probably derived instead from volatile components trapped within the planetesimals that formed the Earth. Both large impacts and the formation of the dense core (Stevenson, 1990) released considerable amounts of gravitational energy that heated the accreting planet and released volatiles to form a secondary atmosphere. This early steam-dominated atmosphere might have been modified by hydrodynamic escape powered by a extreme UV flux from the young sun (Kasting, 1993). Impact erosion, the explosive ejection of a portion of the atmosphere by a large impact, might have removed a substantial fraction of the earliest atmosphere (Walker, 1986; Zahnle et al., 1988). Core formation depleted metallic iron from the upper mantle, allowing the source regions of volcanism to become more oxidized. Thus volcanic and hydrothermal emanations were perhaps only weakly reducing in composition (Kasting et al., 1993). Once the main accretionary and core-forming events were concluded within the first few tens of My, oceans formed and the residual atmosphere was dominated by CO_2, N_2 and H_2O, and included lesser amounts of CO and H_2 (Holland, 1984).

The processes of mantle–crust exchange strongly influenced the crustal and atmospheric volatile inventories. This process was recently modeled for the rare gases, carbon and nitrogen (Zhang and Zindler, 1989; Zhang and Zindler, 1993). Volcanic outgassing very likely reflected the melt–vapor partitioning of volatiles. Zhang and Zindler estimate that amounts roughly equivalent to the present inventory of crustal volatiles were outgassed within the first 0.5 Gy. Therefore, most of Earth's supply of nonradiogenic rare gases and N_2 were partitioned into the atmosphere very early. Interestingly, the crustal and atmospheric inventories of carbon that were achieved by 4.0 Ga might have even been greater than they are today, because subduction of carbon became more efficient with time (McCulloch, 1993) and caused a net rate of return of carbon to the mantle during the past 3 Ga (Des Marais, 1985; Zhang and Zindler, 1993).

It is useful to assess how the carbon cycles between the ancient crust, oceans and atmosphere might have affected global climate. If one assumes that the crustal carbon inventory was the same as it is today (10^{23} g, Holland, 1978), and that approximately 15% of this carbon resided in the atmosphere before carbonate rocks accumulated substantially (Walker, 1985), then the primitive atmosphere would have contained 10 bars of CO_2 and CO during the first few hundred million years of its existence (Kasting, 1993). With this atmosphere, the mean surface temperature is estimated to have been 80 to 90°C (Kasting and Ackerman, 1986).

The lunar cratering record testifies that bodies as large as 100 km in diameter continued to strike the Earth until 3.8 Ga. Even at this late date, cometary or carbonaceous chondritic material would have contributed substantial amounts of water (Chyba, 1990). Both the reduction of atmospheric CO_2 by iron-rich impactors and the oxidation of meteoritic organic matter probably augmented episodically the atmospheric inventories of more reduced gases such as CO and H_2 (Kasting, 1990).

GEOLOGIC EVIDENCE FOR THE COMPOSITION OF THE 3.8–3.0 Ga ATMOSPHERE

Additional constraints about the composition of the early atmosphere can be obtained from the surviving rock record. For example, the presence of sedimentary carbonates in the 3.8 Ga Isua sediments in SW Greenland (Schidlowski, 1988) demonstrates that CO_2 existed in the atmosphere at that time. Clues about ancient global temperatures also are important for estimating atmospheric composition because they can constrain the levels of greenhouse gases required to maintain climate. The absence of evidence for glaciations earlier than 3.0 Ga (Harland, 1983) is at least consistent with the possibility that global temperatures were equal to or greater than today's temperatures. The $^{18}O/^{16}O$ values are lower in Archean than in younger sedimentary cherts, consistent with the interpretation that Archean surface temperatures were warm, perhaps even as high as 70°C (Knauth and

Epstein, 1976). However, surface temperatures inferred from the silica isotopic record must be viewed with some caution, due to the postdepositional recrystallization of silica at elevated temperatures (Hesse, 1990). Also, the silica record might reflect changes during the Precambrian in the style of isotopic exchange between water and Earth's crust (Perry et al., 1978). Still, chemical weathering was very effective during the production of these ancient sediments, which is consistent with warm temperatures. The crust was tectonically and magmatically unstable and produced thick first cycle sediments in the greenstone belts. The weathering of a typical uplifted greenstone sequence produced coarse clastic sediments that are enriched in the most chemically-resistant components of the sequence, such as the cherts and silicified komatiitic and dacitic tuffs (Nocita and Lowe, 1990). These components are derived from silicified sedimentary units that comprised less than 20% of the original rock volume. Thus, despite the rapid uplift and transport of the rocks and their debris, their less chemically-resistant components were efficiently degraded. The highly effective weathering implied by these observations is consistent both with relatively warm, moist conditions and with elevated atmospheric CO_2 concentrations (Lowe, 1994). Altered evaporites also occur in greenstone sequences between 3.5 and 3.2 Ga (e.g., Lowe and Knauth, 1977; Buick and Dunlop, 1990). Their occurrence in such tectonically unstable settings is consistent with high rates of evaporation that would have been favored by elevated temperatures and dry conditions. The deposition of gypsum rather than anhydrite (Barley et al., 1979; Lowe, 1983) indicates that temperatures were probably below 58°C (Walker, 1982).

Several lines of evidence indicate that the atmosphere was more reduced during the Archean than it is today. Some minerals, such as uraninite (UO_2), can be transported as detritus at low O_2 levels, but are oxidized and dissolve readily at high O_2 levels (Grandstaff, 1976). Ore deposits in late Archean sediments of the Witwatersrand Basin in South Africa contain detrital uraninite (Robb and Meyer, 1990) and thus indicate that O_2 levels were low. Ancient soils, or paleosols, were undoubtedly influenced by the composition of the atmosphere. The oxidation state of the atmosphere during weathering affects the mobility of redox sensitive elements such as iron and manganese. Highly oxidized soils retain most, if not all, of the iron in Fe^{3+}-rich minerals that was originally present in the parent rock as a mixture of Fe^{2+} and Fe^{3+} compounds (Holland, 1992). However, iron losses were severe in the upper horizons of Archean and Early Proterozoic paleosols, particularly those developed on more mafic parent rocks (Holland, 1992). A lower atmospheric oxygen level would have permitted Fe^{2+} to be leached from the upper soil horizons as observed in the oldest paleosols.

These features are illustrated in late Archean (2.765–2.715 Ga) weathering profiles in the Fortescue Group, Western Australia (MacFarlane et al., 1994). The investigators established that these were indeed paleosols having well-developed sericitic weathered zones which grade downward into heterogeneous chlorite-rich zones (Fig. 1). The chlorite zones grade downward into unweathered basalts.

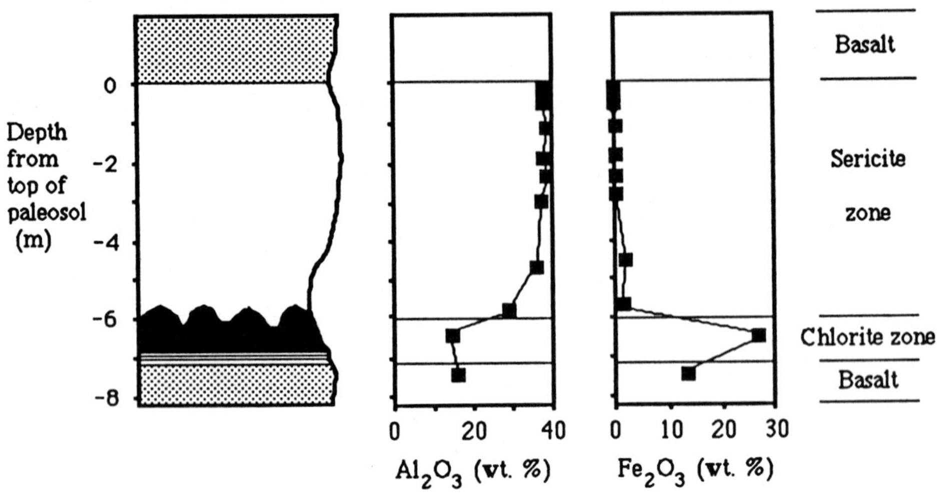

Fig. 1. Stratigraphic section of a Mt. Roe paleosol, showing the relationships between the seric... and chlorite weathered zones to the underlying and overlying basalts (MacFarlane et al., 1994). Also shown are plots of concentration versus depth for an immobile (Al) and mobile (Fe) element (MacFarlane et al., 1994). Concentrations of Al are increased in the sericite zone, relative to the underlying parent rock, due to the removal of mobile constituents during weathering. The depletion of Fe in the sericite zone indicates that Fe^{2+} was removed during weathering under substantially reduced levels of atmospheric O_2.

Elements which were immobile during weathering (Al, Ti, Zr and Th) are enriched in the sericite zone due to removal of more mobile constituents (including Fe, Mn, Mg and Zn). Because Fe is among those constituents depleted in the sericite zone (Fig. 1), it must have been weathered and removed as Fe^{2+}. This scenario indicates that atmospheric O_2 levels during the late Archean were less than 8% of the modern value (MacFarlane et al., 1994).

Shallow water Archean sediments indicate that at least mildly oxidizing conditions prevailed. Silicified or baritized gypsum is widely distributed within both the Barberton and eastern Pilbara greenstone belts (e.g., Lowe and Knauth, 1977; Barley et al., 1979). Gypsum may have been deposited on oceanic volcanic islands (Lowe, 1994), indicating that sulfate was ubiquitous in seawater and therefore that either sunlight (Walker and Brimblecombe, 1985) or O_2 (Ohmoto and Felder, 1987) was available to promote the oxidation of any reduced sulfur that entered surface waters.

Archean BIF was deposited in a stratified ocean at sites where soluble Fe^{2+} upwelling from anoxic deep waters was precipitated after being oxidized in surface waters (Fig. 2a; Klein and Beukes, 1989). Little biological productivity apparently occurred at the site of iron oxidation (Towe, 1983), an observation supported by the very low contents of phosphorus in BIF (Beukes and Klein,

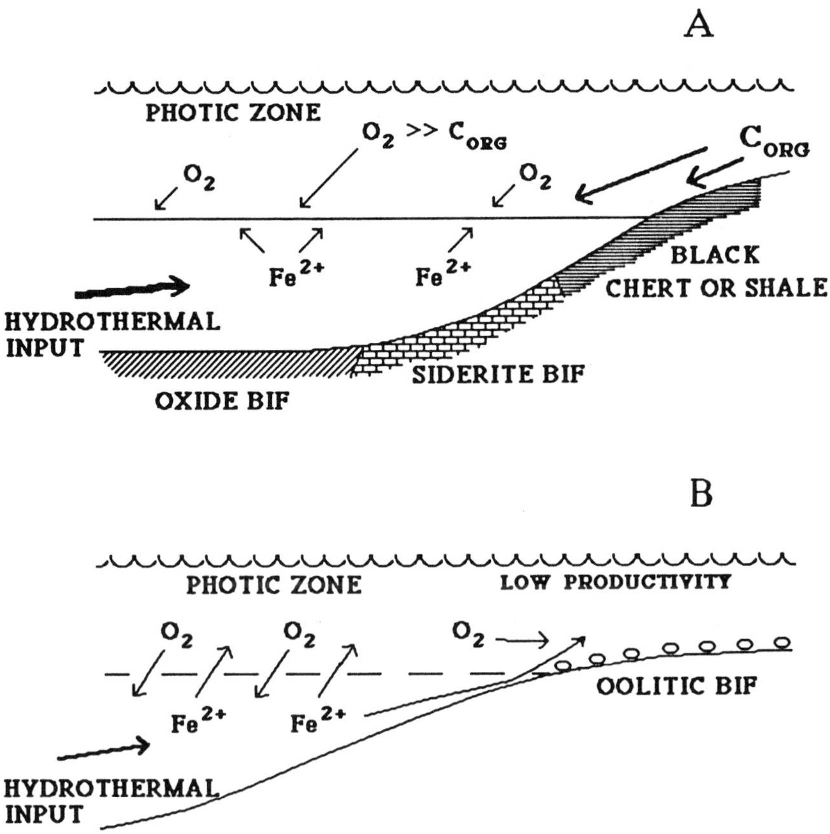

Fig. 2. Models for iron-formation deposition. (a) Archean to Early Proterozoic: stratified ocean with mostly deep water deposition of microbanded iron-formation (Klein and Beukes, 1989). Arrow at upper right depicts organic carbon (C_{org}) imported from shallower water environments. (b) Middle Early Proterozoic: weakening of hydrothermal inputs, breakdown of ocean redox stratification and deposition of hematite-rich oolitic iron-formations. Note that the oxic–anoxic transition in the water column lies below the photic zone (see text). Figure modified from Beukes and Klein (1992).

1992). The absence of BIF deposition on shallow water Archean platforms reveals that the uppermost wind-mixed layer of the ocean contained little Fe^{2+}, indicating that surface seawater was oxidizing (Lowe, 1980; Beukes and Klein, 1992). Although sunlight can cause Fe^{2+} to be oxidized (Braterman et al., 1983), BIF deposition was absent even at water depths some distance below the photic zone (Beukes and Klein, 1992). Thus a mobile oxidant such as O_2 must have penetrated beneath the photic zone to prevent Fe^{2+} from sustaining BIF deposition at those depths. Because the geochemistry of Archean iron-formations resembles the

Proterozoic examples discussed here (Klein and Beukes, 1989), near-surface Archean seawater might have contained at least minor concentrations of dissolved O_2 (Beukes and Klein, 1992). Still, Early Archean atmospheric O_2 concentrations might have been substantially less than 1% of modern levels (Kasting, 1993).

PROCESSES THAT SHAPED THE ARCHEAN ATMOSPHERE

It is useful to evaluate how those processes that add or remove atmospheric constituents might have shaped the atmosphere in the interval of 3.8–3.0 Ga. Although its time of origin is not known, the biosphere would have played an important role. For the purpose of this discussion, it is assumed that the biosphere began at 3.8 Ga. The heavy meteoritic bombardment of Earth ended at about that time, and life's existence at 3.5 Ga is clearly revealed by microfossils, stromatolites and geochemical evidence (Schopf, 1983).

The atmosphere at 3.8 Ga was likely dominated by CO_2, N_2 and H_2O, with lesser amounts of CO, H_2 and reduced sulfur gases (Kasting, 1993). This composition would have been effectively sustained by global rates of volcanic outgassing that exceeded modern rates (Holland, 1984; Veizer et al., 1982 ; Des Marais, 1985). As atmospheric composition ultimately reflects the balance between sources and sinks of the components, it is useful to consider how climate might have affected weathering rates.

The substantial volcanic inputs of CO_2 to the atmosphere ultimately were balanced by the rate of CO_2 removal by weathering (Fig. 3). Because increasing CO_2 concentrations do intensify greenhouse warming that, in turn, increases the rate of CO_2 removal by weathering, CO_2 participates in a negative feedback mechanism that apparently has stabilized Earth's climate (Walker et al., 1981). To sustain the presence of liquid water at 3.8 Ga (Schopf, 1983) despite the predicted lower solar luminosity at that time (Newman and Rood, 1977), the Archean atmosphere must have contained higher-than-modern concentrations of CO_2 and/or other greenhouse gases. An atmospheric CO_2 inventory of approximately 0.2–2 bars could have maintained temperatures in the range 5 to 20°C during the early Archean (Kasting, 1987). This CO_2 inventory declined as the solar constant increased over time, thus stabilizing global temperatures.

The importance of weathering illustrates how the evolution of the continents and land area have also influenced atmospheric composition. The discovery of approximately 4-Ga rocks in NW Canada (Bowring et al., 1989) and a reevaluation of the geochronology of Earth's early crust (e.g., Housh and Bowring, 1993; Jacobsen and Harper, 1993) both imply that the volume of the continents approached modern levels prior to 3.0 Ga. This observation must be reconciled with parallel observations that the global rate of subaerial erosion was considerably lower than it was later in Earth history. Ancient seawater $^{87}Sr/^{86}Sr$ values, recorded in carbonates, typically reflect the balance between strontium inputs from hy-

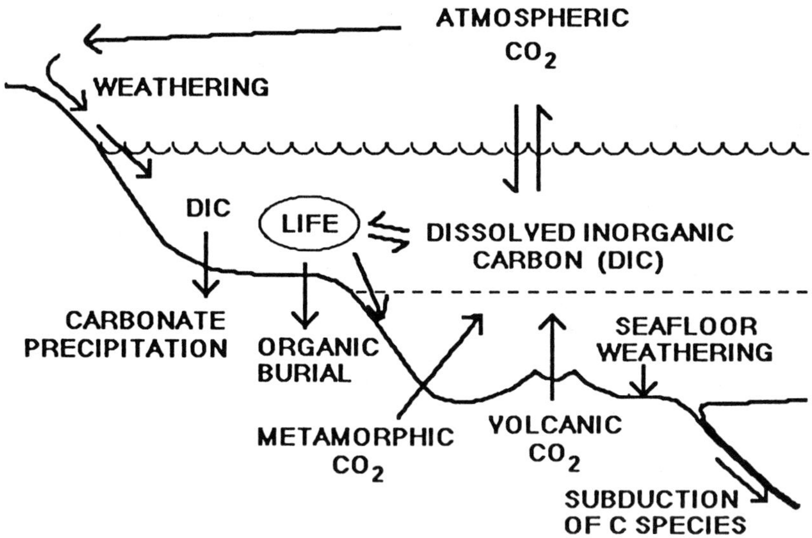

Fig. 3. Schematic diagram of the carbon cycle, depicting carbon reservoirs and processes affecting their abundance and distribution. Thermal emanations are sources of atmospheric carbon dioxide. Both the burial of carbonates and organic carbon and also the subduction of carbon species are sinks.

drothermal and riverine sources. However, $^{87}Sr/^{86}Sr$ values from Archean carbonates indicate that hydrothermal processes dominated the global strontium budget (Veizer et al., 1982; Veizer et al., 1989a; Veizer et al., 1989b). Furthermore, clastic sediments deposited in Early Archean greenstone belts show minimal influence from adjacent continental blocks, in contrast to much more substantial inputs of continental clastics to greenstone sequences during the Late Archean (Lowe, 1992; Thurston and Chivers, 1990, Mueller and Donaldson, 1992; Mueller et al., 1991; Lowe, this volume). The best-preserved sequences indicate that early Archean greenstone belts were derived from mafic volcanic islands and submarine plateaus that were associated with mantle plumes or plate margins, and which formed as large, low-relief simatic shields (Lowe, 1983; Thurston and Chivers, 1990). These unstable, tectonically active platforms were sites for the deposition of mainly volcaniclastic debris and a variety of orthochemical sediments (Lowe, 1982). Much of the carbonate observed in the Barberton and eastern Pilbara greenstone belts formed by hydrothermal alteration and seafloor weathering of submarine volcanic rocks and sediments (de Wit et al., 1982; Veizer et al., 1989a).

Large volumes of Archean crust probably differentiated from the mantle very early in Earth history, yet this crust was prone to extensive alteration and recycling into the mantle (Armstrong, 1991). This early crust required a considerable interval of time to become stabilized by anatexis, metamorphism and underplating (Lowman, 1989). Before stabilization and thickening became widespread, the early Archean "continental" crust might have been almost completely submerged in the global ocean. Thus the areal extent of land, and therefore the global rate of subaerial weathering, would have been reduced.

The prospect that land area was small during the early Archean reconciles observations about weathering rates. If little land was available globally, then no conflict exists between the evidence that rates of subaerial weathering were high in specific localities (Nocita and Lowe, 1990) and the observation that global river discharge exerted a minimal effect upon the $^{87}Sr/^{86}Sr$ values of Archean seawater (Veizer et al., 1982). Minimal land area could also help to explain how Archean temperatures might have been quite warm (see previous section). With smaller land areas, atmospheric CO_2 levels would have increased to higher levels before weathering rates balanced the rate of CO_2 inputs (Fig. 3; Walker, 1985). With higher CO_2 levels and less land area, the removal of CO_2 by weathering of the sea floor (Staudigel and Hart, 1983) should have been more extensive. Pervasive carbonation of Archean submarine basalts did indeed occur (Roberts, 1987). Elevated CO_2 concentrations are compatible with the alkalinity constraints in the global ocean (Walker, 1990), and they certainly would have enhanced the greenhouse effect and sustained a warm climate.

The climatological effects of both land area and Earth's rotation rate have been modelled for the Archean (Jenkins et al., 1993). A faster rotation rate (14 hours per day) decreases the fraction of global cloud coverage by 20%, decreasing planetary albedo. Reduced land area also increases the planet's ability to absorb solar radiation. Thus both increased rotation rate and smaller land area either promoted warmer conditions during the Archean or else the concentrations of atmospheric CO_2 needed to sustain a given temperature would have been lower than previously expected.

The oxidation state of the atmosphere reflects a balance between volcanic, atmospheric, biologic and tectonic processes (Fig. 4). For example, before the origin of life, atmospheric H_2 from volcanoes and H from the photodissociation of water vapor could have been lost to space (Walker, 1977). This loss would have oxidized the crust and the mantle to some unspecified extent (Kasting et al., 1993). However, once bacteria became widespread, they consumed H_2 emanating from the many undersea volcanoes. Virtually all bacteria, including nonphotosynthetic varieties, have a high affinity for H_2 as an electron donor in biosynthesis (Fenchel and Blackburn, 1979). Thus the early biosphere would have captured H_2 from this volcanic source of reducing power, buried much of it as organic matter in sediments, and substantially curtailed its rate of escape to space.

The history of O_2 has been the most extensively studied aspect of the atmospheric redox budget. Abiotic O_2 production by the photodissociation of H_2O and the escape of H to space is very small (Kasting and Walker, 1981). Oxygenic photosynthesis has been a much more robust O_2 source, and it arose certainly by the Late Archean (Buick, 1992; Beukes and Lowe, 1989), and perhaps earlier (Schopf and Packer, 1987). The evidence, summarized earlier, that shallow ocean

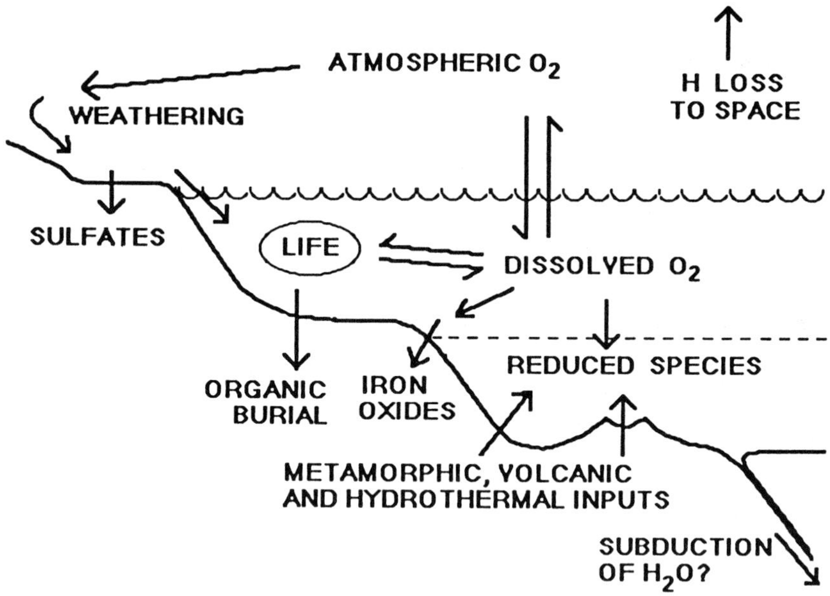

Fig. 4. Schematic diagram of the O_2 cycle, depicting reservoirs and processes that control its abundance in the atmosphere and ocean. The burial of organic carbon (and sedimentary sulfides) allows rates of oxygenic photosynthesis to exceed slightly the rates of biological O_2 consumption, thus creating a net source of O_2. Both the oxidation of volcanic and metamorphic gases and the weathering of rocks are O_2 sinks. Another sink of oxidant is the subduction of H_2O, coupled with the release of reduced gases such as H_2.

waters were at least mildly oxidizing (Fig. 4) is perhaps the most compelling argument that oxygenic photosynthesis existed during the Archean. Shallow seawater would have become anoxic had there not been a source of oxidizing power stronger than the abiotic photodissociation of water. Even at today's lower-than-Archean hydrothermal circulation rates, the modern Earth's substantial inventories of O_2 and seawater sulfate would be consumed in less than 60 million years in the absence of oxygenic photosynthesis (Wollery and Sleep, 1989).

The availability of nutrients to biota living in the shallow seas and coastal environments of the Early to Mid-Archean would have been limited by the combination of a strongly stratified ocean (Beukes and Klein, 1992; Lowe, 1994), and relatively low global rates of subaerial weathering and continental runoff. These conditions favored benthic photosynthetic microbial mat communities over planktonic communities, because microbial mats are highly efficient at recycling and retaining nutrients (Canfield and Des Marais, 1993; Canfield and Des Marais, 1994). However, the organic matter is efficiently recycled in mats (Canfield and Des Marais, 1993), especially if sedimentation rates are low. This would have led to low *net* organic productivity, which is supported by the observation that shallow water Archean sediments are relatively poor in organic matter (Lowe, 1994). Thus, even though Archean deeper-water shales do contain some organic matter (Lowe, 1994), global rates of burial of organic matter from oxygenic photosynthesis were probably low by modern standards. Low rates of organic burial correspond to low net rates of O_2 production (Berner and Canfield, 1989). Furthermore, it has been proposed (Kasting et al., 1993) that the oxidation state of the upper mantle has increased since the Archean. If this occurred, then volcanic emanations during the Archean and Early Proterozoic would have been even more reduced, and atmospheric O_2 levels even lower than currently estimated. The combination of low net O_2 production and higher inputs of reduced volcanic gases seem consistent with the interpretation that Archean O_2 levels were substantially below modern levels.

LATE ARCHEAN ATMOSPHERIC CHANGE

The evolution of Earth's mantle and crust during the Late Archean substantially affected the atmosphere. Following the inevitable decay of radioactive nuclides in the mantle, the heat flow from Earth's interior declined (Turcotte, 1980). This decreased both the seafloor hydrothermal circulation and the volcanic outgassing of reduced species. The style of subduction also changed (McCulloch, 1993). In the Early- to Mid-Archean, subducted slabs were dehydrated, sustained partial melting, and largely disaggregated in the upper 200 km of the mantle. Later, the reduced heat flow and lower temperatures permitted colder, stronger oceanic lithosphere to form. Subducting slabs thus sustained perhaps only partial dehydration and, together with volatiles such as CO_2 and H_2O, penetrated to depths exceeding 600 km (McCulloch, 1993). An increased subduction efficiency of carbon certainly would have affected the crustal and atmospheric carbon budget, but its magnitude is presently unknown. It has been proposed (Kasting et al., 1993) that the upper mantle was oxidized by the subduction of water, followed by the escape of reduced gases. A progressive oxidation of the upper mantle has not been demonstrated, but, if it had occurred, its effect upon atmospheric evolution might have been substantial.

The reworking of Archean continental crust by tectonism, igneous activity and metamorphism also had important consequences for the atmosphere. Through a process termed 'internal differentiation' (Dewey and Windley, 1981), preexisting crust may have become vertically zoned into granitic upper and granulitic lower parts. Also, a subcontinental lithosphere formed perhaps by the extraction of basaltic constituents from the mantle (Jordan, 1988; Hoffman, 1990), and contributed to a thickening and stabilization of continental crust. Thus, crustal evolution during the Late Archean and Early Proterozoic involved the modification, rearrangement and thickening (over- and underplating) of preexisting crust (Lowman, 1989). These processes conceivably contributed to the emergence of vast tracts of continental land area in the interval 3.0 to 2.5 Ga, which led to increased subaerial weathering and the production of abundant clastic sediments (Lowe, 1992). Marine carbonates of this age record a substantial increase in $^{87}Sr/^{86}Sr$ values, indicating greater continental erosion and runoff (Veizer et al., 1992a). New and extensive stable shallow water platforms (Fig. 3) became sites for the deposition and long-term preservation of carbonates (Grotzinger, 1989) and organic matter (Des Marais, 1994).

The long-term increase in solar luminosity (Newman and Rood, 1977) altered the energy balance of the atmosphere. If one assumes that this trend occurred and that long-term temperature changes were relatively minor, due to a negative feedback stabilization involving CO_2 and weathering (Walker et al., 1981), the atmospheric CO_2 concentration is estimated to have declined from perhaps 10 bars or more at 4.6 Ga to less than 0.03 bar at 0.6 Ga (Fig. 5; Kasting, 1992).

Increased continental erosion rates also would have accelerated the rate of CO_2 decline (Berner, 1991; Walker, 1990). Increased weathering would have enhanced the delivery of nutrients to coastal waters, enhancing biological productivity (Betts and Holland, 1991). Greater productivity perhaps removed CO_2 from surface seawater, but, given the high oceanic and atmospheric inorganic carbon contents, the effect of this productivity on the atmosphere should have been minor. However, some authors (Lovelock and Whitfield, 1982; Schwartzman and Volk, 1991) have proposed that soil biota accelerated the weathering process on land, drawing CO_2 levels down even further. In any case, a declining atmospheric CO_2 inventory might have contributed to a late Archean decline in global temperatures. The first well-recorded glaciations occurred in the Late Archean (von Brunn and Gold, 1993) and Early Proterozoic (Harland, 1983). Perhaps these events represent the consequences of this declining CO_2 inventory (Fig. 5).

Although it has not yet been detected in the paleosol record, an increase in atmospheric O_2 levels during the Late Archean (Fig. 6) is consistent with other observations. For example, negative cerium anomalies become more prominent in Proterozoic BIF (Dymek and Klein, 1988; Fryer, 1977), indicating that this element became progressively more oxidized in seawater. The atmospheric O_2 budget (Fig. 4) reflects the balance between its net production by photosynthesis and its consumption by reduced volcanic gases and weathering (Holland, 1984; Berner and Canfield, 1989) . Oxygenic photosynthesis probably evolved prior to the Late Archean (Buick, 1992; Beukes and Lowe, 1989). Rates of continental erosion accelerated during the Late Archean (Veizer, 1994), increasing productivity and sedimentation rates. Higher sedimentation rates increased the burial rate of photosynthetic organic matter (Berner and Canfield, 1989), and the newly-stabilized continental shelves enhanced long-term preservation of these organics (Knoll, 1979; Des Marais, 1994). This increased the net rate of O_2 production. Also, the long-term decline in heat flow decreased the consumption rate of O_2 by volcanic emanations (Holland, 1984). Increased erosion at this time probably increased O_2 consumption somewhat during weathering, but the recently stabilized

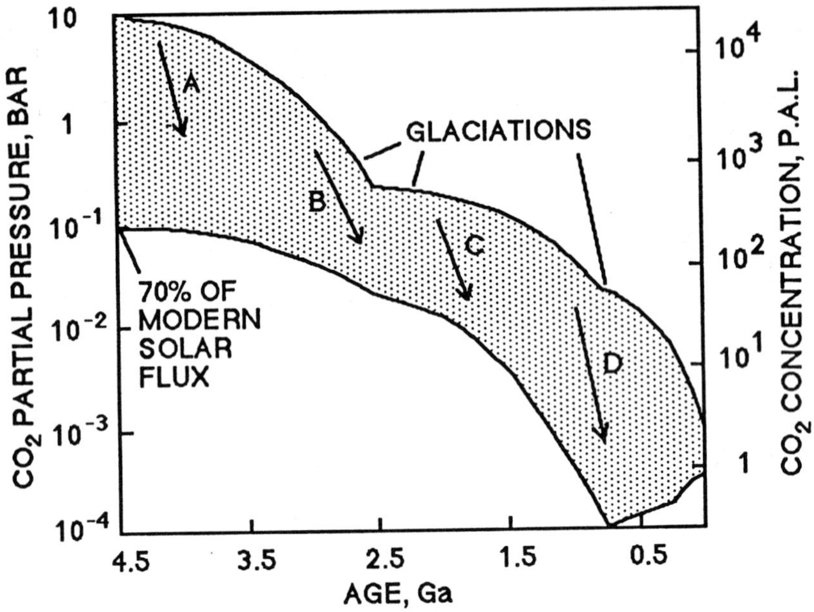

Fig. 5. Estimated change in atmospheric CO_2 levels during Earth history. The long term CO_2 decline is required to decrease the atmospheric greenhouse effect in response to the increase in the solar constant (Walker et al., 1981). The outline of the shaded area and the scales along the vertical axes were adapted from Kasting (1993). Arrows represent those events that likely accelerated the decline in atmospheric CO_2 levels. Letters correspond to the following events: A: cessation of early heavy impactor bombardment; B: stabilization of continental crust (see text); C and D: episodes of continental rifting and orogeny that could have accelerated CO_2 consumption during weathering (Des Marais et al., 1992).

and uplifted landmasses probably were dominated by organic-poor igneous and metamorphic rocks which had survived the generally less-stable Archean crust. The rate of O_2 consumption by the weathering of these rocks would have been exceeded by the rate of O_2 production due to the burial of new, organic-rich aqueous sediments. Thus, atmospheric O_2 levels probably increased, perhaps approaching 10^{-3} bar by 2.5 Ga (Fig. 6; Kasting, 1993). As the Early Proterozoic came to a close, the redox stratification of the global ocean began to break down (Fig. 2b), perhaps due to declining hydrothermal inputs and increased rates of net O_2 production. Deposition of BIF ceased by Middle Proterozoic time (Walker et al., 1983) indicating that the deep ocean had become oxygenated.

Little can be said about the inventories of trace gases in the Archean atmosphere, with the possible exception of methane (CH_4). The presence of kerogens in late Archean rocks with very low $^{13}C/^{12}C$ values is consistent with the hypothesis that CH_4-oxidizing bacteria contributed substantially to the sedimentary carbon inventory (Hayes, 1983; Hayes, 1994). Besides corroborating the view that atmos-

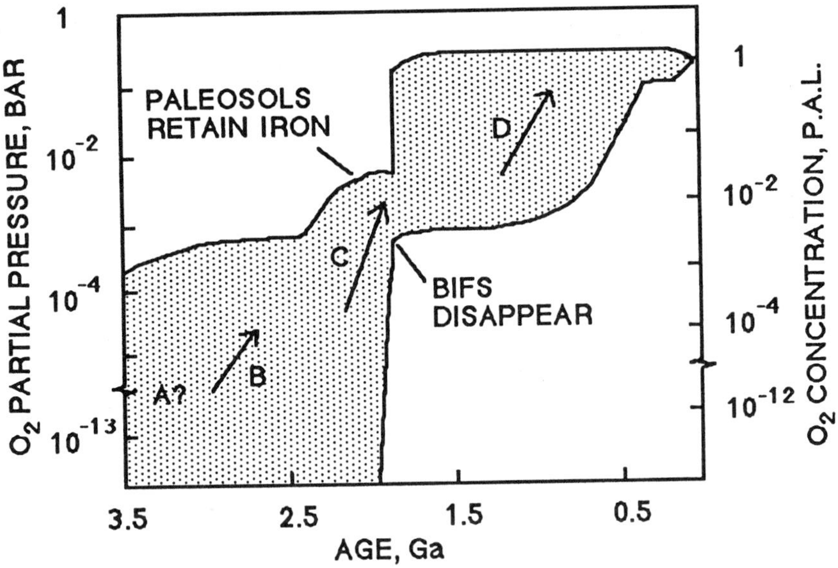

Fig. 6. Estimated changes in atmospheric O_2 levels during the history of the Earth. The long-term increase reflects both the decline in the outgassing rate of reduced volcanic species and also the increase in the net burial of organic matter from oxygenic photosynthesis (e.g., Garrels and Perry, 1974; Holland, 1984, Des Marais et al., 1992). Outline of the shaded area and the scales along the vertical axes were adapted from Kasting (1993). Letters and/or arrows depict those events that would have created episodic increases in atmospheric O_2 levels. Letters depict the following: A: advent of oxygenic photosynthesis ("?" indicates that actual time of origin is unknown); B: stabilization of continents, causing increased preservation of sedimentary organic carbon (see text); C: episode of continental rifting, orogeny and stabilization that enhanced the rates of burial and preservation of organic matter (Des Marais et al., 1992); D: breakup of supercontinent, orogeny and increases in organic burial rates (Des Marais et al., 1992). Event C is recorded by the first appearance of paleosols which retain virtually all of the iron released during the weathering of mafic parent rocks (Holland, 1992). BIF = banded iron formation.

pheric O_2 existed at that time, Hayes' model also estimates that the concentration of atmospheric CH_4 was approximately 20 ppm.

SUMMARY

The general view emerges that the Early Archean prebiotic atmosphere was dominated by endogenic processes, and thus was a weakly reducing mixture composed principally of CO_2, N_2 and H_2O with lesser amounts of H_2 and CO. The development of a pervasive biosphere led to the capture of most of the volcanic H_2 into organic matter in aqueous sediments. The role of subaerial erosion was subordinate to hydrothermal activity, not only because Earth's heat flow was greater at that time, but also because the earliest continental crust was thinner and almost completely submerged, with typically small emergent volcanic islands and microcontinents that sustained very intense weathering. These circumstances are consistent with the notion that Early Archean temperatures were warm. Atmospheric O_2 concentrations were kept low because the net burial of organic matter from oxygenic photosynthesis was probably small, and this O_2 was largely consumed by volcanic emanations.

Atmospheric change in the Late Archean was driven by declining volcanism and hydrothermal emanations, the thickening and stabilization of continental crust, and perhaps also the subduction of volatiles. Increases in land area, weathering and the sedimentation and preservation of significant quantities of photosynthetically-produced organic matter caused CO_2 levels to decline and O_2 levels to increase. Perhaps the weakening of the atmospheric greenhouse effect culminated in Late Archean and Early Proterozoic glaciations. These changes also must have influenced the early evolution of the biosphere, but few details are known at this time.

Future research will lead to a more quantitative understanding of Archean atmosphere and climate. Because the exchange of volatiles between the mantle and crust was important, the oxidation state and volatile inventory of the Archean mantle must be better defined. Did the upper mantle become progressively more oxidized due to the subduction of water? Because weathering and sedimentation strongly affect the atmosphere, the tectonic evolution of Archean continental crust also must be clarified. We must not only quantify the volume of Archean continental crust, we also need to define Archean crustal architecture in order to quantify land area, topography, and rates of erosion. Can specific changes in the Archean atmosphere be linked to specific episodes of tectonic activity and/or biological evolution? Do the 3.8 Ga Isua rocks contain conclusive evidence of a biosphere? How did biota interact with reduced volcanic gases prior to the advent of oxygenic photosynthesis? Did life contribute key trace gases to the Archean atmosphere? What was the inventory and importance of atmospheric trace gases such as CH_4? Was CH_4 an important greenhouse constituent, and did it cause significant amounts of H to be lost to space? A broad consensus on Archean O_2 levels does not yet exist, although most agree that the inventory was substantially below modern levels. When did oxygenic photosynthesis arise? Can carbon isotopic or other records be interpreted to estimate the inventory of photosynthetically-derived organic carbon in the Archean crust? Improved O_2 paleobarometers are needed. Also, paleotemperatures of the Archean surface environment are not well constrained. Can ambiguities be removed from the temperatures which are estimated through $^{18}O/^{16}O$ measurements of silica and other minerals which formed in the surface environment? Can Archean CO_2 levels be estimated more precisely?

Thus a variety of new and improved techniques are needed to interpret the often sparse and poorly-preserved Archean rock record. Such new approaches promise us a clearer view of our origins.

ACKNOWLEDGEMENT

The author is indebted to K. Condie, J. Kasting, K. Klein, J. Walker, and K. Zahnle for constructive criticism of the manuscript. This work was supported by a grant from NASA's Exobiology Program.

REFERENCES

Anders, E. and Owen, T., 1977. Mars and Earth: origin and abundance of volatiles. Science, 198: 453–465.

Armstrong, R.L., 1991. The persistent myth of crustal growth. Aust. J. Earth Sci., 38: 613–630.

Barley, M.E. et al., 1979. Sedimentary evidence for an Archaean shallow-water volcanic-sedimentary facies, eastern Pilbara Block, Western Australia. Earth Planet. Sci. Lett., 43: 74–84.

Berner, R.A., 1991. A model for atmospheric CO_2 over Phanerozoic time. Am. J. Science, 291: 339–376.

Berner, R.A. and Canfield, D.E., 1989. A new model for atmospheric oxygen over Phanerozoic time. Am. J. Sci., 289: 333–361.

Betts, J.N. and Holland, H.D., 1991. The oxygen content of ocean bottom waters, the burial efficiency of organic carbon, and the regulation of atmospheric oxygen. Paleogeog. Palaeoclim. Palaeoecol., 97: 5–18.

Beukes, N.J. and Klein, C., 1992. Models for iron-formation deposition. In: J.W. Schopf and C. Klein (Ed.), The Proterozoic Biosphere, A Multidisciplinary Study. Cambridge University Press, New York, pp. 147–152.

Beukes, N.J. and Lowe, D.R., 1989. Environmental control on diverse stromatolite morphologies in the 3000 Myr Pongola Supergroup, South Africa. Sedimentology, 36: 383–397.

Bowring, S.A. et al., 1989. 3.96 Ga gneisses from the Slave province, Northwest Territories, Canada. Geology, 17: 971–975.

Braterman, P.S. et al., 1983. Photooxidation of hydrated Fe^{2+}: The significance for banded iron formations. Nature, 303: 163–164.

Buick, R., 1992. The antiquity of oxygenic photosynthesis: Evidence from stromatolites in sulphate-deficient Archaean lakes. Science, 255: 74–77.

Buick, R. and Dunlop, J.S.R., 1990. Evaporitic sediments of early Archean age from the Warrawoona Group, North Pole, Western Australia. Sedimentology, 37: 247–277.

Canfield, D.E. and Des Marais, D.J., 1993. Biogeochemical cycles of carbon, sulfur, and free oxygen in a microbial mat. Geochim. Cosmochim. Acta, 57: 3971–3984.

Canfield, D.E. and Des Marais, D.J., 1994. Cycling of carbon, sulfur, oxygen and nutrients in a microbial mat. In: L.J. Stal and P. Caumette (Ed.), Microbial mats: structure, development and environmental significance. NATO ASI Series. Springer, Heidelberg, pp. 255–263.

Chyba, C.F., 1990. Impact delivery and erosion of planetary oceans in the early inner solar system. Nature, 343: 129–133.

de Wit, M.J. et al., 1982. Archean abiogenic and probable biogenic structures associated with mineralized hydrothermal vent systems and regional metasomatism, with implications for greenstone belt studies. Econ. Geol., 77: 1783–1801.

Des Marais, D.J., 1985. Carbon exchange between the mantle and crust and its effect upon the atmosphere, today compared to Archean time. In: E.T. Sundquist and W.S. Broecker (Eds.), The Carbon Cycle and Atmospheric CO_2: Natural Variations Archean to Present. American Geophysical Union, Washington, DC, pp. 602–611.

Des Marais, D.J., 1994. Tectonic control of the crustal organic carbon reservoir during the Precambrian. Chem. Geol., 114: 303–314.

Des Marais, D.J. et al., 1992. Carbon isotope evidence for the stepwise oxidation of the Proterozoic environment. Nature, 359: 605–609.

Dewey, J.F. and Windley, B.F., 1981. Growth and differentiation of the continental crust. Philos. Trans. R. Soc. London, Ser. A, 301: 189–206.

Dymek, R.F. and Klein, C., 1988. Chemistry, petrology and origin of banded iron-formation lithologies from the 3800 Ma Isua supracrustal belt, West Greenland. Precambrian Res., 39: 247–302.

Fenchel, T. and Blackburn, T.H., 1979. Bacteria and Mineral Cycling. Academic Press, New York, 225 pp.

Fryer, B.J., 1977. Rare earth evidence in iron-formations for changing Precambrian oxidation states. Geochim. Cosmochim. Acta, 41: 361–367.

Garrels, R.M. and Perry, E. ., Jr., 1974. Cycling of carbon, sulfur, and oxygen through geologic time. In: E.D. Goldberg (Ed.), The Sea. John Wiley & Sons, New York, pp. 303–336.

Grandstaff, D.E., 1976. A kinetic of the dissolution of uraninite. Econ. Geol., 71: 1493–1506.

Grotzinger, J.P., 1989. Facies and evolution of Precambrian carbonate depositional systems: emergence of the modern platfrom archetype. In: P.D. Crevello et al. (Eds.), Controls on Carbonate Platform and Basin Development. Society of Economic Paleontologists and Mineralogists, Tulsa, OK, pp. 79–106.

Harland, W.B., 1983. The Proterozoic glacial record. Mem. Geol. Soc. Am., 161: 279–288.

Hayes, J.M., 1983. Geochemical evidence bearing on the origin of aerobiosis, a speculative hypothesis. In: J.W. Schopf (Ed.), Earth's Earliest Biosphere. Princeton University Press, Princeton, pp. 291–301.

Hayes, J.M., 1994. Global methanotrophy at the Archean–Proterozoic transition. In: S. Bengtson (Ed.), Early Life on Earth. Nobel Symposium No. 84, Columbia Univ. Press, New York, pp. 220–236.

Hesse, R., 1990. Origin of chert: diagenesis of biogenic siliceous sediments. In: L.A. McIllreath and D.W. Morrow (Eds.), Diagenesis. Geological Association Canada, pp. 227–252.

Hoffman, P.F., 1990. Geological constraints on the origin of the mantle root beneath the Canadian shield. Phil. Trans. R. Soc. Lond. A, 331: 523–532.

Holland, H.D., 1978. The Chemistry of the Atmosphere and Oceans. John Wiley & Sons, Inc., New York, 351 pp.

Holland, H.D., 1984. The Chemical Evolution of the Atmosphere and Oceans. Princeton University Press, Princeton, 582 pp.

Holland, H.D., 1992. Distribution and paleoenvironmental interpretation of Proterozoic paleosols. In: J.W. Schopf and C. Klein (Ed.), The Proterozoic Biosphere, A multidisciplinary Study. Cambridge University Press, New York, pp. 153–155.

Housh, T. and Bowring, S.A., 1993. Geochemical constraints on the formation of the Earth's oldest extant crust. Abstracts with Program, 1993 Annual Meeting of the Geological Society of America, A73.

Jacobsen, S.B. and Harper, C.L.J., 1993. Constraints on the differentiation of the early Earth from the coupled 146,147Sm–142,143Nd systematics. Abstracts with Program, 1993 Annual Meeting of the Geological Society of America, A74.

Jenkins, G.S. et al., 1993. Precambrian climate: the effects of land area and Earth's rotation rate. J. Geophys. Res., 98: 8785–8791.

Jordan, T.H. 1988. Structure and formation of the continental lithosphere. J. Petrol., Special Lithosphere Issue: 11–37.

Kasting, J.F., 1987. Theoretical constraints on oxygen and carbon dioxide concentrations in the Precambrian atmosphere. Precambrian Res., 34: 205–229.

Kasting, J.F., 1990. Bolide impacts and the oxidation state of carbon in the earth's early atmosphere. Origins of Life, 20: 199–231.

Kasting, J.F., 1992. Proterozoic climates: the effect of changing atmospheric carbon dioxide concentrations. In Schopf, J. W. and Klein, C. (Ed.), The Proterozoic Biosphere: A Multidisciplinary Study. Cambridge Univ. Press, Cambridge, 165–168.

Kasting, J.F., 1993. Earth's early atmosphere. Science, 259: 920–926.

Kasting, J.F. and Ackerman, T.P., 1986. Climatic consequences of very high CO2 levels in the earth's early atmosphere. Science, 234: 1383–1385.

Kasting, J.F. et al., 1993. Mantle redox evolution and the oxidation state of the Archean atmosphere. J. Geol., 101: 245–257.

Kasting, J.F. and Walker, J.C.G., 1981. Limits on oxygen concentration in the prebiological atmosphere and the rate of abiotic fixation of nitrogen. J. Geophys. Res., 86: 1147–1158.

Klein, C. and Beukes, N.J., 1989. Geochemistry and sedimentology of a facies transition from limestone to iron-formation deposition in the early Proterozoic Transvaal Supergroup, South Africa. Econ. Geol., 84: 1733–1774.

Knauth, L.P. and Epstein, S., 1976. Hydrogen and oxygen isotope ratios in nodular and bedded chert. Geochim. Cosmochim. Acta, 40: 1095–1108.

Knoll, A.H., 1979. Archean photoautotrophy: Some alternatives and limits. Origins of Life, 9: 313–327.

Lovelock, J.E. and Whitfield, M., 1982. Lifespan of the biosphere. Nature, 296: 561–563.

Lowe, D.R., 1980. Stromatolites 3400 Myr old from the Archaean of Western Australia. Nature, 284: 441–443.

Lowe, D.R., 1982. Comparative sedimentology of the principal volcanic sequences of Archean greenstone belts in South Africa, Western Australia, and Canada: implications for crustal evolution. Precambrian Res., 17: 1–29.

Lowe, D.R., 1983. Restricted shallow-water sedimentation of 3.4 Byr-old stromatolitic and evaporitic strata of the Strelley Pool Chert, Pilbara Block, Western Australia. Precambrian Res., 19: 239–283.

Lowe, D.R., 1992. Major events in the geological development of the Precambrian Earth. In: J.W. Schopf and C. Klein (Ed.), The Proterozoic Biosphere: a Multidisciplinary Study. Cambridge University Press, New York, 67–76.

Lowe, D.R., 1994. Early environments: constraints and opportunities for early evolution. In: S. Bengtson (Ed.), Early Life on Earth. Nobel Symposium No. 84. Columbia Univ. Press, New York, pp. 24–35.

Lowe, D.R. and Knauth, L.P., 1977. Sedimentology of the Onverwacht Group (3.4 billion years), Transvaal, South Africa, and its bearing on the characteristics and evolution of the early earth. J. Geol., 85: 699–723.

Lowman, P.D.J., 1989. Comparative planetology and the origin of continental crust. Precambrian Res., 44: 171–195.

MacFarlane, A.W. et al., 1994. Geology and major and trace element chemistry of late Archean weathering profiles in the Fortescue Group, Western Australia: implications for atmospheric PO_2. Precambrian Res., 65: 297–317.

McCulloch, M.T., 1993. The role of subducted slabs in an evolving earth. Earth Planet. Sci. Lett., 115: 89–100.

Mueller, W. and Donaldson, J.A., 1992. Development of sedimentary basins in the Archean Abitibi belt, Canada: an overview. Can. J. Earth Sci., 29: 2249–2265.

Mueller, W., et al., 1991. The Duparquet Formation: sedimentation in a late Archean successor basin, Abitibi greenstone belt, Quebec, Canada. Can. J. Earth Sci., 28: 1394–1406.

Newman, M.J. and Rood, R.T., 1977. Implications of solar evolution for the Earth's early atmosphere. Science, 198: 1035–1037.

Nocita, B. W. and Lowe, D.R., 1990. Fan-delta sequence in the Archean Fig Tree Group. Precambrian Res., 48: 375–393.

Ohmoto, H. and Felder, R.P., 1987. Bacterial activity in the warmer, sulphate-bearing Archaean oceans. Nature, 328: 244–246.

Perry, E.C.J. et al., 1978. The oxygen isotope composition of 3,800 m.y. old metamorphosed chert and iron formation from Isukasia, West Greenland. J. Geol., 86: 223–239.

Robb, L.J. and Meyer, F.M., 1990. The nature of the Witwatersrand hinterland: Conjectures on the source area problem. Econ. Geol., 85: 511–536.

Roberts, R.G., 1987. Ore deposit models No. 11. Archean lode gold deposits. Geosci. Can., 14: 37–52.

Schidlowski, M., 1988. A 3,800-million-year isotopic record of life from carbon in sedimentary rocks. Nature, 333: 313–318.

Schopf, J.W., 1983. Earth's Earliest Biosphere. Princeton University Press, Princeton, 543 pp.

Schopf, J.W. and Packer, B.M., 1987. Early Archean (3.3 billion to 3.5 billion-year-old) microfossils from the Warrawoona Group, Western Australia. Science, 237: 70–73.

Schwartzman, D.W. and Volk, T., 1991. Biotic enhancement of weathering and surface temperatures on Earth since the origin of life. Global Planet. Change, 90: 357–371.

Staudigel, H. and Hart, S.R., 1983. Alteration of basaltic glass: Mechanism and significance for the oceanic crust–seawater budget. Geochim. Cosmochim. Acta, 47: 337–350.

Stevenson, D.J., 1990. Fluid dynamics of core formation. In: H.E. Newsome and J.H. Jones (Eds.), Origin of the Earth. Oxford Univ. Press, New York, pp. 231–249.

Thurston, P.C. and Chivers, K.M., 1990. Secular variation in greenstone sequence development emphasizing Superior Province, Canada. Precambrian Res., 46: 21–58.

Towe, K.M., 1983. Precambrian atmospheric oxygen and banded iron formations: a delayed ocean model. Precambrian Res., 20: 161–170.

Turcotte, D.L., 1980. On the thermal evolution of the earth. Earth Planet. Sci. Lett., 48: 53–58.

Veizer, J., 1994. The Archean–Proterozoic transition and its environmental implications. In: S. Bengtson (Ed.), Early Life on Earth. Columbia Univ. Press, New York, in press.

Veizer, J. et al., 1992a. Geochemistry of Precambrian carbonates: IV. Early Paleoproterozoic (2.25 $\pm$ 0.25 Ga) seawater. Geochim. Cosmochim. Acta, 56: 875–885.

Veizer, J. et al., 1982. Mantle buffering of the early oceans. Naturwissenschaften, 69: 173–180.

Veizer, J. et al., 1989b. Geochemistry of Precambrian carbonates: II. Archean greenstone belts and Archean sea water. Geochim. Cosmochim. Acta, 53: 859–871.

Veizer, J. et al., 1989a. Geochemistry of precambrian carbonates: I. Archean hydrothermal systems. Geochim. Cosmochim. Acta, 53: 845–857.

von Brunn, V. and Gold, D.J.C., 1993. Diamictite in the Archaean Pongola Sequence of southern Africa. J. Afr. Earth Sci., 16: 367–374.

Walker, J.C.G., 1977. Evolution of the Atmosphere. Macmillan, New York, 318 pp.

Walker, J.C.G., 1982. Climatic factors on the Archean earth. Palaeogeogr. Palaeoclimatol. Palaeoecol., 40: 1–11.

Walker, J.C.G., 1985. Carbon dioxide on the early Earth. Origins of Life, 16: 117–127.

Walker, J.C.G., 1986. Impact erosion of planetary atmospheres. Icarus, 68: 87–98.

Walker, J.C.G., 1990. Precambrian evolution of the climate system. Palaeogeogr. Palaeoclimatol. Palaeoecol., 82: 261–289.

Walker, J.C.G. and Brimblecombe, P., 1985. Iron and sulfur in the prebiologic ocean. Precambrian Res., 28: 205–222.

Walker, J.C.G. et al., 1981. A negative feedback mechanism for the long-term stabilization of Earth's surface temperature. J. Geophys. Res., 86: 9776–9782.

Walker, J. C. G., et al., 1983. Environmental evolution of the Archean–Early Proterozoic Earth. In: J.W. Schopf (Ed.), Earth's earliest biosphere: its origin and evolution. Princeton University Press, Princeton, N.J., pp. 260–290.

Wollery, T.J. and Sleep, N.H., 1989. Interactions of geochemical cycles with the mantle. In: C.B. Gregor et al. (Eds.), Chemical Cycles in the Evolution of the Earth. Wiley, New York, pp. 77–103.

Zahnle, K.J. et al., 1988. Evolution of a steam atmosphere during Earth's accretion. Icarus, 74: 62–97.

Zhang, Y. and Zindler, A., 1989. Noble gas constraints on the evolution of the Earth's atmosphere. J. Geophys. Res., 94: 13719–13737.

Zhang, Y. and Zindler, A., 1993. Distribution and evolution of carbon and nitrogen in Earth. Earth Planet. Sci. Lett., 117: 331–345.

SUBJECT INDEX